Carbon Nanotube–Reinforced Polymers

Carbon Nanotube–Reinforced Polymers

From Nanoscale to Macroscale

Edited by

Roham Rafiee

ELSEVIER

Elsevier
Radarweg 29, PO Box 211, 1000 AE Amsterdam, Netherlands
The Boulevard, Langford Lane, Kidlington, Oxford OX5 1GB, United Kingdom
50 Hampshire Street, 5th Floor, Cambridge, MA 02139, United States

Library of Congress Cataloging-in-Publication Data
A catalog record for this book is available from the Library of Congress

British Library Cataloguing-in-Publication Data
A catalogue record for this book is available from the British Library

ISBN: 978-0-323-48221-9

For information on all Elsevier publications visit our website at
https://www.elsevier.com/books-and-journals

Publisher: Matthew Deans
Acquisition Editor: Simon Holt
Editorial Project Manager: Sabrina Webber
Production Project Manager: Poulouse Joseph
Designer: Greg Harris

Typeset by Thomson Digital

I proudly dedicate this book to my lovely parents, Parvin *and* Bijan
For their devotion, patience, and support during every single moment of my life...
Similar to the meaning of their names in Persian culture:
My mother means to me the sole magnificent, glorious, and shining star in the sky...

and

My father is the only fearless, compassionate, and encouraging hero of my life...
They are endless sources of inspiration and sympathy. Simply, they mean all to me.

Table of Contents

List of Contributors

Héctor Aguilar-Bolados
University of Chile (Universidad de Chile), Santiago, Chile

Omid Akhavan
Sharif University of Technology, Tehran, Iran

Nick K. Anifantis
University of Patras, Patras, Greece

Francis Avilés
Scientific Research Center of Yucatan, AC (CICY), (Centro de Investigación Científica de Yucatán, A.C., Unidad de Materiales), Mérida, Yucatán, Mexico

Naeimeh Bahri-Laleh
Iran Polymer and Petrochemical Institute (IPPI), Tehran, Iran

José N. Canongia Lopes
CQE, Instituto Superior Técnico, University of Lisbon, Lisbon, Portugal

Juan V. Cauich-Rodríguez
Scientific Research Center of Yucatan, AC (CICY), (Centro de Investigación Científica de Yucatán, A.C., Unidad de Materiales), Mérida, Yucatán, Mexico

Aggeliki Chanteli
University of Patras, Patras, Greece

José de Jesús Kú-Herrera
CONACYT-Research Center for Applied Chemistry (CIQA), (Centro de Investigación Científica de Yucatán, A.C., Unidad de Materiales), Saltillo, Mérida, Coahuila, Mexico

Majid Elyasi
Babol Noshirvani University of Technology, Babol, Iran

Bruno Faria
CQE, Instituto Superior Técnico, University of Lisbon, Lisbon, Portugal

Łukasz Figiel

International Institute for Nanocomposites Manufacturing, WMG, University of Warwick, Coventry, United Kingdom; Centre of Molecular and Macromolecular Studies, Polish Academy of Sciences, Łódź, Poland

Vahid Firouzbakht

University of Tehran, Tehran, Iran

Stylianos K. Georgantzinos

University of Patras, Patras, Greece

Roghayeh Ghasempour

University of Tehran, Tehran, Iran

Amin Ghorbanhosseini

University of Tehran, Tehran, Iran

Georgios I. Giannopoulos

University of Patras, Patras, Greece

Tejendra K. Gupta

Institute Center for Microsystems (iMicro), Masdar Institute of Science and Technology, Abu Dhabi, United Arab Emirates

Saeed Herasati

Yasouj University, Yasouj, Iran

Soheil Jafari

University of Sussex, Brighton, Sussex, United Kingdom

Shanmugam Kumar

Institute Center for Energy (iEnergy), Masdar Institute of Science and Technology, Abu Dhabi, United Arab Emirates

Reza Malekimoghadam

University of Tehran, Tehran, Iran

Asimina K. Manta

University of Patras, Patras, Greece

Behzad Mehrafrooz

K.N. Toosi University of Technology, Tehran, Iran

Seyed Amin Mirmohammadi
Islamic Azad University, Tehran, Iran

Abbas Montazeri
K.N. Toosi University of Technology, Tehran, Iran

Hamid Narei
University of Tehran, Tehran, Iran

Andrés I. Oliva-Avilés
Anahuac Mayab University, (Universidad Anáhuac Mayab, División de Ingeniería y Ciencias Exactas), Mérida, Yucatán, Mexico

Ghanshyam Pal
SRM University, Kanchipuram, Tamil Nadu, India

Antonio Pantano
University of Palermo, Palermo, Italy

Timon Rabczuk
Institute of Structural Mechanics, Bauhaus University Weimar, Weimar, Germany

Roham Rafiee
University of Tehran, Tehran, Iran

Samaheh Sadjadi
Iran Polymer and Petrochemical Institute (IPPI), Tehran, Iran

Mohammad Silani
Isfahan University of Technology, Isfahan, Iran

Nuno Silvestre
IDMEC, Instituto Superior Técnico, University of Lisbon, Lisbon, Portugal

Patricio Toro-Estay
University of Chile (Universidad de Chile), Santiago, Chile

Konstantinos I. Tserpes
University of Patras, Patras, Greece

Androniki S. Tsiamaki
University of Patras, Patras, Greece

Nam Vu-Bac

Institute of Structural Mechanics, Bauhaus University Weimar, Weimar, Germany

David Weidt

University of Limerick, Limerick, Ireland

Mehrdad Yazdani-Pedram

University of Chile (Universidad de Chile), Santiago, Chile

Liangchi Zhang

The University of New South Wales, Sydney, NSW, Australia

Xiaoying Zhuang

Tongji University, Shanghai, China; Institute of Continuum Mechanics, Leibniz University Hannover, Hannover, Germany

About the Editor

Dr. Roham RAFIEE received his PhD degree in 2010 in mechanical engineering, focusing on nanocomposites, from Iran University of Science and Technology. He completed both his postgraduate (MSc) and graduate (BSc) studies in the general field of composite materials and structures. He began his industrial experience in different sectors, including wind turbines, composite pipes, strategic planning, and technology transfer projects since 1999.

He joined the University of Tehran in 2011 as an Assistant Professor in the Faculty of New Sciences and Technologies, where he founded the Composites Research Laboratory (www.COMRESLAB.com). He was promoted to the level of Associate Professor in 2015, with unanimous approval of all evaluating committee members. His research interests can be summarized as multiscale modeling of nanocomposites, mechanics of composite materials, design and analysis of composite structures, fatigue modeling of composite structures, and finite element modeling and analysis.

Dr. Rafiee is already the Vice President of the Iranian Composites Association and is also a member of the Iran Composites Scientific Association. He is a senior consultant for different companies and industrial groups and is a member of several graduate student thesis advisory committees. He also collaborates with universities worldwide, including the University of Weimar (Germany) and UPM (Malaysia).

Dr. Rafiee has published 50 ISI papers, 57 international conference papers, and 5 book chapters with Woodhead Publishing, Springer, Taylor & Francis, and World Academic Publishing. He has also organized national and international conferences on composite materials and structures and has registered three patents in the field of nanocomposites and composite structures. He was selected as the Young Distinguished Researcher at the University of Tehran in 2015. He was also selected as the Distinguished Teacher at the University of Tehran in 2017.

Preface

The discovery of carbon nanotubes (CNTs) has established a new era in the field of material science and technology. Exceptional and remarkable mechanical, electrical, and thermal properties of CNTs, as well as their low density and high aspect ratios have rendered them a novel category of multifunctional and unique materials for various potential applications, ranging from nanodevices and/or nanoinstruments to the structural elements in high-tech industries. Incorporating CNT into polymers has developed a new generation of composite materials and has received considerable attention in recent years among the research community, implying their undeniable role in the future of composite science and technology.

A wide spectrum of various industries exploit composite materials either to improve the behaviors of traditional materials/structures or to resolve encountered obstacles and problems. The structure and functions of composites have been diversified due to the advent of new constituents along with manufacturing and design technologies to meet various expectations for different industrial sectors. Therefore, several prospective engineering applications of CNT-reinforced polymers (CNTRP) are anticipated in the form of load-bearing and secondary structural or functional components.

Further development of the CNTRPs and the full utilization of their advantages demand a deep and comprehensive understanding of their behavior. The main objective of this book is to present an overview of CNTs and CNTRPs, covering various aspects from the basic concepts and fundamental issues to sophisticated modeling techniques to estimate mechanical properties.

General topics, including CNT characteristics, synthesis, and applications accompanied with CNTRP processing, enhancing CNT–polymer interaction mechanisms, CNT-grown microfibers, and CNT safety concerns, are introduced. Further, experimental characterization of both isolated CNTs and CNTRPs is presented, followed by the full range of computational approaches for CNT, CNT–polymer interaction, and CNTRPs. A systematic coverage of the computational modeling approaches and techniques employed/developed for the analysis of CNT and CNTRPs at different scales are provided, and corresponding challenges are discussed in an integrated package. Drawing an overall picture of CNTRPs, the current book can potentially be used as a textbook for graduate and postgraduate courses or as a reference book in the field of modeling and analysis of CNTRPs.

Roham Rafiee
Tehran, Iran

Acknowledgments

I would like to take this opportunity to express my sincere appreciation for the invaluable contributions of the authors from different parts of the world in sharing their expertise and devoting their precious time to this book. They all honored me by participating in this project and without them, it would have not been possible to accomplish the current mission of scientific information exchange.

I did know some of the authors before working with them on this project. I was familiar with Timon Rabczuk and Xiaoying Zhuang from my visits to the Structural Mechanic Institute in Bauhaus University of Weimar, Weimar, Germany, and I cooperated with them on some research projects. I met Nuno Silvestre, Konstantinos I. Tserpes, and Antonio Pantano at ICCS events and also followed their research. The contributor to Chapter 2, Soheil Jafari, was my classmate during my PhD program in Iran and we had some very memorable moments together. He undoubtedly fulfills the true meaning of a friend. Abbas Montazeri is a very capable colleague of mine in the field of material science and he is always supportive. It is my extreme pleasure to discuss with him various aspects of the nanoscale. Roghayeh Ghasempour, one of the contributors to Chapters 1 and 7, is my colleague in my school, and she is very active in her own field of research. I am also thankful for my ex-MSc students, Reza Malekimoghadam, Vahid Firouzbakht, and Amin Ghorbanhosseini, who contributed to Chapters 3, 20, and 21 as coauthors. Amin is currently a PhD student in my group, while Vahid and Reza are both research associates of my laboratory (COMRESLAB). It was really an extreme pleasure of mine to work with other authors whom I had not known personally but who are all recognizable researchers in their fields.

Last, but not least, I am grateful for all the support that I have received during this project by the professional staff at Elsevier. My earnest gratitude is extended to Sabrina Webber and, Nicky Carter Joseph Poulouse and Sheela Bernardine Josy for their work throughout the project and also to Simon Holt at the onset of this project.

Roham Rafiee

Tehran, Iran

Introduction

The emergence of carbon nanotubes (CNTs) as an attractive material with superlative properties has been the main motivation of conducted investigations in the fields of nanoscience and nanotechnology in the past few decades. Owing to the unique characterizes of CNTs from different mechanical, electrical, and thermal viewpoints, they are considered to be an ideal and promising candidate for reinforcing polymer, and the potential capabilities of CNT-based composites have received broad attention in recent years. Efficient translation of CNT supreme properties into composites is still an ongoing challenging task, implying the importance of analysis, interpretation, and understanding of CNT-reinforced polymer (CNTRP) behavior.

This book aims to provide a clear insight in the area of CNTRP properties and behavior. Each chapter of this book has been prepared by an independent group of contributors, elaborating different aspects of CNT and CNT/polymer science, with an emphasis on mechanical performance. As this is intended to furnish required information for readers of various levels familiar with this area of science, the first seven chapters of the book focus on preliminary information and basic fundaments of CNTs and CNTRPs. Experimental characterization and modeling procedures for evaluating mechanical properties of CNTs and CNTRPs are discussed in the remaining chapters.

Chapter 1 introduces CNT and its characteristics. A history of CNT is reviewed and also its nanostructure and morphology are elaborated.

Chapter 2 presents engineering applications of CNT, in the fields of coatings, conductor, semiconductor, insulators, biomedicines, sensors, actuators, acoustic actuators, hydrogen storage, water treatment, textiles, photovoltaics, and catalysts. The structural application of CNTs is emphasized in accordance with the main impetus of the book.

Chapter 3 describes processing methods of CNT and highlights advantages/disadvantages of each growth technique.

In Chapter 4, a comprehensive review on different fabrication methods of CNTRP is provided, consisting of conventional methods and recently developed innovative techniques. The important challenges of dispersion and alignment of CNTs in the host matrix during fabrication of CNTRP are also addressed in this chapter. Furthermore, covalent and noncovalent functionalization methods employed to improve the quality of CNT and polymer interaction are briefly explained.

Chapter 5 specifically focuses on chemical modifications through covalent attachment of functional groups to CNT. Moreover, CNT functionalizations interact with thermosettings, thermoplastics, and elastomers, and their influence on mechanical and thermal properties are concisely discussed.

Instead of adding CNT to a polymer, CNTs can be directly grown on the surface of microfibers. The practical procedures of deposition and growth of CNTs in microfibers are introduced in Chapter 6. While the main focus is on glass, carbon, and aramid fibers, some other nonengineered fibers, such as core fibers, are also explained.

In Chapter 7, as the last chapter describing general knowledge on CNTs and CNTRPs, CNT toxicity and CNT-induced toxicity are considered. In parallel with the rapidly growing applications of CNTs, an important concern regarding their potential toxicity to human beings and the environment has arisen. This chapter summarizes that the examination of the corresponding risks and hazards associated with their exposure is an important step during their commercialization.

Prior to estimating mechanical properties of CNT-based composites, mechanical properties of isolated CNTs are required to be extracted. Chapter 8 reviews different experimental and theoretical studies, concentrating on the elastic and inelastic mechanical properties of isolated CNTs. A comparison between obtained results through different modeling techniques employed by researchers and conducted experimental programs is presented. The influence of CNT diameter, chirality, and number of walls on investigated properties is analyzed.

Chapter 9 presents experimentally measured mechanical properties of various CNTRPs with several different polymer matrices. The chapter presents, from a practical point of view, a review on the degree to which CNTs can improve mechanical properties of CNTRP with different classes of host polymers.

CNTRPs are analyzed in terms of electrical and electromagnetic properties in Chapter 10. After introducing key concepts in electrical conductivity and electromagnetic shielding interference, the influence of CNTs on the electrical/electromagnetic properties of CNTRP is discussed and the critical factors are identified.

From a practical point of view, experimental characterization of CNTs and CNT-based composites is a challenging and tedious task. Extremely scattered data obtained from experimental observations that originated from different imposed limitations and obstacles have encouraged many researchers to pursue, alternatively, a variety of theoretical studies. Consequently, computational modeling and simulation techniques play a crucial role not only in facilitating the ongoing applications of CNTs, but also in paving the road toward the future horizon of nanotechnology evolving advancement. Conversely, analysis of the challenges experienced during the implementation of computational tools at different scales of a study is an important issue that will lead to the establishment of novel computational procedures.

Outstanding developments over the past few decades have been seen in the field of modeling and simulation of CNTs and CNT-based composite materials. Concentrating on the mechanical behavior of CNTs and their composites, computational modeling techniques can be divided into three major groups: (1) atomistic modeling; (2) continuum mechanics–based approaches, including both analytical and numerical ones; and (3) multiscale modeling techniques. Different efforts have addressed the mechanical behavior of isolated CNTs and CNT/polymer composites by numerous researchers, and these areas of study are still ongoing. Varied techniques and approaches for assessing the mechanical behavior of CNT and CNT-based composites capturing various effective parameters are outlined in Chapters 11–21. The introduced methods in each chapter of this book reflect the attempts of scientists to capture involved physical phenomena for the purpose of evaluating the real nature of the investigated material more precisely.

Stiffness, strength, and toughness of isolated CNTs under local deformation are evaluated using atomistic simulation in Chapter 11. Three main categories of discrete modeling known as ab initio, tight binding, and molecular dynamics (MD) are explained. Then the fundamental aspects of MD simulation are elaborated on and accompanied with a detailed discussion on the selection of interatomic potentials and time integration schemes. The suitability of continuum shell models in comparison with MD for simulating CNT behavior is investigated. The effect of combined shortening–twisting on CNT local buckling behavior and the chirality and anisotropic effects of CNTs are also discussed.

Chapter 12 presents an integrated computational method for the prediction of Young's modulus and natural frequencies of isolated CNT through 3D finite element (FE) modeling.

The interatomic interactions of CNT nanostructure are simulated using suitable spring and mass elements. The performance of the model is examined by comparing the obtained properties with available results in literature. The developed modeling framework can be applied for both elastic and plastic mechanical responses.

The impact response of an embedded CNT in polymer is investigated in Chapter 13, using a two-scale numerical model. A parametric study is conducted to investigate the influence of different effective parameters on the impact response of an embedded CNT in polymer. The mechanisms of energy absorption and distribution in CNT/polymer composites are analyzed. The constructed model is extended to simulate a nanoindentation test in an upper scale, and the enhancement in energy absorption capability of the polymer is studied.

It is very well known that the interaction between CNT and polymer plays an essential role in defining the reinforcing efficiency, as it is responsible for transferring load from matrix to CNT. It is extremely difficult to obtain reliable experimental data on CNT–polymer interaction and, thus, numerical modeling techniques are preferred methods for this purpose. Chapter 14 performs an extensive review on atomistic-based techniques simulating CNT–polymer interaction present in literature, covering all molecular mechanics, molecular structural mechanics, MD, and coarse-grained MD techniques. The role of CNT–polymer interactions on the load transferring issue is discussed and the capacity of each technique in simulating this important phenomenon is analyzed.

Chapter 15 performs a benchmark study on modeling CNTRPs at different nano-, micro-, meso-, and macroscales, focusing mainly on continuum/FE models. Continuum models are employed to obtain macroscopic properties of CNTRPs wherein associated parameters are fed from a molecular level. Thus, atomic details of nanostructures are incorporated into continuum constitutive models based on the homogenization technique. The developed model can be applied to various materials wherein multiscale modeling is required. The focus is on continuum models for predicting mechanical behavior of CNTRPs.

In Chapter 16, a two-scale continuum model is developed for predicting Young's modulus and yield strength of CNTRPs, taking into account CNT waviness, agglomeration, and CNT–polymer interaction. The model is constructed on the basis of the FE method and is in need of some experimental investigation for collecting required input data. A parametric study is also conducted in this chapter to identify the dominance of effective parameters on the results.

The objective of Chapter 17 is to develop a 3D nonlinear multiscale model for predicting the nonlinear finite deformation of CNT/epoxy composites subjected to compressive loading. A combined experimental–computational approach is suggested. Different qualities of bonding varying from van der Waals interaction to perfect bonding through covalent crosslinks and also CNT waviness are taken into consideration.

The variety of involved length scales and effective parameters of each scale necessitate development of a multiscale modeling technique for accurately analyzing the behavior of CNTRPs. Categorized into hierarchical, sequential, semiconcurrent, and concurrent multiscale methods, an overview on multiscale modeling techniques developed for CNTRPs is summarized in Chapter 18.

Chapter 19 presents a multiscale modeling technique for predicting Young's modulus of CNTRPs. A detailed study is carried out to investigate the influence of CNT–polymer interaction and also CNT waviness on macroscopic elastic properties of CNTRPs. A strategy is suggested for generating and characterizing wavy CNTs and then it is coupled with a commercial FE code to predict mechanical properties of CNTRPs.

Full stochastic multiscale modeling technique has been developed for estimating Young's modulus of CNTRPs in Chapter 20. The developed modeling starts from the nanoscale, passes between the micro- and mesoscales, and ends in the macroscale. A separate representative volume element is defined for each scale, capturing effective parameters of the corresponding scale. Avoiding huge computational efforts at the nanoscale, the lattice structure of CNT is replaced with a discrete space frame structure wherein each C–C bond is replaced with an equivalent beam element. An idea for replacing CNT and CNT–polymer interaction with a virtually equivalent fiber is also developed, which enables the appropriate employment of micromechanical rules. Focusing on the phenomena attributed to the meso- and macroscales, CNT orientations, volume fractions, agglomeration patterns, and nonstraight shapes are treated as random parameters at their effective scale.

The final chapter of the book, Chapter 21, is devoted to multiscale modeling of a carbon fiber coated with CNTs. Both radially aligned and randomly oriented CNTs are taken into account as two available dispersion patterns of CNTs around the core carbon fiber. Involved random parameters are identified for each case and thus a stochastic modeling approach is employed. Effective properties of fuzzy fibers are obtained and compared with available experimental data. After validating the model, a parametric study is also conducted to recognize key parameters influencing estimated properties.

1

CNT Basics and Characteristics

Roghayeh Ghasempour, Hamid Narei
UNIVERSITY OF TEHRAN, TEHRAN, IRAN

CHAPTER OUTLINE

1.1 Introduction to Carbon

Carbon's outstanding capability to bond with a large number of elements makes it the most versatile element in the periodic table. The existence of different structural and geometric isomers, and enantiomers, found in multifarious structures, is a direct consequence of the diverse bonds and their corresponding geometries that carbon can form [1]. Furthermore, the possible configurations of the electronic states of an atom, known as hybridization, are responsible for the geometry of carbon allotropes and their properties. Hence, it is of paramount importance that a brief overview on the physical structure of carbon and the hybridization of atomic orbitals be provided.

Carbon, whose six electrons are shared evenly between the 1s, 2s, and 2p orbitals, is a nonmetallic element with the symbol C. The 1s orbital contains two strongly bound core electrons. Four more weakly bound electrons, known as the valence electrons and involved in chemical bonding, occupy both the 2s and 2p orbitals. Since the energy difference between the 2p energy levels and the 2s level in carbon is far smaller than the binding energy of the chemical bonds, the electronic wave functions for these four electrons can

Carbon Nanotube–Reinforced Polymers. http://dx.doi.org/10.1016/B978-0-323-48221-9.00001-7

readily mix with each other; as a result, the occupation of the 2s and three 2p atomic orbitals changes. The general mixing of 2s and 2p atomic orbitals is called hybridization [2]. Since the type of hybridization is determined by the number of involved p orbitals, it can be categorized into three classes: (1) two hybridized sp^1 orbitals formed by pairing the 2s orbital with one of the 2p orbitals. The linear geometry of the hybridized orbitals provides an angle of 180 degree between them. (2) The 2s orbital is hybridized with two 2p orbitals thereby forming three sp^2 orbitals. These are on the same plane separated by an angle of 120 degree. (3) The 2s orbital hybridizing with the three 2p orbitals, resulting in four sp^3 orbitals separated by an angle of 109.5 degree. sp^3 hybridization results in the tetrahedral arrangements of the bonds. In all three cases, the free energy of forming chemical bonds with other atoms provides the energy required to hybridize the atomic orbitals. The level of hybridization of the carbon orbitals determines the final molecular structure.

While forming a molecule, carbon can bind in sigma (σ) and pi (π) bonds. In sp^n hybridization, $(n + 1)$ σ bonds per carbon atom are formed, so an sp^2 hybridized carbon atom, for instance, is capable of forming three σ bonds and one π bond, so that the number and nature of the bonds determine the geometry and properties of carbon allotropes.

Diamond, graphite, buckminsterfullerene, and nanotubes are four allotropic forms of carbon in the solid phase [1]. Fig. 1.1 presents crystallographic structures of these allotropes.

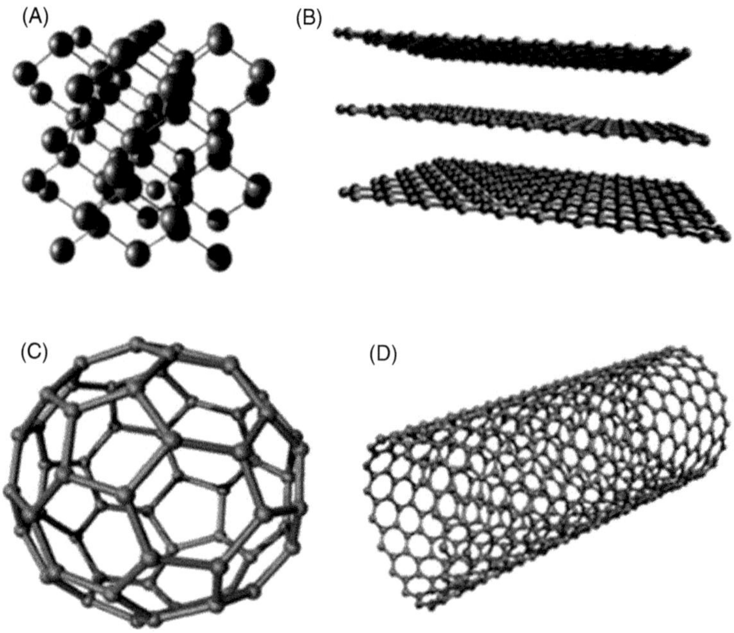

FIGURE 1.1 Allotropes of carbon. (A) Diamond. Diamond has a crystalline structure where sp^3 hybridized carbon atoms are bonded together in a tetrahedral lattice arrangement. (B) Graphite. The carbon atoms are bonded together in sheets of a hexagonal lattice. Van der Waals force bonds the sheets together. The atoms are sp^2 hybridized. (C) Spherical fullerene, C_{60}. The carbon atoms of fullerenes are bonded together in pentagons and hexagons. (D) Tubular fullerene, SWCNT. The carbon atoms are in a tubular formation. *From Z. Ren, et al., Introduction to carbon, Aligned Carbon Nanotubes: Physics, Concepts, Fabrication and Devices, Springer, Berlin, Heidelberg, 2013, p. 2, with permission from Springer [3].*

Diamond comprises pure sp^3 hybridized carbon atoms (Fig. 1.1A). In this material, each carbon atom is surrounded by its neighboring four carbon atoms, forming a tetrahedral lattice arrangement. The crystalline network accounts for the extreme hardness of diamond (hardest known natural material) and its excellent heat conduction properties. Diamond's electrical insulation property and relatively high optical dispersion of the visible spectrum [4] originate from the sp^3 hybridized bonds.

Graphite has a layered and planar structure of sp^2 hybridized carbon atoms bonded in a hexagonal network with a lattice parameter of 0.142 nm while the distance between planes is 0.335 nm (Fig. 1.1B). This results in a two-dimensional (2D) network stacked and loosely bonded through weak van der Waals forces [3]. The opaque, soft, slippery, and electrical conductivity nature of graphite are direct consequences of the different geometries of the chemical bonds. In a graphite sheet, each carbon atom is bonded to only three other atoms thereby allowing electrons to move freely from an unhybridized p orbital to another, forming an endless delocalized π-bond network that leads to its electrical conductivity.

Buckminsterfullerene (or widely known as fullerene), a crystalline formation, comprises a family of spheroidal or cylindrical molecules with a graphite-like structure, while instead of pure hexagons, buckminsterfullerenes may also be formed from pentagonal (or sometimes heptagonal) rings of carbon atoms (Fig. 1.1C). Fullerenes, whose missing (or additional) atoms warp the sheets into spheres, ellipses, or cylinders, are composed of sp^2 hybridized carbon atoms. The tubular form of the fullerenes or carbon nanotubes (CNTs) (Fig. 1.1D) will be the subject of this book, and an elaborate description of their history, structure, and properties will be presented in the subsequent sections of this chapter.

1.2 History

The discovery of CNTs has provoked major controversy even among the scientific communities over recent years. The answer to the question "who should been given credit for the discovery of multiwalled carbon nanotubes (MWCNTs)?" is the main source of the conflict. A review of the literature on the history of CNTs indicates that most scientists had cited Iijima as the discoverer of MWCNTs because he provided the first unambiguous evidence of MWCNTs in 1991 [5], while recent investigations have traced the discovery of MWCNTs back to the early 1950s. Monthioux and Kuznetsov credited Radushkevich and Lukyanovich for the discovery of hollow carbon filaments with nanometer-scale diameter in 1952 by providing a fairly persuasive argument in their paper [6]. Oberlin et al. [7] also synthesized carbon fibers with nanometer-sized diameters using a vapor-growth technique and imaged them via high-resolution transmission electron microscopy (HRTEM) for the first time in 1976. The graphite walls and hollow core of the carbon fiber are clearly discernible from the HRTEM image (Fig. 1.2B). Very few of the pre-1991 researches succeeded in attracting considerable attention among scientific communities, possibly for immaturity of science to consider the nanocharacteristic of these hollow fibers and filaments as well as the limitations of characterization and measurement [6,8].

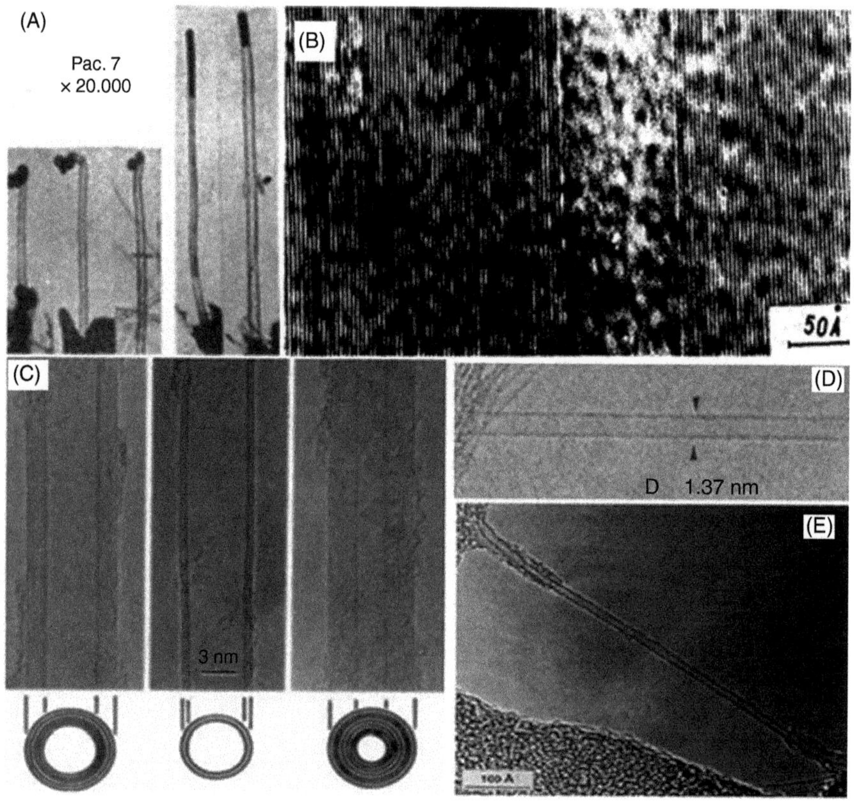

FIGURE 1.2 (A) First TEM images of possible MWCNTs published in 1952. (B) FirstHRTEM image of a hollow fiber consisting of multilayers, published in 1976. (C) HRTEM images of the most well-known MWCNTs published in 1991. (D) Electron micrograph of an SWCNT published in 1993. (E) TEM image of an SWCNT published in 1993. *Part A: Reprinted from M. Monthioux, V.L. Kuznetsov, Who should be given the credit for the discovery of carbon nanotubes?, Carbon 44 (2006) 1621–1623, Copyright 2016, with permission from Elsevier [6]; part B: reprinted from A. Oberlin, M. Endo, T. Koyama, Filamentous growth of carbon through benzene decomposition, J. Cryst. Growth 32 (1976) 335–349, Copyright 2016, with permission from Elsevier [7]; part C: reprinted from S. Iijima, Helical microtubules of graphitic carbon, Nature 354 (1991) 56–58, Copyright 2016, with permission from Macmillan Publishers Ltd. [5]; part D: reprinted from S. Iijima, T. Ichihashi, Single-shell carbon nanotubes of 1-nm diameter, Nature 363 (1993) 603–605, Copyright 2016, with permission from Macmillan Publishers Ltd. [9]; part E: reprinted from D.S. Bethune, C.H. Klang, M.S. de Vries, G. Gorman, R. Savoy, J. Vazquez, et al., Cobalt-catalysed growth of carbon nanotubes with single-atomic-layer walls, Nature 363 (1993) 605–607, Copyright 2016, with permission from Macmillan Publishers Ltd. [10].*

The fortuitous discovery of C_{60} by Kroto et al. [11] in 1985 paved the way for the systematic study of carbon filaments with extremely small diameters. The synthesis of carbon clusters with different sizes, properties, and structure brought fullerenes into the spotlight. Shortly after the discovery of buckminsterfullerene, Iijima, along with other scientists, used the same method of producing C_{60} (DC arc-discharge evaporation) and observed the growth of tube-like graphitic structures with closed ends at the negative end of the electrode using a transmission electron microscope [5]. The observed structures were MWCNTs as shown in Fig. 1.2C. There is no doubt that Iijima's observation of the MWCNTs gave irresistible and increasing momentum on the work on CNTs, but crediting him for the discovery of MWCNTs is neglectful of the scientists who had chanced upon MWCNTs before him.

In contrast with MWCNTs, the discovery history of single-walled carbon nanotubes (SWCNTs) is perfectly clear. The first reports on the formation of SWCNTs were submitted to the June 17th issue of *Nature* in 1993 by two independent groups, one by Iijima and Ichihashi [9] (Fig. 1.2D), the other by Bethune et al. [10] (Fig. 1.2E). The significant aspect of the discovery of SWCNTs is that they are more fundamental than MWCNTs, so they have supported a large body of theoretical studies and predictions.

A great deal of progress has been made in understanding the unique properties of CNTs and in characterizing them over recent years. The most highlighted progress in these areas will be described in the following sections.

1.3 Structure

CNTs can be visualized as rolled-up graphene sheets forming hollow seamless cylinders. As shown in Fig. 1.3, CNTs usually exist in two forms: (1) SWCNTs, comprising a single graphene layer (Fig. 1.3A), and (2) MWCNTs, comprising two or more concentrically rolled graphene sheets (Fig. 1.3B) separated from each other by approximately 0.34 nm as a result of van der Waals forces between adjacent layers. This is slightly larger than the interlayer spacing of graphite. CNTs may be regarded as having a one-dimensional (1D) nanostructure owing to their high aspect ratio, which can be greater than 1000. The chemical bonds in CNTs are essentially sp^2 bonds, similar to those of graphite. The curvature, however, results in quantum confinement and σ–π rehybridization; that is, forcing three σ bonds

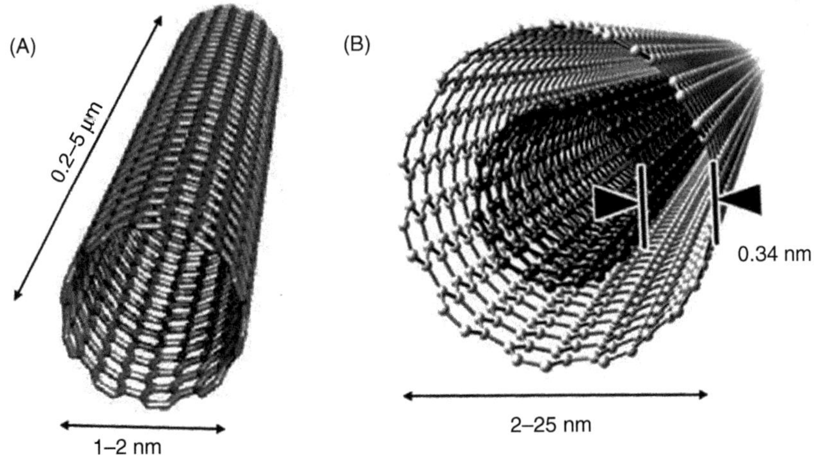

FIGURE 1.3 (A) Schematic diagram of an SWCNT. (B) Schematic diagram of an MWCNT. Typical dimensions of length, diameter, and separation distance between two graphene layers in MWCNTs are shown. The dimensions are taken from Ref. [12]. *Part A: Reprinted from A. Hirsch, Funktionalisierung von einwandigen Kohlenstoffnanoröhren, Angew. Chem. 114 (2002) 1933, Copyright 2016, with permission from Wiley [13]; part B: reprinted from S. Iijima, Carbon nanotubes: past, present, and future, Phys. B Condens. Matter 323 (2002) 1–5, Copyright 2016, with permission from Elsevier [14].*

to be placed slightly out of plane and, for compensation, delocalizing the π orbital more outside the tube. This makes CNTs mechanically stronger, thermally and electrically more conductive, and chemically and biologically more active than graphite. Moreover, they are responsible for the existence of topological defects, such as pentagons and heptagons, incorporated into a hexagonal network to form bent, capped, helical, and toroidal CNTs. All the discussions presented in the following sections concern defect-free CNTs.

An SWCNT is a hollow cylinder of a graphene sheet, so the hexagonal honeycomb lattice of the graphene, shown in Fig. 1.4A, is used to explain the structure of the SWCNT. For this purpose, the chiral vector of the CNT, $\boldsymbol{C}_h$, and translation vector, $\boldsymbol{T}$, are introduced. The vector $\boldsymbol{C}_h$ defines the circumference on the surface of the tube connecting two crystallographically equivalent sites on a 2D graphene sheet. Vector $\boldsymbol{T}$ is parallel to the CNT axis, but perpendicular to the chiral vector. $\boldsymbol{C}_h$ can be uniquely characterized by a pair of

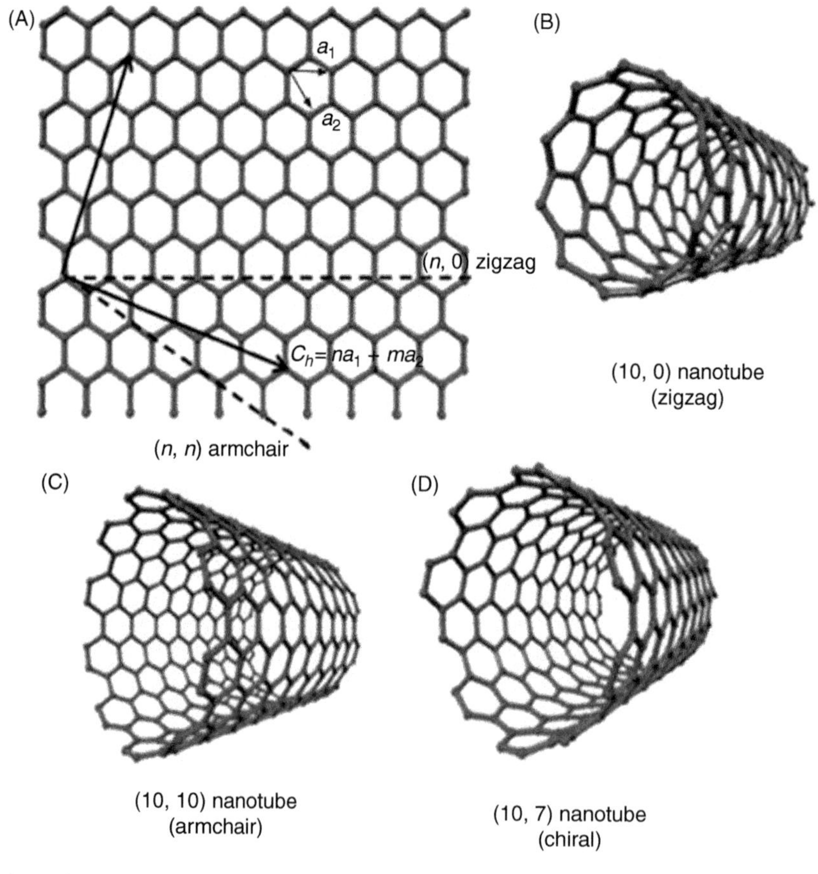

FIGURE 1.4 Schematic representation of the relation between nanotubes and graphene. $C_h = na_1 + ma_2$ is a graphene 2D lattice vector, where $\boldsymbol{a}_1$ and $\boldsymbol{a}_2$ are unit vectors. Integers n and m uniquely define the tube diameter, chirality, and metal versus semiconducting nature. *Reproduced from C.N.R. Rao, R. Voggu, A. Govindaraj, Selective generation of single-walled carbon nanotubes with metallic, semiconducting and other unique electronic properties, Nanoscale 1 (2009) 96–105, with permission of the PCCP Owner Societies [15].*

integers (n, m), known as indices, corresponding to basis vectors $\hat{a}_1$ and $\hat{a}_2$ of the graphene lattice:

$$C_h = n\hat{a}_1 + m\hat{a}_2 \tag{1.1}$$

The nanotube diameter, d_t, can be calculated by

$$d_t = \frac{C_h}{\pi} = \frac{a}{\pi}(m^2 + mn + n^2)^{\frac{1}{2}} \tag{1.2}$$

where C_h is the length of the chiral vector $\boldsymbol{C_h}$, $a = |\hat{a}_1| = |\hat{a}_2| = \sqrt{3}\, a_{C-C}$ is the lattice constant of graphene, and $a_{C-C} = 0.142$ nm is the C—C bond length. The chiral angle θ, the angle between the vector $\boldsymbol{C_h}$ and the basis vector $\hat{a}_1$, is given by

$$\theta = \tan^{-1}\left(\frac{\sqrt{3}\, m}{m + 2n}\right) \tag{1.3}$$

Note that $0 \le |m| \le n$ is used for convention.

Depending on how the graphene sheet is rolled up, three distinct types of CNTs are distinguishable: (1) armchair, (2) zigzag, and (3) chiral. The (n, m) CNTs with $m = n$ correspond to armchair CNTs and $m = 0$ to zigzag CNTs. The zigzag and armchair CNTs are also identified, respectively, by chiral angles of $\theta = 0$ and $\theta = 30$ degree. These two types of achiral CNTs are named from the shape of the cross-sectional ring, as shown at the edge of the nanotubes in Fig. 1.4B–C, respectively. All other (n, m) CNTs with $n \neq m \neq 0$ are commonly referred to as chiral CNTs, which correspond to chiral angle of $0 < \theta < 30$ degree (Fig. 1.4D).

Two models have been established to describe the structure of MWCNTs: (1) the Russian Doll model and (2) the Parchment model. In the Russian Doll (nested shells) model, concentric cylinders of graphitic sheets form MWCNTs, for instance, a (0, 8) SWCNT is placed within a wider (0, 10) SWCNT [3]. The Parchment model, proposed in 1960 [16] to explain the cylindrical structure of carbon fibers, suggests an alternative "scroll" structure for some MWCNTs [17]. In this model a single continuous graphene sheet rolls up or scrolls to form concentric tubes [3].

MWCNTs are a group of coaxial SWCNTs, possibly with different chiralities, so the investigation of their physical properties is more complicated than that of SWCNTs. Combining different diameters and chiralities, which plays a key part in the transport properties of CNTs, notably the electronic ones, results in a large number of individual nanotubes, each with its own distinct electronic, mechanical, and thermal properties that will be discussed briefly in the following sections.

1.4 Physical Properties of CNTs

The peculiar structure of CNTs leads to many exceptional properties such as high electrical and thermal conductivities, high tensile strength, high ductility, and high thermal stability, which make them suitable for various applications, such as gas sensors [17a,17b], energy storage systems and electrodes of batteries and capacitors [17c], catalyst supports [17c] used alone or in composite forms [17d] and thin film. This section provides brief overviews on some of the unique electronic, mechanical, and thermal properties of CNTs.

1.4.1 Electronic Properties

The exceptional electronic properties of CNTs have drawn the greatest attention in CNT research and applications [18]. The nanometer dimensions and the highly symmetric structure of CNTs as well as the unique electronic structure of a 2D graphene sheet are listed as the main reasons for the extraordinary electronic properties of 1D CNT structures [19]. Earlier theoretical calculations [20,21] and later experimental measurements [22–25] have introduced many peculiar electronic properties of CNTs, for example, the metallic and semiconducting characteristics of an SWCNT and the quantum wire feature of an SWCNT, SWCNT bundles, and an MWCNT [18].

As shown in Section 1.3, an SWCNT can geometrically be viewed as a graphene sheet rolled up to form a hollow cylinder; thus, the physics behind the electronic properties of CNTs can be traced back to the electronic structure of graphene. Graphene is a zero-gap semiconductor while early theoretical calculations [20,26] have predicted that SWCNTs can be either metals or semiconductors with different-sized energy gaps. The theory has demonstrated that the electronic properties of SWCNTs depend sensitively on the diameter and helicity of the tubes, - in other words, on the indices (n, m) [19]. To understand the close relationship between the electronic properties and geometric structure of SWCNTs, and the electronic structure of graphene, the following discussion is presented.

The electronic structure of the 2D graphene layer can be analyzed by a simple tight-binding model in which electrons are considered to be a part of atoms forming the solid [27]. The binding model estimates the electronic structure of the 2D graphene layer near the Fermi energy using an antibonding π^* orbital (conduction bands of graphene) and bonding π orbital (valence bands of graphene), shown in Fig. 1.5. The amazing feature of the energy dispersion of graphene is that the upper empty π^* band and the lower occupied π band meet at the Fermi energy, which goes through at the K point in the Brillouin zone. This implies that graphene is a zero band gap semiconductor [12,19].

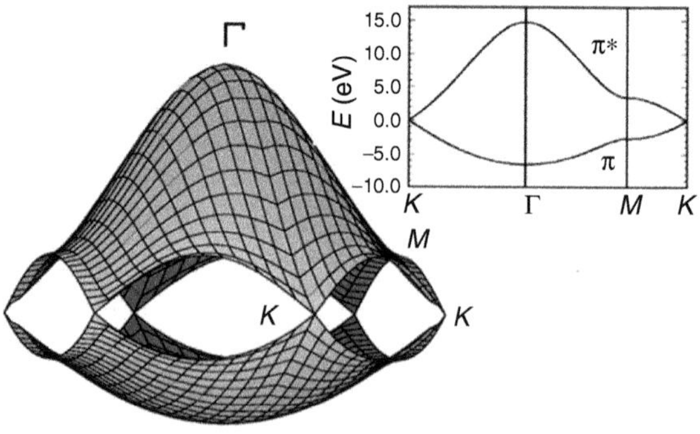

FIGURE 1.5 The energy dispersion relations for the π and π^* bands in graphene are shown throughout the whole region of the Brillouin zone. The inset shows the energy dispersion along the high symmetry points of Γ, *M*, and *K*. *Reprinted from R. Saito, G. Dresselhaus, M. Dresselhaus, Trigonal warping effect of carbon nanotubes, Phys. Rev. B 61 (2000) 2981–2990, Copyright 2016, with permission from American Physical Society [28].*

A simple approximation for the electronic structure of a graphene layer derived from the tight-binding model with the wave vectors (k_x, k_y) is [18]

$$E(k_x, k_y) = \pm\gamma \left\{ 1 + 4\cos\left(\frac{\sqrt{3}\,k_x a}{2}\right)\cos\left(\frac{k_y a}{2}\right) + 4\cos^2\left(\frac{k_y a}{2}\right) \right\}^{\frac{1}{2}} \tag{1.4}$$

where γ is the nearest neighbor-hopping parameter and $a = 0.246$ nm is the lattice constant. $\gamma = 2.5$–3.2 eV from different sources [20–25]. In Eq. (1.4), the negative signs give antibonding π^* orbitals, while the positive signs denote bonding π orbitals.

When the graphene is rolled over to form an SWCNT, a periodic boundary condition, imposed along the tube circumference or the $\boldsymbol{C_h}$ direction, quantizes the 2D wave vector $\boldsymbol{k} = (k_x, k_y)$ along the $\boldsymbol{C_h}$ direction. For this purpose, the $\boldsymbol{k}$ satisfying $\boldsymbol{k} \cdot \boldsymbol{C} = 2\pi q$ (q is an integer) is allowed [18,29]. By cross-sectional cutting of the energy dispersion of graphene with the quantized values of allowed $\boldsymbol{k}$ states, the 1D band structure of an SWCNT is obtained. This method is termed the zone-folding scheme [12]. The allowed $\boldsymbol{k}$ states, shown in Fig. 1.6, depend on the diameter and helicity of the SWCNT. The K point is the place where the π and π^* bands of a graphene sheet meet, defining the Fermi energy [29]. Whenever one of the allowed $\boldsymbol{k}$ vectors crosses the K point, the SWCNT is a 1D metal with a nonzero density of states at the Fermi level; otherwise, the SWCNT is a semiconductor with an energy gap that increases as the tube diameter decreases [19].

Solving the periodic boundary condition, $\boldsymbol{k} \cdot \boldsymbol{C} = 2\pi q$, for varieties of ($n$, m) SWCNTs led to the following general rules: (n, n) SWCNTs, also known as armchair SWCNTs, are always metallic, independent of SWCNT curvature due to their symmetry; (n, m) SWCNTs with $n - m = 3q$, where q is a nonzero integer, are very tiny-gap semiconductors; and

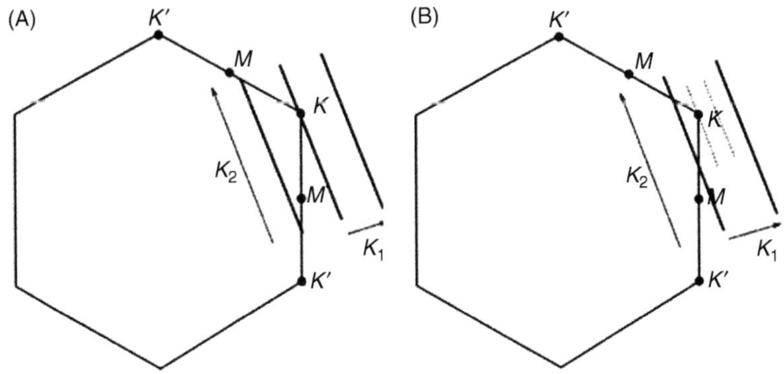

FIGURE 1.6 The wave vector $\boldsymbol{k}$ for 1D carbon nanotubes is shown in the 2D Brillouin zone of graphene as *bold lines* for (A) metallic and (B) semiconducting CNTs. In the direction of $\boldsymbol{K_1}$, discrete $\boldsymbol{k}$ values are obtained by periodic boundary conditions for the circumferential direction of the carbon nanotubes, while in the direction of the $\boldsymbol{K_2}$ vector, continuous $\boldsymbol{k}$ vectors are shown in the 1D Brillouin zone. For metallic nanotubes (A), the *bold line* intersects a K point (corner of the hexagon) at the Fermi energy of graphite. For the semiconductor nanotubes (B), the K point always appears one-third of the distance between two *bold lines*. It is noted that only a few of all the possible *bold lines* are shown near the indicated K point. *Reprinted from R. Saito, G. Dresselhaus, M. Dresselhaus, Trigonal warping effect of carbon nanotubes, Phys. Rev. B 61 (2000) 2981–2990, Copyright 2016, with permission from American Physical Society [28].*

all SWCNTs with $n - m = 3q \pm 1$ are large-gap (~1.0 eV for d_t ~0.7 nm) semiconductors. Strictly within the zone-folding scheme, the $n - m = 3q$ SWCNTs would all be metals, but because of SWCNT curvature effects, a tiny gap opens for the case where q is nonzero. As the diameter of the SWCNT, d_t, increases, the band gaps of the large-gap and tiny-gap SWCNTs decrease with a $1/d_t$ and $1/d_t^2$ dependence, respectively [12]. This means that for most experimentally observed sizes of SWCNTs, the gap in the tiny-gap variety, arising from curvature effects, would be so small that, for most practical purposes, all the $n - m = 3q$ SWCNTs can be considered as metallic at room temperature since their thermal energy is adequate for exciting electrons from the valance to the conduction band [30].

While the zone-folding scheme, based on the tight-binding approach, completely neglects the curvature effects of the SWCNT, it, in general, predicts the electronic properties of SWCNTs accurately. However, strong rehybridization effects between in-plane σ and out-of-plane π orbitals, occurring in small-diameter SWCNTs (<1.5 nm in diameter) owing to the curvature effects, can spoil the results of this relatively simple method. The effect of σ–π rehybridization, which profoundly alters the electronic structure of small-diameter SWCNTs, has been investigated using various approaches, including first-principles or *ab initio* pseudopotential local density functional calculations [31,32]. For instance, the (0, 5) SWCNT, which the zone-folding scheme predicts to be semiconducting, has been characterized as metallic by *ab initio* calculations [33,34].

The electronic states of the CNTs result in the other exceptional properties, such as a strong geometry dependence of the electric polarizability (α) and the quantum wire feature of an SWCNT. Electric polarizability, the relative tendency of a charge distribution, determines the dynamic response of a bound system to external fields. The theoretical calculation [35] showed that the polarizability tensor of an SWCNT were highly anisotropic with $\alpha_{\|} \gg \alpha_{\perp}$, where $\alpha_{\|}$ and $\alpha_{\perp}$ are, respectively, its components parallel and perpendicular to the tube axis. It also predicted that the polarizability of small-gap SWCNTs was greatly enhanced among tubes of similar radii [30]. Experiments have confirmed that, at low temperatures, a single metallic SWCNT [36], an SWCNT rope [22], or an MWCNT [24] intrinsically behave like a quantum wire, in which the conduction appears to occur through well-separated discrete electron states that are quantum-mechanically coherent over distances exceeding many hundreds of nanometers, due to the confinement effect on the tube circumference. The system acts like an elongated quantum dot at adequately low temperatures [18,30].

The unmarked and easily predictable electronic properties and simple structure of CNTs have attracted considerable attention in applications of nanotubes in nanoelectronics. The diameter-dependent energy gap has put a spotlight on the semiconducting SWCNTs. The simple geometry of SWCNT-based field-effect transistors (FETs) gives them a huge advantage over other types of FETs. In fact the SWCNT-based FET is merely composed of two metal electrodes connected by a semiconducting SWCNT on top of a conducting substrate capped by an insulating layer. Semiconducting SWCNTs can be the key component in a single-electron transistor because of their 1D nature. The high conductance sensitivity of SWCNTs to the electrostatic environment, stemming from their considerable surface-to-volume ratio, makes the semiconducting SWCNT a promising candidate for memory applications. Moreover, SWCNTs have also been used for constructing diodes [12].

1.4.2 Mechanical Properties

The mechanical properties of CNTs are a direct consequence of the nature of the chemical bonds between the carbon atoms and of the particular geometrical arrangement of such bonds in CNTs. Since σ bonding is probably the strongest chemical bond known in nature, CNTs, structured with all σ bonding, are expected to possess exceptional mechanical properties. Most of the remarkable mechanical properties of CNTs were first predicted theoretically [37–39] and then confirmed experimentally [40–42]. Most theoretical calculations were conducted on defect-free CNTs and have provided consistent results. In general, the calculations are in agreement with experiments on average. Experimental results reveal a glaring discrepancy, especially for MWCNTs, since MWCNTs contain different amounts of defects from different growth approaches [19].

Young's modulus and the elastic response to deformation are two important parameters that characterize the mechanical properties of CNTs. In the elastic region, experimental measurements and theoretical calculations showed that a CNT is as stiff as diamond with the highest Young's modulus and tensile strength [43]. Researchers have also demonstrated that Young's modulus is independent of tube chirality [44], but dependent on tube diameter. The SWCNTs with diameters between 1 and 2 nm are expected to have the highest value of Young's modulus, approximately 1 TPa. The Young's modulus for MWCNTs is higher than SWCNTs, typically 1.1–1.3 TPa, because the Young's modulus of an MWCNT takes the highest Young's modulus of the SWCNTs contributed in the MWCNT plus contributions from coaxial intertube coupling or van der Waals forces [18].

The extraordinary elastic response of a CNT to deformation has also attracted special attention. Atomic force microscopy (AFM) measurements [45] have revealed that CNTs can be bent to form sharp U-tubes and loops with small curvatures, testifying to their flexibility, toughness, and capacity for reversible deformations [19]. Although most hard materials fail with a strain of 1% or less owing to propagation of dislocations and defects, CNTs can sustain up to 15% tensile strain before fracture. Thus, assuming 1 TPa for Young's modulus of an SWCNT, its tensile strength can be as high as 150 GPa, which is an order of magnitude higher than any other material. Such a high strain is ascribed to an elastic buckling through which high stress gets released [43]. Twisting and bending deformation of CNTs also benefit from elastic buckling. All types of elastic deformations including tensile, twisting, and bending in a CNT are nonlinear, featured by elastic buckling up to 15% strain. This high elastic strain in several deformation modes is attributed to sp^2 rehybridization [18]. Clear evidence of resilience is provided in Fig. 1.7, in which the bends seem fully reversible up to very large bending angles. Table 1.1 presents the calculated Young's modulus (tube axis elastic constant) and tensile strength for a (10, 10) SWCNT and an MWCNT in comparison with other materials.

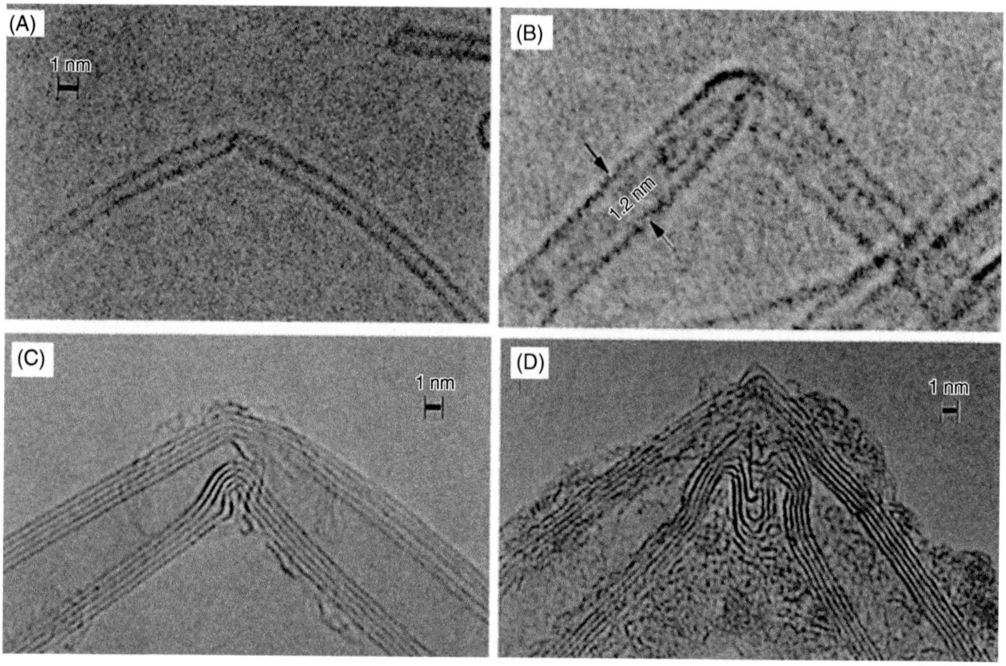

FIGURE 1.7 HRTEM images of bent CNTs under mechanical duress. (A–B) Single kinks in the middle of single-walled nanotubes with diameters of 0.8 and 1.2 nm, respectively. The gap between the tip of the kink and the upper wall is about 0.4 nm in (B). (C–D) A multiwalled tube (diameter ~ 8 nm) showing a single kink and a two-kink complex. The two kinks in (D) are separated by about 3.5 nm. *Reprinted from S. Iijima, C. Brabec, A. Maiti, J. Bernholc, S. Iijima, Structural flexibility of carbon nanotubes, J. Chem. Phys. 104 (1996) 2089–2092, with permission from AIP Publishing [46].*

Table 1.1 Mechanical Properties of Carbon Nanotubes Compared With Some Engineering Materials [43]

Materials	Young's Modulus (GPa)	Tensile Strength (GPa)	Density (g/cm³)
MWCNT	1200	~150	2.6
SWCNT	1054	~150	1.3
Graphite (in-plain)	350	2.5	2.6
Steel	208	0.4	7.8
Wood	16	0.08	0.6

The high stiffness, large tensile strength, and the capability to sustain large deformation without fracture make CNTs attractive for various applications including high-performance composites, high-energy-absorbing materials, hydrogen and VOC absorbers for the environmental applications [17d,46a], and nanoelectromechanical systems.

1.4.3 Thermal Properties

The exceptional heat capacity and thermal conductivity of diamond and graphite have raised researchers' expectations that CNTs might display similar thermal properties at room and elevated temperatures. The thermal properties of CNTs, nonetheless, have not been as extensively studied as their electronic and mechanical properties, in part because the required techniques for such studies are still under development. The CNTs show a broad range of thermal properties stemming from their relation to the corresponding 2D graphene sheet and from their unique structure and nanometer dimensions [19].

The specific heat of individual CNTs should be similar to that of a 2D graphene layer at high temperature. Experimental measurements have shown a temperature-dependent specific heat for MWCNTs, which is consistent with weak interlayer coupling. At lower temperatures, however, the effect of phonon quantization becomes distinct for SWCNTs of small diameter (<2 nm), where a linear temperature dependence of the specific heat is anticipated. In general, CNTs demonstrate quantum confinement effects at low temperatures. For instance, specific heat is 0.3 mJ/(g·K) for a (10, 10) SWCNT, ~0 mJ/(g·K) for an SWCNT bundle, and 2–10 mJ/(g·K) for an MWCNT [18].

The thermal conductivity of graphite, in general, is dominated by the transportation of phonons, and restricted by the small crystallite size [19]. Thus, the long-range crystallinity of CNTs and long phonon mean free path have prompted speculation that CNTs would act like a heat pipe with a longitudinal thermal conductivity that could possibly surpass the in-plane thermal conductivity of graphite [30,47]. Experiments on the thermal conductivity of bulk samples have showed graphite-like behavior for MWCNTs but a different behavior for SWCNTs, particularly a linear temperature dependence at low T, consistent with 1D phonons. Fig. 1.8A shows the thermal conductivity (κ) of three MWCNTs and bulk sample of the laser-vaporization-produced SWCNT for ranges of 4–300K. As shown in Fig. 1.8A, at low temperatures ($T < 100$K), $\kappa(T)$ for MWCNTs increases with $\sim T^2$, which is similar to the $T^{2.3}$

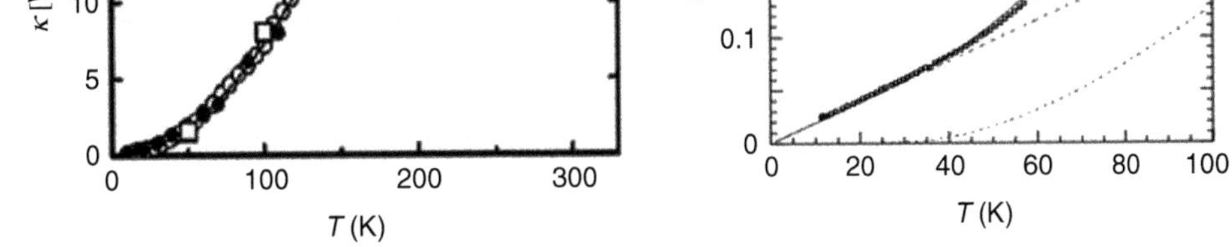

FIGURE 1.8 (A) Thermal conductivity of three distinct MWCNTs. (B) Measured low-temperature thermal conductivity of SWCNTs, compared to acoustic and optical subband. *Part A: Reprinted from W. Yi, L. Lu, Z. Dian-Lin, Z. Pan, S. Xie, Linear specific heat of carbon nanotubes, Phys. Rev. B 59 (1999) R9015–R9018, Copyright 2016, with permission from American Physical Society [48]; part B: from J. Hone, Phonons and thermal properties of carbon nanotubes, In: M.S. Dresselhau, G. Dresselhaus, P. Avouri (Eds.), Carbon Nanotubes: Synthesis, Structure, Properties, and Applications, Springer, Berlin, Heidelberg, 2001, pp. 273–286, with permission from Springer [50].*

behavior in graphite. There are, however, two different behaviors for the thermal conductivity of SWCNT at low temperatures attributed to the competitive impact of acoustic and optical phonon. Fig. 1.8B demonstrates the measured $\kappa(T)$ of SWCNTs, compared with the effect of acoustic and optical phonon. The top dashed line indicates the linear dependency of the acoustic phonon dispersion band and the lower dashed line represents the optical subband. The solid line is the sum of the two types of phonons and it is perfectly consistent with the experimental data below ~100K. As shown in Fig. 1.8B, there is a considerable change in the slope near 35K. Indeed, for $T < 35K$, $\kappa(T)$ has a linear relationship with T and it completely resembles acoustic phonon behavior. However, the optical phonon dispersion band exerts a dominated effect on the thermal conductivity of SWCNTs after 35K. The linear T dependence of $\kappa(T)$ is the representation of the 1D band-structure of individual SWCNTs, which is different from the graphitic behavior of MWCNTs [30,47–49].

Thermal conductivity, the same as electrical conductivity, is 1D for CNTs. Hence, measurements provide a wide range of 200–6000 W/(m·K), showing a great dependence upon the sample quality and alignment [18]. Experimental measurements and theoretical calculations demonstrated that the thermal conductivity of SWCNT ropes and of MWCNTs at room temperature could be in the range of 1800–6000 W/(m·K) [48,49].

In addition to the superb thermal conductivity, CNTs also offer excellent thermal stability. The thermogravimetric analysis of different carbon materials under air flow has proved that CNTs are much more resistant to oxidation than either graphite or C_{60}, indicated by the onset temperature for weight loss and the peak temperature corresponding to maximum oxidation rate. The analysis showed a peak temperature of 695°C for CNTs, which was higher than that of C_{60} (420°C), diamond (630°C), and graphite (645°C) [51]. Thermogravimetric analysis studies of CNTs in argon did not detect any weight change or noticeable thermal transformation at temperatures up to 1000°C, showing no bond breakage by pyrolysis or change in molecular structure [8]. The high thermal stability and the excellent thermal conductivity of CNTs have recently been exploited to develop ideal CNT-based materials for various thermal management applications.

1.5 Characterization of CNTs

Characterization plays a prominent role in the study of emerging materials, including CNTs, to determine their basic physical and chemical properties. Characterization of CNTs is conducted at different levels. Some characterization techniques are performed to study the sizes, shapes, and morphology of CNTs, whereas others are used to derive in-depth structural information. A variety of techniques have been developed and employed to achieve structural information and to probe the fascinating properties of CNTs on the nanoscale. However, each of these techniques exposes just a portion of the full spectrum of the required information. Since the diameters of CNTs are smaller than the wavelengths of visible light (200–600 nm), optical microscopic techniques cannot characterize their morphology and surface features [8]. Hence, electron microscopic techniques have primarily been used for this purpose. This section describes some of the important methods employed for structural and morphological characterization of CNTs.

1.5.1 Electron Microscopy

Electron microscopy employs electrons instead of phonons since the former benefit from the shorter wavelengths, enabling the observation of matter with atomic resolution [52]. Two general types of electron microscopies can be distinguished: scanning electron microscopy (SEM) and transmission electron microscopy (TEM).

SEM produces images by scanning the surface of a sample with a focused beam of high-energy electrons. SEM provides a wealth of information concerning the structural arrangement, spatial distribution, and geometrical features of CNTs [8]. The typical resolution of an SEM is about 2–5 nm; thus, SEM is not appropriate for imaging individual SWCNTs in an SWCNT bundle where it would be more suitable to use field emission scanning electron microscopy [52]. The technique, however, is capable of imaging the ropes of SWCNTs or of observing the strongly oriented forest of MWCNT films [18]. In conjunction with other characterization techniques, SEM can also be exploited to determine the amount of impurities. Fig. 1.9 shows SEM images of CNTs.

For imaging and structural studies of individual CNTs, either SWCNTs or MWCNTs, TEM is the preferred option [55]. TEM is a multifaceted microscopy technique in which a beam of electrons is transmitted through a thin sample at a high-acceleration voltage; the interaction between the sample and the incident beam forms the image [52]. Modern TEMs are capable of imaging at a spatial resolution of less than 0.1 nm, with an electron probe size measuring less than 0.5 nm. In a TEM, if a CNT is oriented perpendicular to the direction of the imaging beam, each shell of the CNT will appear as two parallel lines in the image; thus, it allows one to determine the diameters of CNTs and the number of walls in an MWCNT [55]. TEM imaging identifies the overall perfection of the CNT, and most importantly, if impurities (intentionally or accidentally) are present in the interior of the CNT or on its outer surface. Fig. 1.2 displays typical TEM images of different types of CNTs. If the microscope benefits from an adequately high point resolution, and if the sample is properly oriented along the zone axis, HRTEM images can be achieved.

TEM also offers the advantage of using the electron diffraction and electron energy loss spectroscopy on the nanoscale. Electron diffraction patterns obtained by TEM can be used to determine the local variations of helicity of the individual SWCNT and its average value within the ropes of SWCNTs. Electron energy loss spectroscopy, using the inelastic scattering of the incident electrons on the sample, enables researchers to identify the amount of dopant existing in the CNT structure [18].

TEM suffers from a serious drawback; this method employs intense electron beams (up to 300 keV for most commercial instruments), which can severely damage CNTs [55]. Indeed, a 300 keV electron beam can completely destroy the graphitized structure of an MWCNT within less than 2 min, leaving amorphous carbon behind. It is fortunate that numerous high-resolution TEMs in use today yield satisfactory resolution by applying modest electron energies (100 keV or less). Under these irradiation conditions, a typical CNT can be imaged for much more time, say tens of minutes, with minimal structural damage. Of course, it is possible to take advantage of electron-beam damage. Recent studies have demonstrated that high-energy electron beams in the microscope can also be used

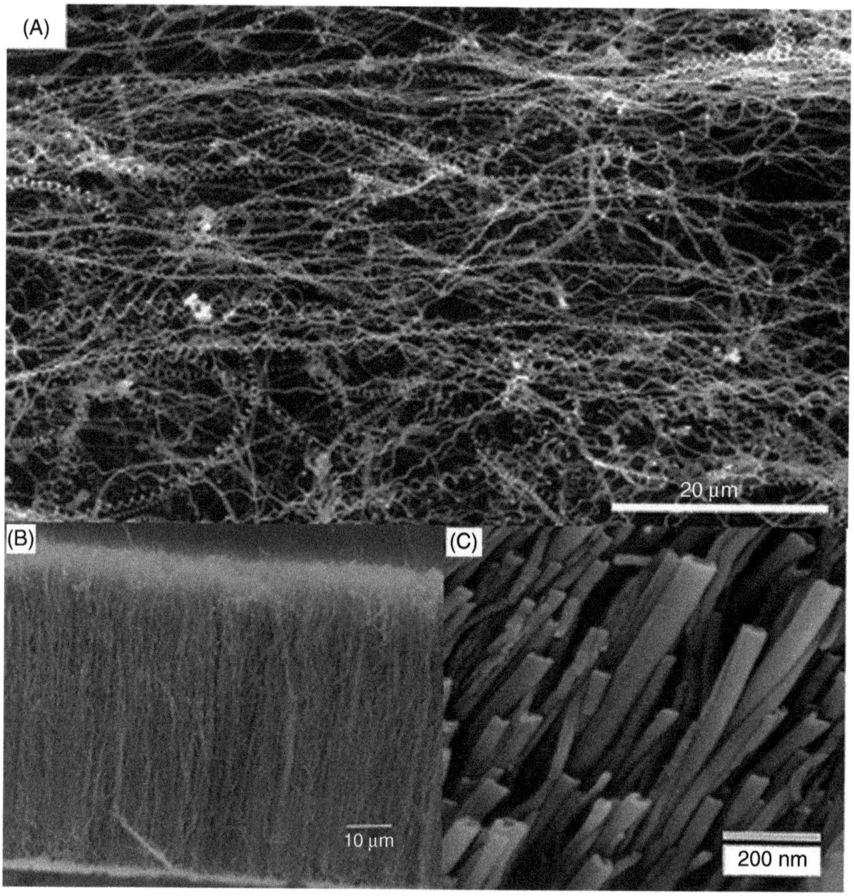

FIGURE 1.9 SEM images. (A) SEM image of a large amount of helically coiled CNTs. (B) Oriented MWCNT film (low resolution) and (C) MWCNTs with open ends revealed under higher magnification. *Part A: Reprinted from M. Zhang, J. Li, Carbon nanotube in different shapes, Mater. Today 12 (2009) 12–18, Copyright 2016, with permission from Elsevier [53]; parts B and C: reprinted from R. Andrews, D. Jacques, A.M. Rao, F. Derbyshire, D. Qian, X. Fan, et al., Continuous production of aligned carbon nanotubes: a step closer to commercial realization, Chem. Phys. Lett. 303 (1999) 467–474, Copyright 2016, with permission from Elsevier [54].*

to modify the structure of CNTs. For example, if the effects of controlled defect density on the structural, transport, and mechanical properties are of interest, TEM can cause a finite amount of defects. The other example is nanowelding of overlapping CNTs using a high-energy electron beam as shown in Fig. 1.10 [56].

1.5.2 Scanning Probe Microscopy

Scanning probe microscopy refers to a family of microscopies that forms images of surfaces using a physical probe, that is, a sharp tip, which scans a sample. The most widely used scanning probe microscopy techniques in the real-mapping of CNTs are AFM and scanning tunneling microscopy (STM) [57].

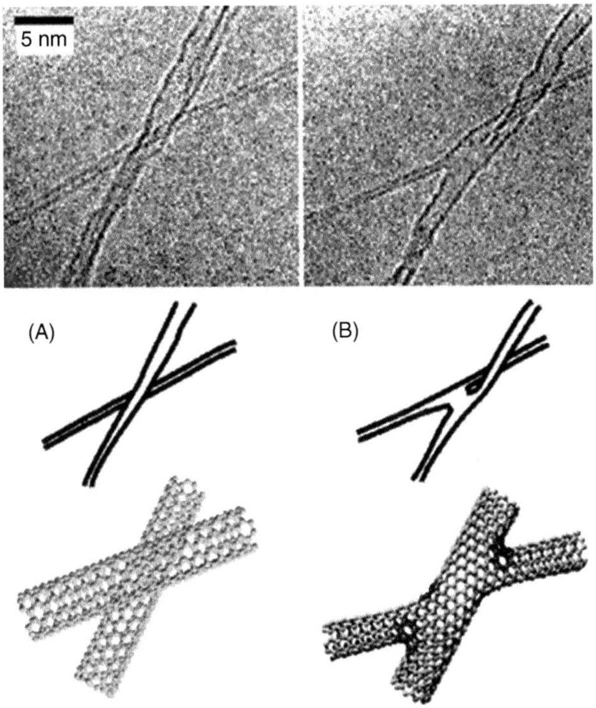

FIGURE 1.10 (A) An SWCNT of ~2.0 nm diameter (running from bottom-left diagonally toward top-right) crossing with an individual SWCNT of ~0.9 nm diameter. (B) 60 s of electron irradiation promotes a molecular connection between the thin and the wide tube, forming an "X" junction. Schematics show that this junction is twisted out of the plane. Molecular models of each image are provided *Reprinted from M. Terrones, F. Banhart, N. Grobert, J.-C. Charlier, H. Terrones, P.M. Ajayan, Molecular junctions by joining single-walled carbon nanotubes, Phys. Rev. Lett. 89 (2002) 75505, Copyright 2016, with permission from American Physical Society [56].*

1.5.2.1 Atomic Force Microscopy

AFM is used to image the surface profile and the real topographical image of CNTs. In a typical AFM, when a microfabricated cantilever-tip assembly scans over the surfaces of a sample, a detector records the deflection of the cantilever by measuring the laser beam reflected from its back. The recorded deflections can be used to provide AFM images of the surface. AFM operation is usually categorized in three modes: contact mode, tapping mode, and noncontact mode [52]. In contact mode, the cantilever remains in contact with the sample, providing high-resolution images [58]. A serious drawback to this mode is that it can damage fragile surfaces. In the tapping mode, also known as intermittent contact mode, the tip oscillates near resonance and intermittently contacts (taps) the surface. In noncontact mode, the tip does not contact the surface, it just oscillates near it. The typical horizontal and vertical distance resolution of an AFM are, respectively, 0.2 and 0.1 nm. Although the resolution of AFM is lower than that of TEM, it can be employed to measure a wide range of surface properties including the hardness of surfaces, surface magnetization, frictional force on the nanoscale, surface charge distribution, yield stress, and the elastic–plastic deformation dynamics of CNTs [52].

The relation between CNTs and AFM is mutually beneficial. As just discussed, AFM provides a wealth of information on the topography and surface features of CNTs. Interestingly, CNTs can be used as AFM tips owing to their 1D nature and high stiffness. Also, the buckling capability of CNTs under mechanical pressure enables a CNT AFM tip to withstand crashes into the sample. Fig. 1.11A–B shows, respectively, SEM and TEM images of an SWCNT surface growth tip consisting of two SWCNTs.

A fascinating application of AFM is the use of the AFM tip as a "finger" to handle CNTs, such as for pushing or rolling CNTs around on a surface. This is extremely useful in the construction of CNT-based devices, or more simply as a tool in studying the properties of

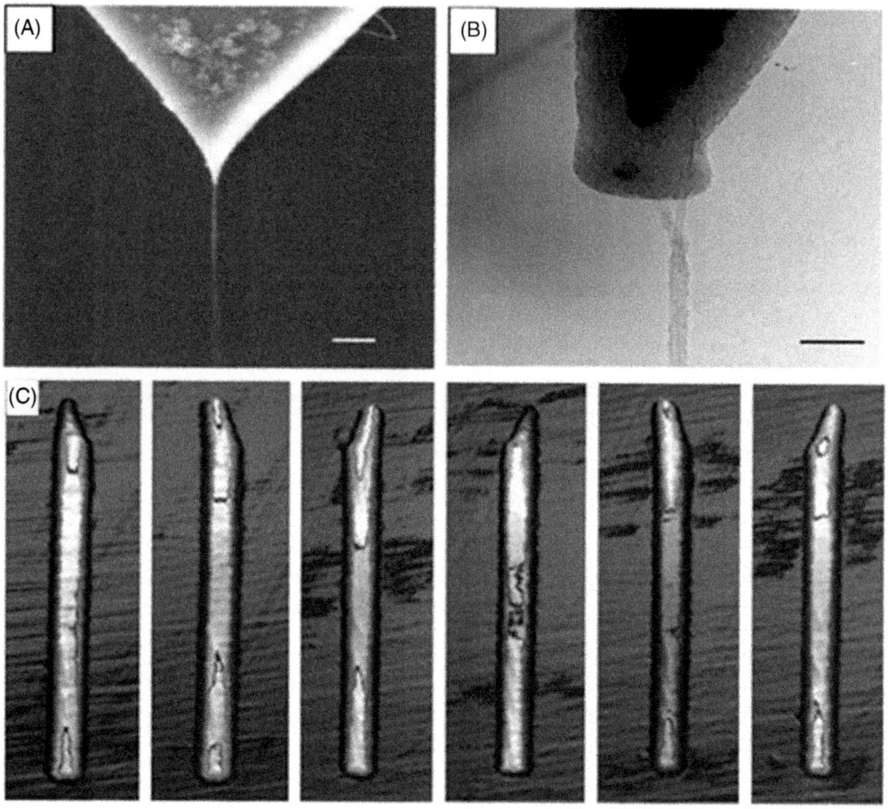

FIGURE 1.11 (A) SEM and (B) TEM images of an SWCNT surface growth tip consisting of two SWCNTs. The scale bars in (A) and (B) are 200 and 20 nm, respectively. (C) AFM images of an MWCNT in different orientations on a graphite surface. The tube was rolled with the AFM tip between images. The tube is 30 nm in diameter and 500 nm long. *Parts A and B: Adapted from J.H. Hafner, C.L. Cheung, C.M. Lieber, Direct growth of single-walled carbon nanotube scanning probe microscopy tips, J. Am. Chem. Soc. 121 (1999) 9750–9751, Copyright 2016, with permission from American Chemical Society [59]; part C: reprinted from J.W. Seo, L. Forró, S. Saito, Carbon Nanotubes: Quantum Cylinders of Graphene, vol. 3, Elsevier Science, Amsterdam, 2008, Copyright 2016, with permission from Elsevier [55].*

CNTs [55]. Fig. 1.11C shows a series of AFM images of an MWCNT in which the MWCNT was rolled on cleaved graphite with the AFM tip.

1.5.2.2 Scanning Tunneling Microscopy

STM is a powerful type of electron microscopy for imaging the surface of a sample at the atomic scale. The principle behind STM is based on the concept of quantum tunneling. When a conducting tip is positioned a few angstroms from a metallic or semiconducting surface, a bias voltage applied between the probe tip and the surface enables electrons to tunnel across the gap. As the tip scans over the surface, the variations in the tunneling current are recorded, which can then be used to produce topographical images of the surface [52]. The typical lateral and depth resolution of STM are 0.1 and 0.01 nm, respectively. STM is known as a surface-sensitive technique and, unlike TEM, it needs an absolutely clean surface, thus its success rate is low.

STM is incredibly useful in the determination of the diameter and chirality of CNTs. Fig. 1.12 shows an STM image of CNTs in which the precise determination of the diameter and chirality of the CNT are perfectly viable. This technique is also able to directly measure the electronic density of states of the CNTs. Moreover, it has been proved that STM is truly valuable in studies related to defects in CNTs and functionalized CNTs. The 1D structure and flexible nature of CNTs have provided the opportunity for CNTs to be used as a probing tip in STM, just as CNTs serve as AFM tips [18].

1.5.3 Raman Spectroscopy

Raman spectroscopy is a vibrational spectroscopy technique with a wide range of applications. It provides vibrational information as well as invaluable information about electronic and phonon properties [60]. Since Raman spectroscopy uses a low-energy photon as a probe, it is nondestructive and noninvasive. There are numerous studies in which Raman spectroscopy was employed to characterize CNTs, particularly SWCNTs. Raman spectroscopy imparts a great deal of information about the configuration, number of walls, and diameters of CNTs, as well as the existence of defects and impurities. The capability of Raman spectroscopy to characterize CNTs at single-nanotube level is largely due to the

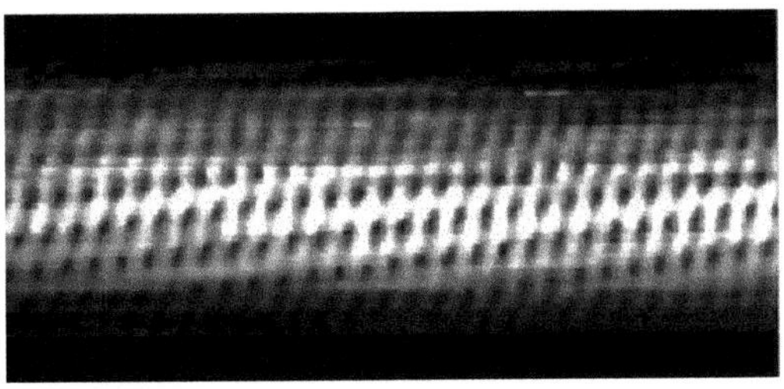

FIGURE 1.12 STM image of a CNT. From the image, the CNT geometrical indices (d, θ) can be deduced. *Reprinted from J.W. Seo, L. Forró, S. Saito, Carbon Nanotubes: Quantum Cylinders of Graphene, vol. 3, Elsevier Science, Amsterdam, 2008, Copyright 2016, with permission from Elsevier [55].*

peculiarly high optical response of CNTs, which is a direct consequence of the 1D confinement of electronic and phonon states.

The basic principle behind Raman spectroscopy relies on the inelastic scattering of light by matter. When the monochromic light, usually from a laser, is shined on a sample, some electrons of its valence energy band are excited to the conduction energy band by absorbing photons. These excited electrons then relax to the valence band by emitting photons. If prior to the relaxation of electrons to the valence band they are scattered by emitting (or absorbing) phonons, the inelastic scattering of light will happen. This phenomenon makes the energy of the scattered photons (light) shift upward or downward. Plotting the intensity of the inelastically scattered light versus the frequency downshift (losing energy) of the scattered light provides the Raman spectrum of the sample. Fig. 1.13 shows Raman spectra from a metallic and a semiconducting SWCNT on the Si/SiO_2 surface. Each peak is assigned to one or more vibrational mode of the SWCNT. The most unique features of Raman spectroscopy for an SWCNT, shown in Fig. 1.13, are the following:

- Radial breathing mode (RBM): The RBM, a unique phonon mode, appears only in CNTs, thus it can be recognized as a fingerprint of the SWCNT. It is a series of low-frequency peaks observed between Raman shift ranges of 100–500 cm^{-1}. The RBM frequency, ω_{RBM}, has an inverse dependence on the diameter of the CNT and is expressed as $\omega_{RBM} = C/d_t$, where $C = 248\ cm^{-1}$ nm for isolated SWCNTs on a SiO_2 substrate [61]. The facile evaluation of CNT diameter has made Raman spectroscopy attractive for real-time monitoring of CNT processing and different separation methods. It can also be used to probe the electronic structure through its intensity [1].
- D-band: D-band of CNTs, located at a Raman shift around 1300 cm^{-1}, can be used to evaluate the defect density existing in the sample. It also offers a proper

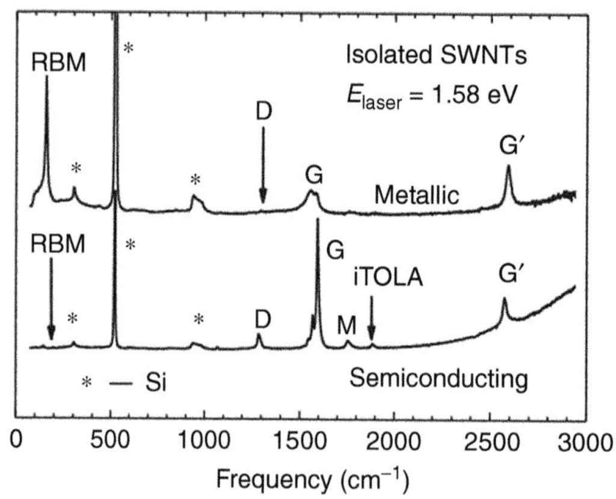

FIGURE 1.13 Raman spectra from a metallic (top) and a semiconducting (bottom) SWCNT at the single nanotube level. The spectra show the RBM, D-band, G-band, and G′-band features. The isolated CNTs are sitting on an oxidized silicon substrate, which provides contributions to the Raman spectra denoted by "*," and these Si features are used for calibration purposes. *Reprinted from M.S. Dresselhaus, G. Dresselhaus, R. Saito, A. Jorio, Raman spectroscopy of carbon nanotubes, Phys. Rep. 409 (2005) 47, Copyright 2016, with permission from Elsevier [61].*

tool for monitoring the extent of covalent sidewall modification in the CNT functionalization [1].

- G-band: SWCNT G-band, corresponding to the planar vibrations of carbon atoms, consists of several peaks in the Raman shift of 1500–1600 cm^{-1}. The most widespread use of G-band frequency is to distinguish between semiconducting and metallic SWCNTs, through strong differences in their Raman line shapes. Semiconducting SWCNTs have a two-peak G-band structure with Lorentzian line shape, whereas metallic SWCNTs produce a G-band structure with a Breit–Wigner–Fano line shape [1].
- G′-band: Measurements have demonstrated that the position of G′-band, located at the Raman shift between 2450 and 2650 cm^{-1}, depends on the chirality and diameter of SWCNTs. Thus its frequency can be used to obtain unique information on the magnitude of curvature effects on the electron and phonon dispersion relations of SWCNTs [61].

Raman spectroscopy also provides other features such as the M-band and the iTOLA second-order modes, which are not discussed here because of their limited importance to the characterization of CNTs.

1.6 Conclusions

Theory and experiment demonstrate the exceptional structures and properties of CNTs. Their small dimensions and extraordinary physical properties have given rise to a wide range of applications.

An SWCNT, depending on its chirality, can be metallic and semiconducting. CNTs also benefit from high stiffness, large tensile strength, and the capability to sustain large deformation without fracture. Moreover, high thermal stability and excellent thermal conductivity are two remarkable thermal properties of CNTs. These properties, along with other structural, chemical, and physical properties, make CNTs one of the most attractive areas to focus on for researchers.

Characterization techniques provide a wealth of information on the structural, morphological, and functional properties of CNTs; therefore, promoting the awareness of researchers and engineers about these techniques is of utmost importance. In the last section of this chapter a brief overview on some of the characterization techniques including electron microscopy, scanning probe microscopy, and Raman spectroscopy was provided.

References

[1] R.J. Grow, Electromechanical properties and applications of carbon nanotubes, in: M.J. O'Connel (Ed.), Carbon Nanotubes: Properties and Applications, CRC Press, Boca Raton, FL, 2006.

[2] R. Saito, G. Dresselhaus, M.S. Dresselhaus, Physical Properties of Carbon Nanotubes, Imperial College Press, London, 1998.

[3] Z. Ren, Y. Lan, Y. Wang, Aligned Carbon Nanotubes: Physics, Concepts, Fabrication and Devices, Springer, Berlin, Heidelberg, 2013.

[4] J. Walker, Optical absorption and luminescence in diamond, Reports Prog. Phys. 42 (1979) 1605–1659.

[5] S. Iijima, Helical microtubules of graphitic carbon, Nature 354 (1991) 56–58.

[6] M. Monthioux, V.L. Kuznetsov, Who should be given the credit for the discovery of carbon nanotubes?, Carbon 44 (2006) 1621–1623.

[7] A. Oberlin, M. Endo, T. Koyama, Filamentous growth of carbon through benzene decomposition, J. Cryst. Growth 32 (1976) 335–349.

[8] P.-C. Ma, J.-K. Kim, Carbon Nanotubes for Polymer Reinforcement, CRC Press, Boca Raton, FL, 2011, p. 224.

[9] S. Iijima, T. Ichihashi, Single-shell carbon nanotubes of 1-nm diameter, Nature 363 (1993) 603–605.

[10] D.S. Bethune, C.H. Klang, M.S. de Vries, G. Gorman, R. Savoy, J. Vazquez, et al. Cobalt-catalysed growth of carbon nanotubes with single-atomic-layer walls, Nature 363 (1993) 605–607.

[11] H.W. Kroto, J.R. Heath, S.C. O'Brien, R.F. Curl, R.E. Smalley, C_{60} buckminsterfullerene, Nature 318 (1985) 162.

[12] Q. Zhang, Carbon Nanotubes and Their Applications, vol. 1, Pan Stanford, Sangapore, 2012.

[13] A. Hirsch, Funktionalisierung von einwandigen Kohlenstoffnanoröhren, Angew. Chem. 114 (2002) 1933.

[14] S. Iijima, Carbon nanotubes: past, present, and future, Phys. B Condens. Matter 323 (2002) 1–5.

[15] C.N.R. Rao, R. Voggu, A. Govindaraj, Selective generation of single-walled carbon nanotubes with metallic, semiconducting and other unique electronic properties, Nanoscale 1 (2009) 96–105.

[16] R. Bacon, Growth, structure, and properties of graphite whiskers, J. Appl. Phys. 31 (1960) 283–290.

[17] D. Qian, G.J. Wagner, W.K. Liu, M.-F. Yu, R.S. Ruoff, Mechanics of carbon nanotubes, Appl. Mech. Rev 55 (6) (2016) 495–533.

[17a] R. Ghasempour, S.Z. Mortazavi, A. Iraji zad, F. Rahimi, Hydrogen sensing properties of multiwalled carbon nanotube films sputtered by Pd, Int. J. Hydrogen Energy 35 (2) (2010) 4445–4449.

[17b] A. Irajizad, R. Ghasempour, Hydrogen sensor based on MWNTs/WO3 hybrid film, 10th Annual IEEE Conference on Sensors, Sensors 2011, Irland, 2011.

[17c] L. Wen, F. Li, H.-M. Cheng, Carbon Nanotubes and Graphene for Flexible Electrochemical Energy Storage: from Materials to Devices, Advance. Material. 28 (2016) 4306–4337.

[17d] R. Ghasempour, A. Irajizad, M.R. Hormozi Nezhad, Design and fabrication of sensitive carbon nanotubes/PMMA film for Acetone vapour detection, Int. J. Nanomanuf. 5 (3–4) (2010) 268–277.

[18] M. Meyyappan, Carbon Nanotubes: Science and Applications, CRC Press, Boca Raton, FL, 2005.

[19] M.S. Dresselhaus, G. Dresselhaus, A. Jorio, Unusual properties and structure of carbon nanotubes, Annu. Rev. Mater. Res. 34 (2004) 247–278.

[20] N. Hamada, S. Sawada, A. Oshiyama, New one-dimensional conductors: graphitic microtubules, Phys. Rev. Lett. 68 (1992) 1579–1581.

[21] J.W. Mintmire, B.I. Dunlap, C.T. White, Are fullerene tubules metallic? Phys. Rev. Lett. 68 (1992) 631–634.

[22] M. Bockrath, D.W. Cobden, P. McEuen, B. Chspra, N.G. Eettl, A.A. Thess, et al., Single-electron transport in ropes of carbon nanotubes, Science 80 (275) (1997) 1922–1925.

[23] T.W. Odom, J.-L. Huang, P. Kim, C.M. Lieber, Atomic structure and electronic properties of single-walled carbon nanotubes, Nature 391 (1998) 62–64.

[24] S. Frank, P. Poncharal, Z.L. Wang, W.A. de Heer, Carbon nanotube quantum resistors, Science 80 (280) (1998) 1744–1746.

[25] J.W.G. Wilder, L.C. Venema, A.G. Rinzler, R.E. Smalley, C. Dekker, Electronic structure of atomically resolved carbon nanotubes, Nature 391 (1998) 59–62.

[26] R. Saito, M. Fujita, G. Dresselhaus, M.S. Dresselhaus, Electronic structure of chiral graphene tubules, Appl. Phys. Lett. 60 (1992) 2204.

[27] S. Reich, C. Thomsen, J. Maultzsch, Wiley, Weinheim, 2004.

[28] R. Saito, G. Dresselhaus, M. Dresselhaus, Trigonal warping effect of carbon nanotubes, Phys. Rev. B 61 (2000) 2981–2990.

[29] J. Bernholc, D. Brenner, M.B. Nardelli, V. Meunier, C. Roland, Mechanical and electrical properties of nanotubes, Annu. Rev. Mater. Res. 32 (2002) 347–375.

[30] M.S. Dresselhaus, G. Dresselhaus, P. Avouris, Carbon Nanotubes: Synthesis, Structure, Properties, and Applications, Springer, Berlin, Heidelberg, 2000.

[31] X. Blase, L.X. Benedict, E.L. Shirley, S.G. Louie, Hybridization effects and metallicity in small radius carbon nanotubes, Phys. Rev. Lett. 72 (1994) 1878–1881.

[32] M.S. Dresselhaus, G. Dresselhaus, J.C. Charlier, E. Hernández, Electronic, thermal and mechanical properties of carbon nanotubes, Philos. Trans. R. Soc. A 362 (2004) 2065–2098.

[33] H.J. Liu, C.T. Chan, Properties of 4 Å carbon nanotubes from first-principles calculations, Phys. Rev. B 66 (2002) 115416.

[34] I. Cabria, J.W. Mintmire, C.T. White, Metallic and semiconducting narrow carbon nanotubes, Phys. Rev. B 67 (2003) 121406.

[35] X. Benedict, S.G. Louie, M.L. Cohen, Static polarizabilities of single-walled carbon nanotubes, Phys. Rev. B 52 (1995) 8541–8549.

[36] S.J. Tans, M.H. Devoret, H. Dai, A. Thess, R.E. Smalley, L.J. Geerligs, et al., Individual single-wall carbon nanotubes as quantum wires, Nature 386 (1997) 474–477.

[37] J.P. Lu, J.I.E. Han, Carbon nonotubes and nanotube-based devices, Int. J. High Speed Electron. Syst. 9 (1998) 101–123.

[38] A. Garg, J. Han, S.B. Sinnott, Interactions of carbon-nanotubule proximal probe tips with diamond and graphene, Phys. Rev. Lett. 81 (1998) 2–5.

[39] J.P. Lu, Elastic properties of carbon nanotubes and nanoropes, Phys. Rev. Lett. 79 (1997) 1297–1300.

[40] E.W. Wong, P.E. Sheehan, C.M. Lieber, Nanobeam mechanics: elasticity, strength, and toughness of nanorods and nanotubes, Science 80 (277) (1997) 1971–1975.

[41] P. Poncharal, Z.L. Wang, D. Ugarte, W.A. de Heer, Electrostatic deflections and electromechanical resonances of carbon nanotubes, Science 80 (283) (1999) 1513–1516.

[42] J.-P. Salvetat, J.-M. Bonard, N.H. Thomson, A.J. Kulik, L. Forró, W. Benoit, et al., Mechanical properties of carbon nanotubes, Appl. Phys. A 69 (1999) 255–260.

[43] J.P. Lu, J. Han, Carbon nanotubes and nanotube-based nano devices, Int. J. High Speed Electron. Syst. 9 (1998) 101–123.

[44] R. Rafiee, M. Heidarhaei, Investigation of chirality and diameter effects on the Young's modulus of carbon nanotubes using non-linear potentials, Compos. Struct. 94 (2012) 2460–2464.

[45] M.R. Falvo, G.J. Clary, R.M. Taylor, V. Chi, F.P. Brooks, S. Washburn, et al., Bending and buckling of carbon nanotubes under large strain, Nature 389 (1997) 582–584.

[46] S. Iijima, C. Brabec, A. Maiti, J. Bernholc, S. Iijima, Structural flexibility of carbon nanotubes, J. Chem. Phys. 104 (1996) 2089–2092.

[46a] R. Ghasempour, A. Iraji zad, Hybrid multiwalled carbon nanotubes and trioxide tungsten nanoparticles for hydrogen gas sensing, J. Phys. D. Appl. Phys. 42 (2009) 165105–165112.

[47] J. Hone, M.C. Llaguno, N.M. Nemes, a.T. Johnson, J.E. Fischer, D.A. Walters, et al., Electrical and thermal transport properties of magnetically aligned single wall carbon nanotube films, Appl. Phys. Lett. 77 (2000) 666–668.

[48] W. Yi, L. Lu, Z. Dian-Lin, Z. Pan, S. Xie, Linear specific heat of carbon nanotubes, Phys. Rev. B 59 (1999) R9015–R9018.

[49] J. Hone, M. Whitney, C. Piskoti, A. Zettl, Thermal conductivity of single-walled carbon nanotubes, Phys. Rev. B 59 (1999) R2514–R2516.

[50] J. Hone, Phonons and thermal properties of carbon nanotubes, in: M.S. Dresselhau, G. Dresselhaus, P. Avouri (Eds.), Carbon Nanotubes: Synthesis, Structure, Properties, and Applications, Springer, Berlin, Heidelberg, 2001, pp. 273–286.

[51] J.D. Saxby, S.P. Chatfield, A.J. Palmisano, A.M. Vassallo, M.A. Wilson, L.S.K. Pang, Thermogravimetric analysis of buckminsterfullerene and related materials in air, J. Phys. Chem. 96 (1992) 17–18.

[52] C.N.R. Rao, K. Biswas, Characterization of nanomaterials by physical methods, Annu. Rev. Anal. Chem. 2 (2009) 435–462.

[53] M. Zhang, J. Li, Carbon nanotube in different shapes, Mater. Today 12 (2009) 12–18.

[54] R. Andrews, D. Jacques, A.M. Rao, F. Derbyshire, D. Qian, X. Fan, et al. Continuous production of aligned carbon nanotubes: a step closer to commercial realization, Chem. Phys. Lett. 303 (1999) 467–474.

[55] J.W. Seo, L. Forró, S. Saito, Carbon Nanotubes: Quantum Cylinders of Graphene, vol. 3, Elsevier Science, Amsterdam, 2008.

[56] M. Terrones, F. Banhart, N. Grobert, J.-C. Charlier, H. Terrones, P.M. Ajayan, Molecular junctions by joining single-walled carbon nanotubes, Phys. Rev. Lett. 89 (2002) 75505.

[57] A. Jorio, G. Dresselhaus, M.S. Dresselhaus, Carbon Nanotubes: Advanced Topics in the Synthesis, Structure, Properties and Applications, vol. 112, Springer, Berlin, Heidelberg, 2007.

[58] J.H. Hafner, C.L. Cheung, A.T. Wooley, C.M. Lieber, Structural and functional imaging with carbon nanotube AFM probes, Prog. Biophys. Mol. Biol. 77 (1) (2001) 73–110.

[59] J.H. Hafner, C.L. Cheung, C.M. Lieber, Direct growth of single-walled carbon nanotube scanning probe microscopy tips, J. Am. Chem. Soc. 121 (1999) 9750–9751.

[60] M.S. Dresselhaus, A. Jorio, R. Saito, Characterizing graphene, graphite, and carbon nanotubes by Raman spectroscopy, Annu. Rev. Condens. Matter Phys. 1 (2010) 89–108.

[61] M.S. Dresselhaus, G. Dresselhaus, R. Saito, A. Jorio, Raman spectroscopy of carbon nanotubes, Phys. Rep. 409 (2005) 47.

2

Engineering Applications of Carbon Nanotubes

Soheil Jafari

UNIVERSITY OF SUSSEX, BRIGHTON, SUSSEX, UNITED KINGDOM

CHAPTER OUTLINE

2.1 Introduction

Nanotechnology, the sixth revolutionary technology introduced in the modern world, has played an important role in research and development areas from the first days of its emergence in the 1990s. Nanotechnology is a branch of technology that deals with dimensions and tolerances of less than 100 nm, and covers a vast variety of products [1].

In this field, the carbon nanotube (CNT), a carbon made into a tube-shaped nanomaterial, is a unique nanoproduct with a very wide application range because of its strong bonding between atoms, as well as the extreme aspect ratio [2,3]. CNTs, introduced by Iijima in 1991 [4], are allotropes of carbon with a cylindrical nanostructure. Nanotubes have been constructed with a length-to-diameter ratio of up to 132 million, significantly

Carbon Nanotube–Reinforced Polymers. http://dx.doi.org/10.1016/B978-0-323-48221-9.00002-9

larger than any other material. These cylindrical carbon molecules have unusual properties that are valuable for nanotechnology, electronics, optics, and other fields of materials science and technology. CNTs have many structures, differing in length, thickness, and number of layers depending on how the graphene sheet has rolled up to form the tube. Their types and characteristics are normally categorized in Table 2.1 and Fig. 2.1 [5,6].

In particular, owing to their extraordinary thermal conductivity, mechanical, and electrical properties, the literature is replete with various engineering applications of CNTs [7]. These applications are discussed in this chapter under the following topics:

- Structural reinforcement
- Coatings and films
- Electromagnetics
- Biotechnology and biomedicine
- Sensors and actuators
- Acoustics and electroacoustics
- Other applications

Table 2.1 General Categorization of CNTs

CNT Type	SWCNTs	MWCNTs
Manufacturing technique	• Arc discharge • Laser ablation method (catalyst is required for synthesis)	• CVD (can be produced without catalyst)
Young's modulus (GPa)	1054	1200
Diameter range (ID) nm	1.5–15	20–50
Most important applications	• Sensors • Electrodes • Nanodevices	• Metal composites for polymers

CNTs, Carbon nanotubes; CVD, chemical vapor deposition; MWCNTs, multiwalled carbon nanotubes; SWCNTs, single-walled carbon nanotubes.

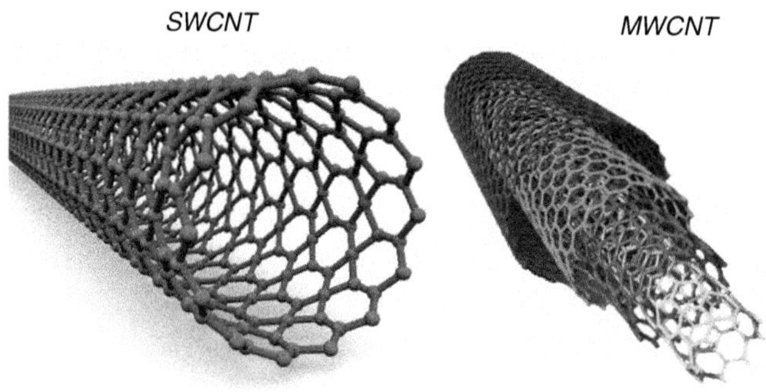

FIGURE 2.1 SWCNTs versus MWCNTs.

It is worthwhile to mention that the applications of CNTs are not limited to the aforementioned fields, and this chapter discusses only the most important engineering applications of CNTs. Worldwide commercial interest in CNTs is reflected in a production capacity that at present exceeds several thousand tons per year.

2.2 Structural Reinforcement

Currently, the most popular use for CNTs is in structural reinforcement. Many structures have been proposed ranging from everyday items (e.g., clothes and sports gear) to combat jackets and space elevators with respect to the CNT's superior mechanical properties [8].

CNTs are excellent for load-bearing reinforcement in composites and also for structural applications because of their mechanical properties. Although thousands of papers on nanotubes are being published every year, commercial applications have been rather slow to develop, primarily because of the high production costs of the best quality nanotubes.

The composite filling is probably the most demanding application of CNTs. CNTs are a promising material as building blocks in hierarchical composite materials given their exceptional mechanical properties. In fact, this application range is vast, ranging from expensive tennis rackets to aircraft body parts and spacecraft.

NASA has invested large amounts of money in developing CNT-based composites for applications, such as the futuristic Mars mission, because experiments have confirmed that nanotubes are one of the stiffest structures ever made [9–13]. In fact, CNTs are considered to be "the ultimate" carbon fiber that has ever been made [14], since recent experiments showed that the nanotubes broke by a sword-in-sheath failure mode (using a nanostressing stage located within a scanning electron microscope) [15]. In addition, the load-bearing reinforcements are very important in composites. Studies on single-walled nanotubes (SWNTs) reveal that they could have a Young's modulus as high as 1 TPa [10], which depends on their plastic behavior [16] and their fracture behavior under tensile loading at high temperature [17].

CNTs are also mixed with polymers and resins [to increase their stiffness, strength, and toughness; e.g., adding ~1 wt.% multiwalled nanotubes (MWNT) to epoxy resin enhances stiffness and fracture toughness by 6 and 23%, respectively, without compromising other mechanical properties [18]] for load-bearing applications [19]. However, this application has some limitations due to the improper interface between the CNTs and the polymer matrix [20–22].

Fig. 2.2 shows some applications of CNT-enhanced resin carbon fiber composites in wind turbine blades and hulls for maritime security boats [23].

In addition, CNT-rich composites have many applications in automotive and aerospace engineering. CNT plastics have been used to enable electrostatically assisted painting of mirror housings, and fuel lines to dissipate electrostatic charge in the automotive industry [24]. Replacing carbon black cathodes with CNTs may have a significant impact on the development of lithium-ion batteries for electric and hybrid vehicles due to an increase in energy density by prevention of electrode degradation, lower impedance, higher discharge capacity, and improved charging cycling.

FIGURE 2.2 (A) Micrograph showing the cross-section of a carbon fiber laminate with CNTs dispersed in the epoxy resin; (B) a lightweight CNT-fiber composite boat hull for maritime security boats. (C) CNT-reinforced polyurethane blade obtained by vacuum bag technique (World's First Carbon Nanotube Reinforced Polyurethane Wind Blades, Case Western Reserve University, Cleveland, Ohio) [23]. *Part A and B, photo courtesy Zyvex Technologies.*

Also, adding CNTs to aerospace materials could protect them against lightning strikes. In a recent research study, MIT experts have strengthened these advanced aerospace materials with what they call "nanostitching." Rows of CNTs perpendicular to the carbon microfibers fill the spaces between them, reinforcing the fiber layers without piercing them. The researchers showed that these materials are not only 10 times stronger than those that don't contain nanotubes, but they are also more than 1 million times more electrically conductive, which suggests that they might protect aircraft from lightning strikes [25].

Also, high tensile strength and electrical conductivity of CNTs are used in developing CNT–metal matrix composites (MMCs). This material is still in the research phase, but the potential areas of application are nanonetworks, nanorobotics, and future soldiers.

Some of the other important structural applications of the CNTs are:

- CNTs can be used in reinforcement of turbine blades to enhance the strength of marine current turbines that must survive in the ocean [26–28].
- CNTs in concrete increase its tensile strength and halt crack propagation.
- CNTs may be able to replace steel in suspension bridges.
- The high strength-to-weight ratios of CNTs enable very high rotational speeds [26].
- Composites that use CNTs as secondary fillers are already commercialized (e.g., cranks, handlebars, front forks, etc., of road racing bikes from various manufacturers).
- Secondary filler CNT composites are also used in ice hockey sticks for durability and controllability [27].

In such applications, MWCNTs are usually used for the increase in compressive strength, and carbon fiber contributes to the tensile strength.

2.3 Coatings and Films Applications of CNTs

CNTs are used widely in marine current turbine coatings to reduce the effect of biofouling due to their nanostructure and hydrophobic nature [12] (Fig. 2.3). Since they can discourage attachment of algae and barnacles, they are a possible alternative to environmentally hazardous biocide-containing paints [28].

CNT-based transparent conducting films can also be used as an alternative to indium tin oxide (ITO) for the following reasons [29]:

- The indium is scarce, and the demand for touch-screen devices is high and growing.
- The CNT transparent conductors are very flexible.

Other coating applications of the CNTs (e.g., thin-film heaters, flame-retardant coating, transparent conductors, and electrostatic dissipative windows) are being pursued industrially [30].

A comprehensive experimental study on efficient coating of transparent and conductive CNT thin films on plastic substrates was done in 2008 in the Institute of Materials Research and Engineering in Singapore to discuss the adhesion promoter role and surfactant in the coating process. The researchers used a dip-coating technique to fabricate SWNT thin films. Their successful results proposed a solution processing technique based on the dip-coating of substrates in a SWNT mixture [31].

CNT-based epoxy primers containing zinc dust were demonstrated in 2009 with the US Army Corps of Engineers at Fort Bragg, NC, as a potential replacement for the traditional three-coat, military-specification systems.

Finally, another industrial application was presented in a white paper in 2013 by Charlie Simpson. He discussed the enhanced strength and electrical conductivity of a newly developed CNT-based anticorrosion coating. The idea was to incorporate sacrificial metal

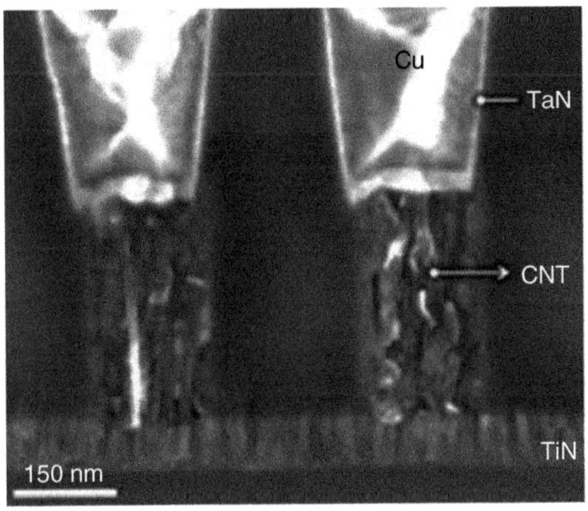

FIGURE 2.3 CMOS-compatible 150 nm vertical interconnects developed by imec and Tokyo Electron Limited.
Photo courtesy of imec.

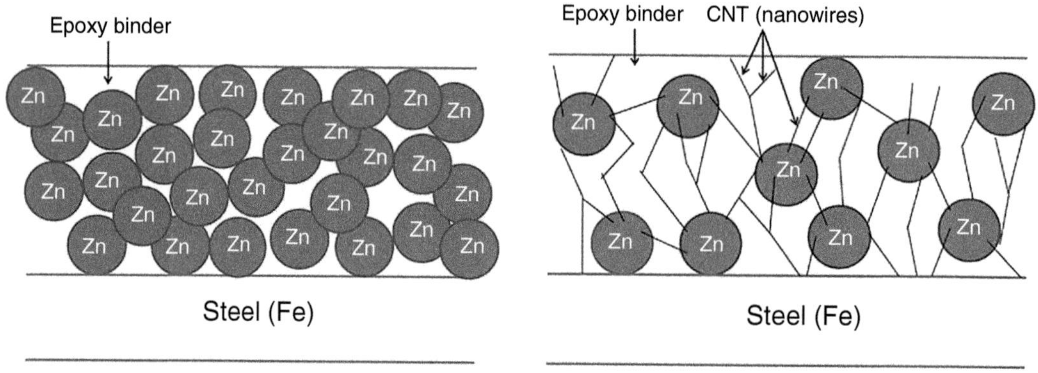

FIGURE 2.4 Conventional zinc-rich epoxy primer CNT-modified zinc-rich epoxy primer [32].

particles (e.g., Zn, Al, Mg) for corrosion inhibition via cathodic protection. The advantage of the described coating was that it could be applied using standard painting equipment (conventional and high-volume, low-pressure [HVLP] air spray, airless spray, brush, roller, etc.). Fig. 2.4 compares the structure of a conventional coating with that of a CNT-modified one [32].

2.4 CNTs in Electromagnetics

Another interesting field of CNT applications is use as electrical conductors, semiconductors, and insulators.

The light weight and remarkable electrical conductivity of CNTs enable them to be used as heat sinks for chipboards, as LCD screen backlights, and even as electrical protection in Faraday cages [33].

The electromagnetic parameters of a high CNT loading film have been characterized in recent years by analyzing the layer number of CNT laminate, CNT alignment, and resin impregnation effects. The results of experiments show:

- the anisotropic electromagnetic characteristic of CNT film,
- increase of complex permittivity and decrease of shielding effectiveness with increasing layer number, and
- decrease of conductivity and shielding effectiveness with resin impregnation into CNT film.

Taking the aforementioned characteristics into account, Akyildiz and Jornet (2010) discussed the possibility of using CNTs in several ways, such as [34]:

- as an alternative to tungsten filaments,
- in magnetic fields development using coated MWCNTs, and
- as antennae for electromagnetic devices.

Due to their light weight, durability, conductivity, and negligible skin effect at high frequencies, it is also reported that CNT diodes are used to replace the ITO in solar cells because of their photovoltaic effect.

In 2009, Park et al. reported on a new principle yielding enhanced electromagnetic shielding, using CNT integrated with a reactive ethylene terpolymer (RET) with the luxury of incorporating reasonable dispersion with low electrical percolation volume fractions (~0.1 vol.%), yielding outstanding microwave shielding efficiency for electromagnetic interference (EMI) applications. They showed experimentally that both single-walled CNTs (SWCNT) and multiwalled CNT (MWCNT) based composites could be used for this purpose. However, their preliminary experiments confirmed that SWNTs were much more effective than were MWNTs [35].

Finally, in 2015 CNT film supported by a Teflon substrate was designed to meet the requirement of wave guide measurement for the characterization of electromagnetic parameters of CNT film by Zhang et al. Their experiments showed extraordinary shielding properties for a five-layer CNT film with shielding effectiveness in the range of 67–78 dB in X-band, which is an interesting area of study for the researchers in the field [36].

2.5 Biotechnological and Biomedical Applications of CNTs

CNTs also have interesting applications in biotechnology and biomedicine thanks to the dimensional and chemical compatibility of CNTs with biomolecules, such as DNA and proteins. Recent studies have shown the selective cancer cell killing ability obtained by hyperthermia due to the thermal conductivity of CNTs [37]. This application is very practical because of the invaluable importance of cancer detection at early stages. The CNTs' potential to enter and analyze single cells could satisfy this requirement [38]. Moreover, some research work confirmed the potential of CNTs for use in drug delivery systems [39].

In 2013, Hau He et al. summarized the main applications of CNTs in pharmacy and medicine as drug, biomolecule, gene delivery to cells or organs, tissue regeneration, and biosensor diagnostics and analysis [40].

The most interesting applications of CNTs in biomedical can be summarized as follows:

- *CNTs for cancer therapy*: This application is usually done by drug delivery, antitumor immunotherapy, and local antitumor hyperthermia therapy. The complete medical details of this application can be found in [41–48].
- *CNTs for infection therapy*: The issue that the CNTs could deal with in this field is the resistance of infectious agents against numerous antiviral substances. Recent studies show that CNTs are able to act as carriers for antimicrobial agents, such as the antifungal amphotericin B. Functionalized CNTs can also act as vaccine delivery procedures [49,50].
- *CNTs for gene therapy by DNA delivery*: Some delivery systems for DNA transfer include liposomes, cationic lipids, and nanoparticles, such as CNTs (recently discovered) [41,51].

- *CNTs for tissue regeneration and artificial implants*: CNTs may be the best tissue engineering candidates for tissue scaffolds because of their biocompatibility and resistance to biodegradation [52–54].
- *CNTs for neurodegenerative diseases and Alzheimer syndrome*: Owing to their tiny dimensions, CNTs are able to cross the blood–brain barrier by various targeting mechanisms for acting as effective delivery carriers for the target brain and many neuroscience applications [55,56].
- *CNTs as antioxidants*: Over the past 2 decades the theory of oxygen-free radicals has been under consideration by researchers. Nevertheless, the potential role of CNTs as free-radical scavengers is still an emerging area of research [57].
- *CNTs as biosensor vehicles for diagnostics and detection*: Many researchers have coupled CNTs with glucose–oxidase biosensors for blood sugar control in diabetic patients with higher accuracy and simpler manipulation than biosensors used alone [43,58].
- *CNTs for solid phase extraction of drugs and biochemicals*: CNTs' surfaces have very good adsorption ability due to their strong interaction with other molecules, so they are investigated as solid phase extraction (SPE) adsorbents used alone or in conjugation with classic SPE sorbents (C18 silica, XAD-2 copolymer) for the analytical extraction of drugs, pesticides, or natural compounds in different media, such as biological fluids, drug preparations, environment, plants, animal organs, and so forth [59,60].
- CNTs also enable fluorescent [61] and photoacoustic imaging [62], as well as localized heating using near-infrared radiation [63]. CNT sensors are used in the food industry and in environmental applications for gas and toxin detection.

Finally, in 2015, a comprehensive and interesting study was done in Poland at Lodz University of Technology and the University of Life Sciences in Lublin on using the magnetic property of CNTs for targeted thermal ablation of cancer cells. The main idea was discussion on the influence of the electromagnetic field on human body tissues by means of a liquid containing magnetic particles. The numerical and experimental results of this study illustrated the potential of CNTs for hyperthermia treatment [61].

2.6 Sensors and Actuators Applications of CNTs

Thanks to their excellent thermal and electrical conductivity, CNTs are widely used in sensors and actuators development. Since the electrical conductivity of SWNTs changes when it encounters oxygen, NH_3, and nitrogen, they could be used in the manufacturing of chemical sensors with fast response time and small dimensions. CNTs are also used in nanoprobe application for different purposes, such as drug delivery, nanoelectrodes, high-resolution imaging, sensors, field-emitting devices, electrostatic force microscopes, and so on [64].

In 2013, Technical University of Munich (TUM) researchers in Germany reported on their new findings to show the way toward low-cost, industrial-scale sensors using CNTs. They introduced a new gas sensor for indicating the freshness level of food packages and for use in wireless air-quality monitors as their leading examples. Moreover, new types of solar cells, and pressure and temperature sensors that could be built into electronic skin for robotic or bionic applications were also discussed in their white paper. The technology used is shown and described in Fig. 2.5.

The initial experiments on the developed gas sensors at TUM show the unique characteristics of this technology, such as ability to rapidly detect and continuously respond to extremely small changes in the concentrations of gases, including ammonia, carbon dioxide, and nitrogen oxide. Such devices can also be fabricated on flexible backing materials through large-area, low-cost processes [65].

In addition, taking the low electron scattering, as well as the band gap of CNTs into account, the high-quality SWNTs are attractive candidates for transistors.

University of Wisconsin–Madison materials engineers have created CNT transistors that outperform state-of-the-art silicon transistors for the first time [66]. Making CNT transistors that are better than silicon transistors is a big milestone toward exploiting CNTs in logic, high-speed communications, and other semiconductor electronics technologies. Since the mobility of CNT thin-film transistors (TFTs) is higher than amorphous silicon, they are used for driving organic light-emitting diodes (OLEDs).

The most recent progress in this field was reported by IBM in 2015 regarding a breakthrough on CNT transistors. IBM researchers say they have overcome the major manufacturing challenge of how to combine nanotubes with the metal contacts that deliver electric current, which has been standing in the way of commercially viable nanotube devices [67]. They addressed the issue by changing the interface between a nanotube and the two metal contacts.

FIGURE 2.5 Flexible, high-performance gas sensors (*left*) were made by spraying a solution of CNTs (*right*) onto a plastic backing. *Image: Heath Group, Caltech.*

2.7 Acoustic and Electroacoustic Applications of CNTs

The natural frequency of CNTs could be tuned widely. This characteristic is used in tunable radio-frequency components, such as nanotransceivers and tunable filters [68]. With respect to this property, sheets of parallel CNTs are used in both acoustic actuators (speakers) and sensors (microphones). These thin-film transducers show excellent acoustic response over a broadband frequency range and have the advantages of being transparent, flexible, extremely thin, and lightweight.

Wearable speakers (nanospeakers) will be one of the future applications of CNTs that could be made from ultrathin, flexible nanotube sheets (which is ongoing by Chinese researchers) [69]. The working mechanism of this device is to heat the air by applying electric current to the tubes. This causes a thermoacoustic effect that results in creating sound waves (like thunder). Unlike conventional speakers, nanospeakers do not emit vibration at all. So will iPods be replaced by nanospeaker-infused jackets? More detail can be found in Ref. [70].

Moreover, recent investigations illustrate that CNTs arrays have an interesting acoustic absorption. The results of experimental studies in the University of Adelaide, Australia, highlight the potential of nanoscopic fibers for use as lightweight acoustic absorbers by quantification of their acoustic characteristics and absorption performance in comparison with conventional porous materials [71]. The researchers showed that a forest of 3 mm CNT arrays can provide as much as 5%–10% acoustic absorption capability within the frequency range 125 Hz–4 kHz. Fig. 2.6 shows the results obtained by that study about the absorption coefficient measurements of CNTs.

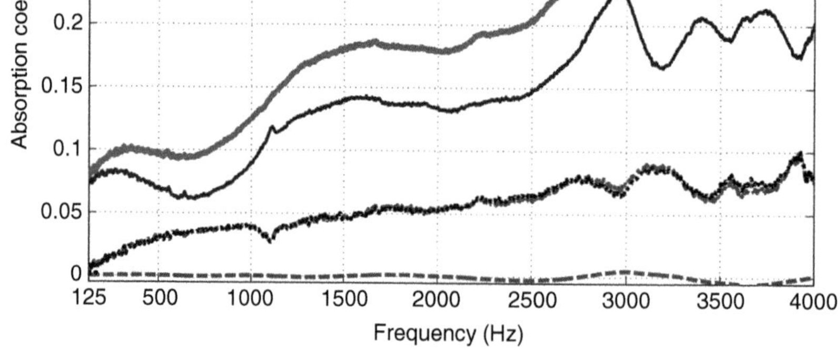

FIGURE 2.6 Measured and corrected normal incidence sound absorption coefficient of CNT forest sample compared with that of tube with rigid termination [71].

The potential use of CNTs in manufacturing higher-speed oscillators (>50 GHz) has been reported by researchers as well [72].

2.8 Other Applications of CNTs

There are many other applications of CNTs in the engineering field. Some of them, along with short descriptions, are presented in this section.

2.8.1 Hydrogen Storage

Methods of hydrogen storage for subsequent use span many approaches, including high pressures, cryogenics, and chemical compounds. These methods are also useful to provide grid energy storage for intermittent energy sources, like wind power, as well as providing fuel for transportation, particularly for ships and airplanes.

There are very promising experiments in the literature clarifying high hydrogen storage capacities in different carbon nanostructures due to their nanosize, as well as cylindrical structure. Some of the reported results are shown in Table 2.2.

In addition to experiments, continuum mechanics models are also generated in the literature for hydrogen storage in SWCNTs and MWCNTs and are validated by atomistic simulations to categorize and analyze CNTs from the hydrogen storage point of view.

2.8.2 Water Treatment

Water treatment/purification is one of the most fascinating applications of CNTs. In recent years, filters containing CNTs have been commercialized because CNTs provide the network with controlled nanoscale porosity [77]. This reduces the energy cost for water desalination in comparison with the commercial filtration by enhancing the permeability [78,79] due to the unfavorable energy barrier that would have to be overcome in order to desolvate the ions, as the hydrated ions are often larger than the diameter of the nanotubes.

Table 2.2 Some of the Reported Storage Capacities of Hydrogen in CNTs

Material	Discharge Capacity (mAh/g)	Stored Hydrogen (wt.%)	References
SWNT	110	0.39	[73]
SWNT	800	2.9	[74]
MWNT	297	1.051	[75]
SWNT	<141	<0.5	[76]
Aligned SWNT	503	1.84	

MWCNTs, Multiwalled nanotubes; SWNT, single-walled nanotubes.

2.8.3 Textiles

In order to transform a cotton fabric surface to be superhydrophobic, CNTs can be used by a dip-coating approach. The fabrics made by this technique have high stability and chemical durability because CNTs are covalently attached on the surface [29]. Recently, researchers have coated conventional cotton thread with highly conductive, biosensing CNTs. The threads can be woven into fabrics that are lightweight and wearable but act as simple, sensitive sensors that can, among other functions, detect human blood.

2.8.4 Catalysts

CNTs can be used as catalysts (to improve the rate of a reaction) because of their large specific surface areas and excellent electron conductivity incorporated with the good chemical inertness and relatively high- oxidation stability [80].

2.8.5 Photovoltaics

Recent studies show that SWCNTs are ideal as thin-film photovoltaics due to their excellent absorption of light across a wide range of wavelengths (from visible to near infrared). Organic photovoltaic devices are fabricated from thin films of organic semiconductors, such as polymers and small molecule compounds and are typically on the order of 100 nm thick [81].

2.9 Conclusions

The main engineering applications of CNTs are discussed in this chapter. As described previously, the growing demand for nanotechnology is penetrating various end-use markets for CNTs. Some applications, such as composite filling and biotechnology are well-established now, whereas some others (e.g., textile applications and acoustic devices) are still in their infancy period and need more research and studies. It seems that in the near future, with new technologies and developments in manufacturing processes and costs, many more engineering applications of CNTs will emerge, as well as expansion of the current applications. Especially, with huge investments in renewable energy technologies all around the world, the role of CNTs in wind turbines, solar panels, electromagnetics, sensors, and actuators would be vital in this regard.

References

[1] R.K. Naidu, N.V. Pulagara, R.S. Dondapati, Carbon nanotubes in engineering applications: a review, Prog. Nanotechnol. Nanomater. 3 (4) (2014) 79–82.

[2] W. Zhu, P.J.M. Bartos, A. Porro, Application of nano technology in construction, Mater. Struct. 37 (2004) 649–658.

[3] N. Barkalina, C. Charalambous, C. Jones, K. Coward, Nanotechnology in reproductive medicine: emerging applications of nanomaterials, Nanomed.Nanotechnol. 10 (5) (2014) 921–938.

[4] S. Iijima, Helical microtubules of graphitic carbon, Nature 354 (1991) 56–58.

[5] G. Mittal, V. Dhand, K.Y. Rhee, S.J. Park, W.R. Lee, A review on carbon nanotubes and graphene as fillers in reinforced polymer, J. Indust. Eng. Chem. 21 (2014) 11–25.

[6] Y.A. Minaev, Thermochemical nanotechnology in metallurgy and machine building, Metallurgist 55 (2012) 771–779 Russian original nos. 11–12, November–December, 2011.

[7] Y.A. Kim, T. Hayashi, M. Endo, M.S. Dresselhaus, Carbon nanofibres, in: R. Vajtai (Ed.), Springer Handbook of Nanomaterials, Springer-Verlag, Berlin, Heidelberg, 2013, pp. 233–262.

[8] B.C. Edwards, E.A. Westling, The Space Elevator, Spageo Inc., San Francisco, (2003).

[9] G. Overney, W. Zhong, D. Tomanek, Structural rigidity and low frequency vibational modes of long carbon tubules, Z. Phys. D 27 (93) (1993) 406.

[10] B.I. Yakobson, C.J. Brabec, J. Bernholc, Nanomechanics of carbon tubes: instabilities beyond linear response, Phys. Rev. Lett. 76 (1996) 2511–2514.

[11] M.M.J. Treacy, T.W. Ebbesen, J.M. Gibson, Exceptionally high Young's modulus observed for individual carbon nanotubes, Nature 381 (1996) 678–680.

[12] E.W. Wong, P.E. Sheehan, C.M. Lieber, Nanobeam mechanics: elasticity, strength, and toughness of nanorods and nanotubes, Science 277 (1997) 1971–1975.

[13] B. Yakobson, Ph. Avouris, Mechanical properties of carbon nanotubes, in: M.S. Dresselhaus, Ph. Avouris (Eds.), Carbon Nanotubes, Springer-Verlag, Berlin, Heidelberg, 2001, pp. 293.

[14] M. Pulickel, M. Ajayan, O.Z. Zhou, Applications of carbon nanotubes, Carbon Nanotubes, Top. Appl. Phys. 80 (2001) 391–425.

[15] M.S. Dresselhaus, G. Dresselhaus, K. Sugihara, I.L. Spain, H.A. Goldberg, Graphite Fibers and Filaments, Springer, Berlin, Heidelberg, (1988).

[16] B.I. Yakobson, Mechanical relaxation and "intramolecular plasticity" in carbon nanotubes, Appl. Phys. Lett. 72 (1998) 918–920.

[17] B.I. Yakobson, C.J. Brabec, J. Bernholc, Structural mechanics of carbon nanotubes: from continuum elasticity to atomistic fracture, J. Comput.-Aided Mater. 3 (1996) 173–182.

[18] R.C. Jatl, K. Arora, N. Anuragi, R. Kaushal, V. Jain, Carbon nanotubes: advancement in science and technology, World J. Pharm. Pharm. Sci. 4 (07) (2015) 1533–1550.

[19] T.W. Chou, L. Gao, E.T. Thostenson, Z. Zhang, J.H. Byun, An assessment of the science and technology of carbon nanotube-based fibers and composites, Compos. Sci. Technol. 70 (2010) 1–19.

[20] P.M. Ajayan, O. Stephan, C. Colliex, D. Trauth, Aligned carbon nanotube arrays formed by cutting a polymer resin–nanotube composite, Science 265 (1994) 1212–1214.

[21] H.D. Wagner, O. Lourie, Y. Feldman, R. Tenne, Stress-induced fragmentation of multiwall carbon nanotubes in a polymer matrix, Appl. Phys. Lett. 72 (1998) 188–190.

[22] L. Jin, C. Bower, O. Zhou, Alignment of carbon nanotubes in a polymer matrix by mechanical stretching, Appl. Phys. Lett. 73 (1998) 1197–1199.

[23] J.N. Coleman, U. Khan, W.J. Blau, Y.K. Gun'ko, Small but strong: a review of the mechanical properties of carbon nanotube–polymer composites, Carbon 44 (2006) 1624.

[24] T.-W. Chou, L. Gao, E.T. Thostenson, Z. Zhang, J.-H. Byun, An assessment of the science and technology of carbon nanotube-based fibers and composites, Compos. Sci. Technol. 70 (1) (2010) 1–19.

[25] Elizabeth A. Thomson, MIT News Office, (March 5, 2009). Available from: http://news.mit.edu/2009/nanostitching-could-lead-much-stronger-airplane-skins-more.

[26] W.P. Gibson, C. Lee, P.F. Ko, D. Reneker, Application of nanofiber technology to non-woven thermal insulation, J. Eng. Fiber. Fabr. 2 (2) (2007) 32–40.

[27] Y. Kong, T. Qiu, J. Qiu, Fabrication of novel micro nano carbonous composites based on self-made hollow activated carbon fibers, Appl. Surf. Sci. 265 (2013) 352–357.

[28] K.-W. Ng, W.-H. Lam, S. Pichiah, A review on potential applications of carbon nanotubes in marine current turbines, Renew. Sust. Energ. Rev. 28 (2013) 331–339.

[29] A. Jorio, M.S. Dresselhaus, G. Dresselhaus, Carbon Nanotubes: Advanced Topics in the Synthesis, Structure, Properties and Applications, Springer-Verlag Berlin Heidelberg, Heidelberg, Germany, (2008) Series Vol. 111.

[30] M. Endo, Y.A. Kim, T. Hayashi, Y. Fukai, K. Oshida, M. Terrones, Structural characterization of cup-stacked-type nanofibers with an entirely hollow core, Appl. Phys. Lett. 80 (2002) 1267–1269.

[31] A. Beigbeder, Preparation and characterisation of silicone-based coatings filled with carbon nanotubes and natural sepiolite and their application as marine fouling-release coatings, Biofouling 24 (2008) 291.

[32] F. Michael, L. De Volder, H. Sameh, R. Tawfick, H. Baughman, A. John Hart, Carbon nanotubes: present and future commercial applications, Science 339 (1) (2013) 535–539.

[33] Z. Wu, Transparent, conductive carbon nanotube films, Science 305 (2004) 1273.

[34] I.F. Akyildiz, J.M. Jornet, Electromagnetic wireless nanosensor networks, J. Nano Commun. Netw. 1 (1) (2010) 3–19.

[35] S.H. Park, P. Theilmann, P. Asbeck, P.R. Bandaru, Enhanced electromagnetic interference shielding through the use of functionalized carbon nanotube-reactive polymer composites, IEEE, TNANO-00013-2009, pp. 1–6.

[36] W. Zhang, H. Xiong, S. Wang, M. Li, Y. Gu, Electromagnetic characteristics of carbon nanotube film materials, Chinese J. Aeronaut. 28 (4) (2015) 1245–1254.

[37] N.W.S. Kam, M. O'Connell, J.A. Wisdom, H.J. Dai, Carbon nanotubes as multifunctional biological transporters and near infrared agents for selective cancer cell destruction, Proc. Natl. Acad. Sci. USA 102 (2005) 11600–11605.

[38] P.R. Bandaru, Electrical properties and applications of carbon nanotube structures, J. Nanosci. Nanotechnol. 7 (2007) 1–29.

[39] W. Sebastien, W. Giorgia, P. Monica, B. Cedric, B. Renato, Targeted delivery of amphotericin b to cells by using functionalized carbon nanotubes, Angew. Chem. 117 (2005) 6516–6520.

[40] H. He, L.A. Pham-Huy, P. Dramou, D. Xiao, P. Zuo, C. Pham-Huy, Carbon nanotubes: applications in pharmacy and medicine, BioMed Res. Int. 2013 (2013) 1–12.

[41] H. Liao, B. Paratala, B. Sitharaman, Y. Wang, Applications of carbon nanotubes in biomedical studies, Methods Mol. Biol. 726 (2011) 223–241.

[42] Z. Liu, S. Tabakman, K. Welsher, H. Dai, Carbon nanotubes in biology and medicine: in vitro and in vivo detection imaging and drug delivery, Nano Res. 2 (2) (2009) 85–120.

[43] M.S. Digge, R.S. Moon, S.G. Gattani, Applications of carbon nanotubes in drug delivery: a review, Int. J. PharmTech Res. 4 (2) (2012) 839–847.

[44] W. Yang, P. Thordarson, J.J. Gooding, S.P. Ringer, F. Braet, Carbon nanotubes for biological and biomedical applications, Nanotechnology 18 (2007) 1–12.

[45] K.T. Al-Jamal, H. Nerl, K.H. Müller, et al. Cellular uptake mechanisms of functionalised multi-walled carbon nanotubes by 3D electron tomography imaging, Nanoscale 3 (6) (2011) 2627–2635.

[46] S.Y. Madani, N. Naderi, O. Dissanayake, A. Tan, A.M. Seifalian, A new era of cancer treatment: carbon nanotubes as drug delivery tools, Int. J. Nanomed. 6 (2011) 2963–2979.

[47] C.L. Lay, J. Liu, Y. Liu, Functionalized carbon nanotubes for anticancer drug delivery, Expert Rev. Med. Devic. 8 (5) (2011) 561–566.

[48] A.M.A. Elhissi, W. Ahmed, I.U. Hassan, V.R. Dhanak, A. D'Emanuele, Carbon nanotubes in cancer therapy and drug delivery, J. Drug Deliv. 2012 (2012) 1–10.

[49] Y. Rosen, N.M. Elman, Carbon nanotubes in drug delivery: focus on infectious diseases, Expert Opin. Drug Deliv. 6 (5) (2009) 517–530.

[50] Y. Rosen, B. Mattix, A. Rao, F. Alexis, Carbon nanotubes and infectious diseases, in: R.J. Hunter (Ed.), Nanomedicine in Health and Disease, Science Publishers, London, UK, 2011, pp. 249–267.

[51] E. Bekyarova, Y. Ni, E.B. Malarkey, et al. Applications of carbon nanotubes in biotechnology and biomedicine, J. Biomed. Nanotechnol. 1 (1) (2005) 3–17.

[52] Y. Usui, H. Haniu, S. Tsuruoka, N. Saito, Carbon nanotubes innovate on medical technology, Med. Chem. 2 (1) (2012) 1–6.

[53] Y. Zhang, Y. Bai, B. Yan, Functionalized carbon nanotubes for potential medicinal applications, Drug Discov. Today 15 (11–12) (2010) 428–435.

[54] B. Kateb, V. Yamamoto, D. Alizadeh, et al. Multi-walled carbon nanotube (MWCNT) synthesis, preparation, labeling, and functionalization, Methods Mol. Biol. 651 (2010) 307–317.

[55] B.G.P. Singh, C. Baburao, V. Pispati, et al. Carbon nanotubes: a novel drug delivery system, Int. J. Res. Pharm. Chem. 2 (2) (2012) 523–532.

[56] R.A. MacDonald, B.F. Laurenzi, G. Viswanathan, P.M. Ajayan, J.P. Stegemann, Collagen–carbon nanotube composite materials as scaffolds in tissue engineering, J. Biomed. Mater. Res. 74 (3) (2005) 489–496.

[57] L.A. Pham-Huy, H. He, C. Pham-Huy, Free radicals, antioxidants in disease and health, Int. J. Biomed. Sci. 4 (2) (2008) 89–96.

[58] J. Wang, Carbon-nanotube based electrochemical biosensors: a review, Electroanalysis 17 (1) (2005) 7–14.

[59] A.H. El-Sheikh, J.A. Sweileh, Recent applications of carbon nanotubes in solid phase extraction and preconcentration: a review, Jordan J. Chem. 6 (1) (2011) 1–16.

[60] D. Xiao, P. Dramou, N. Xiong, et al. Development of novel molecularly imprinted magnetic solid-phase extraction materials based on magnetic carbon nanotubes and their application for the determination of gatifloxacin in serum samples coupled with high performance liquid chromatography, J. Chromatogr. 1274 (2013) 44–53.

[61] D.A. Heller, S. Baik, T.E. Eurell, M.S. Strano, Single-walled carbon nanotube spectroscopy in live cells: towards long-term labels and optical sensors, Adv. Mater. 17 (2005) 2793.

[62] A. De La Zerda, et al. Carbon nanotubes as photoacoustic molecular imaging agents in living mice, Nat. Nanotechnol. 3 (2008) 557.

[63] N.W.S. Kam, M. O'Connell, J.A. Wisdom, H.J. Dai, Carbon nanotubes as multifunctional biological transporters and near-infrared agents for selective cancer cell destruction, Proc. Natl. Acad. Sci. USA 102 (2005) 11600.

[64] H.J. Dai, J.H. Hafner, A.G. Rinzler, D.T. Colbert, R.E. Smalley, Nanotubes as nanoprobes in scanning probe microscopy, Nature 384 (1996) 147–150.

[65] Technische Universitaet Muenchen, With carbon nanotubes, a path to flexible, low-cost sensors, ScienceDaily, (September 25, 2013). Available from: http://phys.org/news/2013-09-carbon-nanotubes-path-flexible-low-cost.html.

[66] D.M. Sun, et al. Flexible high-performance carbon nanotube integrated circuits, Nat. Nanotechnol. 6 (2011) 156.

[67] M. Orcutt, IBM reports breakthrough on carbon nanotube transistors, MIT Technol. Rev. (October 1, 2015). Available from: https://www.technologyreview.com/s/541921/ibm-reports-breakthrough-on-carbon-nanotube-transistors/.

[68] S.W. Lee, E.B. Campbell, Nanoelectromechanical devices with carbon nanotubes, Curr. Appl. Phys. 13 (2013) 1844–1859.

[69] L. Stephens, 3 Futuristic uses of carbon nanotubes (March 24, 2016). Available from: http://interestingengineering.com/3-amazing-uses-carbon-nanotubes/.

[70] L. Xiao, Z. Chen, C. Feng, L. Liu, Z. Bai, Y. Wang, L. Qian, Y. Zhang, Q. Li, K. Jiang, S. Fan, Flexible, stretchable, transparent carbon nanotube thin film loudspeakers, Nano Lett. 8 (12) (2008) 4539–4545.

[71] M. Ayub, A.C. Zande, Q.C. Howard, B.S. Cazzolato, V.N. Shanov, N.T. Alvarez, D.M. Huang, Acoustic absorption behaviour of carbon nanotube arrays, Proceedings of Internoise 2014, Melbourne, Australia.

[72] X. Jia, Controlled formation of sharp zigzag and armchair edges in graphene nanoribbons, Science 323 (2009) 1701–1705.

[73] C. Nutzenadel, A. Zuttel, D. Chartouni, Electrochemical storage of hydrogen in nanotube materials, Electrochem. Solid-State Lett. 2 (1999) 30–32.

[74] N. Rajalakshmi, K.S. Dhathathreyan, A.G. Raj, B.C.S. Kumar, Electrochemical investigation of single-walled carbon nanotubes for hydrogen storage, Electrochim. Acta 45 (2000) 4511–4515.

[75] X.P. Gao, Y. Lan, G.L. Pan, F. Wu, J.Q. Qu, Electrochemical hydrogen storage by carbon nanotubes decorated with metallic nickel, Electrochem. Solid-State Lett. 4 (2001) A173–A175.

[76] G.P. Dai, C. Liu, M. Liu, M.Z. Wang, H.M. Cheng, Electrochemical hydrogen storage behavior of ropes of aligned single-walled carbon nanotubes, Nano Lett. 2 (2002) 503–506.

[77] M.S. Rahaman, C.D. Vecitis, M. Elimelech, Electrochemical carbon nanotube filter performance toward virus removal and inactivation in the presence of natural organic matter, Environ. Sci. Technol. 46 (2012) 1556–1564.

[78] J.K. Holt, H.G. Park, Y. Wang, M. Stadermann, A.B. Artyukhin, Fast mass transport through sub-2-nanometer carbon nanotubes, Science 312 (2006) 1034–1037.

[79] B. Corry, Designing carbon nanotube membranes for efficient water desalination, J. Phys. Chem. B 112 (2008) 1427–1434.

[80] S. Vadukumpully, J. Paul, N. Mahanta, S. Valiyaveettil, Flexible conductive graphene/poly(vinyl chloride) composite thin films with high mechanical strength and thermal stability, Carbon 49 (2011) 198–205.

[81] I. Robel, B.A. Bunker, P.V. Kamat, Single-walled carbon nanotube–CdS nanocomposites as light-harvesting assemblies: photoinduced charge-transfer interactions, Adv. Mater. 17 (20) (2005) 2458–2463.

Further Reading

G. Raniszewski, A. Miaskowski, S. Wiak, The application of carbon nanotubes in magnetic fluid hyperthermia, J. Nanomater. 2015 (2015) 1–8.

3

Carbon Nanotubes Processing

Reza Malekimoghadam, Roham Rafiee
UNIVERSITY OF TEHRAN, TEHRAN, IRAN

CHAPTER OUTLINE

3.1 Introduction

The exploiting carbon nanotubes (CNTs) in all sectors and industries encompassing nanoelectronics and sensors, nanocomposites, aviation, automotive, and so on, require reliable growth techniques of CNTs capable of yielding high-purity material in desirable quantities. Multifarious methods have been exploited and developed to grow CNTs, such as laser ablation [1], arc discharge [2], and chemical vapor deposition (CVD) [3]. Oberlin et al. clearly demonstrated a hollow carbon fiber with nanometer-scale diameters employing a vapor-growth technique [4]. Mats of small fibers were found by Wiles and Abrahamson on one electrode when sparks were passed between two graphite electrodes [5]. Extraordinary contributions to the race for devising methods for the production of carbon nanotubes were made by laser-ablation synthesis of bundles of aligned single-walled nanotubes (SWNTs) with small diameter distribution by Guo et al. at Rice University [6] and by catalytic growth of nanotubes by the CVD method by Yacaman et al. [7]. Methane CVD was developed by Dai's group, which reported the synthesis of bulk amounts of SWNTs by CVD from methane at 900°C [8]. Su et al. significantly improved the yield of this method using Al_2O_3 aerogels impregnated with Fe/Mo nanoparticles as a catalyst [9]. Several groups exploited other hydrocarbons and catalysts to prepare SWNTs. SWNTs using an extremely small amount of C_2H_4 diluted by Ar and an Fe/Mo bimetallic catalyst with an Al_2O_3 support were prepared by Hafner et al. [10]. As a result, single- and double-walled

Carbon Nanotube–Reinforced Polymers. http://dx.doi.org/10.1016/B978-0-323-48221-9.00003-0

nanotubes were observed for reaction temperatures from 700 to 750°C. Two of the earliest techniques for successful CNT growth are synthesis and laser ablation. Of these two, laser ablation is not amenable for large-scale production, but the arc process is suitable for scale-up to meet the material needs of bulk applications, such as composites. The figure of merit for the growth process catering to structural applications is the tons-a-day production ability, in addition to high purity. In contrast, applications, such as nanoelectronics and field emission may require controlled growth on patterned substrates at reasonable rates. This necessity is provided by CVD and related techniques, which are stratified according to the energy source. When a conventional heat source, such as a resistive or inductive heater, furnace, or infrared lamp is used, the technique is called thermal CVD. It should be mentioned that in the literature the term CVD without any prefix commonly refers to thermal CVD, whereas plasma-enhanced CVD (PECVD), refers to the case where a plasma source is invoked to create a glow discharge. It is worth mentioning that carbon nanotubes produced from carbon vapor generated by arc discharge or laser ablation of graphite generally have fewer structural defects than those produced by other known techniques due to the higher growth process temperature that ensures perfect annealing of defects in tubular graphene sheets [11]. Multiwalled nanotubes (MWNTs) produced by these high-temperature methods are perfectly straight, in contrast to defective nanotubes produced at low temperatures in metal-catalyzed CVD processes. Although the quality of low-temperature tubes can be ameliorated by prolonged postsynthesis annealing at temperatures above 2000K, the mechanical and electrical properties of arc-produced MWNTs are far superior [11]. Very high-quality nanotubes are vital for many applications that do not require multiton tube quantities; therefore, high-temperature production methods will not be completely replaced by more productive catalytic methods. On the contrary, the need of industry for the highest quality nanotubes will likely steadily grow, thus making it necessary to scale up the high temperature production methods. SWNTs are almost lacking in structural defects independent of the synthesis process; therefore, alternative methods already strongly compete with arc and laser techniques and may surpass and substitute them in future bulk production. The extent of this substitution will depend on other important qualities of the SWNT product, such as uniformity of tube chirality, size distribution, and ease of purification.

Irrefutably, knowledge of the nanotube growth mechanism is the basis for success in scaling up efforts and in acquiring desirable nanotube qualities. At present, the arc method remains the easiest and cheapest way to obtain significant quantities of SWNTs, whereas the produced nanotubes are less pure than those produced by laser ablation. Hence, many experimental and theoretical studies have attempted to improve the SWNT yield in the arc process and to provide a comprehensive understanding of the growth mechanism. Scaling up, optimizing, and controlling the arc process to make it commercially viable is a significant challenge. On the other hand, it should be taken into account that whatever CNT preparation method is applied, the CNTs are always produced with a number of impurities whose type and amount depend on the technique employed. These impurities interfere with most of the desired properties of CNTs and lead to serious impediment in detailed characterization and applications. One of the most fundamental challenges in CNT science, therefore, is the development of efficient and simple purification methods [12]. Most common purification methods are based on acid treatment of synthesized CNTs [13].

3.2 Arc Discharge

Electric-arc discharge is known as the most common technique and also one of the oldest methods in CNT production. This method is known to be a traditional and simple device to produce high temperatures needed to evaporate carbon atoms in plasma [14]. SWNTs are produced in an arc process invoking covaporization of graphite and metal in a composite anode commonly made by drilling an axial hole in the graphite rod and densely packing it with a mixture of metal and graphite powders [15,16]. Various pure elements and mixtures have been utilized to fill the rod, such as Fe, Co, Ni, Cr, Mn, Cu, and Pd [17–20], but at present only Ni/Y and Co/Ni catalysts, which share many common features, are commonly used in SWNT production. Both perform better in helium atmosphere than in argon, at about 0.6 bar pressure. As can be seen in Fig. 3.1, in the arc discharge technique, two carbon electrodes are invoked to produce an arc by digital current (DC). Initially, the two electrodes are kept independent. The electrodes are kept in a vacuum chamber and an inert gas is supplied to the chamber. The inert gas increases the speed of carbon deposition. Once the pressure is stabilized, the power supply is turned on (about 20 V) and the positive electrode is then gradually brought closer to the negative one to strike the electric arc. In this situation, the electrodes become red-hot and plasma forms. Stabilizing the arc, the rods are kept about a millimeter apart while the CNT deposits on the negative electrode. Once the specific length is achieved, the power supply is cut off and the machine is ready for the cooling phase. Notably, there are some items that should be entirely monitored including the control of arcing current and the optimal pressure of the inert gas in the chamber [21,22]. Moreover, a cylindrical deposit grows at the surface of the cathode, the weight of which constitutes about one half of the weight of the anode consumed in the process and is comprised of a hard gray shell and a soft core.

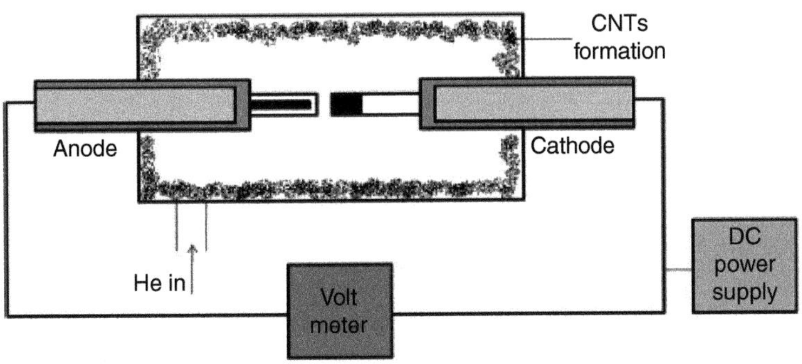

FIGURE 3.1 Schematic of arc discharge to produce CNTs [23].

It has a poorly developed columnar structure and contains MWNTs, multiwall polyhedral particles (MPPs), and graphitic particles. In the Co/Ni system, commonly operating with 2% metals in the anode, the core is diamagnetic and does not contain any metal components. The high temperature prevents expedition of metal vapor on the growth surface, and all metals escape the gap.

However, the outermost layer of the gray shell contains trace amounts of metals, introduced by gas convection onto the exposed side surface. In many other systems, especially with high metal loading of the anode, the deposit may contain metal-filled MWNTs and MPPs or bare spherical metallic particles [1]. In the Ni/Y system the deposit contains large amounts of yttrium [24], whereas metal nanoparticles found in the chamber are enriched in nickel [25]. The mixed carbon and metal vapor that has escaped the gap then condenses onto the product, which moves to the reactor surface and deposits on it. The product is divided into three discrete structural types depending on the deposition area. A spongy soft belt called *collaret* is formed around the cylindrical deposit and represents about 20% of the product weight. A relatively strong cloth-like soot on the chamber walls represents another 70%, and the remaining 10% of the product is a web-like structure suspended in the chamber volume between the cathode and the walls. All three foregoing types contain varying amounts of SWNTs, fullerenes, amorphous carbon, empty and metal-filled MPPs, naked metal particles, and graphitic nano- and microparticles. Thermogravimetric analysis and near-infrared spectroscopy [22] are exploited to precisely determine the metal and SWNT content in the arc material, whereas electron microscopy (tunneling electron microscopy and scanning electron microscopy) and Raman spectroscopy generally prepare only rough assessments of the concentration of product components. These methods appear to be the most useful for analyzing arc-product composition and structure. The collaret contains more SWNTs and metal particles than other components of the product. It is worthwhile to indicate that this difference is about 20% for Co/Ni and 50% for Ni/Y systems. The collaret is formed from an electrostatically deflected flow of soot that is propagating away from the gap. SWNT growth continues in the material that is deflected and captured on the deposit surface for a longer time, leading to enrichment of the collaret in SWNTs. The SWNT yield averaged among all three structural types of the product is about the same in these systems, roughly 20% ± 5% of the total product weight. Fullerenes C_{60} and C_{70} contribute 5–10 wt.% to the total product weight in the Co/Ni system and 1% in the Ni/Y system. The average mass fraction of the metal catalyst is 20% of the total product weight in the Co/Ni system and 35% in the Ni/Y system. Amorphous carbon constitutes 50 wt.% of the total product weight and graphitic particles 5 wt.% in both cases. SWNTs are generally organized in bundles consisting of a few dozen tubes, tightly compounded in a honeycomb lattice with an average separation between tube axes of 1.7 nm. It is worth mentioning that scarce isolated tubes can be observed. The bundles are covered with an amorphous carbon layer 2–5 nm thick, which contains embedded fullerenes. The majority of the tubes have diameters in the range of 1.2–1.5 nm and lengths reaching up to 5 μm in the Ni/Y system and 20 μm in the Co/Ni system. Although in the rare case the individual tubes can be observed, their tips are closed. Furthermore, some bundles terminate with

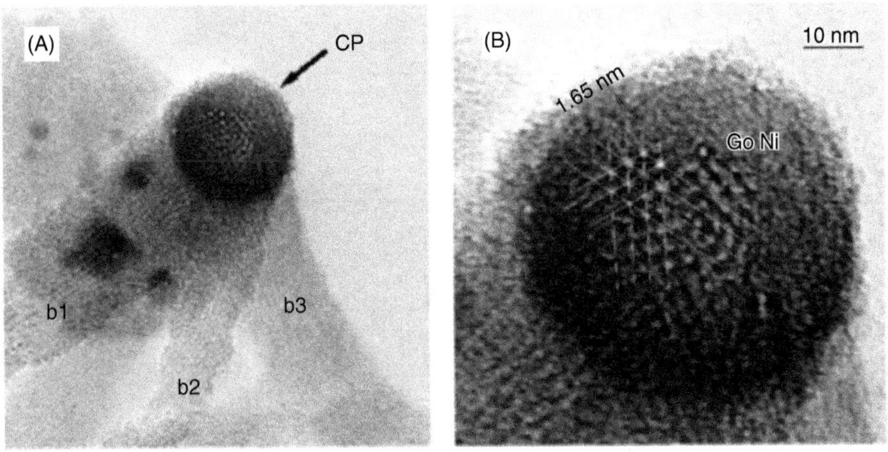

FIGURE 3.2 (A) TEM image of three Co/Ni catalytic metal alloy particle *(CP)* with three single-walled nanotube (SWNT) bundles *(b1, b2, b3)* grown from it. (B) Magnified view of the same particle showing a bundle root with about 20 active sites arranged in a triangular pattern on the surface from which SWNTs of the bundle b1 originate [26].

a spherical metal particle of 10–30 nm in diameter, which exceeds the bundle diameter. Such termination is characteristic of Co/Ni [26] (Fig. 3.2), Ni [27], Ni/La [26], and Ni/Y [21] arc systems, as well as of a nickel catalyst-based laser vaporization system [28]. From the structural properties point of view, it should be emphasized that, SWNT diameter depends on the temperature of the catalytic site at which growth occurs. This temperature is regulated by many factors, including heating of the reaction zone with an externally controlled heat source. The constant temperature thus set by external heating (environment temperature) provides a minimum temperature for the reaction site. It was established that the mean diameter of SWNTs increases with the environment temperature [29]. Other factors affecting reaction-site temperature cause the same kind of dependence, that is, production of thicker SWNTs is favored at higher temperatures [30,31]. Likewise, the environment temperature also affects the yield of SWNTs. Higher SWNT production rate at elevated environment temperatures is generally achieved with arc and laser techniques [29,32].

3.3 Laser Ablation

Pulsed laser vaporization, or laser ablation, is a newer technique currently being developed to create nanotubes. This method, similar to the arc discharge, vaporizes the carbon and later deposits it onto a substrate. In 1995, Smalley's group at Rice University reported that the synthesis of CNTs by laser vaporization becomes an efficient route for the synthesis of bundles of single-walled carbon nanotubes (SWCNTs) with a narrow distribution [6]. In this method, a piece of graphite target is vaporized by laser irradiation under high temperature in an inert atmosphere and the multiwalled carbon nanotubes were formed. Nanotubes produced by laser ablation have higher purity (up to about 90% pure) and their structure is better graphitized than those produced in the arc process.

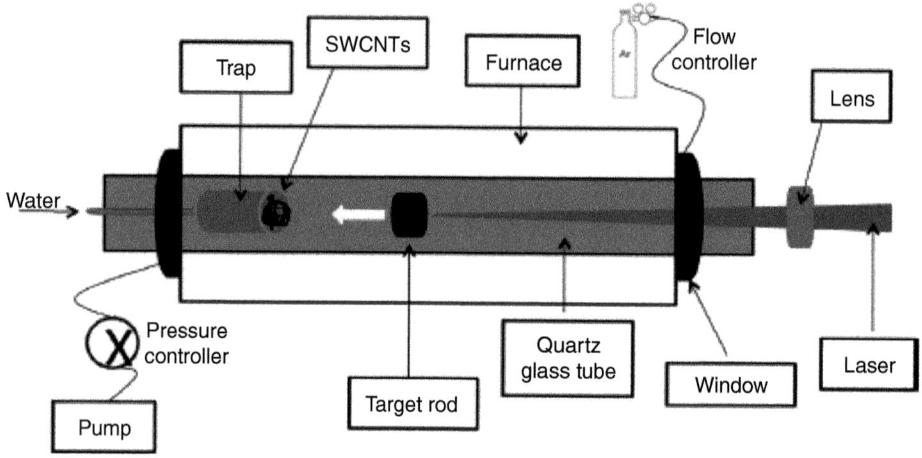

FIGURE 3.3 Diagram of laser furnace apparatus to produce CNTs [6].

The setup of the laser furnace is presented in Fig. 3.3, which comprises of a furnace, a quartz tube with a window, a target carbon composite doped with catalytic metals, a water-cooled trap, and flow systems for the buffer gas to maintain constant pressures and flow rates [33]. A laser beam is introduced through the window and focused on the target located in the center of the furnace. The target is vaporized in high-temperature Ar buffer gas and forms SWCNTs. The Ar flow rate and pressure are typically 1 cm/s and 500 Torr, respectively. The SWCNTs produced are conveyed by the buffer gas to the trap, where they are collected. The vaporization surface is kept as fresh as possible by changing the focus point or moving the target [32].

Evaporative heating of a carbon–metal composite target with laser light pulses, either separate or frequently repeated and with continuous illumination by laser or solar light, can bring about SWNT production under the correct conditions, which has been reported in many literatures [34–36]. The produced SWNT properties are dependent on variations in the process parameters, consisting of light intensity, process temperature and geometry, carrier gas type, pressure, and flow conditions. Typical conditions for producing SWNTs in several laser ablation systems are listed in Table 3.1. A typical 10-ns-long and 300-mJ laser pulse vaporizes 10^{17} carbon and 10^{15} metal atoms, contained in a 0.1-μm-thick layer over a 0.2 cm^2 spot, into argon gas preheated by an external furnace to 1200°C. The surface temperature rises during the laser pulse to about 6000K, with energy continuously deposited into the solid as thermal energy as well as supplying the heat of vaporization for the ablated mass [37]. At the end of the laser pulse, the surface temperature drops quickly to 4000K as a large amount of thermal energy is extracted by the additional after-pulse carbon vaporization and then more slowly cools by thermal conduction reaching 2000K in about 40–50 ns [37]. It was found that the application at this moment of a second equivalent laser pulse increases the SWNT yield [32].

Ejection of graphite particulates is a consequence of thermally induced stress relaxation, as it is occurring later than the single-pulse termination [38]. As a comparison, the

Table 3.1 SWNT Synthesis by Ablation of Carbon–Metal Target Containing Co:Ni = 1:1 Catalyst [11]

N	Ref.	Furnace *T* (°C)	Argon Pressure (kPa)	Pulse Energy (mJ)	Pulse Length (ns)	Pulse Repetition (Hz)	Laser Spot Diameter (mm)	Pulse Fluence (J/cm²)	Pulse Intensity (MW/cm²)	Average Irradiance (W/cm²)	Laser Wavelength (nm)	Total Metal in Target (at.%)	Production Rate (mg/h)	SWNT Yield of Product (wt.%)
1	107	1,200	66	300	10	10	6.5	0.9	100	9	532	1.2	10	50
2[a]	60	1,200	66	250	10	10	5	1.3	130	13	532	1.2	n/s	70
			300	10	10	7	0.8	80	8	1,064				
3[a]	108	1,200	66	490	10	30	6	1.7	170	52	532	2.0	120	65
			550	10	30	6.5	1.7	170	50	1,064				
4[a]	108	1,100	66	930	10	30	7.1	2.4	240	70	1,064	2.0	500	45
			930	10	30	7.1	2.4	240	70	1,064				
5[a]	54, 48	1,050	66	100	7	20	3	1.4	200	28	532	5.2	n/s	65
				100	7	20	3	1.4	200	28	1,064	1.2[b]		
6	112	1,150	105	300	20	20	2	9.5	470	190	1,064	1.2	200	60
7	113	1,150	65	290	20	15	4.5	1.8	90	27	1,079	1.2	100	40
8	104	1,200	92	115	6.5	10	1.5	2.0	310	20	532	9.0	25	55
9	104	1,200	92	115	6.5	0.5	1.5	2.0	310	1	532	9.0	1	55
10	104	1,200	92	115	6.5	0.008	1.5	2.0	310	0.017	532	9.0	0.02	55
11	102	1,000	66	140	8	0.016	1.6	7.0	870	0.1	532[c]	2	n/s	60
12	102	900?	66	140	8	0.016	1.6	7.0	870	0.1	532[c]	2	n/s	15
13	120	400	80	n/s	8	n/s	n/s	3	380	n/s	532	5[d]	n/s	5
14	118	1,200	80	300	6.5	10	3.5	3.1	470	31	532	1.2	100	50
15	58	25	66	1.0	575	24,000	1.0	0.12	0.2	3,000	1,064	1.2[e]	n/s	30
16	58	25	66	2.3	300	10,000	1.0	0.27	0.9	2,800	1,064	1.2[e]	n/s	15
17	58	25	66	3.2	250	6,000	1.0	0.37	1.5	2,300	1,064	1.2[e]	n/s	4
18	58	25	66	5.3	175	3,000	1.0	0.61	3.5	1,800	1,064	1.2[e]	n/s	0
19	105	1,200	80	20,000	2×10^7	s.p.	1	2,550	0.13	1×10^5	10,600	1.2	1,800	60
20	105	25	80	20,000	2×10^7	s.p.	1	2,550	0.13	1×10^5	10,600	1.2	900?	6?
21	125	25	66	∞	∞	∞	1.1	∞	0.025	2×10^4	10,600	4	70	30
22	126	25	53	∞	∞	∞	1.1	∞	0.025	2×10^4	10,600	4	200	50
23	129	2,700	45	∞	∞	∞	5	∞	0.002	2,000	Solar	4	40	10
24	61	3,500	66[f]	∞	∞	∞	8	∞	0.003	3,000	Arc	2[g]	5,000	20

n/s, Nonspecified; s.p., single pulse; ?, denotes rough estimate for a value not specified in the referenced paper; ∞, unidentified.

[a]Dual pulse with 40- to 50-ns gap.

[b]Fe:Ni catalyst.

[c]Both 532 and 1064 nm, simultaneously.

[d]95% C_{60} target.

[e]Porous target.

[f]Helium gas.

[g]Co:Ni = 3:1.

graphite particulate amount in the product is generally smaller with laser ablation than with arc discharge. This explains, in part, larger SWNT yields and relative ease of deep postsynthesis purification of the laser product. During the single-pulse ablation, a flat bubble of 6000K and 400 bar carbon vapor forms near the target surface and starts to expand with supersonic velocity and compresses the background gas [37,38]. Gas pressure and temperature in the plume therefore formed, both quickly decrease until the pressure becomes slightly below that of the compressed background gas [39]. At this moment, less than 300 ns after the pulse, a strong external shock wave is generated and detaches to propagate through the background gas, and an internal shock wave within the plume material commences to propagate in the opposite direction [39]. The internal shock wave engenders backward motion of a portion of the plume toward the target surface. By the time this portion reaches the target (100–600 ns after the pulse), the surface temperature is already below 1000K [39], and carbon and metal components of the plume partially precipitate on the cold surface as amorphous carbon and thin metal films. Debilitating the material, the backward-moving portion of the plume reflects from the surface and begins to move out to join the main body of the plume. Simultaneously, the backward motion in the radial direction causes an axial focusing of the plume.

In about 200–1000 ns this first oscillation of the plume material is completed. Decaying in strength, several consecutive oscillations of the same nature can be observed [39]. Following these oscillations, occurring during less than 0.2 ms after the pulse, the plume material is dissociated into a "smoke ring" or a bubble, depending on conditions, confined within 0.1–1 cm^3 volume, which detaches from the surface. This volume retains carbon for a time interval between a few milliseconds and seconds and then converts into a shapeless turbulent structure. It is commonly agreed that the SWNT growth time in laser ablation lies in this millisecond–second time scale [35,40]. There is no spectral feature directly associated with the growth of SWNTs; therefore, a more accurate determination of the growth time with time-resolved spectral diagnostics of the laser plume is not possible [41–43]. Accurate investigation of the SWNT growth time in laser synthesis emanates from the comparison of the kinetic data on carbon–metal vapor condensation and further evolution in laser and arc ablation processes. This comparison shows close quantitative similarities in the kinetics of laser and arc processes in the time range of SWNT formation under optimal conditions.

The carbon component of the carbon–metal vapor condenses into soot particles and fullerenes during 0.2 ms in the laser and arc process. The amount of fullerenes produced in both systems under similar conditions is the same, about 5%–10% [38]. Condensation of the metal component starts later and is completed in 2.0–2.3 ms in laser systems and in about the same time in the arc process [43,44].

The ephemeral behavior of the laser plume and arc jet flow parameters in the time range of interest, from 0.2 to 30 ms, is very similar. In this range the ablated material propagates by a few centimeters, decelerating from 103 to 10 cm/s linear velocity; temperature declines from 2200 to 1000°C and carbon particle number density from 10^{17} to 10^{16} cm^{-3} in the laser and arc systems [41–43,45]. Under optimal conditions, SWNTs of equal length are formed in both systems. The mentioned comparison absolutely suggests the same growth time for SWNTs produced in the laser and arc processes, which is about 10 ms under optimal conditions. A conservative estimate of 3–30 ms for the SWNT growth time can be obtained with allowance made for the width of the metal catalyst particle diameter distribution and accuracy of cooling-time determination. Unequivocally, the SWNT crop sharply decreases, the SWNTs become much shorter, and the diameters of individual SWNTs and their bundles diminish when the ambient temperature declines from 1200 to 900°C, especially near the lower boundary of this temperature range [46,47]. In this regard, the observation of a short, thin, and low-yield SWNT bundle is quite conceivable and does not contradict the mechanism proposed [39]. The furnace temperature, apparently, is the most critical externally controlled parameter among those studied with pulsed laser ablation, as it primarily determines the temperature of the buffer gas layer of a few centimeters thick near the target, in which the plume expands, cools down, and forms SWNTs. It should be noted that no SWNT formation was observed with the ablation of a graphite–metal target at a furnace temperature below 800°C or above 1300°C with a maximum SWNT crop lying at about 1200°C according to Fig. 3.4. Actually, the existence of the upper temperature limit for SWNT formation, divulged by the abrupt drop in yield at 1300°C furnace temperature, can be ascribed to overheating of the metal catalyst particle with subsequent cessation of SWNT growth by the cold-explosion mechanism. In this respect, the width of the bell-shaped dependence in Fig. 3.4 near the maximum yield (1200°C) should depict the temperature corridor in which the particle works. Thus, the range 1300–1500°C can be specified for the actual temperature of the Ni/Co particle working under optimal conditions in laser and arc systems and using amorphous carbon for the feedstock in SWNT formation by the dissolution–precipitation mechanism. Significantly, the actual temperature of the particle determines the diameter of the SWNTs. With regard to MWNT production by laser

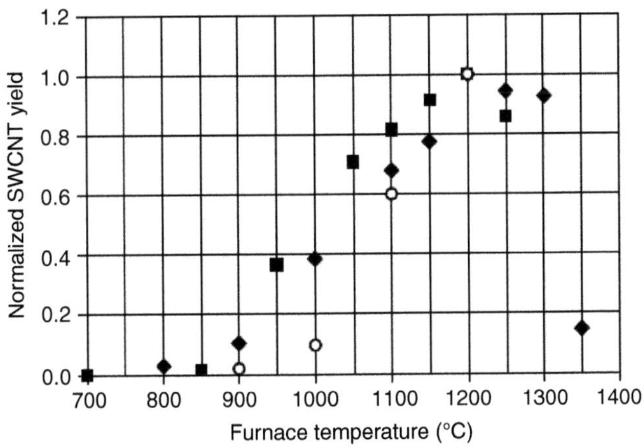

FIGURE 3.4 Influence of the furnace temperature on the relative SWNT yield in the product, obtained under similar laser-pulsing conditions but with different gas-flow velocity near the target. *Solid squares* [48], 1 cm/s linear velocity; *solid diamonds* [47], 1.8 cm/s; *open circles* [49], 4.8 cm/s.

ablation of graphite, during fullerene production experiments employing a laser vaporization apparatus with an ablated graphite sample positioned in an oven, it was found that closed-ended MWNTs were produced in the gas phase through homogeneous carbon-vapor condensation in a hot argon atmosphere [11]. The laser-produced MWNTs are relatively short (300 nm), although the number of layers (ranging between 4 and 24) and the inner diameter (varying between 1.5 and 3.5 nm) are similar to those of arc-produced MWNTs. The yield and quality of MWNTs diminished at oven temperatures below 1200°C, and at 200°C no nanotubes can be observed [11].

As a conclusion for both aforementioned techniques, although arc discharge and laser ablation produce some of the highest quality nanotubes, they suffer from the following disadvantages that limit their utilization as large-scale industrial processes:

1. They are both energy-intensive methods. A large amount of energy is required to provide the arc or laser used for the ablation processes. Such a huge amount of energy is not only unattainable but also unreasonable for large-scale production.
2. Both techniques require solid carbon/graphite as a target, which has to be evaporated to get nanotubes. It is difficult to get such a large amount of graphite to be used as the target in an industrial process, which limits its exploitation in a large-scale production process.
3. Both processes grow nanotubes in highly tangled/unaligned form, mixed with unwanted forms of carbon or catalysts. The CNTs produced by these processes, therefore, require purification to get purged and assembled forms, and designing such refining processes is difficult and expensive. All the aforementioned factors extremely limit the use of arc discharge and laser ablation as large-scale processes for the production of carbon nanotubes.

3.4 Thermal CVD

CVD is essentially a thermal dehydrogenation reaction whereby a transition metal catalyst, such as iron, nickel, or cobalt is invoked to lower the temperature required in order to "crack" a gaseous hydrocarbon feed into carbon and hydrogen. CVD is a versatile process apposite for the manufacturing of coatings, powders, fibers, and monolithic components. This technology is an essential factor in the manufacturing of semiconductors and other electronic components in the coating of tools, bearings, and other wear-resistant parts and in many optical, optoelectronic, and corrosion applications [14].

The CNT samples produced in the reactor by CVD technique is described in Fig. 3.5 [13]. It consists of a quartz tube (1–2 in. in diameter) embedded in a tubular furnace capable of maintaining ±1°C over a 25-cm zone. Thus, it is a hot-wall system at initial atmospheric pressure CVD; therefore, pumping systems are not required. The substrate, typically about or smaller than 1 in. now, is located inside the quartz tube. In thermal CVD, either CO or some hydrocarbon, such as methane, ethane, ethylene, acetylene, or other higher hydrocarbon, is utilized without dilution and the feedstock is measured through

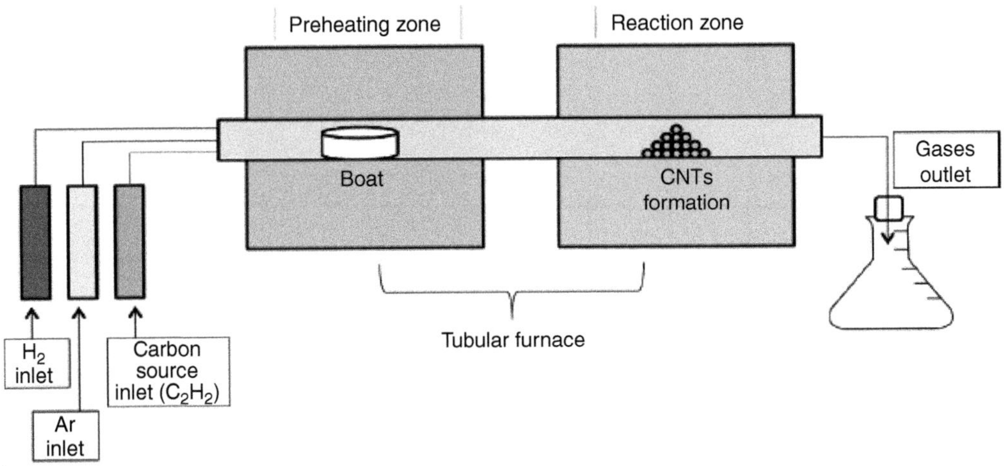

FIGURE 3.5 Diagram of CVD to produce CNTs [13].

a mass flow controller. A typical growth run would involve purging the reactor first with argon or some other inert gas until the reactor attains the desired growth temperature. The gas flow is then switched to the feedstock for the specified growth period. Finally, the gas flow is switched back to the inert gas while the reactor cools down to 300°C or lower before exposing the nanotubes to air. It should be noted that exposure to air at elevated temperatures can cause damage to the CNTs. It is noteworthy that the growth of CNTs takes place at the second furnace and collected in ceramic boats that are placed at the center of the second furnace. Cold-wall reactors common in the semiconductor industry, where the substrate holder is directly heated from below (resistance, inductance, or other heaters), have not been reported in the literature. Since the growth method is catalyst-promoted at temperatures of 500–1000°C and does not depend on precursor dissociation at these temperatures, either hot-wall or cold-wall systems could be designed to be efficient for CNT growth. Admittedly, most batch CVD reactors with well-designed gas inflow (showerhead) and outflow manifolds, wafer holder, and support platform with heating systems underneath should work when growth on large areas becomes desirable. Low-pressure operation is possible if advantages are clearly demonstrated. The catalyst mixture needs to be applied to the substrate first for it to grow on the substrate before loading it inside the reactor, which is the so-called supported catalyst approach. In contrast, CVD can also be used to grow large quantities of nanotubes using a floating-catalyst approach [50–54]. In the latter, a nozzle system may be invoked to inject the vaporized catalyst precursor into the flowing CO or hydrocarbon [51,52]. A second furnace may be required to heat up the catalyst precursor system to its dissociation temperature [55]. The floating catalyst approach is amenable to scale-up for large-scale production to meet the commodity market demands of nanotubes in a variety of structural applications.

Fig. 3.6 depicts MWNTs grown by thermal CVD at 750°C with ethylene feedstock and exploiting a catalyst preparation recipe [11]. In this solution-based approach, gel-based catalysts

FIGURE 3.6 Cylinder-shaped multiwalled nanotube (MWNT) towers grown by thermal CVD [11].

form films, and the characteristics depend on the choice of solvents, concentration, the influence of catalyst ligands, and so on. In Fig. 3.6, a series of cylinder-shaped towers composed of MWNTs is seen, which are fairly uniform, about 200 μm in diameter and 400-μm tall.

3.5 Plasma-Enhanced CVD

The plasma enhancement in CVD first emerged in microelectronics because some processes cannot tolerate the elevated temperatures of thermal CVD. For instance, charring of photoresists on patterned wafers could be a problem at elevated temperatures in integrated circuit manufacturing. Thus, the most common PECVD processes proceed at substantially lower wafer temperatures (room temperature to 100°C); hence, PECVD has become a key step in semiconductor device fabrication.

The carbon source is a hydrocarbon gas, which dissociates either thermally (thermal CVD) or in the presence of a plasma (PECVD) aided by a transition-metal catalyst. The self-assembly of nanotubes is facilitated by catalyst nanoparticles deposited on the substrate's surface, which seed the nanotube growth, defining the location and diameter of the deposited nanotubes. Silicon substrates are typically utilized in electronic applications as it is amenable for device fabrication using standard semiconductor manufacturing tools. The catalyst takes the form of a thin film of transition metal (Fe, Ni, Co, or Mo), which may be applied chemically from a solution containing the catalyst or directly by using techniques, such as thermal evaporation, ion beam sputtering, or magnetron sputtering. The increased surface mobility of the catalyst atoms causes the metal film to coalesce into nanoclusters, when heat is subsequently applied to the substrate [56]. To impede a chemical reaction between the catalyst and the substrate during heat application, a thin diffusion barrier of SiO_2, TiN, or indium tin oxide is deposited onto the Si substrate before the metal catalyst layer is deposited. The catalyst layer can be patterned using conventional lithography, which then leads to highly localized growth

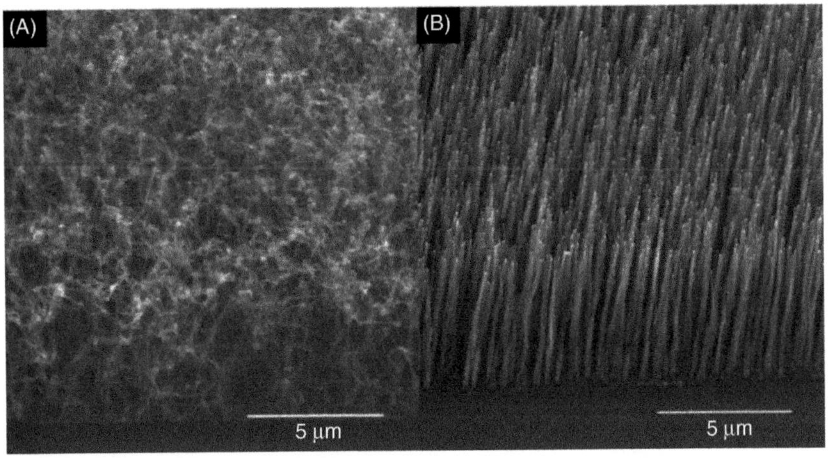

FIGURE 3.7 Deposition (A) without and (B) with plasma.

of carbon nanotubes. Furthermore, to prevent the deposition of amorphous carbon, which interdicts the formation of nanotubes by poisoning the growth catalyst and can also cause short circuits on the substrate surface, the carbon source (typically, C_2H_2 or CH_4) is amalgamated with a hydrogen-rich gas (typically, NH_3 or H_2), which produces reactive species in the plasma to remove any excess carbon. Thermal CVD requires temperatures of 700–1000°C and has two disadvantages: (1) It prevents the use of some desirable substrate materials due to the high temperature, and (2) it produces not randomly oriented and not perfectly straight nanotubes. Therefore, PECVD is introduced to overcome both mentioned problems. Surprisingly, energy in the plasma replaces some of the heat energy, allowing gas dissociation and nanotube formation to take place at lower temperatures (600–700°C), and the electric field in the plasma aligns the nanotubes during growth [57]. There are a number of different techniques available for creating the plasma encompassing radio frequency PECVD, microwave PECVD, inductively coupled PECVD, and DC glow discharge PECVD.

Fig. 3.7 describes the difference between nanotubes deposited with and without plasma [58]. The conditions in the two cases are identical except for the presence of DC glow discharge plasma in the case of Fig. 3.7B. In the purely thermal CVD case, the resultant nanotubes are curly and nonaligned, whereas in the PECVD, the resultant nanotubes are straight and aligned vertically.

A schematic of the PECVD setup is shown in Fig. 3.8 [11]. For CNT growth, the wafer is first loaded in the reactor and the system is pumped down to 10–5 Torr or below to minimize impurities and water vapor. The substrate is then heated to the desired temperature and the feedstock is admitted. Subsequently, the flow rate and chamber pressure can be set to desired levels independent of each other with the aid of a throttle valve. Next, the power from the power source is coupled to the plasma. At the end of the run, the heater, power source, and the gas flow are turned off and the system is purified with argon flow. The wafer is removed after the reactor cools down below 300°C. The reactor can be equipped with

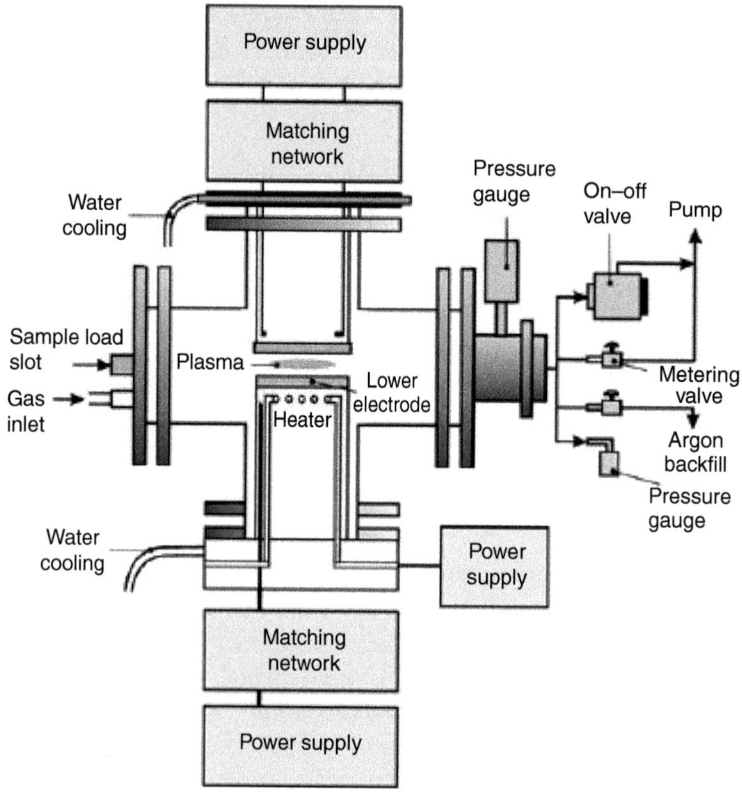

FIGURE 3.8 Schematic of a PECVD setup.

a load-lock system to introduce/withdraw the wafer and reduce the downtime associated with reaching the desired vacuum level.

As a deduction, a comparison of the mentioned techniques according to the synthesis temperature, nanotube graphitization, yield, and relative CNT quantity is shown in Table 3.2 [14,59].

Table 3.2 Comparison of Laser Ablation, Arc Discharge, and CVD Techniques

Process	Laser Ablation	Arc Discharge	CVD
Reaction temperature (°C)	3000	3000–4000	500–1100
Per unit design cost	High	High	Low
Nanotube selectivity	Low	Low	High
Source of carbon	Difficult	Difficult	Easy available
Purification of CNT	High	High	Low
CNTs yield (%)	70	<30	95–99
Process nature	Batch	Batch	Continuous
Process parameter control	Difficult	Difficult	Easy to control
Energy requirement	High	High	Low
Reactor design	Difficult	Difficult	Easy to design reactor
Nanotube graphitization	High	High	Middle

3.6 Catalyst Preparation

Although there have been some studies reporting CNT growth without catalysts in arc discharge processes, it is widely acknowledged that transition-metal catalysts are needed for SWNT and MWNT, growth by CVD and PECVD. Irrefutably, the catalyst on the substrate must be in the form of particles instead of smooth, continuous films since the latter do not appear to yield nanotubes. There have been several studies correlating the catalyst particle size and the diameter of the resulting nanotubes [60–64]. The metals that are employed as catalysts include Fe, Ni, Co, and Mo, which can be applied onto the substrate from solutions containing them, or they can be directly deposited using some physical techniques. These two approaches are different in terms of the needed resources, time, and cost, and the nature of the resulting products.

3.7 Purification

CNTs produced by any of the methods described in this chapter contain impurities, most notably catalyst metal particles and different forms of amorphous carbon. Although it is always a goal to produce CNTs as pure as possible, the research community has developed several postprocessing purification methods [32,65–67], which are explicated here concisely. Several chemical and physical methods are available for postsynthesis processing for the removal of metal catalysts and other impurities, such as amorphous carbon, metal catalyst-containing carbon nanospheres, and polyaromatic carbons including fullerenes. Some common techniques reported in the literature include thermal oxidation in air [32], hydrothermal treatment [65], H_2O-plasma oxidation [66], acid oxidation [67], dispersion separation by microfiltration [68] as chemical purification and high-performance liquid chromatography [69] and filtration [70] as physical purification, and a number of combinations of these techniques. Likewise, Murphy et al. [69] reported the use of a conjugated polymer host system to extract the graphitic particle impurities from CNT soot produced by the arc discharge method. The polymer extraction technique is claimed to be nondestructive due to independence from the difference in the oxidation reaction rates for the crystalline graphitic structures versus the amorphous carbon impurities.

3.8 Conclusions

Although the growth mechanism of CNTs is not exactly known, new synthesis methods for higher yield, higher purity, and low defects of the produced CNTs are the main points of investigation being pushed forward by the burgeoning fields of nanotechnology and nanoscience, which have many ideas of possible applications. In this chapter, different synthesis techniques of CNTs have been reviewed, comprising arc discharge, laser ablation, CVD, and PECVD. Unequivocally, mass production, selective production, and organization of carbon nanotubes continue to be major synthetic challenges toward future applications.

The mechanism of the SWNT synthesis in the arc discharge and laser ablation processes appears to be fairly simple. It is governed mainly by the dynamics of hot vapor cooling and carbon diffusion through the bulk metal catalyst particle. It allows intentional control of the arc and laser processes and is the basis for the successful scaling-up efforts of the arc process. Semiquantitative treatment of the arc and laser processes is possible, which prepares a basis for a detailed mathematical modeling of arc and laser ablation reactors. This is necessary to design practical processes for maximum productivity, yield, and quality of SWNTs. Scaling up the arc discharge processes of SWNT, double-walled carbon nanotubes, and MWNT production is possible and can be realized in practice on an industrial scale.

Significant advances have been achieved during the last years including the development of new CVD methods for mass production and for aligned growth on surfaces, both vertically and horizontally. Mass production and vertically aligned growth by CVD have been significantly boosted by the discovery of "super-growth," adding small amounts of water to the CVD precursor or utilizing alcohol as the carbon feedstock, to proscribe the formation of amorphous carbon, and thus to enhance the formation of clean carbon nanotubes.

Finally, it can be inferred that, among the foregoing techniques, the CVD-based methods appear to be the most suitable for large-scale CNT production because of their scalability. However, arc discharge and laser ablation suffer from many disadvantages for being expensive and uneconomical for large-scale production of CNTs, despite their high quality and yield of CNT produced. Admittedly, a fundamental key to further increase the control of carbon nanotube synthesis is the proper understanding of the different mechanisms of carbon nanotube formation, which determines its structure, electronic properties, and arrangement.

References

[1] C. Journet, P. Bernier, Production of carbon nanotubes, Appl. Phys. A 67 (1) (1998) 1–9.

[2] A. Thess, R.P. Lee, H.J. Nikolaev, P. Dai, J. Petit, C.H. Robert, Crystalline ropes of metallic carbon nanotubes, Science 273 (52) (1996) 483–487.

[3] D.S. Bethune, C.H. Kiang, M.S. Devries, G. Gorman, R. Savoy, J. Vazquez, Cobalt-catalysed growth of carbon nanotubes with single-atomic-layer walls, Nature 363 (1993) 605–607.

[4] A. Oberlin, M. Endo, T. Koyama, Filamentous growth of carbon through benzene decomposition, J. Cryst. Growth 32 (1976) 335–349.

[5] P.G. Wiles, J. Abrahamson, Carbon fibre layers on arc electrodes—I: their properties and cool-down behavior, Carbon 16 (1978) 341–349.

[6] T. Guo, P. Nikolaev, A. Thess, D.T. Colbert, R.E. Smalley, Catalytic growth of single-walled nanotubes by laser vaporization, Chem. Phys. Lett. 243 (1–2) (1995) 49–54.

[7] M.J. Yacaman, M.M. Yoshida, L. Rendon, J.G. Santiesteban, Catalytic growth of carbon microtubules with fullerene structure, Appl. Phys. Lett. 62 (2) (1993) 202–204.

[8] J. Kong, A.M. Cassell, H. Dai, Chemical vapor deposition of methane for single-walled carbon nanotubes, Chem. Phys. Lett. 292 (1998) 567–574.

[9] M. Su, B. Zheng, J. Liu, A scalable CVD method for the synthesis of single-walled carbon nanotubes with high catalyst productivity, Chem. Phys. Lett. 322 (2000) 321–326.

[10] J. Hafner, M. Bronikowski, B. Azamian, P. Nikolaev, D. Colbert, R. Smalley, Catalytic growth of single-wall carbon nanotubes from metal particles, Chem. Phys. Lett. 296 (1998) 195–202.

[11] M. Meyyappan, Carbon Nanotubes Science and Applications, CRC Press, 2005.

[12] I. Kruusenberg, N. Alexeyeva, K. Tammeveski, J. Kozlova, L. Matisen, V. Sammelselg, J. Solla-Gullon, J.M. Feliu, Effect of purification of carbon nanotubes on their electrocatalytic properties for oxygen reduction in acid solution, Carbon 49 (2011) 4031–4039.

[13] N.M. Mubarak, F. Yusof, M.F. Alkhatib, The production of carbon nanotubes using two-stage chemical vapor deposition and their potential use in protein purification, Chem. Eng. J. 168 (2011) 461–469.

[14] N.M. Mubarak, E.C. Abdullah, N.S. Jayakumar, J.N. Sahu, An overview on methods for the production of carbon nanotubes, J. Ind. Eng. Chem. 20 (4) (2013) 1186–1197.

[15] S. Iijima, T. Ichihashi, Single-shell carbon nanotubes of 1-nm diameter, Nature 363 (1993) 603–605.

[16] D.S. Bethune, C.H. Kiang, M.S. De Vries, G. Gorman, R. Savoy, J. Vazquez, R. Beyers, Cobalt-catalysed growth of carbon nanotubes with single-atomic-layer walls, Nature 363 (1993) 605–607.

[17] P.M. Ajayan, J.M. Lambert, P. Bernier, L. Barbedette, C. Colliex, J.M. Planeix, Growth morphologies during cobalt-catalyzed single-shell carbon nanotube synthesis, Chem. Phys. Lett. 215 (1993) 509–517.

[18] S. Seraphin, D. Zhou, J. Jiao, M.A. Minke, S. Wang, T. Yadav, J.C. Withers, Catalytic role of nickel, palladium, and platinum in the formation of carbon nanoclusters, Chem. Phys. Lett. 217 (1994) 191–198.

[19] S. Seraphin, D. Zhou, Single-walled carbon nanotubes produced at high yield by mixed catalysts, Appl. Phys. Lett. 64 (1994) 2087–2089.

[20] C.H. Kiang, W.A. Goddard, R. Beyers, J.R. Salem, D.S. Bethune, Catalytic synthesis of single-layer carbon nanotubes with a wide range of diameters, J. Phys. Chem. 98 (1994) 6612–6618.

[21] J. Gavillet, A. Loiseau, F. Ducastelle, S. Thair, P. Bernier, O. Stéphan, J. Thibault, J.-C. Charlier, Microscopic mechanisms for the catalyst assisted growth of single-wall carbon nanotubes, Carbon 40 (2002) 1649–1663.

[22] M.E. Itkis, D.E. Perea, S. Niyogi, S.M. Rickard, M.A. Hamon, H. Hu, B. Zhao, R,C. Haddon, Purity evaluation of as-prepared single-walled carbon nanotube soot by use of solution-phase near-IR spectroscopy, Nano Lett. 3 (2003) 309–314.

[23] Production and purification of carbon nanotubes, in: T.W. Ebbesen (Ed.), Carbon Nanotubes: Preparation and Properties, CRC Press, Boca Raton, FL, 1997, pp. 139–162.

[24] M. Yudasaka, N. Sensui, M. Takizawa, S. Bandow, T. Ichihashi, S. Iijima, Formation of single-wall carbon nanotubes catalyzed by Ni separating from Y in laser ablation or in arc discharge using a C target containing a NiY catalyst, Chem. Phys. Lett. 312 (1999) 155–160.

[25] C. Journet, W.K. Maser, P. Bernier, A. Loiseau, M. Lamy de la Chapelle, S. Lefrant, P. Deniard, R. Lee, J.E. Fischer, Large-scale production of single-walled carbon nanotubes by the electric-arc technique, Nature 388 (1997) 756–758.

[26] R.O. Loutfy, et al., IWFAC'99, Abstracts, St. Petersburg, 109 (1999).

[27] Y. Saito, M. Okuda, N. Fujimoto, T. Yoshikawa, M. Tomita, T. Hayashi, Single-wall carbon nanotubes growing radially from Ni fine particles formed by arc evaporation, Jpn. J. Appl. Phys. 33 (1994) 526.

[28] D. Golberg, Y. Bando, L. Bourgeois, K. Kurashima, T. Sato, Large-scale synthesis and HRTEM analysis of single-walled B- and N-doped carbon nanotube bundles, Carbon 38 (2000) 2017–2027.

[29] S. Bandow, S. Asaka, Y. Saito, A.M. Rao, L. Grigorian, E. Richter, P.C. Eklund, Effect of the growth temperature on the diameter distribution and chirality of single-wall carbon nanotubes, Phys. Rev. Lett. 80 (1998) 3779.

[30] A.C. Dillon, P.A. Parilla, J.L. Alleman, J.D. Perkins, M.J. Heben, Controlling single-wall nanotube diameters with variation in laser pulse power, Chem. Phys. Lett. 316 (2000) 13–18.

[31] I. Hinkov, et al., Optical plasma control during arc carbon nanotube growth. Proc. 6th ADC/2nd FCT Joint Conference, 816 (2001).

[32] A. Thess, R. Lee, P. Nikolaev, H. Dai, P. Petit, J. Robert, C. Xu, Y.H. Lee, S.G. Kim, A.G. Rinzler, D.T. Colbert, G.E. Scuseria, D. Tomanek, J.E. Fischer, R.E. Smalley, Crystalline ropes of metallic carbon nanotubes, Science 273 (1996) 483–487.

[33] Y. Ando, X. Zhao, T. Sugai, M. Kumar, Growing carbon nanotubes, Mater. Today 7 (2004) 22–29.

[34] F. Kokai, K. Takahashi, M. Yudasaka, R. Yamada, T. Ichihashi, S. Iijima, Growth dynamics of single-wall carbon nanotubes synthesized by CO_2 laser vaporization, J. Phys. Chem. B 103 (1999) 4346–4351.

[35] Deleted in review.

[36] T. Guo, P. Nikolaev, A. Thess, D.T. Colbert, R.E. Smalley, Catalytic growth of single-walled nanotubes by laser vaporization, Chem. Phys. Lett. 243 (1995) 49–54.

[37] E.Y. Lo, et al., AIP Conf. Proc. 288, in: J.C. Miller, D.B. Geohegan (Eds.), 1994, p. 291.

[38] A.A. Puretzky, et al., AIP Conf. Proc. 288, in: J.C. Miller, D.B. Geohegan (Eds.), 1994, p. 365.

[39] A.A. Puretzky, D.B. Geohegan, X. Fan, S.J. Pennycook, Dynamics of single-wall carbon nanotube synthesis by laser vaporization, Appl. Phys. A 70 (2000) 153–160.

[40] Y. Zhang, H. Gu, S. Iijima, Single-wall carbon nanotubes synthesized by laser ablation in a nitrogen atmosphere, Appl. Phys. Lett. 73 (1999) 3827.

[41] S. Arepalli, C.D. Scott, Spectral measurements in production of single-wall carbon nanotubes by laser ablation, Chem. Phys. Lett. 302 (1999) 139–145.

[42] S. Arepalli, Laser ablation process for single-walled carbon nanotube production, J. Nanosci. Nanotechnol. 4 (4) (2004) 317–325.

[43] C.D. Scott, S. Arepalli, P. Nikolaev, R.E. Smalley, Growth mechanisms for single-wall carbon nanotubes in a laser-ablation process, Appl. Phys. A 72 (2001) 573–580.

[44] A.A. Puretzky, D.B. Geohegan, X. Fan, S.J. Pennycook, In situ imaging and spectroscopy of single-wall carbon nanotube synthesis by laser vaporization, Appl. Phys. Lett. 76 (2000) 182–184.

[45] A.V. Krestinin, A.P. Moravsky, Kinetics of fullerene C-60 and C-70 formation in a reactor with graphite rods evaporated in electric arc, Chem. Phys. Rep. 18 (3) (1999) 515–532.

[46] M. Yudasaka, R. Yamada, N. Sensui, T. Wilkins, T. Ichihashi, S. Iijima, Mechanism of the effect of NiCo, Ni and Co catalysts on the yield of single-wall carbon nanotubes formed by pulsed Nd:YAG laser ablation, J. Phys. Chem. B 103 (1999) 6224–6229.

[47] H. Kataura, Y. Kumazawa, Y. Maniwa, Y. Ohtsuka, R. Sen, S. Suzuki, Y Achiba, Diameter control of single-walled carbon nanotubes, Carbon 38 (2000) 1691–1697.

[48] A. Gorbunov, O. Jost, W. Pompe, A. Graft, Solid–liquid–solid growth mechanism of single-wall carbon nanotubes, Carbon 40 (2002) 113–118.

[49] M. Yudasaka, T. Ichihashi, S. Iijima, Roles of laser light and heat in formation of single-wall carbon nanotubes by pulsed laser ablation of $C_xNi_yCo_y$ targets at high temperature, J. Phys. Chem. B 102 (1998) 10201–10207.

[50] B.C. Satishkumar, A. Govindraj, R. Sen, C.N.R. Rao, Single-walled nanotubes by the pyrolysis of acetylene-organometallic mixtures, Chem. Phys. Lett. 293 (1998) 47–52.

[51] P. Nikolaev, M.J. Bronikowski, R. Bradley, F. Rohmund, D.T. Colbert, K.A. Smith, R.E. Smalley, Gas-phase catalytic growth of single-walled carbon nanotubes from carbon monoxide, Chem. Phys. Lett. 313 (1999) 91–97.

[52] M.J. Bronikowski, P.A. Willis, D.T. Colbert, K.A. Smith, R.E. Smalley, Gas-phase production of carbon single-walled nanotubes from carbon monoxide via the HiPco process: a parametric study, J. Vac. Sci. Technol. A 19 (2001) 1800–1805.

[53] H.M. Cheng, F. Li, X. Sun, S.D.M. Brown, M.A. Pimenta, A. Marucci, G. Dresselhaus, M.S. Dresselhaus, Bulk morphology and diameter distribution of single-walled carbon nanotubes synthesized by catalytic decomposition of hydrocarbons, Chem. Phys. Lett. 289 (1998) 602–610.

[54] H.M. Cheng, F. Li, G. Su, H.Y. Pan, L.L. He, X. Sun, M.S. Dresselhaus, Large-scale and low-cost synthesis of single-walled carbon nanotubes by the catalytic pyrolysis of hydrocarbons, Appl. Phys. Lett. 72 (1998) 3282–3284.

[55] R. Sen, A. Govindaraj, C.N.R. Rao, Carbon nanotubes by the metallocene route, Chem. Phys. Lett. 267 (1997) 276–280.

[56] V.I. Merkulov, D.H. Lowndes, Y.Y. Wei, G. Eres, E. Voelkl, Patterned growth of individual and multiple vertically aligned carbon nanofibers, Appl. Phys. Lett. 76 (2000) 3555.

[57] K.B.K. Teo, M. Chhowalla, G.A.J. Amaratunga, W.I. Milne, G. Pirio, P. Legagneux, F. Wyczisk, J. Olivier, D. Pribat, Characterization of plasma-enhanced chemical vapor deposition carbon nanotubes by Auger electron spectroscopy, J. Vac. Sci. Technol. 20 (2002) 116.

[58] M.S. Bell, K.B.K. Teo, R.G. Lacerda, W.I. Milne, D.B. Hash, M. Meyyappan, Carbon nanotubes by plasma-enhanced chemical vapor deposition, Pure Appl. Chem. 78 (6) (2006) 1117–1125.

[59] M.M.A. Rafique, J. Iqbal, Production of carbon nanotubes by different routes—a review, J. Encapsul. Adsorpt. Sci. 1 (2011) 29–34.

[60] Y. Homma, T. Yamashita, P. Finnie, M. Tomita, T. Ogino, Single-walled carbon nanotube growth on silicon substrates using nanoparticle catalysts, Jpn. J. Appl. Phys. 41 (2002) 89.

[61] C.J. Lee, S.C. Lyu, Y.R. Cho, J.H. Lee, K.I. Cho, Diameter-controlled growth of carbon nanotubes using thermal chemical vapor deposition, Chem. Phys. Lett. 341 (2001) 245.

[62] J.I. Sohn, C.J. Choi, S. Lee, T.Y. Seong, Growth behavior of carbon nanotubes on Fe-deposited (001) Si substrates, Appl. Phys. Lett. 78 (2001) 3130.

[63] G.W. Ho, A.T.S. Wee, J. Lin, W.C. Tjiu, Synthesis of well-aligned multiwalled carbon nanotubes on Ni catalyst using radio frequency plasma-enhanced chemical vapor deposition, Thin Solid Films 388 (2001) 73–77.

[64] Deleted in review.

[65] S. Huang, L. Dai, Plasma etching for purification and controlled opening of aligned carbon nanotubes, J. Phys. Chem. B 106 (2002) 3543–3545.

[66] I.W. Chiang, B.E. Brinson, R.E. Smalley, J.L. Margrave, R.H. Hauge, Purification and characterization of single-wall carbon nanotubes, J. Phys. Chem. B 105 (2001) 1157–1161.

[67] S. Bandow, A.W. Rao, K.A. Williams, A. Thess, R.E. Smalley, P.C. Eklund, Purification of single-wall carbon nanotubes by microfiltration, J. Phys. Chem. B 101 (1997) 8839–8842.

[68] S. Niyogi, H. Hu, M.A. Hamon, P. Bhowmik, B. Zhao, S.M. Rozenzhak, J. Chen, M.E. Itkis, M.S. Meier, R.C. Haddon, Chromatographic purification of soluble single-walled carbon nanotubes, J. Am. Chem. Soc. 123 (2001) 733–734.

[69] R. Murphy, J.N. Coleman, M. Cadek, B. McCarthy, M. Bent, A. Drury, R.C. Barklie, W.J. Blau, High yield nondestructive purification and quantification method for multiwalled carbon nanotubes, J. Phys. Chem. B 106 (2002) 3087–3091.

[70] Y. Brhane, T. Gabriel, Production, purification and functionalization of carbon nanotubes for medical application, Int. Res. J. Pharm. 7 (7) (2016) 19–27.

Further Reading

M. Yudasaka, T. Ichihashi, T. Komatsu, S. Iijima, Single-wall carbon nanotubes formed by a single laser-beam pulse, Chem. Phys. Lett. 299 (1999) 91–96.

K. Tohji, H. Takahashi, Y. Shinoda, N. Shimizu, B. Jeyadevan, I. Matsuoka, Purification procedure for single-walled nanotubes, J. Phys. Chem. 101 (1997) 1974–1978.

4

Fabrication of Carbon Nanotube/ Polymer Nanocomposites

Tejendra K. Gupta*, Shanmugam Kumar**

*INSTITUTE CENTER FOR MICROSYSTEMS (IMICRO), MASDAR INSTITUTE OF SCIENCE AND TECHNOLOGY, ABU DHABI, UNITED ARAB EMIRATES;
**INSTITUTE CENTER FOR ENERGY (IENERGY), MASDAR INSTITUTE OF SCIENCE AND TECHNOLOGY, ABU DHABI, UNITED ARAB EMIRATES

CHAPTER OUTLINE

4.1 Introduction

The area of carbon nanotube (CNT)–based polymer nanocomposites (PNCs) has been progressing rapidly in recent years in several fields, such as strain sensing [1–7], supercapacitors [8], electromagnetic interference (EMI) shielding and electrostatic discharge [9–12], aerospace [13–15], and high-end sporting goods, such as tennis rackets, baseball bats, and bicycle frames, etc. [13] due to their excellent mechanical, electrical, piezoresistive, and thermal properties compared to conventional conductive polymer composites. Adding carbon-based fillers in polymers to improve mechanical and functional properties [16,17] is not new. Researchers have already used carbon black widely as reinforcement in thermoplastic and thermoset polymers to improve mechanical, electrical, and thermal properties; for example, it has been extensively used in racing tires to reduce thermal damage [18]. Carbon fiber is also a popular material that is being used for making high-strength and conductive polymer composites for aerospace and automobile applications.

Carbon Nanotube–Reinforced Polymers. http://dx.doi.org/10.1016/B978-0-323-48221-9.00004-2

However, CNTs have a unique place in nanoscience owing to their exceptional electrical, thermal, chemical, and mechanical properties, and have found application in areas as diverse as composite materials, energy storage and conversion, sensors, drug delivery, field emission devices, and nanoscale electronic components. CNTs continue to gain attention and have impacted many fields, and the number of potential applications continues to grow. CNTs have already been used in many sporting goods, such as high-end tennis rackets, baseball bats, and golf clubs. Nowadays, CNTs are becoming easier to produce but cannot be used in bulk form due to their poor translation of outstanding inherent properties of individual CNTs. Therefore, the idea of incorporating CNTs as filler in various polymer matrices, such as thermoplastics, thermosets, and conjugated polymers to fabricate CNT/PNCs has modernized the materials science and technology, and CNT has become one of the major reinforcements for the fabrication of PNCs. For the past 2 decades, a great deal of research has been done on CNT-reinforced PNCs, and the area is still growing.

CNTs are hollow cylinders of graphene with extraordinary electronic and mechanical properties. There are two basic types of CNTs: single-walled carbon nanotubes (SWCNTs), which are the seamless cylinder of a single graphene sheet, and multiwalled carbon nanotubes (MWCNTs), which are made up of coaxial cylinders. CNTs are well-ordered with high aspect ratio; length varies from several hundreds of nanometers to several micrometers or up to millimeters, and diameters are 0.4–2 nm for SWCNTs and 2–100 nm for MWCNTs [19]. The structure of SWCNTs can be seen as wrapping a graphene sheet in a seamless hollow cylinder in which all carbon atoms are densely packed in a honeycomb lattice structure. The structure of MWCNTs can be viewed as several SWCNTs of different radii arranged coaxially with intertube separation of 0.34–0.35 nm, which is close to the interplanar separation of perfectly crystalline graphite [19]. Although the carbon atoms are involved in hexagonal aromatic rings, the C—C bond angles are not planar. This means that the hybridization of carbon atoms is not pure sp^2 but has some contribution of sp^3 character, and this sp^3 character increases as the tube radius of curvature decreases. The effect is the same as for C_{60} fullerene, whose radius of curvature is 0.35 nm and bonds have 10% of sp^3 character [19,20].

Density of SWCNT depends on the rolling of graphene sheet; that is, pure armchair, zigzag, and chiral SWCNTs should have densities of 1.33, 1.34, and 1.40 g/cm^3, respectively. The structure of nanotube influences its properties, including conductance, density, and lattice structure. It is known that electronic structures of SWCNTs determine their electrical properties. Depending on their chiral vector, SWCNTs with small diameter are either metallic or semiconducting, and these differences are caused by the molecular structure that results in a different band structure and thus a different band gap [21]. In this view, the discovery of CNTs and their PNCs is an important addition in the area of nanoscience, playing a key role in modern science and technology. In view of this, research and development on nanoparticles-filled polymers has increased exponentially since the early 1990s with the "discovery" of CNTs [22], the properties of which, particularly modulus, strength, and density, are significantly different from those of traditional microscale fillers (Table 4.1 [23]) and therefore offer exciting possibilities for new generation of composite materials. Structural design and the strength of sp^2 hybridized C—C bond offer a large interest in

Table 4.1 Comparison of Common Materials Versus CNTs in Terms of Strength, Modulus, and Density [23]

Materials	Tensile Strength (MPa)	Tensile Modulus (GPa)	Density (g/cm³)
6061 Aluminum (bulk)	310	69	2.71
4340 Steel (bulk)	1,030	200	7.83
Nylon 6/6 (polymer)	75	2.8	1.14
Polycarbonate (polymer)	65	2.4	1.20
E-glass fiber	3,448	72	2.54
S-2 glass fiber	4,830	87	2.49
Kevlar 49 (aramid fiber)	3,792	131	1.44
T-1000G (carbon fiber)	6,370	294	1.80
CNTs	30,000	1,000	1.90

CNTs, Carbon nanotubes.

mechanical properties of CNTs because sp^2 hybridized C—C bond is the strongest bond [24]. CNTs contain not only sp^2 hybridized C—C bond but also the mixture of σ and π orbitals due to the curvature in graphene sheet after rolling. A great deal of work has been done to study the mechanical properties of CNTs experimentally and theoretically [25–28]. CNTs are the stiffest known fiber yet discovered, with a measured Young's modulus of 1.0 TPa [29]. They have an expected elongation to failure of 20%–30%, which, combined with their stiffness, project to a tensile strength well above 100 GPa in the axial direction. For comparison, the Young's modulus [30] of high-strength steel is around 200 GPa, and its tensile strength is 1–2 GPa. The high stiffness and strength combined with low density implies that nanotubes could serve as an ideal reinforcement in composite materials.

CNTs are attracting much interest for their outstanding properties, which are largely derived from their unusual structure and electronic properties. Since all these properties are concerned directly to the atomic structure of nanotubes, it is quite necessary to have a thorough understanding of the phenomenon to control nanotube size, the number of walls, the helicities, and the structure during growth. The full potential of nanotubes for applications can't be realized until the growth of nanotubes is optimized and well controlled. For utilization of CNTs' properties in real-world applications, such as in composites and hydrogen storage, it is desired to obtain high-quality CNTs in bulk quantity using growth methods that are simple, efficient, and inexpensive. Significant work has been carried out in this field, and various methods have been studied to synthesize CNTs.

Arc discharge [31], laser ablation [32], gas-phase catalytic growth from carbon monoxide [33], and chemical vapor deposition (CVD) from hydrocarbons [34] are different methods for the synthesis of CNTs. The arc discharge and laser ablation are well-established techniques for the production of high-quality and nearly perfect nanotube structures, but due to their scale-up limitations the production cost is too high and hence limits its use. During synthesis, impurities in the form of catalyst particles, amorphous carbon, and fullerenes are also produced. Therefore, purification is required to separate the CNTs from impurities. The gas-phase catalytic growth in which carbon monoxide is used as the carbon source produces high-purity SWCNTs at 1200°C temperature and a pressure of 10 atm. However, this method is not economical due to the high temperature and pressure of the synthesis conditions. The liquid-phase CVD process from hydrocarbons has great potential for the synthesis of large quantity CNTs with fewer impurities and was more amenable to large-scale processing. From the CVD method, we can control the growth of CNT and also their length, diameter, and alignment, and this is a low-cost production method at large scale with high purity. Therefore, the gas phase techniques, such as CVD offer the greatest potential for scaling up CNT production for the processing of composites.

Theoretically, the thermal conductivity of CNTs is very high—5000 W/mK at room temperature [35–37]—while thermal conductivities as high as 3000 W/mK at room temperature have been measured [37], which is equivalent to the thermal conductivity of natural diamond and a single crystal of graphite and almost 10 times higher than that of copper (385 W/mK). Therefore, CNTs are highly useful for making thermally conductive materials for aerospace applications. Due to their one-dimensional nature, charge carriers can travel through nanotubes without scattering, which is called ballistic transport. The absence of scattering means that joule heating is minimized and thus CNT can carry very large current densities up to 100 mA/cm^2 [38–40]. In addition, carrier mobility as high as 10^5 $cm^2/V\,s$ has been observed in semiconducting nanotubes [41]. Superconductivity has also been observed in SWCNTs [42]. The metallic CNTs can carry an electrical current density of 4×10^9 A/cm^2, which is 1000 times greater than that carrried by metals, such as copper [43]. Pure CNTs are nonmagnetic in nature, but most of the prepared CNTs possess some magnetic character that can be attributed to the presence of metal catalyst nanoparticles in the cavities of CNTs.

4.2 Fabrication of CNT/Polymer Nanocomposites

A variety of techniques for the fabrication of PNCs are in practice for incorporation of CNTs into various polymer matrices. The main objective is to debundle the CNTs and to achieve their uniform dispersion in the polymer matrix. At present, there is no single technique that is universally applicable to all situations. Therefore, we here discuss both traditional approaches and new approaches used for the fabrication of CNT/PNCs.

4.2.1 Traditional Approaches

Solution mixing, melt blending, and in situ polymerization are three traditional approaches for the fabrication of CNT/PNCs.

4.2.1.1 Solution Mixing Technique

The solution mixing method is one of the most useful methods for the fabrication of CNT/PNCs on a laboratory scale. This method is also useful to make nanocomposites without losing properties of nanosized fillers [44] because dispersion and interfacial bonding of nanofillers is important to achieve good mechanical and electrical properties. In this

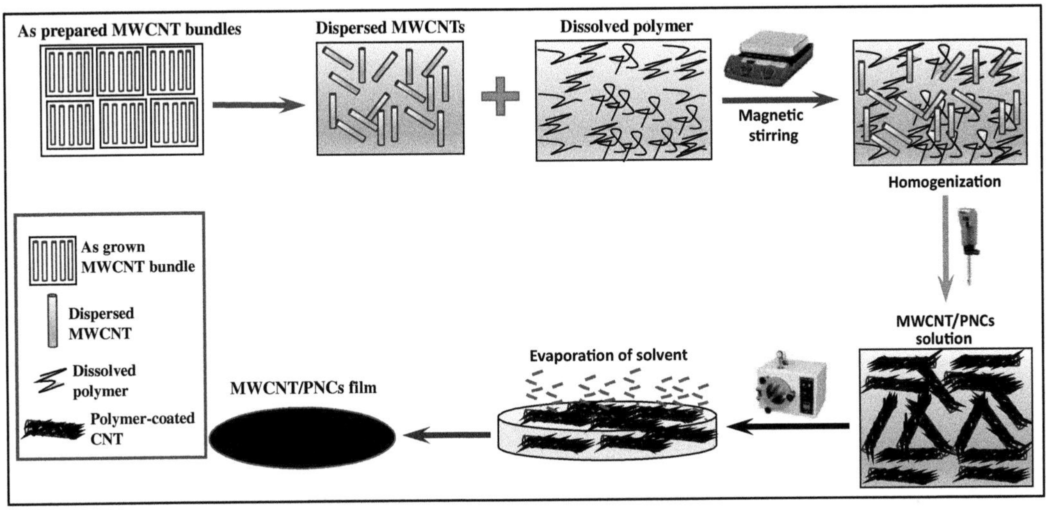

FIGURE 4.1 Schematic diagram of solution mixing technique for CNT/PNCs. *MWCNTs*, Multiwalled carbon naotubes; *PNC*, polymer nanocomposites.

process, the CNTs and polymer are mixed separately in a suitable solvent. The choice of solvent is mainly governed by solubility of the polymer matrix. The solvent for CNTs and polymer matrix may be same or different, but should have good miscibility. The dispersed CNTs and polymer are mixed together, and finally a composite film is produced with and without vacuum-assisted solvent evaporation. In this whole process, ultrasonication, magnetic stirring, and high-speed homogenization techniques are used to disperse and mix CNTs properly into the polymer matrix and are clearly shown in a schematic illustration (Fig. 4.1) [9,10]. In some cases, CNTs are not separately dispersed; rather, they are directly mixed to the polymer solution using high-power ultrasonication or shear mixing. The main disadvantage related to the use of ultrasonication and shear mixer for a long time is that it can shorten the length of CNTs, thereby reducing the properties of PNCs.

To achieve better dispersion and to solve the problem of tube shortening, surface-functionalized CNTs are used. The functionalized CNTs provide good dispersion and improved interfacial interaction with polymer matrix, and participate in the enhancement of electrical, mechanical, thermal, and dielectric properties of the resulting PNCs [9,11,45]. The boiling point of the solvent used for the dispersion of CNTs and polymer matrix also has an impact on the properties of PNCs. In general, low boiling point solvents are preferred due to fast evaporation from the composites. High boiling point solvents are difficult to be removed from the composites, and they also tend to get trapped in the solidifying PNCs. One of the advantages of this technique is the possibility of debundling the CNTs and hence improving the quality of dispersion in polymer matrix [46]. However, this technique cannot be used for insoluble polymers. This technique also contaminates the environment due to use of large amounts of solvents, and therefore its use is limited to only large-scale production of CNT/PNCs. Nevertheless, this technique is still widely used.

4.2.1.2 Melt Blending Technique

Melt processing technique is the most promising approach for the fabrication of thermoplastic PNCs on an industrial scale due to its low cost. This technique is also suitable to insoluble polymers, especially for thermoplastic polymers that cannot be processed with solution techniques due to their insolubility in common solvents. In general, melt processing involves blending of polymer matrix with CNTs by the application of intense shear forces at elevated temperatures, which contributes in partial deagglomeration of CNT bundles and their dispersion into the matrix. The melt blending can be carried out in batch or continuous operation using high shear mixer and twin screw extruder, respectively [47–49]. High shear mixer is often used to prepare master batches of PNCs, which are highly concentrated and used to fabricate the desired amount of CNT-loaded nanocomposites. An extruder consists of single or twin screws, and a twin screw extruder is more effective in terms of uniform mixing. A typical twin screw extruder consists of two rotating screws as shown in Fig. 4.2 [50].

The polymer granules are loaded via hopper into the extruder and the rotating screw catches the polymer granules and pushes them forward. These polymer granules are melted inside the heated melting zone, and shearing of material between the two screws is also taking place. The CNTs are also loaded into the extruder via a separate hopper, and the mixing takes place in the melting zone due to the shearing action and then reaches to homogenization zone [52,53]. Finally, the mixture passes to die before becoming semisolid and cooled by air drying or by passing through a water bath and being chopped into granules for further use, or composites of desired size and shape can be made using an injection molding machine. This is a simple, fast, environmental friendly, and industrially viable technique for large production, and no solvent is required during the preparation of nanocomposites; but it is not very effective in the breaking of bundles of CNTs compared to solution mixing [54]. This technique is not good for those materials that degrade with temperature [46]. Although the melt blending technique is simple, high shear force and high temperature can also break CNTs or even polymer chains. Therefore, the shear force and temperature need to be properly optimized to achieve appropriate dispersion without compromising intrinsic properties of CNTs and polymer matrix.

4.2.1.3 In Situ Polymerization Technique

In situ polymerization is a very effective method to improve the dispersion and interaction of CNTs in polymer matrix and a viable option for the preparation of insoluble and thermally unstable polymer-based composites. This is a very convenient method for preparation of polymer nanocomposites with high CNT loading and good miscibility with almost all polymer matrices. In this process, dispersed CNTs are mixed with a monomer solution and then polymerization takes place in the presence of an initiator under certain conditions [55]. Fig. 4.3 shows all the steps of in situ polymerization for CNT/PNCs.

This method provides good dispersion and compatibility between polymer chains and CNTs as compared to solvent mixing and melt blending techniques. This method also

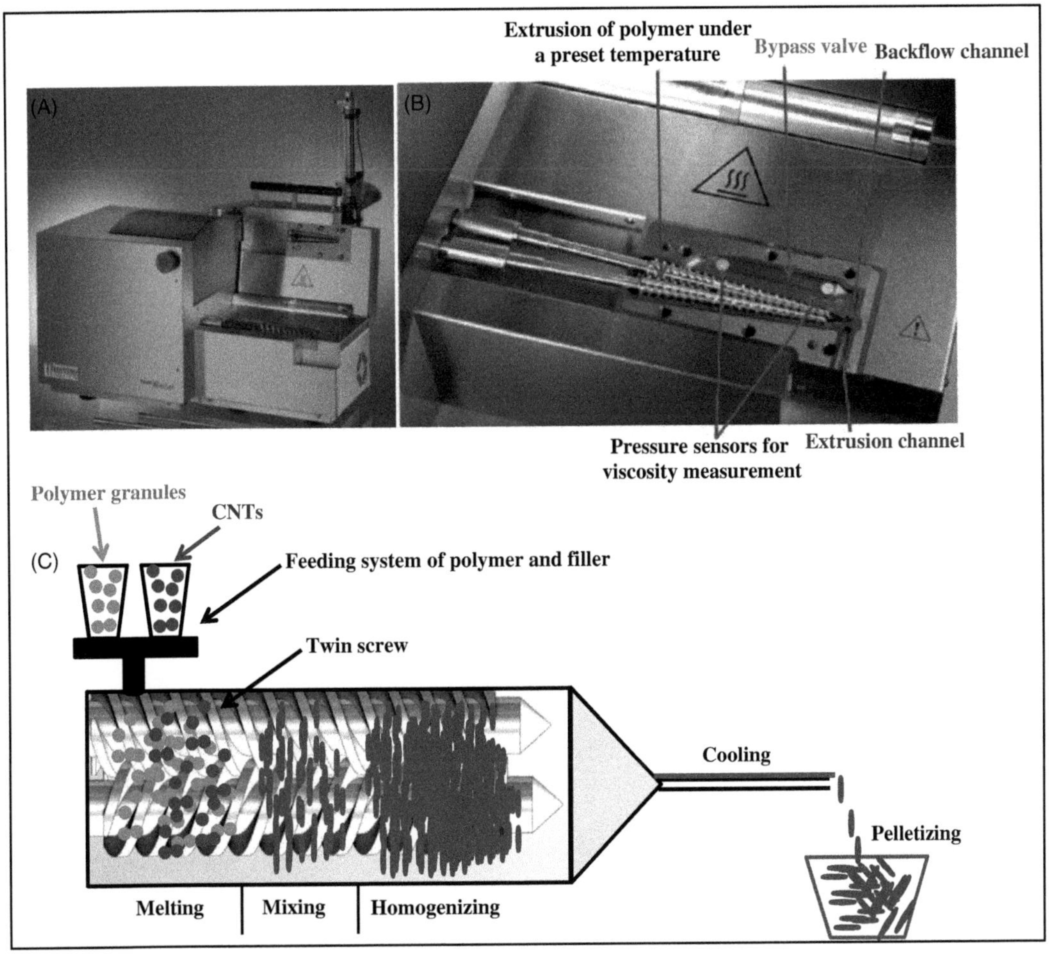

FIGURE 4.2 (A) Twin screw extruder, (B) view of microcompounder showing different valves and channels, and (C) schematic representation of twin screw extruder for melt mixing of CNTs with polymer matrix [51].

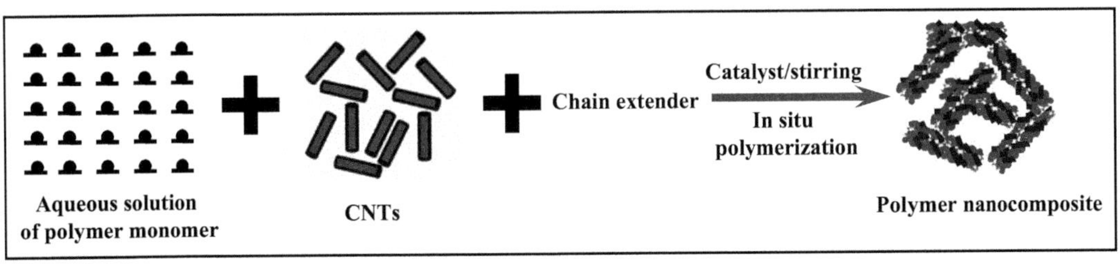

FIGURE 4.3 Schematic representation of in situ polymerization process.

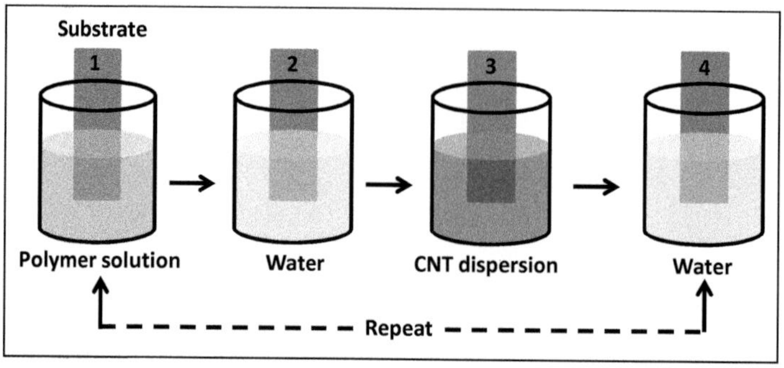

FIGURE 4.4 Schematic representation of LBL film deposition on substrate in which step 1 and step 3 represent the adsorption of positively and negatively charged polyelectrolyte and step 2 and step 4 are washing steps. *From S. Srivastava, N.A. Kotov, Composite layer-by-layer (LBL) assembly with inorganic nanoparticles and nanowires, Acc. Chem. Res. 41 (12) (2008) 1831–1841, with the permission of American Chemical Society [56].*

provides a stronger and more active interface between polymer and CNTs, but it is less applicable as it requires a large amount of solvent for processing.

4.2.2 New Approaches

4.2.2.1 Layer-by-Layer Route

The layer-by-layer (LBL) approach involves building a layered composite film on the substrate by dipping the substrate alternately in a CNT-dispersed polyelectrolyte solution [18,56]. Initially, substrate is dipped in positively charged polyelectrolyte solution; after that the substrate is rinsed with pure water to remove loosely attached polyelectrolytes. To grow the LBL further, the substrate is dipped in negatively charged polyelectrolyte solution; as a result, a double layer of polyelectrolyte is attached on the substrate. The process is shown in Fig. 4.4.

In the same way, this technique can also be used to incorporate inorganic nanoparticles, nanowires, and nanosheets into the organic polymers. In this process, initially the substrate is dipped into the polymer solution so that polymer layer can attach on the surface of substrate; after that the substrate is dipped again in the CNT-dispersed solution. To grow the LBL further, the same process is repeated again. This method has significant advantages as the thickness and CNT ratio can easily be controlled and high nanotube loading can be obtained [18].

4.2.2.2 Bucky Paper–Based Approach

Bucky paper is a thin, porous sheet of CNTs usually formed by filtration of CNT dispersion [57]. Infusing the bucky paper with polymer is a simple way to create PNCs with high loading of CNTs and with improved mechanical and electrical properties. The bucky paper sheet can also be inserted between laminates to enhance the mechanical properties [18].

Teotia et al. [58] reported an approach to make bucky paper–based phenolic resin composites as shown in Fig. 4.5. In this method, CNTs were dispersed in some solvent and added to the dissolved phenolic resin. The suspension was then mixed together and

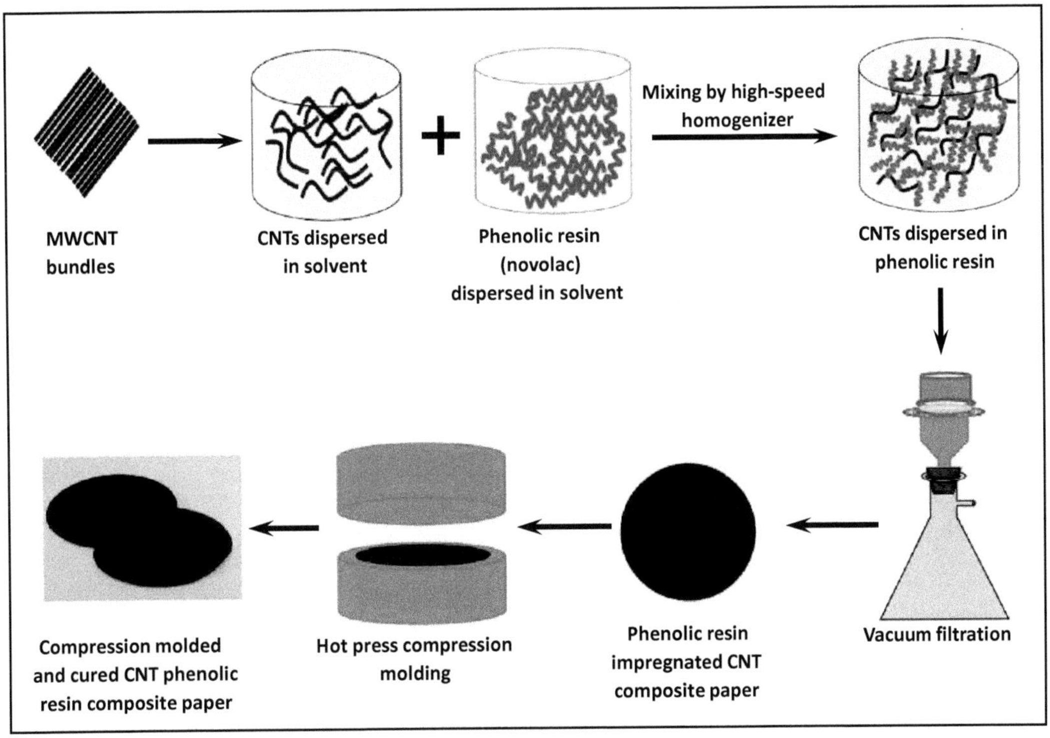

FIGURE 4.5 **Schematic diagram for CNT–phenolic resin paper composite preparation [58].**

homogenized using a high-speed homogenizer at 30,000 rpm. The dispersed CNTs and phenolic resin mixture were then filtered in a specially designed filtration system, and a film of CNT impregnated polymer resin was obtained, which was then dried to get a CNT-phenolic resin prepreg. The prepreg was compression molded using a hydraulic press.

4.3 Dispersion and Alignment of CNTs in Polymer Matrices for Processing of Polymer Nanocomposites

These fabrication processes are good to enhance the mechanical and electrical properties of PNCs, but some problems still remain in the system that hinder the excellent properties of CNTs in PNCs. Poor dispersion and adhesion or interfacial interaction of CNTs in polymer matrix at higher loading is a key challenge because good dispersion provides stabilization of CNTs in polymer matrix. The methods described earlier for the fabrication of PNCs require CNTs to be well dispersed in polymer matrix to maximize the contact surface area with the polymer matrix. CNTs have nanoscale dimensions and therefore synthesis and van der Waals force of interaction create entanglement and force them to agglomerate into bundles, and these bundles make them difficult to dissolve into common solvents.

Therefore, these bundles of CNTs show poor electrical and mechanical properties compared to individual CNTs because of lower surface area of bundled CNTs with imperfect interfacial bonding with polymer matrix. Depending upon the methods used for the fabrication of CNT-reinforced PNCs, various techniques, such as ultrasonication, high-speed magnetic stirring, and homogenization techniques are used to disperse the different amounts of CNTs in polymer matrix to fabricate nanocomposites with enhanced properties. The most effective methods are surface functionalization of CNTs using some chemical treatment or by dispersing CNTs using some dispersing agents, such as surfactants.

Another key challenge is to determine the effect of CNT alignment on the properties of PNCs, because CNTs have asymmetric structure and properties and therefore display highest properties in the oriented reinforcement direction. Therefore, to take the full advantage of CNT properties, CNT alignment is important prior to, during, or after composite fabrication. Several methods, such as electric arc discharge [59], application of electric or magnetic field during composite formation [60–62], composite slicing [63], CVD growth of aligned CNTs [64,65], mechanical stretching of CNT/PNCs [66], melt fiber spinning [54], and electrospinning [67,68] have been reported for the alignment of CNTs.

4.4 Chemical Modifications of CNTs for Processing of Polymer Nanocomposites

Impurities, such as amorphous carbon, fullerenes, and metal catalysts also create difficulties in dispersing CNTs into polymer matrices. Therefore, postsynthesis chemical processing approaches that can purify and separate CNTs or can create different reactive sites on the surface of CNTs are required in order to manipulate the favorable electronic and mechanical properties. The interface between CNTs and polymer matrix should be carefully engineered; otherwise, it will lead to poor load transfer between CNTs when in bundles and may result in interfacial slippage between CNTs and polymer chains. Therefore, to resolve these problems, two major approaches, covalent and noncovalent functionalization, are used as interactions between active materials and CNTs. They are discussed in the following sections.

4.4.1 Covalent Functionalization

Covalent functionalization of CNTs can be achieved by either direct addition reactions of chemical reagents to the sidewall of CNTs or attachment of appropriate surface functional groups, such as carboxylic group or amine group on the CNTs. 1,3-dipolar cycloaddition reaction of CNT with phenol group is shown in Fig. 4.6. This functionalization provides stable dispersion of CNTs in several polar solvents [18].

An advantage of phenol group functionalization is that they allow further functionalization of CNT with some other molecules. However, the most common and facile method for the surface functionalization of CNT is nanotube oxidation using some acid treatment. These types of functionalization of CNTs are much more stable in polar

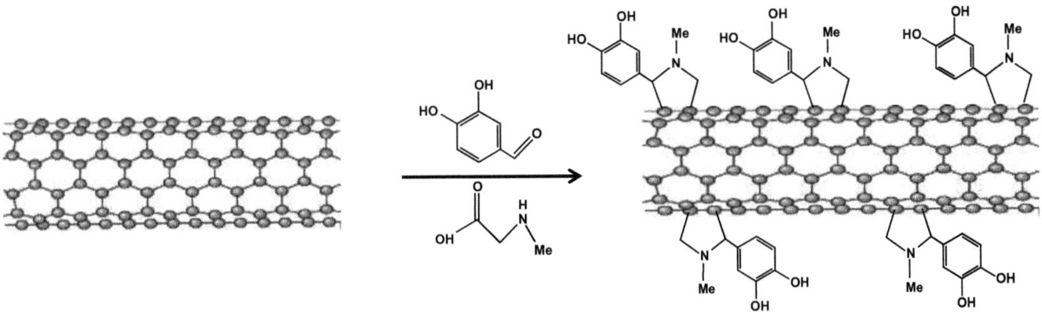

FIGURE 4.6 Schematic representation of covalent functionalization of CNT via 1,3-dipolar cycloaddition reaction [18].

solvents. In a typical reflux setup used for the oxidation of MWCNTs with nitric acid [9,45], oxidation of CNTs was performed by refluxing 1 g of CNTs with 200 mL concentrated nitric acid [60% (v/v)] as shown in Fig. 4.7. The mixture was refluxed for 24 h at 60°C to achieve reasonable surface modification and then cooled to room temperature. After refluxing and cooling, modified MWCNTs were collected by filtration and washed repeatedly with deionized water until neutralization. Finally, they were washed with acetone to make sure that no water molecules are trapped inside the tube bundles. These functionalized CNTs can be easily embedded in polymer matrix to make CNT/PNCs with homogeneous CNT dispersion.

Acid-modified CNTs significantly improve the interfacial bonding between CNT and polymer matrices. The carboxylic group provides stronger CNT–polymer interfacial interaction leading to enhanced value of mechanical properties [9,18,69]. In the same way,

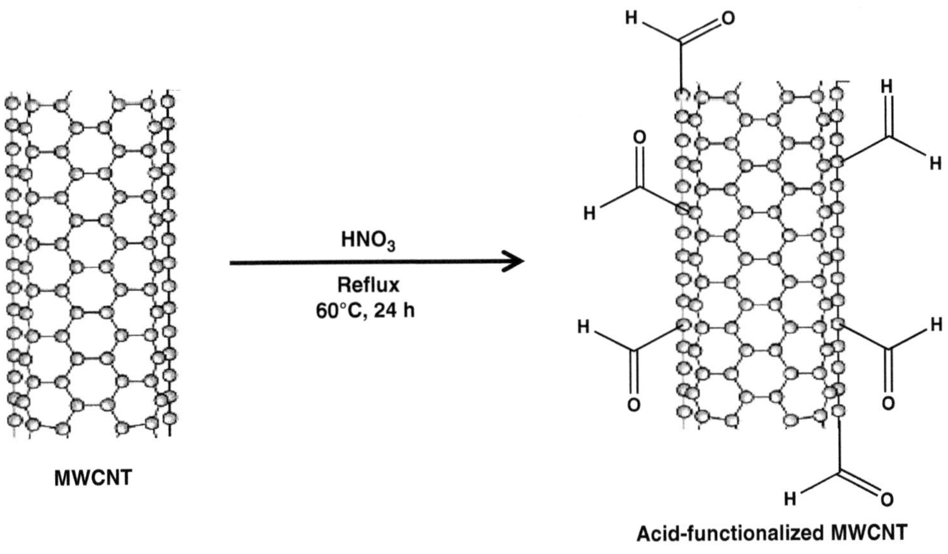

FIGURE 4.7 Schematic diagram of acid functionalization of CNT.

amine functionalization of CNTs can also be done by further treatment of carboxylic group functionalized CNTs with $SOCl_2$ in the presence of dimethylformamide (DMF) for acyl chlorination and then treated again with amine containing solvents, such as triethylenetetraamine (TETA) or hydrazine to create an amine functional group on CNT [45].

The presence of active functional groups, such as the carboxylic group or amine group allows for further covalent functionalization with polymer molecules and is also used to make PNCs. Two main approaches, such as "grafting from" and "grafting to" have been reported for covalent functionalization of CNTs with polymers.

The "grafting from" approach is based on initial immobilization of initiator on the nanotube surface, and then in situ polymerization takes place in order to form polymer molecules onto the CNT surface. The benefit of this technique is that polymer-functionalized CNTs with high grafting density can be prepared, but this method requires strict control of the amount of initiator and substrate used.

The "grafting to" approach is based on the attachment of already functionalized polymer molecules to functional groups on the nanotube surface via appropriate chemical reactions. In this approach, commercially available polymers containing reactive functional groups can be used. However, the main limitation of this technique is that initial binding of polymer chains sterically prevents diffusion of additional macromolecules to the surface, leading to relatively low polymer loading [18].

As shown in Fig. 4.8, several routes have been taken for functionalization, such as sidewall functionalization, defect group functionalization, noncovalent exohedral functionalization with molecules and with polymers, and endohedral functionalization with fullerenes.

4.4.2 Noncovalent Functionalization

The noncovalent functionalization of nanotubes normally involves van der Waals, π–π, CH–π, or electrostatic interactions between polymer molecules and CNT surface [72,73]. The advantage of noncovalent functionalization is that it does not alter the structure of the CNT, and, therefore, both electrical and mechanical properties should also remain unchanged. However, the efficiency of the load transfer might decrease since the forces between the wrapping molecules and the nanotube surface might be relatively weak.

There are several noncovalent approaches for nanotube functionalization, such as surfactant-assisted dispersion, polymer wrapping, plasma polymerization treatment, and polymerization filling technique [74,75]. Surfactant-assisted dispersion is a very common technique that enables the transfer of CNTs to the aqueous phase in the presence of surface-active molecules, such as sodium dodecylsulfate (SDS), *N*-methyl-2-pyrrolidone (NMP), or benzylalkonium chloride. The presence of an aromatic group in the surfactant molecule allows for π–π stacking interactions with the graphitic sidewalls of the CNTs, which results in their effective coating and dispersion. Polymer wrapping involves the utilization of conjugated and aromatic group containing polymers [e.g., polyvinyl pyrrolidone (PVP), poly(phenylenevinylene), pyrene–poly(ethylene glycol) (PEG)], which can wrap around

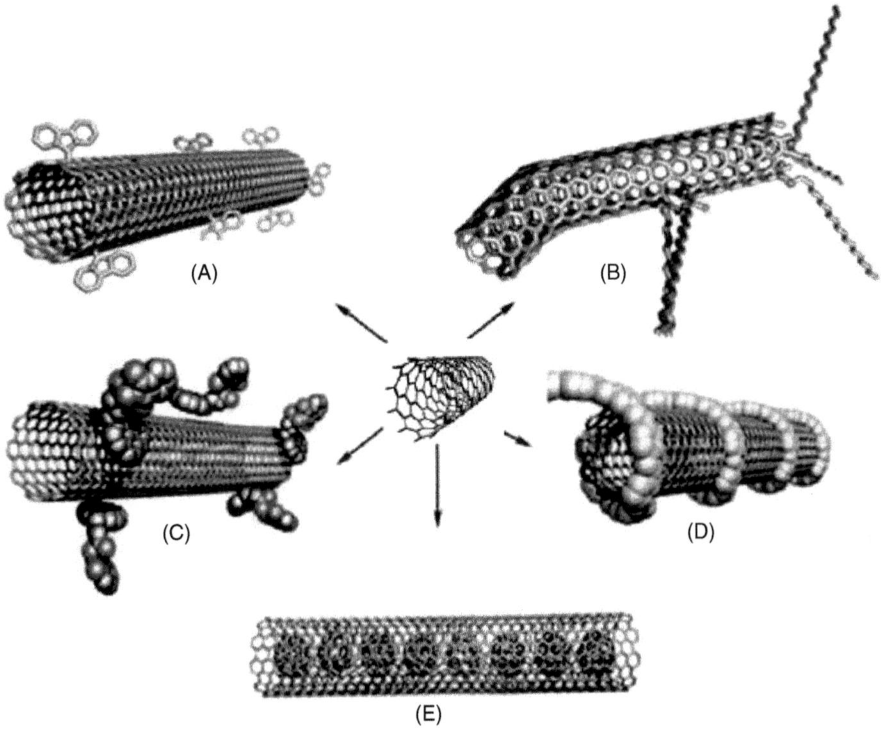

FIGURE 4.8 Possibilities of SWCNT functionalization. (A) Sidewall functionalization, (B) defect group functionalization, (C) noncovalent exohedral functionalization with molecules through π-stacking, (D) noncovalent exohedral functionalization with polymers, and (E) endohedral functionalization with fullerenes [70,71].

CNTs through π–π stacking and van der Waals interactions [75]. Poly(dimethylsiloxane) (PDMS) polymers have also been used for CNT coating via noncovalent CH–π interactions [73,76]. Plasma polymerization treatment enables the coating of CNTs with a very thin (~3 nm) polymer layer. Polymer composites based on these coated nanotubes enhanced interfacial bonding in a PS polymer matrix [77].

Different thermoplastic and thermoset polymer matrices have been tried to realize the superior mechanical properties of CNTs for development of lightweight, strong material. NASA scientists are considering CNT–polymer composites for space elevators. To date, a volume of literature is available on the improvement of mechanical performance of polymers with addition of CNTs. The first study for formation of CNT–polymer composites was carried out by Ajayan et al. [63] in which CNTs were aligned within the epoxy matrix by the shear forces; the overall improvement of 155% in flexural strength with addition of merely 0.3 wt.% amine functionalized MWCNTs in epoxy matrix and an increment of 38% in flexural modulus was achieved. Mathur et al. [78] reported an increment of 158% in flexural strength of phenolic resin with addition of 5 vol.% of MWCNT. Coleman et al. [55] reviewed the mechanical properties of a large number of CNT-reinforced polymer (thermoplastic and thermosetting) composites fabricated by various methods, and reported enhancement in mechanical properties. Du et al. [79] studied the experimental results for mechanical performance of CNTs nanocomposites carried out by different research groups and observed that the improvements are uncertain and far below the simplest theoretical estimates. Haggenmueller [80] applied the Halpen Tsai composite theory to CNT nanocomposites and observed that the experimental elastic modulus is smaller than predicted by more than one order. This is attributed to the lack of perfect load transfer from nanotubes to matrix due to nonuniform dispersion and small interfacial interaction. Although chemical functionalization of CNTs has sorted out those problems to an extent, the best results have yet to be achieved. Also, the aspect ratio is other source of uncertainty in mechanical properties. Defects on the CNT surface are also expected to influence the mechanical properties significantly. The methods of handling nanotubes, including acid treatments and sonication for a long time, are known to shorten nanotube results in decreasing aspect ratio and are detrimental to mechanical properties.

The mechanical properties of CNT-based composites increase up to a certain loading of CNTs, beyond which they start decreasing [81,82]. This may be because of an increase in viscosity of polymers at higher CNTs loading and also cause some surface of CNTs not to be completely covered by polymer matrix due to the large specific surface area of CNTs. Therefore, few studies have been carried out to disperse high loading of CNTs in polymer matrices with improved mechanical properties. Bradford et al. [83] reported 400 MPa (tensile strength) and 22.3 GPa (tensile modulus) for 27 vol.% of MWCNT–epoxy composites. Feng et al. [84] also reported 183 and 408% improvement in tensile strength and tensile modulus respectively at 39.1 wt.% SWCNTs loading compared with those of the neat epoxy. Ogasawara et al. [85] found 50.6 GPa and 183 MPa modulus and ultimate tensile strength, respectively, of CNT (21.4 vol.%)–epoxy composite. These values were 19 and 2.9 times higher than the neat epoxy resin, respectively. Andrews et al. [84] prepared aligned CNT/pitch composites and found the significant improvement in the electrical and mechanical properties, especially due to orientation of CNTs. Du et al. [79] compared the mechanical performance of randomly oriented and aligned CNTs polymer composites. Their study revealed that in aligned CNT–polymer nanocomposites tensile strength and modulus even reached to 3600 MPa and 80 GPa, respectively, which is much higher than the general value of 100 MPa and 6 GPa in cases of randomly oriented CNT–polymer nanocomposites. They also observed that the mechanical properties are always higher for aligned CNTs composites with higher loading whereas the case is different for isotropic CNT–polymer composites.

The extraordinary electrical conductivity of CNTs makes them excellent additives to impart electrical conductivity to PNCs. Many experimental results show that the conductive CNT composites can be constructed at low loading of CNTs due to low percolation threshold originated from the high aspect ratio and conductivity of CNTs [86]. It can be found from almost all the experimental results that CNT nanocomposites exhibit a typical percolation behavior, and CNT reinforcement to polymers can increase the conductivity of resulting composites to several orders of magnitude or even sometimes higher than

10 orders of magnitude. According to percolation theory, the conductivity follows the following power law close to threshold percolation.

$$\sigma = \sigma_0 (\phi - \phi_c)^t$$

where σ and σ_0 are the electrical conductivity of conductive PNCs at different wt.% of CNT and the proportionality constant that is related to the intrinsic conductivity of the CNT, respectively; ϕ is the volume fraction of the CNT, and ϕ_c is the percolation threshold; t is the critical exponent that reflects the dimensionality of the conductive network formed in the conductive PNCs. Theoretical and experimental results have shown that percolation laws are applicable to CNT-based composites, and the enhancement of maximum conductivity and percolation can be achieved with significantly lower filler concentrations than with other carbon and other conductive fillers. But sometimes, the same polymer with the same preparation method and the same concentration of CNT shows different electrical conductivity or percolation threshold, which may be because of some uncertainties, such as type and quality of CNTs, aspect ratios, orientation, purity, and entanglement [87,88]. In CNT-filled PNCs, CNTs are covalently integrated into the matrix and become part of the cross-linked structure rather than just a separated component [89]. When the percolation threshold is reached, a percolated filler network structure is formed at a critical minimum distance between CNTs structures, and electron conduction is facilitated through a "hopping" or "tunneling" mechanism. The formation of a CNT continuous electron path [87,90,91] is the main cause of the electron transport by tunneling or electron hopping, which occurs along CNT interconnects. Pal and Kumar [92] developed a multiscale modeling of short carbon fiber–CNT–polymer matrix hybrid composites to evaluate the effects of electron hopping and the formation of conductive networks on the electrical conductivity of SCF-CNT-PNCs.

Some researchers have suggested that the low percolation concentrations observed in experiments are due to nanoscale phenomena, such as electron hopping [93–95], while many other investigators focus on dispersion arguments, indicating that the low percolation concentrations observed are due to the formation of conducting networks that allow a high percentage of electrons to flow through the composites within the polymer [96–98]. However, Seidel and Lagoudas [99] considered that in the case of MWCNT/PNCs, the formation of conductive networks is seen as the principal cause for the large increases in nanocomposite conductivity observed at very low CNT concentrations, whereas in the case of SWCNT/PNCs, both the formation of conductive networks and electron hopping lead to large increases in nanocomposite conductivity at very low CNT concentrations. The formation of conducting networks is the dominant factor at concentrations below the percolation concentration associated with electron hopping. Ounaies et al. [96] observed nonlinear I–V characteristics when the CNT loadings are in excess of 0.1 vol.%, and this nonlinear behavior is most likely due to a tunneling mechanism. Conduction may occur by electron hopping from a nanotube to an adjacent one when they are close enough and without physically touching each other [100].

As discussed previously, the CNTs have thermal conductivity as high as 5000 W/mK predicted for SWCNTs [101] at room temperature and have experimental value 3000 W/mK for isolated MWCNTs. So it is quite expected that the reinforcement of CNTs can significantly enhance the thermal properties of CNT/PNCs. The improvement in thermal transport properties of CNT/PNCs leads to their applications for usage as printed circuit boards, connectors, thermal interface materials, and heat sinks, and for several other aerospace applications.

4.5 Conclusions and Future Scope

The excellent properties of CNTs have opened up a new age of making advanced multifunctional nanocomposites. Incorporation of CNTs in polymer matrices provides materials that could be used for many high-performance engineering applications. Studies on CNT/PNCs suggest that CNTs have great potential in altering the properties of PNCs. The quality of CNT-reinforced PNCs depends on several factors, such as type of CNTs, chirality, dispersion, alignment, purity, aspect ratio, functionalization, and interfacial interaction between CNTs and polymer matrix. A lot of work has been done on CNT processing and to improve the quality of CNTs. The biggest challenge is to achieve homogeneous dispersion and interfacial adhesion of CNTs with polymer matrix so that maximum filler surface area will be available for load transfer between filler and matrix. The functionalization of CNTs provides an easy way to improve the dispersion and stress transfer between CNTs and polymer matrix, but much improvement is needed in this area. It is important to focus on different techniques of functionalization of CNTs and also find a route that can improve the dispersion and compatibility of CNTs with polymer matrix. Currently, CNTs are widely used in nanocomposites for electronic applications, such as protection from unwanted electromagnetic radiation and also for electrostatic-discharge components. The microwave-absorbing capability of CNTs could be exploited to heat temporary housing structures and may have applications in space exploration. Thin layers of CNTs on polymer sheets might also be used in transparent conducting composites. The high mechanical strength of these PNCs could be utilized to make some high-end sporting goods, such as tennis rackets, baseball bats, and bicycle frames and thus deliver superior performance. In short, the biggest markets for CNT nanocomposites will undoubtedly be for high-value applications that can absorb the added costs; these include commercial sectors, such as electronics, as well as aerospace industries that require lightweight, high-strength, high-temperature-resistant composites. Once the cost of nanotubes becomes comparable to that of carbon fiber or even to that of the much cheaper reinforcing agent carbon black, thermally stable CNT-filled rubber tires could become a reality.

References

[1] X. Wu, Y. Han, X. Zhang, C. Lu, Highly sensitive, stretchable, and wash-durable strain sensor based on ultrathin conductive layer@polyurethane yarn for tiny motion monitoring, ACS Appl. Mater. Interfaces 8 (15) (2016) 9936–9945.

[2] J. Zhao, K. Dai, C. Liu, G. Zheng, B. Wang, C. Liu, et al. A comparison between strain sensing behaviors of carbon black/polypropylene and carbon nanotubes/polypropylene electrically conductive composites, Compos. Part A 48 (2013) 129–136.

[3] H. Liu, W. Huang, J. Gao, K. Dai, G. Zheng, C. Liu, et al. Piezoresistive behavior of porous carbon nanotube–thermoplastic polyurethane conductive nanocomposites with ultrahigh compressibility, Appl. Phys. Lett. 108 (1) (2016) 011904.

[4] L. Lin, S. Liu, S. Fu, S. Zhang, H. Deng, Q. Fu, Fabrication of highly stretchable conductors via morphological control of carbon nanotube network, Small 9 (21) (2013) 3620–3629.

[5] V. Eswaraiah, K. Balasubramaniam, S. Ramaprabhu, Functionalized graphene reinforced thermoplastic nanocomposites as strain sensors in structural health monitoring, J. Mater. Chem. 21 (34) (2011) 12626–12628.

[6] D. Ding, H. Wei, J. Zhu, Q. He, X. Yan, S. Wei, et al. Strain sensitive polyurethane nanocomposites reinforced with multiwalled carbon nanotubes, Energ. Environ. Focus 3 (1) (2014) 85–93.

[7] H. Liu, Y. Li, K. Dai, G. Zheng, C. Liu, C. Shen, et al. Electrically conductive thermoplastic elastomer nanocomposites at ultralow graphene loading levels for strain sensor applications, J. Mater. Chem. 4 (1) (2016) 157–166.

[8] D. Li, Y. Liu, B. Lin, C. Lai, Y. Sun, H. Yang, et al. Synthesis of ternary graphene/molybdenum oxide/poly(p-phenylenediamine) nanocomposites for symmetric supercapacitors, RSC Adv. 5 (119) (2015) 98278–98287.

[9] T.K. Gupta, B.P. Singh, S.R. Dhakate, V.N. Singh, R.B. Mathur, Improved nanoindentation and microwave shielding properties of modified MWCNT reinforced polyurethane composites, J. Mater. Chem. 1 (32) (2013) 9138–9149.

[10] T.K. Gupta, B.P. Singh, S. Teotia, V. Katyal, S.R. Dhakate, R.B. Mathur, Designing of multiwalled carbon nanotubes reinforced polyurethane composites as electromagnetic interference shielding materials, J. Polym. Res. 20 (6) (2013) 1–7.

[11] T.K. Gupta, B.P. Singh, R.B. Mathur, S.R. Dhakate, Multi-walled carbon nanotube-graphene-polyaniline multiphase nanocomposite with superior electromagnetic shielding effectiveness, Nanoscale 6 (2) (2014) 842–851.

[12] B.P. Singh, Prabha, P. Saini, T. Gupta, P. Garg, G. Kumar, et al. Designing of multiwalled carbon nanotubes reinforced low density polyethylene nanocomposites for suppression of electromagnetic radiation, J. Nanopart. Res. 13 (12) (2011) 7065–7074.

[13] M.F. De Volder, S.H. Tawfick, R.H. Baughman, A.J. Hart, Carbon nanotubes: present and future commercial applications, Science 339 (6119) (2013) 535–539.

[14] J. Njuguna, K. Pielichowski, Polymer nanocomposites for aerospace applications: properties, Adv. Eng. Mater. 5 (11) (2003) 769–778.

[15] A.M. Marconnet, N. Yamamoto, M.A. Panzer, B.L. Wardle, K.E. Goodson, Thermal conduction in aligned carbon nanotube–polymer nanocomposites with high packing density, ACS Nano 5 (6) (2011) 4818–4825.

[16] Y. Cui, S. Kundalwal, S. Kumar, Gas barrier performance of graphene/polymer nanocomposites, Carbon 98 (2016) 313–333.

[17] Y. Cui, S. Kumar, B.R. Kona, D. van Houcke, Gas barrier properties of polymer/clay nanocomposites, RSC Adv. 5 (78) (2015) 63669–63690.

[18] M.T. Byrne, Y.K. Gun'ko, Recent advances in research on carbon nanotube–polymer composites, Adv. Mater. 22 (15) (2010) 1672–1688.

[19] W. Cheung, F. Pontoriero, O. Taratula, A.M. Chen, H. He, DNA and carbon nanotubes as medicine, Adv. Drug Deliv. Rev. 62 (6) (2010) 633–649.

[20] Y. Huh, J.Y. Lee, J. Cheon, Y.K. Hong, J.Y. Koo, T.J. Lee, et al. Controlled growth of carbon nanotubes over cobalt nanoparticles by thermal chemical vapor deposition, J. Mater. Chem. 13 (9) (2003) 2297–2300.

[21] V. Mittal, Polymer Nanotube Nanocomposites Synthesis, Properties, and Applications, Scrivener Publishing, Beverly, MA, (2010).

[22] S. Iijima, Helical microtubules of graphitic carbon, Nature 354 (6348) (1991) 56–58.

[23] R.F. Gibson, Principles of Composite Material Mechanics, second ed., CRC Press, Boca Raton, FL, (2007).

[24] D.H. Robertson, D.W. Brenner, J.W. Mintmire, Energetics of nanoscale graphitic tubules, Phys. Rev. 45 (21) (1992) 12592–12595.

[25] M.M.J. Treacy, T.W. Ebbesen, J.M. Gibson, Exceptionally high Young's modulus observed for individual carbon nanotubes, Nature 381 (6584) (1996) 678–680.

[26] A. Krishnan, E. Dujardin, T.W. Ebbesen, P.N. Yianilos, M.M.J. Treacy, Young's modulus of single-walled nanotubes, Phys. Rev. 58 (20) (1998) 14013–14019.

[27] E.W. Wong, P.E. Sheehan, C.M. Lieber, Nanobeam mechanics: elasticity, strength, and toughness of nanorods and nanotubes, Science 277 (5334) (1997) 1971–1975.

[28] M.-F. Yu, O. Lourie, M.J. Dyer, K. Moloni, T.F. Kelly, R.S. Ruoff, Strength and breaking mechanism of multiwalled carbon nanotubes under tensile load, Science 287 (5453) (2000) 637–640.

[29] M.-F. Yu, B.S. Files, S. Arepalli, R.S. Ruoff, Tensile loading of ropes of single wall carbon nanotubes and their mechanical properties, Phys. Rev. Lett. 84 (24) (2000) 5552–5555.

[30] B.I. Yakobson, C.J. Brabec, J. Bernholc, Nanomechanics of carbon tubes: instabilities beyond linear response, Phys. Rev. Lett. 76 (14) (1996) 2511–2514.

[31] C. Journet, W.K. Maser, P. Bernier, A. Loiseau, M.L. de la Chapelle, S. Lefrant, et al. Large-scale production of single-walled carbon nanotubes by the electric-arc technique, Nature 388 (6644) (1997) 756–758.

[32] A.G. Rinzler, J. Liu, H. Dai, P. Nikolaev, C.B. Huffman, F.J. Rodríguez-Macías, et al. Large-scale purification of single-wall carbon nanotubes: process, product, and characterization, Appl. Phys. 67 (1) (1998) 29–37.

[33] P. Nikolaev, M.J. Bronikowski, R.K. Bradley, F. Rohmund, D.T. Colbert, K.A. Smith, et al. Gas-phase catalytic growth of single-walled carbon nanotubes from carbon monoxide, Chem. Phys. Lett. 313 (1–2) (1999) 91–97.

[34] Z. Ren, Z. Huang, J. Xu, J. Wang, P. Bush, M. Siegal, et al. Synthesis of large arrays of well-aligned carbon nanotubes on glass, Science 282 (5391) (1998) 1105–1107.

[35] J.W. Che, T. Cagin, W.A. Goddard, Thermal conductivity of carbon nanotubes, Nanotechnology 11 (2) (2000) 65–69.

[36] M.A. Osman, D. Srivastava, Temperature dependence of the thermal conductivity of single-wall carbon nanotubes, Nanotechnology 12 (1) (2001) 21–24.

[37] S. Berber, Y.K. Kwon, D. Tomanek, Unusually high thermal conductivity of carbon nanotubes, Phys. Rev. Lett. 84 (20) (2000) 4613–4616.

[38] B.Q. Wei, R. Vajtai, P.M. Ajayan, Reliability and current carrying capacity of carbon nanotubes, Appl. Phys. Lett. 79 (8) (2001) 1172–1174.

[39] V.R. Kumar, S. Kumar, G. Pal, T. Shah, High-ampacity overhead power lines with carbon nanostructure–epoxy composites, J. Eng. Mater. Technol. 138 (4) (2016) 041018.

[40] S. Kumar, G. Pal, T. Shah, High performance overhead power lines with carbon nanostructures for transmission and distribution of electricity from renewable sources, J. Clean. Prod. 145 (2017) 180–187.

[41] T. Durkop, B.M. Kim, M.S. Fuhrer, Properties and applications of high-mobility semiconducting nanotubes, J. Phys. 16 (18) (2004) R553–R580.

[42] Z.K. Tang, L.Y. Zhang, N. Wang, X.X. Zhang, G.H. Wen, G.D. Li, et al. Superconductivity in 4 angstrom single-walled carbon nanotubes, Science 292 (5526) (2001) 2462–2465.

[43] S. Hong, S. Myung, Nanotube electronics: a flexible approach to mobility, Nat. Nanotechnol. 2 (4) (2007) 207–208.

[44] A.T. Seyhan, M. Tanoğlu, K. Schulte, Tensile mechanical behavior and fracture toughness of MWCNT and DWCNT modified vinyl-ester/polyester hybrid nanocomposites produced by 3-roll milling, Mater. Sci. Eng. 523 (1) (2009) 85–92.

[45] P. Garg, B.P. Singh, G. Kumar, T. Gupta, I. Pandey, R.K. Seth, et al. Effect of dispersion conditions on the mechanical properties of multi-walled carbon nanotubes based epoxy resin composites, J. Polym. Res. 18 (6) (2011) 1397–1407.

[46] H. Kim, Y. Miura, C.W. Macosko, Graphene/polyurethane nanocomposites for improved gas barrier and electrical conductivity, Chem. Mater. 22 (11) (2010) 3441–3450.

[47] J.-H. Lee, S.K. Kim, N.H. Kim, Effects of the addition of multi-walled carbon nanotubes on the positive temperature coefficient characteristics of carbon-black-filled high-density polyethylene nanocomposites, Scripta Mater. 55 (12) (2006) 1119–1122.

[48] K. Kalaitzidou, H. Fukushima, L.T. Drzal, A new compounding method for exfoliated graphite–polypropylene nanocomposites with enhanced flexural properties and lower percolation threshold, Compos. Sci. Technol. 67 (10) (2007) 2045–2051.

[49] W.-P. Wang, C.-Y. Pan, Preparation and characterization of polystyrene/graphite composite prepared by cationic grafting polymerization, Polymer 45 (12) (2004) 3987–3995.

[50] C.A. Cooper, D. Ravich, D. Lips, J. Mayer, H.D. Wagner, Distribution and alignment of carbon nanotubes and nanofibrils in a polymer matrix, Compos. Sci. Technol. 62 (7) (2002) 1105–1112.

[51] M. Yousfi, S. Alix, M. Lebeau, J. Soulestin, M.-F. Lacrampe, P. Krawczak, Evaluation of rheological properties of non-Newtonian fluids in micro rheology compounder: experimental procedures for a reliable polymer melt viscosity measurement, Polym. Test. 40 (2014) 207–217.

[52] A.S. Babal, B.P. Singh, J. Jyoti, S. Sharma, A.K. Arya, S.R. Dhakate, Synergistic effect on static and dynamic mechanical properties of carbon fiber-multiwalled carbon nanotube hybrid polycarbonate composites, RSC Adv. 6 (72) (2016) 67954–67967.

[53] J. Jyoti, B.P. Singh, S. Rajput, V.N. Singh, S.R. Dhakate, Detailed dynamic rheological studies of multiwall carbon nanotube-reinforced acrylonitrile butadiene styrene composite, J. Mater. Sci. 51 (5) (2016) 2643–2652.

[54] R. Haggenmueller, H. Gommans, A. Rinzler, J.E. Fischer, K. Winey, Aligned single-wall carbon nanotubes in composites by melt processing methods, Chem. Phys. Lett. 330 (3) (2000) 219–225.

[55] J.N. Coleman, U. Khan, W.J. Blau, Y.K. Gun'ko, Small but strong: a review of the mechanical properties of carbon nanotube–polymer composites, Carbon 44 (9) (2006) 1624–1652.

[56] S. Srivastava, N.A. Kotov, Composite layer-by-layer (LBL) assembly with inorganic nanoparticles and nanowires, Acc. Chem. Res. 41 (12) (2008) 1831–1841.

[57] M.F. Arif, S. Kumar, T. Shah, Tunable morphology and its influence on electrical, thermal and mechanical properties of carbon nanostructure-buckypaper, Mater. Design 101 (2016) 236–244.

[58] S. Teotia, B.P. Singh, I. Elizabeth, V.N. Singh, R. Ravikumar, A.P. Singh, et al. Multifunctional, robust, light-weight, free-standing MWCNT/phenolic composite paper as anodes for lithium ion batteries and EMI shielding material, RSC Adv. 4 (63) (2014) 33168–33174.

[59] X. Wang, X. Lin, V. Dravid, J. Ketterson, R.P. Chang, Growth and characterization of buckybundles, Appl. Phys. Lett. 62 (16) (1993) 1881–1883.

[60] E. Camponeschi, R. Vance, M. Al-Haik, H. Garmestani, R. Tannenbaum, Properties of carbon nanotube–polymer composites aligned in a magnetic field, Carbon 45 (10) (2007) 2037–2046.

[61] C. Ma, W. Zhang, Y. Zhu, L. Ji, R. Zhang, N. Koratkar, et al. Alignment and dispersion of functionalized carbon nanotubes in polymer composites induced by an electric field, Carbon 46 (4) (2008) 706–710.

[62] B. Smith, Z. Benes, D. Luzzi, J. Fischer, D. Walters, M. Casavant, et al. Structural anisotropy of magnetically aligned single wall carbon nanotube films, Appl. Phys. Lett. 77 (5) (2000) 663–665.

[63] P. Ajayan, O. Stephan, C. Colliex, D. Trauth, Aligned carbon nanotube arrays formed by cutting a polymer resin–nanotube composite, Science 265 (5176) (1994) 1212–1214.

[64] S. Fan, M.G. Chapline, N.R. Franklin, T.W. Tombler, A.M. Cassell, H. Dai, Self-oriented regular arrays of carbon nanotubes and their field emission properties, Science 283 (5401) (1999) 512–514.

[65] W. Li, S. Xie, L. Qian, B. Chang, Large-scale synthesis of aligned carbon nanotubes, Science 274 (5293) (1996) 1701.

[66] L. Jin, C. Bower, O. Zhou, Alignment of carbon nanotubes in a polymer matrix by mechanical stretching, Appl. Phys. Lett. 73 (9) (1998) 1197–1199.

[67] J. Gao, A. Yu, M.E. Itkis, E. Bekyarova, B. Zhao, S. Niyogi, et al. Large-scale fabrication of aligned single-walled carbon nanotube array and hierarchical single-walled carbon nanotube assembly, J. Am. Chem. Soc. 126 (51) (2004) 16698–16699.

[68] H. Hou, J.J. Ge, J. Zeng, Q. Li, D.H. Reneker, A. Greiner, et al. Electrospun polyacrylonitrile nanofibers containing a high concentration of well-aligned multiwall carbon nanotubes, Chem. Mater. 17 (5) (2005) 967–973.

[69] G. Sui, W. Zhong, X. Yang, Y. Yu, S. Zhao, Preparation and properties of natural rubber composites reinforced with pretreated carbon nanotubes, Polym. Adv. Technol. 19 (11) (2008) 1543–1549.

[70] I.-Y. Jeon, D.W. Chang, N.A. Kumar, J.-B. Baek, Functionalization of carbon nanotubes, InTechOpen, (2011).

[71] W. Khan, R. Sharma, P. Saini, Carbon nanotube-based polymer composites: synthesis, properties and applications, InTechOpen, (2016).

[72] P. Liu, Modifications of carbon nanotubes with polymers, Eur. Polym. J. 41 (11) (2005) 2693–2703.

[73] D. Baskaran, J.W. Mays, M.S. Bratcher, Noncovalent and nonspecific molecular interactions of polymers with multiwalled carbon nanotubes, Chem. Mater. 17 (13) (2005) 3389–3397.

[74] S. Bredeau, S. Peeterbroeck, D. Bonduel, M. Alexandre, P. Dubois, From carbon nanotube coatings to high-performance polymer nanocomposites, Polym. Int. 57 (4) (2008) 547–553.

[75] R. Andrews, D. Jacques, D. Qian, T. Rantell, Multiwall carbon nanotubes: synthesis and application, Acc. Chem. Res. 35 (12) (2002) 1008–1017.

[76] S.-H. Liao, C.-Y. Yen, C.-H. Hung, C.-C. Weng, M.-C. Tsai, Y.-F. Lin, et al. One-step functionalization of carbon nanotubes by free-radical modification for the preparation of nanocomposite bipolar plates in polymer electrolyte membrane fuel cells, J. Mater. Chem. 18 (33) (2008) 3993–4002.

[77] D. Shi, J. Lian, P. He, L. Wang, F. Xiao, L. Yang, et al. Plasma coating of carbon nanofibers for enhanced dispersion and interfacial bonding in polymer composites, Appl. Phys. Lett. 83 (25) (2003) 5301–5303.

[78] R. Mathur, B. Singh, T. Dhami, Y. Kalra, N. Lal, R. Rao, et al. Influence of carbon nanotube dispersion on the mechanical properties of phenolic resin composites, Polym. Compos. 31 (2) (2010) 321–327.

[79] J.H. Du, J. Bai, H.-M. Cheng, The present status and key problems of carbon nanotube based polymer composites, Express Polym. Lett. 1 (5) (2007) 253–273.

[80] R. Haggenmueller, W. Zhou, J. Fischer, K. Winey, Production and characterization of polymer nanocomposites with highly aligned single-walled carbon nanotubes, J. Nanosci. Nanotechnol. 3 (1–2) (2003) 105–110.

[81] S. Kundalwal, S. Kumar, Multiscale modeling of stress transfer in continuous microscale fiber reinforced composites with nano-engineered interphase, Mech. Mater. 102 (2016) 117–131.

[82] G. Pal, S. Kumar, Modeling of carbon nanotubes and carbon nanotube–polymer composites, Prog. Aerosp. Sci. 80 (2016) 33–58.

[83] P.D. Bradford, X. Wang, H. Zhao, J.-P. Maria, Q. Jia, Y. Zhu, A novel approach to fabricate high volume fraction nanocomposites with long aligned carbon nanotubes, Compos. Sci. Technol. 70 (13) (2010) 1980–1985.

[84] R. Andrews, D. Jacques, A.M. Rao, T. Rantell, F. Derbyshire, Y. Chen, et al. Nanotube composite carbon fibers, Appl. Phys. Lett. 75 (9) (1999) 1329–1331.

[85] T. Ogasawara, S.-Y. Moon, Y. Inoue, Y. Shimamura, Mechanical properties of aligned multi-walled carbon nanotube/epoxy composites processed using a hot-melt prepreg method, Compos. Sci. Technol. 71 (16) (2011) 1826–1833.

[86] H. Liu, J. Gao, W. Huang, K. Dai, G. Zheng, C. Liu, et al. Electrically conductive strain sensing polyurethane nanocomposites with synergistic carbon nanotubes and graphene bifillers, Nanoscale 8 (26) (2016) 12977–12989.

[87] C. Min, X. Shen, Z. Shi, L. Chen, Z. Xu, The electrical properties and conducting mechanisms of carbon nanotube/polymer nanocomposites: a review, Polymer 49 (12) (2010) 1172–1181.

[88] A. Dasari, Z.-Z. Yu, Y.-W. Mai, Electrically conductive and super-tough polyamide-based nanocomposites, Polymer 50 (16) (2009) 4112–4121.

[89] J. Zhu, H. Peng, F. Rodriguez-Macias, J.L. Margrave, V.N. Khabashesku, A.M. Imam, et al. Reinforcing epoxy polymer composites through covalent integration of functionalized nanotubes, Adv. Funct. Mater. 14 (7) (2004) 643–648.

[90] A.K. Kota, B.H. Cipriano, M.K. Duesterberg, A.L. Gershon, D. Powell, S.R. Raghavan, et al. Electrical and rheological percolation in polystyrene/MWCNT nanocomposites, Macromolecules 40 (20) (2007) 7400–7406.

[91] E. Kymakis, I. Alexandou, G. Amaratunga, Single-walled carbon nanotube–polymer composites: electrical, optical and structural investigation, Synth. Met. 127 (1) (2002) 59–62.

[92] G. Pal, S. Kumar, Multiscale modeling of effective electrical conductivity of short carbon fiber-carbon nanotube-polymer matrix hybrid composites, Mater. Design 89 (2016) 129–136.

[93] F. Du, R.C. Scogna, W. Zhou, S. Brand, J.E. Fischer, K.I. Winey, Nanotube networks in polymer nanocomposites: rheology and electrical conductivity, Macromolecules 37 (24) (2004) 9048–9055.

[94] S.A. Curran, D. Zhang, W.T. Wondmagegn, A.V. Ellis, J. Cech, S. Roth, et al. Dynamic electrical properties of polymer–carbon nanotube composites: enhancement through covalent bonding, J. Mater. Res. 21 (04) (2006) 1071–1077.

[95] R. Smith, J. Carey, R. Murphy, W. Blau, J. Coleman, S. Silva, Charge transport effects in field emission from carbon nanotube–polymer composites, Appl. Phys. Lett. 87 (26) (2005) 263105.

[96] Z. Ounaies, C. Park, K. Wise, E. Siochi, J. Harrison, Electrical properties of single wall carbon nanotube reinforced polyimide composites, Compos. Sci. Technol. 63 (11) (2003) 1637–1646.

[97] F. Du, C. Guthy, T. Kashiwagi, J.E. Fischer, K.I. Winey, An infiltration method for preparing single-wall nanotube/epoxy composites with improved thermal conductivity, J. Polym. Sci. 44 (10) (2006) 1513–1519.

[98] S.-H. Wu, I. Masaharu, T. Natsuki, Q.-Q. Ni, Electrical conduction and percolation behavior of carbon nanotubes/UPR nanocomposites, J. Reinf. Plast. Comp. 25 (18) (2006) 1957–1966.

[99] G.D. Seidel, D.C. Lagoudas, A micromechanics model for the electrical conductivity of nanotube–polymer nanocomposites, J. Compos. Mater. 43 (9) (2009) 917–941.

[100] L. Chang, K. Friedrich, L. Ye, P. Toro, Evaluation and visualization of the percolating networks in multi-wall carbon nanotube/epoxy composites, J. Mater. Sci. 44 (15) (2009) 4003–4012.

[101] S. Berber, Y.-K. Kwon, D. Tománek, Unusually high thermal conductivity of carbon nanotubes, Phys. Rev. Lett. 84 (20) (2000) 4613.

5

Improving Carbon Nanotube/Polymer Interactions in Nanocomposites

Francis Avilés*, Juan V. Cauich-Rodríguez*, Patricio Toro-Estay**, Mehrdad Yazdani-Pedram**, Héctor Aguilar-Bolados**

*SCIENTIFIC RESEARCH CENTER OF YUCATAN, AC (CICY), (CENTRO DE INVESTIGACIÓN CIENTÍFICA DE YUCATÁN, A.C., UNIDAD DE MATERIALES), MÉRIDA, YUCATÁN, MEXICO; **UNIVERSITY OF CHILE (UNIVERSIDAD DE CHILE), SANTIAGO, CHILE

CHAPTER OUTLINE

5.1 Introduction

Surface modification is the act of modifying the surface of a material by generating physical, chemical, or biological characteristics different from the ones originally found. A common way to achieve this goal is through chemical functionalization, which introduces new functional groups on the surface of the "substrate" material. Besides increasing the chemical reactivity of the surface, these new chemical groups need to be carefully chosen to match the surface groups or molecules of a second material that the substrate is aimed to interact with to promote physicochemical interactions. For carbon nanotubes (CNTs), their strong C–C covalent bonding and ring structure not only render excellent physical properties, but also render low chemical reactivity, which causes difficulties in their

Carbon Nanotube–Reinforced Polymers. http://dx.doi.org/10.1016/B978-0-323-48221-9.00005-4

interaction with polymer molecules. This creates a need for chemical and physical surface modifications on CNTs to improve their physicochemical interactions with host polymers. The selection of the CNT functionalization method and molecule(s) used strongly depends on the chemical nature of the host polymer, and their selection is a topic addressed in this chapter. Although the mechanical properties of CNT/polymer composites may not compete with those of fiber-reinforced polymer composites, the selection of an adequate CNT functionalization method/bridging molecule could greatly assist in creating proper interfaces that improve the mechanical, thermal, electrical, and multifunctional properties of polymer nanocomposites.

5.2 Carbon Nanotube Functionalization Methods

Chemical (or physical) modifications of CNTs are frequently required to achieve their full potential and exploit their outstanding properties. Chemical modifications of CNTs generally occur at their end caps or defective sidewalls. It is generally accepted that with the loss of π-conjugation, there is also a change from sp^2 to sp^3 hybridization. These modifications may allow a good dispersion into a polymer matrix, improve the polymer–CNT load transfer ability in the resulting composite, and also allow proper CNT characterization in solution. Enhanced interactions between CNTs and a polymer matrix can be achieved in several ways. The most common ones are chemical methods for covalent attachment of different organic groups, but there are also alternatives that use defect-group or defect-site functionalization (such as Stone–Wales defects, sp^3-hybridized defects, vacancies, etc.), noncovalent functionalization with surfactants, and noncovalent exohedral and endohedral functionalizations. Although physical and other functionalization methods are touched upon herein, this section focuses on chemical modifications through the covalent attachment of functional groups.

5.2.1 Oxidation

Oxidation is undoubtedly the most common functionalization method for CNTs. It can be used as a final treatment to improve the functionality of single-wall (SWCNT) and multiwall carbon nanotubes (MWCNTs) [1–3], or as the first treatment of a sequence of chemical modifications [4–6]. It is also frequently used to improve the dispersion of CNTs in liquids, organic solvents, and polymers [7,8]. Most of the purification methods used after CNT synthesis employ acids to dilute the metallic catalysts, so a certain degree of oxidation may be inherently conducted even when it is not deliberately attempted [9,10]. For commercial CNTs, this is evident from the physicochemical characterization of the as-received material, showing the presence of oxygen-containing functional groups [11,12]. CNT oxidation can be conducted by physical methods, such as plasma treatments [13], and also by wet-chemical methods [4]. Plasma treatments, such as UV/ozone or oxygen plasma, have the advantage of being gentle to the CNT structure, yielding only minor damage [13–15]. However, the density of oxygen-containing functional groups generated by known plasma treatments is significantly less than that generated by wet-chemical methods; therefore, a compromise exists between CNT damage and the density of functional groups. In spite of these options, the large majority of works dealing with CNT polymer composites in the open literature have preferred wet-chemical methods to promote covalent bonding, albeit with some CNT damage, which seems unavoidable [12,16,17].

CNT oxidation by means of inorganic acids, such as nitric acid (HNO_3), sulfuric acid (H_2SO_4), or a combination of both, are among the most reported methods to date [12,17–20]. The use of potassium permanganate ($KMnO_4$), hydrogen peroxide (H_2O_2), ammonium persulfate ($(NH4)_2S_2O_8$), ozone (O_3), and other similar chemical compounds are also effective oxidation methods [4,5,21]. Successful oxidation methods generate carboxyl, carbonyl, and hydroxyl functional groups, typically as part of carboxylic acids, aldehydes, ketones, or alcohols. The relative contribution of each of these functional groups depends on the oxidation treatment employed. Using X-ray photoelectron spectroscopy, Wepasnick et al. [21] examined six common chemical oxidation methods, and concluded that oxidation with $KMnO_4$ and the mixture of nitric and sulfuric acids are the chemical treatments that yield a higher degree of oxidation (based on O at.%) and generation of COOH functional groups. On the other hand, they report a higher density of C=O groups for treatments using H_2O_2 or O_3. Concomitant with the generation of moieties upon oxidation is the generation of sp^3 hybridization defects in the sp^2 structure of the CNTs [22,23]. For cases where a high concentration of OH groups is desired, chemical reduction can sequentially be used after oxidation [24,25]. For example, Scheibe et al. [25] report the use of sodium tetrahydroborate ($NaBH_4$) to convert carbonyl and lactone groups of oxidized MWCNTs into hydroxyl groups. The generation of these new OH groups, however, may be at the cost of reducing COOH groups [24].

Oxidation may generate structural damage in the form of vacancies, Thrower or Stone–Wales defects and erosion of the CNT walls [26–29]. It is widely accepted that the functional groups generated by oxidative treatments concentrate on the CNT ends and defective sites, as depicted in Fig. 5.1.

The extent of CNT damage generated upon oxidation depends on several factors, such as the oxidant agent used, its concentration, treatment time, and method employed [12,27,29–31]. In that respect, Avilés et al. [12] proposed a sequential combination of 3.0 M HNO_3 and H_2O_2 (30% v/v), both for 2 h, as an oxidation method that reduces CNT damage. Fourier transform infrared spectroscopy (FTIR) analysis of the MWCNTs oxidized by this method (Fig. 5.2A) reveals the presence of O—H, C=O, C—O, and CH/CH_2 functional groups. The N_2 adsorption–desorption isotherms of these MWCNTs correspond to mesoporous materials with a hysteresis loop toward the end of its cycle [6], and, upon oxidation, the adsorbed volume increases (Fig. 5.2B). The increased adsorbed volume upon oxidation is ascribed to new pores coming from the elimination of amorphous carbon and probably from the elimination of impurity traces. Roughening of the external MWCNT walls and a certain extent of MWCNT damage may also contribute to the increased adsorbed volume. Thermogravimetric analysis (TGA) may also be employed to assess oxidation; as seen from Fig. 5.2C, upon mild oxidation; elimination of amorphous carbonaceous material renders increased thermal stability.

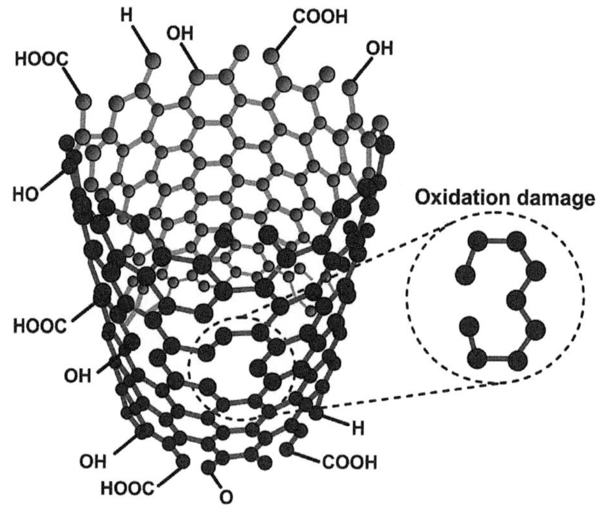

FIGURE 5.1 Schematic of functional groups and structural damage generated by CNT oxidation.

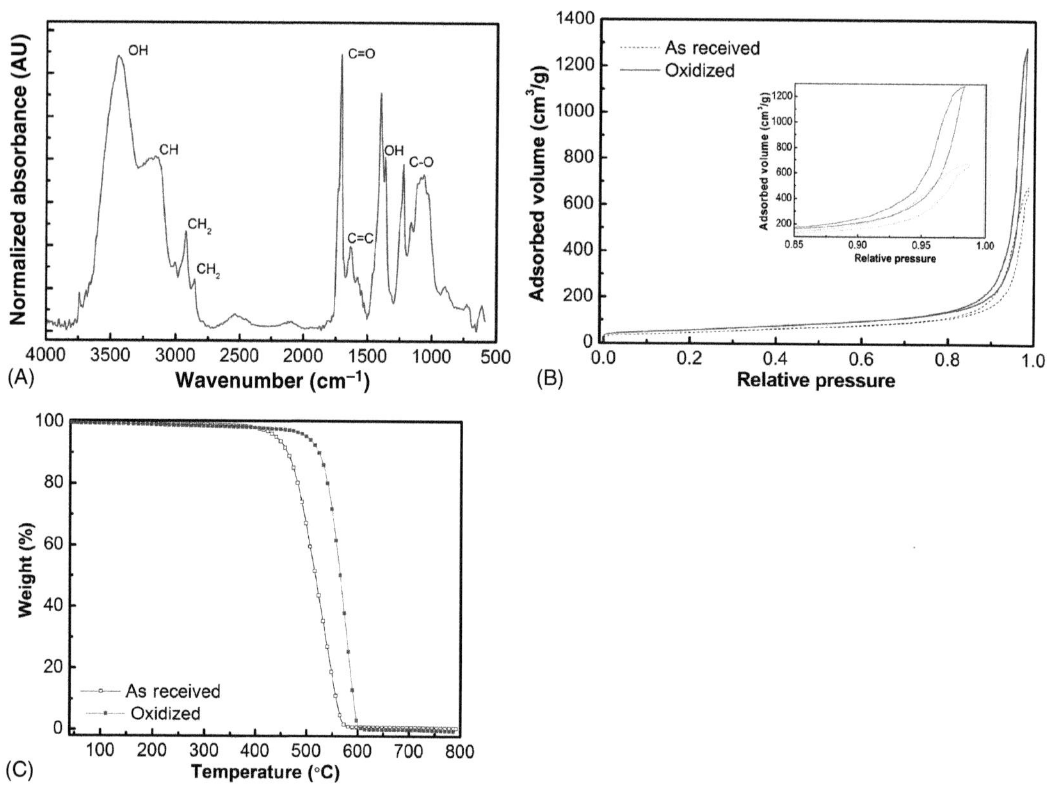

FIGURE 5.2 Characterization of chemically oxidized MWCNTs. (A) Fourier transform infrared spectroscopy (FTIR), (B) adsorption isotherms, and (C) thermogravimetric analysis (TGA).

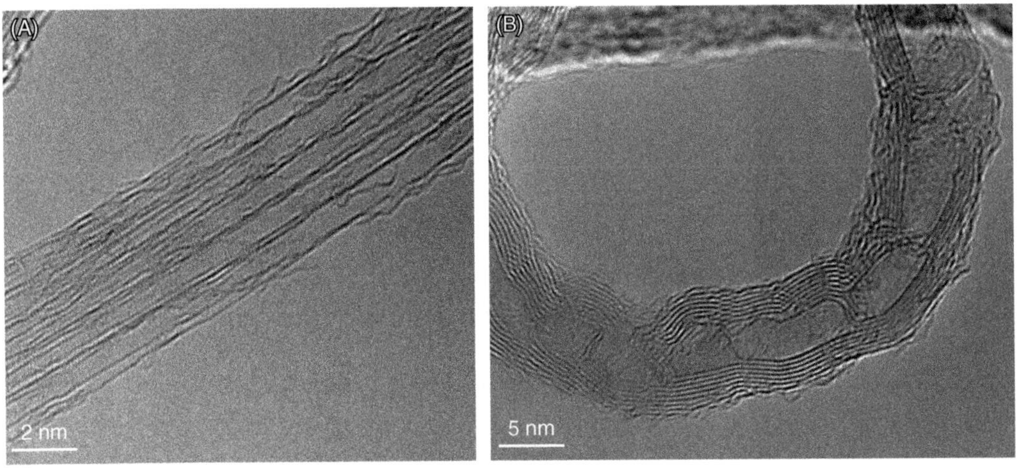

FIGURE 5.3 Transmission electron microscopy (TEM) images of CNTs oxidized under aggressive acid conditions. (A) SWCNTs and (B) MWCNTs.

Other forms to characterize CNT oxidation may employ X-ray photoelectron spectroscopy, UV–Vis transmittance of colloidal dispersions, and quantification of the CNT mobility in solution by zeta (ζ)-potential measurements [6,21,32]. The ζ-potential of CNTs is a strong function of the O/C ratio, pH, and solvent employed [32,33]; ζ-potentials of the order of 30–60 mV have been reported for oxidized MWCNT, depending on the pH and oxidation agent [32,33]; the reported trend is to increase the ζ-potential toward more negative values as the degree of oxidation and/or pH increases.

Oxidations with overly concentrated acids or that are conducted for prolonged time periods may severely damage the surface of CNTs, even causing fragmentation [27,29]. Fig. 5.3 shows transmission electron microscopy (TEM) images of SWCNT bundles (Fig. 5.3A) and an individual MWCNT (Fig. 5.3B), both treated with a 1:1 mixture of H_2SO_4 (99% v/v) and HNO_3 (70% v/v) for 2 h. As seen from this figure, such an aggressive acid condition causes surface roughening, erosion, wall attrition, local bends, wall stripping, thickness reduction (rendering nonuniform wall thickness and diameter), and conversion of part of the graphitic structure of the outermost and innermost layers into disordered carbon, due to the harsh etching action of the acids employed. The situation may be more severe for SWCNTs than for MWCNTs, given their single-layer condition [34]. These defects alter the electronic and vibrational properties of the CNTs, and may greatly impact their mechanical and electrical properties, and those of their polymer composites [22,26].

However, structural defects are active sites for recombination, and have been experimentally and theoretically identified as key locations for further functionalizations [35,36]. It has even been proposed that the control of defects in CNTs could lead to tailored electronic and structural properties, and even control of the CNT geometry [37]. Therefore, the compromise between the generation of a high density of oxygen-containing functional groups and the preservation of length, hybridization, and structural properties of CNTs must be customized depending on the application. For the mechanical and electrical properties of polymer composites, preserving CNT length and structural properties may be the decision driver; however, for applications, such as functionalization with biomolecules, producing a higher density of functional groups may overcome the general desire of structural preservation.

5.2.2 Silanization

One of the most common techniques reported to modify hydrophilicity and improve the chemical compatibility between CNTs and polymer matrices is silanization [24,38,39]. Silanes were originally developed as coupling agents to improve adhesion between organic resins and inorganic surfaces, and are also used as coupling agents for conventional polymer composites made of glass and carbon fibers [40,41]. Organofunctional silanes typically contain organic and inorganic moieties in the same molecule, and therefore they work as coupling agents for polymer composites. Silane-coupling agents are organosilane compounds that have at least two different reactive groups bonded to a silicon atom in a molecule; one of the reactive groups (e.g., methoxy, ethoxy, or acetoxy group) is reactive with the surface of the inorganic material (e.g., glass, carbon, metals, and silica). The other end of the molecule (e.g., epoxy, vinyl, methacryl, or amino group) is reactive with the organic material or synthetic resin, acting in this way as a chemical bonding agent. Therefore, in this functionalization method, the silane-coupling agent is expected to create a bridge at the interface between the CNT and the polymer matrix. A very small amount of an organofunctional silane may not only improve the interfacial strength, but also result in its retention over time. For the case of CNTs, although CNT silanization might be possible directly over as-produced CNTs, it has been mostly achieved by taking advantage of the hydroxyl and/or carboxyl groups previously produced during the CNT oxidation process [24,38,42,43]. Therefore, an oxidation treatment normally precedes CNT silanization. The choice of an adequate silane as a CNT surface treatment is instrumental for the correct compatibility with the polymer matrix and promotion of surface bonding. The selection of a specific silane for a polymer composite is based on matching the organofunctional group of silicon with the resin polymer type to be bonded (Fig. 5.4). For example, CNT functionalization by amino-based silanes (such as 3-aminopropyltriethoxysilane) are adequate for epoxy resins [39,44]. Another silane used for CNTs that matches the functional groups of most epoxy resins is 3-glycidoxypropyltrimethoxysilane [7,45]. 3-Isocyanatopropyltriethoxysilane has been used to strengthen the chemical interface of polyimide and polymethyl methacrylate MWCNT composites [46,47]. A general schematic of possible chemical reactions occurring during MWCNT silanization for three different silanes is depicted in Fig. 5.4. The silanes in this figure are 3-methacryloxypropyltrimethoxysilane (MPS), 3-aminopropyltriethoxysilane, and 3-mercaptopropyltrimethoxysilane (MPTMS), where ethanol or methanol can be produced as by-products.

For example, for polyester or vinyl ester resins, MPS has been reported to promote chemical bonding with MWCNTs, especially when the MWCNTs have been previously

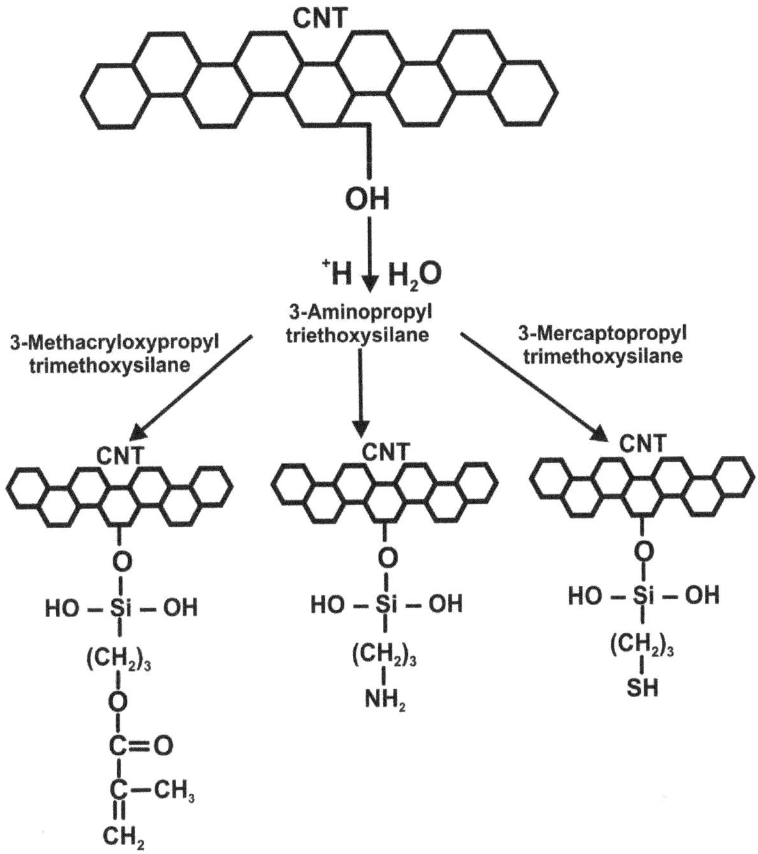

FIGURE 5.4 Different types of silane-coupling agents used for CNT functionalization (after oxidation).

chemically oxidized [43]. For this case, a more detailed schematic of a possible chemical reaction occurring during MWCNT silanization is depicted in Fig. 5.5A. The oxidized MWCNT surface contains (hydrophilic) hydroxyl groups that act as reaction sites (Fig. 5.5A). The MPS contains a trimethoxy group that after hydrolysis also forms hydroxyl groups and methanol (Fig. 5.5B). The hydroxyl groups on the CNT and silane condense to covalently bind a methacrylate group (hydrophobic) on the MWCNT surface (Fig. 5.5C). The methacrylate group (also an ester), matches the chemical structure of several polyester and vinyl ester resins that may be used as matrices for polymer composites. Similar chemical reactions have been proposed for other silanes, such as those shown in Fig. 5.4 [24,38].

It is important to point out that the reaction proposed in Fig. 5.5C rests upon the OH groups present on the (oxidized) CNT surface, but other chemical reactions based on the COOH groups of the CNT are also possible [42]. If needed, additional OH groups can be generated by reducing the COOH groups of the oxidized CNTs [24]. The silanization

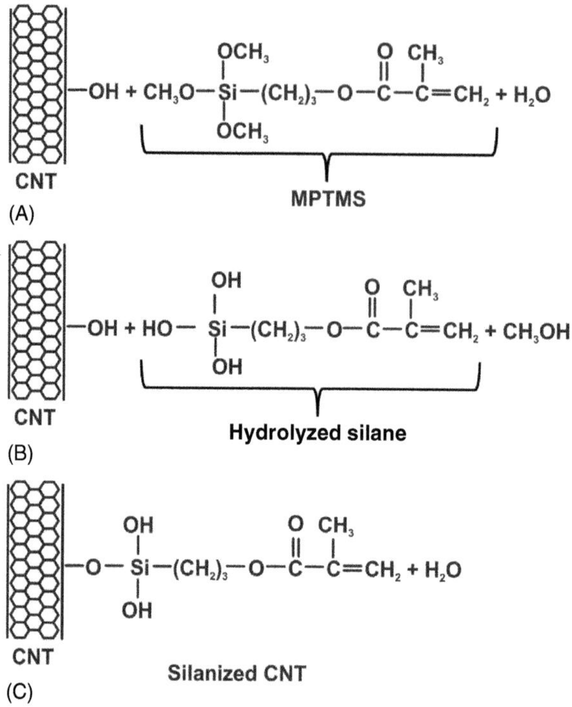

FIGURE 5.5 Schematic representation of a CNT silanization process using 3-methacryloxypropyltrimethoxysilane (MPS). (A) Hydrolysis of silane in presence of oxidized CNTs, (B) reaction between the hydrolyzed silane and the oxidized CNT, and (C) silanized CNT containing functional groups affine to vinyl ester or polyester resins. *MPTMS*, 3-mercaptopropyltrimethoxysilane.

process for CNTs was inherited from the large knowledge accumulated for engineering fibers, such as glass and carbon fibers [40,48]. Although the general procedure used for CNT silanization may be similar to that of conventional (micron-size) engineering fibers, several parameters need to be adjusted for the correct and homogenous silanization of such nanostructures. First of all, for CNTs, the use of ultrasonic waves (or other effective form of dispersion) is highly recommended to achieve homogenous coverage of individual CNTs [6,49]. The concentration of silane in the solution is also quite different to that employed for conventional micron-size fibers. This is because the surface area per unit weight of CNTs is about 2–3 orders of magnitude higher than that of conventional carbon fibers. While conventional engineering fibers use typical silane concentrations (weight of silane with respect to fiber weight) around 0.03–5 wt.% [41,50], successful silane concentrations for MWCNTs have reported the use of 1–2× the weight of silane with respect to the MWCNT weight [6,7,24,44]. On this issue, Avilés et al. [6] reported that the use of MPS at concentrations ranging from 3.5 wt.% to 1000 wt.% (10×) improves the interfacial strength of vinyl ester matrix composites. Based on TGA measurements, they estimated the weight of incorporated silane on the surface of the MWCNTs, finding a decrease in the amount of

FIGURE 5.6 Concentration of silane incorporated on the MWCNT surface as a function of weight concentration of silane used for MWCNT silanization. Inset shows silanized MWCNTs at 1× silane concentrations.

silane that is effectively incorporated to the MWCNT when silane concentrations beyond 2× (200 wt.%) are used (Fig. 5.6).

Field emission scanning electron microscopy (see inset in Fig. 5.6) showed that for 1× and 2× concentrations, the MWCNT surface is homogeneously covered with MPS, but beyond 2× excess of silane occurs; this means that for concentrations beyond 2×, layers of polysiloxanes may react with themselves, modifying the properties of the interface. This overly formation of polysiloxanes may weaken the strength of the interface and degrade the mechanical properties of polymer composites [6].

5.2.3 Fluorination

Halogenation is very common nowadays for the chemical modification of CNTs. Although chlorination and bromination have been reported as means of halogenation [51], fluorination is preferred to enhance the reactivity of the CNT sidewalls [52,53]. Besides improving the chemical reactivity, fluorination facilitates disentanglement of nanotube bundles and improves processability and solubility [52]. Fluorination improves solubility in solvents such as *N,N*-dimethylformamide (DMF), tetrahydrofuran (THF), and alcohols (such as 2-propanol or 2-butanol), but CNTs treated by fluorination do not dissolve in perfluorinated solvents, water, diethyl amine, and acetic acid [52]. This type of functionalization weakens the C–F bonds, increases the chemical reactivity of CNTs, and allows further modifications, particularly the attachment of nucleophilic substituents bearing hydroxyl, amino, and carboxylic as terminal groups or even *N*-alkylamino substitutions or alkylation reactions. CNT fluorination can be reverted by a reaction with hydrazine.

Several methods for fluorination exist, such as direct gas, plasma, and hydrothermal fluorination. Direct fluorination can be achieved by a high-temperature reaction (200–250°C) in the presence of hydrogen fluoride [52], as depicted in Fig. 5.7. Similarly,

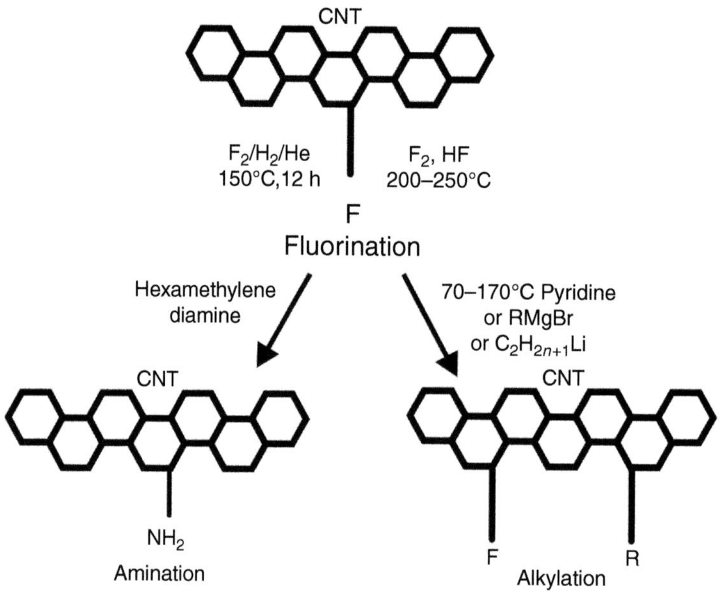

FIGURE 5.7 Common CNT fluorination reactions.

carboxyl-modified CNTs can be treated with a mixture of $F_2/H_2/He$ (2/1/30) at 150°C during 12 h to incorporate F atoms to the CNT surface [54,55]. Fluorine can be identified by attenuated total reflectance FTIR by the presence of a band at 1220–1250 cm^{-1}. In Zhu et al.'s work [55], Raman spectroscopy showed that the appearance of an sp^3 carbon peak at 1301 cm^{-1} indicates that the sidewalls are covalently modified with fluorine, while elemental analysis showed that up to 20 wt.% of F can be incorporated.

Fluorinated CNTs can serve as a reactive substrate for further chemical modifications. Following CNT fluorination, the reaction with alkoxides (e.g., sodium methoxide in methanol) results in methoxylated CNTs [52]. When these reactions are conducted with lithium hydroxide in methanol, a methoxy derivative is also obtained. Organolithium derivatives, such as methyl, *n*-butyl, *n*-hexyl, and phenyl, can also react with fluorinated CNTs to obtain an alkylated derivative that is soluble in chloroform and THF.

By reacting fluorinated CNTs with methyl, ethyl, butyl, hexyl, octyl, and dodecyl lithium derivatives, or alkyl magnesium bromide (Grignard reagent), partial alkylation can be achieved while retaining F atoms [56] (Fig. 5.7). Li, Na, or K hydroxides in the presence of ethylene glycol, 1,3-propanediol, 1,4-butanediol, 1,2-propanediol, 1,2-butanediol, or glycerol yield hydroxyl-terminated CNTs with some remaining atoms of fluorine [57].

5.2.4 Amination

The formation of primary, secondary, tertiary, or even quaternary ammonium salts on CNTs is a common method pursued for different purposes, including solubilization [58], compatibilization with polymeric matrices based on copolyimides [59], heavy metal

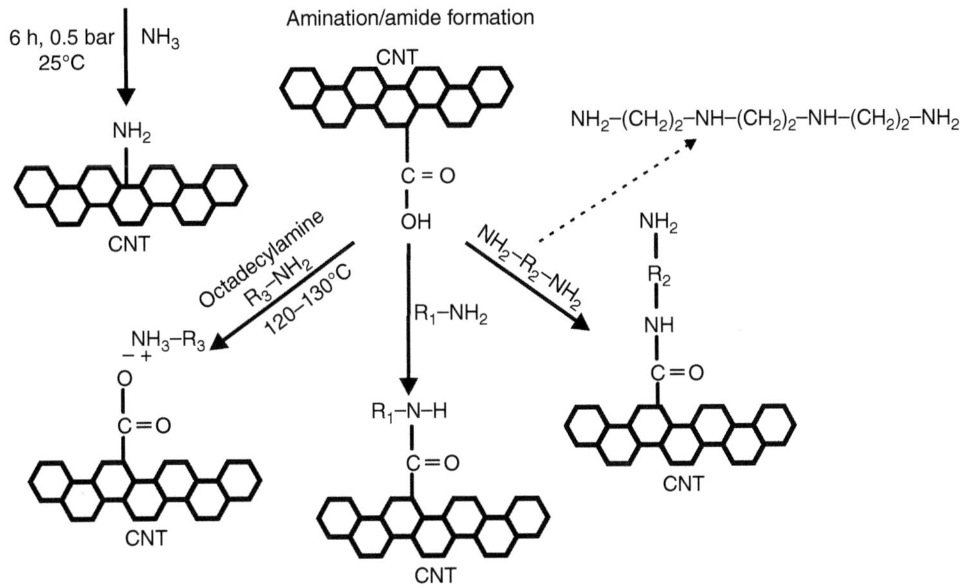

FIGURE 5.8 Typical reactions for CNT–amine/amide formation.

elimination from water effluents [60], crosslinking reactions with epoxy resins [61], and even for delivering genetic material [62,63]. The introduction of amine functional groups has been achieved through a variety of methods comprising gas or liquid phases; identification of amino groups can be achieved by several methods including FTIR, ζ-potential, and TGA. Its quantification by the Kaiser's method or ninhydrin assay (among other techniques) can prove the existence of these chemical moieties. A word of caution is mentioned, as many of these reactions can be reverted (intentionally or unintentionally) during the manufacturing of composites.

Direct amination, as described in Fig. 5.8, has been achieved by ball-mill treatment of MWCNTs in the presence of NH_3 gas flow of 50 mL/min during 6 h at 0.5 bar and 25°C [59,64]. However, oxidation is commonly the first step for the introduction of amino groups. In this regard, CNT amination is achieved after oxidation with nitric acid, combination of nitric and sulfuric acids, or treatments with thionyl chloride to form acyl chloride, followed by subsequent reaction of ethylene diamine [60]. In this case, the linking bridge contains an amide linkage, a hydrolytic unstable moiety, and the end group is a primary amine.

Amine functional groups on the CNT surface are frequently used for improving the chemical compatibilization with epoxy-based composites. For this type of modification not only oxidized CNTs [61,65,66], but also fluorinated CNTs [5,67] can be used. Oxidized CNTs can be treated with triethylenetetramine, diamines (H_2N-R-NH_2), or hydroxyamines [$HNR(CH_2)_nOH$] in the presence of pyridine to yield amino-terminated or hydroxyl-terminated CNTs, respectively, as shown in Fig. 5.8.

Sun et al. [68] covalently introduced amine groups into CNTs for preparing epoxy composites. This was achieved by using carboxylated (COOH) CNTs, followed by conversion into the corresponding acyl chloride, and finally by the reaction of 1,6-diamino hexane. Amination was proved by using FTIR by the presence of a wide band in the range of 2800–3500 cm^{-1}, bands at 2900 cm^{-1} corresponding to C–H stretching of the diamine, and absorptions at 1640, 1490, and 1300 cm^{-1}, corresponding to amide I, amide II, and amide III, respectively. Further evidence arose from the presence of bands related to N–H bending at 1570 cm^{-1}. In their work, ζ-potential was also used as a method to identify the presence of amino groups on the CNTs. In this case, the ζ-potentials of COOH- and NH_2-treated CNTs were quite different at any pH value. For COOH-treated CNTs, the ζ-potential was slightly negative and decreased up to –35 mV when the pH increased from 2 to 9. The behavior was opposite for NH_2-treated CNTs, which present a +30 mV ζ-potential at a pH of 2 that decreases to zero as the pH approaches 9. This later behavior was attributed to the presence of amido groups, which protonate via H^+ capture in the aqueous solution. Finally, by using TGA, decomposition of amino groups was also suggested in the range of 200–430°C, while above 440°C, the decomposition of secondary amidocyanogen was reported. By using TGA, a 3.1 wt.% of amino groups was estimated.

Amination has also been achieved by both cycloaddition reactions [62] and also by direct conversion of the carboxylic groups on the oxidized CNTs into amine groups [63] to deliver either DNA plasmids or siRNA.

Soluble octadecylammonium CNT–carboxylated zwitterions (Fig. 5.8) have been obtained by the reaction of nitric acid–oxidized SWCNTs in the presence of octadecylamine in the molten state (120–130°C) [58]. The authors claim that this ionic functionalization was chosen because this method provides a higher yield than covalent functionalization and is a simpler, cost effective, and can be scaled up.

Amide formation can also be achieved through Claisen reaction using oxidized CNTs treated with *N,N*-dimethylacetamide dimethylacetal [69]. Amide formation has also been suggested by the condensation reaction between carboxylic acid and amino-functionalized CNTs [70]. In Hecker et al.'s work [70], the interconnected CNTs were assisted by using carbodiimide-activated coupling using *N,N*-dicyclohexyl carbodiimide dimethyl amino pyridine. Other possible methods for the formation of primary amines on CNTs make use of the halide derivative and its reaction with an excess of ammonia. Similarly, azides, nitriles, and nitrocompounds can be reduced, leading to amine formation.

Aminated CNTs can be further derivatized. Oxidation of amines with H_2O_2, $KMnO_4$, or peroxyacids can lead to hydroxyl amines, amine oxides, nitro compounds, or nitro derivatives. The reactions of aldehydes and ketones with NH_2-functionalized CNTs can lead to imine formation (Schiff base), which after reduction can result in the production of secondary amines. The reaction with sulfonyl chloride can lead to sulfonamides, a known family of compounds with antibacterial activity. However, in many of these derivatives heat can have a profound effect, such as in the case of Hoffman's elimination, where the ammonium salt can lead to alkene formation. These need to be taken into consideration especially for thermoforming or composite manufacturing at high temperatures.

5.2.5 Other Chemical Modifications

Several organic functional groups other than OH, COOH, F, or NH_2 can be directly introduced to CNTs with specific purposes. Transformation of metallic SWCNTs to semiconducting type with minimal modification of the electronic structure is reported to be achieved through hydrogenation [71]. Hydrogenation can be achieved by the Birch reaction using ammonia as the reaction medium [72] or by flowing hydrogen gas through a hot tungsten filament [71]. Vinylation was used by Kan et al. [73] for molecular-imprinted polymers with the capability of specific adsorption and recognition of the template molecule. For this, carboxylated MWCNTs were treated with thionyl chloride and then allylic alcohol for the selective copolymerization of methacrylic acid and trimethylolpropane trimethacrylate using dopamine as a template. Nitrenes can be thermally generated on CNTs by microwave activation with *p*-toluenesulfonyl, methylsulfonyl, and trimethylsilyl azides, or even isocyanates, yielding the corresponding aziridine [74]. Furthermore, cyclopropanation of CNTs can be achieved by a reaction with diethyl bromomalonate with sodium hydride (Bingel reaction). Direct addition to the unsaturated π-electrons can also be achieved by Diels–Alder addition to render carboxylic groups on the CNT surface [75]. This type of reaction has been extended to various dienes and dienophiles with the aim of introducing alcohol, amine, carboxylic, epoxy, and ester groups [76]. Sidewall functionalization with 12%–23% dichlorocarbene (nucleophilic carbene addition) has generated SWCNTs that are soluble in both THF and dichlorobenzene [77]. The free radical addition of succinic or glutaric acid acyl peroxides in *ortho*-dichlorobenzene at 80–90°C has been reported to introduce 2-carboxyethyl or 3-carboxypropyl groups on the sidewalls of SWCNTs [2]. The introduction of substituted polyrrolidine rings by 1,3-dipolar cycloaddition of azomethine ylides has been reported by Herrero and Prato [78]. Thiol- or mercaptane-functionalized MWCNTs have been prepared by treating hydroxyl-terminated MWCNTs with thiourea in the presence of hydrobromic acid and glacial acetic acid [79]. By using condensation reaction of an *N*-(4-hydroxyphenyl)glycine and formaldehyde, Bae et al. [80] prepared SWCNTs with azomethine ylide groups containing anchored phenol structures. Classic coordination complexes and organometallic compounds have been suggested by Kharisov and Kharissova [81] to increase the solubility of CNT and as a metal precursor for decoration and encapsulation of metals and oxides. Therefore, it is clear that although many chemical reactions have been explored to modify the surface of CNTs to improve their compatibility with polymers, there is plenty of room for further modifications.

5.3 Carbon Nanotube Functionalization for Improved Properties of Polymer Composites

CNT functionalization can be used for a variety of purposes, from improving dispersion, subsequent doping, and sensing biomolecules to improving interfacial adhesion with polymers by the promotion of the formation of covalent bonds. For polymer composite applications, the functional groups created on the surface of the nanotube have to match

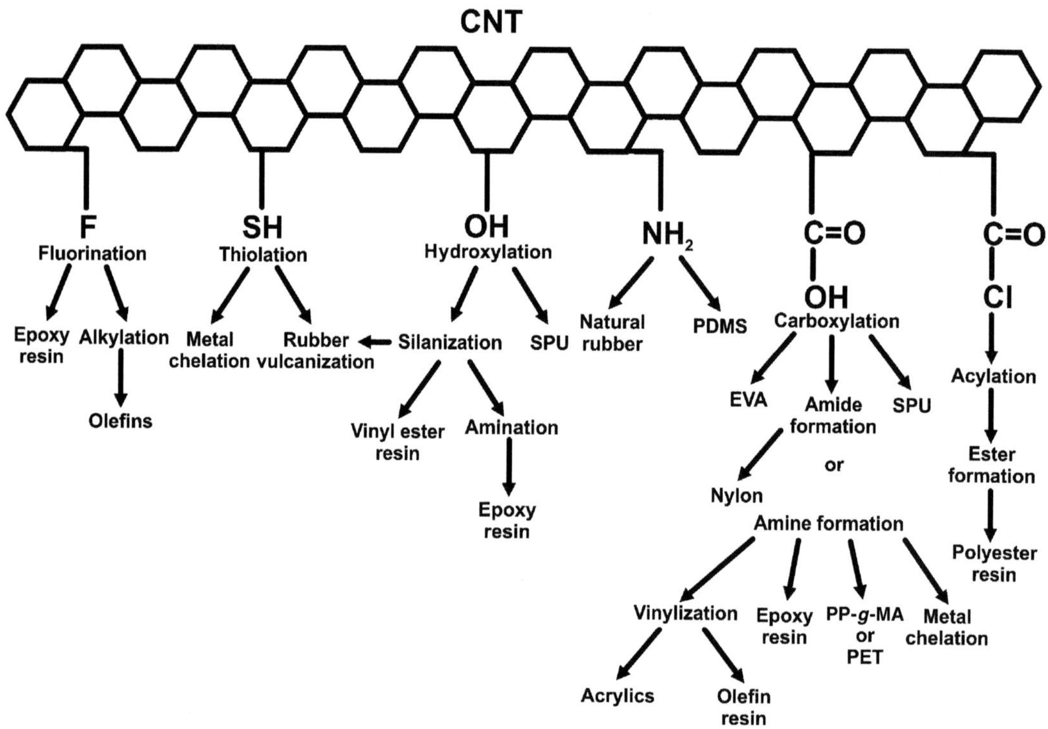

FIGURE 5.9 Routes for CNT functionalization depending on the chemical nature of the polymer matrix. *EVA*, Ethylene–vinyl acetate; *SPU*, segmented polyurethane; *PDMS*, polydimethylsiloxane; *PET*, polyethylene terephthalate; *PP-g-MA*, polypropylene grafted with maleic anhydride.

the active functional groups of the polymer of interest, which can be a challenging task. Fig. 5.9 suggests some functional groups that can interact better with common polymer matrices used for composite preparation. For example, CNT carboxylation can be used directly to fabricate functionalized ethylene–vinyl acetate (EVA) or segmented polyurethane (SPU) composites; however, after further chemical modifications forming amides or amines, they can be used for nylon, epoxy, olefin, or acrylic polymer composites.

This section revises the most common functionalizations that have been conducted to improve interfacial interactions with thermosetting, thermoplastic, and elastomeric polymer matrices, and discusses the effective properties of the resulting functionalized nanocomposites.

5.3.1 Thermosettings

The most common thermosetting resins used for polymer composites are epoxy, vinyl ester, unsaturated polyester, and phenolic resins. The family of epoxy resins is undoubtedly the most studied thermosetting matrix due to its relative high strength and stiffness, low curing shrinkage, and high dimensional and chemical stability; these properties make epoxy resins suitable for high-end applications in diverse fields, such as automotive, aeronautics, and the electronic industry. As for other polymers, the addition of CNTs into epoxy resins may render multifunctional nanocomposites, which can increase its mechanical, electrical, and thermal properties. However, to improve the physicochemical interactions at the interface between CNTs and epoxy, and to improve CNT dispersion, specific chemical functionalizations have to be conducted. Many efforts have been conducted to promote this affinity due to the entangled nature of CNT bundles and the dissimilar polarity between CNTs and epoxies [82]. Regarding CNT dispersion, several techniques, such as ultrasonic dispersion [30], calendaring [83,84], ball-milling [85], stirring, and extrusion [86] have been attempted. An alternative that can tackle the issues of dispersion and interfacial adhesion is functionalization [4]. The most recommended CNT functionalization methods reported in the literature for epoxy polymers are those obtained by fluoration, amination, or silanization (Fig. 5.9). Acylation followed by ester formation is also an option for polyester resins (Fig. 5.9). Direct fluorination of CNTs and their subsequent derivatization provide a versatile tool for the preparation and manipulation of nanotubes with variable noncovalent functionalities [52,55]. Fluorinated CNTs can be integrated into epoxy composites through the formation of covalent bonds in the course of epoxy ring–opening esterification and curing chemical reactions [55]. However, for epoxy resins, CNT functionalization with amino groups or a proper silane (such as 3-glycidoxypropyltrimethoxysilane) have been more commonly reported [39,45,61]. Upon functionalization, the CNT hydrophobicity decreases due to the incorporation of amino or silane groups on the CNT that encourage more reactive sites to covalently bind with thermosetting polar groups [87]. The reaction mechanisms of amino-(left) or carboxylate (right)-functionalized CNTs in the preparation of nanocomposites based on an epoxy matrix are shown in Fig. 5.10.

The final composite is obtained by the reaction between the active functional groups of the epoxy resin and the terminal amine (or carboxyl) functional group of the functionalized CNT. The incorporation mechanism involves one of the epoxy terminal groups as an active or electron acceptor site, and a nucleophilic (amino) terminal group on the functionalized CNT. As a result, a covalent N–C bond is formed joining H atoms of the terminal amino group on the CNTs with the epoxy [88–90].

The preparation of amino CNTs by covalent modification is normally sequential. For example, Iannazzo et al. [91] report functionalization of MWCNTs by three sequential steps, carboxylation, acylation, and, finally, amine modification. Amine-functionalized CNT composites could thus render strong interfacial interactions due to covalent bonding between the amine-treated CNTs and the epoxide resin, improving the mechanical and thermal properties of these composites [61,92]. The interfacial strength is associated with the length, purity, structure, and CNT functionalization, and therefore has a great influence on the final properties of the resulting composites. It is also possible to achieve high thermal stability, as well as a significant increase in strength, using cyclic and aromatic amines. Thus obtaining a composite with high interfacial strength and minimal CNT agglomeration depends on the structure and type of the amine group.

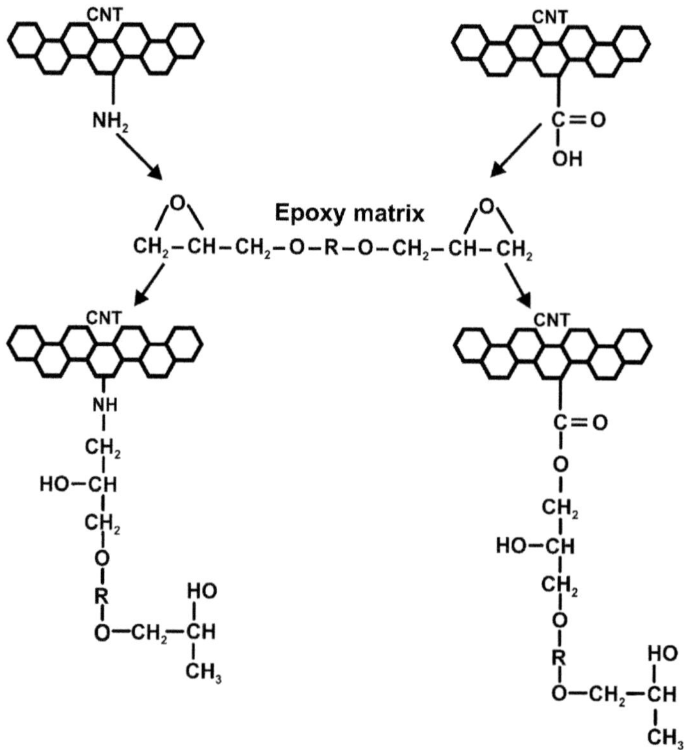

FIGURE 5.10 Reaction mechanism between amino functionalized (left) or carboxylated (right) CNTs and an epoxy polymer.

Silanization is another route for surface modification of CNTs. This covalent functionalization can be carried out by reaction of the required chemical structure of the silane through functional groups at the CNT surface. The most common strategy is to first oxidize the carbonaceous surface by reaction with inorganic acids, leading to the formation of hydroxyl and carboxylic acids [12,38,45]. Once the acid is formed, it can be derivatized by, for example, reaction with thionyl chloride. Acid derivatives have proven to react with alcohols and amines, resulting in ester or amide linkage, respectively, and providing plenty of alternatives for subsequent functionalizations [49]. CNT organophilization with silane-coupling agents has been widely reported [6,38,45,93]. Once the silane is hydrolyzed it may condense with hydroxyl and/or carboxylic acid groups, forming stable covalent bonds. The nonhydrolyzable substituent may remain available for covalent reaction or physical interaction with matrix phases [39]. Achieving optimal surface coating and CNT silanization to improve the properties for thermosetting polymer composites depends on several factors, such as the type of silane, pH, dispersion, and silane concentration [6]. The popularity of this method is its simplicity and the enormous number of silanes available. For a successful linkage, the organofunctional group of the silane has to match the chemical reactivity of the polymer matrix. For epoxy resins, epoxy or amino silanes (such

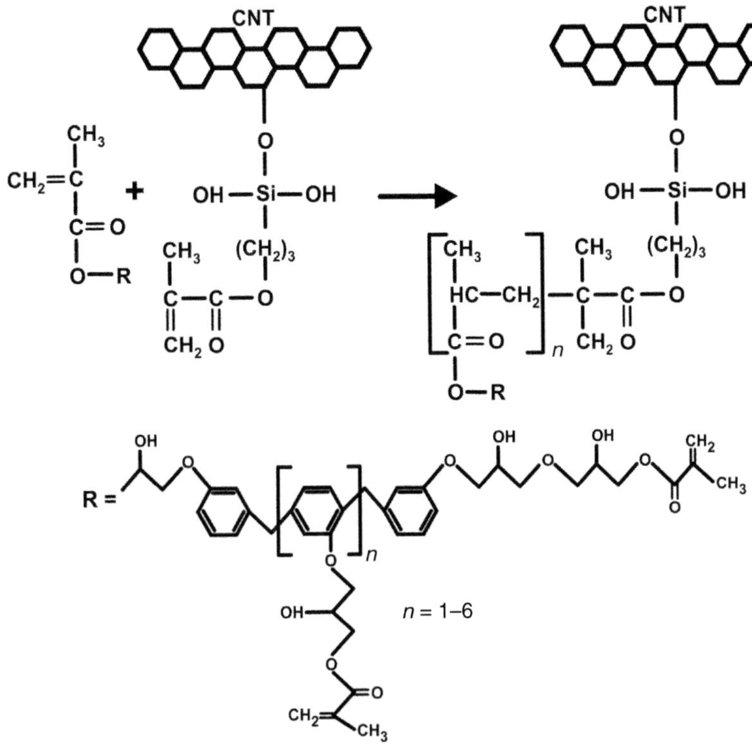

FIGURE 5.11 Reaction mechanism between an MPS-silanized CNT and a vinyl ester resin.

as 3-glycidoxypropyltrimethoxy or 3-aminopropyltriethoxy) are commonly used [39,45]. For vinyl ester and polyester resins, acrylate or vinyl olefin silanes are normally preferred [6]. As an example, Fig. 5.11 shows the interaction mechanism between an MPS-silanized CNT (Fig. 5.5) and a vinyl ester resin. This silane has a methacrylate organoreactive group and trimethoxy hydrolyzable groups, which match the organofunctional groups of the vinyl ester resin employed [43].

The vast majority of published works on mechanical and thermal reinforcement of polymer matrix nanocomposites that use functionalized CNTs are thermosetting matrices. This is probably because of the superior mechanical and thermal performance of thermosettings, as compared to thermoplastics and elastomers. Although both types of CNTs (SWCNTs and MWCNTs) have been extensively addressed, the majority of the reported work focuses on MWCNTs, probably because of their higher availability in large amounts, which is needed to fabricate polymer nanocomposites. It is generally accepted that the mechanical and thermal properties of thermosetting polymer composites increase when CNT weight concentration is increased, but such an increase is nonmonotonic. It has been suggested that the upper bound in CNT concentration to reach mechanical reinforcement is the percolation concentration, which may be easily estimated by electrical conductivity measurements [94]. This is because above certain critical CNT concentration, CNT agglomeration and a large increase in resin viscosity occur, which greatly hinder the mechanical and thermal properties of nanocomposites. Such a critical CNT concentration depends on the CNT type, aspect ratio, functional groups on its surface, resin chemistry, resin viscosity, and dispersion method, among other factors [95–98]. This drawback is particularly relevant for tensile properties, as besides stress concentrations caused by CNT agglomeration, the increased viscosity of the solution may generate trapped bubbles, causing crack tip opening and rapid crack propagation upon tensile loading, greatly hindering their mechanical properties [39,99]. Therefore, if these factors are not carefully controlled, it is not unusual to obtain even lower mechanical/thermal properties with the inclusion of CNTs [95,96,99]. Aside from the large efforts of the scientific/industrial community to improve the CNT dispersion within the polymer, two of the most practical ways to better control this issue is through keeping the CNT concentration significantly low (just above the percolation threshold) and using CNT functionalization to improve the dispersion and mechanical bonding with the host polymer. Both strategies are reflected in Table 5.1, where a few examples of mechanical and thermal properties of CNT/thermosetting polymer nanocomposites using functionalized CNTs have been collected. In this table, the CNT weight concentration is indicated in parenthesis underneath the CNT type (MWCNT or SWCNT) and the increase (+Δ) or decrease (–Δ) in three mechanical and two thermal properties have been collected from the literature. The subscript ``0'' refers to the neat matrix, and the properties addressed are the elastic modulus (E), the ultimate stress (strength, σ^u), the ultimate strain (ε^u), the glass transition temperature (T_g, determined by either differential scanning calorimetry or dynamic mechanical analysis), and the thermal degradation temperature (T_d, determined by TGA).

As seen from this table, among the thermosetting polymers the most reported one is undoubtedly the family of epoxies, likely because of their better mechanical and thermal

Table 5.1 Variation in the Mechanical and Thermal Properties of CNT-Functionalized Thermosetting Polymer Nanocomposites, With Respect to the Neat Matrix

Thermosetting Matrix	CNT Types (wt.%)	CNT Functionalization	$\Delta E/E_0$	$\Delta\sigma^u/\sigma_0^u$	$\Delta\varepsilon^u/\varepsilon_0^u$	ΔT_g (°C)	ΔT_d (°C)	Reference
Epoxy	MWCNT (0.80)	Amino groups	—	—	—	+18	—	[61]
Epoxy	SWCNT (1.0)	Amino groups	+31% (T)	+25%	+31%	—	—	[100]
Epoxy (rubbery)	MWCNT (1.0)	Acid oxidized	+28% (T)	+104%	+60%	—	—	[95]
Epoxy (glassy)	MWCNT (1.0)	Acid oxidized	+4% (T)	−1%	+8%	—	—	
Epoxy	MWCNT (0.25)	Amino groups	+19% (F)	+26%	—	−10	+30	[101]
Epoxy	MWCNT (0.20)	3-Aminopropyltriethoxysilane	+22% (F)	+23%	—	+11	—	[39]
Polyester	MWCNT (0.5)	Amino groups	—	+15% (T)	—	—	—	[102]
Vinyl ester/polyester	MWCNT (0.30)	Amino groups	+31% (T)	+20%	—	—	—	[103]
Vinyl ester	MWCNT (0.2)	MPS	+7% (C)	+5%	+5%	—	—	[6]

C, Compressive; F, flexural; T, tensile.

properties compared to vinyl esters and unsaturated polyesters, as well as their lower curing shrinkage. However, given their lower costs and versatility, vinyl ester and even polyester resins are currently utilized in a large variety of industrial applications, such as automotive, shipbuilding, wind-turbine blade manufacturing, and piping industry, among many others [102,104]. Another distinctive feature observed in Table 5.1 is the very low CNT concentrations used, which are typically around 0.2–0.3 wt.%, that is, just above the electrical percolation threshold [105]. CNT weight concentrations above 0.5 wt.% are deemed high for CNT/thermosetting polymer nanocomposites, and such high concentrations may hinder their mechanical properties. This is significantly different for thermoplastic and elastomeric matrices, where higher CNT concentrations are typically needed, as will be further discussed.

As seen in Table 5.1, amino functional groups are undoubtedly the ones used to functionalize both types of CNTs to improve the mechanical and thermal properties of epoxy nanocomposites [61,89,100,101]; functionalizing CNTs with amines has also been attempted for vinyl ester and unsaturated polyester resins [102,103]. Care needs to be taken, as amine functionalization is expected to modify the kinetics of the epoxy curing reaction, which under certain circumstances, could yield a reduction of the glass transition temperature of the nanocomposite [101]. Depending on the epoxy matrix, carboxylated MWCNTs may also improve the mechanical properties of epoxy composites [95], but is more common that a second, more specific, chemical functionalization is conducted after CNT oxidation. An alternative for a more specific functionalization is the use of an adequate silane, which is frequently used as a sequential treatment after CNT oxidation [6,39]. In this regard, Avilés et al. [6] have pointed out that the concentration of silane in solution, used to functionalize the CNT, plays a paramount role in the resulting mechanical properties of the nanocomposite. As discussed in connection to Fig. 5.6, overly high concentrations of silane may form an excess of polysiloxanes at the interface, decrementing the mechanical (and thermal) properties of polymer composites.

5.3.2 Thermoplastics

Thermoplastic polymers are of great commercial interest because of their ease of processing and variety of applications. Thermoplastic polymers, such as polyethylenes and polypropylenes (commodities), constitute the largest polymer production. The preparation of nanocomposites based on thermoplastic polymers is generally performed by incorporation of reinforcements/fillers using melt-mixing process. Different strategies have been used to facilitate the homogeneous dispersion of nanofillers, such as CNTs in thermoplastic polymers. These strategies include the high-shear melt processing of the thermoplastic polymer to promote the breakdown and dispersion of the nanotubes, or the use of so called “compatibilizers,” among other methods [106]. Compatibilizers are materials that are added to a mixture of incompatible polymers or polymer/fillers to suppress their phase separation by promoting the interaction between the components of the mixture [107]. Nonpolar polymers grafted with polar molecules, such as polyolefins grafted with maleic

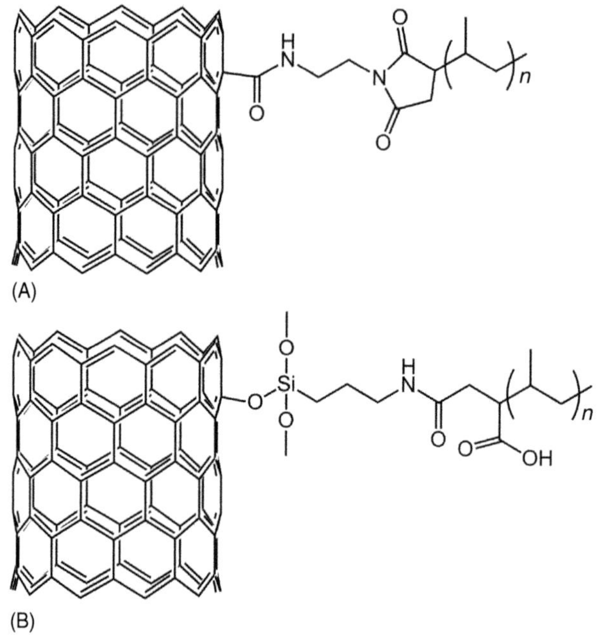

FIGURE 5.12 CNTs grafted with PP obtained from the reaction of functionalized CNTs and PP-*g*-MA. (A) Amine-functionalized CNTs and (B) CNTs functionalized with 3-aminopropyltriethoxysilane.

anhydride, are among the most used compatibilizers. For instance, Lee et al. [108] reported the use of two types of compatibilizers, polypropylene (PP)-grafted maleic anhydride (PP-g-MA) and styrene-ethylene-butylene-styrene-grafted maleic anhydride (SEBS-g-MA), for the preparation of MWCNT/PP composites. MWCNTs were chemically functionalized through acid, amine, and heat treatments. The authors found that the electrical conductivity of the composite was dramatically increased when the heat-treated MWCNTs were added and was further increased with the addition of SEBS-g-MA. The preparation of PP-grafted MWCNTs by melt mixing of PP-g-MA with amine-functionalized MWCNTs (Fig. 5.12A) is also a successful functionalization method that improves the CNT dispersion and mechanical properties of the resulting composite [109]. An alternative is to use an adequate silane, such as 3-aminopropyltriethoxysilane, which results in PP composites with increased flame retardancy and thermal stability [110]; it has been reported that the maleic anhydride moiety reacts with the amine groups of functionalized MWCNTs, yielding amide linkage between MWCNTs and PP-g-MA, Fig. 5.12B.

The grafting method is not only applicable to PP, but also to other polymers such as polyethylene [111], EVA copolymers [112], and polystyrene [113], among many others.

Inclusion of functionalized CNTs in many other thermoplastic polymers, such as those obtained by polycondensation (e.g., polyesters and polyamides), has also been reported. For example, for polyethylene terephthalate, 4-alkyloxybenzoic acids [114],

Table 5.2 Variation in the Mechanical Properties of CNT-Functionalized Thermoplastic Polymer Nanocomposites, With Respect to the Neat Matrix

Thermoplastic Matrix	CNT Type (wt.%)	CNT Functionalization	Processing Method	$\Delta E/E_0$	$\Delta\sigma^u/\sigma_0^u$	Reference
PP	MWCNT (0.10)	Acid oxidized	Melt mixing	+20%	+3%	[119]
	MWCNT (1.0)			+78%	+6%	
PP	MWCNT (1.0)	Acid oxidized	Melt mixing	+31%	+10%	[120]
		Acid oxidized + IA		+40%	+10%	
PP	MWCNT (1.5)	Amino groups	Melt mixing	+1.1	+1.4	[109]
PET	MWCNT (0.5)	Acid oxidized	In situ polymerization	+1.0	+2.0	[115]
		Diamine		+1.09	+2.4	
PET	MWCNT (0.5)	Acid oxidized	Melt mixing	+2.6%	+47%	[116]
PEN	MWCNT (0.5)	Acid oxidized	Melt mixing	+18%	+32%	[121]
Nylon-6	MWCNT (1.0)	Acid oxidized	Melt mixing	+1.15	+1.24	[122]

PEN, Polyethylene naphthalate; IA, itaconic acid.

diamines [115], acid oxidation, and itaconic acid [116] have been used to functionalize MWCNTs. For polyamides, acid treatments [117] and the use of compatibilizers based on styrene–maleic anhydride copolymer [118] have also been addressed. The final aim of this functionalization and/or use of processing aids or compatibilizers is to improve the effective properties of polymer nanocomposite. Table 5.2 presents a brief literature collection listing improved mechanical properties due to the use of functionalized MWCNTs in thermoplastic polymers matrices (numbers are with respect to the neat polymer matrix). As seen from this table, melt mixing is undoubtedly the most used processing method, likely because is ease of use and scalability. However, this process demands a relative large amount of material and this is likely the reason why the use of MWCNTs has been preferred over SWCNTs for thermoplastic melt mixing processing. Larger improvements in mechanical properties can be achieved by using in situ polymerization [115], although this technique may be difficult to scale up. It can also be seen from the Table 5.2 that acid oxidation (commonly by means of nitric and/or sulfuric acids) and amination are common routes for CNT functionalization that result in composites with improved mechanical properties and that grafting organic or inorganic molecules is also of assistance. A common feature (not indicated in the Table 5.2) is that the increments in stiffness (E) and strength (σ^u) in CNT/thermoplastic nanocomposites are concomitant with a significant reduction in ultimate strain (elongation at break) with respect to its neat polymer. This is because CNTs act as physical (or chemical) crosslinking sites that restrict polymer chain motion [115,119,120]. A way to mitigate this effect may be through grafting polymer molecules on the CNTs prior to composite processing [109]. Thermal properties, such as T_g, T_d, and melting/crystallization temperatures of thermoplastic composites, have also been reported to increase by using functionalized CNTs [3,119–121]. It is generally accepted that functionalized CNTs act as nucleating agents that elevate the crystallization temperature of thermoplastics [109].

5.3.3 Elastomers

Elastomers present a broad range of elasticity as their segment links are free enough to assume different conformations of equal energy. This is the result of having a glass transition temperature below room temperature. Also, elastomers may return to their initial shape after being deformed as they have high molecular weights and are crosslinked. The family of elastomers include several types of rubbers, such as natural rubber, butyl rubber, nitrile rubber, polyisoprene, polychloroprene, styrene–butadiene rubber, silicon rubbers, fluoroelastomers, polysulfide rubber, EVA, ethylene propylene diene monomer (EPDM), as well as thermoplastic elastomers, such as SPUs. These polymers, copolymers, or even physical blends have been filled with unmodified and modified CNTs, but several challenges still prevail to improve their processing conditions, and to apply appropriate CNT functionalizations that lead to improved CNT dispersion, mechanical, electrical, thermal, and actuating properties, among others [123–126]

CNTs can be either physically or chemically modified to improve dispersion and compatibility with an elastomeric matrix. For example, regarding physical interactions, MWCNTs can be modified with ionic liquids, such as 1-ethyl-2,3-dimethylimidazolium bis(trifluoromethylsulfonyl), that promote cation π/π–π interactions to significantly improve the mechanical properties of styrene–butadiene rubber composites [127]. In a similar manner, MWCNTs have been modified with ionic liquids, such as 1-butyl-3-methyl imidazolium bis(trifluoromethylsulphonyl)imide, to provide high electrical conductivity, improved tensile modulus, and thermal stability to polychloropene rubber [128]. These improvements strongly depend on the physicochemical features of the CNTs used. The most common groups attached to the CNTs for covalent rubber modification are hydroxyl, carboxyl, amines, and silane coupling agents (after oxidation). Although some improvements have been reported, in some cases the limited amount of functional groups introduced does not guarantee the creation of a proper interface and therefore, marginal improvements in mechanical properties are frequently reported.

Elastomers based on SPUs have been modified with acid-oxidized MWCNTs and SWCNTs, improving CNT dispersion, stiffness, and the strength of the resulting composites [129,130]. For these composites, improvements in stiffness and strength heavily depend on the use of low CNT concentrations (close to the percolation threshold), and these improvements are normally concomitant with significant reductions of elongation at break [129,130]. In the case of segment polyurethanes, it has been observed that the resulting properties of the CNT composite are more influenced by the rigid segment content than by the CNT content [131]. Similarly, highly crystalline (40% rigid segments) thermoplastic polyurethanes based on hexamethylene diisocyanate, butanediol, polycaprolactone, and polycarbonate diol have been mixed with acid-oxidized MWCNTs [132], showing improved mechanical properties. To improve further the interactions between the CNTs and thermoplastic polyurethanes, some attempts have been made to graft the polyol used for the polyurethane synthesis on the CNTs. Jing et al. [133] grafted polycaprolactone onto MWCNTs to improve the dispersion and mechanical properties of the resulting composite. Other routes to functionalize CNTs for polyurethane matrices include the use of toluene diisocyanate, polyethylene glycol, and 3-amino-1,2,4-triazole [134,135].

In the case of polyolefin elastomers, the addition of MWCNTs substantially increases their melt viscosity. Therefore, different works have focused on the addition of thermoplastic polymers to facilitate processing of these elastomeric polyolefins in the presence of MWCNTs [136]. It has been reported that in PP/EPDM blends, SWCNTs act as nucleation centers, which affect the physical and mechanical properties of the composites [137]. In situ incorporation of oxidized MWCNTs during the preparation of a thermoplastic elastomer has resulted in improved MWCNT dispersion in the polymer matrix, as revealed by small-angle X-ray analysis [138]. Likewise, thermoplastic polymers grafted with polar molecules have been used as compatibilizers to favor the dispersion of CNTs in the nanocomposite matrix. This method of preparation has led to a significant increase of the elastic modulus of nanocomposites. For instance, Hemmati et al. [139] reported that nanocomposites based on PP/EPDM (80/20) blends containing 0.5 wt.% SWCNTs and 1 wt.% PP grafted with maleic anhydride as compatibilizer resulted in an increase of tensile modulus of about 60%. The use of compatibilizer prevented the decrease of elongation at break that normally occurs with the addition of CNTs. Another relevant study focused on the functionalization of MWCNTs by acid treatment and the grafting of PP on the MWCNTs [140]. The use of PP-grafted MWCNTs improved the mechanical properties of a PP/EPDM thermoplastic elastomer, achieving a 40% increase of the tensile modulus without decreasing its elongation at break.

Acid oxidation using nitric and/or sulfuric acids are among the preferred CNT functionalization methods used for elastomers. By using 2.8 M nitric acid for 38 h, Bhattacharyya et al. [141] prepared carboxylated MWCNTs and dispersed them in natural rubber assisted by sodium dodecyl sulfate. The resulting composite showed improvements in elastic modulus (10-fold), tensile strength (2-fold), and storage modulus (60-fold) with a concomitant reduction in elongation at break, attaining a low electrical percolation threshold (~1 wt.%). A polydimethylsiloxane matrix was reinforced with alkylamine-modified MWCNTs to obtain dielectric elastomers for capacitive force sensors [142]. For this, a 1:3 v/v ratio of nitric:sulfuric acids was used to oxidize MWCNTs, and subsequently treated with various alkylamines (such as hexylamine, octadecylamine, and dodecylamine) for amide formation. Similarly, when a polyester-modified polydimethylsiloxane solution was use on nitric acid-oxidized MWCNTs, the dispersion, the mechanical and thermal properties of silicone rubber composites were improved [143].

The use of silanes has also been pursued to improve CNT dispersion and the effective properties of elastomeric composites. CNTs with different aspect ratios were modified by a silane-coupling agent and embedded into a silicone elastomer matrix to obtain composites with improved CNT dispersion and enhanced dielectric loss and permittivity [144]. Shanmugharaj et al. [145] oxidized MWCNTs with 10 N sulfuric acid in the

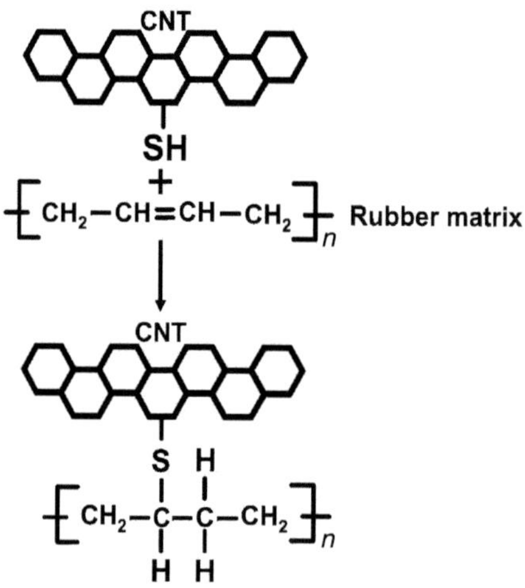

FIGURE 5.13 Thiol crosslinking mechanism for unsaturated rubbery matrices.

presence of potassium dichromate and then conducted a silanization with 3-aminopropyltriethoxysilane, improving the elastic modulus and tensile strength of natural rubber composites. As customary for elastomeric systems, the large final strain of natural rubber was compromised with the introduction of CNTs, but the authors showed that the use of an adequate silane may lessen this negative effect. It has been reported that the use of MWCNTs functionalized with 7-octenyltrichlorosilane favors a significant increase in the elastic modulus of a commercial crosslinked polydimethylsiloxane (Sylgard 184). This was attributed to the fact that 7-octenyl trichlorosilane molecules are able to participate in the crosslinking reaction of polydimethylsiloxane [146]. By further functionalizing oxidized MWCNTs with bis(triethoxysilylpropyl)tetrasulfide, it has also been suggested that in situ silanization (where silane was directly added into the mixing process) may be of higher benefit for the generation of chemical linkages (and consequently, resulting effective properties) of natural rubber composites [147].

Organosulfur compounds can also be used to improve the CNT adhesion to elastomeric matrices. In that respect, Fig. 5.13 shows a possible mechanism for thiol crosslinking of an unsaturated rubbery matrix (polybutadiene) using a thiol-modified CNT.

Other functionalization routes that report improved effective properties of elastomeric composites include reversible addition–fragmentation chain transfer grafting of acrylonitrile and diurethane monoacrylate on MWCNTs for polyurethane composites with actuating properties [148], functionalization with polymers of low glass transition temperature, such as polybutyl acrylate homopolymer for triblock copolymer thermoplastic elastomers [149], and $KMnO_4$ oxidation followed by chemical reduction with diisobutyl aluminum hydride for EVA composites [150].

An important phenomenon that needs to be taken into account when analyzing the elastic behavior of filled rubbers (either functionalized or not) is the Payne effect [151].

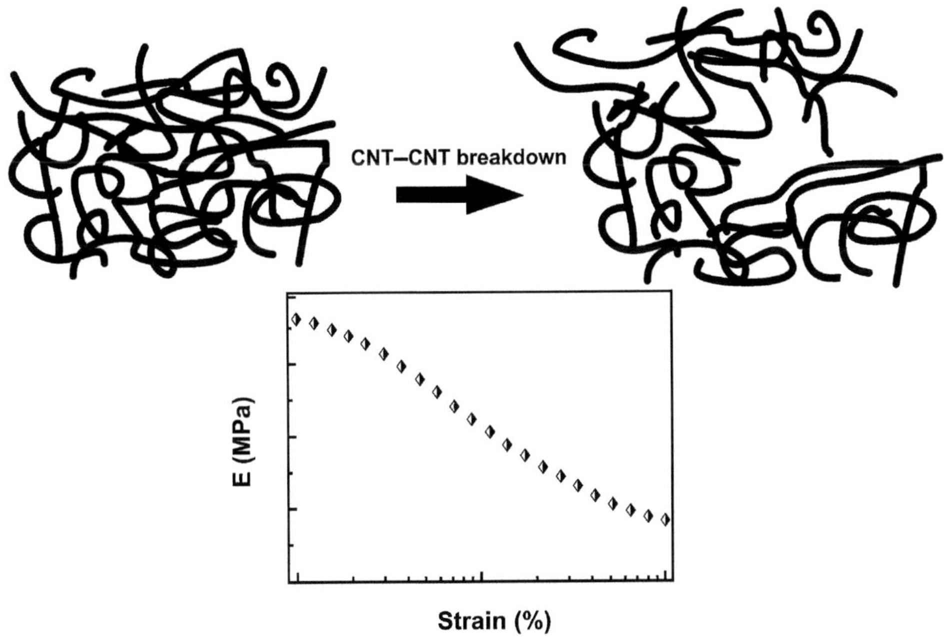

FIGURE 5.14 The strain softening (Payne) effect caused by the loss of CNT-to-CNT interactions with applied strain.

The Payne effect describes an experimentally observed "strain softening" behavior of many filled elastomeric polymers (including those filled with CNTs), where the dynamic elastic modulus decreases with increased applied strain. This effect is strongly related to the material microstructure and occurs for filled rubbery polymers, being minor or inexistent for unfilled rubbers. For the case of CNT-filled rubbers, the Payne effect is attributed to the loss of interactions among CNTs or weakening of the filler–filler interactions as a result of increased strain [141,152,153], as schematically represented in Fig. 5.14. It has been pointed out that the Payne effect is more severe for rubbers filled with MWCNTs than for those filled with carbon black [152]. This behavior has also been observed for nanocomposites based on synthetic rubbers, such as styrene butadiene filled with CNTs [154]. As the Payne effect heavily depends on the filler–filler and filler–matrix interactions [141], it is expected that adequate CNT functionalizations may be able to decrease this nonlinear effect.

Acknowledgments

Economic support from CONACYT (Mexico) through CIAM project No. 188089 and CONICYT (Chile) through project No. 120003 is acknowledged. Dr. MYP also acknowledges support from CONICYT through project FONDECYT 1131139. The authors wish to thank Omar Rodriguez (CICY) for his artistic assistance with the figures.

References

[1] F.H. Gojny, J. Nastalczyk, Z. Roslaniec, K. Schulte, Surface modified multiwalled carbon nanotubes in CNT/epoxy-composites, Chem. Phys. Lett. 370 (2003) 820–824.

[2] H. Peng, L.B. Alemany, J.L. Margrave, V.N. Khabashesku, Sidewall carboxylic acid functionalization of single-walled carbon nanotubes, J. Am. Chem. Soc. 125 (2003) 15174–15182.

[3] K. Yang, H. Han, X. Pan, N. Chen, M. Gu, The effect of chemical treatment on the crystallinity of multi-walled carbon nanotubes, J. Phys. Chem. Solids 69 (2008) 222–229.

[4] K. Balasubramanian, M. Burghard, Chemically functionalized carbon nanotubes, Small 1 (2005) 180–192.

[5] S. Banerjee, T. Hemraj-Benny, S.S. Wong, Covalent surface chemistry of single-walled carbon nanotubes, Adv. Mater. Lett. 17 (2005) 17–29.

[6] F. Avilés, C.A. Sierra-Chi, A. Nistal, A. May-Pat, F. Rubio, J. Rubio, Influence of silane concentration on the silanization of carbon nanotubes, Carbon 57 (2013) 520–529.

[7] H. Chen, O. Jacobs, W. Wu, G. Rüdiger, B. Schädel, Effect of dispersion method on tribological properties of carbon nanotube reinforced epoxy resin composites, Polym. Test. 26 (2007) 351–360.

[8] S.W. Kim, T. Kim, Y.S. Kim, H.S. Choi, H.J. Lim, S.J. Yang, C.R. Park, Surface modifications for the effective dispersion of carbon nanotubes in solvents and polymers, Carbon 50 (2012) 3–33.

[9] P.-X. Hou, C. Liu, H.-M. Cheng, Purification of carbon nanotubes, Carbon 46 (2008) 2003–2025.

[10] S.-H. Su, W.-T. Chiang, C.-C. Lin, M. Yokoyama, Multi-wall carbon nanotubes: purification, morphology and field emission performance, Phys. E 40 (2008) 2322–2326.

[11] R. Schönfelder, M.H. Rümmeli, W. Gruner, M. Löffler, J. Acker, V. Hoffmann, T. Gemming, B. Buchner, et al. Purification-induced sidewall functionalization of magnetically pure single-walled carbon nanotubes, Nanotechnology 18 (2007) 375601–375608.

[12] F. Avilés, J.V. Cauich-Rodríguez, L. Moo-Tah, A. May-Pat, R. Vargas-Coronado, Evaluation of mild acid oxidation treatments for MWCNT functionalization, Carbon 47 (2009) 2970–2975.

[13] T. Xu, J. Yang, J. Liu, Q. Fu, Surface modification of multi-walled carbon nanotubes by O_2 plasma, Appl. Surf. Sci. 253 (2007) 8945–8951.

[14] C. Chen, B. Liang, A. Ogino, X. Wang, M. Nagatsu, Oxygen functionalization of multiwall carbon nanotubes by microwave-excited surface-wave plasma treatment, J. Phys. Chem. C 113 (2009) 7659–7665.

[15] W.-M. Chang, C.-C. Wang, C.-Y. Chen, Plasma treatment of carbon nanotubes applied to improve the high performance of carbon nanofiber supercapacitors, Electrochim. Acta 186 (2015) 530–541.

[16] S.C. Tsang, P.J.F. Harris, M.L.H. Green, Thinning and opening of carbon nanotubes by oxidation using carbon dioxide, Nature 362 (1993) 520–522.

[17] M. Monthioux, B.W. Smith, B. Burteaux, A. Claye, J.E. Fischer, D.E. Luzzi, Sensitivity of single-wall carbon nanotubes to chemical processing: an electron microscopy investigation, Carbon 39 (2001) 1251–1272.

[18] H. Hu, B. Zhao, M.E. Itkis, R.C. Haddon, Nitric acid purification of single-walled carbon nanotubes, J. Phys. Chem. B 107 (2003) 13838–13842.

[19] I.D. Rosca, F. Watari, M. Uo, T. Akasaka, Oxidation of multiwalled carbon nanotubes by nitric acid, Carbon 43 (2005) 3124–3131.

[20] V. Datsyuk, M. Kalyua, K. Papagelis, J. Parthenios, D. Tasis, A. Siokou, I. Kallitsis, C. Galiotis, Chemical oxidation of multiwalled carbon nanotubes, Carbon 46 (2008) 833–840.

[21] K.A. Wepasnick, B.A. Smith, K.E. Schrote, H.K. Wilson, S.R. Diegelmann, H. Fairbrother, Surface and structural characterization of multi-walled carbon nanotubes following different oxidative treatments, Carbon 49 (2011) 24–36.

[22] T. Guo, T.W. Ding, Q.-X. Pei, Y.-W. Zhang, Effect of sp^3-hybridized defects on the oscillatory behavior of carbon nanotube oscillators, Phys. Lett. A 375 (2011) 2400–2404.

[23] J. Zhang, M. Terrones, C.R. Park, R. Mukherjee, M. Monthioux, N. Koratkar, Y.S. Kim, Hurt, et al. Carbon science in 2016: status, challenges and perspectives, Carbon 98 (2016) 708–732.

[24] P.C. Ma, J.K. Kim, B.Z. Tang, Functionalization of carbon nanotubes using a silane coupling agent, Carbon 44 (2006) 3232–3238.

[25] B. Scheibe, E. Borowiak-Palen, R.J. Kalenczuk, Oxidation and reduction of multiwall carbon nanotubes—preparation and characterization, Mater. Charact. 61 (2010) 185–191.

[26] J.-C. Charlier, Defects in carbon nanotubes, Acc. Chem. Res. 35 (2002) 1063–1069.

[27] M.T. Martínez, M.A. Callejas, A.M. Benito, M. Cochet, T. Seeger, A. Ansón, J. Schreiber, C. Gordon, et al. Sensitivity of single wall carbon nanotubes to oxidative processing: structural modification, intercalation and functionalization, Carbon 41 (2003) 2247–2256.

[28] M. Monthioux, J.-C. Charlier, Giving credit where credit is due: the Stone–(Thrower)–Wales designation revisited, Carbon 75 (2014) 1–4.

[29] F. Avilés, A. Ponce, J.V. Cauich-Rodríguez, G.T. Martínez, TEM examination of MWCNTs oxidized by mild experimental conditions, fullerenes, nanotubes and carbon, Nanostructures 20 (2012) 49–55.

[30] K.L. Lu, R.M. Lago, Y.K. Chen, M.L.H. Green, P.J.F. Harris, S.C. Tsang, Mechanical damage of carbon nanotubes by ultrasound, Carbon 34 (1996) 814–816.

[31] A.B. Sulong, C.H. Azhari, R. Zulkifli, M.R. Othman, J. Park, A comparison of defects produced on oxidation of carbon nanotubes by acid and UV ozone treatment, Eur. J. Sci. Res. 33 (2009) 295–304.

[32] P.C. Ma, N.A. Siddiqui, E. Mäder, J.-K. Kim, Correlation between electrokinetic potential, dispersibility, surface chemistry and energy of carbon nanotubes, Compos. Sci. Technol. 71 (2011) 1644–1651.

[33] L. Vanyoreka, R. Meszarosa, S. Baranya, Surface and electrosurface characterization of surface-oxidized multi-walled N-doped carbon nanotubes, Colloids Surf. A 448 (2014) 140–146.

[34] R. Schönfelder, F. Avilés, A. Bachmatiuk, J.V. Cauich- Rodríguez, M. Knupfer, B. Büchner, M.H. Rümmeli, On the merits of Raman spectroscopy and thermogravimetric analysis to assess carbon nanotube structural modifications, Appl. Phys. A 106 (2012) 843–852.

[35] Y. Park, R.J.W.E. Lahaye, Y.H. Lee, Adsorption of Pt on defective carbon nanotube walls: a DFT approach, Comput. Phys. Commun. 177 (2007) 46–146.

[36] E.G. Uc-Cayetano, F. Avilés, J.V. Cauich-Rodríguez, R. Schönfelder, A. Bachmatiuk, M.H. Rümmeli, F. Rubio, et al. Influence of nanotube physicochemical properties on the decoration of multiwall carbon nanotubes with magnetic particles, J. Nanopart. Res. 16 (2014) 2192.

[37] P.T. Araujo, M. Terrones, M.S. Dresselhaus, Defects and impurities in graphene-like materials, Mater. Today 15 (2012) 98–109.

[38] C. Velasco-Santos, A.L. Martínez-Hernández, M. Lozada-Cassou, A. Alvarez-Castillo, V.M. Castaño, Chemical functionalization of carbon nanotubes through an organosilane, Nanotechnology 13 (2002) 495–498.

[39] J. Kathi, K.-Y. Rhee, J.H. Lee, Effect of chemical functionalization of multi-walled carbon nanotube with 3-aminopropyltriethoxysilane on mechanical and morphological properties of epoxy nanocomposites, Compos. A 40 (2009) 800–809.

[40] J.D. Miller, H. Ishida, Quantitative monomolecular coverage of inorganic particulates by methacryl-functional silanes, Surf. Sci. 148 (1984) 601–622.

[41] C. Kaynak, O. Orgun, T. Tincer, Matrix and interface modification of short carbon fiber-reinforced epoxy, Polym. Test. 24 (2005) 455–462.

[42] C. Velasco-Santos, A.L. Martínez-Hernández, V.M. Castaño, Silanization of carbon nanotubes: surface modification and polymer nanocomposites, in: S. Yellampalli (Ed.), Carbon Nanotubes-Polymer Nanocomposites, Intech, Rijeka, Croatia, 2011, pp. 251–280.

[43] F. Avilés, J.V. Cauich-Rodríguez, J.A. Rodríguez-González, A. May-Pat, Oxidation and silanization of MWCNTs for MWCNT/vinylester composites, Express Polym. Lett. 5 (2011) 766–776.

[44] T. Zhou, X. Wang, X.H. Liu, J.Z. Lai, Effect of silane treatment of carboxylic-functionalized multi-walled carbon nanotubes on the thermal properties of epoxy composites, Express Polym. Lett. 4 (2010) 217–226.

[45] P.C. Ma, J.-K. Kim, B.Z. Tang, Effects of silane functionalization on the properties of carbon nanotube/epoxy nanocomposites, Compos. Sci. Technol. 67 (2007) 2965–2972.

[46] S.-M. Yuen, C-C.M. Ma, C.-L. Chiang, J.-A. Chang, S.-W. Huang, S.-C. Chen, C.-Y. Chuang, C.-C. Yang, Silane-modified MWCNT/PMMA composites—preparation, electrical resistivity, thermal conductivity and thermal stability, Compos. A 38 (2007) 2527–2535.

[47] S.-M. Yuen, C-C.M. Ma, C.-L. Chiang, Silane grafted MWCNT/polyimide composites—preparation, morphological and electrical properties, Compos. Sci. Technol. 68 (2008) 2842–2848.

[48] J. Berg, F.R. Jones, The role of sizing resins, coupling agents and their blends on the formation of the interphase in glass fibre composites, Compos. A 29 (1998) 1261–1272.

[49] P.C. Ma, N.A. Siddiqui, G. Marom, J.K. Kim, Dispersion and functionalization of carbon nanotubes for polymer-based nanocomposites: a review, Compos. A 41 (2010) 1345–1367.

[50] J. Gulyás, S. Rosenberger, E. Földes, B. Pukánszky, Chemical modification and adhesion in carbon fiber/epoxy microcomposites; coupling agent and surface coverage, Polym. Compos. 21 (2000) 387–395.

[51] E. Unger, A. Graham, F. Kreupl, M. Liebau, W. Hoenlein, Electrochemical functionalization of multi-wall carbon nanotubes for solvation and purification, Curr. Appl. Phys. 2 (2002) 107–111.

[52] V.N. Khabashesku, W.E. Billups, J.L. Margrave, Fluorination of single-wall carbon nanotubes and subsequent derivatization reactions, Acc. Chem. Res. 35 (2002) 1087–1095.

[53] K.F. Kelly, I.W. Chiang, E.T. Mickelson, R.H. Hauge, J.L. Margrave, X. Wang, G.E. Scuseria, C. Radloff, et al. Insight into the mechanism of sidewall functionalization of single-walled nanotubes: an STM study, Chem. Phys. Lett. 313 (1999) 445–450.

[54] E.T. Mickelson, C.B. Huffman, A.G. Rinzler, R.E. Smalley, R.H. Hauge, J.L. Margrave, Fluorination of single-wall carbon nanotubes, Chem. Phys. Lett. 296 (1998) 188–194.

[55] J. Zhu, J.D. Kim, H. Peng, J.J. Margrave, V.N. Khabashesku, E.V. Barrera, Improving the dispersion and integration of single-walled carbon nanotubes in epoxy composites through functionalization, Nano Lett. 3 (2003) 1107–1113.

[56] P.J. Boul, J. Liu, E.T. Mickelson, C.B. Huffman, L.M. Ericson, I.W. Chiang, K.A. Smith, D.T. Colbert, et al. Reversible sidewall functionalization of buckytubes, Chem. Phys. Lett. 310 (1999) 367–372.

[57] L.V. Zhang, V.U. Kiny, J.L. Margrave, H. Peng, J. Zhu, V.N. Khabashesku, R.F.M. Lobo, Sidewall functionalization of single-walled carbon nanotubes with hydroxyl group-terminated moieties, Chem. Mater. 16 (2004) 2055–2061.

[58] J. Chen, A.M. Rao, S. Lyuksyutov, M.E. Itkis, M.A. Hamon, H. Hu, R.W. Cohn, P.C. Eklund, et al. Dissolution of full-length single-walled carbon nanotubes, J. Phys. Chem. B 105 (2001) 2525–2528.

[59] V. Grosso, D. Vuono, M.A. Bahattab, G.D. Profio, E. Cursio, A.A. Al-Jilil, F. Alsubaie, M. Alfife, et al. Polymeric and mixed matrix polymide membranes, Sep. Purif. Technol. 132 (2014) 684–696.

[60] M. Soleimani, M.G. Afshar, A. Sedghi, Amino-functionalization of multiwall carbon nanotubes and its use for solid phase extraction of mercury ions from fish sample, Nanotechnology 2013 (2013) 1–8.

[61] F.H. Gojny, K. Schulte, Functionalization effect on the thermo-mechanical behavior of multi-wall carbon nanotube/epoxy-composites, Compos. Sci. Technol. 64 (2004) 2303–2308.

[62] D. Pantarotto, R. Singh, D. McCarthy, M. Erhardt, J.P. Briand, M. Prato, K. Kostarelos, A. Bianco, Functionalized carbon nanotubes for plasmid DNA gene delivery, Angew. Chem. Int. Ed. 43 (2004) 5242–5246.

[63] C. Spinato, D. Giust, I.A. Vacchi, C. Menard-Moyon, K. Kostarelos, A. Bianco, Different chemical strategies to aminate oxidised multi-walled carbon nanotubes for siRNA complexation and delivery, J. Mater. Chem. B 4 (2016) 431–441.

[64] C.H. Park, E. Tocci, E. Fontananova, M. Bahattab, S.A. Aljlil, E. Drioli, Mixed matrix membranes containing functionalized multiwalled carbon nanotubes: mesoscale simulation and experimental approach for optimizing dispersion, J. Memb. Sci. 514 (2016) 195–209.

[65] T. Ramanathan, H. Liu, L.C. Brinson, Functionalized SWNT/polymer nanocomposite for dramatic property improvement, J. Polym. Sci. B 43 (2005) 2269–2279.

[66] T. Ramanathan, F.T. Fisher, R.S. Ruoff, L.C. Brinson, Amino-functionalized carbon nanotube for binding to polymer and biological systems, Chem. Mater. 17 (2005) 1290–1295.

[67] J.L. Stevens, A.Y. Huang, H. Peng, I.W. Chiang, V.N. Khabashesku, J.L. Margrave, Sidewall amino-functionalization of single-walled carbon nanotubes through fluorination and subsequent reactions with terminal diamines, Nano Lett. 3 (2003) 331–336.

[68] T. Sun, H. Fan, Q. Zhuo, X. Liu, Z. Wu, Covalent incorporation of aminated carbon nanotubes into epoxy resin network, High Perform. Polym. 26 (2014) 892–899.

[69] P. Lu, R. Zhou, W. Guo, X.C. Zeng, Amide functionalization of graphene and carbon nanotubes: coverage- and pattern-dependent electronic and magnetic properties, J. Phys. Chem. C 116 (2012) 13722–13730.

[70] J.R. Heckert, I.C. Uber, C.M. Whitake, Synthesis of Amide Functionalized Carbon Nanotubes: Year Summary Report, United States Naval Academy (USNA), Annapolis, MD, (2007).

[71] K.S. Kim, D.J. Bae, J.R. Kim, K.A. Park, S.C. Lim, J.-J. Kim, W.B. Choi, C.Y. Park, et al. Modification of electronic structures of a carbon nanotube by hydrogen functionalization, Adv. Mater. 14 (2002) 1818–1821.

[72] S. Pekker, J.P. Salvetat, E. Jakab, J.M. Bonard, L. Forro, Hydrogenation of carbon nanotubes and graphite in liquid ammonia, J. Phys. Chem. B 105 (2001) 7938–7943.

[73] X. Kan, Y. Zhao, Z. Geng, Z. Wang, J.-J. Zhu, Composites of multiwalled carbon nanotubes and molecularly imprinted polymers for dopamine recognition, J. Phys. Chem. C 112 (2008) 4849–4854.

[74] H. Leinonen, J. Rintala, A. Siitonen, M. Lajunen, M. Pettersson, New nitrene functionalizations onto sidewalls of carbon nanotubes and their spectroscopic analysis, Carbon 48 (2010) 2425–2434.

[75] D. Mata, M. Amaral, A.J.F. Fernandes, B. Colaço, A. Gama, M.C. Paiva, P.S. Gomes, R.F. Silva, et al. Diels–Alder functionalized carbon nanotubes for bone tissue engineering: in vitro/in vivo biocompatibility and biodegradability, Nanoscale 7 (2015) 9238–9251.

[76] S. Munirasu, J. Albuerne, A. Boschetti-de-Fierro, V. Abetz, Functionalization of carbon materials using the Diels-Alder reaction, Macromol. Rapid Commun. 31 (2010) 574–579.

[77] H. Hu, B. Zhao, M.A. Hamon, K. Kamaras, M.E. Itkis, R.C. Haddon, Sidewall functionalization of single walled carbon nanotubes by addition of dichlorocarbene, J. Am. Chem. Soc. 125 (2003) 14893–14900.

[78] M.A. Herrero, M. Prato, Recent advances in the covalent functionalization of carbon nanotubes, Mol. Cryst. 483 (2008) 21–32.

[79] M.H. Hsu, H. Chuang, F.Y. Cheng, Y.P. Huang, C.C. Han, K.C. Pao, S.C. Chou, F.K. Shieh, et al. Simple and highly efficient direct thiolation of the surface of carbon nanotubes, RSC Adv. 4 (2014) 14777–14780.

[80] J.H. Bae, A.M. Shanmugharaj, W.H. Noh, W.S. Choi, S.H. Ryu, Surface chemical functionalized single-walled carbon nanotube with anchored phenol structures: physical and chemical characterization, Appl. Surf. Sci. 253 (2007) 4150–4155.

[81] B.I. Kharisov, O.V. Kharissova, Coordination and organometallic compounds in the functionalization of carbon nanotubes, J. Coord. Chem. 67 (2014) 3769–3808.

[82] E.T. Thostenson, Z. Ren, Tsu-Wei, T.-W. Chou, Advances in the science and technology of carbon nanotubes and their composites: a review, Compos. Sci. Technol. 61 (2001) 1899–1912.

[83] E.T. Thostenson, T.-W. Chou, Processing-structure-multi-functional property relationship in carbon nanotube/epoxy composites, Carbon 44 (2006) 3022–3029.

[84] L. Chang, K. Friedrich, L. Ye, P. Toro, Evaluation and visualization of the percolating networks in multi-wall carbon nanotube/epoxy composites, J. Mater. Sci. 44 (2009) 4003–4012.

[85] P.C. Ma, B.Z. Tang, J.-K. Kim, Conversion of semiconducting behavior of carbon nanotubes using ball milling, Chem. Phys. Lett. 458 (2008) 166–169.

[86] T. Villmow, P. Pötschke, S. Pegel, L. Häussler, B. Kretzschmar, Influence of twin-screw extrusion conditions on the dispersion of multi-walled carbon nanotubes in a poly(lactic acid) matrix, Polymer 49 (2008) 3500–3509.

[87] J.N. Coleman, U. Khan, Y.K. Gunko, Mechanical reinforcement of polymers using carbon nanotubes, Adv. Mater. 18 (2006) 679–687.

[88] G. Gabriel, G. Sauthier, J. Fraxedas, M. Moreno-Mañas, M.T. Martínez, C. Miravitlles, J. Casabó, Preparation and characterisation of single-walled carbon nanotubes functionalised with amines, Carbon 44 (2006) 1891–1897.

[89] J. Shen, W. Huang, L. Wu, Y. Hu, M. Ye, The reinforcement role of different amino-functionalized multi-walled carbon nanotubes in epoxy nanocomposites, Compos. Sci. Technol. 67 (2007) 3041–3050.

[90] V. Brancato, A.M. Visco, A. Pistone, A. Piperno, D. Lannazzo, Effect of functional groups of multi-walled carbon nanotubes on the mechanical, thermal and electrical performance of epoxy resin based nanocomposites, J. Compos. Mater. 47 (2012) 3091–3103.

[91] D. Iannazzo, A. Piperno, A. Ferlazzo, A. Pistone, C. Milone, M. Lanza, F. Cimino, A. Speciale, D. Trombetta, A. Saija, S. Galvagno, Functionalization of multi-walled carbon nanotubes with coumarin derivatives and their biological evaluation, Org. Biomol. Chem. 10 (2012) 1025–1031.

[92] S.G. Prolongo, B.G. Melitón, A. Jimenez-Suarez, A. Ureña, Study of efficiency of different commercial carbon nanotubes on manufacturing of epoxy matrix composites, J. Compos. Mater. 48 (2014) 3169–3177.

[93] J.T. Kim, H-C. Kim, S.-K. Kim, J. Kathi, Y. Kyong, K.-Y. Rhee, 3-Aminopropyltriethoxysilane effect on thermal and mechanical properties of multi-walled carbon nanotubes reinforced epoxy composites, J. Compos. Mater. 43 (2009) 2533–2541.

[94] A. Martone, C. Formicola, M. Giordano, M. Zarrelli, Reinforcement efficiency of multi-walled carbon nanotube/epoxy nano composites, Compos. Sci. Technol. 70 (2010) 1154–1160.

[95] L. Liu, H.D. Wagner, Rubbery and glassy epoxy resins reinforced with carbon nanotubes, Compos. Sci. Technol. 65 (2005) 1861–1868.

[96] A. Hernández-Pérez, F. Avilés, A. May-Pat, A. Valadez-González, P.J. Herrera-Franco, P. Bartolo-Pérez, Effective properties of multiwalled carbon nanotube/epoxy composites using two different tubes, Compos. Sci. Technol. 68 (2008) 1422–1431.

[97] B. Krause, T. Villmow, R. Boldt, M. Mende, G. Petzold, P. Pötschke, Influence of dry grinding in a ball mill on the length of multiwalled carbon nanotubes and their dispersion and percolation behaviour in melt mixed polycarbonate composites, Compos. Sci. Technol. 71 (2011) 1145–1153.

[98] F. Avilés, A. May-Pat, G. Canche-Escamilla, O. Rodrıguez-Uicab, J.J. Ku-Herrera, S. Duarte-Aranda, J. Uribe-Calderon, C. Gonzalez, et al. Influence of carbon nanotube on the piezoresistive behavior of multiwall carbon nanotube/polymer composites, J. Intell. Mater. Syst. Struct. 27 (2016) 92–103.

[99] K.-T. Lau, D. Hui, Effectiveness of using carbon nanotubes as nano-reinforcements for advanced composite structures, Carbon 40 (2002) 1597–1617.

[100] J. Zhu, H. Peng, F. Rodriguez-Macias, J.L. Margrave, V.N. Khabashesku, A.M. Imam, K. Lozano, E.V. Barrera, Reinforcing epoxy polymer composites through covalent integration of functionalized nanotubes, Adv. Funct. Mater. 14 (2004) 643–648.

[101] J. Shen, W. Huang, L. Wu, Y. Hu, M. Ye, Thermo-physical properties of epoxy nanocomposites reinforced with amino-functionalized multi-walled carbon nanotubes, Compos. A 38 (2007) 1331–1336.

[102] A.T. Seyhan, F.H. Gojny, M. Tanoglu, K. Schulte, Critical aspects related to processing of carbon nanotube/unsaturated thermoset polyester nanocomposites nanocomposites, Eur. Polym. J. 43 (2007) 374–379.

[103] A.T. Seyhan, M. Tanoglu, K. Schulte, Tensile mechanical behavior and fracture toughness of MWCNT and DWCNT modified vinyl-ester/polyester hybrid nanocomposites produced by 3-roll milling, Mater. Sci. Eng. A 523 (2009) 85–92.

[104] LM Wind Power, Rolles Møllevej, Lunderskov, Denmark, 2016. Available from: http://www.lmwindpower.com/.

[105] J.J. Ku-Herrera, F. Avilés, G.D. Seidel, Self-sensing of elastic strain, matrix yielding and plasticity in multiwall carbon nanotube/vinyl ester composites, Smart Mater. Struct. 22 (2013) 085003.

[106] S. Sathyanarayana, C. Hübner, Thermoplastic nanocomposites with carbon nanotubes, in: I.J. Njuguna (Ed.), Structural Nanocomposites, first ed., Springer, Berlin, Germany, 2013, pp. 19–90.

[107] J.W. Gooch, Compatibilizer, second ed., Encyclopedic Dictionary of Polymers, Springer-Verlag, New York, NY, (2011) (p. 160).

[108] S.H. Lee, E. Cho, S.H. Jeon, J.R. Youn, Rheological and electrical properties of polypropylene composites containing functionalized multi-walled carbon nanotubes and compatibilizers, Carbon 45 (2007) 2810–2822.

[109] B.-X. Yang, J.-H. Shi, K.P. Pramoda, S.H. Goh, Enhancement of the mechanical properties of polypropylene using polypropylene-grafted multiwalled carbon nanotubes, Compos. Sci. Technol. 68 (2008) 2490–2497.

[110] Q. He, T. Yuan, X. Yan, D. Ding, Q. Wang, Z. Luo, T. Shen, S. Wei, D. Cao, Z. Guo, Flame-retardant polypropylene/multiwall carbon nanotube nanocomposites: effects of surface functionalization and surfactant molecular weight, Macromol. Chem. Phys. 215 (2014) 327–340.

[111] B.-X. Yang, K. Pramoda, G. Xu, S.H. Goh, Mechanical reinforcement of polyethylene using polyethylene-grafted multiwalled carbon nanotubes, Adv. Funct. Mater. 17 (2007) 2062–2069.

[112] L. Liu, Y. Wang, F. Xiang, Y. Li, L. Han, Z. Zhou, Effects of functionalized multiwalled carbon nanotubes on the morphologies and mechanical properties of PP/EVA blend, J. Polym. Sci. B 47 (2009) 1481–1491.

[113] J.M. Yuan, Z.-F. Fan, X.-H. Chen, X.-H. Chen, Z.-J. Wu, L.-P. He, Preparation of polystyrene–multiwalled carbon nanotube composites with individual-dispersed nanotubes and strong interfacial adhesion, Polymer 50 (2009) 3285–3291.

[114] H. Lee, S. Oh, J. Choi, J. Kim, J. Han, L. Tan, J. Baek, In situ synthesis of poly(ethylene terephthalate) (PET) in ethylene glycol containing terephthalic acid and functionalized multiwalled carbon nanotubes (MWNTs) as an approach to MWNT/PET nanocomposites, Chem. Mater. 17 (2005) 5057–5064.

[115] S.H. Jin, Y.-B. Park, K.H. Yoon, Rheological and mechanical properties of surface modified multiwalled carbon nanotube-filled PET composite, Compos. Sci. Technol. 67 (2007) 3434–3441.

[116] A. May-Pat, F. Avilés, P. Toro, M. Yazdani-Pedram, J.V. Cauich-Rodríguez, Mechanical properties of PET composites using multi-walled carbon nanotubes functionalized by inorganic and itaconic acids, Express Polym. Lett. 6 (2012) 96–106.

[117] T. Liu, I.Y. Phang, L. Shen, S.Y. Chow, W.-D. Zhang, Morphology and mechanical properties of multiwalled carbon nanotubes reinforced nylon-6 composites, Macromolecules 37 (2004) 7214–7222.

[118] A. Bhattacharyya, P. Potschke, L. Haussler, D. Fischer, Reactive compatibilization of melt mixed PA6/SWNT composites: mechanical properties and morphology, Macromol. Chem. Phys. 206 (2005) 2084–2095.

[119] S.A. Girei, S.P. Thomas, M.A. Atieh, K. Mezghani, S.K. De, S. Bandyopadhyay, A. Al-Juhan, Effect of –COOH functionalized carbon nanotubes on mechanical, dynamic mechanical and thermal properties of polypropylene nanocomposite, J. Thermoplast. Compos. Mater. 25 (2012) 333–350.

[120] M. Yazdani-Pedram, C. Menzel, P. Toro, R. Quijada, A. May-Pat, F. Avilés, Mechanical and thermal properties of multiwalled carbon nanotube/polypropylene composites using itaconic acid as compatibilizer and coupling agent, Macromol. Res. 21 (2013) 153–160.

[121] J.Y. Kim, S.I. Han, S. Hong, Effect of modified carbon nanotube on the properties of aromatic polyester nanocomposites, Polymer 49 (2008) 3335–3345.

[122] W.D. Zhang, L. Shen, I.Y. Phang, T. Liu, Carbon nanotubes reinforced nylon-6 composite prepared by simple melt-compounding, Macromolecules 37 (2004) 256–259.

[123] M. Frogley, D. Ravich, H. Wagner, Mechanical properties of carbon nanoparticle-reinforced elastomers, Compos. Sci. Technol. 63 (2003) 1647–1654.

[124] M. López-Manchado, J. Biagiotti, L. Valentini, J. Kenny, Dynamic mechanical and Raman spectroscopy studies on interaction between single-walled carbon nanotubes and natural rubber, J. Appl. Polym. Sci. 92 (2004) 3394–3400.

[125] L. Bokobza, Multiwall carbon nanotube elastomeric composites: a review, Polymer 48 (2007) 4907–4920.

[126] L. Chen, C. Liu, C. Hu, S. Fan, Electrothermal actuation based on carbon nanotube network in silicone elastomer, Appl. Phys. Lett. 92 (2008) 263104.

[127] J. Abraham, P.M. Arif, L. Kailas, N. Kalarikkal, S.C. George, T. Thomas, Developing highly conducting and mechanically durable styrene butadiene rubber composites with tailored microstructural properties by a green approach using ionic liquid modified MWCNTs, RSC Adv. 6 (2016) 32493–32504.

[128] K. Subramaniam, A. Das, G. Heinrich, Improved oxidation resistance of conducting polychloroprene composites, Compos. Sci. Technol. 74 (2013) 14–19.

[129] J.R. Bautista-Quijano, F. Avilés, J.V. Cauich Rodríguez, Sensing of large strain using multiwall carbon nanotube/segmented polyurethane composites, J. Appl. Polym. Sci. 130 (2013) 375–382.

[130] J.R. Bautista-Quijano, F. Avilés, J.V. Cauich Rodríguez, R. Schönfelder, A. Bachmatiuk, T. Gemming, M.H. Rümmeli, Tensile piezoresistivity and disruption of percolation in single- and multiwall-carbon nanotube/polyurethane composites, Synth. Met. 185-186 (2013) 96–102.

[131] C. Lozano-Pérez, J.V. Cauich-Rodríguez, F. Avilés, Influence of rigid segment and carbon nanotube concentration on the cyclic piezoresistive and hysteretic behavior of multiwall carbon nanotube/segmented polyurethane composites, Compos. Sci. Technol. 128 (2016) 25–32.

[132] B. Fernandez-d'Arlas, M.A. Corcuera, A. Eceiza, Comparison between exfoliated graphite, graphene oxide and multiwalled carbon nanotubes as reinforcing agents of a polyurethane elastomer, J. Thermoplast. Compos. Mater. 28 (2015) 705–716.

[133] Q. Jing, J.Y. Law, L.P. Tan, V.V. Silberschmidt, L. Li, Z.L. Dong, Preparation, characterization and properties of polycaprolactone diol-functionalized multi-walled carbon nanotube/thermoplastic polyurethane composite, Compos. A 70 (2015) 8–15.

[134] J. Lee, E.J. Park, J. Choi, J. Hong, S.E. Shim, Polyurethane/PEG-modified MWCNT composite film for the chemical vapor sensor application, Synth. Met. 160 (2010) 566–574.

[135] B. Cheng, H. Liu, Preparation of multiwalled carbon nanotubes/polyurethane composite, Polym. Prepr. 47 (2006) 229.

[136] A. Kelarakis, K. Yoon, R. Somani, X. Chen, Rheological study of carbon nanofiber induced physical gelation in polyolefin nanocomposite melt, Polymer 46 (2005) 11591–11599.

[137] L. Valentini, J. Biagiotti, K. Kenny, M. López-Manchado, Physical and mechanical behavior of single-walled carbon nanotube/polypropylene/ethylene–propylene–diene rubber nanocomposites, J. Appl. Polym. Sci. 89 (2003) 2657–2663.

[138] G. Broza, Thermoplastic elastomers with multi-walled carbon nanotubes: influence of dispersion methods on morphology, Compos. Sci. Technol. 70 (2010) 1006–1010.

[139] M. Hemmati, A. Narimani, H. Shariatpanahi, A. Fereidoon, M. Ahangari, Study on morphology, rheology and mechanical properties of thermoplastic elastomer polyolefin (TPO)/carbon nanotube nanocomposites with reference to the effect of polypropylene-grafted-maleic anhydride (PP-*g*-MA) as a compatibilizer, Int. J. Polym. Mater. 60 (2011) 384–397.

[140] C. Li, H. Deng, K. Wang, Q. Zhang, F. Chen, Q. Fu, Strengthening and toughening of thermoplastic polyolefin elastomer using polypropylene-grafted multiwalled carbon nanotubes, J. Appl. Polym. Sci. 121 (2011) 2104–2112.

[141] S. Bhattacharyya, C. Sinturel, O. Bahloul, M.L. Saboungi, S. Thomas, J.P. Salvetat, Improving reinforcement of natural rubber by networking of activated carbon nanotubes, Carbon 46 (2008) 1037–1045.

[142] H. Jang, S.S. Lee, H. Yoon, I. Jeon, Y. Ko, J.H. Kim, J. Choi, H. Kim, Enhanced performance in capacitive force sensors using carbon nanotube/polydimethylsiloxane nanocomposites with high dielectric properties, Nanoscale 8 (2016) 5667–5675.

[143] Y. Liu, W. Chi, H. Duan, H. Zou, D. Yue, L. Zhang, Property improvement of room temperature vulcanized silicone elastomer by surface-modified multi-walled carbon nanotube inclusion, J. Alloys Compd. 657 (2016) 472–477.

[144] H. Li, H. Liu, J. Li, D. Wei, Y. Wang, B. Song, Dielectric behaviors of carbon nanotube/silicone elastomer composites, Sci. Eng. Compos. Mater. 22 (2015) 215–221.

[145] A.M. Shanmugharaj, J.H. Bae, K.Y. Lee, W.H. Noh, S.H. Lee, S.H. Ryu, Physical and chemical characteristics of multiwalled carbon nanotubes functionalized with aminosilane and its influence on the properties of natural rubber composites, Compos. Sci. Technol. 67 (2007) 1813–1822.

[146] L. Vast, L. Carpentier, F. Lallemand, J. Colomer, G. Van Tendeloo, A. Fonseca, J. Nagy, Z. Mekhalif, J. Delhalle, Multiwalled carbon nanotubes functionalized with 7-octenyltrichlorosilane and *n*-octyltrichlorosilane: dispersion in Sylgard 184 silicone and Young's modulus, J. Mater. Sci. 44 (2009) 3476–3482.

[147] Y. Nakaramontri, C. Kummerlowe, C. Nakason, N. Vennemann, The effect of surface functionalization of carbon nanotubes on properties of natural rubber/carbon nanotube composites, Polym. Compos. 36 (2015) 2113–2122.

[148] F. Galantini, S. Bianchi, V. Castelvetro, G. Gallone, Functionalized carbon nanotubes as a filler for dielectric elastomer composites with improved actuation performance, Smart Mater. Struct. 22 (2013) 1–10.

[149] M. Ilcikova, M. Mrlik, T. Sedlacek, M. Doroshenko, K. Koynov, M. Danko, J. Mosnacek, Tailoring of viscoelastic properties and light-induced actuation performance of triblock copolymer composites through surface modification of carbon nanotubes, Polymer 72 (2015) 368–377.

[150] H.E. Miltner, S. Peeterbroeck, P. Viville, P. Dubois, B. Van Mele, Interfacial interaction in EVA-carbon nanotube and EVA-clay nanocomposites, J. Polym. Sci. B 45 (2007) 1291–1302.

[151] A.R. Payne, The dynamic properties of carbon black loaded natural rubber vulcanizates. Part II, J. Appl. Polym. Sci. 6 (1962) 368–372.

[152] C. Nah, J. Lim, B. Cho, C. Hong, A. Gent, Reinforcing rubber with carbon nanotubes, J. Appl. Polym. Sci. 118 (2010) 1574–1581.

[153] L. Bokobza, Multiwall carbon nanotube-filled natural rubber: electrical and mechanical properties, Express Polym. Lett. 6 (2012) 213–223.

[154] A. Das, K. Stöckelhuber, R. Jurk, M. Saphiannikova, J. Fritzsche, H. Lorenz, M. Klüppel, G. Heinrich, Modified and unmodified multiwalled carbon nanotubes in high performance solution-styrene–butadiene and butadiene rubber blends, Polymer 49 (2008) 5276–5283.

6

Deposition of Carbon Nanotubes on Fibers

Francis Avilés*, José de Jesús Kú-Herrera**, Andrés I. Oliva-Avilés[†]

*SCIENTIFIC RESEARCH CENTER OF YUCATAN, AC (CICY), (CENTRO DE INVESTIGACIÓN CIENTÍFICA DE YUCATÁN, A.C., UNIDAD DE MATERIALES), MÉRIDA, YUCATÁN, MEXICO; **CONACYT-RESEARCH CENTER FOR APPLIED CHEMISTRY (CIQA), (CENTRO DE INVESTIGACIÓN CIENTÍFICA DE YUCATÁN, A.C., UNIDAD DE MATERIALES), SALTILLO, MÉRIDA, COAHUILA, MEXICO); [†]ANAHUAC MAYAB UNIVERSITY, (UNIVERSIDAD ANÁHUAC MAYAB, DIVISIÓN DE INGENIERÍA Y CIENCIAS EXACTAS), MÉRIDA, YUCATÁN, MEXICO

CHAPTER OUTLINE

6.1 Introduction

Natural and, most recently, engineering fibers have been used for humankind for centuries, from the early uses of natural fibers for textiles, clothing, and ropes, to the modern uses of carbon fibers for structural reinforcement in aircraft, such as the landmark Boeing 787 jet airliner. As for many other materials, the new trends in textiles, fibers, and polymer composites require that, besides fulfilling a primary purpose, the properties of the material allow delivery of one or more secondary functions—in other words, multifunctionality. One approach to reach multifunctionality for the case of engineering fibers, such as

Carbon Nanotube–Reinforced Polymers. http://dx.doi.org/10.1016/B978-0-323-48221-9.00006-6

glass, carbon, or aramid fibers is to deposit carbon nanotubes (CNTs) on their surface, or to directly grow them on the fiber surface. Depositing CNTs on the surface of fibers may greatly modify their surface properties, electrical conductivity, and thermal conductivity, among many other properties. If this CNT deposit/growth is achieved without altering the mechanical and other volumetric properties of the fibers, a multifunctional fiber with improved capabilities and a host variety of applications may be achieved. With this motivation in mind, this chapter reviews the fundamentals of the most common methods used to deposit CNTs on fibers, touching on in situ CNT growth. Progress in this area for glass, carbon, and aramid fibers is the main focus of the chapter.

6.2 Methods of Deposition and Growth of Carbon Nanotubes on Engineering Fibers

The incorporation of CNTs to the surface of conventional (micron-size) engineering fibers represents an important step on the constitution of multiscale hierarchical reinforced composites, with a wide range of applications. Although the larger interest to date relies on polymer matrices, metallic and ceramic matrices are also of interest, depending on the application. The outstanding morphological, mechanical, electrical, and thermal properties of CNTs provide multifunctional characteristics to the composites, improving their mechanical properties, electrical and thermal conductivity, flame retardancy, wear resistance, surface finish, and biological interactions, among many others. As such, the methodology to achieve a homogeneous and stable distribution of CNTs on the surface of micron-size fibers becomes a key factor, and different strategies have been attempted for such a challenging aim.

Among the methods used to cover the surface of engineering fibers with CNTs, dipping deposition (DD), and electrophoretic deposition (EPD), along with in situ growth are nowadays the most used methods reported in the literature. A general overview of the mechanisms underlying each deposition method along with their main advantages and caveats are discussed in this section. Recent literature and reports on the application of such techniques to different fibers are discussed. Other less reported deposition methods are also briefly discussed toward the end of the section.

6.2.1 Dipping Deposition

In the simplest form of the DD method, the CNTs are deposited on the surface of engineering fibers (acting as a substrate) by immersing the fiber into a solution containing dispersed CNTs. A simplified schematic of the DD method is shown in Fig. 6.1. The DD approach is essentially based on the physicochemical interactions between CNTs and the fiber surface and demands adequate CNT dispersion in the dipping solution prior to fiber immersion. Upon fiber immersion, the CNTs approach the fiber due to capillary, van der Waals, and electrostatic forces present between their interfaces. Then, the physicochemical interactions between the chemical groups of the CNTs and those on the fiber surface should provide the desired CNT-to-fiber anchorage, rendering a stable CNT network on

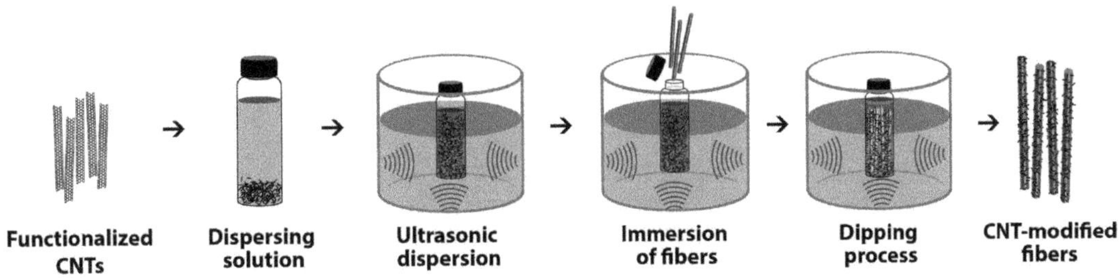

FIGURE 6.1 Schematic of the general process of carbon nanotube (CNT) deposition on the fibers surface by the dipping deposition (DD) approach.

the fiber surface (Fig. 6.1). Therefore, a prerequisite to maintain the CNTs bonded to the fiber after the fiber is pulled-out from the solution is that the CNTs form physical and/or chemical bonds with the fiber surface (or with its surface coating). Because of this, it is customary that the CNTs (and sometimes the fibers) are functionalized prior to DD [1–5]. The functional groups aimed on the CNTs depend on the chemistry of the substrate (fiber), and for conventional glass fibers (GFs) and carbon fibers (CFs) a simple wet chemical oxidation may suffice [5,6]. Several techniques to functionalize both the CNTs and fibers have been reported, and the most important ones are probably covalent attachments, defect-group or defect-site functionalization, and noncovalent functionalization [7–10].

To create chemical groups on the surface of CNTs that enhance the interaction with the fiber surface, oxidation is undoubtedly the most common functionalization technique. CNT oxidation can be conducted by physical methods, such as plasma treatments [11] or by wet chemical methods [9,12], with the latter being more effective in terms of generation of functional groups. Upon oxidation, carboxyl and hydroxyl groups are generated on the CNT surface [9], which may then interact with the functional groups present on the fiber or on its fiber surface coating. Nonionic surfactants are also an option. They are often used as dispersing aids for CNTs, and have functional groups containing hydrophilic and hydrophobic segments. The hydrophobic groups interact with the CNTs through adsorption, improving the CNT wettability. In addition, the surfactants serve as dispersing agents by introducing steric repulsion forces between CNTs, which in turn overcome the van der Waals attractive forces present in the pristine CNT agglomerates [13]. These characteristics promote better conditions for CNT attachment to fiber surfaces. Electrostatic interactions are also used to promote attachment between CNTs and fibers [1,14]. For example, by creating oxygen-containing functional groups on the CNT surface (mostly hydroxyl and carboxylic), they become negatively charged. If the fiber is then positively charged, for example, by protonation of amino groups $(-NH_3^+)$, electrostatic interactions are promoted between the CNT and fiber [1]. Another alternative is the use of cationic polymers with reactive functional groups for CNT treatment, which can then be electrostatically bonded to negatively charged (oxidized) CFs [14]. In this concept, the repulsive forces between the positively charged CNTs are expected to promote good dispersion, and the interactions between negatively charged CFs and the positively charged nanotubes promote a uniform CNT coating. For all these approaches, the CNT dispersion in solution, the type of solution, and the amount of surface charge are paramount parameters to control, and such parameters can be quantified by zeta potential measurements. For example, it is well known that surface oxidized CFs based on polyacrylonitrile (PAN) have a relatively strong negative charge [15–17]. García et al. reported that surface oxidized (PAN-based) CFs in 1 mM potassium chloride in water at pH of 7 present a zeta potential of −50 to −60 mV [15]. Considering this, positively charged CNTs (positive values of zeta potential) will be more attracted to negatively charged CFs [14]. There are several parameters that strongly influence the efficiency of CNT fiber coating by the DD technique, such as the dipping time, the CNT concentration in solution, the pH value of the solution, and the roughness of the fiber surface. Very long dipping times and relatively high concentration of CNTs in the soaking solution may reduce proper distribution of CNTs on the fiber due to formation of CNT agglomerates. A different alternative for polymer fibers consists of swelling the fiber to allow diffusion-controlled CNT ingress. O'Connor et al. used *N*-methylpyrrolidone to first swell Kevlar fibers and then soaked the fibers into a multiwall carbon nanotube (MWCNT) suspension to allow their infiltration [18]. In this case, it was shown that the CNT mass uptake ($m_{\mathrm{CNT}}/m_{\mathrm{Kev}}$) scales with the square root of the soaking time (t), suggesting that the intercalation of CNTs on the fiber surface is diffusion-limited by means of,

$$\frac{m_{\mathrm{CNT}}}{m_{\mathrm{Kev}}} = \left(\frac{m_{\mathrm{CNT}}}{m_{\mathrm{RVE}}}\right)_{\mathrm{sat}} \sqrt{\left(\frac{16D}{b^2\pi}\right)t} \tag{6.1}$$

where the suffix "sat" refers to the saturated value of the mass uptake, D is the diffusion coefficient, and b is the sample thickness.

The pH of the CNT solution also has a very important effect on the final condition of CNT deposition, influencing the amount of negative charges attached to the CNTs, and hence their mobility in solution [19]. By tuning this parameter, the thickness and morphology of the CNT coating can be controlled [1]. The roughness of the fiber surface and its chemical modification are also relevant for a uniform coating. More uniform and stable CNT coating is observed for fibers with rougher surfaces [20]. CNT coverage is improved when the dimensions of the CNTs are similar or below the dimensions of the grooves providing roughness to the fiber surface [20]. Depending on the fiber type, several approaches have been reported to attach reactive functional groups to the fiber surface. One approach takes advantage of the fiber surface coating (also known as "sizing"), commonly used to improve the compatibility of engineering fibers with polymer matrices. Some authors have benefited from the chemical interactions between the CNTs and the commercial sizing of the fibers [4,5,20,21], whereas other works prefer to remove the commercial sizing of the fiber and apply a specific silane (either modified with CNTs or not), to render more control [2,6]. Avilés and coworkers have explored the roles of the (commercial) fiber surface coating and the suspending fluid on the deposition of oxidized MWCNTs on the surface of GFs by the dipping technique [4,5,21]. In a series of works by Avilés and coworkers, GFs were soaked in solutions containing oxidized MWCNTs dispersed within either distilled water or a mixture of water/silane (3-trimethoxysilyl propyl methacrylate in this case). Due to chemical interactions with carboxyl and hydroxyl functional groups on the MWCNT surface, more homogeneous distribution of MWCNTs on the fiber is achieved when the fiber coating is maintained [5]. Also, both water and the hydrolyzed silane were determined to be good dispersive solvents for the CNTs during the dipping process. By an exhaustive physicochemical analysis including infrared, Raman, nuclear magnetic resonance and X-ray photoelectron spectroscopies [4,5], the authors explored the mechanisms and nature of chemical bonding between the oxidized MWCNTs and the commercial sizing of GFs. They report that multiple oxygen-containing functional groups and epoxy groups are present on the complex structure of commercial sizing and are prone to form covalent and hydrogen bonding with the OH and COOH groups of the oxidized CNTs. Another approach for CNT deposition has been dispersing the CNTs in the solutions containing a coupling agent (typically a silane) prior to its application to the fiber. Generally, the CNTs are first dispersed in a solvent, then mixed in the silane solution, and finally the fibers are sized with the CNT/sizing solution [6,22,23]. The type of bonding also determines the quality of CNT coverage onto the fiber surface. A more uniform CNT grafted coating with rather homogeneous surface coverage is observed when covalent bonds are formed, as compared with the case when only physical bonds or simple adsorption occurs [2,3]. When only physical bonds are present between the CNTs and fiber surfaces, MWCNT agglomeration is promoted and noncovered areas can be found along the fiber. A different approach that also involves fiber dipping has been also explored, where the CNTs are dispersed within a nanocomposite solution to firstly form a CNT/polymer coating solution [24–26]. In this approach, CNTs are dispersed in a solvent and later incorporated to a dissolved polymer. The mixed solution is finally used as the dipping solution for the fibers, which are soaked to form a film coating the surface. The functional groups on the fiber surface and the reactivity of the CNT/polymer solution dictate the degree of coating uniformity.

6.2.2 Electrophoretic Deposition

Electrophoresis refers to the motion of a particle with a nonzero net charge produced by Coulombic forces. Biological and nonbiological particles (such as CNTs) generally have a finite surface charge (usually negative for the case of CNTs, due to the presence of acid groups on their surface) and the observation of motion of these particles in uniform fields has been used to characterize and separate them [27]. The Coulomb force ($\boldsymbol{F_C}$) on a particle is given by the product of the electric field ($\boldsymbol{E}$) and the charge on the particle and can be expressed as,

$$\boldsymbol{F_C} = \int_S \sigma_q \boldsymbol{E}\, dS \tag{6.2}$$

where σ_q is the surface charge density and the integral refers to the closed surface S of the particle. As can be understood from Eq. (6.2), electrophoresis is therefore governed by the intensity of the applied electric field and by the effective charge at the surface of the particles. For an alternating current electric field, for example, the behavior of this force is oscillatory with zero time-average. Although early experiments on this area were conducted in polymeric and biological particles suspended on dielectric liquids [28], the range of application of electrophoresis has been tremendously increased and nowadays represents a widely used and successful technique for the placement, manipulation, and deposition of nanoparticles, including CNTs. In EPD, particles suspended in a fluid are attracted and deposited onto a conductive surface of opposite charge, upon application of a direct current (DC) electric field [29]. A typical EPD setup consists of a pair of electrodes (work and counter electrodes) and a power supply that provides the electric field between the electrodes (Fig. 6.2). Thus, two subprocesses can be appreciated on EPD, electrophoretic motion in a liquid phase toward an electrode due to the application of an electric field, and the collection of the particles at the electrode surface and formation of the deposit. There are two types of EPDs depending on which electrode the deposition occurs. When the particles are positively charged, the deposition happens on the cathode and the process is called cathodic EPD. The deposition of negatively charged particles on the positive electrode (anode) is termed as anodic EPD (Fig. 6.2). By a suitable modification of the surface charge on the particles or of the pH of the solution, any of the two modes of deposition are possible. Excellent reviews regarding the fundamentals and applications of EPD can be found elsewhere [30–35]. Among the pioneering efforts on EPD of CNTs over engineering fibers are those by Gao et al. [36]. In their work, single wall carbon nanotubes (SWCNTs) were dispersed in dimethylformamide (DMF) and methanol [36]. A small amount of NaOH (less than 1%) was added during ultrasonic dispersion of CNTs and a DC voltage of 20 V was applied between electrodes located 1 cm apart. When the CNTs were dispersed in DMF with NaOH, they migrated to the positive electrode, evidencing a negative nature of the charge at the CNTs surface. Interestingly, SWCNT deposition was not observed when the SWCNTs were dispersed within DMF without NaOH (i.e., a neutral solution); the SWCNTs migrated toward the negative electrode instead of the positive onewhen NaOH was replaced by salts (such as $MgCl_2$). Therefore, in this case, the migration and subsequent deposition of CNTs under the influence of a DC electric field are attributed to preferential absorption of ions in the solution by the CNTs.

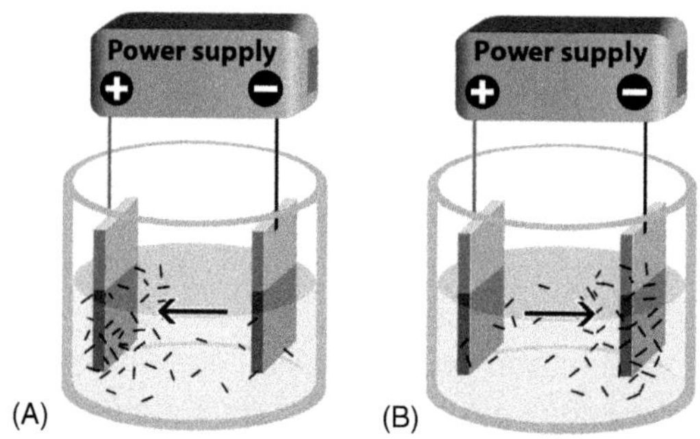

FIGURE 6.2 Schematic illustration of electrophoretic deposition (EPD) of CNTs. (A) Anodic deposition and (B) cathodic deposition.

The use of EPD to deposit CNTs on the surfaces of engineering fibers is nowadays an area of increasing interest. EPD represents a simple, versatile, and very promising technique for deposition CNTs (and other nanostructures) onto fibers that can be automated and utilized for industrial applications, rendering many advantages over other CNT deposition techniques [37]. Generally, CNTs deposited by EPD exhibit high packing density with homogeneous microstructure [38]. There are several strategies and experimental setups designed to use electric fields to deposit CNTs onto micron-size fibers (Fig. 6.3). Fig. 6.3A presents a schematic of an experimental setup designed and implemented by the Aviles' group for the EPD of CNTs onto GFs. The experimental setup consists of an electrophoretic cell comprising a pair of stainless steel electrodes, which are immersed in the CNT/liquid suspension. Piezoelectric transducers are adapted to the electrophoretic cell to apply ultrasonic vibration, which aids CNT dispersion. A dedicated control system allows the GFs to be slowly moved within the cell using a step motor and a winder (Fig. 6.3A), in contact with the positive electrode; the oxidized (negatively charged) CNTs are attracted to the positive electrode, allowing CNT deposition on the nonelectrically conductive GFs. As seen in Fig. 6.3A, both sets of electrodes are slightly shifted in the vertical direction and the order of their polarity is inverted, so both sides of the glass preform are covered upon its horizontal displacement toward the winder. Fig. 6.3B shows a micrograph obtained by scanning electron microscopy (SEM) of a 10 μm diameter GF covered with CNTs by using this EPD system.

Different to other methods of CNT deposition, which use high temperature or excessive chemical treatments, EPD does not induce damage on the fibers. A variety of methods to create stable CNT dispersions in aqueous solutions have been proposed [39], with the aim of creating charged CNTs that will migrate under the influence of an external electric field to the fiber surface coating [40]. The choice of solvent, CNT functionalization method, and voltage can all influence the film formation; for high electric fields, water electrolysis may also affect the film porosity and CNT deposition [40,41].

There are key parameters that control the quality and quantity of CNTs deposited onto fibers by EPD. Some parameters are related to the CNT/liquid suspension, such as the

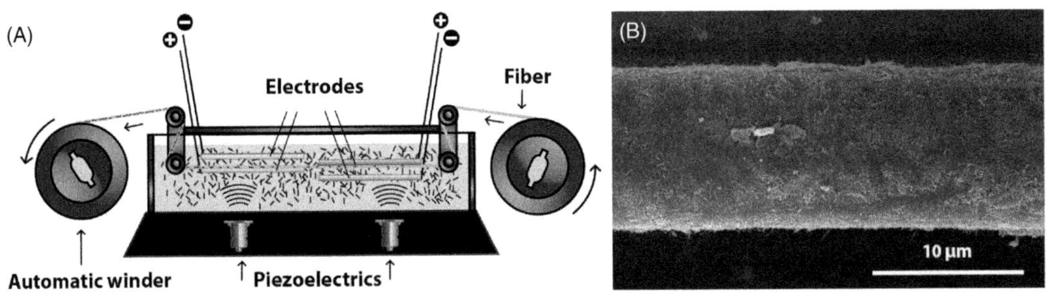

FIGURE 6.3 EPD of CNTs onto nonconductive fibers implemented by Dr. Aviles' group. (A) Illustration of the experimental setup and (B) scanning electron microscopy (SEM) image of a CNT-covered glass fiber with CNTs deposited by EPD.

CNT concentration and the CNT net charge. On the other hand, the applied voltage, the deposition time, the type of electrodes, and the geometry and separation between electrodes are important parameters related to the electric field. Overall, the kinetics of the CNT deposition depends largely on the electric field strength, the stability and concentration of the suspension, the electrophoretic mobility of the particles, and the deposition time [42]. A first approximation to quantify the main governing parameters is summarized by the Hamaker's equation for electrophoretic deposition [43]. The Hamaker's equation is a mathematical model predicting the EPD kinetics of particulate materials. It states that the time-dependent mass of the deposit [$m(t)$] has a linear relationship with the electrophoretic mobility of the particles (μ), the strength of electric field (E), the surface area of the electrodes (A), the coefficient of efficiency of deposition (f), and the particle concentration in suspension (C), that is,

$$m(t) = f\mu EACt \tag{6.3}$$

Eq. (6.3) suggests that the deposition yield follows a linear dependence with the electric field magnitude, although it is known that a minimum value of E is required to achieve the deposition due to the presence of repulsive interparticle interactions. As expected, the deposition yield is proportional to the CNT concentration and to the area of the electrodes. Other factors also indirectly affect the EPD of CNTs. For example, μ depends on the dielectric constant of the CNTs, the viscosity of the medium and the zeta potential of the CNTs, which depends on their surface charge and mobility. However, it is also known that the Hamaker's approach for predicting the EPD kinetics has some caveats under certain conditions; for example, at high EPD voltages and/or long deposition time deviations from the linearity predicted by the Hamaker's law can occur. The increased thickness of the deposited layers of material can negatively affect the electrodes, promoting changes in the applied electric field and eventually causing sharp voltage drops [42]. Deviations of the voltage-deposition yield relationship from linearity at high voltages have been experimentally reported by several authors [44–46]. Other deviations from the linearity predicted by the Hamaker's law are associated to side reactions at specific voltages and/or instability of the suspension during the EPD process [47]. Regarding the electric field magnitude and deposition time, several researchers [31,48,49] have observed that the CNT coverage onto the fibers is higher and more uniform when both parameters increase, in agreement with the kinetic predictions in Eq. (6.3). However, the relationship between electric field and deposition time with the amount of CNT coverage is not linear and after certain values of electric field magnitude and/or deposition time the quality of the deposition decreases, as evidenced by the formation of CNT agglomerates.

It is thus clear that EPD is being increasingly considered as one of the preferred processing methods for the deposition of nanoparticles (including CNTs) over micro- or macroscopic substrates, such as engineering fibers. The success of EPD is based not only on its wide applicability to numerous materials, but also on the fact that it is a versatile cost-effective technique requiring simple instrumentation that is potentially scalable. Research efforts are being increasingly devoted to the exploitation of CNT EPD for a variety of specific applications, such as field emission devices, supercapacitors, bioactive materials, and photocatalytic coatings [31–33]. Arrangements of CNTs exhibiting high aspect ratio and high surface charge are also suitable substrates for entrapment of other functional nanoparticles.

6.2.3 In Situ Growth

Another alternative for coating micron-sized fibers with CNTs is by growing them in situ—that is, by using the fiber surface as a substrate for CNT synthesis. The biggest advantages of in situ growth are the high control of the growth, uniformity, thickness, and density of CNTs, by means of the control of the process parameters. For this approach, chemical vapor deposition (CVD) has been established as the most effective and practical route. The main drawbacks of the CVD technique are related to the usage of high temperatures, which induce thermal damage on the fibers decreasing their mechanical properties [50,51]. Among the first efforts on this area are those reported by Downs and Baker, where a novel concept involving the generation of vapor-grown carbon filaments on the surface of CFs was proposed [52]. The carbon filaments were grown on the fiber by impregnation with copper-nickel catalyst precursors as a product of the decomposition of selected hydrocarbons and carbon monoxide [53]. During this reaction, the hydrocarbons are adsorbed and decomposed at the catalyst particles, followed by diffusion of carbon species through the catalyst particles, producing the filament structure. Since this early study, many research efforts have been devoted to CVD growth of nanostructures on micron-sized fibers.

A schematic of the general process for in situ growth is shown in Fig. 6.4. Unsized fibers are typically used as substrates for CNT growth, being GFs and CFs the most commonly reported ones. To remove the fiber surface coating (sizing) usually present on commercial fibers, preheating treatments (>450°C) under different atmospheres are customarily used [50,54–57]. After sizing removal, metallic catalyst particles are deposited onto the fiber surface. Different means are reported to deposit the metal catalyst required for thermal CVD reactions, including incipient wetness (solution impregnation) methods [58–60],

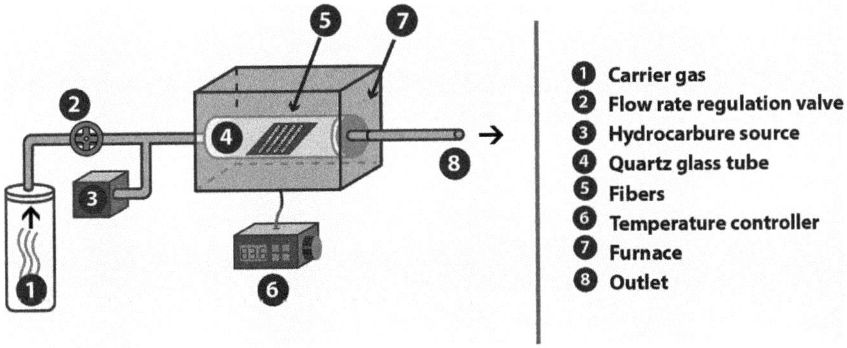

FIGURE 6.4 **Schematic of the process of in situ growth of carbon nanotubes on engineering fibers byCVD.**

polymer-metal complex deposition [61], sputtering [59,62], electron-beam evaporation [63], thermal evaporation [64], and electrodeposition [65,66].

It is also noted that an initial oxidation of the CFs using traditional wet chemistry such as nitric acid oxidation can assist catalyst deposition, particularly via incipient wetness or other solution-based techniques. The polar surface groups, particularly carboxylic groups, encourage fiber wetting by the coating solution and can bind and stabilize catalyst precursor particles [67]. This generates an improved catalyst distribution and therefore a more homogeneous CNT growth. Common deposited catalysts are Fe, SiO_2, Ni, Co, and ferrocene. Once the fibers are catalyst-coated, they are placed in a furnace and CVD growth of CNTs is carried out by using a hydrocarbon source at a high temperature under atmosphere of a carrier gas. Acetylene, ethylene, toluene, or benzene are commonly used carbon sources for CNT growth [56], whereas the carrier gas flowing through the furnace is usually composed by a mixture of N_2, H_2, and Ar gasses, usually above 700°C [54,67,68]. The final quality of CNT coverage on the fiber surface strongly depends on the experimental parameters related to the growth conditions, such as the temperature [58,66,69], flow rate [56,69], type and size of catalyst particles [59,60,70], hydrocarbon source [59,62], and growth time [54–56,69], among the most relevant ones. In general, increase of temperature and carrier flow rate improves the uniformity of the CNT growth and alignment. However, if both properties are considerably high, formation of nonaligned CNTs with different lengths can be observed. This can be associated to the CNT competing mechanisms of nucleation and growth, both dependent on the temperature and flow rate [69]. As such, moderate temperatures and flow rates are favorable for a stable growth—that is, aligned CNTs with a uniform length. The type and size of catalyst play important roles on the CNT growth. The catalyst distribution across the fiber, which can be controlled by the catalyst deposition technique, the oxidation and surface area of the fibers, and the temperature are among the most important factors controlling the homogeneity of the CNT growth on the fiber surface [54,67]. To quantify the CNT growth, nitrogen adsorption can be used to assess the increase of the surface area due to CNT growth [50]. Several studies have explored the effect of the reaction time on the growth and density of CNT coverage. Until certain point, the length and density of CNTs increase with increased growth time, promoting uniform CNT length and coverage. Some studies have also shown that the growth time decreases if the fibers are oxidized [69]. However, if the reaction time is too long, the CNTs bend and become no longer normal to the fiber surface [54]. Such a hoop symmetry disruption can be attributed to the circumferential tension exceeding the inherent lateral strength of the CNTs [55]. The CNT-to-fiber bonding also represents an important characteristic of the in situ growth technique and a very relevant feature. An alternative to characterize the CNT-to-fiber bonding is to monitor the CNT network morphology surrounding the fiber by SEM imaging before and after exposing the CNT/fiber system to solvents under shear mixing or ultrasonic agitation [69]. Distilled water, acetone, methanol and isopropanol are common solvents used for this aim. Compared to other deposition technique, such as DD and EPD, the in situ growth technique generally shows very strong bonding between the CNTs and fiber surface, as observed from the unperturbed morphology observed after the mixing or agitation process. Due to the nature of the CVD process, the catalyst particles can be embedded into the fiber by etching catalysis, which results in CNT roots strongly embedded into the fiber surface. Fracture tests performed on individual CNTs attached to CFs by the CVD technique have suggested that the CNT-to-fiber bonding are essentially composed by C—C covalent bonding [57]. CNT growth has been found to be much more uniform for inorganic fibers such as GFs. This effect has been attributed to the different adhesion energy and diffusion rates of iron on the substrate surface [71]. Injection CVD typically allows the growth of longer CNTs than CVD using predeposited catalysts, often with a higher degree of orientation; this is driven by steric interactions within the dense growing layer [71].

Although the CVD in situ growth process is an efficient technique for covering micron-sized fibers with CNTs, the use of high temperatures and predeposited catalysts, together with the difficulties in scaling up the process, impose serious limitations on the practical application of this technique [41]. Moreover, the properties of the pristine fiber may be seriously degraded because of the high temperature, catalyst, and gas environment nedeed.

6.2.4 Other Deposition Methods

Although DD, EDP, and CVD in situ growth are the most reported methods to cover the surface of engineering fibers with CNTs, other techniques have also been explored with important results. For example, a spray-up process has been implemented to disperse thin layers of CNTs on the surface of GFs [72]. In such a process, the CNTs are dispersed by ultrasound in organic solvents (ethanol or toluene). Then, a certain concentration of high temperature initiators is added and mixed into the solution. A mist spray gun is used to evenly coat the fiber surface with CNTs. Finally the solvents are completely evaporated. A similar approach has also been reported for the deposition of SWCNTs onto Kevlar fibers, previously covered by polyurethane [73]. The fibers are placed on a hot plate at 200°C, and once the solution reaches the hot fibers, the solvent is evaporated leaving the SWCNTs on the fiber surface. Under a constant spray rate, SWCNT layers with thickness of about 2 μm have been obtained by this method in 30 s. A different technique based on a chemical process of esterification of surface groups to anchor CNTs on the fiber surface has also been reported by Vivet et al. [74]. In this process, CNT anchoring is achieved by an esterification between the carboxylic and hydroxyl groups on the CFs and the CNT surface. The esterification follows the Steglich method [75] and uses dicyclohexylcarbodiimide as a coupling agent and 4-dimethylamino-pyridine as the catalyst. After the thermal treatment, the CF tow is spread on a glass substrate. A CNT solution is first deposited drop by drop over the entire length of the tow and the solution with the catalyst is then deposited in the same way. The tow is maintained at 20°C for 1 h and then dried in an oven at 100°C for another 1 h. Intensive washing is then used to eliminate nonanchored CNTs and residues. In a recent report, a novel mechanical approach consisting of a melt extrusion method to coat polyester yarns with polypropylene (PP) and MWCNTs has been reported [76]. In this approach, MWCNTs and PP pellets are blended and then dried in an oven. The MWCNT/PP pellets are then melt mixed. Polyester yarns (which serve as the substrate) are fed into a yarn coating machine through its die with MWCNT/PP pellets being melt extruded to coat the yarns, and then cooled. In this manner, the authors are able to produce electrically conductive wires of polyester covered with a MWCNT/PP mixture with a very attractive scalable process.

6.3 Carbon Nanotube-Modified Fibers for Multiscale Polymer Composites

The use of fiber-reinforced polymers (FRPs) for aerospace, marine, automotive, and renewable energy industries has progressively increased in the last decades, mainly because of their high specific strength and stiffness [77]. To date, CFs, GFs, and aramid fibers (AFs) dominate the market for structural applications where lightweight is a requisite. Table 6.1 summarizes the main properties of commercial GFs, CFs, and AFs. CFs may be considered electrically conductive, but GFs and AFs are insulating materials.

A recently explored approach aimed to impart (or increase) electrical conductivity, enhance mechanical properties and add multifunctional properties to FRPs, consists in bonding CNTs to the surface of these engineering fibers [81]. For multiscale composites, the approach of placing (or growing) CNTs on the fibers may overcome processing limitations associated with CNT agglomeration and increased viscosity, compared to the approach in which CNTs are dispersed within the matrix [81,82]. From the mechanical point of view, the multiscale composites manufactured using those CNT-modified fibers may exhibit enhanced out-of-plane properties [83–85]. For nonelectrically conductive fibers (GFs and AFs), an electrically percolated CNT network can be formed along the fibers upon CNT deposition/growth. The new electrical features provided by the CNT network on the

Table 6.1 Properties of Glass, Carbon, and Aramid Fibers [77–80]

Fiber Type	d (μm)	E (GPa)	σ_u (GPa)	ε_u (%)	ρ (g/cm³)	v	α_L	k_t (W/m/°C)	ρ_e (μΩm)
Glass fibers									
E-glass	8–14	72.3	3.45	4.4	2.55	0.20	5.0–5.4	1.05	10^{19}
S-glass	10	85.0	4.8	5.3	2.49	0.22	1.6–2.9	1.05	10^{20}
Carbon fibers									
Torayca T300 (PAN)	7	230	3.53	1.5	1.75	0.2	−0.6	3.06	18
Hexcel AS4 (PAN)	7	241	4.13	1.6	1.77	—	−0.9	8.1–9.3	15–18
Thornel P55 (pitch)	10	379	1.9	0.5	2.0	—	−1.3	120	8.5
Thornel P100 (pitch)	10	758	2.41	0.32	2.16	—	−1.45	520	2.5
Aramid fibers									
Kevlar 49	12	131	3.62	2.8	1.45	0.35	−2	0.04	—
Kevlar 149	12	179	345	1.9	1.47	—	—	—	—

d, Fiber diameter; E, elastic modulus; σ_u, tensile strength; ε_u, ultimate strain; ρ, density; v, Poisson's ratio, α_L, coefficient of longitudinal thermal expansion; k_t, thermal conductivity; ρ_e, electrical resistivity.

fibers can be exploited to self-monitor temperature [86], resin crosslinking [3], humidity [87], strain [19], and structural damage [88,89], among other external stimuli. Therefore, besides the possibility of improving mechanical properties of multiscale composites, depositing/growing CNTs onto engineering fibers would render additional functionalities. This section reviews the state of the art on this topic, and discusses the challenges and opportunities of the resulting CNT-modified fibers, toward the development of multiscale hierarchical composites with multifunctional properties.

6.3.1 CNT-Modified Glass Fibers

The successful history of FRPs as lightweight materials in a wide range of engineering structures over the last decades was basically triggered by the mass production of GF-reinforced plastics [90]. Given their high specific strength and stiffness, thermal stability, and low flammability, in combination with its corrosion resistance, GF-reinforced polymers are widely used for engineering structures [79]. Among all engineering fibers, the vast majority of FRPs to date are made of GFs. GFs are the reinforcement of choice in more than 90% of all FRPs, and it is estimated that the global demand in 2015 was more than five million tons [91]. One of the main reasons is probably because of their cost-effectiveness [90]. Commercial GFs typically contain a proprietary coating known as “sizing,” whose purpose is to protect them from handling and to promote adhesion between the fibers and matrix. Typically, the GF sizing contains an adhesion promoter (coupling agent), a protective film former, lubricants, surfactants, and antistatic agents.

Among the CNT deposition/growth methods discussed in Section 6.1, DD and EPD are the preferred methods for GFs. In situ growth largely detriments the GF strength because of the high temperatures required to grow CNTs on their surface, and thus it is not recommended for GFs (unless CNT synthesis at low temperature is someday achieved). Regardless of the deposition method, a requisite is that the CNTs remain bonded to the fiber surface forming physical or chemical bonds, to properly exploit the multifunctional properties of the resulting hierarchically structured fiber.

The concentration and dispersion of CNTs achieved on the fiber varies depending on the method and conditions used. Since the CNT content over GFs upon CNT deposition is typically small, an accurate quantification of the amount of CNTs is challenging. On the other hand, for in situ growth, the CNT content with respect to the fiber mass can be as high as ~8 wt.% [92], although CNT contents in the order of 0.12%–0.18% has also been reported using CVD [1]. For CNT deposition using DD or EPD, additional chemical agents or surfactants may be added to the suspending medium, which increases the complexity of this quantification. Therefore, SEM is most commonly used to assess the dispersion and density of CNTs on GFs.

Regarding the chemical interactions, GFs have been used as substrates for CNT deposition using either as-received fibers—that is, those maintaining their proprietary surface coating [93] or by first removing the fiber sizing [38,94]. Taking advantage of the various functional groups present in the fiber coating, chemical bonding between oxidized CNTs

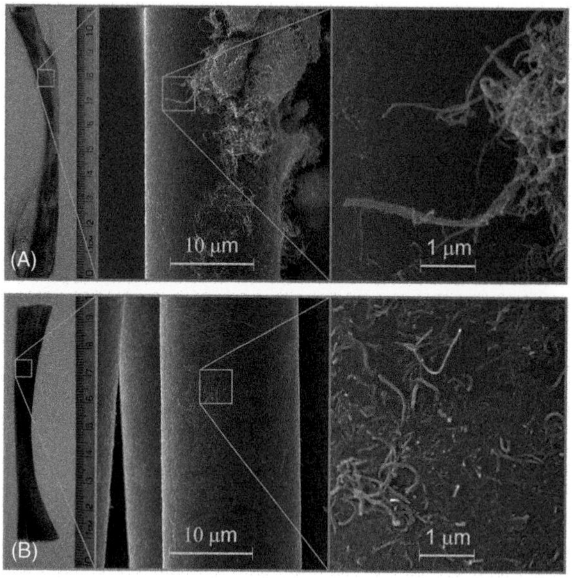

FIGURE 6.5 Photographs and SEM micrographs of MWCNT-modified GFs. (A) SR-CNT and (B) SC-CNT.

and the surface of GFs can be achieved [5]. In a series of CNT deposition experiments by our group, the role of the fiber sizing and the use of a silane coupling agent on the properties of CNT-modified GFs have been explained [4,5,21]. To investigate the role of the suspending fluid, GFs dip-coated by MWCNTs were used as-received (denoted as "SC" herein, for "sizing-containing") and after removing the sizing (denoted "SR" for "sizing removed"). Distilled water was employed as suspending fluid for CNT deposition. Fig. 6.5 shows photographs and SEM micrographs of CNT-modified fibers using GFs whose coating has been removed (SR-CNT) or kept (SC-CNT). From Fig. 6.5, it is evident that when MWCNTs are deposited onto GFs whose coating has been removed (Fig. 6.5A), a low density of nonuniformly distributed MWCNTs is achieved. In contrast, when MWCNTs are deposited onto GFs that maintained their sizing (Fig. 6.5B), the resulting GFs achieve a high density of MWCNTs forming a uniform layer on the GF surface.

Raman spectroscopy can also be exploited to investigate the spatial distribution and density of MWCNTs on the fiber. The use of Raman spectroscopy to such an aim relies on the fact that the GFs do not present Raman features (or they are too weak) in the region where the MWCNTs show their stronger Raman features (1100–1800 cm^{-1}). Therefore, the Raman signal of the CNT-modified GFs is associated solely to the presence of MWCNTs on their surface [21]. In an experiment designed for this aim, 20 Raman spectra were collected for each fiber conditions every 500 μm along the fiber, thus mapping the fiber length [21]. By comparing the intensities of the G (~1585 cm^{-1}) or D (~1330 cm^{-1}) bands of the Raman maps shown in Fig. 6.6, the CNT-modified fibers that maintained their sizing (SC-CNT, Fig. 6.6B) exhibit higher G (or D) band intensities (≈10×) with respect to the CNT-modified

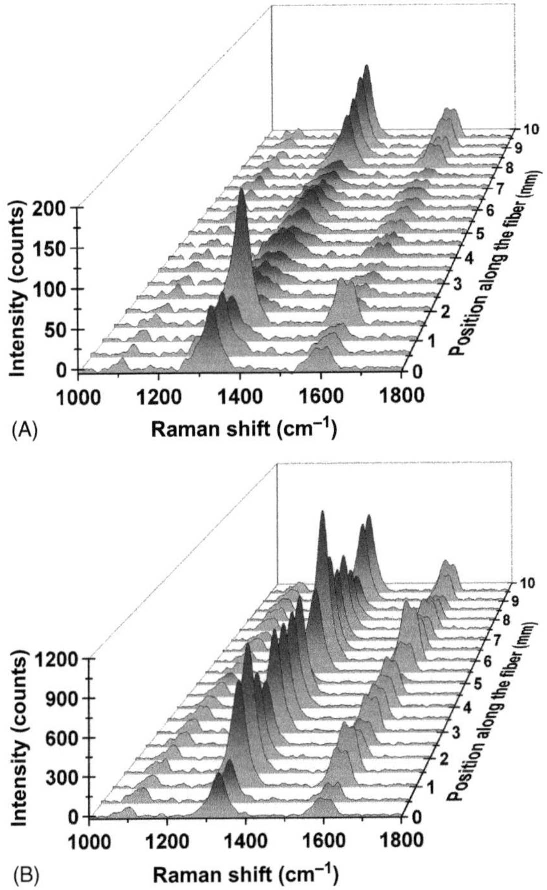

FIGURE 6.6 Raman spectra of MWCNT-modified GFs. (A) SR-CNT and (B) SC-CNT.

fiber whose sizing was removed (SR-CNT, Fig. 6.6A). The significantly more homogeneous distribution of the G band intensities of the CNT-fiber which maintained its coating is interpreted as a clear sign of a more homogeneous CNT deposition on the fiber [21]. Thus, Raman analysis agrees very well with the SEM observations of Fig. 6.5, providing an excellent tool to map the MWCNT spatial distribution onto GFs at a larger scale.

Surface roughness is another feature that is greatly modified when CNTs are incorporated onto the fiber surface. Atomic force microscopy (AFM) images of the surfaces of neat GFs and MWCNT-modified fibers are shown in Fig. 6.7. In Fig. 6.7, Rr and CV stand for root square mean and coefficient of variation, respectively. CV was obtained from 25 measurements taken from different locations on the fiber surface. Regarding the surface roughness of neat GFs (without CNTs), SR fibers (Fig. 6.7A) exhibit a rougher surface than that of SC fibers (Fig. 6.7B), as a consequence of the partial removal of the fiber sizing. Upon MWCNT deposition, GFs achieve an evident increase in surface roughness (Figs. 6.7C–D). The higher content of MWCNTs deposited on the surface of SC-CNT (Fig. 6.7D), with respect to

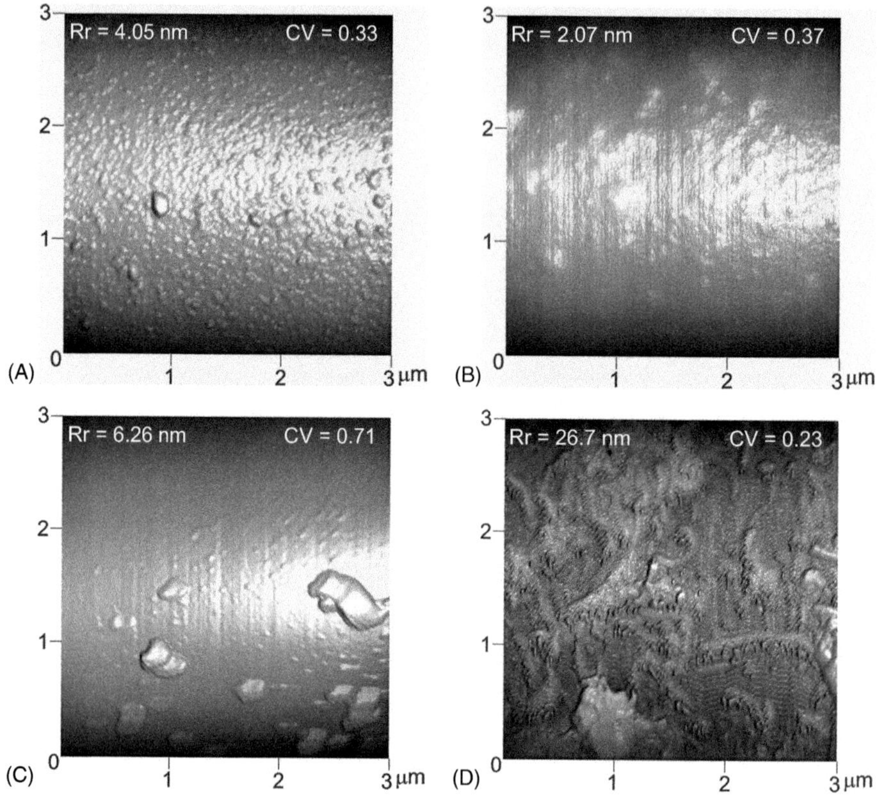

FIGURE 6.7 Atomic force microscopy (AFM) micrographs of GFs before and after MWCNT-deposition. (A) SR, (B) SC, (C) SR-CNT, and (D) SC-CNT.

that of SR-CNT (Fig. 6.7C), points out again that the fiber coating plays an important role for modification of the GF surface due to CNT deposition.

In other studies, in-house synthesized (without fiber sizing) E-glass fibers were dip-coated with commercial MWCNTs [2]. Such GFs are deprived of surface coating and were functionalized with 3-aminopropyltriethoxysilane, whereas MWCNTs were functionalized with carbonyl chloride groups. The surface of the silane-functionalized GFs exhibited an average surface roughness of ~1.18 nm, and upon CNT deposition their surface roughness reached an average value of ~6.14 nm [2]. For multiscale fiber-reinforced composites, the increase in fiber roughness is desired since it can increase mechanical interlocking between GFs and the polymer matrix [2,21].

Upon CNT deposition, GFs may form electrically percolated conductive networks on their surface, becoming electrically conductive. The electrical resistance of GFs decreases several orders of magnitude depending on the technique and conditions used for CNT deposition/growing [2,3,21,49]. This feature opens new opportunities toward the development of hierarchically structured multifunctional smart composites. Table 6.2 summarizes the electrical resistance of CNT-modified fibers obtained using different deposition

Table 6.2 Electrical Resistance of CNT-Modified Glass Fibers (1 mm Long)

Deposition Method	CNT Treatment	Deposition/ Growing Time	Concentration of MWCNTs in the Suspending Fluid	Electrical Resistance (Ω)	Reference
DD (assisted by ultrasound)	Chemical oxidation	3 h	0.125 mg/mL in water	0.3–7 × 10^6	[21]
			0.125 mg/mL in MPS/ water	1.3–8 × 10^6	
DD	Commercial MWCNT-COOH	15 min	0.5 wt.% in Igepal Co-970/water	0.5–300 × 10^6	[38]
DD	MWCNT-COCl	30 min	0.1 mg/mL in THF	3–4.5 × 10^6	[3]
EPD	Commercial MWCNT-COOH	10 min	0.05 wt.% in Igepal Co-970/water	15–80 × 10^6	[38]
EPD	Commercial CNT-COOH	10 min	0.05 wt.% in Igepal Co-970/Glymo/water	50–100 × 10^6	[38]
EPD	CNT-PEI	15 min	0.5 mg/mL in water	1.5–28 × 10^6	[49]
EPD	As received	3.5 min	0.05 wt.% in SDS/water	0.12–0.33 × 10^6	[95]
In situ growth	NA	25 min	NA	0.1–1.2 × 10^3	[92]

CNT, Carbon nanotube; DD, dipping deposition; EPD, electrophoretic deposition; Glymo, (3-glycidyloxypropyl)trimethoxysilane; Igepal Co-970, nonylphenoxy poly(ethyleneoxy)ethanol, branched; MPS, 3-(trimethoxysilyl)propyl methacrylate; MWCNT, multiwall carbon nanotube; NA, not applicable; PEI, polyethyleneimine; SDS, sodium dodecyl sulfate; THF, tetrahydrofuran.

methods. Most of the electrical resistances shown in Table 6.2 were originally conducted for an electrode separation of 1 mm; for comparative purposes in the cases where the measurements were made over larger gaps the values were linearly interpolated to obtain equivalent resistance for 1 mm of electrode separation. For the data collected in Table 6.2, the lowest electrical resistance per unit length (0.1×10^3 Ω/mm) was achieved by using in situ growth [92], and the highest (3×10^8 Ω/mm) was obtained by DD [38]. In general, CNT-coated GFs by EPD exhibit lower electrical resistance per unit length than those covered by DD, since it is more feasible to achieve a more homogenous distribution and higher density of CNTs onto GFs by using EPD than DD [38,49]. It has also been shown that by using EPD it is possible to tailor the electrical conductivity of the fibers by controlling the thickness of the CNT layer deposited on their surface [49]. Even though an evident decrease in electrical resistance of the GFs is achieved upon CNT deposition, it is important to point out that the longer the electrode separation the more difficult it is to maintain electrical continuity along the fiber [49]. As seen from Table 6.2, for all CNT deposition methods, the resulting CNT-modified fibers achieve large variations in electrical resistance values, and this effect is more drastic for DD [38]. However, when fibers are arranged into tows, variations in electrical resistance among specimens are reduced as a consequence of formation of multiple contacts among adjacent fibers [21].

Inspired by the efficient hierarchical architecture of natural materials, such as bone and bamboo (which exhibit excellent strength, stiffness, and toughness), CNT-modified fibers have been explored to manufactured multiscale composites mimicking such natural materials. Improvements in the interfacial shear strength between CNT-coated fibers and polymer matrices of ~40 and ~58% have been reported for DD [21] and EPD [49], respectively.

These improvements may result from synergistic effects of different toughening mechanism including generations of new rigid/tough fiber/nanotube/matrix interfaces, nanotube pull-out, and interfacial crack bridging due to the presence of CNTs; a chemical component rendered by the functional groups on the CNTs and their reaction with the fiber surface may also play an important role. Using CNT-modified GFs, improved interlaminar properties of multiscale composites have been achieved; improvements have been as small as 3% for the case of GFs with CNTs deposited by spray coating [72], and up to 69% for CNTs grown on the surface of GFs [63].

Regarding the multifunctional properties that GFs achieve upon CNT deposition, it has been shown that when CNT-modified fibers are stretched, the Raman G' band shifts to lower wavenumbers [25,96], which can be used to self-track the fiber deformation. Taking advantage of the electrical network formed onto GFs, it is possible to self-sense strain and damage in advanced polymer composites. The deformation of the CNT network developed on the fiber surface when is loaded yields changes in the electrical resistance of the composite. This feature has been explored using single CNT-modified GFs [2,86,94], tows [97], and laminates [89]. It has been shown that composites using CNT-modified fibers have the potential not only to track damage progression of the composite but also to provide information on the dominant failure mechanism [89]. Other potential applications based on this multifunctional feature are related to monitoring of resin cross-linking, UV, and thermal sensing [3].

6.3.2 CNT-Modified Carbon Fibers

CFs are lightweight high-performance fibers with excellent chemical resistance; they are lighter and stiffer than GFs, but significantly more expensive. For CFs, incorporation of CNTs onto their surface has been conducted through spray coating [98], electrospray coating [99], DD [14,20], EPD [83,100,101], and in situ growth [50,54,57]. The morphology of CVD-grown CNTs onto CFs can render radially or randomly oriented CNTs depending on the processing conditions used. As for GFs, the main drawback of using in situ growth is the low viability to extend the process for mass production and the decrease of the fiber strength [67,102–105]. The reduction of tensile strength of CFs can be as high as 55% when the CVD process reaches the carbon-iron eutectic temperature (723°C) [67]. Recent attempts have been made to improve this situation through fiber oxidation, lowering the manufacturing temperature below the eutectic temperature and controlling catalytic growth [103,106].

Taking advantage of the surface reactivity of commercially available CFs due to the presence of their fiber sizing, attachment of oxidized CNTs on CFs is viable through DD or EPD [20]. Fig. 6.8 shows representative SEM micrographs of CFs before (Fig. 6.8A) and after (Fig. 6.8B) CNT deposition using DD. CNT-modified CFs using DD exhibit a surface with randomly oriented CNTs as shown in Fig. 6.8B.

Among the methods described in Section 6.1, EPD is one of the preferred methods used to attach CNTs onto CFs. This is because CFs are electrically conductive and they can be

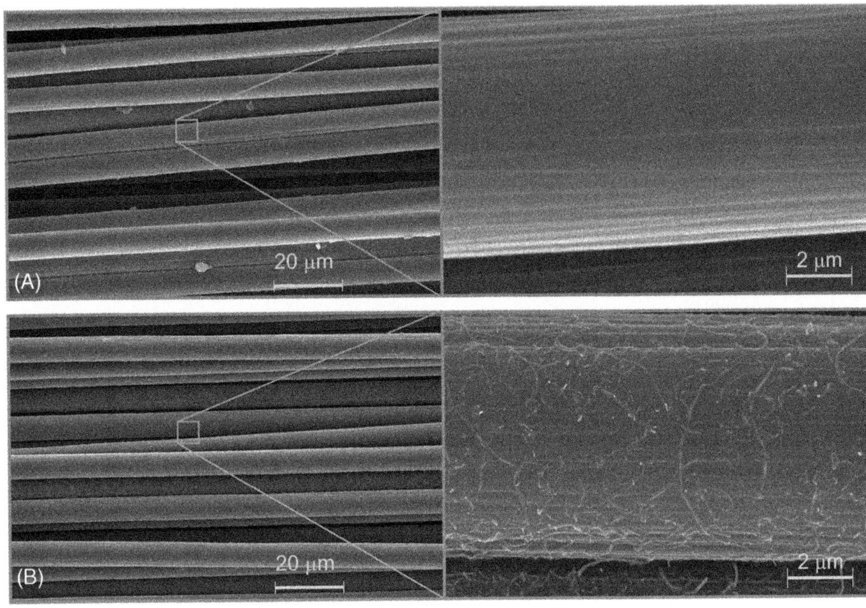

FIGURE 6.8 Carbon fibers before and after CNT diping deposition. (A) As-received and (B) CNT-modified fibers.

used as electrodes, which is not possible for GFs. Using EPD, a denser and more homogenous deposition of CNTs can be achieved in comparison to DD. The CNT content deposited onto CFs using EPD can be has high as 10%–13% [100,107]. Furthermore, EPD has a better control of the amount of CNTs deposited onto CFs [108].

Upon CNT deposition, CFs exhibit important modifications in their surface properties, such as increased surface area and roughness, albeit bulk or volumetric properties are meant to be preserved. Table 6.3 summarizes reported changes in roughness, surface area, and specific capacitance (capacitance normalized by mass) of CFs upon CNT deposition. Depending on the method of synthesis, CFs can exhibit a relatively smooth (average roughness 4.6 ±1.5 nm) or rough (35.1 ± 4.7 nm) surface [20]. The presence of CNTs on the CF surface increases its tortuosity, increasing the average fiber surface roughness. Li et al. reported that for rough CFs their roughness increases from 35.1 ± 4.7 to 49.6 nm (+40%, based on averages), whereas for smooth fibers the presence of CNTs yields more drastic changes, from 4.6 ± 1.5 to 39.6 ± 6.3 nm (+760%) [20]. Given this increased surface tortuosity, the surface area of the CNT-modified CFs is also greatly increased due to the presence of CNTs on their surface, as shown in Table 6.3. Increments in surface area as high as ~100× have been reported for in situ growth [55]; this is possible since this method produces a denser population of CNTs attached to the fibers. Different from GFs, CFs are electrically conductive (Table 6.1), and incorporation of CNTs on their surface does not yield a considerable impact on the CFs electrical conductivity. However, their specific capacitance (capacitance normalized by the sample mass) is largely improved upon CNT deposition (Table 6.3). Increments of specific capacitance of up to three orders of magnitude

Table 6.3 Carbon fiber Surface Property Modification upon CNT Deposition or Growth

Property (Increase)	Measurement Technique	CNT Deposition Method	Fiber Type (Treatment)	CNT Treatment	Reference
Surface roughness (120%)	AFM	Electrospray coating	Panex 35 PAN-based (oxidized)	As-received	[99]
Surface roughness (760%)	AFM	DD	T700SC PAN-based (as received)	COOH	[20]
Surface roughness (40%)	AFM	DD	T300B PAN-based (as received)	OH	[20]
Surface area (100×)	Adsorption isotherms	CVD	T300 PAN-based (thermally desized)	NA	[55]
Surface area (2.5×)	Adsorption isotherms	Electrospray coating	Panex 35PAN-based (oxidized)	As-received	[99]
Surface area (73×)	Adsorption isotherms	EPD	PAN-based (thermally desized)	HCl	[109]
Specific capacitance (~3.5×)	Cyclic voltammetry	CNT solution dropping	T300 PAN-based (OH-CF)	COOH	[110]
Specific capacitance (~3000×)	Cyclic voltammetry	EPD	PAN-based (thermally desized)	HCl	[109]

AFM, Atomic force microscopy; CNT, carbon nanotube; CVD, chemical vapor deposition; DD, dipping deposition; EPD, electrophoretic deposition; NA, not applicable; PAN, polyacrylonitrile.

compared to raw CFs have been reported using EPD and cetyltrimethylammonium bromide as surfactant [109].

Another motivation of placing (or growing) CNTs on CFs is related to improving the mechanical properties of polymer composites, especially the interfacial and interlaminar properties. For instance, improvements in the interfacial shear strength of the order of 1.5 times have been reported using CNT-modified CFs using EPD [99] and DD [111]. The interlaminar shear strength of FRP composites is another property that can be improved by the incorporation of CNTs onto CFs; improvements up to 36% in the short beam shear strength have been found for CF-CNT/epoxy composites by using CFs with CNTs deposited using EPD [112]. In general, carbon reinforced polymer composites manufactured using CNT-modified CFs exhibit higher interlaminar and interfacial shear strengths than those made of conventional CFs (without CNTs). These interlaminar and interfacial improvements result from synergic effects related to the increased surface roughness of the CNT-modified CFs [99], increased surface area, as well as (in some cases) the use of coupling agents [100,113] and/or chemically functionalized CFs [99] and/or CNTs [114,115].

6.3.3 CNT-Modified Aramid Fibers

AFs are synthetic organic fibers of the aromatic polyamide family. Although their elastic modulus and strength are not as high as those made of carbon, AFs are tough, flexible, and render high-energy absorption, making them attractive for ballistic protection. Given their high crystallinity, the surface of AFs is chemically inert and smooth, which inhibits proper adhesion to most polymer matrices [116]. Chemical activation, application of surface coatings, and surface roughening are some approaches that the scientific community has been exploring to overcome the poor fiber/matrix adhesion for aramid polymer composites [117]. Attachment of CNTs onto their surface is another approach recently explored to improve fiber/matrix adhesion [118]. Although CNT deposition/growth on CFs and GFs have been extensively explored, for AFs only a few research works exist. In situ growth of CNTs typically requires high temperatures (>700°C) that can decompose AFs [119]. Thus, this approach is not commonly used to cover AFs with CNTs. However, a recent patent claims a method that allows growing CNTs on AFs that involve applying a barrier coating to protect the fibers from thermal degradation, using low growth temperatures (450°C) and short processing times [120]. More common techniques to cover AFs with CNTs are probably DD and spray coating. A sequential spray coating, first by a polyurethane solution and then by a CNT suspension, has been reported as an effective approach for CNT deposition onto AFs that does not necessarily require chemical functionalization of the fibers [73]. However, given their inertness, for most of the reported cases, AFs have been functionalized to promote physicochemical interactions with CNTs [118,119,121].

AFs, as GFs, are insulators and upon CNT deposition a drastic decrease in electrical resistance is achieved. As an example, a bundle of CNT-modified AFs has been used as an electric wire to power on a light-emitting diode [73]. Electrical resistance of bundles (~1000 filaments) of CNT-modified AFs obtained using DD aided by ultrasonic dispersion render electrical resistances on the order of 0.2–50 MΩ/cm [121]. AFs coated by CNTs have also shown a piezoresistive response upon mechanical loading [119]. From the mechanical viewpoint, it has been reported that impact performance of aramid fabric increased up to 50% upon CNT deposition using DD in *N*-methylpyrrolidone [122]. These CNT-modified AFs with enhanced surface properties have a potential use as multifunctional structural and sensing elements in lightweight structures with high-energy absorption characteristics.

6.3.4 Other Fibers Modified by Carbon Nanotubes

The majority of research works attempting to bond CNTs to the surface of fibers of micrometric diameter have been conducted using CFs, GFs, or (to a less extend) AFs; however, few research works have also focused on modifying other fibers made of silica [123,124], alumina [85,124], titanium [125], jute [87,126], and polyethylene terephthalate [76,127]. CNT-modified silica fibers have been obtained by in situ growth of CNTs and exhibited a reduction in tensile strength of 35% compared to the pristine fibers [123]. Composites made of CNT-modified silica fibers/poly(methyl methacrylate) showed an increase in their interfacial shear strength of 78% with respect to the pristine fiber [123]. Similar to silica fibers, CNT-modified alumina fibers have also been manufactured by direct growth of CNTs using CVD [85,124]. Composites made of CNT-modified alumina fibers and an epoxy matrix exhibited improvements in their fracture toughness in the range of 67%–76%, when compared to their pristine counterparts [85]. Promising results have also been obtained using CNT-modified jute fibers to sense humidity, temperature, and strain in jute/epoxy composites [87]. Therefore, it is clear that modifications of organic and inorganic fibers by depositing (or growing) CNTs on their surface have great potential applications, since it imparts enhanced mechanical and multifunctional properties to conventional composites. The multiscale and hierarchical organization of CNTs, surrounding the micrometric fibers that reinforce polymer matrices, is a novel topic that has a great impact on the effective properties of the composite. However, several unknowns, caveats, and processing challenges still need to be resolved before unraveling the full potential of these CNT-coated fibers, especially for multiscale hierarchical composites.

Acknowledgments

Authors wish to thank Alejandro Can (CICY) for the SEM micrograph of the CNT-coated fiber and for his input regarding the experimental EPD setup. Economic support from CONACYT through projects No. 220513 (FA) and 235905 (AIOA) is acknowledged.

References

[1] L. Zhang, D. Su, L. Jin, C. Li, Polyamide 6 composites reinforced with glass fibers modified with electrostatically assembled multiwall carbon nanotubes, J. Mater. Sci. 47 (2012) 5446–5454.

[2] L. Tzounis, M. Kirsten, F. Simon, E. Mäder, M. Stamm, The interphase microstructure and electrical properties of glass fibers covalently and non-covalently bonded with multiwall carbon nanotubes, Carbon 73 (2014) 310–324.

[3] L. Tzounis, M. Liebscher, E. Petinakis, A.S. Paipetis, E. Mäder, M. Stamm, CNT-grafted glass fibers as a smart tool for the epoxy cure monitoring, UV-sensing and thermal energy harvesting in model composites, RSC Adv. 6 (2016) 55514–55525.

[4] J.J. Ku-Herrera, J.V. Cauich, F. Avilés, On the role of fiber coating in the deposition of multiwall carbon nanotubes onto glass fibers, Nanosci. Nanotechnol. Lett. 6 (2014) 932–935.

[5] J.J. Ku-Herrera, F. Avilés, A. Nistal, J.V. Cauich-Rodríguez, F. Rubio, J. Rubio, P. Bartolo-Pérez, Interactions between the glass fiber coating and oxidized carbon nanotubes, Appl. Surf. Sci. 330 (2015) 383–392.

[6] A. Warrier, A. Godara, O. Rochez, L. Mezzo, F. Luizi, L. Gorbatikh, S.V. Lomov, A.W. VanVuure, et al. The effect of adding carbon nanotubes to glass/epoxy composites in the fibre sizing and/or the matrix, Compos. Part A 41 (2010) 532–538.

[7] S.C. Tsang, P.J.F. Harris, M.L.H. Green, Thinning and opening of carbon nanotubes by oxidation using carbon dioxide, Nature 362 (1993) 520–522.

[8] M. Monthioux, B.W. Smith, B. Burteaux, A. Claye, J.E. Fischer, D.E. Luzzi, Sensitivity of single-wall carbon nanotubes to chemical processing: an electron microscopy investigation, Carbon 39 (2001) 1251–1272.

[9] F. Avilés, J.V. Cauich-Rodríguez, L. Moo-Tah, A. May-Pat, R. Vargas-Coronado, Evaluation of mild acid oxidation treatments for MWCNT functionalization, Carbon 47 (2009) 2970–2975.

[10] F. Avilés, C.A. Sierra-Chi, A. Nistal, A. May-Pat, F. Rubio, J. Rubio, Influence of silane concentration on the silanization of multiwall carbon nanotubes, Carbon 57 (2013) 520–529.

[11] T. Xu, J. Yang, J. Liu, Q. Fu, Surface modification of multi-walled carbon nanotubes by O2 plasma, Appl. Surf. Sci. 253 (2007) 8945–8951.

[12] K. Balasubramanian, M. Burghard, Chemically functionalized carbon nanotubes: review, Small 2 (2005) 180–192.

[13] X. Gong, J. Liu, S. Baskaran, R.D. Voise, J.S. Young, Surfactant-assisted processing of carbon nanotube/polymer composites, Chem.Mater. 12 (2000) 1049–1052.

[14] T. Kamae, L.T. Drzal, Carbon fiber/epoxy composite property enhancement through incorporation of carbon nanotubes at the fiber–matrix interphase—part I: the development of carbon nanotube coated carbon fibers and the evaluation of their adhesion, Compos. Part A 43 (2012) 1569–1577.

[15] A.B. García, A. Cuesta, M.A. Montes-Moran, A. Martinez-Alonso, J.M.D. Tascon, Zeta potential as a tool to characterize plasma oxidation of carbon fibers, J.Colloid Interface Sci. 192 (1997) 363–367.

[16] A. Bismarck, C. Wuertz, J. Springer, Basic surface oxides on carbon fibers, Carbon 37 (1999) 1019–1027.

[17] A. Bismarck, J. Springer, Characterization of fluorinated PAN-based carbon fibers by zeta-potential measurements, Colloids Surf. A 159 (1999) 331–339.

[18] I. O'Connor, H. Hayden, J.N. Coleman, Y.K. Gun'ko, High-strength, high-toughness composite fibers by swelling kevlar in nanotube suspensions, Small 5 (2009) 466–469.

[19] S. Gao, R.-C. Zhuang, J. Zhang, J.-W. Liu, E. Mäder, Glass fibers with carbon nanotube networks as multifunctional sensors, Adv.Funct. Mater. 20 (2010) 1885–1893.

[20] M. Li, Y. Gu, Y. Liu, Y. Li, Z. Zhang, Interfacial improvement of carbon fiber/epoxy composites using a simple process for depositing commercially functionalized carbon nanotubes on the fibers, Carbon 52 (2013) 109–121.

[21] J.J. Ku-Herrera, A. May-Pat, F. Avilés, An assessment of the role of fiber coating and suspending fluid on the deposition of carbon nanotubes onto glass fibers for multiscale composites, Adv. Eng. Mater. 18 (2016) 963–971.

[22] P. Sureeyatanapas, R.J. Young, SWNT composite coatings as a strain sensor on glass fibres in model epoxy composites, Compos. Sci.Technol. 69 (2009) 1547–1552.

[23] G.J. Gallo, E.T. Thostenson, Electrical characterization and modeling of carbon nanotube and carbon fiber self-sensing composites for enhanced sensing of microcracks, Mater. Today Commun. 3 (2015) 17–26.

[24] N.A. Siddiqui, M.L. Sham, B.Z. Tang, A. Munir, J.K. Kim, Tensile strength of glass fibres with carbon nanotube–epoxy nanocomposite coating, Compos. Part A 40 (2009) 1606–1614.

[25] L. Liu, P.C. Ma, M. Xu, S.U. Khan, J.K. Kim, Strain-sensitive Raman spectroscopy and electrical resistance of carbon nanotube-coated glass fibre sensors, Compos. Sci. Technol. 72 (2012) 1548–1555.

[26] P.C. Ma, J.W. Liu, S.L. Gao, E. Mäder, Development of functional glass fibres with nanocomposite coating: a comparative study, Compos. Part A 44 (2013) 16–22.

[27] H. Morgan, N.G. Green, AC Electrokinetics: Colloids and Nanoparticles, Research Studies Press Ltd, Baldock, Hertfordshire, United Kingdom, (2003).

[28] T.B. Jones, Electromechanics of Particles, Cambridge University Press, New York, NY, (1995).

[29] L. Besra, M. Liu, A review on fundamentals and applications of electrophoretic deposition (EPD), Prog. Mater. Sci. 52 (2007) 1–61.

[30] A.R. Boccaccini, I. Zhitomirsky, Application of electrophoretic and electrolytic deposition techniques in ceramics processing, Curr. Opin. Solid State Mater. Sci. 6 (2002) 251–260.

[31] A.R. Boccaccini, K. Cho, J.A. Roether, B.J.C. Thomas, E.J. Minay, M.S.P. Shaffer, Electrophoretic deposition of carbon nanotubes, Carbon 44 (2006) 3149–3160.

[32] A.R. Boccaccini, J. Cho, T. Subhani, C. Kayab, F. Kaya, Electrophoretic deposition of carbon nanotube-ceramic nanocomposites, J.Euro. Ceramic Soc. 30 (2010) 1115–1129.

[33] I. Corni, M.P. Ryan, A.R. Boccaccini, Electrophoretic deposition: from traditional ceramics to nanotechnology, J.Euro. Ceramic Soc. 28 (2008) 1353–1367.

[34] J.H. Dickerson, A.R. Boccaccini, Electrophoretic Deposition of Nanomaterials, Springer, New York, NY, (2012).

[35] O. Van der Biest, L.J. Vandeperre, Electrophoretic deposition of materials, Ann. Rev. Mater. Sci. 29 (1999) 327–352.

[36] B. Gao, G.Z. Yue, Q. Qiu, Y. Cheng, H. Shimoda, L. Fleming, O. Zhou, Fabrication and electron field emission properties of carbon nanotube films by electrophoretic deposition, Adv. Mater. 13 (2001) 1770–1773.

[37] A.J. Rodriguez, M.E. Guzman, C.-S. Lim, B. Minaie, Synthesis of multiscale reinforcement fabric by electrophoretic deposition of amine-functionalized carbon nanofibers onto carbon fiber layers, Carbon 48 (2010) 3256–3259.

[38] J. Zhang, R. Zhuang, J. Liu, E. Mäder, G. Heinrich, S. Gao, Functional interphases with multi-walled carbon nanotubes in glass fibre/epoxy composites, Carbon 48 (2010) 2273–2281.

[39] M.S.P. Shaffer, X. Fan, A.H. Windle, Dispersion and packing of carbon nanotubes, Carbon 36 (1998) 1603–1612.

[40] J. Cho, K. Konopka, K. Rozniatowski, E. Garcıa-Lecina, M.S.P. Shaffer, A.R. Boccaccini, Characterisation of carbon nanotube films deposited by electrophoretic deposition, Carbon 47 (2009) 58–67.

[41] E. Bekyarova, E.T. Thostenson, A. Yu, H. Kim, J. Gao, J. Tang, H.T. Hahn, T.-W. Chou, et al. Multiscale carbon nanotube-carbon fiber reinforcement for advanced epoxy composites, Langmuir 23 (2007) 3970–3974.

[42] M. Diba, D.W.H. Fam, A.R. Boccaccini, M.S.P. Shaffer, Electrophoretic deposition of graphene-related materials: a review of the fundamentals, Prog. Mater. Sci. 82 (2016) 83–117.

[43] H.C. Hamaker, Formation of a deposit by electrophoresis, Trans. Faraday Soc. 35 (1940) 279–287.

[44] Y. Su, I. Zhitomirsky, Cataphoretic assembly of cationic dyes and deposition of carbon nanotube and graphene films, J.Colloid Interface Sci. 399 (2013) 46–53.

[45] Y. Su, I. Zhitomirsky, Electrophoretic deposition of graphene, carbon nanotubes and composite films using methyl violet dye as a dispersing agent, Colloids Surf. A 436 (2013) 97–103.

[46] K. Shi, I. Zhitomirsky, Electrophoretic nanotechnology of graphene–carbon nanotube and graphene–polypyrrole nanofiber composites for electrochemical supercapacitors, J.Colloid Interface Sci. 407 (2013) 474–481.

[47] M. Diba, A. García-Gallastegui, R.N.K. Taylor, F. Pishbin, M.P. Ryan, M.S.P. Shaffer, A.R. Boccaccini, Quantitative evaluation of electrophoretic deposition kinetics of graphene oxide, Carbon 67 (2014) 656–661.

[48] Q. An, A.N. Rider, E.T. Thostenson, Hierarchical composite structures prepared by electrophoretic deposition of carbon nanotubes onto glass fibers, ACS Appl. Mater. Interfaces 5 (2013) 2022–2032.

[49] S. Tamrakar, Q. An, E.T. Thostenson, A.N. Rider, B.Z. Haque, J.W. Gillespie Jr., Tailoring interfacial properties by controlling carbon nanotube coating thickness on glass fibers using electrophoretic deposition, ACS Appl. Mater. Interfaces 8 (2016) 1501–1510.

[50] E.T. Thostenson, W.Z. Li, D.Z. Wang, Z.F. Ren, T.W. Chou, Carbon nanotube/carbon fiber hybrid multiscale composites, J.Appl. Phys. 91 (2002) 6034–6037.

[51] Y. Wang, Z. Xu, L. Chen, Y. Jiao, X. Wu, Multi-scale hybrid composites-based carbon nanotubes, Polym. Compos. 32 (2011) 159–167.

[52] W.B. Downs, R.T.K. Baker, Novel carbon fiber-carbon filament structure, Carbon 29 (1991) 1173–1179.

[53] E. Fitzer, K.H. Geigl, W. Huttner, R. Weiss, Chemical interactions between the carbon fiber surface and epoxy resins, Carbon 18 (1980) 389–393.

[54] P. Agnihotri, S. Basu, K.K. Kar, Effect of carbon nanotube length and density on the properties of carbon nanotube-coated carbon fiber/polyester composites, Carbon 49 (2011) 3098–3106.

[55] P. Lv, Y.-Y. Feng, P. Zhang, H.-M. Chen, N. Zhao, W. Feng, Increasing the interfacial strength in carbon fiber/epoxy composites by controlling the orientation and length of carbon nanotubes grown on the fibers, Carbon 49 (2011) 4665–4673.

[56] S. Rahmanian, K.S. Thean, A.R. Suraya, M.A. Shazed, M.A.M. Salleh, H.M. Yusoff, Carbon and glass hierarchical fibers: Influence of carbon nanotubes on tensile, flexural and impact properties of short fiber reinforced composites, Mater. Des. 43 (2013) 10–16.

[57] C. Wang, Y. Li, L. Tong, Q. Song, K. Li, J. Li, Q. Peng, X. He, et al. The role of grafting force and surface wettability in interfacial enhancement of carbon nanotube/carbon fiber hierarchical composites, Carbon 69 (2014) 239–246.

[58] S. Zhu, C.H. Su, S.L. Lehoczky, I. Muntele, D. Ila, Carbon nanotube growth on carbon fibers, Diamond Relat. Mater. 12 (2003) 1825–1828.

[59] S.H. Jo, J.Y. Huang, S. Chen, G.Y. Xiong, D.Z. Wang, Z.F. Ren, Field emission of carbon nanotubes grown on carbon cloth, J. Vac. Sci. Technol. B 23 (2005) 2363–2368.

[60] F. Cesano, S. Bertarione, D. Scarano, A. Zecchina, Conecting carbon fibers by means of catalytically grown nanofilaments: formation of carbon-carbon composites, Chem. Mater. 17 (2005) 5119–5123.

[61] O. Shafranska, A. Voronov, A. Kohut, X.F. Wu, I.S. Akhatov, Polymer–metal complexes as a catalyst for the growth of carbon nanostructures, Carbon 47 (2009) 3137–3139.

[62] T. Susi, A.G. Nasibulin, H. Jiang, E.I. Kauppinen, CVD synthesis of hierarchical 3D MWCNT/carbon fiber nanostructures, J. Nanomat. 2008 (2008) 425195.

[63] E.J. Garcia, B.L. Wardle, A.J. Hart, N. Yamamoto, Fabrication and multifunctional properties of a hybrid laminate with aligned carbon nanotubes grown in situ, Compos. Sci. Technol. 68 (2008) 2034–2041.

[64] K.L. Kepple, G.P. Sanborn, P.A. Lacasse, K.M. Gruenberg, W.J. Ready, Improved fracture toughness of carbon fiber composites functionalized with multi walled carbon nanotubes, Carbon 46 (2008) 2026–2033.

[65] T.D. Makris, R. Giorgi, N. Lisi, L. Pilloni, E. Salernitano, M.F. De Riccardis, D. Carbone, Carbon nanotube growth on PAN- and pitch-based carbon fibres by HFCVD, Fullerenes Nanotub. Car. Nanostruct. 13 (2005) 383–392.

[66] K.H. Hung, S.S. Tzeng, W.S. Kuo, B.Q. Wei, T.H. Ko, Growth of carbon nanofibers on carbon fabric with Ni nanocatalyst prepared using pulse electrodeposition, Nanotechnol. 19 (2008) 295602 8 pp.

[67] H. Qian, A. Bismarck, E.S. Greenhalgh, G. Kalinka, M.S.P. Shaffer, Hierarchical composites reinforced with carbon nanotube grafted fibers: The potential assessed at the single fiber level, Chem. Mater. 20 (2008) 1862–1869.

[68] S.T. Patton, Q. Zhang, L. Qu, L. Dai, A.A. Voevodin, J. Baur, Electromechanical characterization of carbon nanotubes grown on carbon fiber, J.Appl. Phys. 106 (2009) 104313 9 pp.

[69] Q. Zhang, J. Liu, R. Sager, L. Dai, J. Baur, Hierarchical composites of carbon nanotubes on carbon fiber: influence of growth condition on fiber tensile properties, Compos. Sci. Technol. 69 (2009) 594–601.

[70] N. Yamamoto, A.J. Hart, E.J. Garcia, S.S. Wicks, H.M. Duong, A.H. Slocum, B.L. Wardle, High-yield atmospheric-pressure growth of aligned carbon nanotubes on ceramic fibers for multifunctional enhancement of structural composites, Carbon 47 (2009) 551–560.

[71] H. Qian, E.S. Greenhalgh, M.S.P. Shaffer, A. Bismarck, Carbon nanotube-based hierarchical composites: a review, J.Mater. Chem. 20 (2010) 4751–4762.

[72] J. Zhu, A. Imam, R. Crane, K. Lozano, V.N. Khabashesku, E.V. Barrera, Processing a glass fiber reinforced vinyl ester composite with nanotube enhancement of interlaminar shear strength, Compos. Sci. Technol. 67 (2007) 1509–1517.

[73] C. Xiang, W. Lu, Y. Zhu, Z. Sun, Z. Yan, C.-C. Hwang, J.M. Tour, Carbon nanotube and graphene nanoribbon-coated conductive Kevlar fibers, ACS Appl. Mater. Interfaces 4 (2012) 131–136.

[74] A. Vivet, B.B. Doudou, C. Poilane, J. Chen, M. Ayachi, A method for the chemical anchoring of carbon nanotubes onto carbon fibre and its impact on the strength of carbon fibre composites, J. Mater. Sci. 46 (2011) 1322–1327.

[75] B. Neises, W. Steglich, Simple method forthe esterification of carboxylic acids, Angewandte Chemie Int. Ed. 17 (1978) 522–524.

[76] J.H. Lin, Z.I. Lin, Y.J. Pan, C.T. Hsieh, M.C. Lee, C.W. Lou, Manufacturing techniques and property evaluations of conductive composite yarns coated with polypropylene and multi-walled carbon nanotubes, Compos. Part A 84 (2016) 354–363.

[77] E.J. Barbero, Introduction to Composite Materials Design, second ed., CRC Press, Boca Raton, FL, (2010).

[78] F.T. Wallenberger, Commercial and experimental glass fibers, in: F.T. Wallenberger, P.A. Bingham (Eds.), Fiberglass and Glass Technology: Energy-Friendly Compositions and Applications, Springer, New York, NY, 2010.

[79] F.R. Jones, Glass fibres, in: J.W. Hearle (Ed.), High Performance Fibres, CRC Press, Boca Raton, FL, 2001.

[80] I.M. Daniel, O. Ishai, Engineering Mechanics of Composite Materials, Oxford University Press, New York, NY, (2006).

[81] A.D.B.L. Ferreira, P.R.O. Nóvoa, A.T. Marques, Multifunctional material Systems: a state-of-the-art review, Compos. Struct. 151 (2016) 3–35.

[82] M.L. Shofner, Hierarchical composites containing carbon nanotubes, in: C.-S. Kim, C. Randow, T. Sano (Eds.), Hybrid and Hierarchical Composite Materials, Springer, New York, NY, 2015.

[83] J. Guo, C. Lu, F. An, Effect of electrophoretically deposited carbon nanotubes on the interface of carbon fiber reinforced epoxy composite, J. Mater. Sci. 47 (2012) 2831–2836.

[84] W. Lee, S.-B. Lee, O. Choi, J.-W. Yi, M.-K. Um, J.-H. Byun, E.T. Thostenson, T.-W. Chou, Formicary-like carbon nanotube/copper hybrid nanostructures for carbon fiber-reinforced composites by electrophoretic deposition, J.Mater. Sci. 46 (2011) 2359–2364.

[85] S.S. Wicks, R.G. de Villoria, B.L. Wardle, Interlaminar and intralaminar reinforcement of composite laminates with aligned carbon nanotubes, Compos. Sci. Technol. 70 (2010) 20–28.

[86] J. Zhang, J. Liu, R. Zhuang, E. Mäder, G. Heinrich, S. Gao, Single MWNT-glass fiber as strain sensor and switch, Adv. Mater. 23 (2011) 3392–3397.

[87] R.-C. Zhuang, T.T.L. Doan, J.-W. Liu, J. Zhang, S.-L. Gao, E. Mäder, Multi-functional multi-walled carbon nanotube-jute fibres and composites, Carbon 49 (2011) 2683–2692.

[88] J.J. Ku-Herrera, B. Pinto, V. La Saponara, R.H.R. Castro, F. Avilés, Tailored self-sensing of failure mechanisms in glass fiber/carbon nanotube/vinyl ester multiscale hierarchical composites loaded in tension, J.Multifunct. Compos. 2 (2014) 171–181.

[89] J.J. Ku-Herrera, O.F. Pacheco-Salazar, C.R. Ríos-Soberanis, G. Domínguez-Rodríguez, F. Avilés, Self-sensing of damage progression in unidirectional multiscale hierarchical composites subjected to cyclic tensile loading, Sensors 16 (2016) 400.

[90] S. Thomas, K. Joseph, S.K. Malhotra, K. Goda, M.S. Sreekala (Eds.), Polymer Composites, Wiley, Baden-Württemberg, Weinheim, 2012.

[91] J. Thomason, P. Jenkins, L. Yang, Glass fibre strength: a review with relation to composite recycling, Fibers 4 (2016) 18–42.

[92] A. Rahaman, K.K. Kar, Carbon nanomaterials grown on E-glass fibers and their application in composite, Compos. Sci. Technol. 101 (2014) 1–10.

[93] A. Godara, L. Gorbatikh, G. Kalinka, A. Warrier, O. Rochez, L. Mezzo, F. Luizi, A.W. van Vuure, et al. Interfacial shear strength of a glass fiber/epoxy bonding in composites modified with carbon nanotubes, Compos. Sci. Technol. 70 (2010) 1346–1352.

[94] J. Zhang, R. Zhuang, J. Liu, C. Scheffler, E. Mäder, G. Heinrich, S. Gao, A single glass fiber with ultrathin layer of carbon nanotube networks beneficial to in-situ monitoring of polymer properties in composite interphases, Soft Mater. 12 (2014) 115–120.

[95] B. Hao, Q. Ma, S. Yang, E. Mäder, P.-C. Ma, Comparative study on monitoring structural damage in fiber-reinforced polymers using glass fibers with carbon nanotubes and graphene coating, Compos. Sci. Technol. 129 (2016) 38–45.

[96] A.H. Barber, Q. Zhao, H.D. Wagner, C.A. Baillie, Characterization of E-glass–polypropylene interfaces using carbon nanotubes as strain sensors, Compos. Sci. Technol. 64 (2004) 1915–1919.

[97] J. Sebastian, N. Schehl, M. Bouchard, M. Boehle, L. Li, A. Lagounov, K. Lafdi, Health monitoring of structural composites with embedded carbon nanotube coated glass fiber sensors, Carbon 66 (2014) 191–200.

[98] K. Almuhammadi, M. Alfano, Y. Yang, G. Lubineau, Analysis of interlaminar fracture toughness and damage mechanisms in composite laminates reinforced with sprayed multi-walled carbon nanotubes, Mater. Des. 53 (2014) 921–927.

[99] Q. Li, J.S. Church, M. Naebe, B.L. Fox, Interfacial characterization and reinforcing mechanism of novel carbon nanotube-carbon fibre hybrid composites, Carbon 109 (2016) 74–86.

[100] Q. An, A.N. Rider, E.T. Thostenson, Electrophoretic deposition of carbon nanotubes onto carbon-fiber fabric for production of carbon/epoxy composites with improved mechanical properties, Carbon 50 (2012) 4130–4143.

[101] E. Moaseri, M. Karimi, M. Baniadam, M. Maghrebi, Enhancement in mechanical properties of multi-walled carbon nanotube–carbon fiber hybrid epoxy composite: effect of electrostatic repulsion, Appl. Phys. A 122 (2016) 1–8.

[102] R.J. Sager, P.J. Klein, D.C. Lagoudas, Q. Zhang, J. Liu, L. Dai, J.W. Baur, Effect of carbon nanotubes on the interfacial shear strength of T650 carbon fiber in an epoxy matrix, Compos. Sci. Technol. 69 (2009) 898–904.

[103] H. Qian, A. Bismarck, E.S. Greenhalgh, M.S.P. Shaffer, Carbon nanotube grafted carbon fibres: a study of wetting and fibre fragmentation, Compos. Part A 41 (2010) 1107–1114.

[104] F. An, C. Lu, J. Guo, S. He, H. Lu, Y. Yang, Preparation of vertically aligned carbon nanotube arrays grown onto carbon fiber fabric and evaluating its wettability on effect of composite, Appl. Surf. Sci. 258 (2011) 1069–1076.

[105] F. An, C. Lu, Y. Li, J. Guo, X. Lu, H. Lu, S. He, et al. Preparation and characterization of carbon nanotube-hybridized carbon fiber to reinforce epoxy composite, Mater. Des. 33 (2012) 197–202.

[106] K.J. Kim, J. Kim, W.-R. Yu, J.H. Youk, J. Lee, Improved tensile strength of carbon fibers undergoing catalytic growth of carbon nanotubes on their surface, Carbon 54 (2013) 258–267.

[107] E. Moaseri, M. Karimi, M. Maghrebi, M. Baniadam, Two-fold enhancement in tensile strength of carbon nanotube-carbon fiber hybrid epoxy composites through combination of electrophoretic deposition and alternating electric field, Int. J. Solids. Struct. 51 (2014) 774–785.

[108] A. Battisti, D. Esqué-de los Ojos, R. Ghisleni, A.J. Brunner, Single fiber push-out characterization of interfacial properties of hierarchical CNT-carbon fiber composites prepared by electrophoretic deposition, Compos. Sci. Technol. 95 (2014) 121–127.

[109] H. Chen, Y. Li, Y. Feng, P. Lv, P. Zhang, W. Feng, Electrodeposition of carbon nanotube/carbon fabric composite using cetyltrimethylammonium bromide for high performance capacitor, Electrochim. Acta 60 (2012) 449–455.

[110] M.S. Islam, Y. Deng, L. Tong, S.N. Faisal, A.K. Roy, A.I. Minett, V.G. Gomes, Grafting carbon nanotubes directly onto carbon fibers for superior mechanical stability: towards next generation aerospace composites and energy storage applications, Carbon 96 (2016) 701–710.

[111] F.-H. Zhang, R.-G. Wang, X.-D. He, C. Wang, L.-N. Ren, Interfacial shearing strength and reinforcing mechanisms of an epoxy composite reinforced using a carbon nanotube/carbon fiber hybrid, J. Mater. Sci. 44 (2009) 3574–3577.

[112] F. Zhao, Y. Huang, L. Liu, Y. Bai, L. Xu, Formation of a carbon fiber/polyhedral oligomeric silsesquioxane/carbon nanotube hybrid reinforcement and its effect on the interfacial properties of carbon fiber/epoxy composites, Carbon 49 (2011) 2624–2632.

[113] B. Yu, Z. Jiang, X.-Z. Tang, C.Y. Yue, J. Yang, Enhanced interphase between epoxy matrix and carbon fiber with carbon nanotube-modified silane coating, Compos. Sci. Technol. 99 (2014) 131–140.

[114] J. Guo, C. Lu, F. An, S. He, Preparation and characterization of carbon nanotubes/carbon fiber hybrid material by ultrasonically assisted electrophoretic deposition, Mater. Lett. 66 (2012) 382–384.

[115] L. Dong, Y. Li, L. Wang, Z. Wan, F. Hou, J. Liu, Combination effect of physical drying with chemical characteristic of carbon nanotubes on through-thickness properties of carbon fiber/epoxy composites, J. Mater. Sci. 49 (2014) 4979–4988.

[116] P. Lee-Sullivan, K.S. Chian, C.Y. Yue, H.C. Looi, Effects of bromination and hydrolysis treatments on the morphology and tensile properties of Kevlar-29 fibres, J. Mater. Sci. Lett. 13 (1994) 305–309.

[117] S. Thomas, K. Joseph, S.K. Malhotra, K. Goda, Meyyarappallil, S. Srekala (Eds.), Polymer Composites, Wiley-VCH Verlag GmbH & Co. KGaA, Weinheim, Baden-Württemberg, 2012.

[118] W. Chen, X.M. Qian, X.Q. He, Z.Y. Liu, J.P. Liu, Surface modification of Kevlar by grafting carbon nanotubes, J. Appl. Polym. Sci. 123 (2012) 1983–1990.

[119] G.J. Ehlert, H.A. Sodano, Fiber strain sensors from carbon nanotubes self-assembled on aramid fibers, J. Intell. Mater. Syst. Struct. 25 (2014) 2117–2121.

[120] T.K. Shah, H.C. Malecki, M.R. Alberding, S.H. Gardner, CNT-infused aramid fiber materials and process therefore. US patent application US2,011,171,469A1 (2011).

[121] O. Rodriguez-Uicab, F. Aviles, P.I. Gonzalez-Chi, G. Canche-Escamilla, S. Duarte-Aranda, M. Yazdani-Pedram, P. Toro, F. Gamboa, et al. Deposition of carbon nanotubes onto aramid fibers using as-received and chemically modified fibers, Appl. Surf. Sci. 385 (2016) 379–390.

[122] E.D. LaBarre, X. Calderon-Colon, M. Morris, J. Tiffany, E. Wetzel, A. Merkle, M. Trexler, Effect of a carbon nanotube coating on friction and impact performance of Kevlar, J. Mater. Sci. 50 (2015) 5431–5442.

[123] H. Qian, A. Bismarck, E.S. Greenhalgh, M.S.P. Shaffer, Carbon nanotube grafted silica fibres: characterising the interface at the single fibre level, Compos. Sci. Technol. 70 (2010) 393–399.

[124] N. Lachman, E. Wiesel, R. Guzman de Villoria, B.L. Wardle, H.D. Wagner, Interfacial load transfer in carbon nanotube/ceramic microfiber hybrid polymer composites, Compos. Sci. Technol. 72 (2012) 1416–1422.

[125] J. Li, S. Srivastava, J.G. Ok, Y. Zhang, M. Bedewy, N.A. Kotov, A.J. Hart, Multidirectional hierarchical nanocomposites made by carbon nanotube growth within layer-by-layer-assembled films, Chem. Mater. 23 (2011) 1023–1031.

[126] L. Tzounis, S. Debnath, S. Rooj, D. Fischer, E. Mäder, A. Das, M. Stamm, G. Heinrich, High performance natural rubber composites with a hierarchical reinforcement structure of carbon nanotube modified natural fibers, Mater. Des. 58 (2014) 1–11.

[127] F. Ruan, L. Bao, Mechanical enhancement of UHMWPE fibers by coating with carbon nanoparticles, Fibers Polym. 15 (2014) 723–728.

7

Toxicity and Safety Issues of Carbon Nanotubes

Hamid Narei*, Roghayeh Ghasempour*, Omid Akhavan**

*UNIVERSITY OF TEHRAN, TEHRAN, IRAN; **SHARIF UNIVERSITY OF TECHNOLOGY, TEHRAN, IRAN

CHAPTER OUTLINE

7.1 Introduction

Nanotechnology is a rapidly emerging discipline resulting in the discovery and production of unique nanomaterials [1,1a]. Among these diverse nanomaterials, carbon nanotubes (CNTs) can be considered as the perfect archetype for nanotechnology since they present exceptional mechanical, electrical, optical, thermal, and chemical properties. The extraordinary physicochemical properties of CNTs have offered a wide range of potential applications in numerous areas including ultrastiff materials, electronic devices, reinforced rods, quantum wires, mechanical memories, wastewater treatment, gas sensors, and microfabrication of conjugated polymers [1b,1c,1d,1e]. There are also several fascinating prospects for the

Carbon Nanotube–Reinforced Polymers. http://dx.doi.org/10.1016/B978-0-323-48221-9.00007-8

application of engineered CNTs in medicine. With the considerable progress in the functionalization of carbon nanostructure, especially CNTs, they have been proposed as versatile excipients for drug and vaccine delivery and imaging of disease processes [2,3,3a,3b]. Furthermore, the exceptional absorption in the near-infrared region unlocks the potential of CNTs to be utilized for biological sensing [4,5]. However, such a widespread use of CNTs is fraught with concerns for the high exposure of workers in occupational settings and consumers, as well as the release of an increasing amount into the environment. Thus, it is imperative that systematic studies be conducted to examine the toxic potential of CNTs.

CNTs can enter living organisms through the respiratory tract (inhalation and pharyngeal aspiration), gastrointestinal tract (ingestion), dermal and ocular absorption (skin and eyes), and injection (blood circulation) [6]. Airway exposure to CNTs and/or associated contaminants or catalytic materials that arise during the production process may cause acute and chronic inflammation, oxidative stress, granuloma formation, fibrosis, lung tumor and mesothelioma, and genotoxic effects in the lungs [7]. The dermal exposure to CNTs could be responsible for oxidative stress by the formation of free radicals and inflammation by the induction of redox-sensitive transcription factors [1]. The CNTs entering living organisms through different routes can translocate to the bloodstream, distribute rapidly in other organs including the heart, kidney, spleen, liver, and bone marrow, and cause diverse systemic toxicity [3].

The extremely high aspect ratio and low solubility of CNTs, as well as their structural characteristic resembling fibers have aroused considerable concerns that exposure to CNTs may lead to some disease associated with exposure to asbestos, which caused a worldwide pandemic of disease in the 20th century that continues up to the present [8]. To understand the biological and physiological responses following CNT exposure, it is necessary to determine the parameters affecting their toxicity. Studies have shown that their physicochemical characteristics are key determinants of CNT interaction with living organisms, and thus determine their toxicity. Hence, the extent of their toxicity depends on their length, diameter, surface area, tendency to agglomerate, biopersistence, and the presence and nature of catalyst residues, as well as the chemical functionalization of the CNTs. In fact, the same properties that make CNTs attractive also make them harmful to human health and the environment [3].

After CNTs enter the body via different routes, they interact with cellular macromolecules to induce specific pathologic effects. It is of utmost importance that the underlying mechanisms elucidating how these CNT–biological interactions occur at the molecular level be identified since these mechanisms would be employed to form a better understanding of CNT-induced toxicity. Different mechanisms have been proposed for the toxicity of CNTs in living organisms including oxidative stress, physical interference, proliferation, fibroblastic response, altered gene expression, and impairment of DNA [9,10].

Although, given the numerous expected applications, it is very likely that some of the CNTs will be released into the environmental ecosystems during each step of their lifecycle (i.e., at production, use, and disposal) [11], their potential environmental impacts have not been well characterized. Thus, the effects of CNTs on the environment must be taken into full consideration as new, unknown, and potentially hazardous materials [12].

In the end, it should be noted that this chapter is not supposed to provide an exhaustive review on CNT-induced toxicity for pathologists or other related professions; rather, it is aimed at prompting awareness of researchers and engineers about the deleterious effects of CNTs on human beings and the environment to encourage them to adopt safety requirements in the production of CNTs and CNT-containing materials, and/or CNT-related experimental studies.

7.2 Effects of CNTs on Systems and Organs of the Human Body

The nanodimension and biopersistence of CNTs and their physical characteristics resembling pathogenic fibers have given rise to widespread speculations that exposure to CNTs may induce numerous diseases in different organs and systems of the human body (Fig. 7.1). Although toxicity from exposure to CNTs is yet to be identified in humans, a large number of toxicology studies in animals as well as in vitro studies have concluded that exposure to CNTs can elicit a series of consistent or transient pathogenic responses. However, it should be noted that this by no means implies that results provided by a specific study, with specific samples, can be qualitatively or quantitatively extrapolated to all types of CNTs [13]. Since inhalation is recognized as the most relevant route of exposure, especially in an occupational setting, the earliest studies have focused on toxicity in lung cells and tissues. This section provides a brief overview on some of the pathogenic responses associated with exposure to CNTs.

7.2.1 Pulmonary Toxicity of CNTs

The respiratory tract consists of a series of branching airways. These begin with the extrathoracic region (nasal and oral cavities) and lead via the larynx, trachea, bronchi, bronchioles, and alveolar ducts to the alveoli (Fig. 7.2). The alveoli, saucer- or dome-shaped structures with a smooth inner surface, expand on inspiration. Gas exchanges only across alveolar walls. It is believed that the total surface area of the alveoli exceeds 100 m^2 [14]. Owing to this large surface area, the lung is a common target of many toxicants including respirable fibers [15].

The World Health Organization defines a respirable fiber, that is, a fiber that can be inhaled into the alveolar/gas exchange regions of the lungs, as a particle with a minimum aspect ratio of 3:1 with a length greater than 5 μm and a diameter less than 3 μm [17]. Exposure to pathogenic fibers, especially asbestos, causes human diseases, such as asbestosis, bronchogenic carcinoma, mesothelioma, pleural fibrosis, and pleural plaque [8]. Fig. 7.3 depicts situations where the mentioned diseases cause the pleural pathologies.

Considering the low density of CNTs and their physical characteristics, respirable CNTs are likely to be generated during manufacturing as a result of synthesis, transfer,

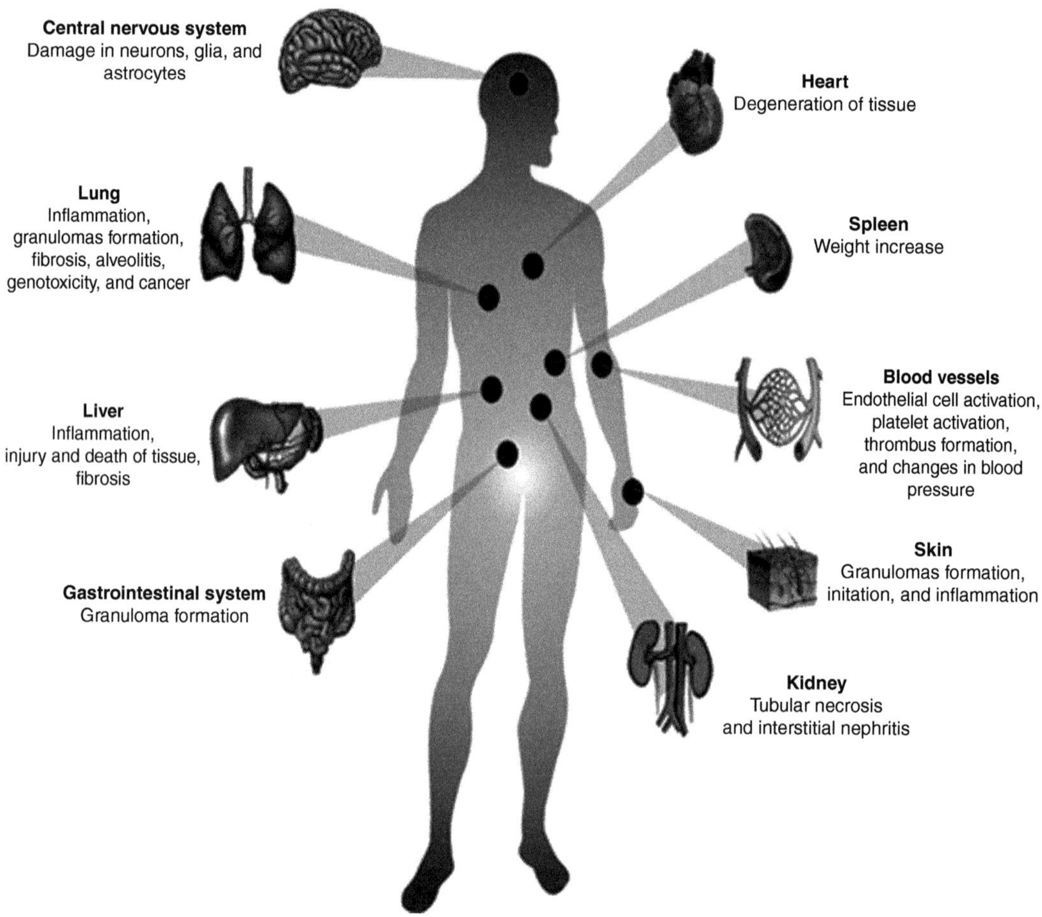

FIGURE 7.1 Main affected organs and systems from exposure to nanotubes. Nanotubes may interact through airways, the gastrointestinal tract, or skin, while reaching several organs including the liver, kidney, lungs, brain, and blood vessels. Nanotubes bioaccumulate in these sites and induce responses, such as oxidative stress, DNA damage, and cytotoxicity. *Reprinted from Y. Rodriguez-Yanez, B. Munoz, A. Albores, Mechanisms of toxicity by carbon nanotubes. Toxicol. Mech. Methods 23 (2013) 178–195, Copyright 2016, with permission from Taylor & Francis Ltd. [9].*

weighting, agitation, and mixing. Although pulmonary toxicity of CNTs is yet to be identified in humans, rodent studies associated with CNT exposure have disclosed a pattern of effects similar to those observed following exposure to asbestos including inflammation, granuloma formation, fibrosis, and mesothelioma.

Inflammation, an adaptive response, is triggered in the presence of noxious stimuli and conditions including infection, trauma, and exogenous inducers of nonmicrobial origin (allergens, irritants, foreign bodies, and toxic compounds) [18]. Therefore, it can occur in any tissue [19]. In general, the purpose of a controlled inflammatory response is to provide protection against offending agents and to allow the host to adapt. However, chronic inflammation

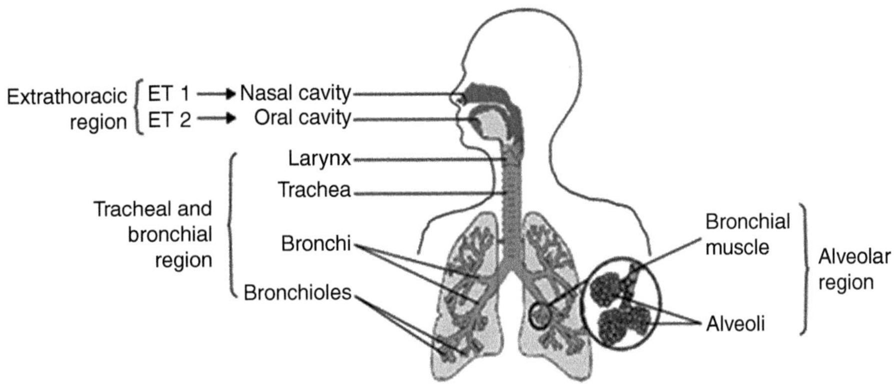

FIGURE 7.2 Different anatomical regions of the respiratory tract. *Reprinted from J-M. Lopez-Cuesta, C. Longuet, C. Chivas-Joly, Thermal degradation, flammability, and potential toxicity of polymer nanocomposites. J. Njuguna, K. Pielichowski, H. Zhu (Eds.), Health and Environmental Safety of Nanomaterials, Woodhead Publishing, 2014, pp. 278–310, Copyright 2016, with permission from Elsevier [16].*

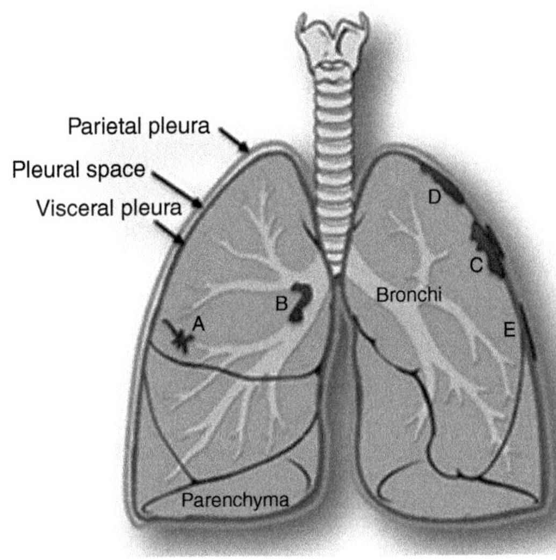

Label	Pathology	Site
A	Asbestosis	Parenchyma
B	Bronchogenic carcinoma (lung cancer)	Bronchi
C	Mesothelioma	Pleurae
D	Pleural fibrosis	Pleurae
E	Pleural plaque	Parietal pleura

FIGURE 7.3 Diagram showing the situation of the types of pathology caused by fibers. *Reprinted from K. Donaldson, C.A. Poland, F.A. Murphy, M. MacFarlane, T. Chernova, A. Schinwald, Pulmonary toxicity of carbon nanotubes and asbestos—similarities and differences. Adv. Drug. Deliv. Rev. 65 (2013) 2078–2086, Copyright 2016, with permission from Elsevier [8].*

may cause adaptive changes that become detrimental and lead to persistent tissue damage [18]. Elimination of exogenous agents followed by a resolution and repair phase is highly expected of a successful acute inflammatory response, which is mediated mainly by tissue-resident and recruited macrophages. Macrophages are a part of the mononuclear phagocytic system, that is, a class of cells having in common the property of phagocytosis, whereby the cells engulf and eliminate viruses, bacteria, and other foreign agents, and ingest abnormal or worn-out body cells. Failure to eliminate these intruders leads to chronic inflammatory response. Unsuccessful attempts by macrophages to encapsulate the intruders may result in the formation of granulomas, in which the offending agents are walled off by layers of fused macrophages as a final attempt to protect the host [18].

Numerous in vivo lung toxicology studies, as well as in vitro cell cultures have reported that exposure to single-walled carbon nanotubes (SWCNTs) and multiwalled carbon nanotubes (MWCNTs) can elicit acute and chronic inflammatory responses and induce granuloma. For instance, Lam et al. [20] conducted one of the first studies in this area in which they investigated the toxicity of SWCNTs in mice. They reported rapid and persistent granulomas associated with particle aggregates, as well as inflammation in the lungs, most frequently observed in high doses. In another study, Ma-Hock et al. [21] performed a 90-day inhalation toxicity study with MWCNTs in rats. They documented an increase in lung weights, pronounced multifocal granulomatous inflammation, and diffuse histiocytic and neutrophilic inflammation as some consequences of head–nose exposure to MWCNTs. He et al. [22,23] demonstrated in vitro that SWCNTs and MWCNTs stimulated secretion of inflammatory mediators from a variety of cell types, such as macrophages, alveolar and bronchial epithelial.

Pulmonary fibrosis, a disease in which dense fibrous connective tissue replaces lung tissue, is the end result of a variety of inflammatory diseases of the lungs. The fibrous tissue makes it more difficult for the lungs to work properly through stiffening the lungs and reducing space available for inhaled air, and thereby interfering with gas exchange; the condition may progress to respiratory failure. Fibrosis can involve the alveolar region, the bronchi, and the pleura [24]. It is believed that exposure to CNTs can be a reason for an early onset of fibrogenic response in vivo. Shvedova et al. [25] demonstrated that mice exposed by pharyngeal aspiration to purified SWCNTs suffered from a robust but acute inflammation with early onset yet progressive fibrosis and granulomas. They associated granulomas with the deposition of agglomerates in the terminal bronchioles and proximal alveoli, whereas onset of fibrosis was attributed to the deposition of more dispersed SWCNTs in the distal alveoli. Ryman-Rasmussen et al. [26] exposed male mice to aerosols of high-dose (30 mg/m^3) and low-dose (1 mg/m^3) MWCNTs for 6 h. They reported that inhaled MWCNTs had migrated to the subpleural region and thereby increasing subpleural fibrosis.

Increasing evidence indicates that individuals with preexisting respiratory diseases, such as asthma, are the most susceptible populations to nanoparticles exposure [27]. In vivo studies have proposed that mice with allergic asthma may exhibit more potent immune responses associated with exposure to CNTs [28]; moreover, CNTs may serve as allergens upon repeated exposure [29].

Fig. 7.4 shows a possible pathway leading to inflammation and longer-term effects after inhalation exposure to CNTs. Some of the related mechanisms will be discussed in Section 7.4. Mesothelioma, another disease related to the respiratory tract, will be explained in Section 7.2.3.

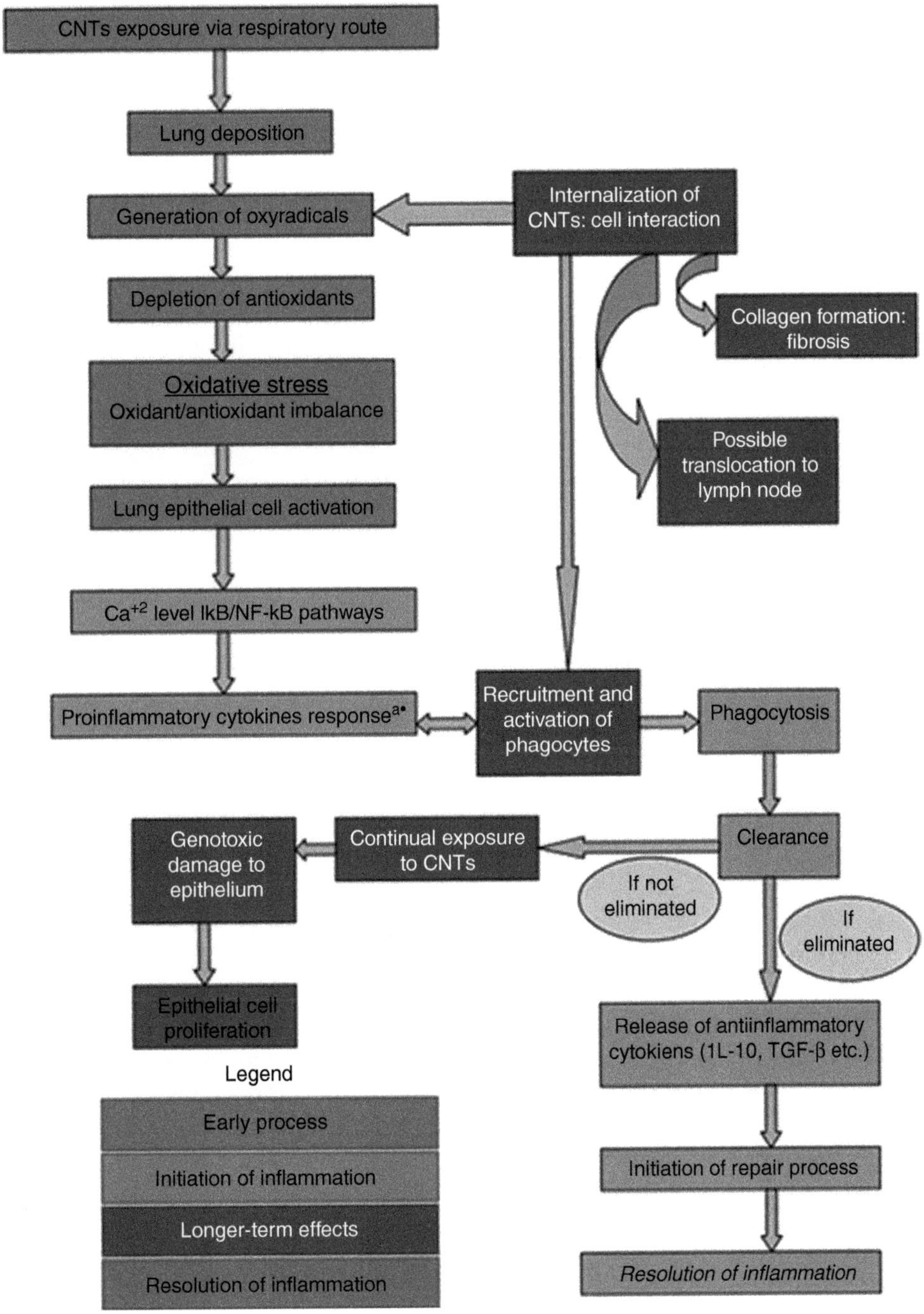

FIGURE 7.4 Possible pathways after the inhalation of carbon nanotubes *(CNTs)*. *Reprinted from A.K. Jain, N. Kumar Mehra, N. Lodhi, V. Dubey, D.K. Mishra, P.K. Jain, et al., Carbon nanotubes and their toxicity. Nanotoxicology 1 (2007) 167–197, Copyright 2016, with permission from Taylor & Francis Ltd. [44].*

7.2.2 Genotoxicity of CNTs

Genotoxicity means damage to the genomic DNA, well known to be implicated in considerable pathophysiological effects on cells, tissues, and organisms. Carcinogenicity and genetic diseases are major implications of DNA damage; thus, genotoxicity testing should be a dispensable facet of regularity risk assessment [13]. Genotoxic studies of CNTs, conducted in vivo and in vitro, have justified the assumption that CNTs may induce various genotoxic effects, as discussed in the ensuing paragraphs.

The potential of carbon nanostructures, particularly SWCNTs and MWCNTs, to cause single-strand breaks is indicated by a variety of studies. Such implications have been reported in different cell types including macrophages [30], lung epithelial cells [31,32], fibroblasts [33,34], and stem cells [34a]. In addition, Pacurari et al. [35] observed that SWCNTs had the potential to inflict DNA damage in normal and malignant human mesothelial cells. Furthermore, CNTs (and CNT-like forms) have been shown to cause more severe lesions, that is, DNA double-strand breaks [35a]. It was documented that exposing human umbilical vein endothelial cells, as well as mouse embryonic fibroblasts to MWCNTs led to the formation of γ-H2AX foci, a sensitive indicator of DNA double-strand breaks [36,37]. SWCNTs and MWCNTs were also shown to cause H2AX phosphorylation in normal human dermal fibroblasts [38].

CNTs have been shown to cause oxidative DNA damage. Enhanced oxidation of purines (bicyclic compounds with the chemical formula of $C_5H_4N_4$) and pyrimidines (heteroaromatic compounds with the chemical formula of $C_4H_4N_2$) were detected after treatment with both types of CNTs in mouse macrophages [39]. Moreover, treatment of human lung epithelial cells and rat lung epithelial cells with SWCNTs and MWCNTs, respectively, caused increased micronucleus frequencies [40,41]. SWCNTs and MWCNTs have also been reported to generate micronuclei in human lymphocyte cultures [38] and in mouse macrophages [30]. Di Giorgio et al. [30] also reported that SWCNTs and MWCNTs might induce chromosomal aberrations in mouse macrophage cells. However, there are reports indicating that MWCNTs failed to cause micronucleus and chromosomal aberrations in Chinese hamster lung fibroblasts [42] and in Chinese hamster lung cells [43].

A number of available studies suggest that CNTs can cause aneuploidy, the presence of an abnormal number of chromosomes in a cell. Muller et al. [45] showed that the treatment of human epithelial cells with MWCNTs, in addition to clastogenic effects (DNA strand breaks), led to numerical chromosome alterations. In the treatment of mouse macrophages with MWCNTs and SWNCTs, chromosomal breakage and decondensation, as well as numerical chromosome changes could be detected [30].

In support of in vitro findings, a number of in vivo studies also indicate that CNTs are potentially capable of eliciting genotoxic responses. For instance, Jacobsen et al. [46] investigated the toxicity of SWCNTs after 3 h intratracheal instillation in mice, and they reported that SWCNTs could cause DNA single-strand breakage in cells obtained from their bronchoalveolar lavage. In another study, SWCNTs were found to stimulate increases in k-ras mutations in the lung tissue DNA, after a single pharyngeal aspiration in mice [47].

Patlolla et al. [48] also investigated the genotoxic potential of MWCNTs in mice and they detected significant genotoxic effects including increased DNA strand breaks in leukocyte DNA and enhanced micronucleus frequencies and chromosomal aberrations in bone-marrow cells. Nevertheless, it should be mentioned that in vivo studies could also be addressed [49,50] in which specific genotoxic responses were shown to be absent.

7.2.3 Carcinogenicity of CNTs

Cancer is well accepted as a disease of the genome, that is, the complete set of genes or genetic material presents in a cell or an organism. Genomic changes in the critical portion of the genome are generally acknowledged to be responsible for carcinogenesis. The alterations may result in the activation of oncogenes, genes that in certain circumstances can transfer a cell into a tumor cell, and/or the inactivation of tumor suppressor genes [51]. To design studies investigating carcinogenic potential, it is of paramount importance to understand whether xenobiotics are primary or secondary carcinogens. Primary carcinogens are considered to interact directly with target cells and cause DNA damage by their intrinsic physicochemical or chemical reactivity, while secondary carcinogens act via cellular mechanisms as well as activating inflammatory cells, which may lead to genotoxicity by the production of reactive oxygen species (ROS). Identifying xenobiotics as secondary carcinogens would make it possible to define thresholds in the sense of no-effect levels for exposure regulations [52].

Owing toCNTs' physical characteristic and biopersistence, besides inflammation and fibrosis, the induction or the promotion of lung tumors and mesothelioma is anticipated from the pulmonary exposure to CNTs. Mesothelioma is a rare, malignant cancer that develops in the thin, protective layer of tissue that covers the lungs (pleural mesothelioma) or the abdomen (peritoneal mesothelioma). Takagi et al. [53] investigated the carcinogenicity of a certain type of MWCNTs (Mitsui MWCNT-7) in the p53 hetero-knockout mice after intraperitoneal injection. The tumor-suppressor gene p53 is most frequently inactivated in human cancers and is recognized as being responsible for the Li–Fraumeni syndrome, the cancer-prone hereditary disease [51]. Although they found that MWCNTs might cause malignant mesothelioma, their work was criticized for the (relatively high) dose, as well as the experimental system of heterogeneous knockout animals [52]. Nonetheless, they repeated their study using much lower exposures and reported that the MWCNTs were as carcinogenic to mesothelial cells as asbestos [54]. In a small-scale study, Sakamoto et al. [55] observed that injection of the same MWCNTs in the scrotum of Fischer 344 rats, a strain that is prone to show spontaneous mesothelioma, caused mesothelioma. It should be noted that due to the presence of a mesothelial cell layer in the scrotum, mesothelioma might arise in this organ. The aforementioned studies have been discussed controversially since they used very sensitive models to investigate the carcinogenicity of MWCNTs [51]. Hence, Nagai et al. [56] used noncancer-prone rodents to demonstrate the mesothelial carcinogenicity of MWCNTs. Their study showed that the time required for the development of mesothelioma after intraperitoneal injection of MWCNTs with 50 nm diameter was much less than that required for asbestos-developed mesothelioma at the same weight [56,57]. However, there are studies reporting no carcinogenic effects [58,59].

Investigations suggesting that SWCNTs and MWCNTs may act as a strong cancer promoter have also been presented. Shvedova et al. [60] exposed mice to SWCNTs by pharyngeal aspiration. Two days after aspiration, Lewis lung carcinoma was inoculated in the mice. They reported that inoculation with cancer cells led to lung tumors 21 days after injection in vehicle control mice, and preexposure to SWCNTs resulted in the significantly greater lung tumor growth. In another study, Sargent et al. [61] employed a two-stage initiation/promotion protocol to examine whether inhaled MWCNTs could act as a cancer initiator and/or promoter. First, they intraperitoneally injected mice with either corn oil (vehicle) or methylcholanthrene (a highly carcinogenic polycyclic aromatic as the initiator). One week later, they then exposed mice to either filtered air or MWCNTs by inhalation. Their results demonstrated that MWCNTs did not significantly increase the incidence of tumor formation in mice treated with corn oil, while they induced a pronounced increase in tumor incidence in mice into which the initiator had been injected [62].

7.3 Determinants of CNT-Induced Toxicity

7.3.1 Physical Characteristics

An ideal SWCNT (MWCNT) can be considered as a sheet (a series of sheets) of hexagonally shaped carbon atoms rolled up to make a seamless, hollow cylinder. The diameter of individual CNTs typically ranges from 0.4 to 3 nm for SWCNTs and from 1.4 to 100 nm for MWCNTs. However, due to van der Waals forces between carbon atoms, seldom do SWCNTs exist as individual nanotubes and they mostly agglomerate to form ropes or bundles of 20–50 individual nanotubes with diameters that could be as high as 30 nm and lengths exceeding 1 μm. In comparison with SWCNTs, van der Waals forces are much weaker for MWCNT, and, as a consequence, they often exist as single tubes or form a few bundles [44].

Electron micrographs of aerosol CNTs have demonstrated that they can be classified into two general types: (1) fiber-like and (2) broadly spherical. Fiber-like CNTs consist of dispersed or lightly agglomerated individual tubes or ropes of tubes with diameters typically around 50 nm (Fig. 7.5A), while broadly spherical CNTs comprise near-spherical, open agglomerated clumps of tubes with a wide range of diameters, from 100 nm up to hundreds of micrometers (Fig. 7.5B). In the limited observation made to date, the fiber-like CNTs were only observed for MWCNTs [14]. Although CNT particles, in general, are divided into two categories to simplify discussion of their transport and deposition behavior, it should be noted that it is likely that aerosol CNTs exist in a mix of such particle forms and sizes, as well as intermediate forms, such as fiber-like particles with significant branches, or broadly spherical particles with tails of single tubes or ropes protruding from the bulk [14].

As demonstrated for high aspect ratio materials, the toxicity of CNTs is heavily dependent on their length. In one of the first studies investigating the role of length in

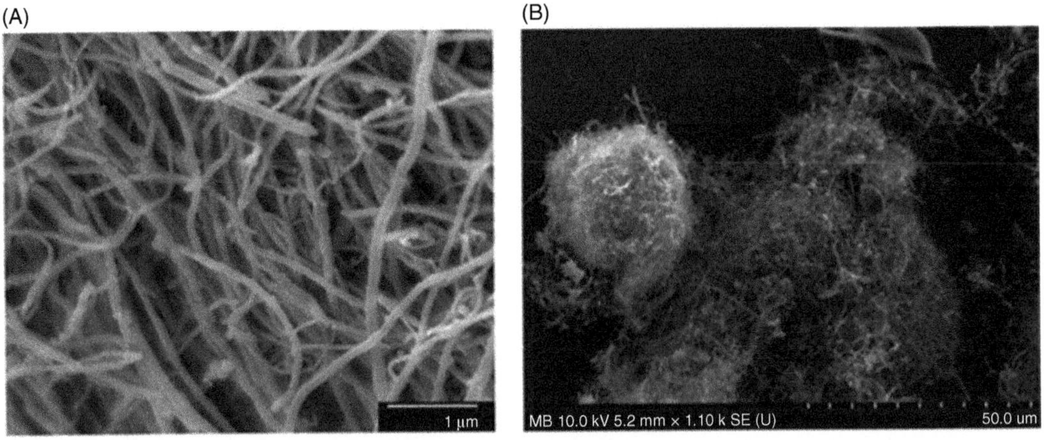

FIGURE 7.5 MWCNTs imaged with scanning electron microscopy. (A) Relatively straight MWCNT sample, and (B) entangled MWCNT sample. *Reprinted from H.J. Johnston, G.R. Hutchison, F.M. Christensen, S. Peters, S. Hankin, K. Aschberger, et al., A critical review of the biological mechanisms underlying the in vivo and in vitro toxicity of carbon nanotubes: the contribution of physico-chemical characteristics. Nanotoxicology 4 (2010) 207–246, Copyright 2016, with permission from Taylor & Francis Ltd. [63].*

inflammation, Poland et al. [64] demonstrated that fiber-shape MWCNTs with lengths of 5–20 μm, injected into the abdominal cavity of mice, exhibited asbestos-like, length-dependent, pathogenic behavior and induced mesothelial granulomatous inflammation. In contrast to fiber-shaped CNTs, the short tangle ones did not cause any significant inflammation. Murphy et al. [65] confirmed the aforementioned findings through direct instillation of long and short CNTs into the pleural cavity of mice. The authors suggested that short fibers and CNT tangles readily cleared off from the cavities by negotiating through stomatal openings on the surface of the parietal pleural wall and the diaphragm (Fig. 7.6A), while long CNT fibers cannot navigate the peritoneal and pleural cavities, leading to their accumulation on the parietal surface (Fig. 7.6B). Then macrophages would try to phagocytose the retained CNTs, but the long CNTs could not be effectively engulfed by gathering macrophages, resulting in frustrated phagocytosis that could further stimulate recruitment of inflammatory cells and mesothelial cell damage, leading to chronic inflammation, granuloma development, fibrosis, and malignancy [66] (Fig. 7.7).

Although the diameter plays an important role as the length of CNTs in their toxicological assessment, it has been far less evaluated. Furthermore, based on the existing literature, drawing a firm conclusion on whether thinner or thicker tubes exhibit higher toxicity is difficult as there are several glaring inconsistencies. On the one hand, Fenoglio et al. [67] investigated the toxicity of two samples of highly pure MWCNTs similar in hydrophobicity, surface reactivity, and length (<5 μm), but markedly different diameters (9.4 vs. 70 nm). They analyzed the toxicity of MWCNTs in vitro and in vivo in murine alveolar macrophages and rats, respectively. The results showed that the thinner MWCNTs were significantly more toxic than their thicker counterparts, both in vitro and in vivo, when compared on a mass–dose basis. In a similar study, thin MWCNTs (diameter ~50 nm)

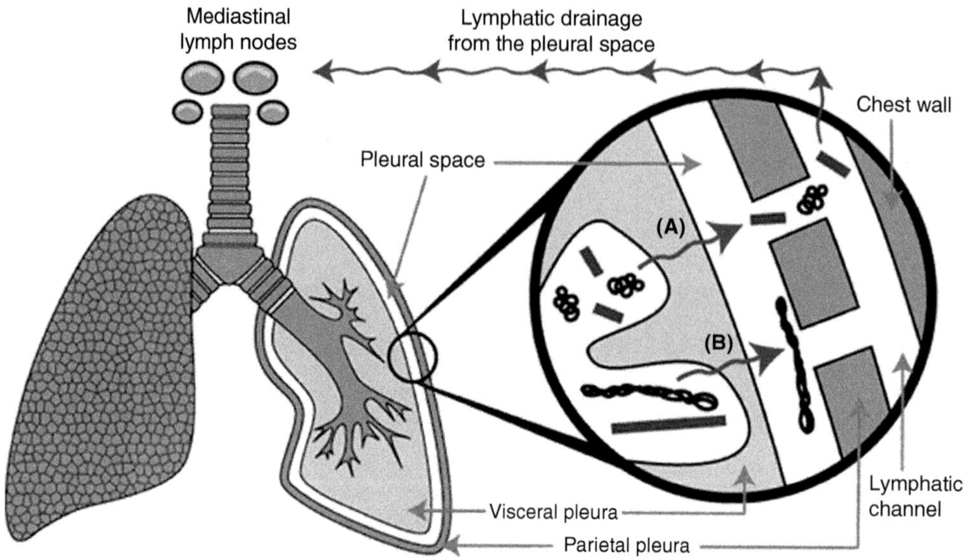

FIGURE 7.6 Diagrammatic representation of length-dependent clearance from the pleural space. (A) Short fibers and small CNT tangles that deposit in alveoli that are situated subpleurally and migrate to the pleural space and exit in the flow of pleural fluid through the stomata, where they follow the lymphatic drainage to the mediastinal lymph nodes. (B) Long fibers and long CNTs also reach the pleural space from subpleural alveoli but they cannot negotiate through the stomata and are retained, where they cause inflammation and potentially long-term disease. *Reprinted from F.A. Murphy, C.A. Poland, R. Duffin, K.T. Al-Jamal, H. Ali-Boucetta, A. Nunes, et al., Length-dependent retention of carbon nanotubes in the pleural space of mice initiates sustained inflammation and progressive fibrosis on the parietal pleura. Am. J. Pathol. 178 (2011) 2587–2600, Copyright 2016, with permission from Elsevier [65].*

with high crystallinity caused mesothelial cell membrane piercing and cytotoxicity in vitro and subsequent inflammogenicity and mesotheliomagenicity in vivo. Conversely, thick (diameter ~150 nm) or tangled (diameter ~2–20 nm) MWCNTs appeared to be less toxic, inflammogenic, and carcinogenic [56]. On the other hand, Wang et al. [68] reported a higher toxicity of short CNTs (1–5 μm) with diameters of 40–100 nm, compared to CNTs with diameters in the range between 10 and 20 nm to alveolar macrophages. This discrepancy about the effects of diameter may be attributed to the inconsistencies in other factors, such as the variability of CNTs employed in different studies and the difference in the concentration/dose.

The intrinsic biological activity of CNTs is closely correlated with their specific surface area. From a biological perspective, high surface area provides high contact area with biological structures, leading to pronounced biological activity. Indeed, CNTs have a highly increased surface area owing to their high surface-to-volume ratio. Comparing the toxicity of SWCNTs with MWCNTs clearly demonstrated the correlation between surface area and toxicity. It was estimated that SWCNTs were approximately 8.5-fold more fibrogenic than MWCNTs per microgram of dose through assessing the fibrotic potential of CNTs in the lungs using the target site dose. Presumably SWCNTs are lighter and have a larger surface area than MWCNTs and, consequently, are more toxic on an equal weight basis [10]

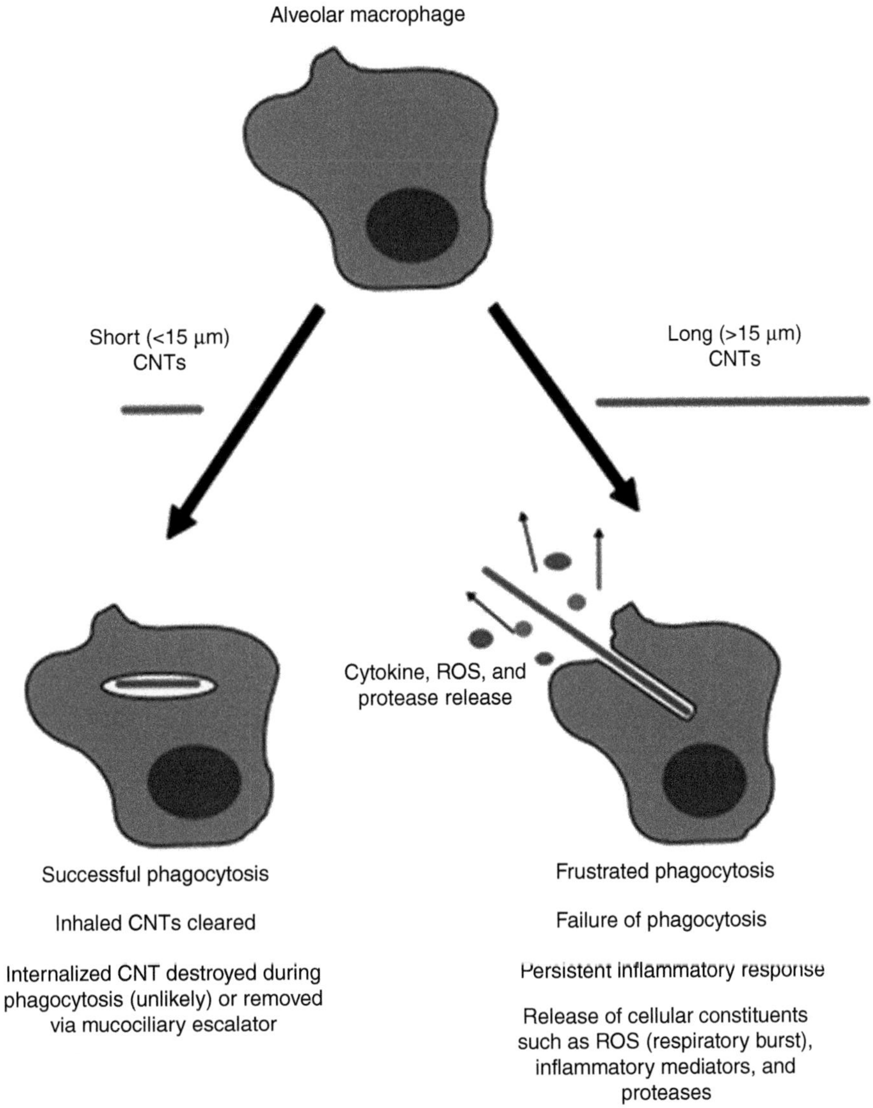

FIGURE 7.7 Importance of CNT length to their clearance by alveolar macrophages. Short CNTs (<15 μm) are easily cleared by macrophages, whereas longer CNTs (>15 μm) are not engulfed by macrophages, and frustrated phagocytosis ensues. *ROS*, reactive oxygen species. *Reprinted from H.J. Johnston, G.R. Hutchison, F.M. Christensen, S. Peters, S. Hankin, K. Aschberger, et al., A critical review of the biological mechanisms underlying the in vivo and in vitro toxicity of carbon nanotubes: The contribution of physico-chemical characteristics. Nanotoxicology 4 (2010) 207–246, Copyright 2016, with permission from Taylor & Francis Ltd. [63].*

The aggregation state of CNTs may alter their shape and the overall surface area available for interaction with cells and molecules, and thus, may have significant implications for their biodistribution and toxicity. There is evidence prompting speculation that dispersed CNTs would not be readily recognized by macrophages, while large agglomerates of CNTs would be easily recognized [29]. However, despite singlet or small bundles of CNTs, large, agglomerated CNTs are more difficult to be cleared off, resulting in the formation of granulomas [10]. Furthermore, agglomerated CNTs may release singlet CNT fibers over time, leading to alteration of the internal dose in different organ systems in the body, and thereby, be implicated in extended and, potentially, unexpected systematic toxicity [10,69].

7.3.2 Biopersistence

Another critical determinant in CNT-induced toxicity is their biopersistent behavior after depositing on the targeted organs. The particle biopersistence is defined as the likelihood of the deposited particle to resist changes in number, dimension, chemical composition, surface area, surface chemistry, or other physicochemical characteristics. Any or all of these parameters may be changed during the residence time of a particle in the organ where it deposits, and such alterations affect the potential, long-term responses to the particle [17]. It is well acknowledged that pristine CNTs are highly biopersistent for several reasons. First, given that graphene, the building block of CNTs, is an extremely strong material made of strong sp^2 carbon—carbon bonds, CNTs are generally considered to be resistant to biological degradation mechanisms. Second, as mentioned in the previous section, CNTs with a large aspect ratio are more difficult to be engulfed and cleared off by macrophages and through size-dependent mechanisms, for instance stomatal openings. Last, their propensity to aggregate and form large bundles makes it more difficult for macrophages to remove them from the site of deposition [10]. As a consequence of the CNTs' biopersistence, continued presence and accumulation in the organ and, consequently, chronic inflammation, fibrosis, and formation of granuloma are anticipated [70].

7.3.3 Surface Chemistry

The surface chemistry of CNTs potentially affects their behavior, and is relevant to consider as a determinant of toxicity since it is the surface of the CNTs with which cells and molecules interact [63]. Modification of the surface of CNTs has the ability to alter their toxic responses. It has been demonstrated that surface modification can enhance and reduce toxicity, depending on the particular modification taken. On the one hand, acid treatment would oxidize CNTs to introduce carboxyl and hydroxyl groups on the surface of CNTs, changing the bioactivity and interaction of CNTs with other molecules [10]. Bottini et al. [71] reported that acid-oxidized MWCNTs induced higher toxicity toward Jurkat cells than their pristine counterparts. On the other hand, compared with pure MWCNTs, nitrogen-doped MWCNTs showed significantly reduced toxicity and increased tolerance in exposed mice [72].

Functionalization potentially affects CNT-induced toxicity. For instance, Ursini et al. [73] exposed lung cells to pristine or carboxylated MWCNTs (MWCNT or MWCNT–COOH) and analyzed a panel of toxic responses including cell survival, cytokine expression, and DNA damage. They reported that MWCNT–COOH had increased cytotoxicity in bronchial cells compared with its pristine counterparts, while the pristine MWCNTs had showed higher toxicity toward alveolar cells than MWCNT–COOH. In another study, atomic layer deposition was employed to coat MWCNTs with a thin film of aluminum oxide (Al_2O_3). It has been documented that the Al_2O_3-coated MWCNTs induced lower fibrosis in mice than pristine MWCNTs [74].

7.3.4 Impurities

Impurities present in a CNT may contribute to their toxicity. Potential impurities range from residual, metal catalysts including iron and nickel to amorphous carbon and diverse hydrocarbons, primarily introduced during their manufacture [63]. Transition metals, such as iron, are well known to induce toxicity via redox cycling and stimulation of ROS production. CNTs may be processed to improve the purity of manufactured CNTs by utilizing postproduction processes, such as the use of acid and heat treatments, to remove most of the metal catalysts from the CNT structure. These processes, however, have been shown to often affect other properties of CNTs including length and surface chemistry, and thereby making it difficult to determine the role of impurities in toxicity [63,75].

A number of studies have investigated the role of metals in toxicity induced by CNTs. In one of the early studies in this area, Shvedova et al. [76] demonstrated that unpurified SWCNTs (iron content, 30%) were able to induce higher toxicity within skin cells in vitro than acid-treated counterparts (iron content, 0.23%), and the addition of metal chelators could reduce the observed toxicity, indicating that the iron content played a significant part in SWCNT-induced toxicity. In another study, Kagan et al. [77] reported that SWCNTs containing high iron content (~26%) stimulated increased ROS production compared with the iron-depleted counterparts (iron content, ~0.23%) in cultured cells and cell-free systems. In contrast to these findings, some different CNT samples with iron content ranging from 0.53% to 26.9% did not appear to impact their potential to produce granulomas in mice, suggesting that granuloma formation could not be simply and solely related to the total iron content of samples. Accordingly, it is possible that the tendency of CNTs to aggregate stimulated a foreign body response in the lungs [20].

7.4 Mechanisms of CNT-Induced Toxicity

Studies have proposed several different mechanisms for CNT-induced toxicity in living organisms including oxidative stress, physical interference, proliferation, fibroblastic response, altered gene expression, and impairment of DNA [9,78]. In this section, due to the importance of oxidative stress, these mechanisms will be reviewed in two general categories: (1) oxidative stress-related mechanism and (2) nonoxidative stress-related mechanisms.

7.4.1 Oxidative Stress-Related Mechanism of CNT-Induced Toxicity

Oxidative stress is one of the most consistent observations that would associate CNT exposure with toxicity. It is a cellular stress state caused by an imbalance between the production of ROS and antioxidant defense, which favors ROS presence, resulting in harming of the cell. Internal metabolism and exogenous exposures, such as bacterial infection and exposure to particles, fibers, and transition metals, can stimulate the production of ROS [10]. Oxygen radicals ($O^{\cdot -}$, $^{\cdot}OH$, $RO_2^{\cdot}$, and $RO^{\cdot}$) and oxygen species that are easily converted to radicals and/or are oxidizing agents (H_2O_2, HOCl, O_3, and $^1O_2$) are known biologically relevant ROS. The definition of ROS implies that the generation of ROS is inevitable during cell life, and it has been proven that a moderate level of ROS is essential for some crucial cellular functions, such as regulation of cell proliferation and immune response, gene expression, and killing of invading microbes during phagocytosis. However, excessive ROS would overwhelm the body's antioxidant capacity, resulting in damages to macromolecules, such as DNA mutation, DNA strand break, lipid peroxidation, and protein peptide chain break, and eventually cell death. Oxidative stress is the common pathogenic pathway of quite different diseases including aging, neurodegeneration, cancer, and chronic inflammatory pathology, such as asbestosis and silicosis [79,80]. Hence, substances capable of inducing oxidative stress may contribute to develop several serious disorders in humans.

SWCNTs and MWCNTs have been found to induce an increase in ROS production, often accompanied with high levels of oxidative markers, and depletion of antioxidants in vivo [81,82], while in vitro oxidative stress has been observed in models of the skin [76] and lung [83]. The contribution of oxidative stress has beenproven by the fact that the pretreatment of animals/cells with antioxidants has the potential to alleviate oxidative stress and inflammatory cytokine expression induced by CNTs [84].

CNTs may induce ROS via a direct effect on the cells, such as macrophages, bronchial and alveolar epithelial cells, and fibroblasts [22,23,81,83]. Transition metals used as catalysts during the production of CNTs, redox-active groups, such as quinones, resulting from functionalization, and relatively stable free radical intermediates existing in the reactive surfaces of carbonaceous particles are major sources that induce oxidative stress [85,86].

CNTs can stimulate excessive production of ROS in several ways. First, long fiber-like CNTs can be responsible for frustrated phagocytosis in which activated macrophages are not able to effectively encase the long fibers that exceed their body diameter. Frustrated macrophages release the contents of their phagosomes including antimicrobial agents, digestive enzymes, and extremely toxic reactive nitrogen species and ROS, causing destruction of surrounding tissues and local inflammation [87]. Second, mitochondria, the powerhouse and a major source of ROS production in mammalian cells under physiologic conditions, are prone to be damaged by pathogenic fibers resulting in oxidative stress. He et al. reported mitochondrial damage by SWCNTs [23] and MWCNTs [22] and thereby elevated ROS production in lung cells. Last, Fenton ($Fe^{2+} + H_2O_2 \rightarrow Fe^{3+} + {}^{\cdot}OH + OH^-$) or Fenton-like ($Mn^{n+} + H_2O_2 \rightarrow Mn^{(n+1)+} + {}^{\cdot}OH + OH^-$) reactions can be triggered by the presence of impurities in CNTs, such as iron and other transition metals from CNT production, which result in the overproduction of ROS and thus induction of oxidative stress in biological systems [10].

In addition to directly damaging macromolecules and cellular structures leading to cytotoxicity and cell death [79], ROS can prompt the activation of signaling pathways, such as nuclear factor NF-κB, responsible for activating the increased expression of a variety of proinflammatory and profibrotic genes implicated in the induction of inflammation and fibrosis [22,23]. Moreover, ROS are capable of serving as signaling molecules to regulate growth factor-induced proliferation and differentiation of fibroblasts, which are crucial in the fibrogenic response to CNT. Considering the biopersistence of CNTs in the body, the continuous production of ROS and thereby prolonged oxidative stress in CNT-deposited tissues would be expected [10].

7.4.2 Nonoxidative Stress-Related Mechanism of CNT-Induced Toxicity

While oxidative stress has been reported as the main mechanism of CNT-induced toxicity in numerous studies, there is ample evidence that CNTs can induce cellular damage in the absence of oxidative stress. One of the most common nonoxidative stress-related mechanisms is physical interference of nanostructures with cellular and extracellular constituents, having great potential to alter vital cellular processes and induce a range of cellular injury, including cell death [87a]. Studies pertaining to this mechanism have investigated the behavior of both phagocytes, such as macrophages, and nonphagocytic cells. Hirano et al. [88] reported that uptake of MWCNTs by macrophage cell lines led to mechanical rupture of the cell membrane. The authors assumed that MWCNTs bound with macrophages through the macrophage receptor with collagenous structure or MARCO receptors, which encouraged the plasma membrane to extend along the CNTs, resulted in damage to the membrane and thereby cell death. Kostarelos et al. [89] reported that various types of functionalized CNTs (SWCNTs and MWCNTs) could be taken up by a wide range of cells. The authors suggested that cellular internalization of functionalized CNTs did not solely depend on endocytosis; rather, their penetration of the cell membrane could occur owing to their ability to act like a "nanosyringe" and thus enabling them to puncture the cell membrane and be internalized. They also mentioned that the capacity of CNTs to enter cells might depend on the cell type, dimensions of CNTs, and their tendency to form larger aggregated structures [63].

Physical interference of CNTs may be encountered at more subtle levels than the aforementioned, overt physical damage. It has been proposed that due to their similarities to the microtubules of the spindle, CNTs have the potential to enter the nucleus, be incorporated into the centrosome, and interfere with the mitotic spindle to result in clastogenic (DNA strand break) and aneuogenic responses (chromosomal loss), which are the conditions strongly predisposing to cancer [90,91]. In addition, CNT may stimulate cells to produce ROS with the capability to damage DNA, leading to DNA base oxidation and DNA strand breaks. It is also possible that elevated inflammatory response to CNT causes damage to DNA. Furthermore, CNT may suppress DNA repair and, as a consequence, facilitate genotoxic processes [10].

Cell proliferation has been perceived as an underlying molecular and cellular mechanism of some pathologic outcomes of CNT exposure in animals including tumorigenesis, fibrosis, and inflammation. There is overwhelming evidence establishing that CNTs stimulate the proliferation of several types of cells in animals and in vitro. For instance, Vietti et al. [92] investigated the proliferative response to MWCNTs assessed on mouse primary lung fibroblasts, human fetal lung fibroblasts, mouse embryonic fibroblasts, and mouse lung fibroblasts. They reported that long MWCNTs directly modulated fibroblast proliferation in vitro, in a dose-dependent manner, and induced lung fibrosis in vivo (the fibroblast proliferation is further discussed in the ensuing paragraphs). In another study, Xu et al. [93] documented that exposure to MWCNTs induced hyperplastic proliferative lesions of the visceral mesothelium in rats. They recorded that proliferating cell nuclear antigen indices of MWCNT-treated rats were approximately 10-fold that of the vehicle control. The authors associated the hyperplastic lesions with inflammatory cell infiltration and inflammation-induced fibrotic lesions of the pleural tissues. Furthermore, elevated mesothelial cell proliferation observed in vitro supported the notion that CNT-stimulated proliferation of the mesothelium may result in the formation of mesotheliomas [10]. Hence, MWCNTs clearly elicit pathological responses in certain cells by triggering cell proliferation.

Fibroblasts, the principal active cells of connective tissues, play a prominent part in tissue fibrosis [94]. Exposure to CNT causes fibroblasts to increase in number and transform into myofibroblasts, the primary collagen-producing cells. Myofibroblasts secrete numerous matrix proteins including collagens and fibronectin, as well as matrix-modulating enzymes, such as metalloproteinases, a known fibrogenic mediator in the lung, resulting in the deposition and processing of collagen proteins. Eventually, near CNT deposits, the parenchyma would be replaced with collagen fibers resulting in lung fibrosis. Pulmonary interstitial fibroblasts may evolve from several sources during fibrosis, conceivably including the following: (1) proliferation of resident fibroblasts and thereby increased production of collagen matrix, which are hallmarks of lung fibrosis in the lungs; (2) recruitment of bone marrow-derived fibrocytes from blood into the lung interstitial space; and (3) epithelial–mesenchymal transition of airway and alveolar epithelial cells [10]. SWCNTs and MWCNTs have been reported to trigger fibroblast proliferation in vitro [92,94,95], transformation of fibroblasts into myofibroblasts [22,23], and induction of epithelial–mesenchymal transition [96,97]. The role of fibrocyte recruitment in the toxicity of CNTs has not yet been documented [10].

Studies suggest two mechanisms for the effects of CNT on fibroblasts. First, CNTs may mimic endogenous collagen fibers or other fibrous structures to stimulate fibroblasts to adhere to their surface, resulting in the activation and thereby proliferation and differentiation of fibroblasts. Second, CNTs may stimulate the secretion of profibrogenic mediators, for example, the inflammatory cytokines TNF-α and IL-1β, as well as the growth factors TGF-β1 and PDGF, exerting powerful mitogenic effects on fibroblasts, which lead to the proliferation and differentiation of the cells [98]. While some in vitro and in vivo studies support these ideas, there is no definitive proof demonstrating which mechanism can explain the fibrotic effects of CNT in the lungs in vivo [10].

7.5 Ecotoxicological Effects of CNTs

CNTs are acknowledged as a novel class of nanomaterials providing exceptional structural, physical, and chemical properties with numerous applications. A minute amount of CNTs, in general, is produced and dispersed in the environment by natural phenomena including lightning and fire. Although this tiny amount is not considered as an environmental pollutant, their increasing production and usage have resulted in the release of a larger amount of CNTs into the environment and thereby disrupting the perfect balance of nature and ecotoxicity. CNTs may be released into the environment through gradual erosion and abrasion, as well as improper burial or obliteration of CNT-containing materials. Burning materials containing CNTs, such as polymers, is a method of obliteration. The presence of CNTs in polymers results in enhanced flame retardancy because of char layer formation. During burning in combustion, nanomaterials present in the composition (dry, dispersible single particles, or agglomerates) may be transformed into gaseous phase and eventually enter the air and the environment with the exhaust gas and cause soil, air, and water pollution [16,99]. Based on nanotechnology development and the elevated production rate of nanomaterials, a large number of studies, particularly on the toxicity of CNTs, have been conducted on their toxicity assessment and deleterious effects on the environment.

As mentioned in previous sections, the available literature has revealed that CNTs are able to elicit various toxic responses in different cells and organisms, which often occur through the direct exposure of these cells and organisms to CNTs. However, questions, such as "May transfer of these materials from the environment to organisms be possible?," "Does the environmental changes resulting from their existence in nature affect the different organisms' health?," and "How much ecotoxicity can be hazardous to health?," have been recently addressed. While numerous researches have been conducted on the ecotoxicity of CNTs, due to the lack of a globally defined standard, there are different viewpoints on the degree of their ecotoxicity, and obtained results contain different and somehow controversial information [100,101]. Hence, the lack of adequate awareness about the risk and toxicity potential of nanomaterials is one of the most serious limitations on their commercialization. This can be investigated in two areas: (1) nanomaterials properties and their detrimental effects on the environment and (2) identifying and tracking nanomaterials in the environment.

To conduct a thorough risk assessment, it is essential to recognize how CNTs may induce toxicity and the degree of their toxicity in the environmentally relevant media. Physical characteristic of CNTs, potential for release of metals from catalyst particles, and dispersion methods employed to disperse CNTs are considered as the most important determinants, which were discussed in Section 7.3.

Quantification of CNTs in environmentally relevant media is a challenging task for two main reasons. First, carbon, oxygen, and hydrogen (owing to the inevitable presence of defects), which are the main compounds of nature, are elements of which CNTs are composed. Hence, it is not possible to utilize detection methods based on elemental detection including inductively coupled plasma mass spectrometry, instrumental neutron activation analysis, and atomic absorption spectroscopy. Second, CNTs within a sample typically have a wide range of diameters and lengths, which imposes a major limitation to the application of chromatographic methods, such as gas and liquid chromatography mass spectrometry. However, there are a limited number of techniques that take advantage of the difference in the physicochemical properties of CNTs and other materials to quantify

CNTs in environmentally relevant media including thermal gravimetric analysis, thermal optical transmittance, and chemothermal oxidation at 375°C (CTO-375), which use the higher thermal stability of CNTs compared with that of the biological or the environmental media in which the CNTs are dispersed [102]. Relatively poor detection limits widespread application of these techniques in soils and sediments, while the limits are lower in water. Radioactive labeling of CNTs is a technique that has been successfully employed with both types of CNTs in different media including soils, sediments, liquids, and organisms [103–106].

Studies have reported two general trends with regard to the ecotoxicity of CNTs: (1) A lack of uptake of CNTs by multicellular organisms and (2) a lack of toxicity in soils and sediments except at extremely high concentrations [102].

Studies with earthworms [104,107,108] and a range of sediment organisms [100,104,106,109] have not reported CNT absorption across the gut tract and systemic circulation throughout organisms. For instance, Ferguson et al. [106] reported that SWCNTs were not observed in estuarine invertebrates after a depuration period. Indeed, they observed the same SWCNT concentration in the organism pellets as in the sediment, which indicated that CNT uptake and elimination had occurred. A study conducted on oligochaeta indicated that CNTs measured in the organisms were from sediment not yet purged from the organisms' guts. After depuration for two days, complete elimination from the worms was detected in clean sediment [109]. Aquatic organisms in the water column, such as daphnia [105,110,111] and amphibians [11,12] have also demonstrated no absorption of CNTs into their tissues. However, a number of studies have documented that some plants absorbed CNTs. For example, Wild and Jones [112] reported that CNTs pierced root cell walls of wheat plants but complete internalization was not observed. MWCNT uptake into all parts of tomato plants including the food crops were observed in a study conducted by Khodakovskaya et al. [113].

Investigations have demonstrated that ecotoxicity in soils and sediments would usually occur in the g/kg range or even higher concentrations of CNTs. It has been reported that acute toxicity was not observed at CNT concentrations up to 3 mg/g in soils with earthworms [104,105,107,108]. It has also been shown that MWCNT and SWCNT concentrations of 0.3 and 0.003 mg/g, respectively, might not cause toxicity in sediments [104,109]. Similarly, Chung et al. [114] indicated that only 0.5 mg/g of MWCNTs caused toxicity to soil microorganism communities, whereas concentrations of 0.05 mg/g and lower rarely induced toxic effects.

7.6 Conclusions

As a novel class of nanomaterials, CNTs provide unique structural, physical, and chemical properties leading to the widespread and rapidly growing use of CNTs in a great variety of applications. However, owing to their physical characteristic of resembling pathogenic fibers and biopersistence, considerable concerns have been raised that exposure to CNTs may elicit some toxic responses. The currently available literature suggests that exposure to CNTs may induce toxic effects in vivo and in vitro. Effects that have been identified include acute and chronic inflammation, oxidative stress, granuloma formation, fibrosis, lung tumor and mesothelioma, genotoxic effects, and diverse systemic toxicities. The inconsistent results observed among different studies offer clear evidence that the effects may be extremely variable in its dependence on their heterogeneous physicochemical properties. In light of the pulmonary and systemic responses reported in vitro and in animal studies, it is prudent to minimize exposure to CNTs.

References

[1] H. Narei, R. Ghasempour, Y. Noorollahi, The effect of employing nanofluid on reducing the bore length of a vertical ground-source heat pump, Energy Convers Manag. 123 (2016) 581–591.

[1a] S. Ghasempour, M. Ali, R. Ghasempour, M. Alipour, Investigating the cytotoxicity of iron oxide nanoparticles in in vivo and in vitro studies, Exp. Toxicol. Pathol. 67 (2015) 509–515.

[1b] A.R. Murray, E. Kisin, S.S. Leonard, S.H. Young, C. Kommineni, V.E. Kagan, et al. Oxidative stress and inflammatory response in dermal toxicity of single-walled carbon nanotubes, Toxicology 257 (2009) 161–171.

[1c] R. Ghasempour, S.Z. Mortazavi, A. Iraji zad, F. Rahimi, Hydrogen sensing properties of multi-walled carbon nanotube films sputtered by Pd, Int. J. Hydrogen Energy 35 (2) (2010) 4445–4449.

[1d] R. Ghasempour, A. Iraji zad, Hybrid multiwalled carbon nanotubes and trioxide tungsten nanoparticles for hydrogen gas sensing, J. Phys. D 42 (2009) 165105–165112.

[1e] R. Ghasempour, A. Iraji zad, M.R. Hormozi Nezhad, Design and fabrication of sensitive carbon nanotubes/PMMA film for Acetone vapour detection, Int. J. Nanomanuf. 5 (3–4) (2010) 268–277.

[2] M. Foldvari, M. Bagonluri, Carbon nanotubes as functional excipients for nanomedicines: I. pharmaceutical properties, Nanomed. Nanotechnol. Biol. Med. 4 (2008) 173–182.

[3] A.A. Shvedova, E.R. Kisin, D. Porter, P. Schulte, V.E. Kagan, B. Fadeel, et al. Mechanisms of pulmonary toxicity and medical applications of carbon nanotubes: two faces of Janus?, Pharmacol. Ther. 121 (2009) 192–204.

[3a] O. Akhavan, E. Ghaderi, H. Emamy, Nontoxic concentrations of PEGylated graphene nanoribbons for selective cancer cell imaging and photothermal therapy, J. Mater. Chem. 22 (2012) 20626–20633.

[3b] O. Akhavan, E. Ghaderi, S. Aghayee, Y. Fereydooni, A. Talebi, The use of a glucose-reduced graphene oxide suspension for photothermal cancer therapy, J. Mater. Chem. 22 (2012) 13773–13781.

[4] P. Cherukuri, S.M. Bachilo, S.H. Litovsky, R.B. Weisman, Near-infrared fluorescence microscopy of single-walled carbon nanotubes in phagocytic cells, J. Am. Chem. Soc. 126 (2004) 15638–15639.

[5] M.J. O'Connell, S.M. Bachilo, C.B. Huffman, V.C. Moore, M.S. Strano, E.H. Haroz, et al. Band gap fluorescence from individual single-walled carbon nanotubes, Science 297 (2002) 593–596.

[6] J. Kayat, V. Gajbhiye, R.K. Tekade, N.K. Jain, Pulmonary toxicity of carbon nanotubes: a systematic report, Nanomed. Nanotechnol. Biol. Med. 7 (2011) 40–49.

[7] C.-W. Lam, J.T. James, R. McCluskey, S. Arepalli, R.L. Hunter, A review of carbon nanotube toxicity and assessment of potential occupational and environmental health risks, Crit. Rev. Toxicol. 36 (2006) 189–217.

[8] K. Donaldson, C.A. Poland, F.A. Murphy, M. MacFarlane, T. Chernova, A. Schinwald, Pulmonary toxic ity of carbon nanotubes and asbestos—similarities and differences, Adv. Drug. Deliv. Rev. 65 (2013) 2078–2086.

[9] Y. Rodriguez-Yanez, B. Munoz, A. Albores, Mechanisms of toxicity by carbon nanotubes, Toxicol. Mech. Methods 23 (2013) 178–195.

[10] J. Dong, Q. Ma, Advances in mechanisms and signaling pathways of carbon nanotube toxicity, Nanotoxicology 9 (2015) 658–676.

[11] F. Mouchet, P. Landois, P. Puech, E. Pinelli, E. Flahaut, L. Gauthier, Carbon nanotube ecotoxicity in amphibians: assessment of multiwalled carbon nanotubes and comparison with double-walled carbon nanotubes, Nanomedicine 5 (2010) 963–974.

[12] F. Mouchet, P. Landois, E. Sarremejean, G. Bernard, P. Puech, E. Pinelli, et al. Characterisation and in vivo ecotoxicity evaluation of double-wall carbon nanotubes in larvae of the amphibian *Xenopus laevis*, Aquat. Toxicol. 87 (2008) 127–137.

[13] R.P.F. Schins, C. Albrecht, K. Gerloff, D. van Berlo, Genotoxicity of carbon nanotubes, in: K. Donaldson, C. Poland, R. Duffin, J. Bonner (Eds.), The Toxicology of Carbon Nanotubues, Cambridge University Press, Cambridge, 2012.

[14] A. Buckley, R. Smith, R. Maynard, The aerodynamic behaviour and pulmonary deposition of carbon nanotubes, in: K. Donaldson, C. Poland, R. Duffin, J. Bonner (Eds.), The Toxicology of Carbon Nanotubes, Cambridge University Press, Cambridge, 2012, pp. 38–62.

[15] J. Muller, F. Huaux, D. Lison, Respiratory toxicity of carbon nanotubes: how worried should we be?, Carbon 44 (2006) 1048–1056.

[16] J.-M. Lopez-Cuesta, C. Longuet, C. Chivas-Joly, Thermal degradation, flammability, and potential toxicity of polymer nanocomposites, in: J. Njuguna, K. Pielichowski, H. Zhu (Eds.), Health and Environmental Safety of Nanomaterials, Woodhead Publishing, 2014, pp. 278–310.

[17] D.B. Warheit, M.P. DeLorme, CNT biopersistence and the fibre paradigm, in: K. Donaldson, C. Poland, R. Duffin, J. Bonner (Eds.), The Toxicology of Carbon Nanotubes, Cambridge University Press, Cambridge, 2012.

[18] R. Medzhitov, Origin and physiological roles of inflammation, Nature 454 (2008) 428–435.

[19] K. Bhattacharya, F.T. Andón, R. El-Sayed, B. Fadeel, Mechanisms of carbon nanotube-induced toxicity: focus on pulmonary inflammation, Adv. Drug. Deliv. Rev. 65 (2013) 2087–2097.

[20] C.-W. Lam, J.T. James, R. McCluskey, R.L. Hunter, Pulmonary toxicity of sing-wall carbon nanotubes in mice 7 and 90 days after intracheal instillation, Toxicol. Sci. 77 (2004) 126–134.

[21] L. Ma-Hock, S. Treumann, V. Strauss, S. Brill, F. Luizi, M. Mertler, et al. Inhalation toxicity of multiwall carbon nanotubes in rats exposed for 3 months, Toxicol. Sci. 112 (2009) 468–481.

[22] X. He, S.H. Young, D. Schwegler-Berry, W.P. Chisholm, J.E. Fernback, Q. Ma, Multiwalled carbon nanotubes induce a fibrogenic response by stimulating reactive oxygen species production, activating NF-κB signaling, and promoting fibroblast-to-myofibroblast transformation, Chem. Res. Toxicol. 24 (2011) 2237–2248.

[23] X. He, S.-H. Young, J.E. Fernback, Q. Ma, Single-walled carbon nanotubes induce fibrogenic effect by disturbing mitochondrial oxidative stress and activating NF-κB signaling, J. Clin. Toxicol. s5 (2012) 395–401.

[24] W.A.H. Wallace, P.M. Fitch, A.J. Simpson, S.E.M. Howie, Inflammation-associated remodelling and fibrosis in the lung—a process and an end point, Int. J. Exp. Pathol. 88 (2) (2007) 103–110.

[25] A.A. Shvedova, E.R. Kisin, R. Mercer, A.R. Murray, V.J. Johnson, A.I. Potapovich, et al. Unusual inflammatory and fibrogenic pulmonary responses to single-walled carbon nanotubes in mice, Am. J. Physiol. Lung Cell. Mol. Physiol. 289 (2005) L698–708.

[26] J.P. Ryman-Rasmussen, M.F. Cesta, A.R. Brody, J.K. Shipley-Phillips, J.I. Everitt, E.W. Tewksbury, et al. Inhaled carbon nanotubes reach the subpleural tissue in mice, Nat. Nanotechnol. 4 (2009) 747–751.

[27] J.C. Bonner, Nanoparticles as a potential cause of pleural and interstitial lung disease, Proc. Am. Thorac. Soc. 7 (2010) 138–141.

[28] J.P. Ryman-Rasmussen, E.W. Tewksbury, O.R. Moss, M.F. Cesta, B.A. Wong, J.C. Bonner, Inhaled multiwalled carbon nanotubes potentiate airway fibrosis in murine allergic asthma, Am. J. Respir. Cell. Mol. Biol. 40 (2009) 349–358.

[29] J.P. Ryman-Rasmussen, M.E. Andersen, J.C. Bonner, Fate and effects of carbon nanotubes following inhalation, in: K. Donaldson, C. Poland, R. Duffin, J. Bonner (Eds.), The Toxicology of Carbon Nanotubes, Cambridge University Press, Cambridge, 2012.

[30] M.L. Di Giorgio, S. Bucchianico, A.M. Di Ragnelli, P. Aimola, S. Santucci, A. Poma, Effects of single and multi walled carbon nanotubes on macrophages: cyto and genotoxicity and electron microscopy, Mutat. Res., Genet. Toxicol. Environ. Mutagen. 722 (2011) 20–31.

[31] K. Yamashita, Y. Yoshioka, K. Higashisaka, Y. Morishita, T. Yoshida, M. Fujimura, et al. Carbon nanotubes elicit DNA damage and inflammatory response relative to their size and shape, Inflammation 33 (2010) 276–280.

[32] N.R. Jacobsen, G. Pojana, P. White, P. Møller, C.A. Cohn, K. Smith Korsholm, et al. Genotoxicity, cytotoxicity, and reactive oxygen species induced by single-walled carbon nanotubes and C 60 fullerenes in the FE1-Muta™Mouse lung epithelial cells, Environ. Mol. Mutagen. 49 (2008) 476–487.

[33] A. Patlolla, B. Knighten, P. Tchounwou, Multi-walled carbon nanotubes induce cytotoxicity, genotoxicity and apoptosis in normal human dermal fibroblast cells, Ethn. Dis. 20 (2010) 1–17.

[34] H. Yang, C. Liu, D. Yang, H. Zhang, Z. Xi, Comparative study of cytotoxicity, oxidative stress and genotoxicity induced by four typical nanomaterials: the role of particle size, shape and composition, J. Appl. Toxicol. 29 (2009) 69–78.

[34a] O. Akhavan, E. Ghaderi, H. Emamy, F. Akhavan, Genotoxicity of graphene nanoribbons in human mesenchymal stem cells, Carbon 54 (2013) 419–431.

[35] M. Pacurari, X.J. Yin, J. Zhao, M. Ding, S.S. Leonard, D. Schwegler-Berry, et al. Raw single-wall carbon nanotubes induce oxidative stress and activate MAPKs, AP-1, NF-κB, and Akt in normal and malignant human mesothelial cells, Environ. Health Perspect. 116 (2008) 1211–1217.

[35a] O. Akhavan, E. Ghaderi, A. Akhavan, Size-dependent genotoxicity of graphene nanoplatelets in human stem cells, Biomaterials 33 (2012) 8017–8025.

[36] Y.-Y. Guo, J. Zhang, Y.-F. Zheng, J. Yang, X.-Q. Zhu, Cytotoxic and genotoxic effects of multi-wall carbon nanotubes on human umbilical vein endothelial cells in vitro, Mutat. Res. 721 (2011) 184–191.

[37] L. Zhu, D.W. Chang, L. Dai, Y. Hong, DNA damage induced by multiwalled carbon nanotubes in mouse embryonic stem cells, Nano. Lett. 12 (2007) 3592–3597.

[38] J. Cveticanin, G. Joksic, A. Leskovac, S. Petrovic, A.V. Sobot, O. Neskovic, Using carbon nanotubes to induce micronuclei and double strand breaks of the DNA in human cells, Nanotechnology 21 (2010) 15102.

[39] L. Migliore, D. Saracino, A. Bonelli, R. Colognato, M.R. D'Errico, A. Magrini, et al. Carbon nanotubes induce oxidative DNA damage in RAW 264.7 cells, Environ. Mol. Mutagen. 303 (2010) 294–303.

[40] H.K. Lindberg, G.C.M. Falck, S. Suhonen, M. Vippola, E. Vanhala, J. Catalán, et al. Genotoxicity of nanomaterials: DNA damage and micronuclei induced by carbon nanotubes and graphite nanofibres in human bronchial epithelial cells in vitro, Toxicol. Lett. 186 (2009) 166–173.

[41] J. Muller, A. Fonseca, J.B. Nagy, N. Moreau, M. Delos, E. Raymundo-pin, et al. Structural defects play a major role in the acute lung toxicity of multiwall carbon nanotubes: toxicological aspects, Chem. Res. Toxicol. 21 (9) (2008) 1698–1705.

[42] U. Wirnitzer, B. Herbold, M. Voetz, J. Ragot, Studies on the in vitro genotoxicity of baytubes, agglomerates of engineered multi-walled carbon-nanotubes (MWCNT), Toxicol. Lett. 186 (2009) 160–165.

[43] M. Asakura, T. Sasaki, T. Sugiyama, M. Takaya, S. Koda, K. Nagano, et al. Genotoxicity and cytotoxicity of multi-wall carbon nanotubes in cultured Chinese hamster lung cells in comparison with chrysotile a fibers, J. Occup. Health 52 (2010) 155–166.

[44] A.K. Jain, N. Kumar Mehra, N. Lodhi, V. Dubey, D.K. Mishra, P.K. Jain, et al. Carbon nanotubes and their toxicity, Nanotoxicology 1 (2007) 167–197.

[45] J. Muller, I. Decordier, P.H. Hoet, N. Lombaert, L. Thomassen, F. Huaux, et al. Clastogenic and aneugenic effects of multi-wall carbon nanotubes in epithelial cells, Carcinogenesis 29 (2008) 427–433.

[46] N.R. Jacobsen, P. Møller, K.A. Jensen, U. Vogel, O. Ladefoged, S. Loft, et al. Lung inflammation and genotoxicity following pulmonary exposure to nanoparticles in ApoE−/− mice, Part. Fibre Toxicol. 6 (2009) 2.

[47] A.A. Shvedova, E. Kisin, A.R. Murray, V.J. Johnson, O. Gorelik, S. Arepalli, et al. Inhalation vs. aspiration of single-walled carbon nanotubes in C57BL/6 mice: inflammation, fibrosis, oxidative stress, and mutagenesis, AJP Lung Cell. Mol. Physiol. 295 (2008) L552–L565.

[48] A.K. Patlolla, S.M. Hussain, J.J. Schlager, S. Patlolla, P.B. Tchounwou, Comparative study of the clastogenicity of functionalized and nonfunctionalized multiwalled carbon nanotubes in bone marrow cells of Swiss–Webster mice, Environ. Toxicol. 25 (2010) 608–621.

[49] J.K. Folkmann, L. Risom, N.R. Jacobsen, H. Wallin, S. Loft, P. Møller, Oxidatively damaged DNA in rats exposed by oral gavage to C60 fullerenes and single-walled carbon nanotubes, Environ. Health Perspect. 117 (2009) 703–708.

[50] K. Szendi, C. Varga, Lack of genotoxicity of carbon nanotubes in a pilot study, Anticancer Res. 28 (2008) 349–352.

[51] S. Toyokuni, Genotoxicity and carcinogenicity risk of carbon nanotubes, Adv. Drug Deliv. Rev. 65 (2013) 2098–2110.

[52] K. Unfried, Experimental carcinogenicity of carbon nanotubes in the context of other fibres, in: K. Donaldson, C. Poland, R. Duffin, J. Bonner (Eds.), The Toxicology of Carbon Nanotubes, Cambridge University Press, Cambridge, 2012.

[53] A. Takagi, A. Hirose, T. Nishimura, N. Fukumori, A. Ogata, N. Ohashi, et al. Induction of mesothelioma in p53+/− mouse by intraperitoneal application of multi-wall carbon nanotube, J. Toxicol. Sci. 33 (2008) 105–116.

[54] A. Takagi, A. Hirose, M. Futakuchi, H. Tsuda, J. Kanno, Dose-dependent mesothelioma induction by intraperitoneal administration of multi-wall carbon nanotubes in p53 heterozygous mice, Cancer Sci. 103 (2012) 1440–1444.

[55] Y. Sakamoto, D. Nakae, N. Fukumori, K. Tayama, A. Maekawa, K. Imai, et al. Induction of mesothelioma by a single intrascrotal administration of multi-wall carbon nanotube in intact male Fischer 344 rats, J. Toxicol. Sci. 34 (2009) 65–76.

[56] H. Nagai, Y. Okazaki, S. Chew, N. Misawa, Y. Yamashita, S. Akatsuka, et al. Diameter and rigidity of multiwalled carbon nanotubes are critical factors in mesothelial injury and carcinogenesis, PNAS 108 (2011) 1330–1338.

[57] L. Jiang, S. Akatsuka, H. Nagai, S.H. Chew, H. Ohara, Y. Okazaki, et al. Iron overload signature in chrysotile-induced malignant mesothelioma, J. Pathol. 228 (2012) 366–377.

[58] C. Varga, K. Szendi, Carbon nanotubes induce granulomas but not mesotheliomas, In Vivo 24 (2010) 153–156.

[59] J. Muller, M. Delos, N. Panin, V. Rabolli, F. Huaux, D. Lison, Absence of carcinogenic response to multiwall carbon nanotubes in a 2-year bioassay in the peritoneal cavity of the rat, Toxicol. Sci. 110 (2009) 442–448.

[60] A.A. Shvedova, A.V. Tkach, E.R. Kisin, T. Khaliullin, S. Stanley, D.W. Gutkin, et al. Carbon nanotubes enhance metastatic growth of lung carcinoma via up-regulation of myeloid-derived suppressor cells, Small 9 (2013) 1691–1695.

[61] L.M. Sargent, D.W. Porter, L.M. Staska, A.F. Hubbs, D.T. Lowry, L. Battelli, et al. Promotion of lung adenocarcinoma following inhalation exposure to multi-walled carbon nanotubes, Part. Fibre Toxicol. 11 (2014) 3.

[62] G. Oberdörster, V. Castranova, B. Asgharian, P. Sayre, Inhalation exposure to carbon nanotubes (CNT) and carbon nanofibers (CNF): methodology and dosimetry, J. Toxicol. Environ. Health 18 (2015) 121–212.

[63] H.J. Johnston, G.R. Hutchison, F.M. Christensen, S. Peters, S. Hankin, K. Aschberger, et al. A critical review of the biological mechanisms underlying the in vivo and in vitro toxicity of carbon nanotubes: the contribution of physico-chemical characteristics, Nanotoxicology 4 (2010) 207–246.

[64] C.A. Poland, R. Duffin, I. Kinloch, A. Maynard, W.A.H. Wallace, A. Seaton, et al. Carbon nanotubes introduced into the abdominal cavity of mice show asbestos-like pathogenicity in a pilot study, Nat. Nano. 3 (2008) 423–428.

[65] F.A. Murphy, C.A. Poland, R. Duffin, K.T. Al-Jamal, H. Ali-Boucetta, A. Nunes, et al. Length-dependent retention of carbon nanotubes in the pleural space of mice initiates sustained inflammation and progressive fibrosis on the parietal pleura, Am. J. Pathol. 178 (2011) 2587–2600.

[66] K. Donaldson, F.A. Murphy, R. Duffin, C.A. Poland, Asbestos, carbon nanotubes and the pleural mesothelium: a review of the hypothesis regarding the role of long fibre retention in the parietal pleura, inflammation and mesothelioma, Part. Fibre Toxicol. 7 (2010) 5.

[67] I. Fenoglio, E. Aldieri, E. Gazzano, F. Cesano, M. Colonna, D. Scarano, et al. Thickness of multiwalled carbon nanotubes affects their lung toxicity, Chem. Res. Toxicol. 25 (2012) 74–82.

[68] X. Wang, G. Jia, H. Wang, H. Nie, L. Yan, X.Y. Deng, et al. Diameter effects on cytotoxicity of multi-walled carbon nanotubes, J. Nanosci. Nanotechnol. 9 (2009) 3025–3033.

[69] R.R. Mercer, J.F. Scabilloni, A.F. Hubbs, L.A. Battelli, W. McKinney, S. Friend, et al. Distribution and fibrotic response following inhalation exposure to multi-walled carbon nanotubes, Part. Fibre Toxicol. 10 (2013) 33.

[70] R.R. Mercer, A.F. Hubbs, J.F. Scabilloni, L. Wang, L.A. Battelli, S. Friend, et al. Pulmonary fibrotic response to aspiration of multi-walled carbon nanotubes, Part. Fibre Toxicol. 8 (2011) 21.

[71] M. Bottini, S. Bruckner, K. Nika, N. Bottini, S. Bellucci, A. Magrini, et al. Multi-walled carbon nanotubes induce T lymphocyte apoptosis, Toxicol. Lett. 160 (2006) 121–126.

[72] J.C. Carrero-Sánchez, A.L. Elías, R. Mancilla, G. Arrellín, H. Terrones, J.P. Laclette, et al. Biocompatibility and toxicological studies of carbon nanotubes doped with nitrogen, Nano. Lett. 6 (2006) 1609–1616.

[73] C.L. Ursini, D. Cavallo, A.M. Fresegna, A. Ciervo, R. Maiello, G. Buresti, et al. Differences in cytotoxic, genotoxic, and inflammatory response of bronchial and alveolar human lung epithelial cells to pristine and COOH-functionalized multiwalled carbon nanotubes, Biomed. Res. Int. (2014) 2014.

[74] A.J. Taylor, C.D. McClure, K.A. Shipkowski, E.A. Thompson, S. Hussain, S. Garantziotis, et al. Atomic layer deposition coating of carbon nanotubes with aluminum oxide alters pro-fibrogenic cytokine expression by human mononuclear phagocytes in vitro and reduces lung fibrosis in mice in vivo, PLoS One (2014) 9.

[75] P.M.V. Raja, J. Connolley, G.P. Ganesan, L. Ci, P.M. Ajayan, O. Nalamasu, et al. Impact of carbon nanotube exposure, dosage and aggregation on smooth muscle cells, Toxicol. Lett. 169 (2007) 51–63.

[76] A.A. Shvedova, V. Castranova, E.R. Kisin, D. Schwegler-Berry, A.R. Murray, V.Z. Gandelsman, et al. Exposure to carbon nanotube material: assessment of nanotube cytotoxicity using human keratinocyte cells, J. Toxicol. Environ. Health 66 (2003) 1909–1926.

[77] V.E. Kagan, Y.Y. Tyurina, V.A. Tyurin, N.V. Konduru, A.I. Potapovich, A.N. Osipov, et al. Direct and indirect effects of single walled carbon nanotubes on RAW 264.7 macrophages: role of iron, Toxicol. Lett. 165 (2006) 88–100.

[78] J. Du, S. Wang, H. You, X. Zhao, Understanding the toxicity of carbon nanotubes in the environment is crucial to the control of nanomaterials in producing and processing and the assessment of health risk for human: a review, Environ. Toxicol. Pharmacol. 36 (2013) 451–462.

[79] Q. Ma, Transcriptional responses to oxidative stress: pathological and toxicological implications, Pharmacol. Ther. 125 (2010) 376–393.

[80] V. Costa, P. Moradas-Ferreira, Oxidative stress and signal transduction in Saccharomyces cerevisiae: insights into ageing, apoptosis and diseases, Mol. Aspects Med. 22 (2001) 217–246.

[81] B. Rothen-Rutishauser, D.M. Brown, M. Piallier-Boyles, I.a. Kinloch, A.H. Windle, P. Gehr, et al. Relating the physicochemical characteristics and dispersion of multiwalled carbon nanotubes in different suspension media to their oxidative reactivity in vitro and inflammation in vivo, Nanotoxicology 4 (2010) 331–342.

[82] S.G. Han, R. Andrews, C.G. Gairola, Acute pulmonary response of mice to multi-wall carbon nanotubes, Inhal. Toxicol. 22 (2010) 340–347.

[83] T. Thurnherr, C. Brandenberger, K. Fischer, L. Diener, P. Manser, X. Maeder-Althaus, et al. A comparison of acute and long-term effects of industrial multiwalled carbon nanotubes on human lung and immune cells in vitro, Toxicol. Lett. 200 (2011) 176–186.

[84] A.A. Shvedova, E.R. Kisin, A.R. Murray, O. Gorelik, S. Arepalli, V. Castranova, et al. Vitamin E deficiency enhances pulmonary inflammatory response and oxidative stress induced by single-walled carbon nanotubes in C57BL/6 mice, Toxicol. Appl. Pharmacol. 221 (2007) 339–348.

[85] A.M. Knaapen, P.J.A. Borm, C. Albrecht, R.P.F. Schins, Inhaled particles and lung cancer. Part A: mechanisms, Int. J. Cancer 109 (2004) 799–809.

[86] P. Kovacic, R. Somanathan, Biomechanisms of nanoparticles (toxicants, antioxidants and therapeutics): electron transfer and reactive oxygen species, J. Nanosci. Nanotechnol. 10 (2010) 7919–7930.

[87] D.M. Brown, I.A. Kinloch, U. Bangert, A.H. Windle, D.M. Walter, G.S. Walker, et al. An in vitro study of the potential of carbon nanotubes and nanofibres to induce inflammatory mediators and frustrated phagocytosis, Carbon N Y 45 (2007) 1743–1756.

[87a] O. Akhavan, E. Ghaderi, Toxicity of graphene and graphene oxide nanowalls against bacteria, ACS Nano 4 (2010) 5731–5736.

[88] S. Hirano, S. Kanno, A. Furuyama, Multi-walled carbon nanotubes injure the plasma membrane of macrophages, Toxicol. Appl. Pharmacol. 232 (2008) 244–251.

[89] K. Kostarelos, L. Lacerda, G. Pastorin, W. Wu, S. Wieckowski, J. Luangsivilay, et al. Cellular uptake of functionalized carbon nanotubes is independent of functional group and cell type, Nat. Nanotechnol. 2 (2007) 108–113.

[90] L.M. Sargent, A.F. Hubbs, S.H. Young, M.L. Kashon, C.Z. Dinu, J.L. Salisbury, et al. Single-walled carbon nanotube-induced mitotic disruption, Mutat. Res.—Genet. Toxicol. Environ. Mutagen. 745 (2012) 28–37.

[91] L.M. Sargent, S.H. Reynolds, V. Castranova, Potential pulmonary effects of engineered carbon nanotubes: in vitro genotoxic effects, Nanotoxicology 4 (2010) 396–408.

[92] G. Vietti, S. Ibouraadaten, M. Palmai-Pallag, Y. Yakoub, C. Bailly, I. Fenoglio, et al. Towards predicting the lung fibrogenic activity of nanomaterials: experimental validation of an in vitro fibroblast proliferation assay, Part. Fibre Toxicol. 10 (2013) 52.

[93] J. Xu, M. Futakuchi, H. Shimizu, D.B. Alexander, K. Yanagihara, K. Fukamachi, et al. Multi-walled carbon nanotubes translocate into the pleural cavity and induce visceral mesothelial proliferation in rats, Cancer Sci. 103 (2012) 2045–2050.

[94] L. Wang, R.R. Mercer, Y. Rojanasakul, A. Qiu, Y. Lu, J.F. Scabilloni, et al. Direct fibrogenic effects of dispersed single-walled carbon nanotubes on human lung fibroblasts, J. Toxicol. Environ. Health 73 (2010) 410–422.

[95] X. Wang, T. Xia, S.A. Ntim, Z. Ji, S. George, H. Meng, et al. Quantitative techniques for assessing and controlling the dispersion and biological effects of multiwalled carbon nanotubes in mammalian tissue culture cells, ACS Nano. 4 (2010) 7241–7252.

[96] W.-W. Cheng, Z.-Q. Lin, B.-F. Wei, Q. Zeng, B. Han, C.-X. Wei, et al. Single-walled carbon nanotube induction of rat aortic endothelial cell apoptosis: reactive oxygen species are involved in the mitochondrial pathway, Int. J. Biochem. Cell Biol. 43 (2011) 564–572.

[97] T. Chen, H. Nie, X. Gao, J. Yang, J. Pu, Z. Chen, et al. Epithelial-mesenchymal transition involved in pulmonary fibrosis induced by multi-walled carbon nanotubes via TGF-beta/Smad signaling pathway, Toxicol. Lett. 226 (2014) 150–162.

[98] J. Dong, D.W. Porter, L.A. Batteli, M.G. Wolfarth, D.L. Richardson, Q. Ma, Pathologic and molecular profiling of rapid-onset fibrosis and inflammation induced by multi-walled carbon nanotubes, Arch. Toxicol. 89 (2015) 621–633.

[99] K. Pielichowski, S. Michalowski, Nanostructured flame retardants: performance, toxicity, and environmental impact, in: J. Njuguna, K. Pielichowski, H. Zhu (Eds.), Health and Environmental Safety of Nanomaterials, Woodhead Publishing, 2014, pp. 251–277.

[100] T. Galloway, C. Lewis, I. Dolciotti, B.D. Johnston, J. Moger, F. Regoli, Sublethal toxicity of nano-titanium dioxide and carbon nanotubes in a sediment dwelling marine polychaete, Environ. Pollut. 158 (2010) 1748–1755.

[101] M. Allegri, D.K. Perivoliotis, M.G. Bianchi, M. Chiu, A. Pagliaro, M.A. Koklioti, et al. Toxicity determinants of multi-walled carbon nanotubes: the relationship between functionalization and agglomeration, Toxicol. Rep. 3 (2016) 230–243.

[102] E.J. Petersen, Ecotoxicological effects of carbon nanotubes: test methods and current research, in: J. Njuguna, K. Pielichowski, H. Zhu (Eds.), Health and Environmental Safety of Nanomaterials, Woodhead Publishing, 2014, pp. 175–199.

[103] E.J. Petersen, R.A. Pinto, P.F. Landrum, W.J. Weber, Influence of carbon nanotubes on pyrene bioaccumulation from contaminated soils by earthworms, Environ. Sci. Technol. 43 (2009) 4181–4187.

[104] E.J. Petersen, Q. Huang, W.J. Weber, Relevance of octanol-water distribution measurements to the potential ecological uptake of multi-walled carbon nanotubes, Environ. Toxicol. Chem. 29 (2010) 1106–1112.

[105] E.J. Petersen, J. Akkanen, J.V.K. Kukkonen, W.J. Weber, Biological uptake and depuration of carbon nanotubes by *Daphnia magna*, Environ. Sci. Technol. 43 (2009) 2969–7295.

[106] P.L. Ferguson, G.T. Chandler, R.C. Templeton, A. DeMarco, W.A. Scrivens, B.A. Englehart, Influence of sediment-amendment with single-walled carbon nanotubes and diesel soot on bioaccumulation of hydrophobic organic contaminants by benthic invertebrates, Environ. Sci. Technol. 42 (2008) 3879–3885.

[107] E.J. Petersen, R.A. Pinto, L. Zhang, Q. Huang, P.F. Landrum, W.J. Weber, Effects of polyethyleneimine-mediated functionalization of multi-walled carbon nanotubes on earthworm bioaccumulation and sorption by soils, Environ. Sci. Technol. 45 (2011) 3718–3724.

[108] E.J. Petersen, Q. Huang, W.J. Weber, Bioaccumulation of radio-labeled carbon nanotubes by *Eisenia foetida*, Environ. Sci. Technol. 42 (2008) 3090–3095.

[109] E.J. Petersen, Q. Huang, W.J. Weber, Ecological uptake and depuration of carbon nanotubes by *Lumbriculus variegatus*, Environ. Health Perspect. 116 (2008) 496–500.

[110] A.J. Edgington, A.P. Roberts, L.M. Taylor, M.M. Alloy, J. Reppert, A.M. Rao, et al. The influence of natural organic matter on the toxicity of multiwalled carbon nanotubes, Environ. Toxicol. Chem. 29 (2010) 2511–2518.

[111] E.J. Petersen, R.A. Pinto, D.J. Mai, P.F. Landrum, W.J. Weber, Influence of polyethyleneimine graftings of multi-walled carbon nanotubes on their accumulation and elimination by and toxicity to *Daphnia magna*, Environ. Sci. Technol. 45 (2011) 1133–1138.

[112] E. Wild, K.C. Jones, Novel method for the direct visualization of in vivo nanomaterials and chemical interactions in plants, Environ. Sci. Technol. 43 (2009) 5290–5294.

[113] M.V. Khodakovskaya, K. de Silva, D.A. Nedosekin, E. Dervishi, A.S. Biris, E.V. Shashkov, et al. Complex genetic, photothermal, and photoacoustic analysis of nanoparticle-plant interactions, Proc. Natl. Acad. Sci. USA 108 (2011) 1028–1033.

[114] H. Chung, Y. Son, T.K. Yoon, S. Kim, W. Kim, The effect of multi-walled carbon nanotubes on soil microbial activity, Ecotoxicol. Environ. Saf. 74 (2011) 569–575.

8

Mechanical Properties of Isolated Carbon Nanotube

Ghanshyam Pal*, Shanmugam Kumar**

*SRM UNIVERSITY, KANCHIPURAM, TAMIL NADU, INDIA; **INSTITUTE CENTER FOR ENERGY (IENERGY), MASDAR INSTITUTE OF SCIENCE AND TECHNOLOGY, ABU DHABI, UNITED ARAB EMIRATES

CHAPTER OUTLINE

8.1 Introduction

Among all the allotropes of carbon, carbon nanotubes (CNTs) have gathered tremendous attention from engineers and scientific researchers in a variety of disciplines. Structurally, CNTs are composed of a single cylinder (single-walled carbon nanotubes, SWCNTs) or multiple concentric cylinders (multiwalled carbon nanotubes, MWCNT) of rolled-up graphene sheets ranging over a few nanometers in diameter and hundreds of microns in length. CNTs have (almost) perfect atomic structure comprising sp^2-hybridized carbon atoms periodically arranged in a benzene-like hexagonal lattice along the nanotube axis. The covalent bonding between these sp^2-hybridized carbon atoms provides CNTs with exceptionally high modulus (~1 TPa) and tensile strength (100–150 GPa) and other superior mechanical properties. In addition, the network of CNTs has been shown to possess high

Carbon Nanotube–Reinforced Polymers. http://dx.doi.org/10.1016/B978-0-323-48221-9.00008-X

electrical and thermal conductivities, promoting their multifunctional applications as structural health monitoring sensors [1–8], biomedical actuators [9–15], material for flexible and wearable sensors [16–22], and reinforcement in lightweight structural composites with improved electrical conductivity and electromagnetic shielding [23–28].

The CNT multifunctionality along with its exceptional elastic modulus, lightweight, and high aspect ratio are primary motivating factors for incorporating it as the reinforcement or filler for modern composite materials. Hence, tremendous research emphasis is being placed on structural and mechanical characterization of CNTs through experimental and computational methods. The elastic and inelastic responses of CNTs are the function of its diameter, graphene rolling direction (chirality), number of concentric walls, aspect ratio and presence of various structural defects, impurities, functional groups, and so forth. Generally, most of the nanosystems are too small for direct measurements, but too large to be described by current rigorous first principles calculation, the critical issue for nanotechnology is to develop the ability to understand, model, and simulate the behavior of these low-dimensional structures and to make the connection between their structural properties and functions [29]. The nanoscale diameter coupled with microscopic length and uniform morphology facilitates the modeling of mechanical response of CNTs using quantum/atomistic and continuum-mechanics-based methods. Due to longitudinal invariance of the atomic structure, fracture of CNTs, although influenced by the morphological factors, is a complex phenomenon.

The remaining chapter has been arranged as follows. Section 8.2 provides the details about structure of CNTs, including their classification based on the chirality. Following which, we discuss the elastic properties of SWCNTs and MWCNTs in Section 8.3. This section is divided into two main parts; Section 8.3.1 compares the Young's modulus of CNTs obtained through different experiments, and Section 8.3.2 reports the Young's modulus of the same determined through various simulation techniques ranging from quantum mechanics-based methods to continuum mechanics-based approaches. Proceeding further, we report the effect of different CNT parameters on its Young's modulus, where Section 8.3.2.1 discusses the effects of CNT chirality, Section 8.3.2.2 discusses the effects of CNT functionalization, Section 8.3.2.3 outlines the effects of vacancy and Stone–Wales defects, and Section 8.3.2.4 comments on the effects of number of CNT sleeves on its Young's modulus. After the linear elasticity, Section 8.4 provides details about large deformation elasticity observed in CNTs, including the radial elasticity (Section 8.4.1) and buckling of CNTs (Section 8.4.2). The tensile strength and fracture of CNTs are discussed in Section 8.5; Section 8.5.1 provides an insight about the effects of high temperature annealing on the defect density and tensile strength of CNTs. The chapter closes with a brief overview of mechanical properties of CNTs as provided in Section 8.6.

8.2 Structure of Carbon Nanotubes

It is reported that CNTs, the tubular carbon nanostructures, were first reported by L. V. Radushkevich and V. M. Lukyanovich in 1952 using transmission electron microscopy (TEM) [30]. In 1991, while studying the deposits on graphitic negative-electrode in arc-discharge evaporation process, Iijima reported similar concentric nanotube structures

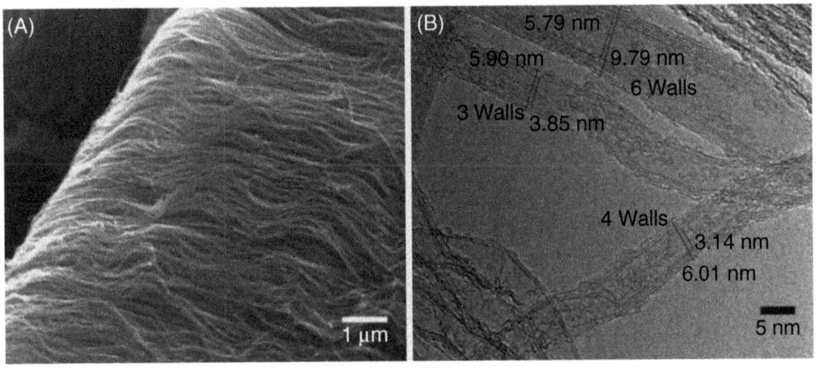

FIGURE 8.1 (A) Scanning electron microscope (SEM) image of aligned multiwalled carbon nanotubes (MWCNTs) [33]. (B) High-resolution TEM (HRTEM) image of MWCNTs showing the interior concentric walls [33].

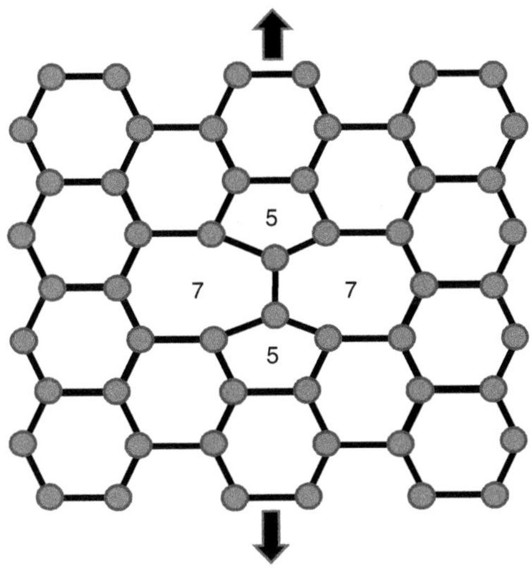

FIGURE 8.2 Atomic structures of graphene sheets and chiral vectors for carbon nanotubes (CNTs).

with sp^2-hybridized carbon with fullerene-like end caps [31,32]. In practice, CNTs are fabricated using chemical vapor deposition (CVD), arc-discharge (AD) evaporation, and thermal plasma (TP) methods. The nanocylinder made of a rolled-up single graphene sheet is called a single-walled carbon nanotube (SWCNT) and the concentric cylinders comprising multiple graphene sheets are called multiwalled carbon nanotubes (MWCNTs) (Fig. 8.1). Depending upon the rolling direction in the plane of a graphene sheet, CNTs are classified into three unique geometries: (1) armchair, (2) zigzag, and (3) chiral (Fig. 8.2). Armchair and zigzag CNTs form the two ends of the CNT microstructure while the chiral CNT structure lies in between these two extremes. The geometric differentiation is formally defined in terms of chiral vectors, a_1 and a_2 (Fig. 8.2). The chiral vector in the rolled-up configuration forms the top/bottom end of the CNT and is given as follows:

$$v_c = m.a_1 + n.a_2 \tag{8.1}$$

For zigzag CNT, n = integer, m = 0; for example, (n, m) = (10, 0).
For armchair CNTs, $n = m$ = integer; for example, (n, m) = (8, 8).
For chiral CNTs, n, m = integer, nm; for example, (n, m) = (7, 9).
The circumference of a SWCNT, Φ, and its radius, R_{CNT}, are given as follows.

$$\Phi = (a_{C-C}\sqrt{3})\sqrt{n^2 + m^2 + mn} \tag{8.2}$$

$$R_{cnt} = \Phi / 2\pi \tag{8.3}$$

where a_{c-c} is the bond length for C—C bond in CNT ($\approx 1.41 - 1.44$Å).

For large diameter CNT ($1/R_{cnt} \ll 1$), the electrical nature depends upon the chiral vector. When $n = m$, the CNT is metallic in nature, and if $(n-m)$ is a multiple of 3, the CNT is semiconducting with a very small band gap; otherwise, the CNTs are moderately semiconducting.

8.3 Elastic Properties of CNTs

Pristine CNTs have well-defined atomic positions and a stable chemical structure (bond strength, bond length, bond angles, etc.), which depends upon the rolling direction of the graphene sheet, or chirality. In practice, the hexagonal structure of CNTs contains imperfections, such as chemical impurities, due to the presence of catalyst atoms, atomic level vacancies (single or multiple atoms), and Stone–Wales defects that affect the elastic and inelastic response of the CNTs.

8.3.1 Experimentally Determined Young's Modulus of CNTs

Elastic modulus, or Young's modulus, is a primary (intensive) material property required for mechanical and structural design. For continuum level, the elastic or Young's modulus is defined as the following:

$$Y = \frac{1}{V}\frac{\partial^2 \varepsilon}{\partial \epsilon^2} \tag{8.4}$$

where Y is the Young's modulus, $\in$ is the direct uniform strain, and ε is the total strain energy. A number of experimental methods have been applied to evaluate the Young's modulus of CNTs. Due to the nanoscale size of CNTs, the difficulties associated with their handling in experiments give rise to uncertainty in the measurements. Most of the experimental approaches to measure the elastic properties of CNTs involve the bending or vibrational response of the cantilever or fixed beam configuration of isolated CNTs. Treacy et al. [34] attached the bundles of carbon nanotubes to a 3-mm nickel ring and studied the vibrating free end of the CNTs using transmission electron microscope (TEM) at 300 and 600 K. The

CNTs on the ring are treated as cantilever beams vibrating due to thermal loading. The vibration energy of the CNT is given as following:

$$W_n = \frac{1}{2} c_n u_n^2 \tag{8.5}$$

where u_n is the vibration amplitude of CNT in nth mode. The term, c_n, defined as effective spring constant, is given as follows:

$$c_n = \frac{\pi \beta_n^4 Y(a^4 b^4)}{16L^3} \tag{8.6}$$

where β_n represent a constant related to the mode of vibration, Y is the Young's modulus, a, b, and L are the inner diameter, outer diameter, and the length of CNTs. The TEM measurements plot between mean-square vibrational amplitude and temperature shows that the effects of other nonequilibrium perturbation was negligible. The average reported Young's modulus of CNTs, as reported in this study, is 1.8 TPa with a wide variation between 0.4 and 4.1 TPa. Krishnan et al. used similar approach involving stochastically driven oscillator and reported the average Young's modulus equal to 1.3–0.4/+0.6 TPa. Demczyk et al. deposited MWCNTs on a gold-plated tensile test fixture in the fixed-fixed boundary condition and stretched in TEM to calculate the MWCNT Young's modulus [35]. An atomic force microscope tip of known spring constant and a piezoelectric drive were used to apply the force on MWCNTs deposited on the tensile test device. Fig. 8.3 shows the tensile test device and the MWCNTs spanning the device.

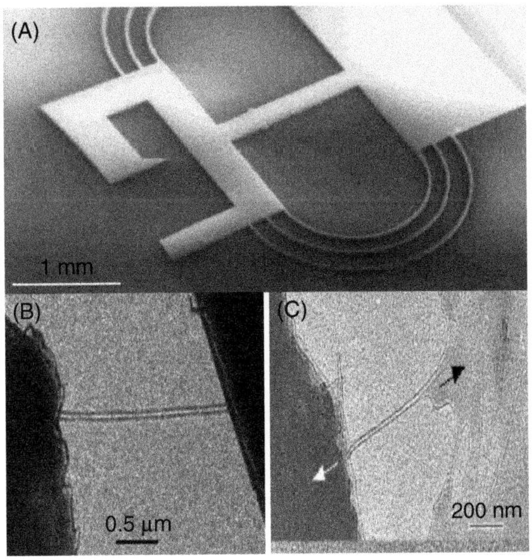

FIGURE 8.3 (A) Microscale tensile test device used in transmission electron microscopy (TEM) for MWCNT Young's modulus measurement [35]; (B) fixed-fixed MWCNT deposited on the tensile (stretching) test device [35]; and (C) application of stretching load on the MWCNT. The *arrows* show the direction of the loading [35].

Another experimental approach to determine the Young's modulus of CNTs involves the application of atomic force microscope (AFM). Citing the restriction related to size of samples and amplitude of thermally induced CNT vibration and limits on achievable mechanical parameters, Wong et al. used AFM to calculate the Young's modulus and tensile strength of CNTs [36]. In their study, they plotted the bending force-displacement response of cantilever (rod-like) CNTs using AFM tip of known spring constant. The recorded force-displacement data was used to calculate the Young's modulus of CNTs using the beam theory as following [36].

$$P(x,y)=3YI\frac{y}{x^3}+\frac{f}{8}(x-L-6\frac{L^2}{x}) \tag{8.7}$$

where P is the applied load, Y is the CNT Young's modulus, I is the moment of inertia, L is the beam length, f is the tip friction force (negligible), and $k(x) = dP/dy$ is the nano beam spring constant. A similar approach using MWCNT deposited on well-polished alumina ultrafiltration membrane was used by Salvetat et al. to determine the Young's modulus and other mechanical properties of MWCNTs [37–39]. Table 8.1 contains the records of experimentally determined values of CNT Young's modulus.

8.3.2 Computationally Evaluated Young's Modulus of CNTs

Several computational approaches, ranging from highly sophisticated quantum mechanics–based density functional theory (DFT) and tight binding (TB) methods, molecular dynamics to molecular-mechanics (MM) supported quasicontinuum and equivalent continuum methods, have been applied to study the mechanical behavior of CNTs. Due to its well-defined and translational structural invariance along the length, the structural equivalency between SWCNTs and engineering structures, where C—C chemical bonds are replaced by truss or beam elements, has enabled researchers to apply the aforementioned

Table 8.1 Experimentally Determined Young's Modulus of SWCNTs and MWCNTs

Types of CNTs	Experimental Approachs	Young's Modulus (TPa)	References	Years
MWCNT	TEM	1.8	Treacy et al. [34]	1996
MWCNT	AFM	1.28	Wong et al. [36]	1997
MWCNT	SEM	1.0	Muster et al. [40]	1998
SWCNT	TEM	1.25	Krishnan et al. [41]	1998
MWCNT	AFM	0.81	Selvetat et al. [37,38]	1999
SWCNT	AFM	1.002	Yu et al. [42]	2000
MWCNT	AFM	1.2	Tombler et al. [43]	2000
MWCNT	TEM	1.25	Demczyk et al. [35]	2002
MWCNT	TEM	0.35 ± 0.11	Guhados et al. [44]	2007
MWCNT	TEM	0.62–1.2	Ding et al. [45]	2007
SWCNT	EMA	0.97	Wu et al. [46]	2008

AFM, Atomic force microscopy; CNTs, carbon nanotubes; EMA, electro-magnetic actuation; MWCNT, multiwalled carbon nanotubes; SEM, scanning electron microscopy; SWCNT, single-walled carbon nanotube; TEM, transmission electron microscopy.

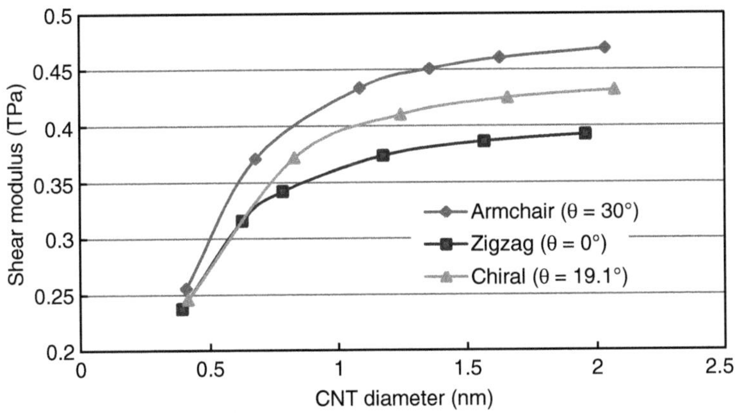

FIGURE 8.4 Variation of shear modulus of single-walled carbon nanotube (SWC NT) with respect to nanotube chirality and diameter [47].

continuum level approaches with acceptable level of success in studying the mechanical properties of CNTs. Some of the early attempts of characterizing Young's modulus and strength of CNTs were carried out utilizing the in-plane properties of graphene sheet. These studies have reported that the mechanical properties of CNTs are function of the physical properties, such as the nanotube diameter, chirality, number of concentric nanotubes in MWCNTs, presence of the defects, and adsorption of other chemical moieties (functionalization).

Lu et al. developed an MM–based finite element (FE) model to study the elastic properties of SWCNTs. Modified Morse potential is utilized to establish the physical parameters of the beam elements representing the covalent C—C bonds in SWCNTs. Their results showed a plateaued variation of shear modulus with the SWCNT diameter (Fig. 8.4). However, it is also reported that the armchair SWCNTs possess 15%–20% higher shear modulus than zigzag nanotube with comparable diameter [47]. Several authors used a similar FE-based modeling approach to compare the elastic properties of SWCNTs and MWCNTs [48–50]. According to their results, armchair SWCNTs have higher Young's modulus than the zigzag SWCNTs of equivalent diameter. The zigzag SWCNTs are relatively more brittle (lower strain to failure) than their armchair SWCNT counterparts [49]. Generally, the results of these CNT modeling studies have been verified and compared with the experimental observations. Table 8.2 consists of the summary of computationally obtained Young's modulus of CNTs calculated using different physics-based and numerical approaches.

8.3.2.1 Effect of CNT Chirality

Sánchez-Portal compared the mechanical and vibrational properties of SWCNTs and MWCNTs using quantum mechanics/quasicontinuum hybrid multiscale approach [77]. The average Young's modulus of MWCNTs is reported to be equal to 1.28 TPa with a wide variation of 0.59 TPa [77]. According to their results, zigzag SWCNTs have higher Poisson's ratio than the corresponding armchair SWCNTs [77]. In another study, Chang applied

Table 8.2 Computationally Determined Young's Modulus of SWCNTs

Types of CNTs	Modeling Approachs	Young's Modulus (TPa)	References	Years
SWCNT	Empirical force model	0.97–1.11	Lu [51]	1998
SWCNT	Car-Parrinello MD and DFT	1.08	Zhang et al. [52]	1998
SWCNT	Tight binding method	1.22	Hernandez et al. [53]	1998
SWCNT	Molecular dynamics	0.97–1.12	Yao and Lordi [54]	1998
SWCNT	Molecular dynamics model	1.24–1.36	Jin and Yuan [55]	2003
SWCNT	Equivalent continuum frame model	1.05	Li [56]	2003
SWCNT	Quantum mechanical model	0.82	Troya et al. [57]	2003
SWCNT	Equivalent continuum plate model	0.6–0.8	Wang [58]	2004
SWCNT	Tight binding method	0.95	Cai et al. [59]	2004
SWCNT	Quasi-continuum method	0.7	Arroyo and Belytschko [60]	2004
SWCNT	Atomistic continuum model	1.08	Zhang et al. [61]	2005
SWCNT	Nonlinear and rotation springs	0.91	Meo and Rossi [62]	2006
SWCNT	MM/continuum thin cylinder model	1.0–1.2	Wu et al. [63]	2006
SWCNT	Quasi-continuum model	0.7	Guo et al. [64]	2006
SWCNT	Molecular mechanics supported finite element model	0.97–1.08	Kalamkarov et al. [65]	2006
SWCNT	Equivalent continuum Truss model	2.52	Selmi et al. [66]	2007
SWCNT	Equivalent continuum shell model	0.83	Hu et al. [67]	2008
SWCNT	Vibrational frequency method	0.96	Wernik and Meguid [68]	2008
SWCNT	Atomistic continuum modeling	1.2–1.4	Cheng et al. [69]	2009
SWCNT	Molecular mechanics supported finite element model	0.91–0.92	Rossi and Meo [70]	2009
SWCNT	Pseudo potential DFT	1.06–1.16	Cai et al. [71]	2009
SWCNT	Nonlocal rod model	1.08	Chandraseker et al. [72]	2009
SWCNT	Nanoscale MM/continuum model	1.04	Shokrieh and Rafiee [73]	2010
SWCNT	Finite element beam model	0.95	Wernik and Meguid [68]	2010
SWCNT	Finite element spring model	1.3	Rafiee and Heidarhaei [74]	2012
SWCNT	Finite element spring model	0.99–1.06	Lu and Hu [47]	2012
SWCNT	Nonlocal rod model	0.93	Malagu et al. [75]	2015
SWCNT	3D-finite element model	1.08	Sakharova et al. [76]	2015

CNTs, Carbon nanotubes; DFT, density functional theory; MD, molecular dynamics; MM, molecular-mechanics; SWCNT, single-walled carbon nanotube.

MM–based anisotropic shell model to evaluate the mechanical response of SWCNTs and also reported the similar variation of Poisson's ratio with respect to nanotube diameter (Fig. 8.5) [78]. A similar trend with respect to Poisson's ratio is observed in the case of oxygenated-SWCNT [79].

However, they reported a reverse trend in the case of Young's modulus for pristine and oxygenated-CNTs (Fig. 8.6) [74,75,79–81].

In a tight-binding study by Hernández et al., the Young's modulus of SWCNTs of different radius and chirality varies in a rather narrow range as 1.24 ± 0.02 TPa [82]. The chirality of SWCNT also affects the monomer-SWCNT reactivity. According to DFT study by Ahangari et al., the physiosorption of epoxy molecule over zigzag SWCNT is favorable over the same with armchair SWCNT [83]. Bogar et al. studies the effect of Hookean axial tensile

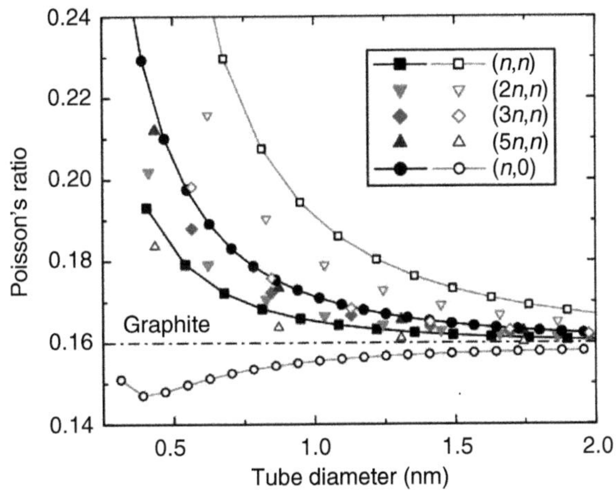

FIGURE 8.5 Variation of Poisson's ratio of SWCNT with respect to nanotube chirality and diameter [78].

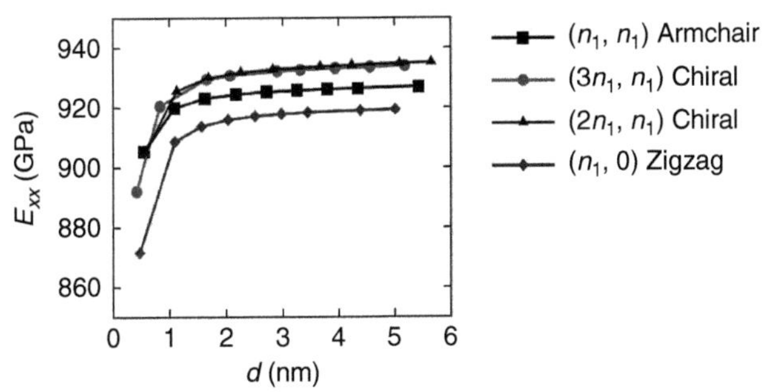

FIGURE 8.6 Young's modulus of pristine SWCNT as a function of diameter and chirality [75].

stress on the electronic band gap structure and reported the semiconductor to metal transition at small (<5%) deformation [84].

8.3.2.2 Effects of CNT Functionalization

Covalent binding of chemical groups on the surface of a CNT is defined as the functionalization of CNTs. Functionalization of CNTs facilitates the dispersion of the CNT in the matrix and improves the binding of polymer chains with CNTs and changes the circular cross-section to an elliptical section. Milowska et al. studied the effects of functionalization of zigzag CNTs with hydroxyl, amine, and carboxyl groups [85,86]. While comparing the effects of the functionalization on the pristine and defective SWCNTs, Milkowska et al. reported the functionalization of CNTs above a threshold concentration might reduce the Young's modulus of defective CNTs (Fig. 8.7) [86]. In general, the functionalization of a CNT reduces its Young's, bulk and shear moduli [86]. In a similar MD study involving hydrogen, hydroxyl, carboxyl, and primary amine as the functionalization groups, Shah et al. reported a 30%–34% drop in SWCNT Young's modulus and 24%–42% drop in shear

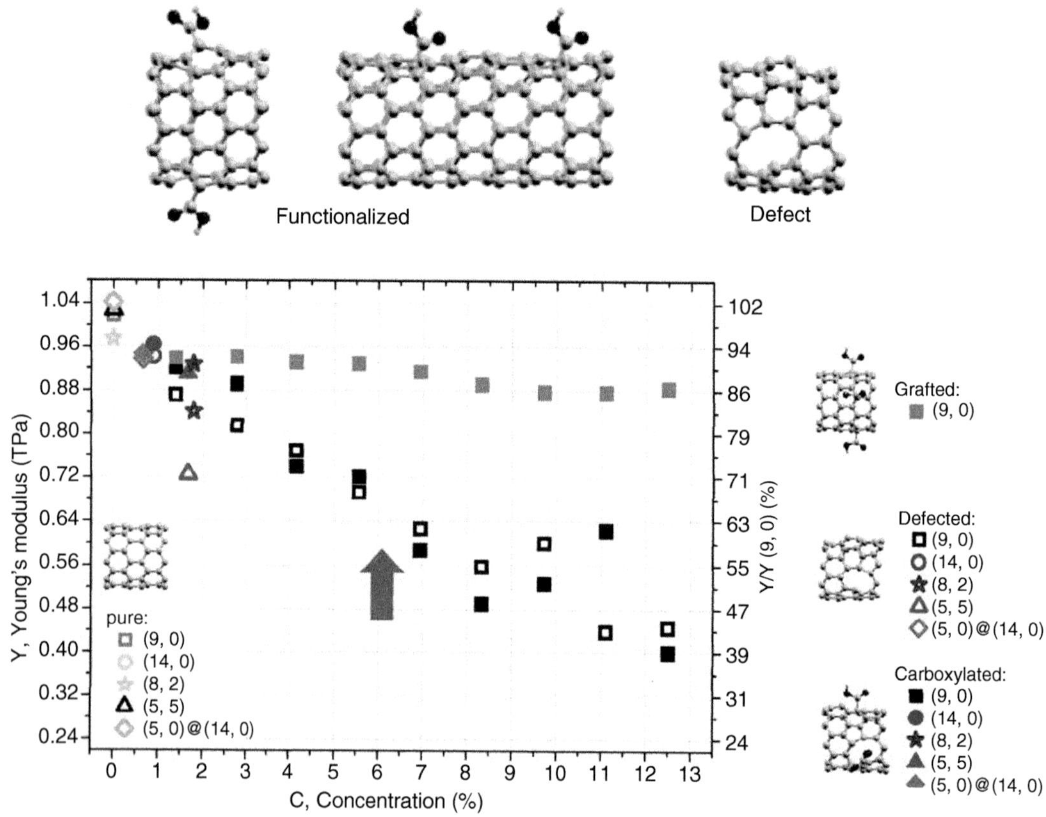

FIGURE 8.7 Molecular dynamics (MD) simulation results for the variation of SWCNT Young's modulus as a function of carboxyl group functionalization and single atom vacancy [86].

modulus at 20% functionalization [87]. Coto et al. reported a similar trend in the drop in SWCNT Young's modulus [88]. The deleterious effects of functionalization are more pronounced in SWCNTs than MWCNTs.

Huang et al. studied another promising application of covalent functionalization of MWCNT. In this multiscale study, the authors proposed to increase the twisting and bending stability of MWCNTs by providing a covalent bridge among the concentric nanotubes of the MWCNT [89]. They reported improved twisting stability of sp^3-bridged MWCNT over sp^2 bridged MWCNT in terms of postrippling stiffness and maximum allowable torque. Both sp^3- and sp^2-bridged MWCNTs show improved response in terms of maximum allowable bending moment and compressive strength over pristine MWCNTs. If fabricated at a strategic location along the length of the MWCNT, the proposed covalent bridge may improve the interwall shear resistance and delay the formation of the defects under the cyclic loading [90–92].

8.3.2.3 Effects of CNT Defects

According to quantum molecular dynamics simulation results by Nardelli et al. [93–94] and experimental verification by Bozovic et al. [95], CNTs subjected to more than 5%–10%

axial strain release the strain energy through an irreversible local rearrangement of chemical bonds, termed as "5—7—7—5" or Stone–Wales defect (Fig. 8.8). These local defects function as the nucleation site for dislocations and the plastic deformation of the CNT.

Presence of Stone–Wales and vacancy defects in SWCNTs diminishes the overall mechanical performance of nanotubes and facilitates the fracture by functioning as a fracture initiation site [86,96–104]. Sakharova et al. developed a FE model to study the variation in the mechanical properties of SWCNTs as a function of the amount of vacancy defects. The Young's modulus of SWCNT with 10% defects drops by 43% and the shear modulus drops by 33% [104]. Fefey et al. used MD modeling to study the effects of vacancy defects on the Young's modulus of a (6, 6) SWCNT. According to their results, the presence of six vacancy defects, equivalent to a defect percentage of 2.5%, reduced the Young's modulus by 13.7% (Fig. 8.9) [105].

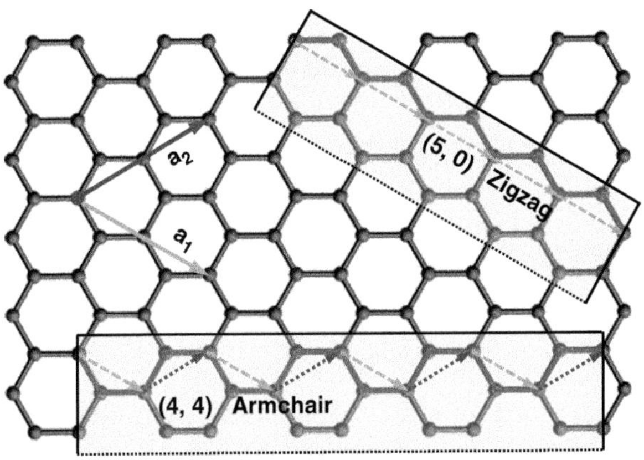

FIGURE 8.8 Formation of imperfection in CNT lattice under tensile loads.

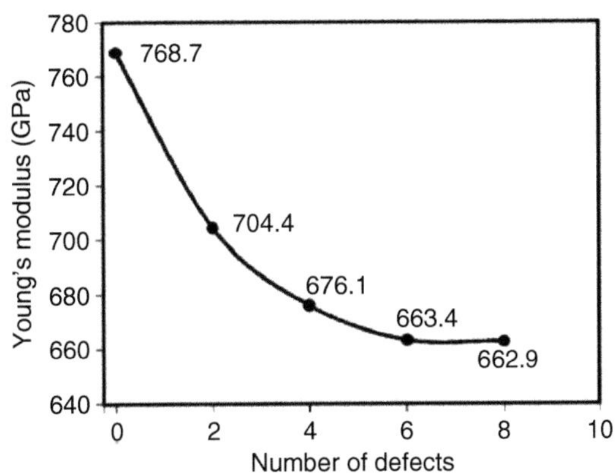

FIGURE 8.9 Variation of SWCNT Young's modulus as a function of density of single vacancy defects [105].

Presence of vacancy defects also diminishes the tensile strength and critical strain of SWCNTs. However, the extent of strength reduction due to presence of defects is chirality dependent [96,103]. Size of vacancy defect is another deteriorating factor for mechanical properties of SWCNT. SWCNTs with a single bigger vacancy cluster have inferior buckling properties than SWCNTs with multiple smaller vacancy clusters with equal number of missing atoms [106]. Among the two types (vacancy and Stone–Wales) defects, Stone–Wales defect are more deleterious, especially in cases of buckling [101] and SWCNT transverse modulus [100]. In addition, defects located near the end of SWCNT and deflection points produce more reduction in the buckling load than those appearing at the inflection points [107]. During the fabrication of nanocomposites, defect sites (Stone–Wales, one atom, and two atom vacancies) are preferred by the monomer molecules for physiosorption. In fact, the monomer-SWCNT interaction energy decreased as the number of defects increased [83].

8.3.2.4 Effect of Number of Concentric Sleeves

Ganji et al. performed a DFT-based study to show the effect of number of walls on the Young's and shear moduli of SWCNT, double-walled CNTs (DWCNTs), and triple-walled CNTs (TWCNTs). According to their calculations, for a fixed outer diameter, the CNT mechanical properties decrease as the number of walls increase [108] (Fig. 8.10). However, if the number of nanotube walls is increased while keeping the fixed inner radius, the Young's modulus of the MWCNT increases slightly the increase in the size of the CNT [99]. Doh et al. performed MM-based FE study of the mechanical properties of multiple DWCNTs. An interesting observation that came out of their results is that for a comparable nanotube outer diameter, DWCNTs with armchair SWCNTs as outer and zigzag SWCNT as inner tube have exhibited higher ultimate strength than those with zigzag SWCNT as outer and armchair SWCNT as inner tube [81]. This observation further supports the fact that zigzag SWCNTs are more brittle than armchair SWCNTs.

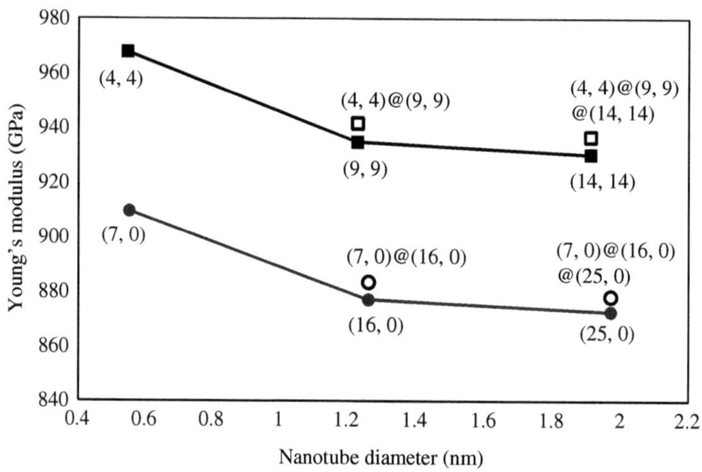

FIGURE 8.10 Effect of number of concentric tubes on the Young's modulus of MWCNT [108].

8.4 Large Elastic Deformation in CNTs

In the case of open structures, such as trusses or frames, the buckling behavior is a localized phenomenon where the failure of structure under compressive or bending load is controlled by the sectional out-of-plane buckling of the individual members of the structure. CNTs are long and slender open structures composed of C—C covalent bonds as the individual members and undergo large elastic deformation, beyond Hookean limits, during the process of buckling. The structural mechanics-based analogy, presented earlier, forms the basis of molecular mechanics supported equivalent continuum or finite element methods, which, in addition to classical molecular dynamics, are applied for the analysis of large deformation elastic behavior of CNTs (Section 8.3.2).

8.4.1 Radial Elasticity of CNTs

From the Atomistic point of view, the out-of-plane bending behavior of local C—C covalent bonds can be directly related to bond inversion energy [109–111]. In fact, for both SWCNTs and MWCNTs, the radial elastic modulus is much lower than their Young's modulus. Chen et al. used modified molecular structural mechanics (MSM) to show that the depending upon the diameter (0.39–2.35 nm) of zigzag SWCNTs, the radial modulus of elasticity varies between 60 and 0.4 GPa, whereas the same of different armchair SWCNTs (0.48 nm < SWCNT diameter < 2.38 nm) varies between 30 and 0.3 GPa [109]. However, various MD studies reported an even lower range for the radial modulus of elasticity for both armchair and zigzag SWCNTs (Fig. 8.11) [110,112,113].

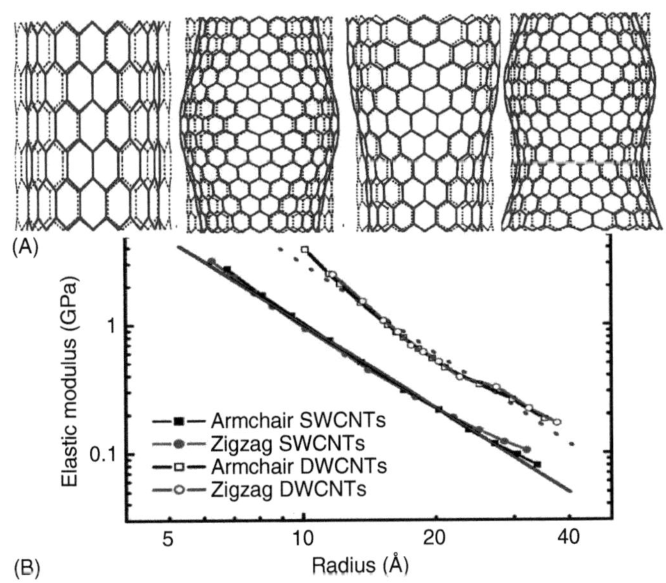

FIGURE 8.11 (A) Radial breathing (like) modes of SWCNT [110]; and (B) variation of radial modulus of elasticity with respect to CNT diameter [113].

Experimental studies report a similar range of radial elastic modulus between 0.4 and 57 GPa [114–118]. Although these experimental results vary over a much larger range, they all reinforce the following two observations. The first observation is that the radial modulus of elasticity for both, SWCNTs or MWCNTs, is at least two to three orders of magnitude smaller that the corresponding axial modulus of elasticity or Young's modulus; and secondly the CNT radial modulus of elasticity decreases rapidly as the diameter of the CNT increases. Muthaswamy et al. compared the radial elasticity of CNTs grown using arc-discharge (AD) method and chemical vapor deposition (CVD) method. They reported that, due to higher degree of radial microstructural uniformity along the tube axis, AD CNTs possess 20% higher average radius modulus than CVD CNTs.

8.4.2 Buckling of CNTs

The buckling behavior of a structural element is primarily governed by length to diameter (i.e., slenderness) ratio. When subjected to large deformation due to axial compressive loading (strains ~5%–10%), CNTs undergo sideways buckling (low aspect ratio) or longitudinal collapse (high slenderness). Specifically, CNTs with less aspect ratio (less than or equal to six) undergo minor compression prior to final buckling [119]. In general, when compressed beyond critical strain, buckling or collapse is the energetically preferred mode of CNT deformation as it promotes an instantaneous release of strain energy. Due to their large aspect ratio and hollow structure, CNTs have shown to accommodate extremely large bending strains (>30%) by forming ripples or kinks, a wave-like local rearrangement of atoms, on the compressive side of the bent tube (Fig. 8.12) [106,120,121]. Lourie et al. studied the bending and collapse of weakly embedded CNTs in epoxy matrix using TEM and reported similar formation of ripples and kinks in CNTs

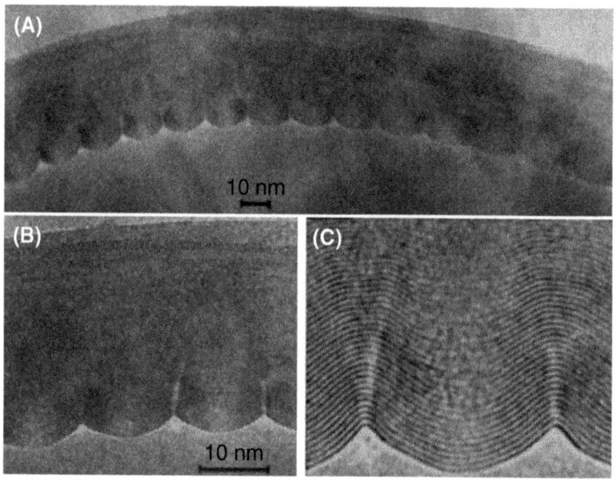

FIGURE 8.12 (A) High-resolution transmission electron microscopy (TEM) image of MWCNT subjected to bending load [120]; (B–C) magnified portion of bent tube TEM image showing the ripple pattern on the compressive side of the MWCNT [120].

under high bending loads [122]. Wang et al. showed that beyond critical bending strain, ripples originate in the central portion of CNT surface and propagate toward the nanotube ends covering the entire surface and upon continued straining, these ripples get deeper and get converted in to kinks. The presence of ripples or kinks on the CNT surface may affect the bending stiffness of the CNT deleteriously [123]. They also show that MWCNTs are more susceptible to ripple formation during buckling than SWCNTs. Duan et al. used SEM to perform bending of SWNCTs and MWCNTs till the point of buckling [124]. In their experiments, they observed that beyond critical bending, SWCNTs with small diameter experience abrupt buckling while MWCNTs show higher resilience and undergo gradual bending.

In their MD study, Yakobson et al. showed that these large strain morphological patterns (ripples and kins) are reversible in nature [119]. Cao et al. used MD simulations to study the buckling behavior of SWCNTs under bending and compression loading [125]. According to their results, for SWNCTs with same dimensional parameters, the critical buckling strain does not depend upon the nanotube chirality; however, it is different in bending and compression for the same nanotube. On the other hand, critical buckling bending curvature for the SWCNTs with fixed length is inversely proportional to the square of the nanotube diameter (Fig. 8.13). The buckling behavior of SWCNTs can be strengthened through developing internal hydrostatic pressure by infilling the SWCNT with small molecules [125]. Pantano et al. [126] used FE modeling to calculate the buckling point of SWCNT and their results compare well with the molecular dynamics simulation results reported by Yakobson et al. [119].

Recently, continuum mechanics–based nonlocal higher-order beam and generalized shell theories have been applied to study the buckling behavior of CNTs [123,127–132]. Amara et al. developed an analytical expression for the critical buckling strain of DWCNT

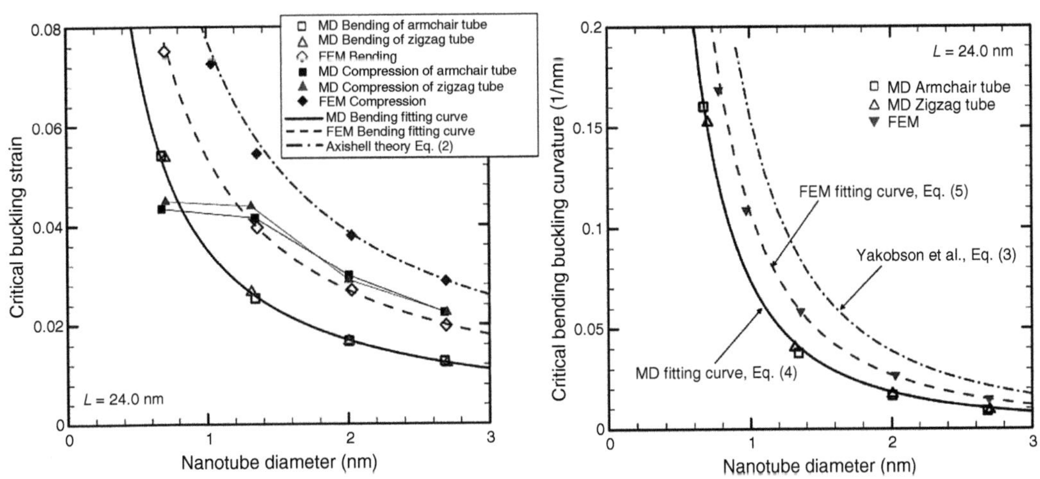

FIGURE 8.13 (A) Comparison between the critical buckling strains of bending and axial compression of SWCNTs with same length [125]; and (B) variation in critical bending buckling curvature with respect to SWCNT diameter [125].

while incorporating the nanolength scale effects using nonlocal elasticity theory [127]. According to their results, the local (classical) beam model overestimates the critical buckling strain. Similar results were reported by Khademolhosseini et al. for torsional buckling load for SWCNTs [129]. It is further reported that, among others, CNT chirality (armchair versus zigzag) and tube diameter play an important role in determining the appropriate value of the nonlocal constants. The incorporation of interlayer interaction through nonlocal van der Waals forces increases the critical buckling load for MWCNTs [130].

8.5 Tensile Strength of CNTs

In general, fracture is a dynamic process where strain or loading rate affect the material behavior and strength. The inelastic failure or fracture of CNTs is an interesting area of contemporary research. The fracture of CNTs involves energy relaxations by bond breaking, formation of defects, and propagation of defects by coalescence. During the low strain rate loading, 90 degree rotation of C—C bond to form a "5—7—7—5" Stone–Wales defect is the preferred mechanism of defect formation. As the loading continues, the more such defects are nucleated and promote plastic flow of the CNTs by coalescing (Fig. 8.14). It is observed that the formation of these defects along the length of CNTs takes place in a helical fashion and the overall movement of the "Stone–Wales slip" is chirality dependent.

In a MD study, Nardelli [94] reported that although CNTs break in brittle fashion at low temperatures, at higher temperature and low strains the type of CNT fracture depends upon its diameter and the symmetry. While most of the armchair and zigzag CNTs with smaller diameter (for $n < 14$) behave in a ductile manner at high temperatures and low strains, zigzag CNTs with larger diameter show brittle behavior [94]. Baykasoglu et al. [134] applied molecular mechanics (MM) approach to study the

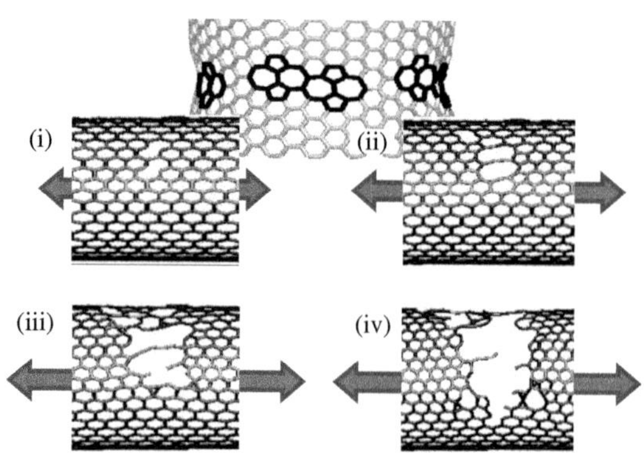

FIGURE 8.14 Formation and coalescence of Stone–Wales defect through different stages under the axial tensile loading, resulting in CNT fracture [133].

effects of CNT chirality of its fracture behavior. Their findings corroborated Nardelli's results that all CNTs (chiral and nonchiral) undergo brittle fracture at low temperature. However, presence of structural defects in CNTs reduces the failure stress values by 27%–52%.

Typically, CVD-grown or otherwise fabricated CNTs are a few nanometers in diameter and have high aspect ratio. Due to its nanoscale diameter, there are no tensile fixtures to clamp a single CNT and perform the direct measurement of tensile strength and produce reliable results. The issue related to proper clamping of CNTs is further complicated by the possibility that, in case of MWCNTs, the externally applied load may primarily be borne by the outer shells. During fabrication, CNTs grow in the form of bundles due to mutual nonbonding (Coulombic and non-Coulombic) interactions. The purification and separation of a single CNT from the bundle is an extremely cumbersome and very low-yield process. Therefore, most of the reported tensile strength values for CNTs are the average outcome of either bending experiments or pullout experiments performed on the bundle of CNTs. Yu et al. attached a bundle of MWCNT to two opposing AFM tips by electron beam deposition method [135]. The tensile load was applied by pulling the bottom cantilever and the resulting tensile strength is between 11 and 63 GPa. However, these results may not be the true representative of tensile strength as the failure of load-carrying capacity did not occur due to yield or fracture of MWCNTs; rather it was due to the pulling and separation of concentric tubules or "sword-in-sheath" failure. To alleviate the problem related to the clamping of the CNTs, Wagner et al. fabricated a polymer thin film containing CNTs and subjected it to tensile loads [136]. They studied the fragmentation of CNTs due to tensile loading using TEM and estimated the tensile strength of CNT equal to 45 GPa. On the other hand, CNT-bending experiments seem to offer more reliable estimation of load-bearing capacity of CNTs due to better experimental controls. Falvo et al. performed bending of a CNT and measured the maximum surface strain. Based on their measurement, they estimated the tensile strength in the range of 100–150 GPa [137]. Table 8.3 contains the records of experimentally determined values for the tensile strength ($\sigma_{t,max}$) and Table 8.4 consists of the same determined through computational approaches.

Table 8.3 Experimentally Determined Tensile Strength of SWCNTs and MWCNTs

References	Types of CNTs	$\sigma_{t,max}$ (GPa)	Years
Falvo et al. [137]	MWCNT	100–150	1997
Wagner et al. [136]	MWCNT	55	1998
Walters et al. [138]	MWCNT	45 ± 7	1999
Demczyk et al. [35]	MWCNT	150	2000
Yu et al. [135]	MWCNT	11–63	2000
Ding et al. [45]	MWCNT	10–66	2007
Khandoker et al. [139]	MWCNT	48	2011

CNTs, Carbon nanotubes; MWCNT, multiwalled carbon nanotubes.

Table 8.4 Computationally Determined Tensile Strength of CNTs

References	Methods	Types of CNTs	$\sigma_{t,max}$ (GPa)	$\varepsilon_{t,max}$	Years
Belytschko et al. [140]	MD	(20, 0)	84	0.16	2002
		(8, 8)	112	0.187	
Ogata and Shibutani [141]	TB	(8, 0)	86.8	0.211	2003
		(8, 8)	93	0.207	
Mielke et al. [142]	DFT/MM	(10, 0)	105	0.2	2004
		(5, 5)	110	0.3	
Liew et al. [133]	MD	(10, 10)	140–148	0.28	2004
Lu and Bhattacharya [143]	MD	(17, 0)	95	0.15	2006
Tserpes and Papanikos [144]	ABC	(20, 0)	98	0.16	2007
		(12, 12)	122	0.19	
Duan et al. [145]	MM	(10, 0)	105.6	0.333	2007
		(10, 10)	134	0.435	
Khare et al. [146]	QM/MM	(10, 0)	106	0.181	2007
		(5, 5)	147	0.247	
Baykasoglu et al. [134]	MM/FE	(10, 0)	95	0.16	2007
		(5, 5)	122	0.21	
Rossi and Meo [70]	MD	(5, 0)	93.8	0.172	2009
		(5, 5)	117.1	0.20	
Chen and Chen [147]	MD	—	80	0.21	2014
Jhon et al. [148]	MD	(9, 0)	94	0.164	2016
		(5, 5)	123	0.216	
Zhu et al. [103]	MD	(8, 0)	101.8	0.273	2016
		(7, 7)	105.9	0.235	
Yang et al. [149]	MD	(17, 0)	200	0.20	2016
		(10, 10)	175	0.20	

ABC, Atomistic-based continuum method; CNTs, carbon nanotubes; DFT, density functional theory; FE, finite element; MD, molecular dynamics; MM, molecular mechanics; QM, quantum mechanics; TB, tight binding method.

8.5.1 Effects of CNT Annealing

As reported in the preceding sections that presence of vacancy or Stone–Wales defects in CNT affect the elastic properties and promote CNT fracture. Yamamoto et al. annealed CVD-grown MWCNT over a series of temperature (1200, 1800, 2200, and 2600°C) and performed tensile tests on annealed CNT using in situ SEM approach [150]. The MWCNT annealed at 1800, 2200, and 2600°C have approximately ~5.3, ~5.2, and ~1.8 times the tensile strength of MWCNT annealed 1200°C. The Raman scattering spectrum and TEM images of the annealed MWCNT indicate that the improvement in MWCNT tensile strength is due to conversion of defective carbon network into stable graphitic planar structure and improved crystallinity in MWCNTs (Fig. 8.15). A similar trend of improvement in mechanical properties of MWCNTs is also reported by [151,152].

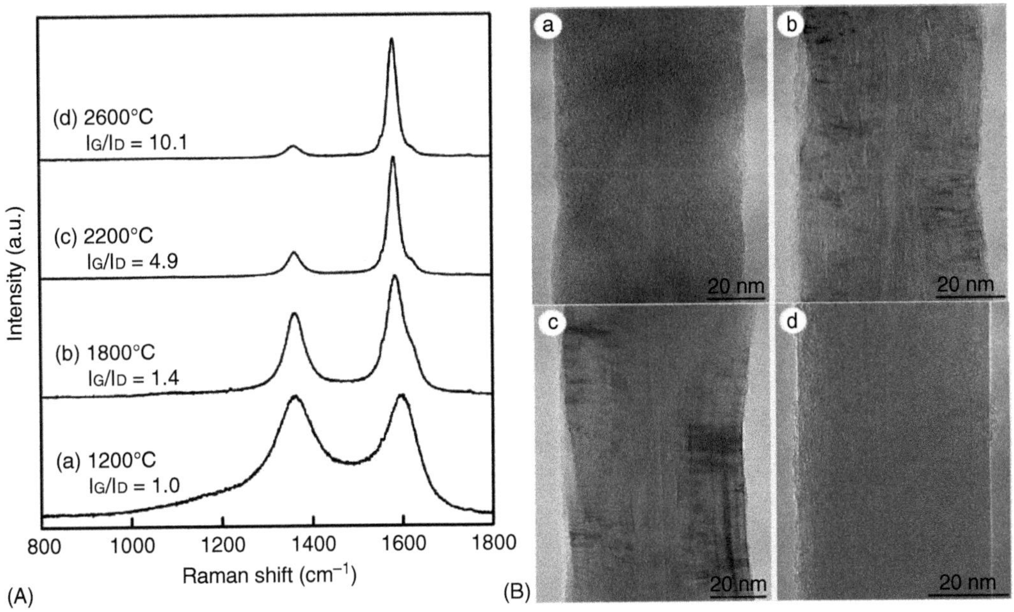

FIGURE 8.15 (A) Raman scattering spectrum of CVD-grown MWCNT annealed at (a) 1200°C, (b) 1800°C, (c) 2200°C, and (d) 2600°C. The spectrum shows a pair of bands around 1360 cm^{-1} (D-band) and 1590 cm^{-1} (G-band). The ratio of intensity of these bands, I_G/I_D or *R*-value, in inversely proportional to the number of defects in CNTs; (B) TEM images of CVD-grown MWCNT annealed at (a) 1200°C, (b) 1800°C, (c) 2200°C, and (d) 2600°C. Slide (A) shows MWCNT annealed at 1200°C, showing irregularly spaced fringes, representing highly disordered MWCNT nanostructure. The degree of fringe alignment, and hence nanostructure crystallinity, increases as annealing temperature is increased from 1800 to 2600°C [slide (B)–slide (D)] [150]. *Used with permission from G. Yamamoto, K. Shirasu, Y. Nozaka, Y. Sato, T. Takagi, T. Hashida, Structure-property relationships in thermally-annealed multi-walled carbon nanotubes, Carbon 66 (2014) 219–226.*

8.6 Epilogue

Because of their small size, determination of elastic and inelastic properties of CNTs is still an active area of research. The reported values of these properties are not only affected by the physical parameters of the CNTs but also by the method and process parameters of the fabrication procedure. In general, it is well accepted that CNTs show exceptionally large Young's modulus (~1 TPa), large elastic strains in buckling ($\gtrsim$ 30%) and high tensile strength. Under the large axial strains (~5%–10%), CNTs start to deteriorate due to formation of various defects. However, the reversibility of the buckling deformation at large strains shows that the defect formation in bending occurs at a much higher strain values than those in axial loading.

The nanoscale size and structural configurations of CNTs challenge the limits of testing equipments and the expertise of experimentalist. A variety of experimental tools have been utilized to ascertain the mechanical behavior of CNTs with great success. In case of modeling, the fundamental investigations related to the effects of chirality, defects, and functional groups on the mechanical performance of CNTs are performed using modeling methods that include electron/atomic level characteristics. These methods include quantum mechanics–based DFT, TB, and MD approaches. However, due to limited spatial and time scales of these methods and exorbitant computational cost, the methods are not suited for complete virtual design and nanosystems optimization process. On the other hand, continuum mechanics-based methods (atomistic-based quasicontinuum or molecular mechanics supported finite element analysis with truss, beam, or shell elements) are computationally efficient and can incorporate input from atomistic modeling under the framework of hierarchical or concurrent multiscale methods. Many excellent review articles offer a comprehensive overview of modeling techniques applicable for CNT and their nanocomposites [29,153–155]. Each carefully performed experiment and simulation provides designers with extra confidence for applying CNTs in numerous multifunctional applications.

References

[1] J.W. Lee, G.R. Kirikera, I. Kang, M.J. Schulz, V.N. Shanov, Structural health monitoring using continuous sensors and neural network analysis, Smart Mater. Struct. 15 (5) (2006) 1266.

[2] W. Zhang, V. Sakalkar, N. Koratkar, In situ health monitoring and repair in composites using carbon nanotube additives, Appl. Phys. Lett. 91 (13) (2007) 133102.

[3] E.T. Thostenson, T.-W. Chou, Carbon nanotube-based health monitoring of mechanically fastened composite joints, Compos. Sci. Technol. 68 (12) (2008) 2557–2561.

[4] N.D. N.D. Alexopoulos, C. Bartholome, P. Poulin, Z. Marioli-Riga, Structural health monitoring of glass-fiber-reinforced composites using embedded carbon nanotube (CNT) fibers, Compos. Sci. Technol. 70 (2) (2010) 260–271.

[5] J. Sebastian, N. Schehl, M. Bouchard, M. Boehle, J. Li, A. Lagounov, K. Lafdi, Health monitoring of structural composites with embedded carbon nanotube-coated glass-fiber sensors, Carbon 66 (2014) 191–200.

[6] A. Baltopoulos, N. Polydorides, L. Pambaguian, A. Vavouliotis, V. Kostopoulos, Exploiting carbon nanotube networks for damage assessment of fiber-reinforced composites, Compos. Part B 76 (2015) 149–158.

[7] L. Vertuccio, L. Guadagno, G. Spinelli, P. Lamberti, V. Tucci, S. Russo, Piezoresistive properties of resin reinforced with carbon nanotubes for health-monitoring of aircraft primary structures, Compos. Part B 107 (2016) 192–202.

[8] G.J. Gallo, E.T. Thostenson, Spatial damage detection in electrically anisotropic fiber-reinforced composites using carbon nanotube networks, Compos. Struct. 141 (2016) 14–23.

[9] M. Musameh, J. Wang, A. Merkoci, Y. Lin, Low-potential stable (NADH) detection at carbon-nanotube-modified glassy carbon electrodes, Electrochem. Commun. 4 (10) (2002) 743–746.

[10] Y. Lin, F. Lu, Y. Tu, R. Zhifeng, Glucose biosensors based on carbon nanotube nanoelectrode ensembles, Nano Lett. 4 (2) (2004) 191–195.

[11] J. Wang, G. Liu, R.M. Jan, Ultrasensitive electrical biosensing of proteins and DNA: carbon-nanotube derived amplification of the recognition and transduction events, J. Amer. Chem. Soc. 126 (10) (2004) 3010–3011.

[12] E.N. Primo, M.B. Oviedo, C.G. Snchez, M.D. Rubianes, G.A. Rivas, Bioelectrochemical sensing of promethazine with bamboo-type multiwalled carbon nanotubes dispersed in calf-thymus double stranded (DNA), Bioelectrochem. 99 (2014) 8–16.

[13] N. Yang, X. Chen, T. Ren, P. Zhang, D. Yang, Carbon nanotube based biosensors, Sens. Actuatators B 207 (2015) 690–715 Part A.

[14] M. Ramanathan, M. Patil, R. Epur, Y. Yun, V. Shanov, M. Schulz, W.R. Heineman, M.K. Datta, P.N. Kumta, Gold-coated carbon nanotube electrode arrays: immunosensors for impedimetric detection of bone biomarkers, Biosens. Bioelectron. 77 (2016) 580–588.

[15] P.A. Danoglidis, M.S. Konsta-Gdoutos, E.E. Gdoutos, S.P. Shah, Strength, energy absorption capability and self-sensing properties of multifunctional carbon nanotube reinforced mortars, Construct. Build. Mater. 120 (2016) 265–274.

[16] Y. Wang, Z. Yang, Z. Hou, D. Xu, L. Wei, E.S.-W. Kong, Y. Zhang, Flexible gas sensors with assembled carbon nanotube thin films for (DMMP) vapor detection, Sens. Actuators. B 150 (2) (2010) 708–714.

[17] T. Yamada, Y. Hayamizu, Y. Yamamoto, Y. Yomogida, A. Izadi-Najafabadi, D.N. Futaba, K. Hata, A stretchable carbon nanotube strain sensor for human-motion detection, Nat. Nanotechnol. 6 (5) (2011) 296–301.

[18] S. Ryu, P. Lee, J.B. Chou, R. Xu, R. Zhao, A.J. Hart, S.-G. Kim, Extremely elastic wearable carbon nanotube fiber strain sensor for monitoring of human motion, ACS Nano 9 (6) (2015) 5929–5936.

[19] J. Shi, X. Li, H. Cheng, Z. Liu, L. Zhao, T. Yang, Z. Dai, Z. Cheng, E. Shi, L. Yang, Z. Zhang, A. Cao, H. Zhu, Y. Fang, Strain sensing: graphene-reinforced carbon nanotube networks for wearable strain sensors, Adv. Funct. Mater. 26 (13) (2016) 2038–12038.

[20] J. Foroughi, G.M. Spinks, S. Aziz, A. Mirabedini, A. Jeiranikhameneh, G.G. Wallace, M.E. Kozlov, R.H. Baughman, Knitted carbon-nanotube-sheath/spandex-core elastomeric yarns for artificial muscles and strain sensing, ACS Nano 10 (10) (2016) 9129–9135.

[21] T. Dinh, H.-P. Phan, T.-K. Nguyen, A. Qamar, M.A.R. Foisal, C.-D. Nguyen Viet Tand Tran, Y. Zhu, N.-T. Nguyen, D.V. Dao, Environment-friendly carbon nanotube based flexible electronics for noninvasive and wearable healthcare, J. Mater. Chem. C 4 (2016) 10061–10068.

[22] J. Zhou, X. Xu, H. Yu, G. Lubineau, Deformable and wearable carbon nanotube microwire-based sensors for ultrasensitive monitoring of strain, pressure and torsion, Nanoscale 9 (2) (2017) 604–612.

[23] N. Li, Y. Huang, F. Du, X. He, X. Lin, H. Gao, Y. Ma, F. Li, Y. Chen, P.C. Eklund, Electromagnetic interference (EMI) shielding of single-walled carbon nanotube epoxy composites, Nano Lett. 6 (6) (2006) 1141–1145.

[24] C.-S. Zhang, Q.-Q. Ni, S.-Y. Fu, K. Kurashiki, Electromagnetic interference shielding effect of nanocomposites with carbon nanotube and shape memory polymer, Compos. Sci. Technol. 67 (14) (2007) 2973–2980.

[25] S.H. Park, P.T. Theilmann, P.M. Asbeck, P.R. Bandaru, Enhanced electromagnetic interference shielding through the use of functionalized carbon-nanotube-reactive polymer composites, IEEE Trans. Nanotechnol. 9 (4) (2010) 464–469.

[26] P. Verma, P. Saini, V. Choudhary, Designing of carbon nanotube/polymer composites using melt recirculation approach: effect of aspect ratio on mechanical, electrical and EMI shielding response, Mater. Des. 88 (2015) 269–277.

[27] A.P. Singh, M. Mishra, D.P. ashim, T.N. Narayanan, M.G. Hahm, P. Kumar, J. Dwivedi, G. Kedawat, A. Gupta, B.P. ingh, A. Chandra, R. Vajtai, S.K. Dhawan, P.M. Ajayan, B.K. Gupta, Probing the engineered sandwich network of vertically aligned carbon nanotube reduced graphene oxide composites for high performance electromagnetic interference shielding applications, Carbon 85 (2015) 79–88.

[28] G. Pal, S. Kumar, Multiscale modeling of effective electrical conductivity of short carbon fiber-carbon nanotube-polymer matrix hybrid composites, Mater. Des. 89 (2016) 129–136.

[29] G. Pal, S. Kumar, Modeling of carbon nanotubes and carbon nanotubepolymer composites, Prog. Aerospace Sci. 80 (2016) 33–58.

[30] M. Monthioux, V.L. Kuznetsov, Who should be given the credit for the discovery of carbon nanotubes?, Carbon 44 (9) (2006) 1621–1623.

[31] S. Iijima, Helical microtubules of graphitic carbon, Nature 354 (6348) (1991) 56–58.

[32] S. Iijima, T. Ichihashi, Single-shell carbon nanotubes of 1-nm diameter, Nature 363 (6430) (1993) 603–605.

[33] M.F. Arif, S. Kumar, T. Shah, Tunable morphology and its influence on electrical, thermal and mechanical properties of carbon nanostructure-buckypaper, Mater. Des. 101 (2016) 236–244.

[34] M.M.J. Treacy, T.W. Ebbesen, J.M. Gibson, Exceptionally high Young's modulus observed for individual carbon nanotubes, Nature 381 (6584) (1996) 678–680.

[35] B.G. Demczyk, Y.M. Wang, J. Cumings, M. Hetman, W. Han, A. Zettl, R.O. Ritchie, Direct mechanical measurement of the tensile strength and elastic modulus of multiwalled carbon nanotubes, Mater. Sci. Eng. A 334 (12) (2002) 173–178.

[36] E.W. Wong, P.E. Sheehan, C.M. Lieber, Nanobeam mechanics: elasticity, strength, and toughness of nanorods and nanotubes, Science 277 (1997) 1971–1975.

[37] J.-P. Selvetat, A.D. Briggs, J.-M. Bonard, R.R. Bacsa, A.J. Kulik, T. Stockli, N.A. Burnham, L. Forro, Elastic and shear moduli of single-walled carbon nanotube ropes, Phys. Rev. Lett. 82 (1999) 944–947.

[38] J.-P. Salvetat, J.-M. Bonard, N.H. Thomson, A.J. Kulik, L. Forró, W. Benoit, L. Zuppiroli, Mechanical properties of carbon nanotubes, Appl. Phys. A 69 (3) (1999) 255–260.

[39] J.-P. Salvetat, G.A.D. Briggs, J.-M. Bonard, R.R. Bacsa, A.J. Kulik, T. Stöckli, N.A. Burnham, L. Forró, Elastic and shear moduli of single-walled carbon nanotube ropes, Phys. Rev. Lett. 82 (1999) 944–947.

[40] J. Muster, M. Burghard, S. Roth, G.S. Duesberg, E. Hernndez, A. Rubio, Scanning force microscopy characterization of individual carbon nanotubes on electrode arrays, J. Vac. Sci. Technol. B 16 (5) (1998) 2796–2801.

[41] A. Krishnan, E. Dujardin, T.W. Ebbesen, P.N. Yianilos, M.M.J. Treacy, Young's modulus of single-walled nanotubes, Phys. Rev. B 58 (1998) 14013–14019.

[42] M.-F. Yu, B.S. Files, S. Arepalli, R.S. Ruoff, Tensile loading of ropes of single wall carbon nanotubes and their mechanical properties, Phys. Rev. Lett. 84 (2000) 5552–5555.

[43] T.W. Tombler, C. Zhou, L. Alexseyev, J. Kong, H. Dai, L. Liu, C.S. Jayanthi, M. Tang, S.-Y. Wu, Reversible electromechanical characteristics of carbon nanotubes under local-probe manipulation, Nature 405 (6788) (2000) 769–772.

[44] G. Guhados, W.K. Wan, X.L. Sun, J.L. Hutter, Simultaneous measurement of Young's and shear moduli of multiwalled carbon nanotubes using atomic force microscopy, J. Appl. Phys. 101 (2007) 033514.

[45] W. Ding, L. Calabri, K.M. Kohlhaas, X. Chen, D.A. Dikin, R.S. Ruoff, Modulus, fracture strength, and brittle vs. plastic response of the outer shell of arc-grown multiwalled carbon nanotubes, Exp. Mech. 47 (1) (2007) 25–36.

[46] Y. Wu, M. Huang, F. Wang, X.M.H. Huang, S. Rosenblatt, L. H, H. Yan, S.P. OBrien, J. Hone, T.F. Heinz, Determination of the Young's modulus of structurally defined carbon nanotubes, Nano Lett. 8 (12) (2008) 4158–4161.

[47] X. Lu, Z. Hu, Mechanical property evaluation of single-walled carbon nanotubes by finite element modeling, Compos. Part B 43 (4) (2012) 1902–1913.

[48] J.R. Xiao, B.A. Gama, J.W. Gillespie Jr., An analytical molecular structural mechanics model for the mechanical properties of carbon nanotubes, Int. J. Solid. Struct. 42 (1112) (2005) 3075–3092.

[49] E. Mohammadpour, M. Awang, Nonlinear finite-element modeling of graphene and single- and multiwalled carbon nanotubes under axial tension, Appl. Phys. A 106 (3) (2012) 581–588.

[50] R. Merli, C. Lzaro, S. Monlen, A. Domingo, Geometrical nonlinear formulation of a molecular mechanics model applied to the structural analysis of single-walled carbon nanotubes, Int. J. Solid. Struct. 58 (2015) 157–177.

[51] J.P. Lu, Elastic properties of carbon nanotubes and nanoropes, Phys. Rev. Lett. 79 (1997) 1297–1300.

[52] P. Zhang, P.E. Lammert, V.H. Crespi, Plastic deformations of carbon nanotubes, Phys. Rev. Lett. 81 (1998) 5346–5349.

[53] E. Hernandez, C. Goze, P. Bernier, A. Rubio, Elastic properties of c and $b_x c_y n_z$ composite nanotubes, Phys. Rev. Lett. 80 (1998) 4502–4505.

[54] N. Yao, V. Lordi, Young's modulus of single-walled carbon nanotubes, J. Appl. Phys. 84 (4) (1998) 1939–1943.

[55] Y. Jin, F.G. Yuan, Simulation of elastic properties of single-walled carbon nanotubes, Compos. Sci. Technol. 63 (11) (2003) 1507–1515.

[56] C. Li, T.-W. Chou, Elastic moduli of multiwalled carbon nanotubes and the effect of van der Waals forces, Comp. Sci. Technol. 63 (11) (2003) 1517–1524.

[57] D. Troya, S.L. Mielke, G.C. Schatz, Carbon nanotube fracture differences between quantum mechanical mechanisms and those of empirical potentials, Chem. Phys. Lett. 382 (12) (2003) 133–141.

[58] Q. Wang, Effective in-plane stiffness and bending rigidity of armchair and zigzag carbon nanotubes, Int. J. Solids Struct. 41 (20) (2004) 5451–5461.

[59] J. Cai, R.F. Bie, X.M. Tan, C. Lu, Application of the tight-binding method to the elastic modulus of (C60) and carbon nanotube, Phys. B 344 (14) (2004) 99–102.

[60] M. Arroyo, T. Belytschko, Finite crystal elasticity of carbon nanotubes based on the exponential cauchy-born rule, Phys. Rev. B 69 (2004) 115415.

[61] H.W. Zhang, J.B. Wang, X. Guo, Predicting the elastic properties of single-walled carbon nanotubes, J. Mech. Phys. Solid. 53 (9) (2005) 1929–1950.

[62] M. Meo, M. Rossi, Prediction of Young's modulus of single wall carbon nanotubes by molecular-mechanics based finite element modelling, Compos. Sci. Technol. 66 (1112) (2006) 1597–1605.

[63] Y. Wu, A.Y.T. Xiaochun Zhang, Leung, Weifang, Zhong, An energy-equivalent model on studying the mechanical properties of single-walled carbon nanotubes, Thin-Walled Struct. 44 (6) (2006) 667–676.

[64] X. Guo, J.B. Wang, H.W. Zhang, Mechanical properties of single-walled carbon nanotubes based on higher order cauchyborn rule, Int. J. Solids Struct. 43 (5) (2006) 1276–1290.

[65] A.L. Kalamkarov, A.V. Georgiades, S.K. Rokkam, V.P. Veedu, M.N. Ghasemi-Nejhad, Analytical and numerical techniques to predict carbon nanotubes properties, Int. J. Solids Struct. 43 (2223) (2006) 6832–6854.

[66] A. Selmi, C. Friebel, I. Doghri, H. Hassis, Prediction of the elastic properties of single walled carbon nanotube reinforced polymers: a comparative study of several micromechanical models, Compos. Sci. Technol. 67 (10) (2007) 2071–2084.

[67] Y.-G. Hu, K.M. Liew, Q. Wang, X.Q. He, B.I. Yakobson, Nonlocal shell model for elastic wave propagation in single- and double-walled carbon nanotubes, J. Mech. Phys. Solid. 56 (12) (2008) 3475–3485.

[68] J.M. Wernik, S.A. Meguid, Atomistic-based continuum modeling of the nonlinear behavior of carbon nanotubes, Ada Mechanica 212 (1–2) (2010) 167–179.

[69] H.-C. Cheng, Y.-L. Liu, Y.-C. Hsu, W.-H. Chen, Atomistic-continuum modeling for mechanical properties of single-walled carbon nanotubes, Int. J. Solids Struct. 46 (78) (2009) 1695–1704.

[70] M. Rossi, M. Meo, On the estimation of mechanical properties of single-walled carbon nanotubes by using a molecular-mechanics based (FE) approach, Compos. Sci. Technol. 69 (9) (2009) 1394–1398.

[71] J. Cai, C.Y. Wang, T. Yu, S. Yu, Wall thickness of single-walled carbon nanotubes and its Young's modulus, Phys. Scr. 79 (2) (2009) 025702.

[72] K. Chandraseker, S. Mukherjee, J.T. Paci, G.C. Schatz, An atomistic-continuum cosserat rod model of carbon nanotubes, J. Mech. Phys. Solid. 57 (6) (2009) 932–958.

[73] M.M. Shokrieh, R. Rafiee, Prediction of Young's modulus of graphene sheets and carbon nanotubes using nanoscale continuum mechanics approach, Mater. Des. 31 (2) (2010) 790–795.

[74] R. Rafiee, M. Heidarhaei, Investigation of chirality and diameter effects on the Young's modulus of carbon nanotubes using non-linear potentials, Compos. Struct. 94 (8) (2012) 2460–2464.

[75] M. Malag, E. Benvenuti, A. Simone, One-dimensional nonlocal elasticity for tensile single-walled carbon nanotubes: a molecular structural mechanics characterization, Eur. J. Mech. A 54 (2015) 160–170.

[76] N.A. Sakharova, A.F.G. Pereira, J.M. Antunes, C.M.A. Brett, J.V. Fer-nandes, Mechanical characterization of single-walled carbon nanotubes: numerical simulation study, Compos. Part B 75 (2015) 73–85.

[77] D. Sánchez-Portal, E. Artacho, J.M. Soler, A. Rubio, P. Ordejón, Ab initio structural, elastic, and vibrational properties of carbon nanotubes, Phys. Rev. B 59 (1999) 12678–12688.

[78] T. Chang, A molecular based anisotropic shell model for single-walled carbon nanotubes, J. Mech. Phys. Solid. 58 (9) (2010) 1422–1433.

[79] R. Ansari, M. Mirnezhad, F. Sadeghi, Elastic properties of chiral carbon nanotubes under oxygen adsorption, Phys. E 70 (2015) 129–134.

[80] T. Chang, H. Gao, Size-dependent elastic properties of a single-walled carbon nanotube via a molecular mechanics model, J. Mech. Phys. Solid. 51 (6) (2003) 1059–1074.

[81] J. Doh, J. Lee, Prediction of the mechanical behavior of double walled-CNTs using a molecular mechanics-based finite element method: effects of chirality, Comput. Struct. 169 (2016) 91–100.

[82] E. Hernández, C. Goze, P. Bernier, A. Rubio, Elastic properties of *c* and $B_xC_yN_z$ composite nanotubes, Phys. Rev. Lett. 80 (1998) 4502–4505.

[83] M. Ghorbanzadeh Ahangari, A. Fereidoon, M. Jahanshahi, M.D. Ganji, Electronic and mechanical properties of single-walled carbon nanotubes interacting with epoxy: a (DFT) study, Phys. E 48 (2013) 148–156.

[84] F. Bogár, J.W. Mintmire, F. Bartha, T. Mező, C. Van, Alsenoy, Density-functional study of the mechanical and electronic properties of narrow carbon nanotubes under axial stress, Phys. Rev. B 72 (2005) 085452.

[85] K.Z. Milowska, M. Birowska, J.A. Majewski, Ab-initio study of structural, mechanical and electronic properties of functionalized carbon nanotubes, AIP Conf. Proc. 1566 (1) (2013) 147–148.

[86] K.Z. Milowska, Influence of carboxylation on structural and mechanical properties of carbon nanotubes: composite reinforcement and toxicity reduction perspectives, J. Phys. Chem. C 119 (47) (2015) 26734–26746.

[87] P.H. Shah, R.C. Batra, In-plane elastic moduli of covalently functionalized single-wall carbon nanotubes, Compos. Mater. Sci. 83 (2014) 349–361.

[88] B. Coto, I. Antia, M. Blanco, I. Martinez de Arenaza, E. Meaurio, J. Barriga, J.-R. Sarasua, Molecular dynamics study of the influence of functionalization on the elastic properties of single and multiwall carbon nanotubes, Comput. Mater. Sci. 50 (12) (2011) 3417–3424.

[89] X. Huang, H. Yuan, W. Liang, S. Zhang, Mechanical properties and deformation morphologies of covalently bridged multi-walled carbon nanotubes: multiscale modeling, J. Mech. Phys. Solids 58 (11) (2010) 1847–1862.

[90] A. Kis, G. Csanyi, J.-P. Salvetat, T.-N. Lee, E. Couteau, A.J. Kulik, W. Benoit, J. Brugger, L. Forro, Reinforcement of single-walled carbon nanotube bundles by intertube bridging, Nat. Mater. 3 (3) (2004) 153–157.

[91] S. Zhang, S.L. Mielke, R. Khare, D. Troya, R.S. Ruoff, G.C. Schatz, T. Belytschko, Mechanics of defects in carbon nanotubes: atomistic and multiscale simulations, Phys. Rev. B 71 (2005) 115403.

[92] S. Zhang, T. Zhu, T. Belytschko, Atomistic and multiscale analyses of brittle fracture in crystal lattices, Phys. Rev. B 76 (2007) 094114.

[93] M.B. Nardelli, B.I. Yakobson, J. Bernholc, Brittle and ductile behavior in carbon nanotubes, Phys. Rev. Lett. 81 (1998) 4656–4659.

[94] M.B. Nardelli, B.I. Yakobson, J. Bernholc, Mechanism of strain release in carbon nanotubes, Phys. Rev. B 57 (1998) 4277–4280.

[95] D. Bozovic, M. Bockrath, J.H. Hafner, C.M. Lieber, H. Park, M. Tinkham, Plastic deformations in mechanically strained single-walled carbon nanotubes, Phys. Rev. B 67 (2003) 033407.

[96] M. Sammalkorpi, A. Krasheninnikov, A. Kuronen, K. Nordlund, K. Kaski, Mechanical properties of carbon nanotubes with vacancies and related defects, Phys. Rev. B 70 (2004) 245416.

[97] Q. Lu, B. Bhattacharya, Effect of randomly occurring Stone–Wales defects on mechanical properties of carbon nanotubes using atomistic simulation, Nanotechnology 16 (4) (2005) 555.

[98] J.R. Xiao, J. Staniszewski, J.W. Gillespie Jr., Tensile behaviors of graphene sheets and carbon nanotubes with multiple Stone–Wales defects, Mater. Sci. Eng. A 527 (3) (2010) 715–723.

[99] A. Ghavamian, M. Rahmandoust, A. chsner, A numerical evaluation of the influence of defects on the elastic modulus of single and multi-walled carbon nanotubes, Comput. Mater. Sci. 62 (2012) 110–116.

[100] K. Sharma, K.K. Saxena, M. Shukla, Effect of multiple Stone–Wales and vacancy defects on the mechanical behavior of carbon nanotubes using molecular dynamics, Procedia Eng. 38 (2012) 3373–3380.

[101] M. Eftekhari, S. Mohammadi, A.R. Khoei, Effect of defects on the local shell buckling and post-buckling behavior of single and multi-walled carbon nanotubes, Comput. Mater. Sci. 79 (2013) 736–744.

[102] H. Torabi, M. Shariati, E. Sedaghat, A.L. Zadeh, Buckling behavior of perfect and defective dwcnts under axial, bending and torsional loadings via a structural mechanics approach, Meccanica 48 (8) (2013) 1959–1974.

[103] L. Zhu, J. Wang, F. Ding, The great reduction of a carbon nanotubes mechanical performance by a few topological defects, ACS Nano 10 (6) (2016) 6410–6415.

[104] N.A. Sakharova, A.F.G. Pereira, J.M. Antunes, J.V. Fernandes, Numerical simulation study of the elastic properties of single-walled carbon nanotubes containing vacancy defects, Compos. Part B 89 (2016) 155–168.

[105] E.G. Fefey, R. Mohan, A. Kelkar, Computational study of the effect of carbon vacancy defects on the Young's modulus of (6, 6) single wall carbon nanotube, Mater. Sci. Eng. B 176 (9) (2011) 693–700.

[106] D.D.T.K. Kulathunga, K.K. Ang, J.N. Reddy, Molecular dynamics analysis on buckling of defective carbon nanotubes, J. Phys. 22 (34) (2010) 345301.

[107] R.H. Poelma, H. Sadeghian, S. Koh, G.Q. Zhang, Effects of single vacancy defect position on the stability of carbon nanotubes, Microelectron. Reliab. 52 (7) (2012) 1279–1284.

[108] M.D. Ganji, A. Fereidoon, M. Jahanshahi, M.G. Ahangari, Investigation of the mechanical properties of multiwalled carbon nanotubes using density functional theory calculations, J. Comput. Theor. Nanosci. 9 (7) (2012) 980–985.

[109] W.-H. Chen, H.-C. Cheng, Y.-L. Liu, Radial mechanical properties of single-walled carbon nanotubes using modified molecular structure mechanics, Comput. Mater. Sci. 47 (4) (2010) 985–993.

[110] H.-C. Cheng, Y.-L. Liu, C.-H. Wu, W.-H. Chen, On radial breathing vibration of carbon nanotubes, Comput. Methods Appl. Mech. Eng. 199 (4548) (2010) 2820–2827.

[111] M. Zheng, X. Chen, I.-T. Bae, C. Ke, C. Park, M.W. Smith, K. Jordan, Radial mechanical properties of single-walled boron nitride nanotubes, Small 8 (1) (2012) 116–121.

[112] C. Zhu, W. W Guo, T.X. Yu, C.H. Woo, Radial compression of carbon nanotubes: deformation and damage, super-elasticity and super-hardness, Nanotechnology 16 (8) (2005) 1035.

[113] X.-F. Wang, Z.-J. Xu, Z.-Y. Zhu, Reversible mechanical bistability of carbon nanotubes under radial compression, Chem. Phys. 334 (13) (2007) 144–147.

[114] W. Shen, B. Jiang, B.S. Han, S.S. Xie, Investigation of the radial compression of carbon nanotubes with a scanning probe microscope, Phys. Rev. Lett. 84 (2000) 3634–3637.

[115] M.-F. Yu, T. Kowalewski, R.S. Ruoff, Investigation of the radial deformability of individual carbon nanotubes under controlled indentation force, Phys. Rev. Lett. 85 (2000) 1456–1459.

[116] I. Palaci, S. Fedrigo, H. Brune, C. Klinke, M. Chen, E. Riedo, Radial elasticity of multiwalled carbon nanotubes, Phys. Rev. Lett. 94 (May 2005) 175502.

[117] L. Muthaswami, Y. Zheng, R. Vajtai, G. Shehkawat, P. Ajayan, R.E. Geer, Variation of radial elasticity in multiwalled carbon nanotubes, Nano Lett. 7 (12) (2007) 3891–3894.

[118] Y.H. Yang, W.Z. Li, Radial elasticity of single-walled carbon nanotube measured by atomic force microscopy, Appl. Phys. Lett. 98 (4) (2011) 041901.

[119] B.I. Yakobson, C.J. Brabec, J. Bernholc, Nanomechanics of carbon tubes: instabilities beyond linear response, Phys. Rev. Lett. 76 (1996) 2511–2514.

[120] Philippe Poncharal, Z.L. Wang, Daniel Ugarte, Walt A. de Heer, Electrostatic deflections and electromechanical resonances of carbon nanotubes, Science 283 (5407) (1999) 1513–1516.

[121] Y.Y. Zhang, Y. Xiang, C.M. Wang, Buckling of defective carbon nanotubes, J. Appl. Phys. 106 (11) (2009) 113503.

[122] O. Lourie, D.M. Cox, H.D. Wagner, Buckling and collapse of embedded carbon nanotubes, Phys. Rev. Lett. 81 (1998) 1638–1641.

[123] C.G. Wang, Y.P. Liu, J. Al-Ghalith, T. Dumitric, M.K. Wadee, H.F. Tan, Buckling behavior of carbon nanotubes under bending: from ripple to kink, Carbon 102 (2016) 224–235.

[124] X. Duan, C. Tang, J. Zhang, W. Guo, Z. Liu, Two distinct buckling modes in carbon nanotube bending, Nano Lett. 7 (1) (2007) 143–148.

[125] G. Cao, X. Chen, Buckling of single-walled carbon nanotubes upon bending: molecular dynamics simulations and finite element method, Phys. Rev. B 73 (2006) 155435.

[126] A. Pantano, D.M. Parks, M.C. Boyce, Mechanics of deformation of single- and multi-wall carbon nanotubes, J. Mech. Phys. Solids 52 (4) (2004) 789–821.

[127] K. Amara, A. Tounsi, I. Mechab, E.A. Adda-Bedia, Nonlocal elasticity effect on column buckling of multiwalled carbon nanotubes under temperature field, Appl. Math. Model. 34 (12) (2010) 3933–3942.

[128] K. Kiani, Nonlocal and shear effects on column buckling of single-layered membranes from stocky single-walled carbon nanotubes, Compos. Part B 79 (2015) 535–552.

[129] F. Khademolhosseini, R.K.N.D. Rajapakse, A. Nojeh, Torsional buckling of carbon nanotubes based on nonlocal elasticity shell models, Comput. Mater. Sci. 48 (4) (2010) 736–742.

[130] X.Q. He, S. Kitipornchai, K.M. Liew, Buckling analysis of multiwalled carbon nanotubes: a continuum model accounting for van der Waals interaction, J. Mech. Phys. Solid. 53 (2) (2005) 303–326.

[131] U. Gven, A comparative study on estimation of critical buckling temperatures of single-walled carbon nanotubes, Compos. Part B 99 (2016) 521–527.

[132] A.R. Setoodeh, M. Khosrownejad, P. Malekzadeh, Exact nonlocal solution for postbuckling of single-walled carbon nanotubes, Phys. E 43 (9) (2011) 1730–1737.

[133] K.M. Liew, X.Q. He, C.H. Wong, On the study of elastic and plastic properties of multi-walled carbon nanotubes under axial tension using molecular dynamics simulation, Ada Mater. 52 (9) (2004) 2521–2527.

[134] C. Baykasoglu, M. Kirca, A. Mugan, Nonlinear failure analysis of carbon nanotubes by using molecular-mechanics based models, Compos. Part B 50 (2013) 150–157.

[135] M.-F. Yu, O. Lourie, M.J. Dyer, K. Moloni, T.F. Kelly, R.F. Ruoff, Strength and breaking mechanism of multiwalled carbon nanotubes under tensile load, Science 287 (5453) (2000) 637–640.

[136] H.D. Wagner, O. Lourie, Y. Feldman, R. Tenne, Stress-induced fragmentation of multiwall carbon nanotubes in a polymer matrix, Appl. Phys. Lett. 72 (2) (1998) 188–190.

[137] G.J. Falvo, G.J. Clary, R.M. Taylor, V. Chi, F.P. Brooks, S. Washburn, R. Superfine, Bending and buckling of carbon nanotubes under large strain, Nature 389 (1997) 582–584.

[138] D.A. Walters, L.M. Ericson, M.J. Casavant, J. Liu, D.T. Colbert, K.A. Smith, R.E. Smalley, Elastic strain of freely suspended single-wall carbon nanotube ropes, Appl. Phys. Lett. 74 (25) (1999) 3803–3805.

[139] N. Khandoker, S.C. Hawkins, R. Ibrahim, C.P. Huynh, F. Deng, Tensile strength of spinnable multiwall carbon nanotubes, Procedia Eng. 10 (2011) 2572–2578.

[140] T. Belytschko, S.P. Xiao, G.C. Schatz, R.S. Ruoff, Atomistic simulations of nanotube fracture, Phys. Rev. B 65 (2002) 235430.

[141] S. Ogata, Y. Shibutani, Ideal tensile strength and band gap of single-walled carbon nanotubes, Phys. Rev. B 68 (2003) 165409.

[142] S.L. Mielke, D. Troya, S. Zhang, J.-L. Li, S. Xiao, R. Car, R.S. Ruoff, G.C. Schatz, T. Belytschko, The role of vacancy defects and holes in the fracture of carbon nanotubes, Chem. Phys. Lett. 390 (46) (2004) 413–420.

[143] Q. Lu, B. Bhattacharya, Fracture resistance of zigzag single walled carbon nanotubes, Nanotechnol. 17 (5) (2006) 1323.

[144] K.I. Tserpes, P. Papanikos, The effect of Stone–Wales defect on the tensile behavior and fracture of single-walled carbon nanotubes, Compos. Struct. 79 (4) (2007) 581–589.

[145] W.H. Duan, Q. Wang, K.M. Liew, X.Q. He, Molecular mechanics modeling of carbon nanotube fracture, Carbon 45 (9) (2007) 1769–1776.

[146] R. Khare, S.L. Mielke, J.T. Paci, S. Zhang, R. Ballarini, G.C. Schatz, T. Belytschko, Coupled quantum mechanical/molecular mechanical modeling of the fracture of defective carbon nanotubes and graphene sheets, Phys. Rev. B 75 (2007) 075412.

[147] H. Chen, S. Chen, The fracture behaviors of carbon nanotube and nanoscroll reinforced silicon matrix composites, Carbon 67 (2014) 344–351.

[148] Y.I. Jhon, C. Kim, M. Seo, W.J. Cho, S. Lee, Y.M. Jhon, Tensile characterization of single-walled carbon nanotubes with helical structural defects, Scien. Rep. 50 (2016) 20324.

[149] Lin Yang, Israel Greenfeld, H. Daniel Wagner, Toughness of carbon nanotubes conforms to classic fracture mechanics, Sci. Adv. 2 (2) (2016) el500969.

[150] G. Yamamoto, K. Shirasu, Y. Nozaka, T. Takagi, T. Hashida, Measurement of carbon nanotube tensile strength and mechanical properties of nanotube/alumina composite prepared by pressureless sintering, In: Proceedings of ECCM15—15th European Conference on Composite Materials, 2012, pp. 1–8.

[151] G. Yamamoto, K. Shirasu, Y. Nozaka, W. Wang, T. Hashida, Microstructure property relationships in pressureless-sintered carbon nanotube/alumina composites, Mater. Sci. Eng. A 617 (2014) 179–186.

[152] B. Zhang, L. Zhao, Y. Cheng, D. Golberg, M.-S. Wang, Reversible tuning of individual carbon nanotube mechanical properties via defect engineering, Nano Lett. 16 (8) (2016) 5221–5227.

[153] E.T. Thostenson, T.-W. Chou, On the elastic properties of carbon nanotube-based composites: modelling and characterization, J. Phys. D 36 (5) (2003) 573.

[154] C. Li, T.-W. Chou, Modeling of Carbon Nanotubes and Their Composites, Springer, Netherlands, Dordrecht, (2006) pp. 55–65.

[155] R. Rafiee, R.M. Moghadam, On the modeling of carbon nanotubes: a critical review, Compos. Part B 56 (2014) 435–449.

9

Mechanical Properties of CNT/Polymer

Antonio Pantano
UNIVERSITY OF PALERMO, PALERMO, ITALY

CHAPTER OUTLINE

9.1 Introduction

This chapter focuses on the mechanical properties of CNTs reinforced composites made with several different polymer matrixes.

One way to take advantage of the marvelous properties of the carbon nanotubes (CNTs) consists in incorporating them into a matrix to build composite materials. The best candidates for this task are undoubtedly polymers, which thanks to their strength, toughness, low weight, and easy processing have been used in a broad variety of industrial application. The extraordinary mechanical properties, together with high ratios (100-10,000) of geometric aspect, stiffness-to-weight, and strength-to-weight, all point to CNTs as potentially ideal reinforcing agents in advanced composites. But not only stiffness and strength

Carbon Nanotube–Reinforced Polymers. http://dx.doi.org/10.1016/B978-0-323-48221-9.00009-1

of the polymers can be improved by adding CNTs, also thermal and electrical conductivities, optical properties, toughness, fatigue resistance, and damping characteristics of formed composites can be enhanced. However, before seeing an extensive use of CNTs enhanced polymer composites there are a few difficult challenges that need to be addressed, in particular it is important to develop inexpensive mass production techniques for CNTs, to be able to accurately control their geometrical features (like diameter, length, and chirality), to achieve the ability to disperse the CNTs homogeneously throughout the matrix, and to efficiently transfer mechanical load from the matrix to the CNTs.

Table 9.1 shows available data in this field. The table is organized as follows: the type of polymer used as matrix is reported in column 1. CNT type, single-walled carbon nanotubes (SWCNTs) or multiwalled carbon nanotubes (MWCNTs), is given in column 2. In column 3 the processing methods are indicated. Column 4 shows at what weight percentage of CNTs the properties reported in the following columns of the row have been measured. In column 5 the improvement in the composite modulus due to the reinforcing action of the CNTs is given in percentage with respect to the modulus of the neat polymer. In a similar way, the improvement in the composite tensile strength, elongation at break, and toughness are reported, respectively, in columns 6, 7, and 8. The reference numbers are listed in column 9.

9.2 Polyethylene–Carbon Nanotube Composites

El Achaby et al. [1] synthesized high-density polyethylene (HDPE)/MWCNT nanocomposites by the melt mixing method and measured their mechanical properties. In the HDPE/MWCNTs nanocomposite the tensile strength and modulus were improved, with respect to neat HDPE, by 57 and 58%, respectively.

Ruan et al. [2] reported a hardness drastically improved in high-strength and ultrahigh molecular weight form polyethylene (UHMWPE) films for the addition of 1% by weight of MWCNTs. A combination of tensile and Raman spectroscopic measurements have shown that the presence of MWCNTs in the composite can lead to an increase of 150% in the strain energy density in comparison with the film to pure UHMWPE reports of similar stretching. This is accompanied by an increase of 140% in the ductility and up to 25% of the tensile strength. Ruan et al. attributed the observations for the improvement of the chain mobility UHMWPE induced by MWCNTs.

Tang et al. [3] used HDPE as matrix material for a CNT composites. In order to evaluate the mechanical properties of the HDPE–nanotube composite, they constructed and implemented a small punch test technique. The punch test results demonstrated that the stiffness, the yield strength, and the fracture toughness of MWCNT/HDPE composite films all increase with growing percentage of MWCNTs.

Bin et al. [4] prepared MWCNTs and UHMWPE composites by gelation/crystallization from solutions. Several experiment allowed to determine that for their composites 15 wt.% is the optimum content of MWCNTs. The composite film could be elongated up to 100-fold; the Young's modulus reached 58 GPa.

Table 9.1 Mechanical Properties of CNT-Based Polymer Composites

Polymer Matrix	Nanotube Type	Sample Type	CNT Weight Fraction (%)	Composite Modulus Improvement (%)	Composite Tensile Strength Increase (%)	Composite Elongation at Break ε (%)	Toughness Improvement (%)	References
UHMWPE	MWNTs	Melt mixing	3	58	57			1
UHMWPE	MWNTs	Solution casting	1	38	49	300		2
HDPE	MWNTs	Melt mixing	5	10	12		5	3
UHMWPE	MWNTs	Gelation/crystallization from solution	15			−250		4
PE	MWNTs	Polyethylene-grafted	0.5	50	28	4	60	5
PMMA	MWNTs	Melt mixing	1	0				6
PMMA	MWNTs	Solvent blending–melt mixing—fiber spinning	8	100	26	−2300		7
PMMA	MWNTs	Melt mixing	1	0	0		170	8
PMMA	MWNTs	Polymer grafting	0.10	103	85	3.8		9
PP	MWNTs	Ultrasonication	3	15		−240		10
PP	MWNTs	Pan milling–melt mixing	1	36	0	6.7		11
PP	MWNTs	Shear mixing	0.75	38	15			12
PP	MWNTs	Ultrasonication and stir	2.1	37				13
PVA	MWNTs	Wet spinning	5	16	36	−19		14
PVA	MWNTs	Casting and water evaporation	1	100	9		30	15
PVA	MWNTs	Solution casting	60					16
PVA	MWNTs	Soaking nanotube sheets (Buckypaper)		300	900		3000	17
PS	MWNTs	In situ emulsion	3	50	200	200		19
PS	MWNTs	Solution mixing	0.26	55	44			20
PS	MWNTs	Melt mixing	25	236	113			21
PS	MWNTs	Spin and film casting	5	120	50			22
PVC	MWNTs	Bath sonication	0.5	696	−10	−60		23
PVC	MWNTs	Solution mixing	0.33	60				20
PVC	MWNTs	Solution mixing	0.20	40	74		145	24
SBA	MWNTs	Solution mixing	3	300				25
SBA	MWNTs	Solution mixing	15	600				26

(continued)

Table 9.1 Mechanical Properties of CNT-Based Polymer Composites (*cont.*)

Polymer Matrix	Nanotube Type	Sample Type	CNT Weight Fraction (%)	Composite Modulus Improvement (%)	Composite Tensile Strength Increase (%)	Composite Elongation at Break ε (%)	Toughness Improvement (%)	References
Epoxy	MWNTs	Solution mixing	0.5	20	14	16		27
Epoxy	MWNTs	Solution mixing	1.5	17	16	−23		28
Epoxy	MWNTs	Solution mixing	2	26	50			31
Nylon-6	MWNTs	Melt extrusion	1	15	46			32
Nylon-6	MWNTs	Melt blending	2	213	161	−27		33
Nylon-6	MWNTs	Melt extrusion	5	29				34
Nylon-6	MWNTs	Melt compounding	1	115	124			35
Nylon-610	MWNTs	In situ polymerization	1.5	170	40	25		36
PI	MWNTs	In situ polymerization	0.4	27	47			37
PI	MWNTs	In situ polymerization	1.89	6				38
PI	SWNTs	Melt extrusion	1	10	9.6	12	105	39
PI	MWNTs	Ultrasonication	7	61	31			40
SBBS	MWNTs	Melt extrusion	3	102	22.9	−17		41
PEI	SWNTs	Layer-by-layer deposition	50	3000	2700			42

Yang et al. [5] created a composite adding MWCNTs to a polyethylene (PE) polymer matrix. A small amount of maleic anhydride group has been grafted to MWCNT-NH_2 by a simple melt blending process. PE-*g*-MWCNTs are well dispersed in the PE matrix, leading to a highly effective reinforcement effect. The storage modulus, Young's modulus, tensile strength, yield stress, ultimate strain, and toughness of PE are all improved upon the addition of PE-*g*-MWCNTs. The retention of ductility and the improvement of toughness of PE by PE-*g*-MWCNTs are particularly remarkable. On the other hand, the addition of as-received MWCNTs to PE leads to marginal increases in the Young's modulus and yield stress, but significant decreases in the ductility and toughness. SEM images of the two composites are shown in Fig. 9.1. The stress–strain curves of composites of PE reinforced with PE-*g*-MWCNTs are shown in Fig. 9.2A. For comparison, composites of PE reinforced with as-received MWCNTs were also studied, and their stress–strain curves and

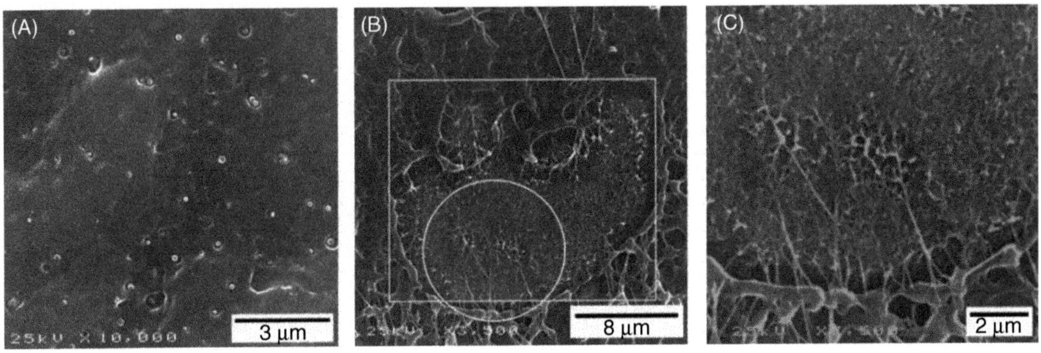

FIGURE 9.1 SEM images of composites. (A) PE/PE-*g*-MWCNTcomposite; (B) PE/MWCNT composite; (C) enlarged section of (B). *Reproduced with permission from B.X. Yang, K.P. Pramoda, G.Q. Xu, S.H. Goh, Adv. Funct. Mater. 17 (2007) 2062 [5]. Copyright 2007, Wiley.*

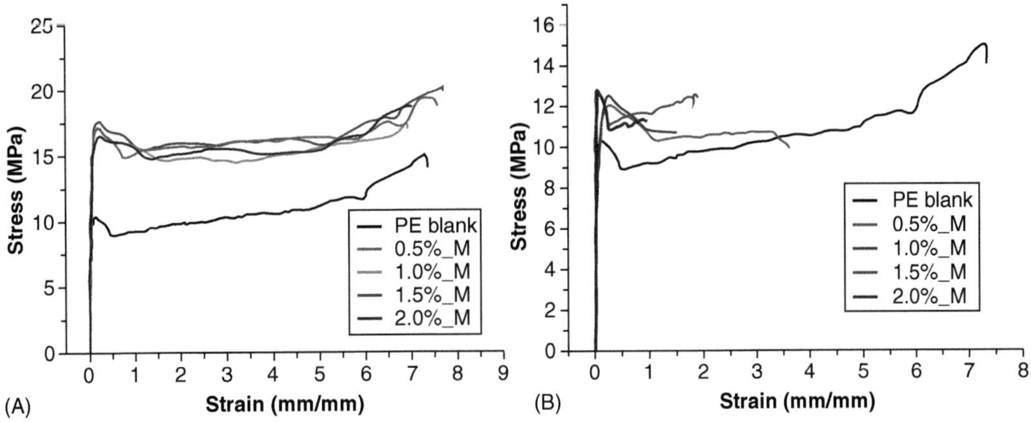

FIGURE 9.2 Stress–strain curves. (A) PE/PE-*g*-MWCNTs composites; (B) PE/MWCNTs composites. *Reproduced with permission from B.X. Yang, K.P. Pramoda, G.Q. Xu, S.H. Goh, Adv. Funct. Mater. 17 (2007) 2062 [5]. Copyright 2007, Wiley.*

mechanical properties are shown in Fig. 9.2B. The addition of as-received MWCNTs improves the Young's modulus and yield stress. However, the enhancement is not as good as that brought by PE-*g*-MWCNTs.

9.3 Polymethyl Methacrylate–Carbon Nanotube Composites

Zamfirova et al. [6] produced nanocomposites based on poly (methyl methacrylate) (PMMA) and 1% in weight MWCNTs, both functionalized and not. The samples were studied by the depth sensing indentation methods based on measurement of load–displacement curves at a constant loading speed. The dynamic hardness, Martens hardness, indentation hardness, indentation modulus, and the elastic part of the deformation work were measured. The material characteristics were determined revealing basic mechanical behavior of the samples. Elasticity does not differ significantly for all samples. The deformability reduction after the addition of the CNTs is due to the reduction of the plastic deformation component. This macroscopic result is a consequence of MWCNTs constraining effect on some segments of the backbone chain.

Haggenmueller et al. [7] used combinations of solvent casting and melt mixing to disperse SWCNT materials in PMMA. Composite fibers were melt spun. The elastic modulus and yield strength of SWCNT–PMMA composite fibers increased with nanotube loading and draw ratio. For all fibers, the modulus increased somewhat as the draw ratio increases, while the yield stress shows a stronger dependence on draw ratio. For example, the yield stress in nanocomposite fibers with 5 wt.% purified soot approximately doubles as the draw ratio increases from 40 to 300. Nanocomposite fibers with >1 wt% nanotubes failed at strains only slightly greater than the yield point.

Gorga and coworkers [8] melt-mixed SWCNT/PMMA nanocomposites. The orientation of MWCNTs in PMMA proved to be the only way to substantially toughen the nanocomposite. The mechanical properties of MWCNT/PMMA nanocomposites were studied as a function of nanotube orientation, length, concentration, and type. A level of 1 wt.% MWCNTs in PMMA (oriented nanocomposite) exhibited the largest increase in tensile toughness with a 170% improvement over oriented PMMA. Growths in the modulus and yield strength were notable only at a high loading of 10 wt.% with increases of 38% and 25%, respectively. None of the nanotube/PMMA composites showed mechanical properties close to the values expected from simple rule of mixture and orientation considerations. Orientation of MWCNTs in PMMA proved to be the only way to substantially toughen the nanocomposite.

Kim and coworkers [9] added MWCNTs to PMMA using poly(3-hexylthiophene) (P3HT)-graft-PMMA as a compatibilizer. P3HT-*g*-PMMA proved to be very effective for dispersing MWCNT in PMMA matrix and in improving mechanical properties of the resultant composite. The fluorescence emission and Raman spectra pinpoint the strong π–π interaction between P3HT-*g*-PMMA and MWCNT in the composite. The resulting composite has increased both Young's modulus and tensile strength of PMMA matrix without sacrifice of elongation at break. Fig. 9.3 shows stress–strain curves for neat PMMA, PMMA/MWCNT, and PMMA/MWCNT/P3HT-*g*-PMMA with various MWCNT contents.

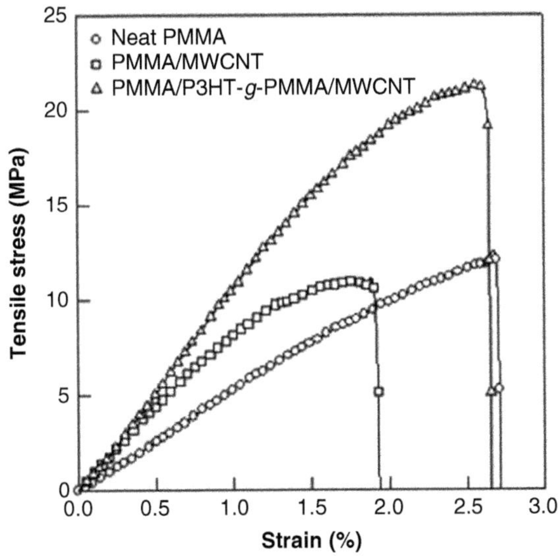

FIGURE 9.3 Stress–strain curves for neat PMMA, PMMA/MWCNT, and PMMA/MWCNT/P3HT-*g*-PMMA with various MWCNT contents. *MWCNT*, Multiwalled carbon nanotubes; *PMMA*, poly (methyl methacrylate). *Reproduced with permission from K.H. Kim, W.H. Jo, Compos. Sci. Technol. 68 (2008) 2120 [9]. Copyright 2008, Elsevier.*

9.4 Polypropylene–Carbon Nanotube Composites

Zhong et al. [10] studied the effect of ultrasonic treatment on dispersion of CNTs, graphene nanoplates, and carbon black in polypropylene (PP) using twin screw extrusion. The efficiency of the ultrasonic treatment was verified through measurements of rheological, electrical, and mechanical properties. The untreated PP/CNT composites have shown a worse dispersion because of the presence of large primary agglomerates of CNT. The yield stress and Young's modulus of all the composites increase with filler concentration with a saturation of the reinforcement at high concentrations. At a 3 wt.% loading of MWCNTs the Young's modulus improved by 15%, while the elongation at break reduced by 240%.

Xia et al. [11] prepared a PP/MWCNT composite in which good dispersion is obtained using a novel solid-state mechanochemical pulverizing process. A laboratory-made pan mill was used, it was then melt-mixed. The process had the following three functions: first, it improved the cohesion between the polymer and CNTs by mechanochemical effects; second, it cut the CNTs and reduced the number of defects and entanglements of long CNTs; and third, it improved the dispersion of the CNTs in PP. During pan milling, the original CNTs, a few micrometers long, were cut to 0.4–0.5 μm. For the unmilled composite a modest 8.8% increase in the elastic modulus was observed in a composite with a 3 wt.% loading of MWCNTs, while when the pan mill used a 36% improvement in the modulus was obtained with a 1 wt.% loading of MWCNTs.

Manchado et al. [12] dispersed SWCNTs in isotactic polypropylene (iPP) by shear mixing. From their results, it is deduced that the reinforcing effect of SWCNTs is very marked as compared to carbon black. As the SWCNT content in the polymer increases, the Young's modulus gradually increases. Moreover, the tensile modulus goes from 0.85 GPa for pristine PP to 1.19 GPa for the 0.75 wt.% nanocomposite. However, it is worth mentioning that a further increase in SWCNT proportion in the composite (1 wt.%) provides a marked decrease in the tensile modulus, whereas when carbon black is used as a filler, the property gradually increases with an increase in filler content. This is probably due to the fact that at the highest SWCNT concentration in the composite the formation of aggregates is higher diminishing the interfacial filler–polymer adhesion. It is also pointed out that this improvement on the stiffness and strength of the composite by adding SWCNTs occurs without any reduction of the deformation at break of the material.

Koval'chuk et al. [13] functionalized MWCNT via attachment of $(CH_2)_{10}CH_3$ alkyl chains in order to optimizing PP/MWCNT nanocomposite properties. They obtained good nanotube dispersion and nanocomposites combining enhanced stiffness and improved plasticity with respect to unmodified MWCNTs. However, the interfaces between nonpolar polyolefin matrices and CNTs is not strong enough. For the authors the most encouraging approach for improving the interfacial interactions between polymer chains and filler particles is the incorporation of polar functional groups in PP macromolecules. A clear mechanical reinforcement was observed in the iPP and syndiotactic polypropylene (sPP) composites even at low loading levels, for example, at 0.1 wt.% MWCNT loading Young's modulus of iPP increases by 22%, and modulus of sPP grows by 34% as shown in Fig. 9.4. Modulus keeps growing up at higher filler concentrations. Ultimate Young's modulus improvement of iPP is 37% at 2.1 wt.% MWCNT content. For the sPP/MWCNT composites maximal mechanical reinforcement is achieved at 0.4 wt.% nanotube loading with a 66% increase in the Young's modulus. Posterior modulus reduction at higher nanotube contents signifies considerable impact of MWCNT aggregation that takes place in this system. Testing on the ultimate tensile strength of the iPP/MWCNT and iPP/C_{11}–MWCNT nanocomposites prove that it is little affected by nanotube loading. This reveals limited load transfer between the polymer matrix and modified nanotubes (Fig. 9.5).

9.5 Polyvinyl Alcohol–Carbon Nanotube Composites

Lai et al. [14] produced by wet-spinning process PVA composite fibers with a large fraction of MWNTs modified by both covalent and noncovalent functionalization. The Young's modulus of composite fibers enhanced with increasing the content of MWNTs, and it raised gradually from 6.7 GPa for the pure PVA fiber to 12.8 GPa for the composite fibers with 27.6 wt.% MWNTs. Meanwhile, the tensile strength increased gradually from 0.39 GPa for the pure PVA fiber to 0.74 GPa for the composite fibers with 14.4 wt.% MWNTs. Nevertheless, the tensile strength of the composite fibers decreased as the MWNTs content up to 27.6 wt.%.

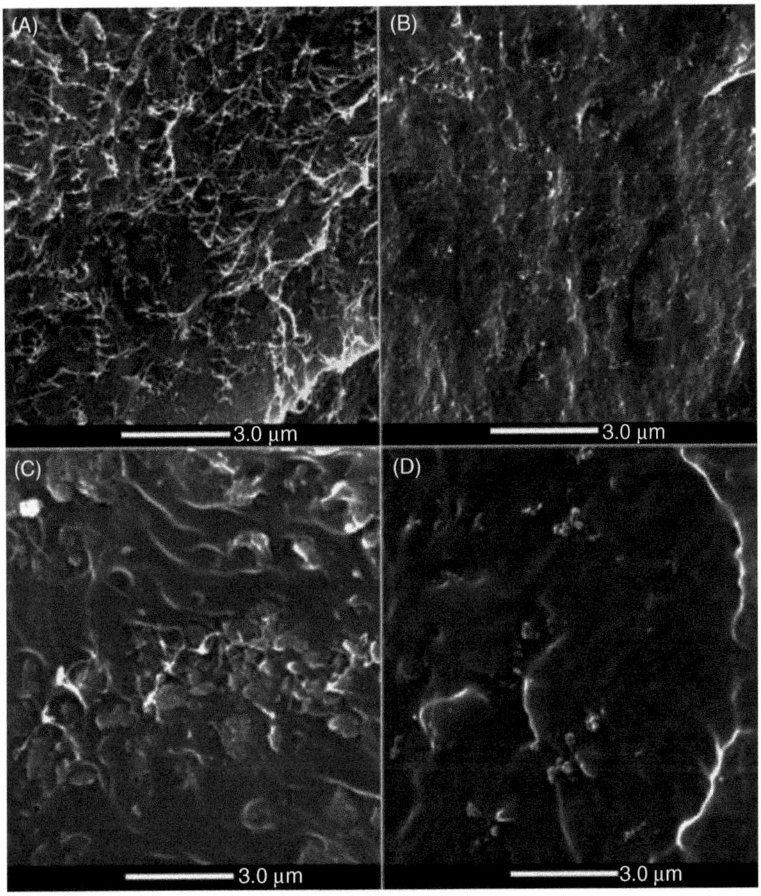

FIGURE 9.4 SEM images of the iPP and sPP nanocomposites. iPP/1.4 wt.% MWCNT (A), iPP/1.2 wt.% C_{11}-MWCNT (B), sPP/1.5 wt.% MWCNT (C), and sPP/1.5 wt.% C_{11}-MWCNT (D). *iPP*, Isotactic polypropylene; *sPP*, syndiotactic polypropylene. *Reproduced with permission from A.A. Koval'chuk, V.G. Shevchenko, A.N. Shchegolikhin, P.M. Nedorezova, A.N. Klyamkina, A.M. Aladyshev, Macromolecules 41 (2008) 7536 [13]. Copyright 2008, American Chemical Society.*

Cadek et al. [15] manufactured composite films of polyvinyl alcohol (PVA) and six different types of CNTs for different nanotube loading levels. Tensile tests were carried out on freestanding composite films of PVA and six different types of CNTs for different nanotube loading levels. Significant increases in Young's modulus by up to a factor of 2 were observed in all cases. Theories, such as the rule-of-mixtures or the Halpin–Tsai theory could not explain the relative differences between composites made from different tube types. They found that the reinforcement scales linearly with the total nanotube surface area in the films, indicating that low diameter multiwall nanotubes are the best tube type for reinforcement. In addition, in all cases crystalline coatings around the nanotubes were detected by calorimetry, suggesting compatible polymer–nanotube interfaces. Thus, the reinforcement appears to be critically dependent on the polymer–nanotube interfacial interaction.

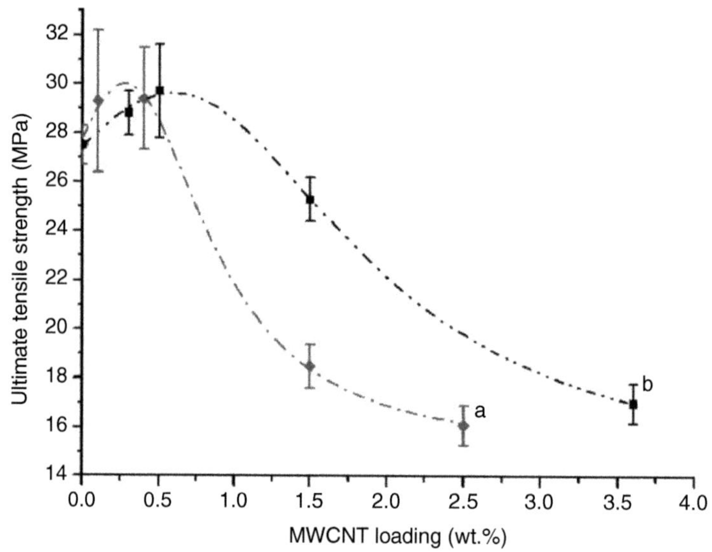

FIGURE 9.5 Young's modulus of iPP/MWCNT (a) and iPP/C_{11}–MWCNT (b) nanocomposites as a function of MWCNT loading. *Reproduced with permission from A.A. Koval'chuk, V.G. Shevchenko, A.N. Shchegolikhin, P.M. Nedorezova, A.N. Klyamkina, A.M. Aladyshev, Macromolecules, 41 (2008) 7536 [13]. Copyright 2008, American Chemical Society.*

Shaffer and coworkers [16] fabricated composite films containing nanotubes, based on the formation of a stable colloidal intermediate. The densities of the composites followed a linear law of mixtures, and the electrical conductivities showed typical percolation behavior. The presence of the nanotubes stiffened the material, particularly at high temperature, and in some cases retards the onset of thermal degradation. The indications from this work are that nanotubes may not ideally be suited to a straightforward reinforcing role, but that they could find application as a modifier for polymers, particularly as an improved matrix for conventional fiber composites for service at high temperatures.

Coleman et al. [17] intercalated organic polymers, such as PVA, polyvinyl pyrrolidone (PVP), and polystyrene (PS), into SWCNT sheets by soaking the sheets in polymer solutions. In all cases the Young's modulus, tensile strength, and toughness were calculated. The mechanical properties of the pristine sheet were: Young's modulus = 2.3 GPa, tensile strength = 6.29 MPa, and toughness = 1.41×10^4 J/m^3. The properties of the polymer intercalated sheets resulted significantly increased in all cases. The modulus was increased monotonically with loading level up to 6.9 GPa for the PVA. Strength for the PVA sample was increased up to 57 MPa. PS was clearly the best reinforcing agent, followed by PVP, with PVA as a close third. This result is reasonable in terms of the relative hydrophobicities of these polymers. A very hydrophobic polymer like PS is expected to interact more strongly with SWCNT than such amphiphilic materials as PVP or PVA, resulting in good stress transfer.

Vigolo et al. [18] produced SWCNTs in macroscopic fibers by a spinning process. Stress versus strain curve of a raw and a stretched fiber are made from the same batch of CNTs. Young's modulus and the tensile strength are substantially higher for the stretched fiber. They are, respectively, of about 10 GPa and 125 MPa for the raw fiber and 40 GPa and 230 MPa for the stretched fiber. The fiber has been stretched by 145% in water.

9.6 Polystyrene–Carbon Nanotube Composites

Lahelin et al. [19] incorporated CNTs into PS and PMMA matrices via in situ emulsion and emulsion/suspension polymerization methods. The loading of CNTs in the composites varied from 0 to 15 wt.%, depending on the CNTs used. The dispersion of MWCNT in the composites was excellent. By adding a 3.0 wt.% of pristine MWCNTs to the PS, the mechanical properties were improved; the Young's modulus was 50% higher, and the strain at break values doubled.

Blake et al. [20] functionalized CNTs with chlorinated PP. PS polymer–composite films were prepared by a solution casting technique. Good improvement in the mechanical properties were found, but modulus and strength could not be sustained to high-nanotube volume fractions. PS-CNT composites reached their maximum for both modulus and strength at $V_f = 0.26$ vol.%. Along a fracture surface it was possible to discover nanotube pullout rather than nanotube failure. Pullout of the functionalized nanotubes happened without any detachment of the functional groups from the nanotube (Figs. 9.6 and 9.7).

Andrews et al. [21] used shear mixing to produce PS/MWCNT composites. Low concentrations of MWCNT in polymer matrices generally resulted in an increase in Young's modulus and reduction in the tensile strength. At higher concentrations, both stiffness and strength were significantly improved. Removal of defects and impurities from the nanotubes by graphitization further improves the tensile properties of the composites. Observation of the tensile failure mechanism suggests that surface treatment of the nanotubes could be used to improve interfacial bonding and increase tensile strength.

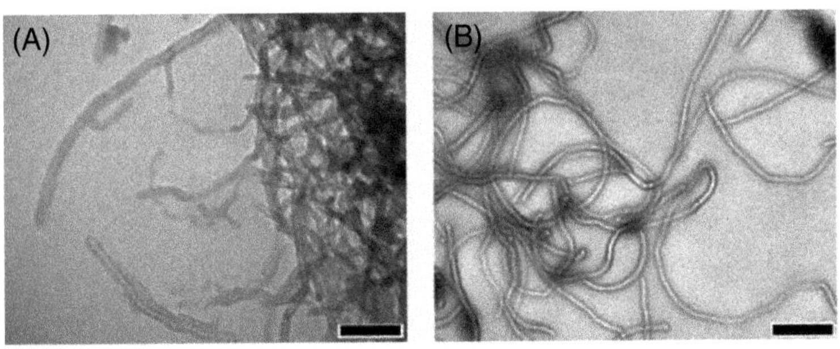

FIGURE 9.6 TEM images of CPP grafted MWCNTs (A) and pristine MWCNTs (B). *Reproduced with permission from R. Blake, J.N. Coleman, M.T. Byrne, J.E. McCarthy, T.S. Perova, W.J. Blau, A. Fonseca, J.B. Nagy, Y.K. Gun'ko, J. Mater. Chem. 16 (2006) 4206 [20]. Copyright 2006, RSC Publishing.*

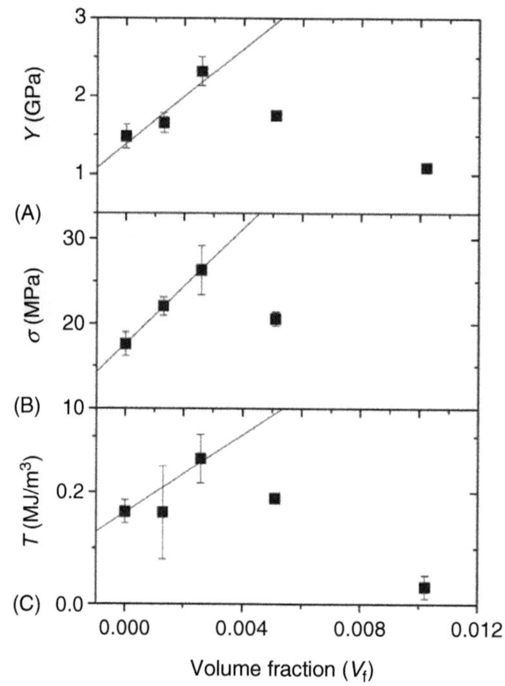

FIGURE 9.7 Young's modulus (A), strength (B), and toughness (C) as functions of volume fraction for CPP grafted nanotubes in polystyrene (PS). *Reproduced with permission from R. Blake, J.N. Coleman, M.T. Byrne, J.E. McCarthy, T.S. Perova, W.J. Blau, A. Fonseca, J.B. Nagy, Y.K. Gun'ko, J. Mater. Chem. 16 (2006) 4206 [20]. Copyright 2006, RSC Publishing.*

When interfacial adhesion is poor, the nanotubes pull out of the matrix. Controlling this failure mechanism will be instrumental to achieving true superstrong materials based on CNTs.

Safadi et al. [22] studied the basic relationships between processing conditions and the mechanical and electrical properties of MWCNT reinforced polymer composites. The tensile modulus and strain to failure of samples compared well to classical micromechanical models, increasing with MWCNT loading. The composite films presented lower strains at the yield stress than neat PS films. The presence of MWCNTs at 2.5 vol.% fraction approximately doubles the tensile modulus.

9.7 Polyvinyl Chloride–Carbon Nanotube Composites

Yazdani et al. [23] examined the mechanical properties of MWCNT-filled polyvinyl chloride (MWCNT/PVC) composites at 0.5 wt.%. The influence of MWCNT on the failure strain is found to be significant, because the failure strain of the pristine matrix is significantly reduced, up to 60%, when mixed with MWCNT regardless of the dispersion quality.

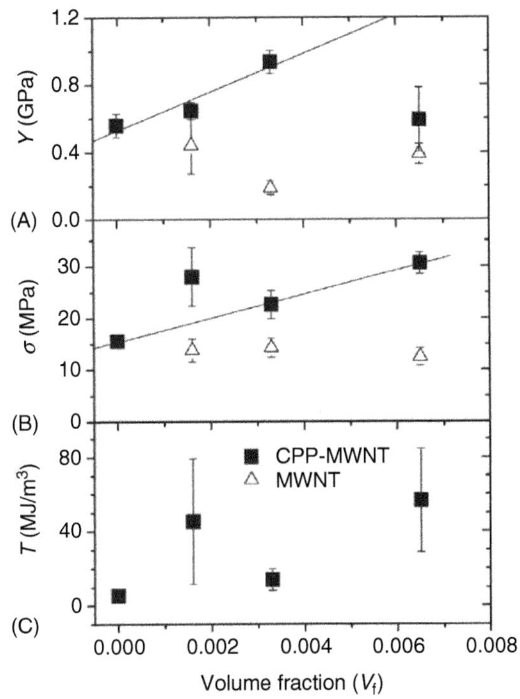

FIGURE 9.8 Young's modulus (A), strength (B), and toughness (C) as functions of volume fraction for CPP grafted nanotubes in PVC. The open triangles represent data for unfunctionalized MWCNT in PVC. *Reproduced with permission from R. Blake, J.N. Coleman, M.T. Byrne, J.E. McCarthy, T.S. Perova, W.J. Blau, A. Fonseca, J.B. Nagy, Y.K. Gun'ko, J. Mater. Chem. 16 (2006) 4206 [20]. Copyright 2006, RSC Publishing.*

A nonuniform dispersion of CNT particles and bundles has been observed to reduce the tensile strength of an otherwise pristine matrix, at least for the range of concentrations examined in the study. Reduction in tensile strength in the latter case may be attributed to the formation of stress concentration zones at the polymer–CNT interface around CNT bundles, promoting the failure of the composite.

In addition to PS-CNT composites, Blake et al. [20] have also produced PVC-based composites. Even in this case, significant mechanical reinforcement with increases in Young's modulus, strength and toughness was obtained. Fig. 9.8 shows Young's modulus, strength and toughness as functions of volume fraction.

Shi and coworkers [24] grafted poly *n*-butyl methacrylate (PBMA) onto MWCNTs using atom transfer radical polymerization. Thanks to the grafted PBMA, MWCNTs were better dispersed in PVC and the efficiency of load transfer from the matrix to the nanotubes was improved. The mechanical performances of PVC were greatly enhanced, it was found that the storage modulus, Young's modulus, yield stress, tensile strength, ultimate strain, and toughness were significantly improved even at low MWCNTs content. The miscibility between PVC and PBMA improving the load transfer from PVC to MWCNTs, produced an enhanced mechanical performance.

9.8 Polystyrene-*co*-Butyl Acrylate–Carbon Nanotube Composites

Dalmas et al. [25] prepared MWCNTs reinforced nanocomposite with an amorphous polystyrene-*co*-butyl acrylate [P(S-BuA)] latex. Good dispersion was obtained using an anionic surfactant. An improvement in the mechanical properties was recorded. Fig. 9.9 shows low-magnification TEM micrographs of a microsection from a P(S-BuA) film filled with 3 vol.% CNTs. Fig. 9.10 shows the true stress–strain curves for tensile tests performed on

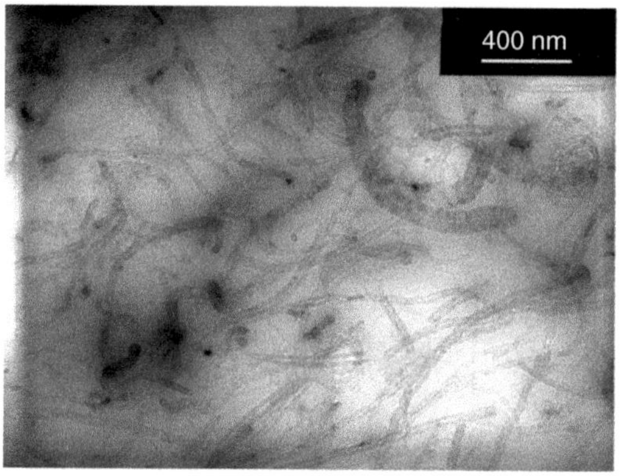

FIGURE 9.9 Low-magnification TEM micrographs of a microsection from a P(S-BuA) film filled with 3 vol.% CNTs. CNTs, Carbon nanotubes; P(S-BuA), polystyrene-co-butyl acrylate. *Reproduced with permission from F. Dalmas, L. Chazeau, C. Gauthier, K. Masenelli-Varlot, R. Dendievel, J.Y. Cavaillé, J. Polym. Sci. B 43 (2005) 1186 [25]. Copyright 2005, Wiley.*

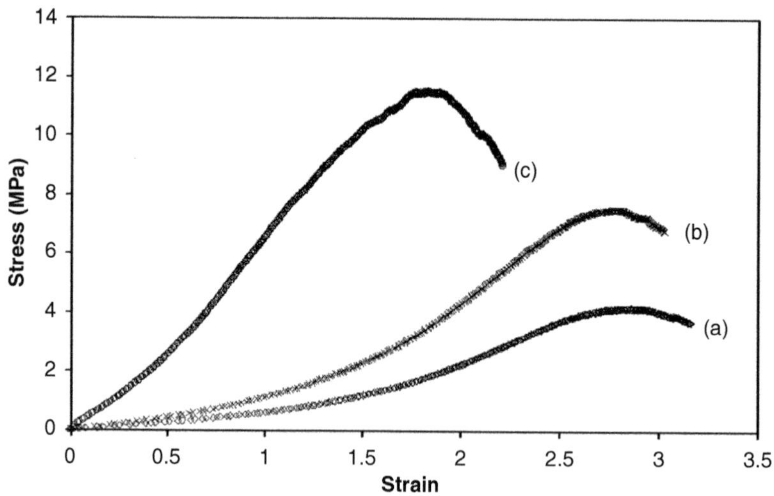

FIGURE 9.10 True stress–strain curves for tensile tests performed on (a) the unfilled P(S-BuA) matrix, (b) the related composite filled with 1.5 vol.% CNTs, and (c) the related composite filled with 3 vol.% CNTs. *Reproduced with permission from F. Dalmas, L. Chazeau, C. Gauthier, K. Masenelli-Varlot, R. Dendievel, J.Y. Cavaillé, J. Polym. Sci. B 43 (2005) 1186 [25]. Copyright 2005, Wiley.*

(1) the unfilled P(S-BuA) matrix, (2) the related composite filled with 1.5 vol.% CNTs, and (3) the related composite filled with 3 vol.% CNTs.

Dufresne et al. [26] mixed amorphous polystyrene-*co*-butyl acrylate latex and an aqueous suspension of CNTs. The mechanical characterization displayed a continuous reinforcing effect of the CNTs without lowering of the elongation at break up to 3 wt.%.

9.9 Epoxy–Carbon Nanotube Composites

Zhang et al. [27] prepared MWCNT-NH_2 with surface covalently linked ethylenediamine through an amidation reaction process. The reinforcement efficiencies of MWCNT-NH_2 and MWCNT-COOH in epoxy matrix and the mechanical and thermomechanical properties of the resulting carbon fiber/epoxy composites were studied. At a 0.5 wt.% loading of MWCNT, the Young's modulus improved by 20%, the strain at break by 16% and the tensile strength by 14%.

Song et al. [28] found that the epoxy composites embedded with poorly dispersed CNTs have higher tensile modulus. However, as the CNTs loading increases, tensile strength of the composite filled with well-dispersed CNTs increases but that of the composite filled with poorly dispersed CNTs decreases. The agglomerates present in poorly dispersed composite cause cracks to initiate and propagate easily. They found a 17% increase in tensile modulus of the composite for 1.5 wt.% MWCNTs. The limited improvements were attributed to the weak bonding between the CNTs and the surrounding matrix. If the interfacial bonding between the CNTs and the polymer resin is weak and the load transfer from the polymer to the CNTs is not large enough for the CNTs to be broken under tensile loading. Instead, the CNTs tend to be pulled out. Fig. 9.11 shows field-emission scanning electron microscopy (FESEM) images of CNTs/epoxy nanocomposites prepared without using the solvent. In Fig. 9.12 it is possible to notice the improvements in dispersion when the nanocomposites are prepared by using the solvent. Fig. 9.13 plots the graphs of the (1) tensile modulus and (2) tensile strength of CNTs/epoxy composites with respect to CNTs loading.

Rajoria and coworkers [29] studied damping properties of CNT–epoxy composites for use in structural vibration applications. Both single-walled and multiwalled nanotube-epoxy composites were prepared with different proportion of nanotubes. Free and forced vibration tests are conducted on cantilevered beams. Multiwalled nanotubes are observed to be a better reinforcement than single-walled nanotubes. Up to a 700% increase in damping ratio is observed for multiwalled a nanotube–epoxy beam as compared to the plain epoxy beam. Another important observation is the maximum increase in damping ratio versus nanotube weight ratio curve for both multiwalled nanotube and single-walled nanotube reinforcements. This indicates a limit on the enhancement in damping ratio that can be achieved by utilizing CNTs in epoxy.

Biercuk et al. [30] loaded an epoxy resin with 1 wt.% unpurified SWCNTs. The Vickers hardness raised monotonically with SWCNT loading up to a factor of 3.5 at 2 wt.%. These results suggest that the mechanical properties of SWCNT–epoxy composites are improved, without the need to chemically functionalize the nanotubes.

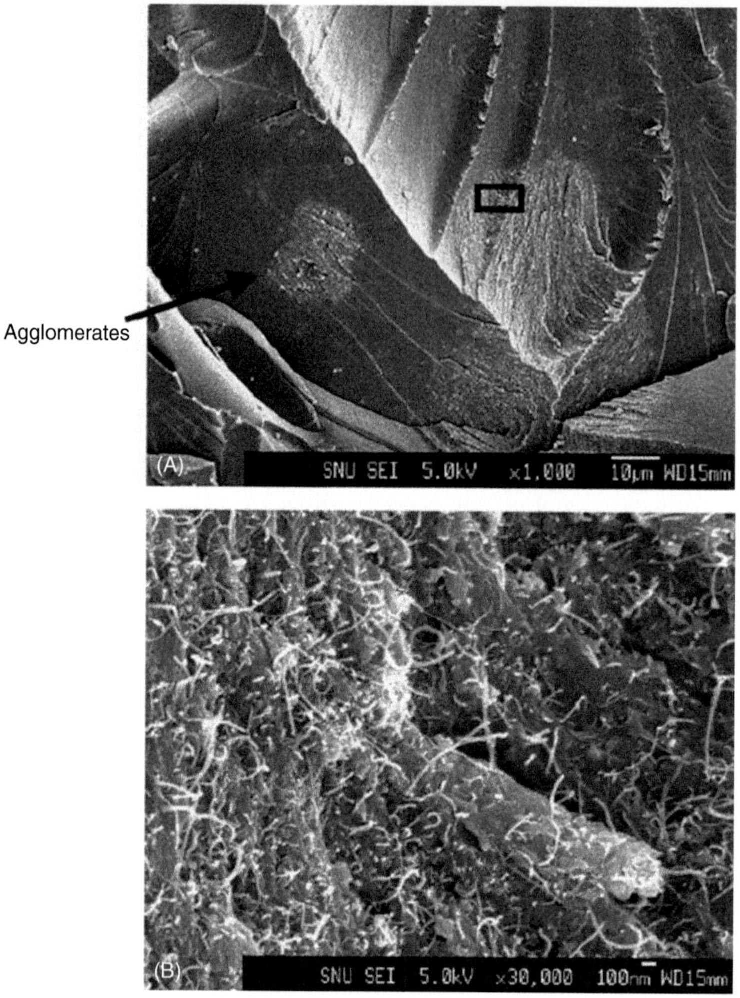

FIGURE 9.11 FESEM images of CNTs/epoxy nanocomposites prepared without using the solvent. *FESEM*, Field-emission scanning electron microscopy. *Reproduced with permission from Y.S. Song, J.R. Youn, Carbon 43 (2005) 1378 [28]. Copyright 2005, Elsevier.*

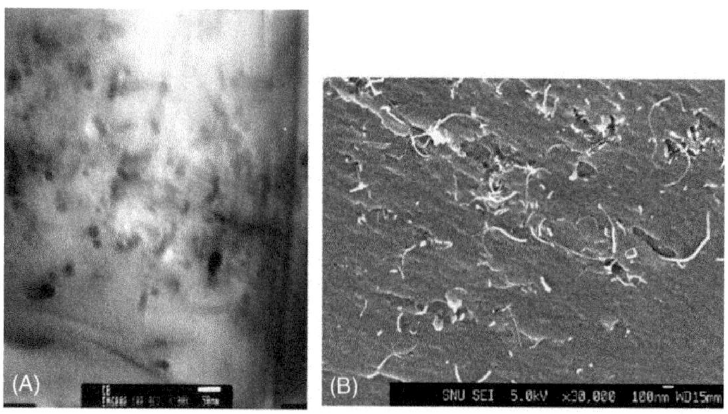

FIGURE 9.12 (A) TEM and (B) FESEM images of the fracture surface for nanocomposites prepared by using the solvent. *Reproduced with permission from Y.S. Song, J.R. Youn, Carbon 43 (2005) 1378 [28]. Copyright 2005, Elsevier.*

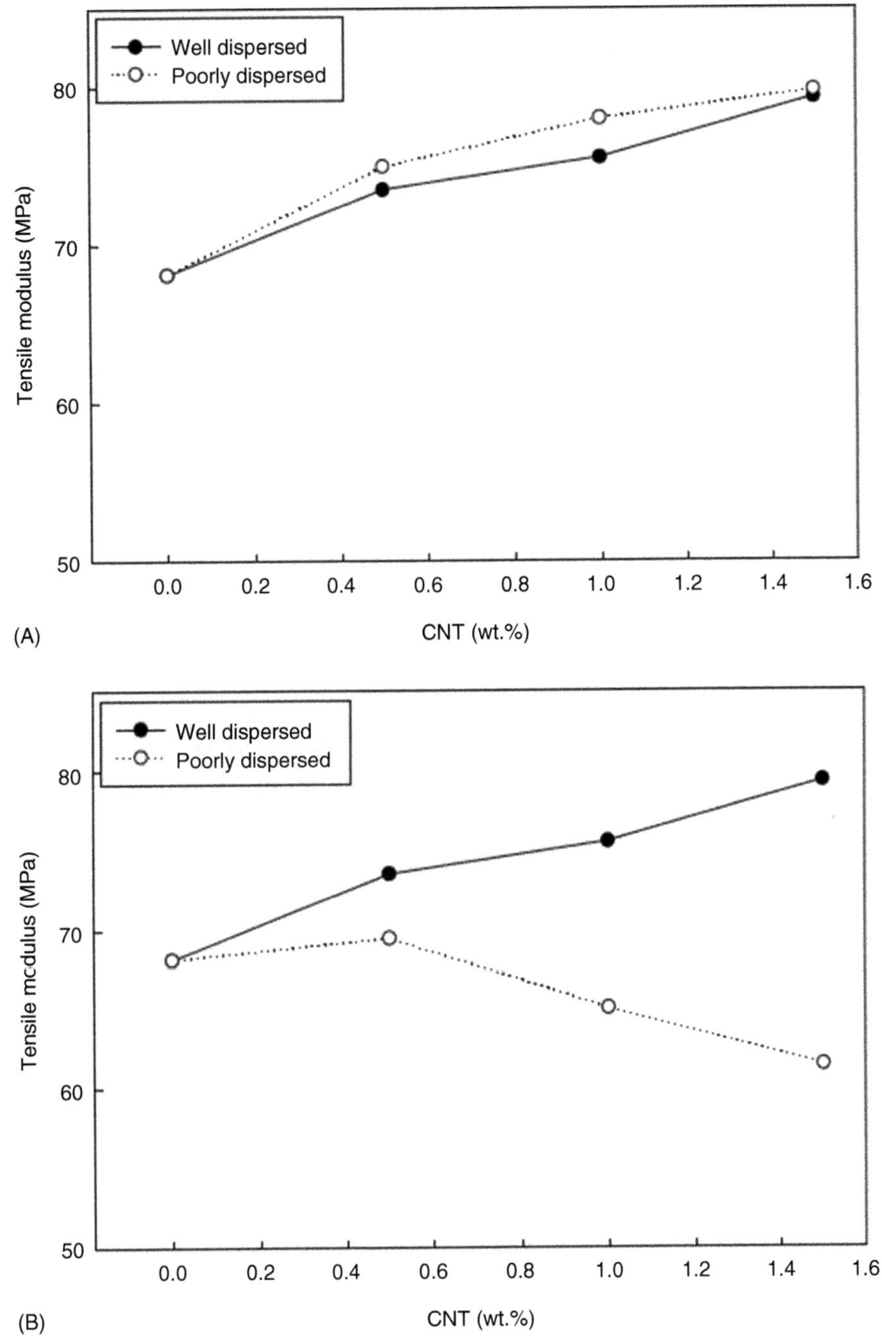

FIGURE 9.13 Mechanical properties of CNTs composites. (A) Tensile modulus and (B) tensile strength of CNTs/epoxy composites with respect to CNTs loading. *Reproduced with permission from Y.S. Song, J.R. Youn, Carbon 43 (2005) 1378 [28]. Copyright 2005, Elsevier.*

Park et al. [31] investigated the fiber damage and the reinforcing effect of carbon nanocomposites with their content. The fracture of carbon fiber was detected by nondestructive acoustic emission relating to electrical resistivity under double-matrix composites test. Tests proved that there was a good sensitivity for fiber damage, such as fiber fracture, matrix deformation, and fiber tension. Damage sensitivity by electrical resistance measurement might be related to closely three-dimensional network structure, percolation structure. A significant reinforcing effect was recorded and increased with CNT content. The best sensitivity for the testing technique were obtained for the 2.0 vol.% CNT case where a percolation structure could be observed.

9.10 Nylon–Carbon Nanotube Composites

Choi et al. [32] used pyrene derivatives as interfacial agents to improve the mechanical strength of the PA66/MWCNT composites. Among the composites examined, the PA66 composite fabricated with PBC-adsorbed MWCNTs exhibited the best interfacial adhesion between MWCNT and PA66 and the highest level of MWCNT dispersion in PA66 matrix. As the composite interfacial adhesion increased, so did the mechanical strength of the composite. The Young's modulus improved by 15% and the tensile strength by 46%.

Liu et al. [33] prepared MWCNTs/nylon-6 (PA6) nanocomposites by a melt-compounding approach. The mechanical properties improved quickly with the loading level of the nanotubes. A composite with only 2 wt.% MWCNTs reached an elastic modulus of 1242 MPa, about 213% higher than PA6 that has a modulus of 396 MPa. Yield strength was improved by about 161% from 18 to 47 MPa. There was a reduction in the elongation at break that decreased of about 27%, indicating that the composite became somewhat brittle compared with neat PA6. Toughness was strongly enhancement with respect to neat PA6. Nanoindentation testing revealed a significant difference in the loading–hold–unloading curves of neat PA6 and its nanocomposites as a function of CNTs content. During loading the forces are incremented at constant velocities and they keep getting higher with increasing the CNTs concentration, indicating that the nanocomposite's resistance to indentation increases with nanotube concentration. Typical stress–strain curves for neat PA6 and MWCNTs/PA6 composites with different CNTs loadings revealed a pronounced yield and postyield drop for neat PA6, which was not present for MWCNTs-reinforced PA6 composites.

Meincke et al. [34] have prepared a composite of polyamide-6 and CNTs. The polyamide-6 was extruded with the CNT using a corotating twinscrew extruder to produce the samples PA-NT-x (x being the weight percentage of nanotubes in the composite) and PA-CB-x (x being the weight percentage of carbon black in the composite), see Fig. 9.14. The investigated samples showed an increased modulus with increasing filler content, see Fig. 9.15. The CNTs were able to reinforce the polymer matrix much more than carbon blacks. The elongation at break was lowered, adding CNTs or carbon black. In both cases the material becomes brittle even at low filler contents.

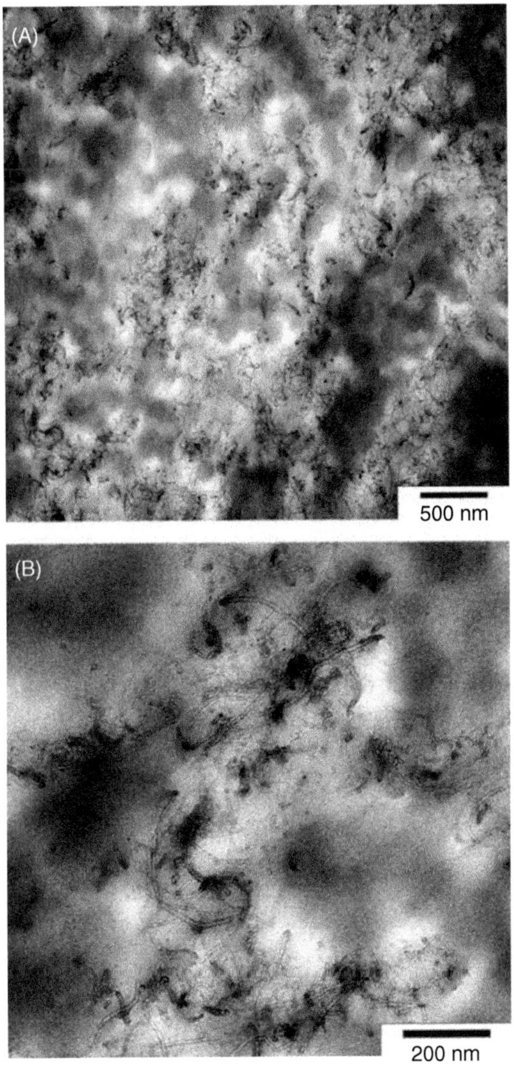

FIGURE 9.14 TEM-micrographs of the PA-ABS-NT-5 composite. (A) 500 nm scale bar, (B) 200 nm scale bar. The carbon nanotubes are selectively located in the polyamide-6 phase. *Reproduced with permission from O. Meincke, D. Kaempfer, H. Weickmann, C. Friedrich, M. Vathauer, H. Warth, Polymer 45 (2004) 739 [34]. Copyright 2004, Elsevier.*

De Zhang et al. [35] manufactured MWCNTs/nylon-6 nanocomposites by melt-compounding. Upon incorporation of only 1 wt.% MWCNTs, the elastic modulus of PA6 is improved by about 115% from 396.0 to 852.4 MPa, and the tensile strength at yield is improved by about 124% from 18.0 to 40.3 MPa. The elongation at break slightly decreases, indicating that the composite becomes somewhat brittle compared with neat PA6.

Kang et al. [36] produced polyhexamethylenesebacamide (nylon 610) nanocomposites containing well-dispersed MWCNTs via the in situ interfacial polymerization.

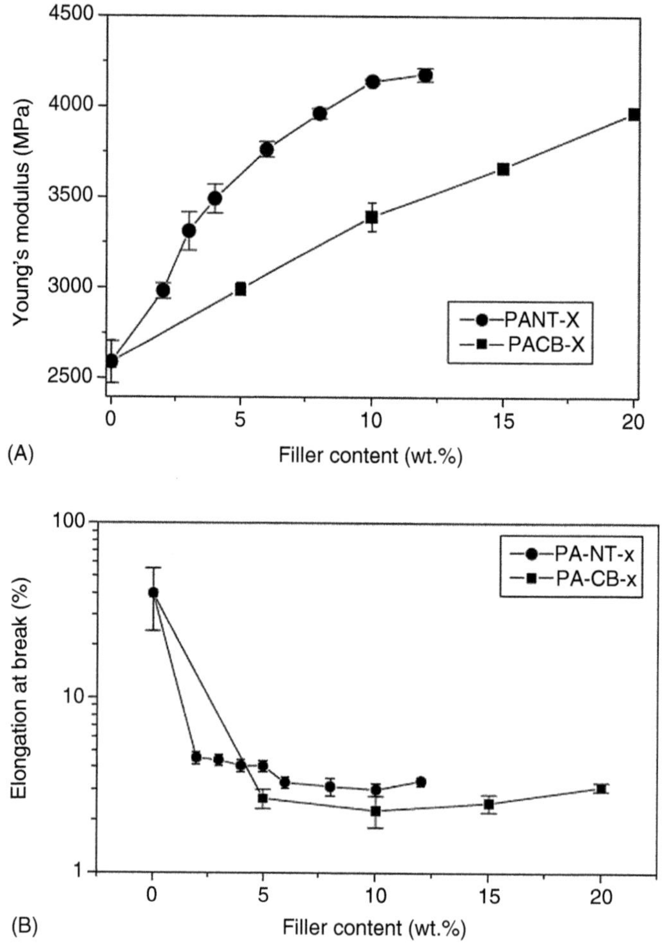

FIGURE 9.15 Young's moduli (A) and elongation at break (B) of PA-NT composites and PA-CB-composites plotted versus the filler content. *Reproduced with permission from O. Meincke, D. Kaempfer, H. Weickmann, C. Friedrich, M. Vathauer, H. Warth, Polymer 45 (2004) 739 [34]. Copyright 2004, Elsevier.*

Tensile tests of the composite sheet showed a 170% increase in the Young's modulus with slight increases in the tensile strength and the elongation at break (about 40% and 25%, respectively). This suggests an interaction between the acid-treated MWCNTs and nylon 610.

9.11 Polyimide–Carbon Nanotube Composites

Dong et al. [37] fabricated polyimide (PI)/NH_2–MWCNTs composite fibers. Different works suggested that there exist interfacial interactions between NH_2-MWCNTs and PIs, including the hydrogen bonding and π–π interaction. The resulting composite fiber containing

0.4 wt.% NH_2-MWCNTs presented a tensile strength of 2.41 GPa, a 47% increase over neat PI fiber, and a modulus of 99 GPa, a 27% raises compared with neat PI fiber.

Jiang et al. [38] fabricated composite films of PI-CNTs by in situ polymerization process of PI with MWCNTs. MWCNTs with different shapes and distribution in diameter, MWCNT-I and MWCNT-II were used as fillers. The MWCNT-I were synthesized by catalytic decomposition method and had a diameter in the range 10–20 nm with a length 10–20 μm, while the MWCNT-II had an average diameter around 20 nm and were produced by vapor ablation technique. The Young's modulus for the composite films of PI was measured by tensile tests for different MWCNTs contents. It was found to increase about 6% with a MWCNT content in the range 0.077–1.89 vol.%. The Young's modulus decreased slightly when MWCNT content further increase. Fig. 9.16 shows the SEM patterns of two kinds of CNTs. Fig. 9.17 shows the SEM pattern of the cross-section of a PI–MWCNT-II composite film.

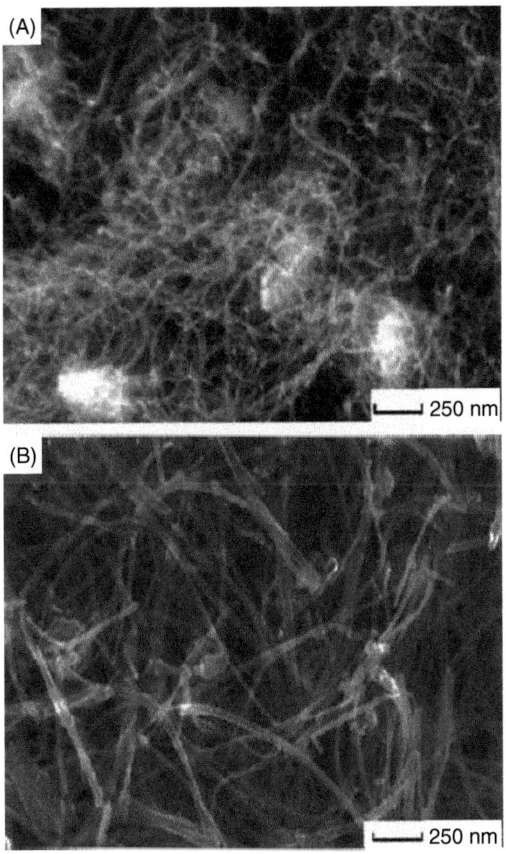

FIGURE 9.16 SEM patterns of two kinds of carbon nanotubes. (A) MWCNT-I, (B) MWCNT-II. *Reproduced with permission from X. Jiang, Y. Bin, M. Matsuo, Polymer 46 (2005) 7418 [38]. Copyright 2005, Elsevier.*

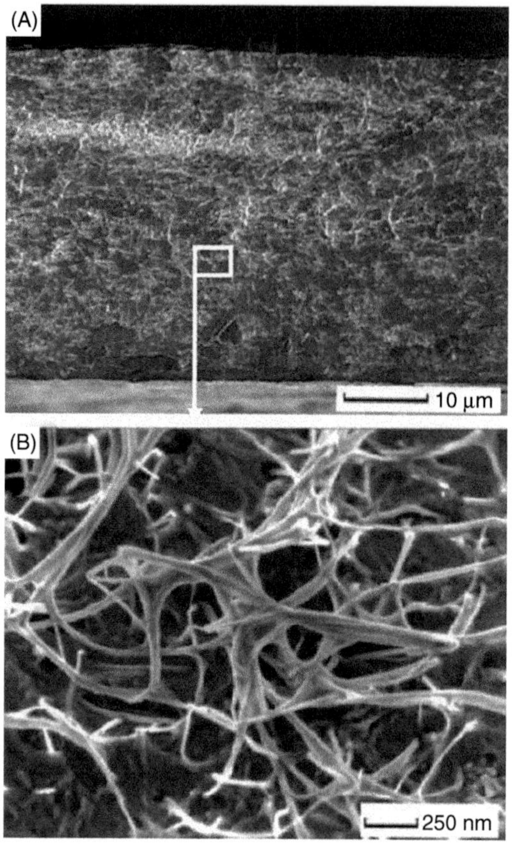

FIGURE 9.17 SEM pattern of the cross-section of PI–MWCNT-II (MWCNT = 1.14 vol.%) composite film. *PI,* Polyimide. *Reproduced with permission from X. Jiang, Y. Bin, M. Matsuo, Polymer 46 (2005) 7418 [38]. Copyright 2005, Elsevier.*

Siochi et al. [39] melt extruded small quantities of high temperature SWCNT reinforced nanocomposite with a PI matrix. Both films and rods were manufactured. When the composite was produced as rods, SWCNTs alignment in the fiber direction was induced by shear forces present during the melt extrusion and fiber drawing processes. This alignment resulted in improved tensile modulus and yield stress in SWCNTs/PI nanocomposite fibers with respect to unoriented nanocomposite films having the same SWCNTs concentration, where SWCNTs hadn't any preferred orientation. The Young's modulus was 40% higher than for pure PI when a quantity of 1 wt.% of SWCNT was added.

Yuen et al. [40] produced acid and amine modified MWCNT/PI composites. Microphotographs proved that unmodified MWCNT aggregated while MWCNTs modified by acid and by amine can be dispersed uniformly in the PI matrix. As the modified MWCNTs have a greater affinity to the polymer matrix than that of unmodified MWCNTs, modified MWCNTs significantly improve the tensile properties of the PI. Tensile strength was increased from 102 to 121 MPa when 6.98 wt.% of unmodified MWCNTs were added, while with 6.98 wt.% of acid-modified MWCNTs it reached 134 MPa. Amine-modified MWCT/PI composites had tensile strength equal to 124 MPa for 0.99 wt.% of amine-modified MWCNTs. However, as the amine-modified MWCNT content increased, the tensile strength decreased to 110 MPa at an amine-modified MWCNT content of 6.98 wt.%. The Young's modulus of the MWCNT/PI composite increased from 2.3 GPa of the neat PI to 3.2 GPa when 6.98 wt.% of unmodified MWCNTs were added, while with 6.98 wt.% of acid-modified MWCNTs it reached 3.7 GPa. Amine-modified MWCT/PI composites had Young's modulus equal 3.4 GPa for 0.99 wt.% of amine-modified MWCNTs.

9.12 Polystyrene-*b*-Butadiene-*co*-Butylene-*b*-Styrene-Carbon Nanotube Composites

Li and coworkers [41] used an high-shear processing technology that can strongly improve the dispersion of unmodified MWCNTs in polystyrene-*b*-butadiene-*co*-butylene-*b*-styrene (SBBS). The rotation speeds of the screw used were 300 rpm, 1000 rpm, and 2000 rpm and the processed composites were denoted as CP300, CP1000, and CP2000, respectively. The mechanical properties of the composites improved significantly with the screw rotation speed used during processing. Probably higher screw rotation speed resulted in better dispersion of MWCNTs in the matrix (Figs. 9.18–9.20).

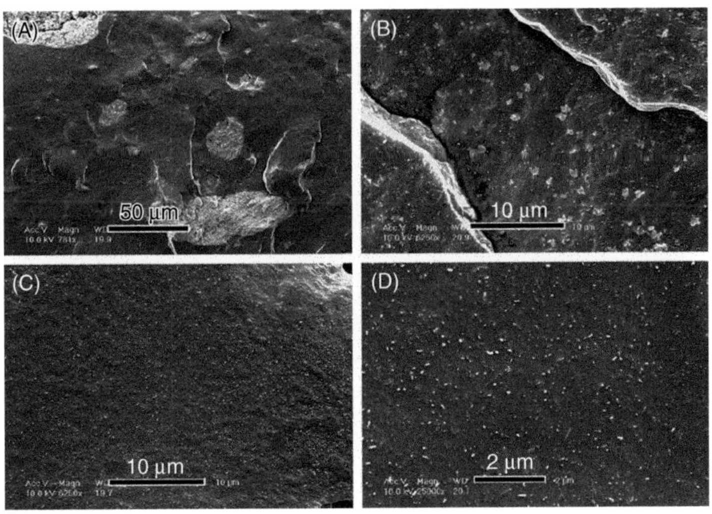

FIGURE 9.18 SEM images for the fracture surface of the MWCNT/SBBS composites. (A) CP300, (B) CP1000, (C) CP2000 with low magnification, and (D) CP2000 with high magnification. SBBS, Polystyrene-*b*-butadiene-*co*-butylene-*b*-styrene. *Reproduced with permission from Y. Li, H. Shimizu, Polymer 48 (2007) 2203 [41]. Copyright 2007, Elsevier.*

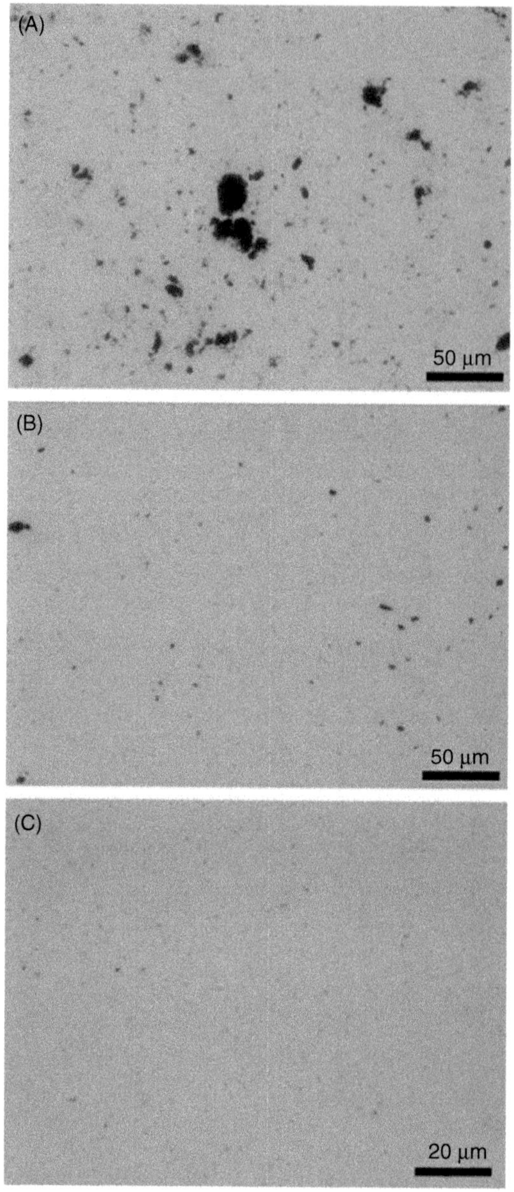

FIGURE 9.19 Optical micrographs of the film casting from toluene solutions of MWCNT/SBBS composites. (A) CP300, (B) CP1000, and (C) CP2000. *Reproduced with permission from Y. Li, H. Shimizu, Polymer 48 (2007) 2203 [41]. Copyright 2007, Elsevier.*

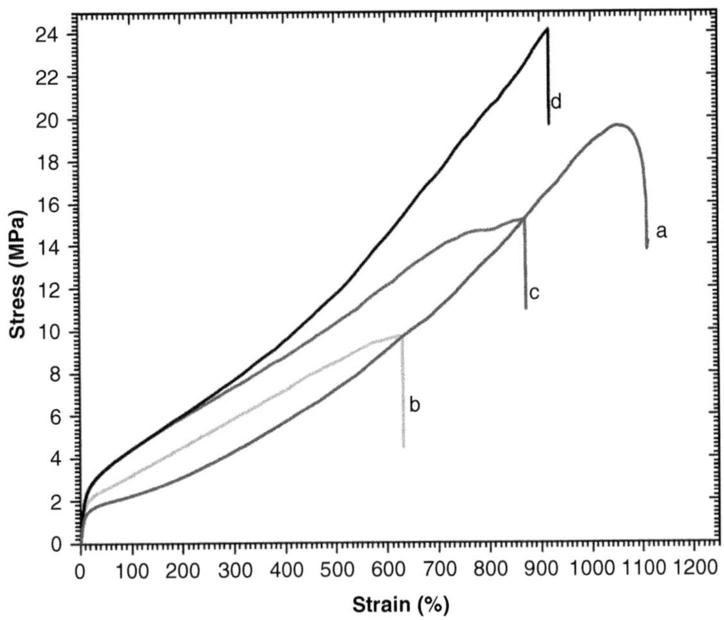

FIGURE 9.20 Stress–strain curves of (a) neat SBBS, (b) CP300, (c) CP1000, and (d) CP2000. *Reproduced with permission from Y. Li, H. Shimizu, Polymer 48 (2007) 2203. Copyright 2007, Elsevier.*

9.13 Methyl-Ethyl Methacrylate–Carbon Nanotube Composites

Velasco-Santos et al. [42] prepared composites by using unmodified MWCNTs and methyl-ethyl methacrylate copolymer. The MWCNTs used had been obtained by the arc discharge approach and had 10–20 nm outer diameter and lengths ranging from 1 to 10 μm. The storage modulus (E') was obtained by DMA, and it was found that a sample with 1 wt.% without additives increased the modulus with respect to the neat polymer by more than 200% at 40°C. Results seems to reveal that a small quantity of MWCNTs in a polymer matrix is able of enhancing the storage modulus meaningfully, but large quantities of nanotubes make the storage modulus to decreases.

Marrs et al. [43] tested fatigue performance of MWCNT/polymer composites. MWCNTs were added to methyl methacrylate–styrene copolymer (MMA-*co*-Sty), a chief component of commercial bone cement. MWCNTs were synthesized by using a chemical vapor deposition process dispersed throughout the molten matrix of MMA-*co*-Sty with two heated stainless steel, counter-rotating sigma rotors. The composites were tested to fatigue failure in tension–compression cycles in a saline environment. For a loading with a 20 MPa peak stress amplitude it was obtained that 2 and 5 wt.% MWCNTs produced the largest increases in Weibull minimum fatigue life value, respectively, 565% and 592% greater fatigue lives than neat polymer matrix. The smallest improvement, equal to 307%,

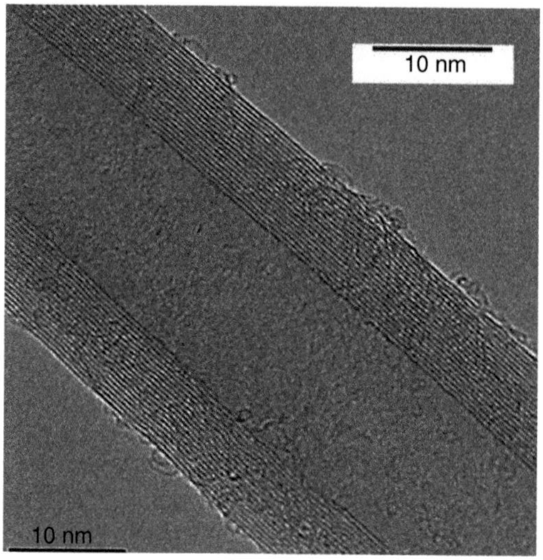

FIGURE 9.21 This transmission electron micrograph shows the highly ordered walls of a MWCNT whose outer diameter is about 20 nm. *Reproduced with permission from B. Marrs, R. Andrews, D. Pienkowski, Carbon 45 (2007) 2098 [43]. Copyright 2007, Elsevier.*

was recorded for the 0.5 wt.% MWCNT loading. The fact that the 2 wt.% MWCTs sample generated the optimal enhancement in the Weibull minimum fatigue life may be due to the increased difficulty in mixing higher loadings of MWCNTs. Thus, MWCNTs seem to be able to enhance the fatigue performance of this polymer much better than traditional graphite fibers (Figs. 9.21–9.23).

9.14 Polyethyleneimine–Carbon Nanotube Composites

Mamedov et al. [44] made SWCNT/polymer composite following the protocol of layer-by-layer (LBL) assembly. This technique is based on the alternating adsorption of monolayers of SWCNTs and polyelectrolyte chains, polyethyleneimine (PEI), attracted to each other by electrostatic and van der Waals interactions. In order to further enhance the structural integrity of the film, crosslinking was induced by adding a layer of polyacrylic acid (PAA) to introduce carboxyl functionalities for crosslinking and heating the film. The high temperature during heating creates covalent bonds between PEI and CNTs. The comparison with stretching curves for polyelectrolytes showed that the incorporation of nanotubes in the LBL structure resulted in the transfer of the SWCNT strength to the entire assembly. The ultimate tensile strength was found to be around 220 MPa with the highest value recorded equal to 325 MPa. According to the mixing law it is clear that a polyelectrolyte matrix, having a tensile strength of 9 MPa, makes a negligible contribution to the strength of the composite, even if it occupies about 50% of its volume fraction (Figs. 9.24–9.26).

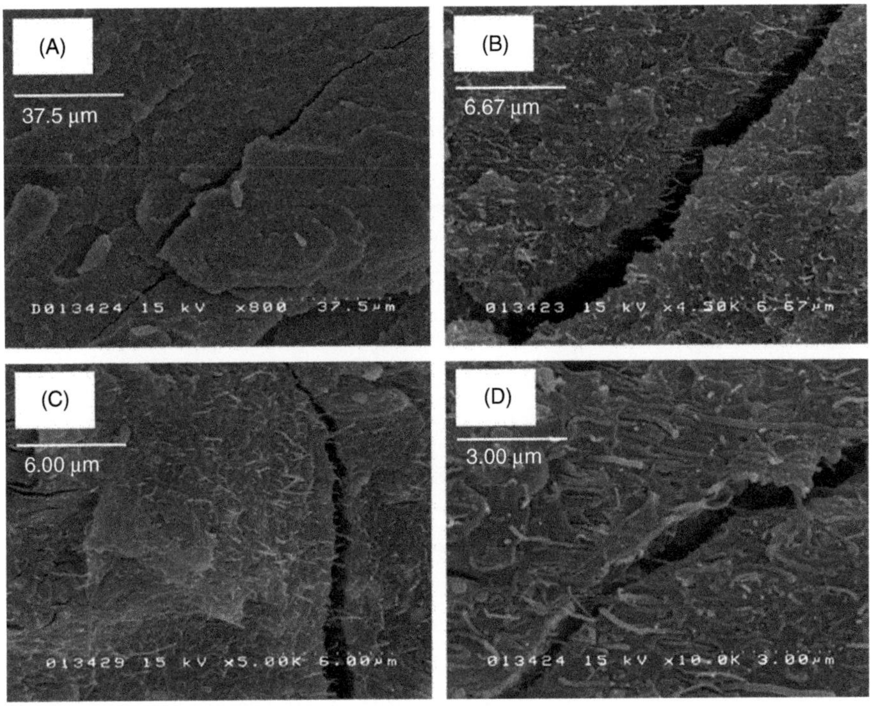

FIGURE 9.22 A series of SEM images documenting one microcrack (A) observed on the freeze-fractured surface of a test specimen loaded with 5 wt.% MWCNTs and stressed at 20 MPa. Magnification of the crack in images (B), (C), and (D) revealed the matrix-reinforcing behavior of the MWCNTs in response to crack growth. The MWCNTs are clearly observed as finger-like projections protruding into the wake of the crack normal to the direction of crack growth. *Reproduced with permission from B. Marrs, R. Andrews, D. Pienkowski, Carbon 45 (2007) 2098 [43]. Copyright 2007, Elsevier.*

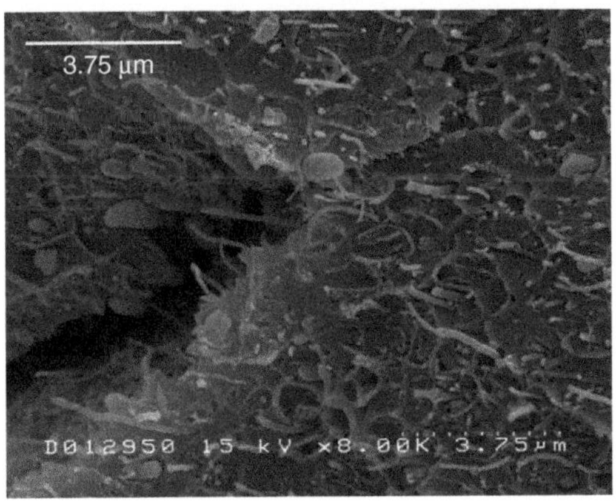

FIGURE 9.23 SEM image of a specimen loaded with 0.5 wt.% MWCNTs and stressed at 20 MPa. This clearly shows MWCNTs in the crack wake that have reoriented perpendicularly to the crack face. Particles of barium sulfate, a radio pacifier commonly used in bone cement, are also seen. Barium sulfate is inert in the MMA-*co*-Sty matrix and, therefore, the particles offer very little mechanical reinforcement. In fact, agglomerations of barium sulfate are known sources of crack initiation. *Reproduced with permission from B. Marrs, R. Andrews, D. Pienkowski, Carbon 45 (2007) 2098. Copyright 2007, Elsevier.*

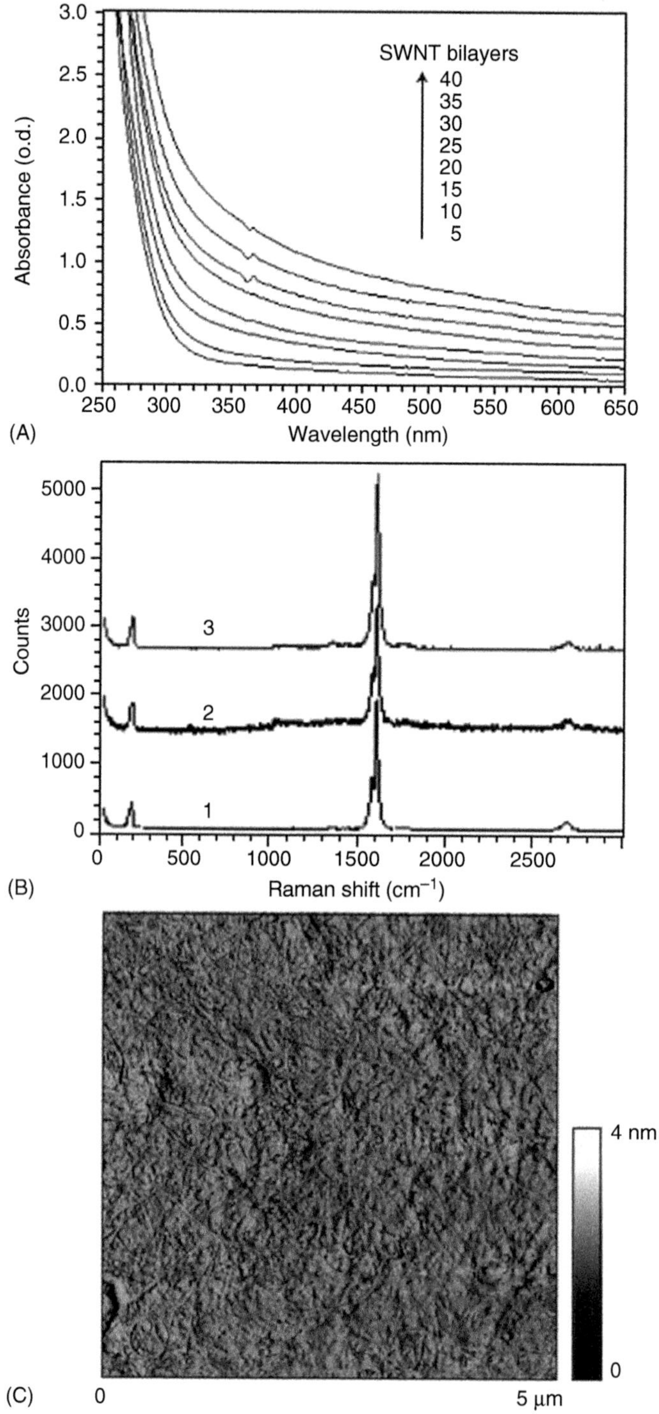

FIGURE 9.24 Structural characterization of SWCNT multilayers. (A) Sequential ultraviolet–visible optical density (o.d.) spectra of a glass substrate in the course of the LBL deposition of SWCNT. The spectra were taken for a total number of (PEI/SWCNT) bilayers indicated in the graph. (B) Raman scattering spectra of SWCNT dispersion (1), LBL film on a glass substrate (2), and free-standing film (3). (C) Tapping-mode AFM image (DI, Multimode IIIA) of a silicon wafer bearing $(PEI/PAA)(PEI/SWCNT)_5$. *LBL*, Layer-by-layer; *PAA*, polyacrylic acid; *PEI*, polyethyleneimine; *SWCNT*, single-walled carbon nanotube. *Reproduced with permission from A.A. Mamedov, N.A. Kotov, M. Prato, D.M. Guldi, J.P. Wicksted, A. Hirsch. Nat. Mater. 1 (2002) 190 [44]. Copyright 2002, Nature Publishing Group.*

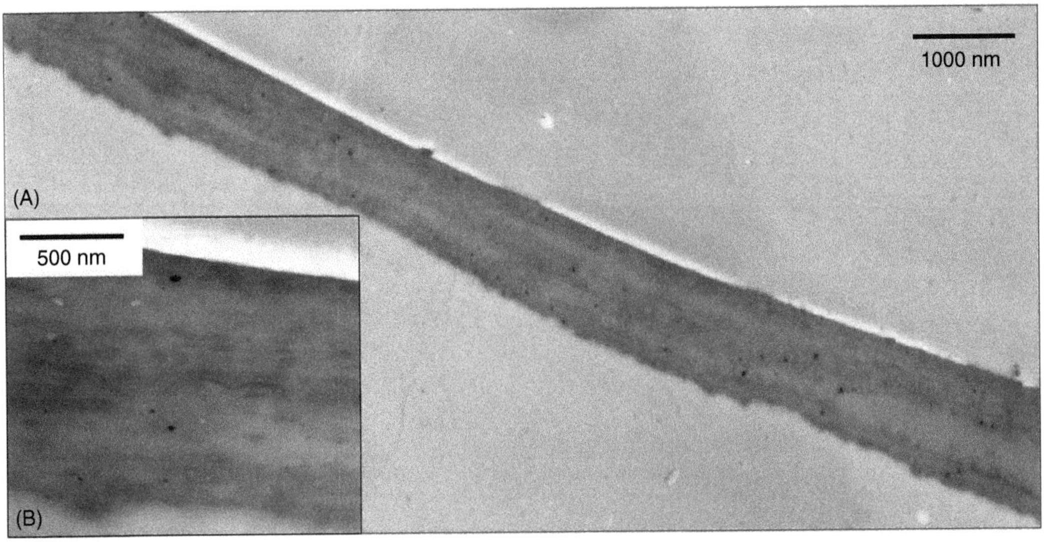

FIGURE 9.25 Examination by TEM of the homogeneity of the SWCNT LBL film. (A) Survey and (B) close-up TEM images of SWCNT film cross-sections. The top and bottom sides of the film are slightly different in roughness: the one that was adjacent to the flat substrate is smoother than the "growth" surface of the film. *Reproduced with permission from A.A. Mamedov, N.A. Kotov, M. Prato, D.M. Guldi, J.P. Wicksted, A. Hirsch. Nat. Mater. 1 (2002) 190 [44]. Copyright 2002, Nature Publishing Group.*

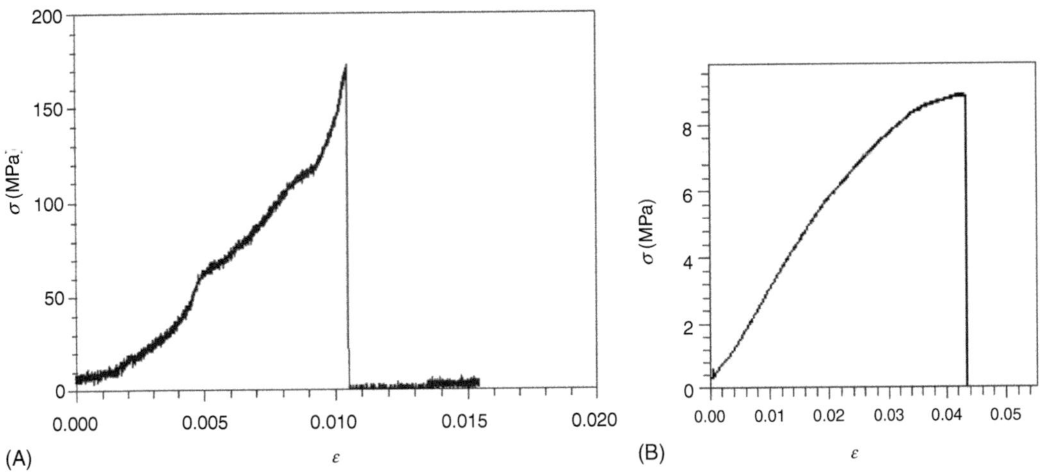

FIGURE 9.26 Typical tensile strength curves of the SWCNT LBL films. Stress (σ)–strain (ε) dependence for: (A) ((PEI/PAA)(PEI/SWCNT)$_5$)$_8$ and (B) a similar freestanding multilayer film made solely from polyelectrolytes (PEI/PAA)$_{40}$. The dependence of the mechanical properties of the SWCNT LBL composites on humidity was tested in the range of relative humidity of 30%–100%, at a temperature of 298°C, and was found to be negligible. *Reproduced with permission from A.A. Mamedov, N.A. Kotov, M. Prato, D.M. Guldi, J.P. Wicksted, A. Hirsch. Nat. Mater. 1 (2002) 190 [44]. Copyright 2002, Nature Publishing Group.*

References

[1] M. El Achaby, A. Qaiss, Processing and properties of polyethylene reinforced by graphene nanosheets and carbon nanotubes, Mater. Des. 44 (2013) 81.

[2] S.L. Ruan, P. Gao, X.G. Yang, T.X. Yu, Toughening high performance ultrahigh molecular weight polyethylene using multiwalled carbon nanotubes, Polymer 44 (2003) 5643.

[3] W. Tang, M.H. Santare, S.G. Advani, Melt processing and mechanical property characterization of multi-walled carbon nanotube/high density polyethylene (MWNT/HDPE) composite films, Carbon 41 (2003) 2779.

[4] Y. Bin, M. Kitanaka, D. Zhu, M. Matsuo, Development of Highly Oriented Polyethylene Filled with Aligned Carbon Nanotubes by Gelation/Crystallization from Solutions, Macromolecules 36 (2003) 6213.

[5] B.X. Yang, K.P. Pramoda, G.Q. Xu, S.H. Goh, Mechanical Reinforcement of Polyethylene Using Polyethylene-Grafted Multiwalled Carbon Nanotubes, Adv. Funct. Mater. 17 (2007) 2062.

[6] G. Zamfirova, V. Gaydarov, F. Faraguna, E. Vidović, A. Jukić, Influence of carbon nanotubes functionalization on the mechanical properties of polymethacrylate nanocomposites, Colloids Surf. A 510 (2016) 169.

[7] R. Haggenmueller, H.H. Gommans, A.G. Rinzler, J.E. Fischer, K.I. Winey, Aligned single-wall carbon nanotubes in composites by melt processing methods, Chem. Phys. Lett. 330 (2000) 219.

[8] R.E. Gorga, R.E. Cohen, Toughness enhancements in poly(methyl methacrylate) by addition of oriented multiwall carbon nanotubes, J. Polym. Sci. B 42 (2004) 2690.

[9] K.H. Kim, W.H. Jo, Improvement of tensile properties of poly(methyl methacrylate) by dispersing multi-walled carbon nanotubes functionalized with poly(3-hexylthiophene)-graft-poly(methyl methacrylate), Compos. Sci. Technol. 68 (2008) 2120.

[10] J. Zhong, A.I. Isayev, X. Zhang, Ultrasonic twin screw compounding of polypropylene with carbon nanotubes, graphene nanoplates and carbon blac, Eur. Polym. J. 80 (2016) 16.

[11] H. Xia, Q. Wang, K. Li, G.H. Hu, Preparation of polypropylene/carbon nanotube composite powder with a solid-state mechanochemical pulverization process, J. Appl. Polym. Sci. 93 (2004) 378.

[12] M. Manchado, L. Valentini, J. Biagiotti, J.M. Kenny, Thermal and mechanical properties of single-walled carbon nanotubes–polypropylene composites prepared by melt processing, Carbon 43 (2005) 1499.

[13] A.A. Koval'chuk, V.G. Shevchenko, A.N. Shchegolikhin, P.M. Nedorezova, A.N. Klyamkina, A.M. Aladyshev, Effect of carbon nanotube functionalization on the structural and mechanical properties of polypropylene/MWCNT composites, Macromolecules 41 (2008) 7536.

[14] D. Lai, Y. Wei, L. Zou, Y. Xu, H. Lu, Wet spinning of PVA composite fibers with a largefraction of multi-walled carbon nanotubes, Prog. Nat. Sci. 25 (2015) 445.

[15] M. Cadek, J.N. Coleman, K.P. Ryan, V. Nicolosi, G. Bister, A. Fonseca, Reinforcement of polymers with carbon nanotubes: the role of nanotube surface area, Nano Lett. 4 (2004) 353.

[16] M.S.P. Shaffer, A.H. Windle, Fabrication and characterization of carbon nanotube/poly(vinyl alcohol) composites, Adv. Mater. 11 (1999) 937.

[17] J.N. Coleman, W.J. Blau, A.B. Dalton, E. Munoz, S. Collins, B.G. Kim, J. Raazal, M. Selvidge, G. Vieiro, R.H. Baughman, Improving the mechanical properties of single-walled carbon nanotube sheets by intercalation of polymeric adhesive, Appl. Phys. Lett. 82 (2003) 1682.

[18] B. Vigolo, P. Poulin, M. Lucas, P. Launois, P. Bernier, Improved structure and properties of single-wall carbon nanotube spun fibers, Appl. Phys. Lett. 81 (2002) 1210.

[19] M. Lahelin, M. Annala, A. Nykänen, J. Ruokolainen, J. Seppälä, In situ polymerized nanocomposites: polystyrene/CNT and poly(methyl methacrylate)/CNT composites, Compos. Sci. Technol. 71 (2011) 900.

[20] R. Blake, J.N. Coleman, M.T. Byrne, J.E. McCarthy, T.S. Perova, W.J. Blau, A. Fonseca, J.B. Nagy, Y.K. Gun'ko, Reinforcement of poly(vinyl chloride) and polystyrene using chlorinated polypropylene grafted carbon nanotubes, J. Mater. Chem. 16 (2006) 4206.

[21] R. Andrews, D. Jacques, M. Minot, T. Rantell, Fabrication of carbon multiwall nanotube/polymer composites by shear mixing, Macromol. Mater. Eng. 287 (2002) 395.

[22] B. Safadi, R. Andrews, E.A. Grulke, Multiwalled carbon nanotube polymer composites: synthesis and characterization of thin films, J. Appl. Polym. Sci. 84 (2002) 2660.

[23] H. Yazdani, B.E. Smith, K. Hatami, Multi-walled carbon nanotube-filled polyvinyl chloride composites: influence of processing method on dispersion quality, electrical conductivity and mechanical properties, Composites A 82 (2016) 65.

[24] J.H. Shi, B.X. Yang, K.P. Pramoda, S.H. Goh, Enhancement of the mechanical performance of poly(vinyl chloride) using poly(n-butyl methacrylate)-grafted multi-walled carbon nanotubes, Nanotechnology 18 (2007) 375704.

[25] F. Dalmas, L. Chazeau, C. Gauthier, K. Masenelli-Varlot, R. Dendievel, J.Y. Cavaillé, Multiwalled carbon nanotube/polymer nanocomposites: processing and properties, J. Polym. Sci. B 43 (2005) 1186.

[26] A. Dufresne, M. Paillet, J.L. Putaux, R. Canet, F. Carmona, P. Delhaes, S. Cui, Processing and characterization of carbon nanotube/poly(styrene-co-butyl acrylate) nanocomposites, J. Mater. Sci. 37 (2002) 3915.

[27] Q. Zhang, J. Wu, L. Gao, T. Liu, W. Zhong, G. Sui, G. Zheng, W. Fang, X. Yang, Dispersion stability of functionalized MWCNT in the epoxy–amine system and its effects on mechanical and interfacial properties of carbon fiber composites, Mater. Des. 94 (2016) 392.

[28] Y.S. Song, J.R. Youn, Influence of dispersion states of carbon nanotubes on physical properties of epoxy nanocomposites, Carbon 43 (2005) 1378.

[29] H. Rajoria, N. Jalili, Passive vibration damping enhancement using carbon nanotube-epoxy reinforced composites, Compos. Sci. Technol. 65 (2005) 2079.

[30] M.J. Biercuk, M.C. Llaguno, M. Radosavljevic, J.K. Hyun, A.T. Johnson, J.E. Fischer, Carbon nanotube composites for thermal management, Appl. Phys. Lett. 80 (2002) 2767.

[31] J.M. Park, D.S. Kim, J.R. Lee, T.W. Kim, Nondestructive damage sensitivity and reinforcing effect of carbon nanotube/epoxy composites using electro-micromechanical technique, Mater. Sci. Eng. C 23 (2003) 971.

[32] E.Y. Choi, S.C. Roh, C.K. Kim, Noncovalent functionalization of multi-walled carbon nanotubes with pyrene-linked nylon66 for high performance nylon66/multi-walled carbon nanotube composites, Carbon 72 (2014) 160.

[33] T.X. Liu, I.Y. Phang, L. Shen, S.Y. Chow, W.-D. Zhang, Morphology and mechanical properties of multiwalled carbon nanotubes reinforced nylon-6 composites, Macromolecules 37 (2004) 7214.

[34] O. Meincke, D. Kaempfer, H. Weickmann, C. Friedrich, M. Vathauer, H. Warth, Mechanical properties and electrical conductivity of carbon-nanotube filled polyamide-6 and its blends with acrylonitrile/butadiene/styrene, Polymer 45 (2004) 739.

[35] W. De Zhang, L. Shen, I.Y. Phang, T.X. Liu, Carbon nanotubes reinforced nylon-6 composite prepared by simple melt-compounding, Macromolecules 37 (2004) 256.

[36] M. Kang, S.J. Myung, H.J. Jin, Nylon 610 and carbon nanotube composite by in situ interfacial polymerization, Polymer 47 (2006) 3961.

[37] J. Dong, Y. Fang, F. Gan, J. An, X. Zhao, Q. Zhang, Enhanced mechanical properties of polyimide composite fibers containing amino functionalized carbon nanotubes, Compos. Sci. Technol. 135 (2016) 137.

[38] X. Jiang, Y. Bin, M. Matsuo, Electrical and mechanical properties of polyimide–carbon nanotubes composites fabricated by in situ polymerization, Polymer 46 (2005) 7418.

[39] E.J. Siochi, D.C. Working, C. Park, P.T. Lillehei, J.H. Rouse, C.C. Topping, A.R. Bhattacharyya, S. Kumar, Melt processing of SWCNT-polyimide nanocomposite fibers, Composites B 35 (2004) 439.

[40] S.M. Yuen, C.C.M. Ma, Y.Y. Lin, H.C. Kuan, Preparation, morphology and properties of acid and amine modified multiwalled carbon nanotube/polyimide composite, Compos. Sci. Technol. 67 (2007) 2564.

[41] Y. Li, H. Shimizu, High-shear processing induced homogenous dispersion of pristine multiwalled carbon nanotubes in a thermoplastic elastomer, Polymer 48 (2007) 2203.

[42] C. Velasco-Santos, A.L. Martinez-Hernandez, F. Fisher, R. Ruoff, V.M. Castano, Dynamical–mechanical and thermal analysis of carbon nanotube–methyl-ethyl methacrylate nanocomposites, J. Phys. D 36 (2003) 1423.

[43] B. Marrs, R. Andrews, D. Pienkowski, Multiwall carbon nanotubes enhance the fatigue performance of physiologically maintained methyl methacrylate–styrene copolymer, Carbon 45 (2007) 2098.

[44] A.A. Mamedov, N.A. Kotov, M. Prato, D.M. Guldi, J.P. Wicksted, A. Hirsch, Molecular design of strong single-wall carbon nanotube/polyelectrolyte multilayer composites, Nat. Mater. 1 (2002) 190.

10

Electrical and Electromagnetic Properties of CNT/Polymer Composites

Seyed Amin Mirmohammadi*, Samaheh Sadjadi**, Naeimeh Bahri-Laleh**

*ISLAMIC AZAD UNIVERSITY, TEHRAN, IRAN; **IRAN POLYMER AND PETROCHEMICAL INSTITUTE (IPPI), TEHRAN, IRAN

CHAPTER OUTLINE

10.1 Introduction

Carbon nanotube (CNT) was discovered by Iijima in 1991 and its first polymeric nanocomposite was fabricated by Ajayan in 1994 [1,2]. CNT was brought up as ultimate carbon fiber owing to its excellent electrical (10^4–10^6 S/cm) and thermal attributes (3000–6000 W/mK). The higher aspect ratio and intrinsic electrical conductivity of CNTs allow them to form electrically conductive nanocomposites with tunable electromagnetic response at low concentration of CNTs, thanks to its low percolation threshold. In fact, CNTs make a 3D

Carbon Nanotube–Reinforced Polymers. http://dx.doi.org/10.1016/B978-0-323-48221-9.00010-8

electrically conductive lattice within the polymer matrix; therefore electrons can cross from one nanoparticle to another and consequently overcome the high electrical resistivity of polymeric matrix [3–6].

It is well known that nanocomposites, an enhanced class of materials, possess a suitable combination of mechanical, thermal, electrical, magnetic, and dielectric attributes [7–9] and are useful for efficient electromagnetic interference (EMI) shielding materials [10–15]. Metallic nanoparticles are probably the most general nanofiller for EMI shielding and other related applications, because of their high electrical and electromagnetic properties, and exceptional mechanical and thermal attributes, with versatile shapes, such as tubes, spheres, wires, and rods. Nonetheless, they introduce some substantial disadvantages into the ultimate nanocomposite, such as corrosion perceptivity, high reflectivity, high density, and extravagant high-priced processing [16–20]. Owing to these disadvantages, alternative nanofillers were immediately replaced with metallic ones and carbon-based nanoparticles, for example, carbon black, carbon fibers, graphene, graphite, and CNTs were simultaneously found to be appropriate nanofillers due to their adequate electrical conductivity and remarkable corrosion resistance. Ultimately carbon-based nanocomposites, because of their facile processability, lower density, suitable mechanical properties, have attracted researchers' attentions in the past few decades. Among mentioned carbon-based nanofillers, application of high aspect ratio, ones such as CNTs, graphene, and carbon fibers reveals instantaneous success to lower the percolation threshold, in addition to superior thermal and mechanical attributes and excellent electromagnetic responses [5,21–23].

Polymer-matrix composites (PMCs) based on thermoplastic or thermosetting polymers as matrices and CNTs as reinforcer component are extremely attractive case study for many scientists and engineers in the field of electronics and electromagnetism, due to natural versatility, acceptable corrosion resistance, low density, easy processability, potential of different filler incorporation, and flexibility of their polymeric matrices; and significant aspect ratio ($>10^2$), high electrical conductivity (sometimes more than copper), optoelectronic property, and low density (sometimes, compared to aluminum) of CNT reinforcer components. Some examples of their important applications are: radio-frequency identification (RFID) systems, circuits, antennas, absorbing media, EMI-shielding materials, electrostatic dissipative (ESD) composites, coatings for enclosures, antistatic materials, conductive adhesives, 3D batteries, super capacitors, fuel cells, and so forth [24–31].

In fact, polymeric materials are generally applied as insulating materials attributed to their negligible electromagnetic property and low electrical conductivity (about 10^{-15} S/m). An insulating polymer can be transformed to a conductive polymeric composite by physical or chemical incorporation of conductive CNTs above a specific concentration threshold when the CNTs create an interconnected structure or network (called percolation threshold), which is well defined by percolation theory. Because of low percolation threshold, small CNT quantities are required and consequently economical nanocomposites can be obtained. It is obvious that merely small amounts of single or multiwall CNTs are required to acquire adequately high values of electromagnetic or electrical conductivity. Not unexpectedly, this result alone can ensure the future commercial applicability of CNT nanocomposites. Moreover, the replacement of carbon black, the most routinely industrially applicable conductive filler, or other conductive fillers with CNTs to prepare electromagnetic and electrically conducting polymer composites is anticipated to have a remarkable effect on their wide-ranging industrial and scientific applications. A major disadvantage of using carbon black or other fillers is the decline of the mechanical properties of the composite because of high required content (up to 30–40 wt.%) attributed to their low aspect ratio, in order to gain appropriate electrical conductivity levels. In these high concentrations aggregates and agglomerates of filler are formed, which can result in some defect points (stress concentration point) within the matrix and subsequently deteriorated mechanical properties will occur [11,26–29,32–34].

It is clear that making a perfect outline of all aspects of this extensive subject is not possible in the framework of this introduction. Thus, we did our best to cover the most characteristic, important, and recently discussable aspects in the context of electromagnetic and electrical properties of single or multiwall CNT–based polymeric composites in the following sections.

10.2 Basic Concepts in Electromagnetism and Electrical Conductivity

10.2.1 Electrical Conductivity

As previously mentioned, polymers (both thermoplastic and thermoset types) are versatile choices as matrix materials to fabricate CNTs- or conductive fillers–based nanocomposites because of their significant properties, such as flexibility, low density, corrosion resistance, facile processability, and so forth [35,36]. Nevertheless, for electrical or electromagnetic applications, the most important necessity is the potential of electrical conductivity [37,38]. Hence, a broad range and different types of conductive fillers have been applied to fabricate electrically conductive composite materials with a wide range of values of electrical conductivity and/or electromagnetic properties, like permeability or permittivity [39–43]. Concerning this, although polymeric nanocomposites have preferable advantages compared to metals, fabrication of a polymeric composite with electrical and/or magnetic properties includes comparatively difficult and complex processes due to the high aspect ratio (specific surface area) of fillers, especially nanofillers that contribute to the formation of their aggregation or agglomeration, which results in unreliable properties in the fabricated material. Thus, the progress in advanced and commercially reasonable production of these materials requires coherent tactics and plans, and appropriate endeavors to gain accurate control over electromagnetic properties [10,12,14].

In most composites, the matrix chiefly plays the role of preserving the fillers and inducing processability to the system. In addition, many polymers have poor electroconductivity and are transparent to electromagnetic radiations. So, most of these essential properties are supplied by the character of the filler component and thus, to induce dielectric/magnetic characters, fillers in the scales of macro, micro, or nano with electric/magnetic dipoles, such as graphite, carbon black, graphene, CNTs, and magnetic (ferrites) or dielectric (titanates) particles are required [10–12,15,37,42–45].

Experiments on the electrical behavior of polymer/CNT nanocomposites is undertaken by either direct current (DC) resistance (without frequency-dependent sources) or alternative current (AC) impedance measurements (dependent on frequency) on bulk or thin film samples [24]. In general, AC electrical conductivity variation with frequency shows two distinct regions; at low nanotube content, conductivity increases with frequency while at high nanotube content (above percolation concentration) frequency independent conductivity behavior is observed [46,47]. While the DC conductivity (σ) data follow percolation behavior described by Eq. (10.1) (see Section 10.2.2), AC data at low nanotube content shows some divergence at low nanotube loadings and good agreement at high nanotube loadings [48]. Explanation for the difference in conductivity at low nanotube concentrations is based on hopping conduction mechanisms for AC and DC conductivity; DC conductivity is limited by the maximum range of activation energies that charge carriers have to hop across, while AC conductivity is limited by the highest barrier experienced as the charge travels for one half period of the field [49]. Below the percolation threshold, where a poor CNT network dominates, the higher barrier energy for DC conductivity means that AC conductivity in a small part of the network is more favorable, giving higher AC conductivity.

10.2.2 Percolation Threshold

Percolation threshold is defined by percolation theory. The essence of this theory is to define the interconnectivity between particles for conductive fillers, regularly or randomly positioned within the matrix space, of a specific composite. The base of this theory is that for each conductive composite, there is a critical filler concentration known as percolation threshold. In this specific concentration an interconnected network of fillers is formed, which spans inside the composite and causes the system to "percolate" [45,50–52].

The term *percolation theory* was introduced by Broadbent and Hammersley in 1957 [53] and was used as an approach to explain and obviate the problems of fluid flow through a static medium. Initial research focused on determining the percolation thresholds in two- and 3D geometries. Two types of percolations were explained: site percolation, in which sites in a matrix are either filled or empty (good or bad filler distribution and dispersion), and bond percolation, in which the sites in the matrix are completely filled, but are either connected or not.

The percolation theory determines the behavior of a composite containing conducting and nonconducting components. In such materials, the sudden transition from insulator to conductor is the indication of a percolation threshold. The relation between electrical conductivity of composite and concentration of conductive filler in the neighborhood of the percolation threshold is described by a simple power law formula [45,50–52,54]:

$$\sigma_c = \sigma_0 (\varphi_f - \varphi_c)^t \tag{10.1}$$

where σ_c, composite electrical conductivity (S/cm); σ_o, filler electrical conductivity (S/cm); ϕ_f, fraction of conductive filler (v/v or wt.%); ϕ_c, percolation threshold (v/v or wt.%); *t*, fitting parameter called critical exponent.

It is clear that this equation can be used at concentrations above percolation threshold, that is, when $\phi_f > \phi_c$. Fitting parameter value, *t*, depends mainly on the lattice dimension, aspect ratio of the filler, modality or quality of dispersion, and orientation of the filler. For example, because of greater aspect ratio of the fibers than other fillers, this value is higher for the composites reinforced with fibers. The predictions from this equation are appertained on the ϕ_c and *t* values. The value of ϕ_c can be experimentally specified while the *t* value is determined by fitting of the data. If the host polymer and filler have the same values of density, mass fraction is equal to volume fraction and can be applied in this formula. The formula described by percolation theory has been applied by most of the researchers to proportionate experimental data and achieve the values of percolation threshold. The percolation threshold value is typically calculated by plotting the electrical conductivity of the composite (σ_c) versus filler reduced mass fraction and fitting with a power law function. Numerous reports demonstrate that the percolation threshold depends decisively on the polymer type, preparation method of composite, CNT aspect ratio, disentanglement of CNTs aggregates and agglomerates, uniform and regular spatial distribution of CNTs, and their alignment or orientation degree [45,52]. Effect of such variables on the percolation threshold values of the CNT/polymer composites will be discussed briefly in Section 10.3.

10.2.3 Shielding Phenomenon

EMI shielding is required to safeguard the environment and working area from irradiation produced by computers and electromagnetic equipment, as well as to ensure sensitive circuits. EMI shielding is generally interpreted as the debarment of propagation of the magnetic and electric waves from one area to another by applying magnetic or conducting materials. In EM terminology, in other words, the term *shield* is commonly assigned to an enclosure that thoroughly surrounds an electronic product or a part of it and works as a barrier against transmission of EM emissions. Briefly, shielding is a technique of attaining a determined level of attenuation using a well-designed shield.

The EMI shielding effectiveness (SE), of a material is defined as the ratio between the outgoing power (P_t) and incident power (P_i) of an electromagnetic wave [25,36,55]. SE is represented in decibels (dB) and expressed by:

$$\mathrm{SE}\,(\mathrm{dB}) = -10\log(P_t / P_i) \tag{10.2}$$

The total shielding effectiveness (SE_{total}) has two mechanisms to minimize emitting the electromagnetic beams via a system: (1) by reflection (SE_R), the main mechanism for metallic shields and (2) absorption (SE_A) of the wave and subsequent dissipation of the irradiation power within the material [16–18,20,56]. The total EMI SE (SE_{total}) is the

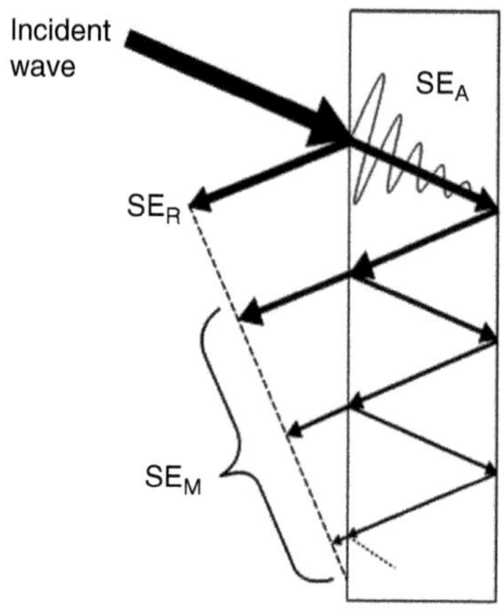

FIGURE 10.1 Multiple reflections (MR), reflection loss (R), and absorption loss (A) within a slab.

summation of the SE due to absorption (SE_A), reflection (SE_R), and multiple reflection (SE_M) [57] (Fig. 10.1), that is,

$$SE_{total} = SE_A + SE_R + SE_M \tag{10.3}$$

For a shielding material containing only a single layer, when SE_A is ≥ 10 dB, then $SE_M \rightarrow 0$ and can be overlooked.

The transmittance T is calculated by the ratio of P_t to P_i:

$$T = (P_t / P_i) \tag{10.4}$$

Thus, the SE_{total} can be written as formula (10.5) for a shielding material:

$$SE_{total} = -10 \log T \tag{10.5}$$

Then, A_{eff}, effective absorbance with respect to the power of the incident electromagnetic wave inside the shielding material is described as:

$$A_{eff} = (1 - R - T)/(1 - R) \tag{10.6}$$

where, R is the reflectance. Then SE due to effective absorbance and reflectance can be represented as:

$$SE_R = -10 \log(1 - R) \tag{10.7}$$

$$SE_A = -10 \log(1 - A_{eff}) = -10 \log[T/(1 - R)] \tag{10.8}$$

The reflection loss is attributed to the disparity between the incident signal and the surface blockage of the shield. The absorption loss is related to the physical characteristics of shield material and, unlike reflection loss, it is independent of the type of source field. When electromagnetic waves cross through an area, their intensity or amplitude is reduced. This absorption loss or dissipation of energy takes place due to ohmic losses of medium, and bulk heating will occur So, a well-designed absorbing shield should have high electromagnetic attributes (low values of permittivity/permeability) and appropriate thickness.

10.3 Electrical/Electromagnetic Behavior of CNT/Polymer Composites

One of the major attributes of CNTs is to provide electrical conduction at a lower filler loading compared to other carbon materials [58]. Percolation threshold of CNTs for enhanced electrical or electromagnetic attributes depends upon plenty of factors, such as aspect ratio, loading level, surface functions, intrinsic electrical conductivity, orientation, dispersion, and type of CNTs, and also, type of host polymer, sample thickness, and processing method [59]. Almost all of these parameters will be discussed in the following sections.

10.3.1 Effect of CNT Type, Morphology, and Dimension

As mentioned before, CNTs possess distinguished properties, such as intrinsically high mechanical, thermal, and electrical properties [60]. The electrical conductivities of SWCNTs range between 10^2 and 10^6 S/cm, while this value for MWCNTs is in the range of 10^3 and 10^5 cm [61]. Mostly, MWCNTs are considered as a preferred candidate for industrial uses due to their availability in bulk and lower costs, and hence most research efforts have been devoted to MWCNT dispersion in polymer matrix and there are few reports on SWCNT/polymer composites.

Interestingly, type of the nanotubes (SWCNT or MWCNT) does not have a clear impact on the percolation threshold and maximum conductivity [62]. Maximum conductivity of 10,000 S/m has been reported for polymethylmetacrylate (PMMA), filled with 10 wt.% $SOCl_2$-treated SWCNT [63] (with percolation threshold of 0.17 wt.%), of 3000 S/m for polyaniline containing 15 wt.% SWCNT (with percolation threshold of 0.3 wt.%) [64], of 2000 S/m for polyurethane containing 15 wt.% MWCNT (with percolation threshold of 1 wt.%) [65], and of 1000 S/m for PS filled with 2 wt.% MWCNT (with percolation threshold of 0.15–0.2 wt.%) [66]. From these data one can conclude the higher conductivity of SWCNT-containing composites; however, it is not consistent in all polymer/CNT systems. For example, by comparing electrical conductivity versus filler content in PANi/MWCNT [67] and PANi/SWCNT [68] composites (Fig. 10.2) one can see that higher conductivities are achieved in MWCNT/PANi systems.

The filler effectiveness is also influenced by the morphology of CNT. To elucidate this parameter, an example provided by Lee et al. [69] is discussed in detail. The authors prepared MWCNT/silicone composites with two different types of MWCNT bundles grown by catalysts with different morphologies. It was established that the shape of the catalyst particles significantly affects the order of agglomeration of MWCNTs. Though the same precursor composition was used, catalyst particles made from precursors gelation followed by flame synthesis (FS) consisted of chunk-type particles, while those from spraying of

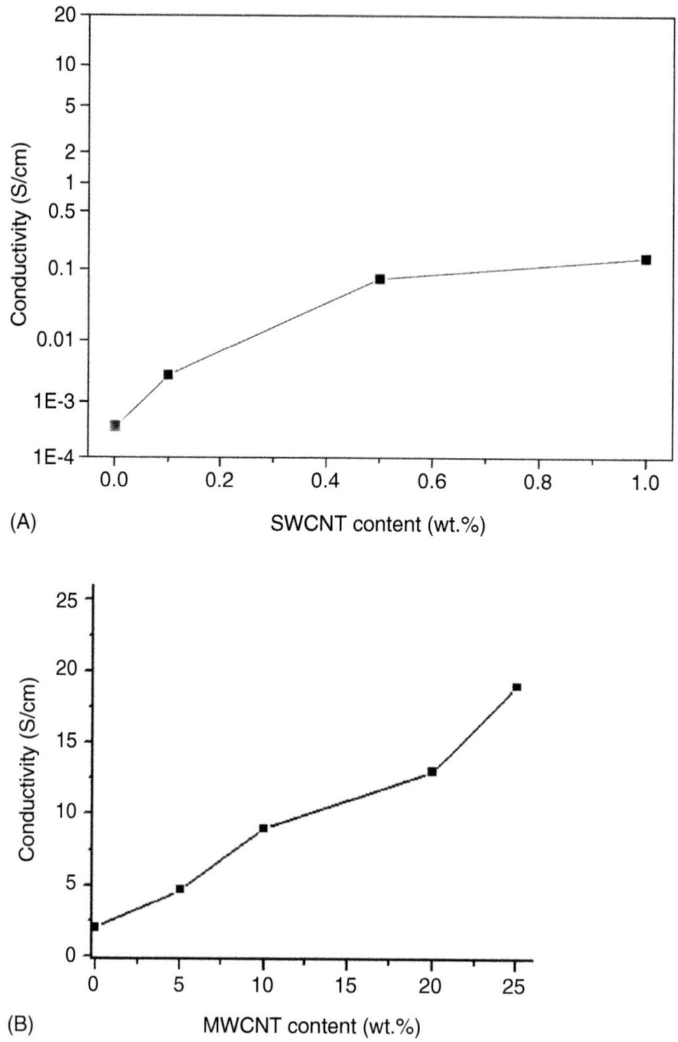

FIGURE 10.2 Electrical conductivity versus carbon nanotubes (CNT) content in (A) PANi/SWCNT [68] and (B) PANi/MWCNT [67] composites.

the precursor solution followed by thermal decomposition (STD) were fabricated with the shape of thin sheets. After CVD growth, the MWCNT bundles maintained rod-like morphology for STD catalysts (Fig. 10.3B and D) but they entangled to form large aggregates for FS catalysts (Fig. 10.3A and C).

Single STD-MWNTs bundles also possessed a smaller MWCNT population with more room inside, which finally rendered the MWCNT/silicone composite significantly conductive. This emerged from the effective dispersion of the MWCNTs. Actually, using state-of-the-art catalyst for the synthesis of MWCNT, MWCNT bundles with low

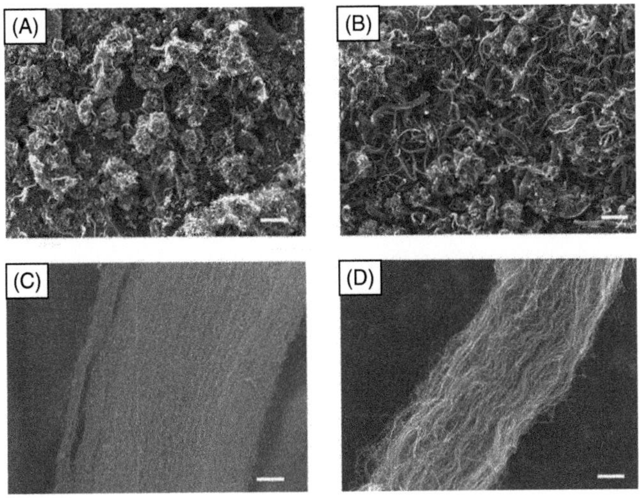

FIGURE 10.3 Morphologies of CVD grown MWCNTs from (A) FS and (B) STD catalysis (scale bar: 100 mm). Morphologies of single MWCNT bundle grown from (C) FS and (D) STD catalysis (scale bar: 1 mm) [69].

agglomeration can be furnished, which resulted in high conductivity of 1407 S/m at relatively low MWCNT loading (21.3 wt.%) due to effective MWCNT debundling.

Another effective parameter on percolation threshold of the nanocomposite is the aspect ratio (l/d) of the filler [66]. In this regard, Celzard and Munson–McGee introduced an advanced percolation theory [70,71], which includes the aspect ratio of particles in determining the critical volume fraction required to achieve a percolated, conducting network for rod-like particles with aspect ratios of >100, the percolation threshold was estimated to be 0.24–1.35 vol.% [71]. The experimental findings showed lower percolation thresholds, compared to theoretical results. This can be justified by the limitation of Celzard and Munson–McGee model to volumetric effects. For example, any cohesive forces (particle–matrix or particle–particle interaction) and Brownian motion of the fillers, induced by dipole–dipole and van der Waals interactions (Coulomb-forces) were not considered in the so-called model.

Fig. 10.4 shows the distribution of percolation thresholds in epoxy-MWCNT nanocomposites. The percolation threshold reduces in a hyperbolic manner as a function of the aspect ratio of the filler according to the several published analytical models [48].

The diameter of CNT also has a significant effect on the electrical conductivity of CNT/polymer composites. To study such effects, Ding et al. [72] have prepared PET/MWCNT nanocomposites containing neat MWCNT of different aspect ratio and diameter by coagulation method. They have shown that high aspect ratio of MWCNT can facilitate filler overlapping and the formation of conductive networks. By controlling the aspect ratios of MWCNT, as well as the thickness of interfacial polymer layer, the electrical feature of PET/MWCNT nanocomposites can be adjusted.

Similar results were achieved by Ram et al. [73], who studied effect of aspect ratio of short carbon fibers (SCF) on the electrical conductivity of PVDF. They have also found that

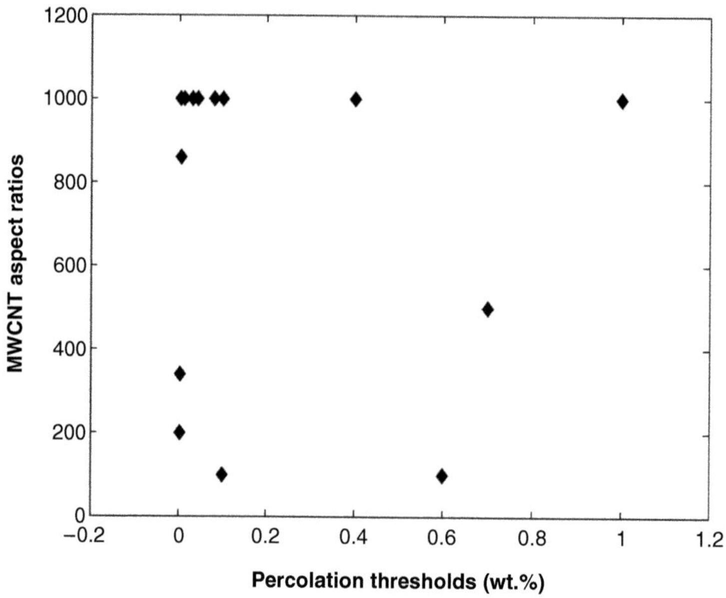

FIGURE 10.4 Effect of filler aspect ratio on the percolation threshold of MWCNT–epoxy composites. *From W. Bauhofer, J.Z. Kovacs, A review and analysis of electrical percolation in carbon nanotube polymer composites, Compos. Sci. Technol. 69 (2009) 1486–1498* [62].

percolation threshold of PVDF/SCF composites decreases with an increase in the aspect ratio of the SCFs in the PVDF matrix.

Applying an electrically conductive CNT foam with low density that can be filled or infiltrated with the polymer matrix can be considered as an interesting alternative approach for designing conductive polymer composites [74]. With this method, the uniformity of the dispersed phase, and hence the conductivity of the composite, is established by the preformed CNT network of the scaffold. To produce CNT scaffolds, a suspension of purified SWNTs in deionized water was prepared and then sonicated so that it can be completely dispersed. Upon dispersion of the SWNTs, the cross-linking of CNT bundles was performed via organic sol–gel chemistry. Typically, organic sol–gel chemistry includes the polymerization of organic precursors to fabricate highly cross-linked organic gels that can be dried and pyrolyzed to furnish porous carbon framework. Conductive SWNT-based nanofoams (aerogels) are employed as scaffolds to produce PDMS composites [75]. The obtained composites exhibited electrical conductivities over 100 S/m and showed a ~300% increase in the elastic modulus with as little as 1 vol.% nanotube loading.

10.3.2 Effect of Polymer Type

Effect of matrix type on the electrical conductivity of MWCNT nanocomposites was well studied by Thomassin and coworkers [76]. In this regard, PCL (polycaprolactone), PS (polystyrene), PEtOC (polyethylene-*co*-octene), PVC (polyvinyl chloride), PMMA (polymethylmetacrylate), PC (polycarbonate), and PP (polypropylene) were selected as polymer

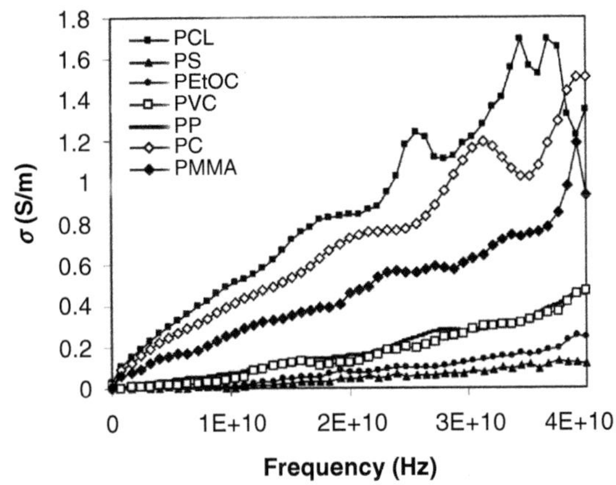

FIGURE 10.5 Effect of polymer type on the electrical conductivity of MWCNT/polymer nanocomposites with 2 wt.% of filler content [76].

matrix, and 2 wt.% of MWCNT was dispersed by melt-mixing in an extruder. Results showed higher conductivity in the composites based on PC, PMMA, and PCL matrix (Fig. 10.5) due to better nanotube dispersion. A common characteristic feature of these three polymers is a carbonyl group in the constitutive monomer unit. It is therefore proposed that π–π interactions between the carbonyl groups in the chains and aromatic structure of the CNT are very beneficial for the nanocomposites morphology and subsequently their electrical properties.

To provide more insight into the effect of matrix type on the final properties of the resulted composites, the reported experimental observations for the percolation threshold and electrical conductivity of the CNT/polymer composites with 2–3 wt.% of the filler are tabulated in Table 10.1.

As is clear from Table 10.1, polymer type and processing method play important roles in the final properties of fabricated material. For example, conductivity of 3200 S/m was obtained for nafion-coated SWCNT network (18 wt.% of the filler) [84], while relatively low

Table 10.1 Effect of Polymer Type on the Conductivity and Percolation Threshold of Different MWCNT Composites

Polymer Types	CNT Types	CNT Weight Fraction	CNT Dispersion Method	Conductivity (S/m)	Percolation Threshold (%)	References
PU	SWCNT	2	Solution	0.022	0.2	[77]
PET	MWCNT	2.5	In situ polymerization	0.3	0.06	[78]
PMMA	MWCNT	2	Electrospinning	0.053	0.05	[79]
PC	MWCNT	3	Extrusion	5	1	[80]
VE[a]	MWCNT	2	Sonication	0.04	0.5	[81]
PS	MWCNT	2	—	1000	0.2	[66]
PVDF	MWCNT	2	In situ polymerization	10	0.07	[82]
Epoxy	MWCNT	2	Milling	17.6	0.2	[83]

[a]Vinylester.

conductivity (21.5 S/m) was observed for polypyrrole-coated CNT networks [85]. The in situ polymerized CNT/polypyrrole composite exhibited the conductivity of 1600 S/m [86]. Peng and Sun [87] used dropcasting of the polymers onto CNT arrays. In this approach, the composites based on sulfonated poly(ether ether ketones) and polystyrene matrices showed conductivity of 6670 and 1330 S/m, respectively.

Using compression molding, MWNT/poly(phenylene sulfide) composite [88] and SWNT/poly(*para*-phenylene vinylene) composite with conductivities of ~1000 S/m and [89] ~1000 S/m at SWNT content of 64 wt.% were obtained, respectively. For this type of CNT/polymer composites, maximum value of ~3000 S/m was achieved for the SWNT/PANi composite [64] and the MWNT/PMMA composite containing 40 wt.% of filler [90].

Some studies report the use of CNTs in silicone-based elastomers [58]. CNTs show outstanding affinity toward poly(dimethylsiloxane) (PDMS). It was shown that even tiny amounts of MWCNTs bring significant changes in the thermal, mechanical, and electrical properties of PDMS. In fact, insulator-conductor transition, corresponding to the formation of a conductive path throughout the sample reaching the so-called percolation threshold, occurs at a filler loading of 0.05 phr, which is very low with regard to hydrocarbon rubbers, where it was shown that CNTs form the interconnecting filler network between 2 and 3 phr [91]. This large reduction in the amount of conductive inclusions is particularly valuable for the elaboration of electrically conductive silicones in commercial applications [58].

10.3.3 Effect of Dispersion Conditions/Degree of Homogeneity

The main challenge facing the commercial success of CNT-based composites is the intrinsic strength of their agglomerates, which prevents good filler dispersion in the matrix, a key attribute for any reinforcement. In order to benefit from the outstanding properties of CNTs in polymer composite, various approaches for circumventing the tube–tube interactions and optimization of the dispersion in the host matrix, such as surface functionalization of the tube and the use of surfactants, have been developed [92,93]. Chemical functionalization can be achieved either on the sidewalls of the nanotubes or on the defects generated on CNT surfaces through acid treatments. It is expected to result in a very strong CNT-matrix interaction with a trade-off on the structural characteristics of CNTs. It is worth noting that covalent functionalization of CNTs can change the H-bonds and consequentially modify CNT stacking morphologies and its solubility in solvents [46]. Bose et al. reported that by NH_2 functionalization of MWCNT, the electrical percolation threshold increases in related MWCNT/PA6 composites [94]. The same results were obtained by Gojny et al. [95] for different CNT/epoxy composites. They have found that the amino-functionalized CNTs have higher percolation thresholds compared to the untreated nanotubes, with the threshold ranging between 0.3 and 0.5 wt.% for MWCNT–NH_2 and between 0.1 and 0.3 wt.% for DWCNT–NH_2. Upon functionalization, π conjugation bonds on the outer walls are destroyed, accompanied with the transforming of sp^2 hybridization of carbon into sp^3 to create active sites for effective interaction with the polymer matrix.

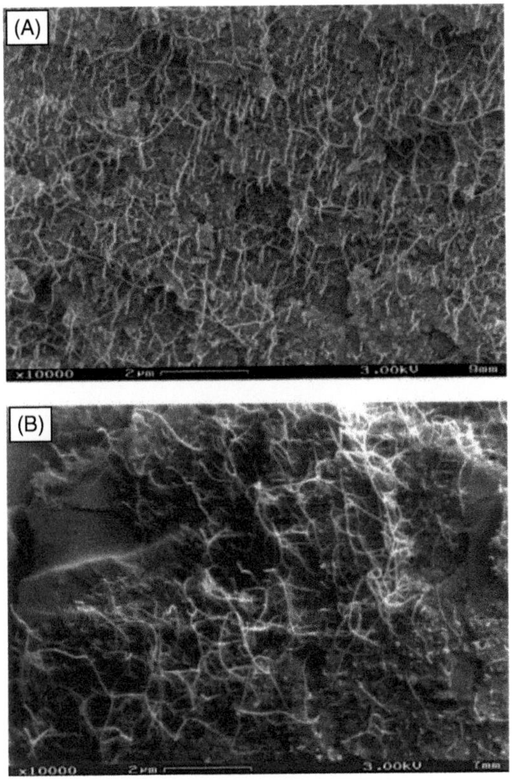

FIGURE 10.6 SEM image of CNTs on fracture surfaces of the epoxy/CNT composites: (A) with and (B) without surfactant [92].

Interfacial interaction is one of the most critical issues in CNT/polymer composites. The role of surfactants as processing aids in CNT/polymer composites was studied by Gong et al. [92]. The authors used polyoxyethylene 8 lauryl as a nonionic surfactant to aid CNT dispersion in epoxy matrix. The surfactant introduces a steric-hindrance repulsive force between the nanotubes, which overcomes the van der Waals attractive force between the carbon surfaces in a poor solvent. The morphologies of the CNTs on the fracture surface of the CNT/epoxy nanocomposites were compared in the presence (Fig. 10.6A) and absence (Fig. 10.6B) of the surfactant. As shown, in the absence of the surfactant, the CNTs in the fractured sample are wavy, very long, and lumped together, illustrating highly agglomerated CNTs with weak matrix-CNT interactions. In the presence of surfactant however, the CNTs in the fractured sample are aligned along one direction and more uniformly distributed. These founding proved that the dispersion of the nanotubes is better, therefore the load can efficiently be transferred to the tubes during the fracture process [92].

It was found that modification of CNT by surfactants reduces electrical conductivity and increases the percolation threshold. In a study, Chen and coworkers [96] investigated the effect of addition of up to 5 wt.% of CNTs modified by 3-aminopropyltriethoxysilane to a silicone matrix. The results indicated that the electrical conductivity of the composite remained unchanged. In another example, the effect of incorporation of CNTs modified by 7-octenyltrichlorosilane was examined by Vast [97]. It was reported that oxidation of CNTs induces defects, which prevents from obtaining conducting composites. Vilcacova et al. [98] reported that modification of CNTs by dodecylbenzene sulfonic acid (DBSA), due to the separation/isolation of individual CNTs, increases the percolation threshold from 2 to 4 vol.%. Concluding this section, although surfactants modify dispersion of CNT fibers in the polymer composites, they induce negative effect on the conductivity properties of the final fabricate.

10.3.4 Effect of Loading Amount

The electrical percolation threshold is the critical filler concentration at which the insulator to conductor transition occurs. At this juncture, the emergence of conductive pathway results in leap of electrical conductivity of the composite.

The electrical conductivity of a filled composite does not have a linear relationship with the filler content. In the case of using a low amount of filler, the electrical properties are dominated by the host matrix since the conductive particles are separated from each other. At a certain amount of filler, the formation of a conducting path is achieved throughout the sample, reaching the so-called percolation threshold and the insulator–conductor transition [91].

The reflection and absorption components of PMMA/MWCNT composites as a function of CNT loading at a particular frequency of 12 GHz were investigated by Pande et al. [25]. The results established that using more content of the MWCNT led to SE_R and SE_A increases. Furthermore, SE_R increase was much slower than SE_A. Fig. 10.7 illustrates the variation of EMI SE due to absorption and reflection at 12 GHz with increasing MWCNT loadings. The difference between SE_A and SE_R becomes more pronounced as CNT content increases, possibly due to the fact that the increase in the content of CNT results in increase of the absorption contribution to electromagnetic shielding.

10.3.5 Geometrical Characteristics of Composites

It has been shown that composite thickness plays an important role in obtaining electromagnetic properties. In fact, more interface area would bring higher EMI shielding, since the entering power is reflected, scattered, and absorbed for many times at the interface between the polymer and the conductive layer [99]. Then, one can expect the thicker samples would own higher EMI shielding values. In order to further validate this conjecture, Wang et al. [99] investigated the EMI shielding effectiveness for PVDF/MWCNTs containing 7 wt.% of the nanofiller with different thicknesses. As shown in Fig. 10.8, SE_{total} and SE_A increased with sample thickness and the growth trend maintained consistent with each other. The increased EMI shielding effectiveness primarily came from the increase of SE_A. It demonstrated that the absorption rather than reflection dominated the shielding mechanism.

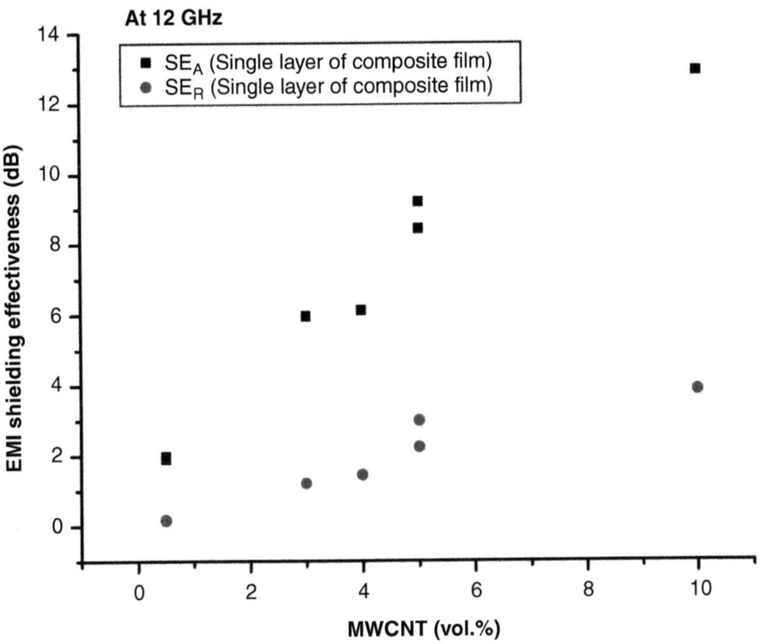

FIGURE 10.7 Effect of MWCNT amount on SE_A and SE_R of a single layer PMMA–MWCNT composite film at 12 GHz [25].

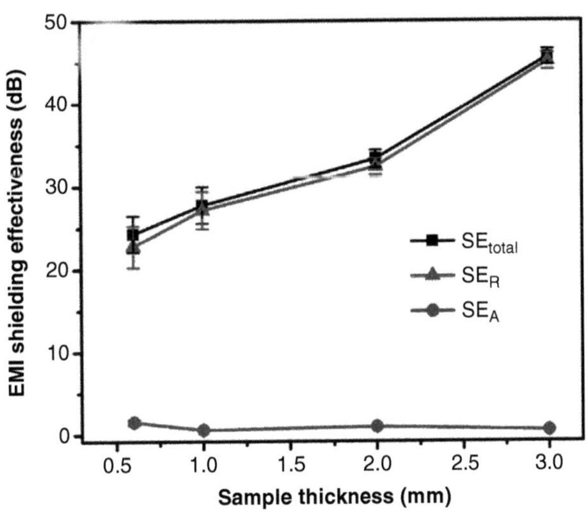

FIGURE 10.8 Effect of sample thickness on the EMI shielding effectiveness of PVDF/MWCNT 7 wt.% composites at the frequency of 9.5 GHz [99].

Mentioning of a study by Pande et al. [25] on the electromagnetic evaluation of MW-CNT-PMMA nanocomposite (10%v/v MWCNT) may be interesting here. They prepared stacked PMMA-MWCNT (10 vol.%) composites by stacking n (n = 1–7) layers of 0.3 mm thick composite films to investigate the effect of sample thickness in EMI shielding in stacked composites. The stacked composite films are referred to as SCFn (n: number of the layers of composite films with n = 1–7). Effect of number of layers on the SE_A, SE_R, and SE_{total} are demonstrated in Fig. 10.9A–C, respectively. As clear from Fig. 10.9, while SE_A and SE_{total} increased with the increase in number of layers, SE_R was almost the same for all layers (Fig. 10.9B). In fact, they found out that remarkable electromagnetic shielding could be obtained with seven layers of stacked composite films (total thickness = 2.1 mm). In this sample, EMI shielding effectiveness (SE) up to 40 dB in the X-band, that is, frequency range of 8.2–12.4 GHz, was achieved, which was much higher than the 30 dB achieved by stacking two layers of 1.1-mm thick MWCNT–PMMA bulk composite with total 2.2 mm thickness.

Another parameter that affects the electrical properties of CNT-polymer nanocomposites is CNTs orientation in the polymeric matrix. Du et al. [100] studied the effect of alignment on the electrical conductivity of SWCNT/PMMA nanocomposites made by a coagulation method with CNTs aligned using melt fiber spinning and compared to Monte Carlo simulation results. In this work, it was reported that slightly aligned rather than perfectly aligned nanotubes gave the highest conductivity. Wang et al. [101] fabricated aligned SW-CNT/epoxy nanocomposites by stretching semi-dried samples in one direction. Electrical conductivity was then tested in the longitudinal and transverse directions (Fig. 10.10).

Higher conductivity (up to three orders of magnitude higher) and a lower percolation threshold (0.05 wt.% SWCNT) were obtained in the alignment direction. However, the solution casting method used impacted negatively on the mechanical properties of the composite due to excessive porosity. These results contradict those obtained by Du et al. [100], who assert that there are poor contacts between the CNTs, when nanotubes are highly aligned, which lead to a higher percolation threshold and a decrease in the electrical conductivity in comparison with those in a composite with randomly oriented CNTs. Further evidence of higher conductivity in partially, rather than perfectly, aligned CNTs is provided in the Monte Carlo simulations of Benham et al. [102] who studied the effect of measurement direction and alignment on the electrical conductivity of CNT thin films.

10.3.6 Mixture of Carbon Nanotubes With Other Fillers

Simultaneous incorporating of CNTs together with second type of filler can benefit from the possible synergistic effects that could arise between the two different fillers. It is expected that the combination of two kinds of fillers promote the properties of the hybrid composite compared to single-filler materials. This improvement can be assigned to the combination of benefits of each kind of filler.

In this context, combination of CNT fibers with carbon black (CB) attracted immense attention. Combination of these two types of fillers is based on promoting the interactions

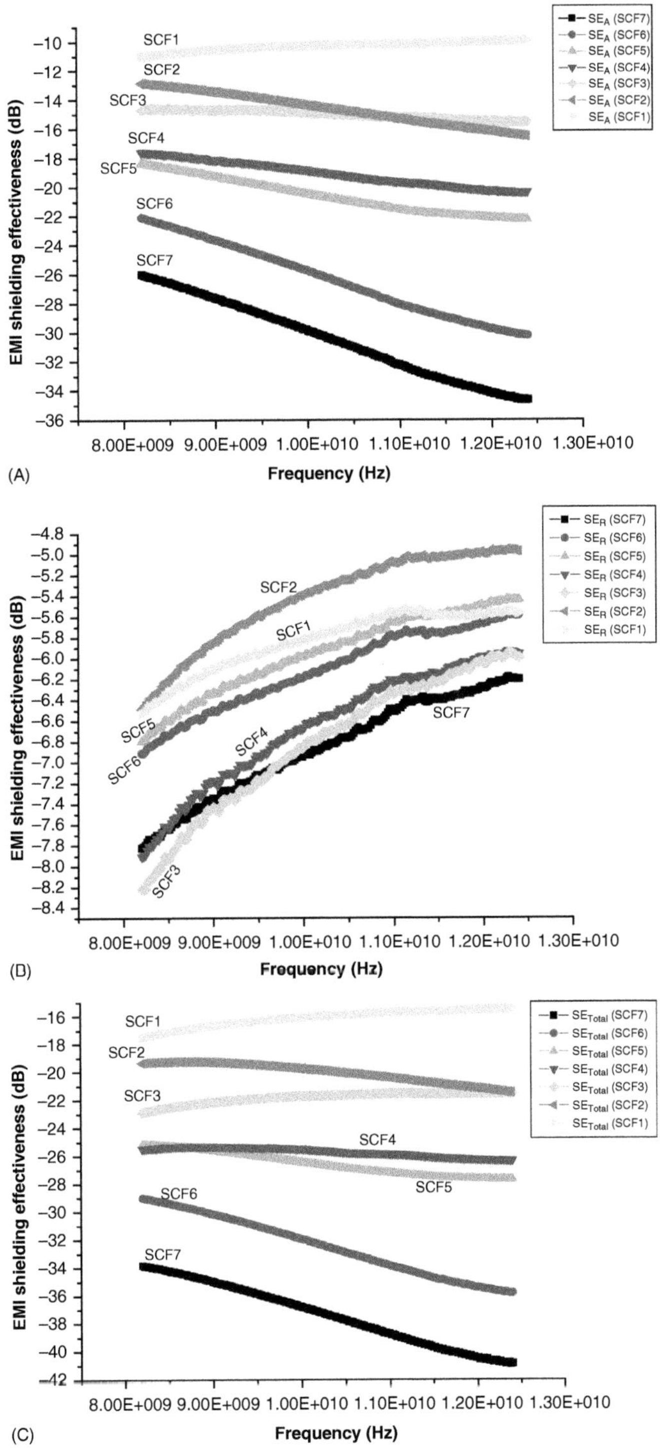

FIGURE 10.9 Effect of number of layers on (A) SE$_A$, (B) SE$_R$, and (C) SE$_{total}$ in PMMA–MWCNT stacked composites [25].

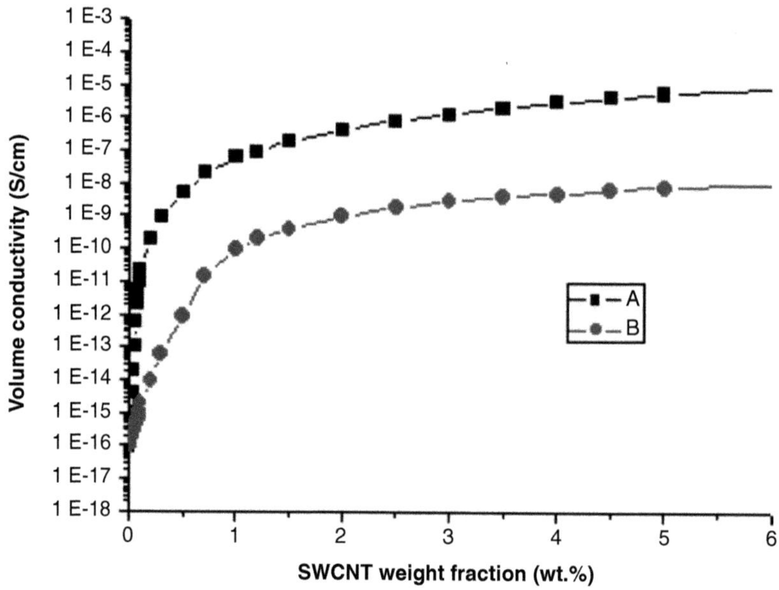

FIGURE 10.10 DC volume conductivity as a function of SWCNT weight fraction in epoxy/SWCNT composite. (A) Along the SWNT alignment and (B) perpendicular to the SWCNT alignment [101].

between the two types of fillers and creating bridges between local clusters thus reducing percolation threshold for electrical conductivity [91]. Indeed, CNTs form a filler network interconnecting together between 2 and 3 phr, and addition of nanotubes in composites containing a constant CB content of 10 phr reduces the percolation threshold and resistivity for CB. These effects are the outcome of better filler dispersion and make clear that the incorporation of nanotubes, by bridging the uncontacted CB particles, improves the formation of connected structures.

In another attempt, CB and carbon fibers were used as fillers of low-density polyethylene in order to move the percolation threshold to a decreased filler loading [103]. The combination of the fillers resulted in a synergy effect and continuous conductive pathways within the CB/CF-filled composite were generated. The authors believed that the introduction of carbon fibers into CB/PE composites provides long conductive pathways to connect unabridged CB aggregates.

Chen et al. investigated the effect of graphite nanosheets (GNS) on the EMI SE and electrical conductivity of the SWCNTs filled polyaniline (PANi) composites [68]. The authors found that the increase in filler loading leads to improvement of the electric conductivity, and the loading amount of 1.0 wt.% SWCNTs/GNS results in the highest electric conductivity of 16.2 S/cm, five orders of magnitude higher than that of pure PANi (Fig. 10.11). Lower conductivity of SWCNT/PANi composites is probably due to nonuniform distribution of SWCNTs in the PANi matrix.

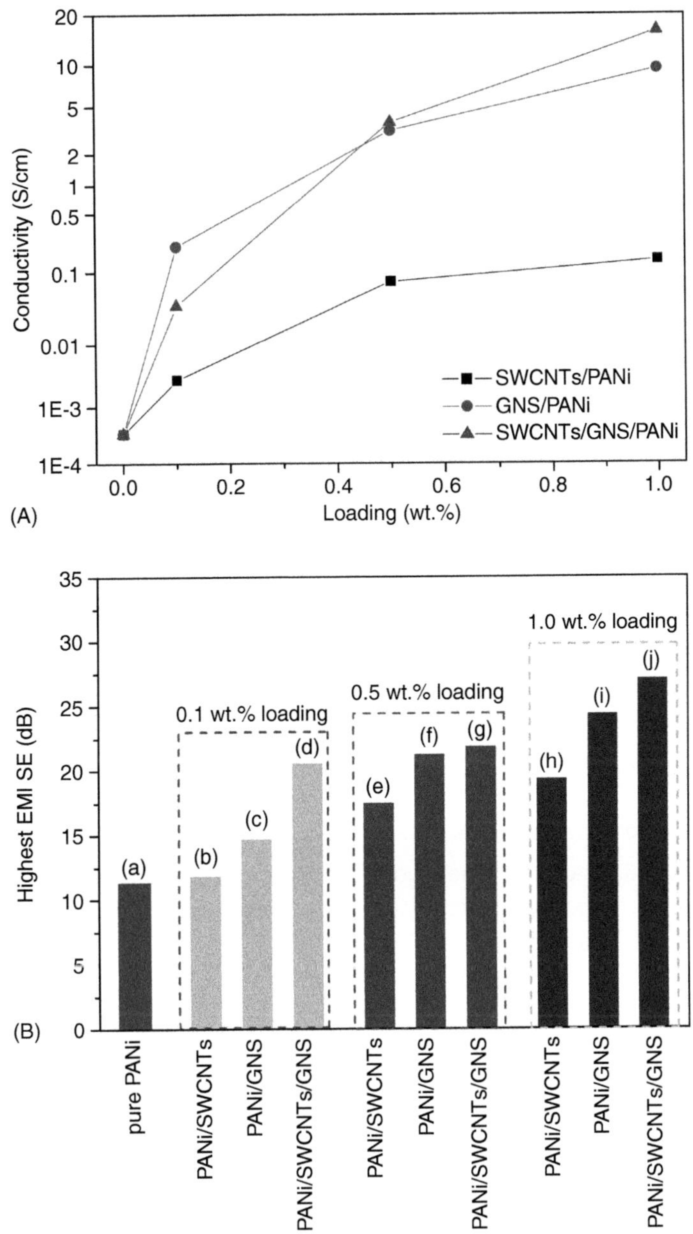

FIGURE 10.11 (A) Electric conductivity and (B) EMI SE of pure PANi and PANi/SWCNTs and PANi/SWCNTs/GNS nanocomposites [68].

It is noteworthy that the order of addition of the fillers remarkably affects the electrical properties of the polymer/filler1/filler2 composites. As an example, Xiao et al. [104] prepared two different composites, that is, PVDF/CNT/Graphene and PVDF/Graphene/CNT composites, which differed only in the order of addition of the fillers. PVDF/CNT/Graphene composite was obtained by adding trivial amount of CNTs (2 wt.%) into the PVDF/Graphene composites while PVDF/Graphene/CNT composite was prepared by adding a small quantity of Graphene (2 wt.%) into the PVDF/CNT composites. Compared with the PVDF/Graphene composites, which exhibit percolated network structure of filler at extremely high filler content (20 wt.%), the addition of CNTs greatly promotes the formation of the 3D hybrid network structure in the PVDF/CNT/Graphene composites. However, addition of Graphene into the PVDF/CNT composites did not alter the percolated network structure. This observation can be attributed to that the dense network structure of CNTs was present in the composites.

10.4 Conclusions

A lot of research has been done, and still is being performed, to induce electrical conductivity in the insulating polymer matrixes, however, a higher aspect ratio and lower percolation threshold of CNTs made CNT/polymer fabricates an interesting issue. In this chapter, main factors affecting electrical/electromagnetic properties of polymer/CNT composites have been discussed in brief. In summary:

1. CNT type, as SWCNT or MWCNT, does not have a clear effect on the electrical conductivity.
2. By increasing CNT aspect ratio, the percolation threshold reduces in a hyperbolic manner.
3. Polymer type and processing method affect the final properties of fabricated material. In this regard, with increasing π–π interactions between the matrix and CNT, conductivity will increase.
4. The presence of surfactant, although modifies CNT alignment and dispersion, it decreases sample efficiency.
5. Electrical and electromagnetic behavior of a composite are significantly affected by CNT amount. In fact, electrical conductivity increases significantly until percolation threshold is reached; after that, changes become far less pronounced.
6. By increasing interface area in the composite, higher EMI shielding will be obtained, since the entering power is reflected, scattered, and absorbed many times at the interface between the polymer and the conductive layer.
7. Electrical conductivity and shielding effectiveness increase by increasing sample thickness. It demonstrated that the absorption rather than reflection dominated the shielding mechanism.
8. Effect of CNT alignment on the electrical conductivity and percolation threshold is not well understood. Both increase and decrease in electrical conductivity were seen in the aligned samples.

9. The combination of CNT with another filler promotes the properties of the hybrid composite compared to single-filler material. The order of fillers addition does also affect the electrical properties of the composite. In order to get a sample with improved properties, beside filler amount, the order of addition should be adjusted as well.

In conclusion, considerable work is still needed to improve further the long-standing electrical/electromagnetic properties of the final CNT/polymer fabricates due to the formation of large aggregates or agglomerates, which results unreliable properties in the fabricated material. Thus, the progression in advanced and commercially reasonable production of these materials needs coherent tactics and plans, and appropriative endeavors should be performed to control accurately the dispersion geometry in the final polymer/CNT fabricates.

References

[1] S. Iijima, Helical microtubules of graphitic carbon, Nature 354 (1991) 56.

[2] P. Ajayan, O. Stephan, C. Colliex, D. Trauth, Aligned carbon nanotube arrays formed by cutting a polymer resin-nanotube composite, Science 265 (1994) 1212–1214.

[3] M. Moniruzzaman, K.I. Winey, Polymer nanocomposites containing carbon nanotubes, Macromolecules 39 (2006) 5194–5205.

[4] R. Ramasubramaniam, J. Chen, H. Liu, Homogeneous carbon nanotube/polymer composites for electrical applications, Appl. Phys. Lett. 83 (2003) 2928–2930.

[5] Y. Gogotsi, V. Presser, Carbon nanomaterials, CRC Press, Boca Raton, FL, United States, (2006).

[6] R.H. Baughman, A.A. Zakhidov, W.A. de Heer, Science 297 (2002) 787–792.

[7] S.A. Mirmohammadi, M. Nekoomanesh-Haghighi, S. Mohammadian Gezaz, N. Bahri-Laleh, Polybutadiene/polyhedral oligomeric silsesquioxane nanohybrid: investigation of various reactants in polyesterification reaction, Polym. Int. 65 (2016) 516–525.

[8] S. Mirmohammadi, M. Imani, H. Uyama, M. Atai, In situ photocrosslinkable nanohybrids based on poly(ε-caprolactone fumarate)/polyhedral oligomeric silsesquioxane: synthesis and characterization, J. Polym. Res. 20 (2013) 1–13.

[9] S.A. Mirmohammadi, M. Nekoomanesh-Haghighi, S.M. Gezaz, N. Bahri-Laleh, M. Atai, In-situ photocrosslinkable nanohybrid elastomer based on polybutadiene/polyhedral oligomeric silsesquioxane, Mater. Sci. Eng. 68 (2016) 530–539.

[10] E.T. Thostenson, C. Li, T.-W. Chou, Nanocomposites in context, Compos. Sci. Technol. 65 (2005) 491–516.

[11] Z. Spitalsky, D. Tasis, K. Papagelis, C. Galiotis, Carbon nanotube–polymer composites: chemistry, processing, mechanical, and electrical properties, Prog. Polym. Sci. 35 (2010) 357–401.

[12] B.A. Rozenberg, R. Tenne, Polymer-assisted fabrication of nanoparticles and nanocomposites, Prog. Polym. Sci. 33 (2008) 40–112.

[13] A. Brydon, G.G. Cameron, Chemical modification of unsaturated polymers, Prog. Polym. Sci. 4 (1975) 209–249.

[14] M. Alexandre, P. Dubois, Polymer-layered silicate nanocomposites: preparation, properties and uses of a new class of materials, Mater. Sci. Eng. 28 (2000) 1–63.

[15] K. Didehban, E. Yarahmadi, F. Nouri-Ahangarani, S.A. Mirmohammadi, N. Bahri-Laleh, Radar absorption properties of Ni0. 5Zn0. 5Fe2O4/PANI/epoxy nanocomposites, J. Chin. Chem. Soc. 62 (2015) 826–831.

[16] P. Saini, V. Choudhary, B. Singh, R. Mathur, S. Dhawan, Polyaniline–MWCNT nanocomposites for microwave absorption and EMI shielding, Mater. Chem. Phys. 113 (2009) 919–926.

[17] P. Saini, V. Choudhary, K.N. Sood, S.K. Dhawan, Electromagnetic interference shielding behavior of polyaniline/graphite composites prepared by in situ emulsion pathway, J. Appl. Polym. Sci. 113 (2009) 3146–3155.

[18] C.R. Paul, Electromagnetics for Engineers, Wiley International, Hoboken, NJ, United States, (2004) 279.

[19] H.W. Ott, Shielding, Electromagnetic Compatibility Engineering, John Wiley & Sons, Hoboken, NJ, United States, 2009, pp. 238–301.

[20] J. Joo, A. Epstein, Electromagnetic radiation shielding by intrinsically conducting polymers, Appl. Phys. Lett. 65 (1994) 2278–2280.

[21] H. Marsh, E.A. Heintz, F. Rodriguez-Reinoso, Introduction to Carbon Technologies, Universidad de Alicante, Secretariado de Publicaciones, Alicante, Spain, (1997).

[22] D.D.L. Chung, Comparison of submicron-diameter carbon filaments and conventional carbon fibers as fillers in composite materials, Carbon 39 (2001) 1119–1125.

[23] J. Wu, D.D.L. Chung, Increasing the electromagnetic interference shielding effectiveness of carbon fiber polymer–matrix composite by using activated carbon fibers, Carbon 40 (2002) 445–447.

[24] A. Nikfarjam, R. Rafiee, M. Taheri, Electrical and electromagnetic properties of isolated carbon nanotubes and carbon nanotube-based composites, Polyolefins J. 4 (2016) 43–68.

[25] S. Pande, B. Singh, R. Mathur, T. Dhami, P. Saini, S. Dhawan, Improved electromagnetic interference shielding properties of MWCNT–PMMA composites using layered structures, Nanoscale Res. Lett. 4 (2009) 327.

[26] F. Nanni, M. Valentini, Electromagnetic properties of polymer–carbon nanotube composites, Polymer–Carbon Nanotube Composites, Woodhead Publishing, Cambridge, United Kingdom, (2011) pp. 329–346.

[27] S. Kara, E. Arda, F. Dolastir, Ö. Pekcan, Electrical and optical percolations of polystyrene latex–multi-walled carbon nanotube composites, J. Colloid Interf. Sci. 344 (2010) 395–401.

[28] H. Ha, K. Ha, S.C. Kim, An empirical equation for electrical resistivity of thermoplastic polymer/multi-walled carbon nanotube composites, Carbon 48 (2010) 1939–1944.

[29] R.R. Kohlmeyer, M. Lor, J. Deng, H. Liu, J. Chen, Preparation of stable carbon nanotube aerogels with high electrical conductivity and porosity, Carbon 49 (2011) 2352–2361.

[30] X. Gao, S. Zhang, F. Mai, L. Lin, Y. Deng, H. Deng, Q. Fu, Preparation of high-performance conductive polymer fibres from double percolated structure, J. Mater. Chem. 21 (2011) 6401–6408.

[31] C.-L. Huang, C. Wang, Rheological and conductive percolation laws for syndiotactic polystyrene composites filled with carbon nanocapsules and carbon nanotubes, Carbon 49 (2011) 2334–2344.

[32] B.P. Singh, Prabha, P. Saini, T. Gupta, P. Garg, G. Kumar, I. Pande, S. Pande, R.K. Seth, S.K. Dhawan, R.B. Mathur, Designing of multiwalled carbon nanotubes reinforced low density polyethylene nanocomposites for suppression of electromagnetic radiation, J. Nanopart. Res. 13 (2011) 7065–7074.

[33] E. Messina, N. Leone, A. Foti, G. Di Marco, C. Riccucci, G. Di Carlo, F. Di Maggio, A. Cassata, L. Gargano, C. D'Andrea, B. Fazio, O.M. Maragò, B. Robba, C. Vasi, G.M. Ingo, P.G. Gucciardi, Double-wall nanotubes and graphene nanoplatelets for hybrid conductive adhesives with enhanced thermal and electrical conductivity, ACS Appl. Mater. Interf. 8 (2016) 23244–23259.

[34] Y. Liu, S. Kumar, Polymer/carbon nanotube nanocomposite fibers: a review, ACS Appl. Mater. Interf. 6 (2014) 6069–6087.

[35] P. Saini, V. Choudhary, S. Dhawan, Electrical properties and EMI shielding behavior of highly thermally stable polyaniline/colloidal graphite composites, Polym. Advan. Technol. 20 (2009) 355–361.

[36] K. Singh, A. Ohlan, P. Saini, S. Dhawan, Poly (3, 4-ethylenedioxythiophene) γ-Fe2O3 polymer composite: super paramagnetic behavior and variable range hopping 1D conduction mechanism—synthesis and characterization, Polym. Adv. Technol. 19 (2008) 229–236.

[37] P. Saini, V. Choudhary, B.P. Singh, R.B. Mathur, S.K. Dhawan, Enhanced microwave absorption behavior of polyaniline-CNT/polystyrene blend in 12.4–18.0 GHz range, Synthetic Met. 161 (2011) 1522–1526.

[38] D.D.L. Chung, Electromagnetic interference shielding effectiveness of carbon materials, Carbon 39 (2001) 279–285.

[39] N.E. Kazantseva, Y.I. Bespyatykh, I. Sapurina, J. Stejskal, J. Vilčáková, P. Sáha, Magnetic materials based on manganese–zinc ferrite with surface-organized polyaniline coating, J. Magn. Magn. Mater. 301 (2006) 155–165.

[40] G. Li, S. Yan, E. Zhou, Y. Chen, Preparation of magnetic and conductive NiZn ferrite-polyaniline nanocomposites with core-shell structure, Colloid Surf. 276 (2006) 40–44.

[41] O. Yavuz, M.K. Ram, M. Aldissi, P. Poddar, S. Hariharan, Synthesis and the physical properties of MnZn ferrite and NiMnZn ferrite-polyaniline nanocomposite particles, J. Mater. Chem. 15 (2005) 810–817.

[42] J. Deng, Y. Peng, C. He, X. Long, P. Li, A.S.C. Chan, Magnetic and conducting Fe3O4–polypyrrole nanoparticles with core-shell structure, Polym. Int. 52 (2003) 1182–1187.

[43] Y. Yang, M.C. Gupta, K.L. Dudley, R.W. Lawrence, Conductive carbon nanofiber–polymer foam structures, Adv. Mater. 17 (2005) 1999–2003.

[44] P. Saini, M. Arora, G. Gupta, B.K. Gupta, V.N. Singh, V. Choudhary, High permittivity polyaniline–barium titanate nanocomposites with excellent electromagnetic interference shielding response, Nanoscale 5 (2013) 4330–4336.

[45] G. Pandey, E.T. Thostenson, Carbon nanotube-based multifunctional polymer nanocomposites, Polym. Rev. 52 (2012) 355–416.

[46] M. Bumsuk Kim, M. Jongjin Lee, M. Insuk Yu, Electrical properties of single-wall carbon nanotube and epoxy composites, J. Appl. Phys. 94 (2003) 6724–6728.

[47] A. Allaoui, S. Bai, H.M. Cheng, J.B. Bai, Mechanical and electrical properties of a MWNT/epoxy composite, Compos. Sci. Technol. 62 (2002) 1993–1998.

[48] M.K. Njuguna MK, Characterization of multiwall carbon nanotube-polymer composites for strain sensing applications, School of Chemistry, Physics and Mechanical Engineering Science and Engineering Faculty, Queensland University of Technology, Brisbane, Australia, 2012.

[49] B.E. Kilbride, J.N. Coleman, J. Fraysse, P. Fournet, M. Cadek, A. Drury, S. Hutzler, S. Roth, W.J. Blau, Experimental observation of scaling laws for alternating current and direct current conductivity in polymer-carbon nanotube composite thin films, J. Appl. Phys. 92 (2002) 4024.

[50] R.H. Cruz-Estrada, M.J. Folkes, Structure formation and modelling of the electrical conductivity in SBS-polyaniline blends, Part I: Percolation theory approach, J. Mater. Sci. Lett. 21 (2002) 1427–1429.

[51] R.H. Cruz-Estrada, M.J. Folkes, Structure formation and modelling of the electrical conductivity in SBS-polyaniline blends, Part II: Generalized effective media theories approach, J. Mater. Sci. Lett. 21 (2002) 1431–1434.

[52] M. Weber, M.R. Kamal, Estimation of the volume resistivity of electrically conductive composites, Polym. Comp. 18 (1997) 711–725.

[53] S.R. Broadbent, J.M. Hammersley, Percolation processes: I. Crystals and mazes. In: Mathematical Proceedings of the Cambridge Philosophical Society, Cambridge University Press, Cambridge, United Kingdom, 1957.

[54] A. Aharony, D. Stauffer, Introduction to Percolation Theory, Taylor & Francis, London, United Kingdom, (2003).

[55] T.A. Skotheim, R.L. Elsenbaumer, J.R. Reynolds, Handbook of Conducting Polymers, second ed., CRC Press, Boca Raton, FL, United States, (1997).

[56] J.-M. Thomassin, C. Jérôme, T. Pardoen, C. Bailly, I. Huynen, C. Detrembleur, Polymer/carbon based composites as electromagnetic interference (EMI) shielding materials, Mater. Sci. Eng. 74 (2013) 211–232.

[57] Y. Yang, M.C. Gupta, K.L. Dudley, R.W. Lawrence, Novel carbon nanotube-polystyrene foam composites for electromagnetic interference shielding, Nano Lett. 5 (2005) 2131–2134.

[58] L. Bokobza, M. Rahmani, Carbon nanotubes: exceptional reinforcing fillers for silicon rubbers, Kautsch Gummi Kunstst 62 (2009) 112–117.

[59] M. Zhang, J. Li, Carbon nanotube in different shapes, Materials Today 12 (2009) 12–18.

[60] N. Saifuddin, A.Z. Raziah, A.R. Junizah, Carbon nanotubes: a review on structure and their interaction with proteins, J. Chem. 2013 (2013) 676815.

[61] S. Sathyanarayana, C. C. Hubner, Thermoplastic nanocomposites with carbon nanotubes, in: J. Njuguna (Ed.), Structural Nanocomposites: Perspectives for Future Applications, Springer, Berlin Heidelberg, Heidelberg, Germany, 2013, pp. 19–60.

[62] W. Bauhofer, J.Z. Kovacs, A review and analysis of electrical percolation in carbon nanotube polymer composites, Compos. Sci. Technol. 69 (2009) 1486–1498.

[63] V. Sk, U. Dettlaff-Weglikowska, S. Roth, Electrical and mechanical properties of nanocomposites of single-wall carbon nanotubes with PMMA, Synth. Metals 152 (2005) 349–352.

[64] G. Blanchet, C. Fincher, F. Gaio, Polyaniline nanotube composites: a high-resolution printable conductor, Appl. Phys. Lett. 82 (2003) 1290–1292.

[65] H. Koerner, W. Liu, M. Alexander, P. Mirau, H. Dowty, R.A. Vaia, Deformation-morphology correlations in electrically conductive carbon nanotube—thermoplastic polyurethane nanocomposites, Polymer 46 (2005) 4405–4420.

[66] N. Grossiord, J. Loos, L. van Laake, M. Maugey, C. Zakri, C.E. Koning, A.J. Hart, High-conductivity polymer nanocomposites obtained by tailoring the characteristics of carbon nanotube fillers, Adv. Funct. Mater. 18 (2008) 3226–3234.

[67] P. Saini, V. Choudhary, B.P. Singh, R.B. Mathur, S.K. Dhawan, Polyaniline–MWCNT nanocomposites for microwave absorption and EMI shielding, Mater. Chem. Phys. 113 (2009) 919–926.

[68] Y.-J. Chen, N.D. Dung, Y.-A. Li, M.-C. Yip, W.-K. Hsu, N.-H. Tai, Investigation of the electric conductivity and the electromagnetic interference shielding efficiency of SWCNTs/GNS/PAni nanocomposites, Diamond Relat. Mater. 20 (2011) 1183–1187.

[69] S.-E. Lee, S. Cho, H. Kim, I. Han, Y. Sohn, Advanced catalyst design induced enhancement of multi-walled nanotube debundling and electrical conductivity of multi-walled nanotube/silicone composites, RSC Adv. 6 (2016) 48120–48128.

[70] S.H. Munson-McGee, Estimation of the critical concentration in an anisotropic percolation network, Phys. Rev. 43 (1991) 3331–3336.

[71] A. Celzard, E. McRae, C. Deleuze, M. Dufort, G. Furdin, J. Mareché, Critical concentration in percolating systems containing a high-aspectratio filler, Phys. Rev. 53 (1996) 6209–6214.

[72] C. Ding, K. Liu, C. Guo, D. Jia, B. Cheng, Effects of diameter and aspect ratio of carbon nanotubes on crystalline and electrical properties of poly(ethylene terephthalate) nanocomposites, Polym. Eng. Sci. 56 (2016) 408–417.

[73] R. Ram, M. Rahaman, A. Aldalbahi, D. D. Khastgir, Determination of percolation threshold and electrical conductivity of polyvinylidene fluoride (PVDF)/short carbon fiber (SCF) composites: effect of SCF aspect ratio, Polym. Int. 66 (4) (2017) 573–582.

[74] Y. Yu, H. Chen, Y. Liu, V.S.J. Craig, L.H. Li, Y. Chen, A. Tricoli, Porous carbon nanotube/polyvinylidene fluoride composite material: superhydrophobicity/ superoleophilicity and tunability of electrical conductivity, Polymer 55 (2014) 5616–5622.

[75] M.A. Worsley, S.O. Kucheyev, J.D. Kuntz, A.V. Hamza, J.J.H. Satcher, T.F. Baumann, Stiff and electrically conductive composites of carbon nanotube aerogels and polymers, J. Mater. Chem. 19 (2009) 3370–3372.

[76] J.-M. Thomassin, X. Lou, C. Pagnoulle, A. Saib, L. Bednarz, I. Huynen, C. Detrembleur, Multiwalled carbon nanotube/poly(ε-caprolactone) nanocomposites with exceptional electromagnetic interference shielding properties, J. Phys. Chem. 111 (2007) 11186–11192.

[77] Z. Liu, G. Bai, Y. Huang, Y. Ma, F. Du, F. Li, T. Guo, Y. Chen, Reflection and absorption contributions to the electromagnetic interference shielding of single-walled carbon nanotube/polyurethane composites, Carbon 45 (2007) 821–827.

[78] E. Logakis, P. Pissis, D. Pospiech, A. Korwitz, B. Krause, U. Reuter, P. Potschke, Low electrical percolation threshold in poly(ethylene terephthalate)/multi-walled carbon nanotube nanocomposites, Eur. Poly. J. 46 (2010) 928–936.

[79] B. Sundaray, V. Subramanian, T. Natarajan, K. Krishnamurthy, Electrical conductivity of a single electrospun fiber of poly(methyl methacrylate) and multiwalled carbon nanotube nanocomposites, Appl. Phys. Lett. 88 (2006) 143114.

[80] P. Pötschke, M. Abdel-Goad, I. Alig, S. Dudkin, D L, Rheological and dielectrical characterization of melt mixed polycarbonate multiwalled carbon nanotube composites, Polymer 45 (2004) 8863–8870.

[81] O. Gryshchuk, J. Karger-Kocsis, R. Thomann, Z. Knya, I. Kiricsi, Multiwall carbon nanotube modified vinylester and vinylester-based hybrid resins, Compos. Part A 37 (2006) 1252–1259.

[82] Z. Zhao, W. Zheng, W. Yu, B. Long, Electrical conductivity of poly(vinylidene fluoride)/carbon nanotube composites with a spherical substructure, Carbon 47 (2009) 2118–2120.

[83] Z. He, X. Zhang, M. Chen, M. Li, Y. Gu, Z. Zhang, Q. Li, Effect of the filler structure of carbon nanomaterials on the electrical, thermal, and rheological properties of epoxy composites, J. Appl. Polym. Sci. 129 (2013) 3366–3372.

[84] B.J. Landi, R.P. Raffaelle, M.J. Heben, J.L. Alleman, W. VanDerveer, T. Gennett, Single-wall carbon nanotube–nafion composite actuators, Nano Lett. 2 (2002) 1329–1332.

[85] X. Zhang, J. Zhang, R. Wang, T. Zhu, Z. Liu, Surfactant-directed polypyrrole/CNT nanocables: synthesis, characterization, and enhanced electrical properties, Chem. Phys. Chem. 5 (2004) 998–1002.

[86] J. Fan, M. Wan, D. Zhu, B. Chang, Z. Pan, S. Xie, Synthesis, characterizations, and physical properties of carbon nanotubes coated by conducting polypyrrole, J. Appl. Polym. Sci. 74 (1999) 2605–2610.

[87] H. Peng, X. Sun, Highly aligned carbon nanotube/polymer composites with much improved electrical conductivities, Chem. Phys. Lett. 471 (2009) 103–105.

[88] J. Yang, T. Xu, A. Lu, Q. Zhang, Q. Fu, Electrical properties of poly(phenylene sulfide)/multiwalled carbon nanotube composites prepared by simple mixing and compression, J. Appl. Polym. Sci. 109 (2008) 720–726.

[89] H. Aarab, M. Baïtoul, J. Wéry, R. Almairac, S. Lefrant, E. Faulques, J.L. Duvail, M. Hamedoun, Electrical and optical properties of PPV and single-walled carbon nanotubes composite films, Synth. Met. 155 (2005) 63–67.

[90] H.M. Kim, K. Kim, C.Y. Lee, J. Joo, S.J. Cho, H.S. Yoon, D.A. Pejaković, J.W. Yoo, A.J. Epstein, Electrical conductivity and electromagnetic interference shielding of multiwalled carbon nanotube composites containing Fe catalyst, Appl. Phys. Lett. 84 (2004) 589–591.

[91] L. Bokobza, M. Rahmani, C. Belin, J.-L. Bruneel, N.-E. El Bounia, Blends of carbon blacks and multiwall carbon nanotubes as reinforcing fillers for hydrocarbon rubbers, J. Polym. Sci. 46 (2008) 1939–1951.

[92] X. Gong, J. Liu, S. Baskaran, R.D. Voise, J.S. Young, Surfactant-assisted processing of carbon nanotube/polymer composites, Chem. Mater. 12 (2000) 1049–1052.

[93] V.D. Punetha, S. Rana, H.J. Yoo, A. Chaurasia, J.T. McLeskey Jr., M.S. Ramasamy, N.G. Sahoo, J.W. Cho, Functionalization of carbon nanomaterials for advanced polymer nanocomposites: a comparison study between CNT and graphene, Prog. Polym. Sci. 67 (2016) 1–47.

[94] S. Bose, A.R. Bhattacharyya, A.P. Bondre, A.R. Kulkarni, P. Pötschke, Rheology, electrical conductivity, and the phase behavior of cocontinuous PA6/ABS blends with MWNT: correlating the aspect ratio of MWNT with the percolation threshold, J. Polym. Sci. 46 (2008) 1619–1631.

[95] F.H. Gojny, M.H.G. Wichmann, B. Fiedler, I.A. Kinloch, W. Bauhofer, A.H. Windle, K. Schulte, Evaluation and identification of electrical and thermal conduction mechanisms in carbon nanotube/epoxy composites, Polymer 47 (2006) 2036–2045.

[96] G.X. Chen, H.S. Kim, B.H. Park, J.S. Yoon, Highly insulating silicone composites with a high carbon nanotube content, Carbon 44 (2006) 3373–3375.

[97] L. Vast, Z. Mekhalif, A. Fonseca, J.B. Nagy, J. Delhalle, Preparation and electrical characterization of a silicone elastomer composite charged with multi-wall carbon nanotubes functionalized with 7-octenyltrichlorosilane, Compos. Sci. Technol. 67 (2007) 880–889.

[98] J. Vilčáková, R. Moučka, P. Svoboda, M. Ilčíková, N. Kazantseva, M. Hrˇibová, M. Mičušík, M. Omastová, Effect of surfactants and manufacturing methods on the electrical and thermal conductivity of carbon nanotube/silicone composites, Molecules 17 (2012) 13157–13174.

[99] H. Wang, K. Zheng, X. Zhang, T. Du, C. Xiao, X. Ding, C. Bao, L. Chen, X. Tian, Segregated poly(vinylidene fluoride)/MWCNTs composites for high-performance electromagnetic interference shielding, Compos. Part A 90 (2016) 606–613.

[100] F. Du, J.E. Fischer, K.I. Winey, Effect of nanotube alignment on percolation conductivity in carbon nanotube/polymer composites, Phys. Rev. 72 (2005) 121404.

[101] Q. Wang, J. Dai, W. Li, Z. Wei, J. Jiang, The effects of CNT alignment on electrical conductivity and mechanical properties of SWNT/epoxy nanocomposites, Compos. Sci. Technol. 68 (2008) 1644–1648.

[102] A. Behnam, J. Guo, A. Ural, Effects of nanotube alignment and measurement direction on percolation resistivity in single-walled carbon nanotube films, J. Appl. Phys. 102 (2007) 044313.

[103] W. Di, G. Zhang, J. Xu, Y. Peng, X. Wang, Z. Xie, Positive-temperature-coefficient/negative-temperature-coefficient effect of low-density polyethylene filled with a mixture of carbon black and carbon fiber, J. Polym. Sci. 41 (2003) 3094–3101.

[104] Xiao Y-j, Wang W-y, Chen X-j, T. Lin, Zhang Y-t, Yang J-h, Y. Wang, Z-w. Zhou, Hybrid network structure and thermal conductive properties in poly(vinylidene fluoride) composites based on carbon nanotubes and graphene nanoplatelets, Compos. Part A 90 (2016) 614–625.

11

Atomistic Simulations of Carbon Nanotubes: Stiffness, Strength, and Toughness of Locally Buckled CNTs

Nuno Silvestre*, Bruno Faria**, José N. Canongia Lopes**

*IDMEC, INSTITUTO SUPERIOR TÉCNICO, UNIVERSITY OF LISBON, LISBON, PORTUGAL;
**CQE, INSTITUTO SUPERIOR TÉCNICO, UNIVERSITY OF LISBON, LISBON, PORTUGAL

CHAPTER OUTLINE

11.1 Introduction

From an engineering-based definition, a *prismatic* tube possesses a longitudinal axis (denote by x) orthogonal to its circular cross-section plane (centroidal system denoted by s–z) and the dimensions and circular shape of its cross-section do not vary with x. Furthermore, a *prismatic thin-walled* tube is a prismatic member with two dimensions (in directions x and s) much higher than the third dimension, which is generally denoted as the tube thickness (t). Fig. 11.1A depicts a prismatic thin-walled tube. In mechanical engineering, there are many applications of prismatic thin-walled tubes. Due to external actions, either static

Carbon Nanotube–Reinforced Polymers. http://dx.doi.org/10.1016/B978-0-323-48221-9.00011-X

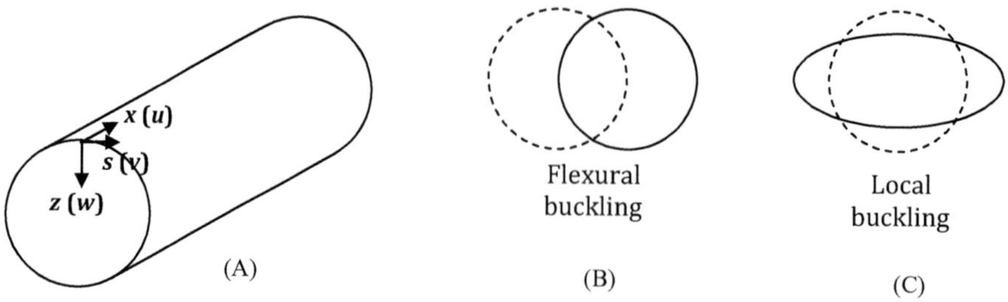

FIGURE 11.1 Thin-walled tube (A) coordinate and displacement system, (B) flexural mode, and (C) local mode.

or dynamic, these tubes are often subjected to compressive stresses. The simultaneous action of these compressive stresses and the inherent reduced thickness of thin-walled tubes make these tubes sensitive to loss of stability and ensuing collapse.

In accordance with the theory of instability of thin-walled structures, it is common to distinguish between two types of buckling modes: (1) flexural modes and (2) local modes. Flexural buckling takes place when the member axis (x) deforms but the cross-section only shows a rigid-body displacement in its own plane (Fig. 11.1B). Thus, if the displacement in the thickness direction is designated as w, it is possible to mention that flexural buckling takes place if

$$w_{,ss} = 0, \tag{11.1}$$

where $(-)_{,s} = \partial(-)/\partial s$.

Inversely, local buckling occurs when the cross-section deforms in its own plane (s–z) but the structure axis remains straight (Fig. 11.1C). In other words, local buckling is kinematically characterized by

$$w_{,ss} \neq 0, \tag{11.2}$$

and nonnull bending curvature of the circular cross-section midline arises. When both modes occur simultaneously, mixed (or coupled) instability takes place. Nowadays, flexural buckling of thin-walled members is very well understood. Unfortunately, local buckling is not at the same stage as flexural buckling. Due to circular cross-section deformation, local buckling is kinematically more complex than flexural buckling.

Thin-walled tubes typically used in mechanical applications and systems possess a size that is easily measurable and observable at our length scale, denoted as macroscale. The material that constitutes these members is treated as a continuous medium (continuum). At macroscale, an elastic material is characterized by the generalized Hooke's law

$$\sigma_{ij} = C_{ijkl}\varepsilon_{kl}, \tag{11.3}$$

where C_{ijkl} is the fourth order for a general elastic material (with symmetry properties $C_{ijkl} = C_{jikl} = C_{ijlk} = C_{klij}$), and σ_{ij} and ε_{kl} are stress and strain tensors, respectively. Due to their nature, prismatic thin-walled members are almost always subjected to plane stress state in their walls. Usually, the normal stress component in the thickness direction is always much lower than the stress components in the wall midplane. Thus, the simplest stress–strain elastic constitutive equation of a wall under a plane stress state reads

$$\begin{aligned}\sigma_{xx} &= C_{xxxx}\varepsilon_{xx} + C_{xxss}\varepsilon_{ss}\\ \sigma_{ss} &= C_{xxss}\varepsilon_{xx} + C_{ssss}\varepsilon_{ss},\\ \sigma_{xs} &= C_{xsxs}\varepsilon_{xs}\end{aligned} \tag{11.4}$$

where σ_{xx}, σ_{ss}, and σ_{xs} are normal and shear stresses and ε_{xx}, ε_{ss}, and ε_{xs} are normal and shear strains. Eq. (11.3) is often called the local stress–strain relation because the stress state at a given material point depends solely on the strain components at that point. On the other hand, the member dimensions, such as the length and thickness, are measurable at macroscale, that is, they have physical meaning to the "human eye." The structural analysis of isotropic thin-walled members considers two mechanical variables: (1) the membrane stiffness $\bar{A}$ and (2) the bending stiffness $\bar{D}$, which depend on the longitudinal elastic modulus E ($C_{xxxx} = E/(1 - \nu^2)$) for an isotropic material and wall thickness t by means of the relations

$$\begin{aligned}\bar{A} &= C_{xxxx}t\\ \bar{D} &= C_{xxxx}t^3/12.\end{aligned} \tag{11.5}$$

Thus, the wall thickness is easily measured at macroscale and the traditional methods of structural analysis, such as the finite element method and the finite strip method, can be used with confidence to model the entire volume of the member and simulate its local buckling behavior.

Due to the advent of nanotechnology in the 1990s, the manipulation of matter at an atomic and molecular scale, denoted as nanoscale, became possible. Generally, nanotechnology works with materials, devices, and other structures with at least one dimension sized from one to several nanometers. A new branch of nanotechnology is molecular self-assembly, which comprises the development of new materials with dimensions on the nanoscale and the direct control of matter at the atomic scale. At nanoscale, new atomic structures (nanostructures) can be thought of and "built." Examples of these nanostructures are well known nowadays, mainly those made of carbon, such as graphene, nanotubes, and fullerenes. The discovery of fullerenes greatly expanded the number of known carbon allotropes, which until recently were limited to graphite, diamond, and amorphous carbon. These carbon-based structures have two dimensions much higher than the third one, that is, they are thin-walled, and display remarkable electrical, magnetic, optical, and mechanical properties. Due to the latter, graphene and nanotubes are often used in nanodevices (as structural members) and nanomaterials (as reinforcement). At their scale, and depending on the intended use, thin-walled nanostructures are also acted upon by several types of loads and may also buckle in a similar way as their macroscale counterparts, either in flexural or local modes (see the local buckling of a multiwalled nanotube in Fig. 11.2).

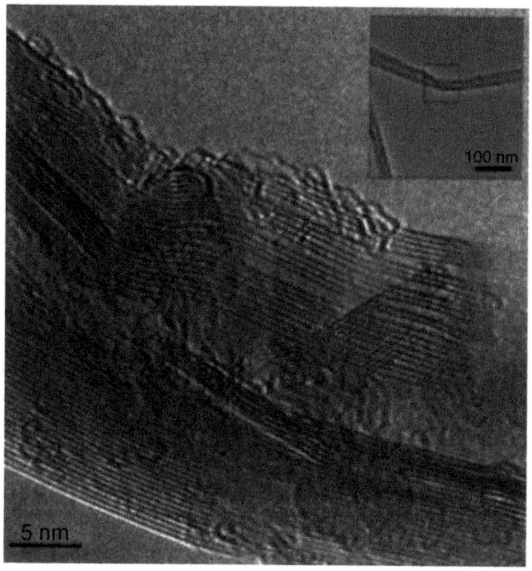

FIGURE 11.2 Local buckling of a multiwalled carbon nanotube.

However, unlike the thin-walled macroscale members, the modeling of matter at nanoscale as continuous media is far from consensus. The nanostructure components are the atoms, which are singular entities, and they are linked by means of interatomic forces, both covalent and noncovalent. Thus, the material of these thin-walled nanostructures displays an inherently discrete nature. To make it more complex, the quantum effects become dominant in these nanostructures—one atom is linked to their adjacent atoms through the covalent bonds but other forces emerge from all the nonbonded atoms in the neighboring environment. Much of the fascination with nanotechnology stems from these quantum and surface phenomena that matter exhibits at the nanoscale. Multiple recent experimental results have shown a significant size-effect in mechanical properties when the dimensions of the specimen or the probed material volume become small.

In contrast to the conventional continuum mechanics, the mechanics at the nanoscale should consider not only the stress at a given point but also the stress state at the environment of that point. The stress–strain relation at nanoscale is designated as nonlocal because the stress at a point in an elastic continuum is dependent on the strain at all the points of the domain. In the absence of body forces, the nonlocal stress relation for a linear elastic and homogenous elastic body at a point x reads

$$t_{ij}(x)=\int \lambda(|x-x'|)\cdot\sigma_{ij}(x')\cdot dV(x'), \quad \vee x\in V, \tag{11.6}$$

where t_{ij} is the stress tensor of the nonlocal elasticity and σ_{ij} is the local stress tensor of classical elasticity theory at any point x' in the body and satisfies the constitutive relation given by Eq. (11.3). In Eq. (11.6), V is the region occupied by the body, $|x - x'|$ is the Euclidean distance between the point and all the points in its environment, and λ $(|x - x'|)$ is the nonlocal kernel function. The properties of the nonlocal kernel function have been discussed in detail by Eringen [1]. By matching the dispersion curves with lattice models, Eringen [1,2] proposed a nonlocal model with the linear differential operator defined by

$$\mathcal{L}=1-(e_0a)^2\nabla^2, \tag{11.7}$$

where the Laplace operator in the coordinate system reads $\nabla^2=\partial^2/\partial x^2+\partial^2/\partial s^2$ and the nonlocal constitutive relation given by Eq. (11.6) is reduced to the differential equation

$$\mathcal{L}\, t_{ij}=\sigma_{ij}. \tag{11.8}$$

In Eq. (11.7), a is an internal characteristic length (lattice parameter, granular size, or molecular diameters) and e_0 is a constant appropriate to each material for adjusting the model to match some reliable results by experiments or other theories. As the product e_0a tends to zero, the nonlocal theory approaches the local theory. Introducing Eq. (11.7) into Eq. (11.8) and using Eq. (11.3), the following is obtained:

$$\begin{aligned} t_{xx}-(e_0a)^2\nabla^2 t_{xx} &= C_{xxxx}\varepsilon_{xx}+C_{xxss}\varepsilon_{ss} \\ t_{ss}-(e_0a)^2\nabla^2 t_{ss} &= C_{xxss}\varepsilon_{xx}+C_{ssss}\varepsilon_{ss}. \\ t_{xs}-(e_0a)^2\nabla^2 t_{xs} &= C_{xsxs}\varepsilon_{xs} \end{aligned} \tag{11.9}$$

This set of nonlocal constitutive relations has been widely applied to study the local buckling of thin-walled nanostructures. However, it might be questioned that it is always necessary to know the reliable values of e_0a to get an accurate local buckling behavior. The set of available e_0a values is still questionable and quantum mechanics principles are needed to calibrate the parameters e_0 and a. The classical continuum theories, on the other hand, lack the capability of representing such effects since they do not include any internal length scale. Consequently, these theories are expected to fail when the specimen size or the wavelength of the perturbation become comparable with the internal length scale(s) of the material.

Another issue that is still unsolved in the modeling of thin-walled nanostructures is to know the real value of their thickness. Due to their discrete nature and absence of a continuum, the modeling of nanostructures using traditional numerical methods of structural analysis, such as the shell finite element method fails. Usually, the thickness t and the elastic modulus E of carbon nanostructures (graphene and nanotubes) are obtained by fitting the atomistic simulation results of the membrane stiffness $\bar{A}$ and the bending stiffness $\bar{D}$, using Eq. (11.5). This gives the thickness

$$t=\sqrt{\frac{12\bar{D}}{\bar{A}}}, \tag{11.10}$$

and elastic modulus

$$E=\frac{\bar{A}}{t}(1-\nu^2)=\frac{12\bar{D}}{t^3}(1-\nu^2). \tag{11.11}$$

The plethora of results available in the literature displays a large scattering; t varies from 0.6 to 6.0 Å and E also varies by a similar order of magnitude [3–12]. Several researchers attributed this huge scatter to the dependence of these parameters on the chirality of the carbon hexagonal lattice and type of loading [13]. This is called the "Yakobson's paradox" in tribute to Boris Yakobson, who was the first researcher to unveil this odd behavior of carbon nanostructures [14]. Nowadays, the scientific community is confident that, unlike macroscale structures, the membrane stiffness $\bar{A}$ and the bending stiffness $\bar{D}$ of nanostructures do not depend on only the thickness but also on other quantum effects. This way, several shell theories not depending on the thickness t and elastic modulus E have recently been proposed, but full consensus and agreement between the technical and scientific communities have not yet been reached [15].

During the last decade, the authors of this chapter have been working on the analysis and characterization of local buckling phenomena in carbon nanotubes (CNTs). This chapter summarizes some of the contributions of the authors to the current state of the art of atomistic modeling of CNTs affected by local buckling, namely, the study of their stiffness, strength, and toughness.

11.2 Atomistic Modeling and Molecular Dynamics

At nanoscale, atomic structures, such as graphene, nanotubes, and fullerenes display a behavior that sometimes mimics their macroscale counterpart. The components of these nanostructures are the atoms and interatomic bonds. However, the experimental measurement of mechanical properties at nanoscale is difficult to obtain because the straightforward manipulation of the nanostructure is rather complex to achieve. Currently, there are still few appropriate measuring techniques at the nanoscale and it is really difficult to obtain some reliable data from indirect measurements. Thus, the theoretical and numerical techniques play an important role in investigating the mechanical properties of nanostructures. There are two main groups of methods: (1) discrete (or atomistic) and (2) continuum. The discrete (atomistic) methods are subdivided into three main types:

1. Ab initio methods Latin term meaning "from the beginning" and is derived from the Latin ab (from) + initio, ablative singular of *initium* (beginning), sometimes designated as first-principles methods, enable the calculation of atomic structures directly from the first principles of quantum mechanics. The main objective of ab initio methods is the rigorous determination of the solution of the Schrödinger equation. The ab initio method treats separately the electronic and ionic degrees of freedom, thus needing a wave function description for the electrons [16]. In the framework of first-principles methods, such as the Born–Oppenheimer and the Car–Parrinello molecular dynamics (MD), the nuclei move on a potential energy surface given by the electronic Schrödinger equation [17]. Due to the *curse of dimension* of the electronic Schrödinger equation for many-body systems, these approaches are only useful for a small number of particles. Many high-dimensional problems are difficult to solve for any numerical method (algorithm). Their complexity, that is, the cost of an optimal algorithm, increases exponentially with the dimension. Many problems suffer from the curse of dimension. The term "curse of dimensionality" seems to go back

to R. Bellman. Since the ab initio method is too expensive and time-consuming, it is common practice to incorporate some approximations, such as the density functional theory, which is based on the fact that the ground-state electronic energy is a unique functional of the electronic density, instead of many-body interacting wave functions.

2. The tight-binding method might suggest a quantum mechanical model describing the properties of tightly bound electrons in solids. However, the electrons in this model should be tightly bound to the atom to which they belong and they should have limited interaction with states and potentials on surrounding atoms of the solid. Therefore, the wave function of the electron will be rather similar to the atomic orbital of the free atom to which it belongs. This is the reason why the tight-binding method is a class of a one-electron model, allowing the calculation of electronic states using a set of wave functions for isolated atoms located at each atomic site. The name "tight-binding" refers to the fact that the electrons in this model should be regarded as tightly bound to the atom to which they belong. Therefore, the wave function of the electron will be rather similar to the atomic orbital of the free atom where it belongs to. The advantage of the tight-binding method is that it can handle a much larger system than the ab initio method.
3. The MD method treats the atoms as classical entities (position and momentum). It is based on Newton's second law or the equation of motion, depending on the force exerted on the atom, its mass, and its acceleration. From knowledge of the force on each atom, it is possible to determine the acceleration of each atom in the system. Integration of the equations of motion then yields a trajectory that describes the positions, velocities, and accelerations of the particles as they vary with time. From this trajectory, the average values of properties can be determined. The method is deterministic; once the positions and velocities of each atom are known, the state of the system can be predicted at any time in the future or the past. MD simulations can be time consuming and computationally expensive if a high number of atoms is involved, but they are much faster than the ab initio and tight-binding methods. The underlying assumption of classical MD is that the electrons are treated as a single and classical entity. If this is not a reasonable approximation, it is mandatory to resort to ab initio MD and consider electron motions. This work is concerned only with classical MD. The use of ab initio methods is addressed elsewhere [18].

11.2.1 Molecular Dynamics Fundamentals

In the simplest terms, MD is a method for generating the trajectories of a system of N atoms by direct numerical integration of Newton's equations of motion, with appropriate specification of an interatomic potential and suitable initial and boundary conditions. At any instance of time t, the atom coordinates and momenta are labeled as $3N$-dimensional vectors:

$$\begin{aligned} \boldsymbol{r}(t) &= \{r_1(t);\ r_2(t);\ldots;\ r_N(t)\} \\ \boldsymbol{p}(t) &= \{p_1(t);\ p_2(t);\ldots;\ p_N(t)\} \end{aligned} \tag{11.12}$$

where r_i and p_i represent the coordinates and linear momenta of atom i. The simulation proceeds with the system in a prescribed initial configuration $\boldsymbol{r}(t_0)$ and velocity $\boldsymbol{v}(t_0)$ at time $t = t_0$. As the simulation proceeds, the atoms evolve through a sequence of time steps $t_0, t_1, \ldots, t_L$, where $t_k = t_0 + k\Delta t$ and Δt is the time step of the MD simulation. The simulation runs for L number of time steps and covers a time interval $L\Delta t$. Typical values of L can range from 10^4 to 10^8 and $\Delta t \sim 10^{-15}$ s (10^{-15} of a second is called a femtosecond. That is one quadrillionth, or one millionth of one billionth, of a second. For context, a femtosecond is to a second what a second is to about 31.7 million years. The word femtosecond is formed by the SI prefix femto and the SI unit second. Its symbol is fs.).

The simulation system has a total energy E (or Hamiltonian $\mathcal{H}$), which is the sum of the kinetic K and potential U energies of the atoms. Here, for simplicity, it is assumed that all atoms have the same mass m.

$$\mathcal{H}(\boldsymbol{r},\boldsymbol{p}) = E = K + U = \sum_{j=1}^{N} \frac{\boldsymbol{p}_j \cdot \boldsymbol{p}_j}{2m} + U(\boldsymbol{r}_1, \ldots, \boldsymbol{r}_N) \quad (11.13)$$

and $U = U(\boldsymbol{r})$ is a prescribed interatomic interaction potential (often designated “force field” in chemistry), which is a function of all the particle coordinates in the system if each particle is allowed to interact with all the others without restriction. Thus, the dependence of U on the particle coordinates can be as complicated as the system under study demands. However, for the present explanation, an approximation is introduced—the assumption of a two-body or pair-wise additive interaction, which is sufficient to illustrate the essence of the MD simulation.

The dynamics of the system is described by Hamilton’s equations of motion

$$v = \frac{\partial r}{\partial t} = \nabla_{\mathrm{p}} \mathcal{H}(\boldsymbol{r},\boldsymbol{p}) \quad F = \frac{\partial p}{\partial t} = -\nabla_{\mathrm{r}} \mathcal{H}(\boldsymbol{r},\boldsymbol{p}) \quad (11.14)$$

It is furthermore assumed that the interaction of the atoms is given by a conservative potential $U(\boldsymbol{r})$ with $\partial U / \partial t = 0$. Moreover, Hamilton’s equations of motion [Eq. (11.13)] are then equivalent to Newton’s equations of motion together with the impulse relation $p_j(t) = m v_j(t)$. Newton’s equations of motion are given by Newton’s second law

$$m\, \boldsymbol{a}_j = \boldsymbol{F}_j \quad j = 1, \ldots, N \quad (11.15)$$

where the force acting on atom j is defined as

$$\boldsymbol{F}_j = -\nabla_{\mathrm{r}} U(\boldsymbol{r}) \quad j = 1, \ldots, N \quad (11.16)$$

A Hamiltonian system that consists of the equations of motion [Eq. (11.14)] and the Hamiltonian [Eq. (11.13)] gives the general basis of MD. If Eq. (11.13), together with Eq. (11.14), is used to calculate the total time derivative of the Hamiltonian

$$\frac{d}{dt}\mathcal{H}(r,p) = \nabla_{\mathrm{r}}\mathcal{H}\frac{dr}{dt} + \nabla_{\mathrm{p}}\mathcal{H}\frac{dp}{dt} = \nabla_{\mathrm{r}}\mathcal{H} \cdot \nabla_{\mathrm{p}}\mathcal{H} + \nabla_{\mathrm{p}}\mathcal{H}(-\nabla_{\mathrm{r}}\mathcal{H}) = 0 \quad (11.17)$$

it is seen that the total energy E is conserved over time. In this case, the so-called microcanonical (NVE) ensemble is obtained. Here, the number of particles (N), the volume (V), and the total energy (E) are constant over time. Temperature and pressure are not constant over time. They, however, are thermodynamic quantities anyway, which are meaningful only at equilibrium [19]. Due to the so-called ergodic hypothesis the *ergodic hypothesis* is that the ensemble averages used to compute expectation values can be replaced by time averages over the simulation [20], if a microscopic dynamic variable α takes on instantaneous values $\alpha(t)$ along a trajectory, then the time average results in the related thermodynamic value

$$\langle\alpha\rangle_t = \lim_{t\to\infty}\frac{1}{t}\int_0^t \alpha(\tau)d\tau \tag{11.18}$$

This dynamic variable can be any function of the coordinates and momenta of the particles of the system. The thermodynamic temperature T is related to the average of the kinetic energy K by the equipartition theorem

$$\left\langle \frac{1}{2}\sum_{j=1}^{N}\frac{\boldsymbol{p}_j\cdot\boldsymbol{p}_j}{m}\right\rangle_t = \langle K\rangle_t = \frac{N_f k_B T}{2} \tag{11.19}$$

where N_f denotes the number of degrees of freedom of the physical system, and k_B is Boltzmann's constant. In an unrestricted system with N atoms, N_f is $3N$ because each atom has three velocity components. In particular, the thermodynamic temperature T is given by the average over time of the instantaneous temperature

$$T_{\text{inst}} = \frac{2K}{N_f k_B} \tag{11.20}$$

Furthermore, the thermodynamic pressure P is related to the thermodynamic temperature T, the volume V, and the internal virial $W = \langle W_{\text{inst}}\rangle_t$, by the equation of state

$$PV = \frac{1}{3}N_f k_B T + \langle W_{\text{inst}}\rangle_t \tag{11.21}$$

which is derived from the classical virial theorem [19]. Here, the instantaneous internal virial W_{inst} is related to the fluctuations of the potential energy U with respect to the volume V by $W_{\text{inst}}/V = -(dU/dV)$. If the instantaneous pressure is defined in the form

$$P_{\text{inst}} = \frac{2}{3V}K + \frac{1}{Z}W_{\text{inst}} \tag{11.22}$$

then the thermodynamic pressure is also given by the average $P = \langle P_{\text{inst}}\rangle_t$. This way, MD simulations for an NVE ensemble can be performed to obtain equilibrium thermodynamic properties like temperature and pressure, as the volume, the number of atoms, and the energy are controlled. However, in "reality," the user controls the temperature instead of the energy. The constant energy dynamics (NVE) does rarely represent the experimental conditions for the system simulated. Mechanical properties like elastic constants can be calculated using microscopic expressions in the form of fluctuations of total energy E keeping the temperature T uniform. Because of the nature of the simulation, it is not possible to switch from the microcanonical ensemble to the canonical ensemble (or NVT ensemble) in which the temperature T is controlled instead of the energy E. Several methods have been introduced to keep the temperature constant while using the microcanonical ensemble. Popular techniques to control temperature include the Nosé–Hoover thermostat and the Berendsen thermostat (scaling of velocities to obtain an exponential relaxation of the temperature to T). In the Nosé–Hoover thermostat, a heatbath is introduced in the Hamiltonian with an extra degree of freedom s. The NVT Hamiltonian now reads

$$\mathcal{H}_{\mathrm{NVT}}(\boldsymbol{r},\boldsymbol{p},p_s,s)=\sum_{j=1}^{N}\frac{\boldsymbol{p}_j\cdot\boldsymbol{p}_j}{2ms^2}+U(\boldsymbol{r}_1,\ldots,\boldsymbol{r}_N)+\frac{p_s^2}{2Q}+g\ k_{\mathrm{B}}\ T\ ln(s) \tag{11.23}$$

where the first/second terms are the physical terms and the third/forth terms are the thermostat terms (often called friction terms), which are used to control the temperature T of the system. Like the Hamiltonian $\mathcal{H}$ of the NVE ensemble [Eq. (11.13)], the function $\mathcal{H}_{\mathrm{NVT}}$ remains constant over time but now turns non-Hamiltonian. But the nonfictitious physical energy (total energy E), given by the first and second terms of Eq. (11.23), is not conserved over time. This way, it becomes possible to determine the variation of the total energy of the system, keeping the temperature constant.

11.2.2 Selection of Potentials for MD

There are many types of potentials (force fields) to describe the forces between atoms. In particular, the reactive empirical bond order (REBO) potential model of Brenner [21] has been widely used to study the formation of carbon fullerenes and their properties, the energetics of CNTs, the growth of CNTs, carbon nanotube fracture, and the mechanics of CNTs and filled CNTs. In particular, Brenner's potential has been successfully applied to examine CNT composites, such as CNT-reinforced polymer and functionalized CNT-reinforced polymer. The REBO potential model is given as a sum over bonds

$$U=\sum_{i=1}^{N}\sum_{\substack{j=1\\j>i}}^{N}\left[u_B(r_{ij})+u_W(r_{ij})\right] \tag{11.24}$$

where r_{ij} is the bond length between atoms i and j. Here, the term u_B consists of a pair potential term u_R to model interatomic core–core repulsive interactions and a pair potential term u_A to describe the attractive interactions between valence electrons and cores, given by

$$u_B(r_{ij})=f_{ij}(r_{ij})\left[u_R(r_{ij})-\bar{B}_{ij}u_A(r_{ij})\right] \tag{11.25}$$

To be precise, the repulsive and the attractive pair terms u_R and u_A are defined as

$$u_R(r_{ij})=\frac{D_{ij}^{(e)}}{S_{ij}-1}e^{-\sqrt{2S_{ij}}\beta_{ij}(r_{ij}-R_{ij}^{(e)})} \tag{11.26}$$

$$u_A(r_{ij})=\frac{D_{ij}^{(e)}}{S_{ij}-1}S_{ij}e^{-\sqrt{(2/S_{ij})}\beta_{ij}(r_{ij}-R_{ij}^{(e)})} \tag{11.27}$$

with the values of parameters $D_{ij}^{(e)}$, S_{ij}, $R_{ij}^{(e)}$, and β_{ij} being specified by the type of the atoms i and j. The factor $\bar{B}_{ij}$ is a so-called empirical bond-order term, which modulates valence electron densities and depends on the bond lengths and the angles. Due to the fact that the factor $\bar{B}_{ij}$ varies subjected to the local configuration of atoms, Brenner's potential is a reactive many-body potential model. The term u_B is restricted to nearest neighbor interactions or bonded atoms by a smooth truncation function f_{ij} given by

$$f_{ij}(r_{ij}) = \begin{cases} 1 & \text{if } r_{ij} \le R_{ij}^{(1)} \\ \frac{1}{2}\left[1+\cos\left(\pi \frac{r_{ij}-R_{ij}^{(1)}}{R_{ij}^{(2)}-R_{ij}^{(1)}}\right)\right] & \text{if } R_{ij}^{(1)} < r_{ij} < R_{ij}^{(2)} \\ 0 & \text{if } r_{ij} \ge R_{ij}^{(2)} \end{cases} \tag{11.28}$$

where the cutoff distances $R_{ij}^{(1)}$ and $R_{ij}^{(2)}$ are only dependent on the type of atoms i and j. This function is also called the cut-off function and aims to restrict the range of interactions among only the nearest-neighbor carbon atoms. The parameter $\bar{B}_{ij}$ represents the multibody coupling, which depends on neighbor atoms through the bond angle, and it is given by

$$\begin{aligned} \bar{B}_{ij} &= \frac{1}{2}(B_{ij}+B_{ji}) \\ B_{ij} &= \left[1+\sum\nolimits_{k=1(\neq i,j)}^{N} G_i(\theta_{ijk}) f_{ik}(r_{ik}) e^{\alpha_{ijk}(r_{ij}-R_{ij}^{(e)}-r_{ik}+R_{ik}^{(e)})}\right]^{-\delta_i} \end{aligned} \tag{11.29}$$

The parameter δ_i depends on the type of atom i and the parameter α_{ijk} depends on the types of atoms i, j, and k, respectively. Furthermore, θ_{ijk} defines the bond angle between carbon bonds i–j and i–k, and G_i is the angular function given by

$$G(\theta) = a\left[1+\frac{c^2}{d^2}-\frac{c^2}{d^2+(1+\cos\theta)^2}\right] \tag{11.30}$$

where a, c, and d are the potential parameters used in the Brenner potential [21]. Note that in the case of $\bar{B}_{ij}=1$, $f_{ij}=1$, and $S_{ij}=2$, the first term in Eq. (11.23) reduces to the usual diatomic Morse potential model.

The second term u_W in Eq. (11.24) denotes the contribution from van der Waals energy modeled by the Lennard–Jones potential. It simulates the interatomic forces arising from noncovalent bonds and it is given by

$$u_W(r_{ij}) = 4\varepsilon_{ij}\left[\left(\frac{\sigma_{ij}}{r_{ij}}\right)^{12}-\left(\frac{\sigma_{ij}}{r_{ij}}\right)^{6}\right] \tag{11.31}$$

where ε_{ij} is the minimum energy, σ_{ij} is the finite distance at which the interatomic potential is zero, and r_{ij} is the distance between the atoms (Fig. 11.3A). At r_m (distance at which the potential reaches its minimum), the potential function has the value $-\varepsilon_{ij}$. These parameters can be fitted to reproduce experimental data or accurate quantum chemistry calculations.

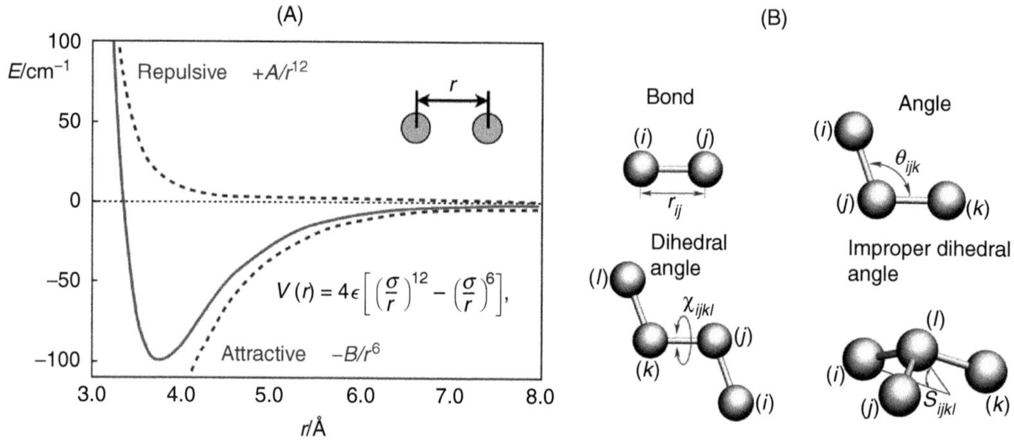

FIGURE 11.3 (A) Lennard–Jones energy curve and (B) components of pair-wise potentials.

REBO is a many-body bond order potential. The bond order term is a function of local atomic hybridization and conjugation. It allows for covalent bond breaking and forming associated with the change of atomic hybridization. Therefore, it can model chemical reactions. Other more simple potentials exist (e.g., Amber and Charmm), but they all consider harmonic approximations to pair-wise bond potential.

The harmonic approximation of the intramolecular potential is acceptable when the bond deformation is not particularly high, but other functions must be used to take into account more extreme conditions:

$$U=\sum_{\text{bonds}} k_b(r-r_0)^2+\sum_{\text{angles}} k_a(\theta-\theta_0)^2+\sum_{\text{dihedrals}} k_d[1-\cos(n\omega-\gamma)] + \sum_{\text{improper dihedrals}} k_i(S-S_0)^2+\sum_{i=1}^{N}\sum_{\substack{j=1\\ j>i}}^{N} u_W(r_{ij})+\frac{q_i q_j}{4\pi\varepsilon_0 r_{ij}} \tag{11.32}$$

The meanings of the right-hand side terms of Eq. (11.32) are as follows:

- The first term (summing over bonds) represents the energy between covalently bonded atoms.
- The first term in Eq. (11.32) describes the component of the potential energy of the system arising due to stretching of the bonds between pairs of atoms in the system. Here, k_b is the stiffness parameter and r_0 is the equilibrium distance. r is the actual distance between two atoms with indices i and j (Fig. 11.3B). The summation in this term is performed over all topologically defined bonds, the total number of which is equal to N. This harmonic (ideal spring) force is a good approximation near the equilibrium bond length, but becomes increasingly poor as atoms separate.
- The second term (summing over angles) represents the energy due to the geometry of electron orbitals involved in covalent bonding. It describes the component of the potential energy of the system arising due to the variation of angles between every

topologically defined triple of atoms in the system (Fig. 11.3B), denoted as θ. Here, k_a is the stiffness parameter and θ_0 is the equilibrium value of an angle in a triple of atoms. The summation in the second term of Eq. (11.32) is performed over different topologically distinct triples of atoms in the system.

- The third term (summing over torsions) represents the energy for twisting a bond due to bond order (double bonds) and neighboring bonds. Note that a single bond may have more than one of these terms, such that the total torsional energy is expressed as a Fourier series. It is known as the torsion energy. It is characterized through a dihedral angle formed by every four atoms (denoted as i, j, k, and l) connected via chemical bonds (Fig. 11.3B). Since this term represents the energy of rotation around a covalent bond, which is the source of most conformational flexibility in biomolecules, they must provide a smooth energy for 360 degree. This is done in most cases with a single sinusoid, as in Eq. (11.32). Here, ω is the angle between the plane containing the first three atoms (ijk) and the plane containing the last three atoms (jkl) (Fig. 11.3B). The "multiplicity" n is typically 1, 2, or 3. k_d is the dihedral spring constant.
- The fourth term in Eq. (11.32) describes the so-called improper dihedral angles, which are used in the molecular topology to maintain planarity. As such, the harmonic form with a large spring constant k_{id} and equilibrium value S_0 (typically equal to zero) is used to restrain the configuration of an atom and three atoms bonded to it (Fig. 11.3B). Similar to the case of dihedral angles, S is the angle between the plane containing the first three atoms with indices (ijk) and the plane containing the last three atoms with the indices (jkl), which define this dihedral angle.
- The fifth term (double summation over i and j) represents the nonbonded energy between all atom pairs, which can be decomposed into van der Waals (first term of summation) and electrostatic (second term of summation) energies.

Based on these terms, which possess physical and mechanical meaning, it is possible to extract the bond stiffness components (axial, bending, and torsion), to model the covalent bonds by means of bar elements and noncovalent bonds by means of springs, and to sim ulate molecular systems using standard macroscale methods. These are called molecular mechanics (MM) methods. The MM simulations of molecular systems are purely static and can only simulate the mechanical behavior. Other effects, such as temperature, pressure, kinetic energy, velocities, and accelerations are not dealt with in MM simulations. Thus, their scope is much more limited than MD simulations.

11.2.3 Time Integration in Molecular Dynamics

Now that the equations of motion and potential are defined, it is necessary to find the numerical solution of the system of equations of motion. A time integration scheme must be employed. For simplicity, the following Hamilton's equations of motion are considered:

$$v = \frac{\partial r}{\partial t} = \nabla_p \mathcal{H}(\boldsymbol{r}, \boldsymbol{p}) = m^{-1}\boldsymbol{p} \qquad F\frac{\partial p}{\partial t} = \nabla_r \mathcal{H}(\boldsymbol{r}, \boldsymbol{p}) = -\nabla_r \mathrm{U}(\boldsymbol{r}) \tag{11.33}$$

with generalized coordinates $\boldsymbol{r}$ and momenta $\boldsymbol{p}$. Despite this equation looking deceptively simple, actually, it is complicated due to the fact that the N-body problem generally cannot be solved exactly for $N > 2$. As a system of coupled second-order, nonlinear ordinary differential equations, Eq. (11.33) is solved numerically in MD simulations. For the numerical solution of this system of first-order ordinary differential equations, time must be discretized. Thus, the system Eq. (11.33) is considered only at time points, $t_k = t_0 + k\Delta t$ and Δt is the time step of the MD simulation. By a Taylor expansion up to third order of the coordinates $\boldsymbol{r}\,(t+\Delta t)$ and $\boldsymbol{r}\,(t-\Delta t)$ around time t, it is obtained as

$$\begin{aligned} \boldsymbol{r}(t+\Delta t) &= \boldsymbol{r}(t)+\boldsymbol{v}(t)\Delta t+\frac{1}{2}\boldsymbol{a}(t)\Delta t^2+\frac{1}{6}\dddot{\boldsymbol{r}}(t)\Delta t^3+O(\Delta t^4)= \\ &= \boldsymbol{r}(t)+m^{-1}\boldsymbol{p}(t)\Delta t+\frac{1}{2}m^{-1}\boldsymbol{F}(t)\Delta t^2+\frac{1}{6}\dddot{\boldsymbol{r}}(t)\Delta t^3+O(\Delta t^4) \end{aligned} \tag{11.34}$$

$$\begin{aligned} \boldsymbol{r}(t-\Delta t) &= \boldsymbol{r}(t)-\boldsymbol{v}(t)\Delta t+\frac{1}{2}\boldsymbol{a}(t)\Delta t^2-\frac{1}{6}\dddot{\boldsymbol{r}}(t)\Delta t^3+O(\Delta t^4)= \\ &= \boldsymbol{r}(t)-m^{-1}\boldsymbol{p}(t)\Delta t+\frac{1}{2}m^{-1}\boldsymbol{F}(t)\Delta t^2-\frac{1}{6}\dddot{\boldsymbol{r}}(t)\Delta t^3+O(\Delta t^4) \end{aligned} \tag{11.35}$$

The sum of Eqs. (11.35) and (11.34) gives

$$\boldsymbol{r}(t+\Delta t)+\boldsymbol{r}(t-\Delta t)=2\boldsymbol{r}(t)+m^{-1}\boldsymbol{F}(t)\Delta t^2+O(\Delta t^4) \tag{11.36}$$

The difference between Eqs. (11.35) and (11.34) gives

$$\boldsymbol{r}(t+\Delta t)-\boldsymbol{r}(t-\Delta t)=2m^{-1}\boldsymbol{p}(t)\Delta t+O(\Delta t^3) \tag{11.37}$$

Eq. (11.36) leads to an approximation for the coordinates

$$\boldsymbol{r}(t+\Delta t)\approx 2\boldsymbol{r}(t)-\boldsymbol{r}(t-\Delta t)+m^{-1}\boldsymbol{F}(t)\Delta t^2 \tag{11.38}$$

with an error of $O(\Delta t^4)$. With the help of Eq. (11.34), Eq. (11.37) leads to the approximation for the momenta

$$\boldsymbol{p}(t+\Delta t)\approx \boldsymbol{p}(t)+\frac{\Delta t}{2}[\boldsymbol{F}(t)+\boldsymbol{F}(t+\Delta t)] \tag{11.39}$$

with an error of $O(\Delta t^2)$. Eqs. (11.38) and (11.39) correspond to the so-called velocity–Verlet time–integration scheme, given by

$$\begin{aligned} \boldsymbol{r}^0 &:= \boldsymbol{r}(t_0) \\ \boldsymbol{p}^0 &:= \boldsymbol{p}(t_0) \\ \boldsymbol{r}^{n+1} &:= 2\boldsymbol{r}^n-\boldsymbol{r}^{n-1}+m^{-1}\Delta t^2\boldsymbol{F}^n \\ \boldsymbol{p}^{n+1} &:= \boldsymbol{p}^n+\frac{\Delta t}{2}(\boldsymbol{F}^n+\boldsymbol{F}^{n+1}) \end{aligned} \tag{11.40}$$

Because the momenta $\boldsymbol{p}$ are not used in the third expression in Eq. (11.40), there is comparatively low accuracy of the order $O(\Delta t^2)$ has no impact on the accuracy of the coordinates **r**. It however affects the calculation of the temperature T and the kinetic energy K. To improve the accuracy for these quantities, several other time-integration schemes to compute the equations of motions related to Hamiltonian [22] or non-Hamiltonian [23] dynamics can be employed [24,25]. The accuracy can be improved if either a smaller time step Δt is used, or if a scheme of higher order is employed.

11.3 Behavior of Locally Buckled Carbon Nanotubes

11.3.1 Suitability of Continuum Shell Models

After the discovery of CNTs, much research has been done to characterize their behavior. Due to their hollow configuration, it is well known that CNTs under compression, bending, and torsion are sensitive to local buckling [26,27]. MD simulations have shown that the buckling behavior of a given CNT depends on its length and radius [28]. It is also known that MD analyses are computationally expensive and time-consuming and are often limited to a maximum number of atoms. To avoid this problem, a large amount of investigations has been carried out using continuum shell analyses. One issue that remains less clear concerns the adoption of CNT thickness to use in shell models. It is known that graphene (CNT wall) possesses a membrane stiffness $C = Yh = 360\ \mathrm{J/m^2}$, where Y is the Young's modulus. Thus, it is possible to adopt either (1) $Y = 1.06$ TPa and $h = 3.40$ Å (thick-walled CNT) or (2) $Y = 5.5$ TPa and $h = 0.66$ Å (thin-walled CNT) or any other Y and h values that satisfy $Yh = 3.60$ TPa Å. The first option has a clear physical meaning, since $h = 3.40$ Å is the equilibrium interlayer spacing of adjacent CNTs. The second option ($h = 0.66$ Å) has less physical meaning, but it is widely accepted that it gives better results in CNT analysis using shell models. In this section, the suitability of simple shell models to assess the local buckling behavior of CNTs are presented, trying to clarify if they should be modeled as thin ($h = 0.66$ Å) or thick ($h = 3.40$ Å) shells. Thus, an MD study on the linear and buckling behavior of CNTs under either shortening or twisting is presented [29].

Molecular dynamics simulations have been carried out using the DL-POLY_2 simulation package [30] and the interatomic forces were derived from the Tersoff–Brenner potential (similar to REBO potential). Displacement control is adopted in the MD simulations since kinematical quantities are imposed at the atoms located at the edges of the CNT. The origin of the cylindrical axis system (x, θ, z) is located at the center of the CNT midsection, where x is the coordinate along the longitudinal axis, θ is the angular coordinate, and z is the radial coordinate (Fig. 11.4).

For a given CNT with diameter D and length L, both end sections are located at $x = \pm L/2$. The displacements in the system (x, θ, z) are denoted by u, ϕ, and w, respectively. For shortening deformations, incremental axial displacements were imposed in opposite directions, +0.025 Å for atoms located at $x = -L/2$ and −0.025 Å for atoms located at $x = +L/2$. For twisting deformations, incremental rotations were imposed in opposite directions, +0.5 degree = $+\pi/360$ rad for the atoms located at $x = -L/2$ and −0.5 degree = $-\pi/360$ rad for

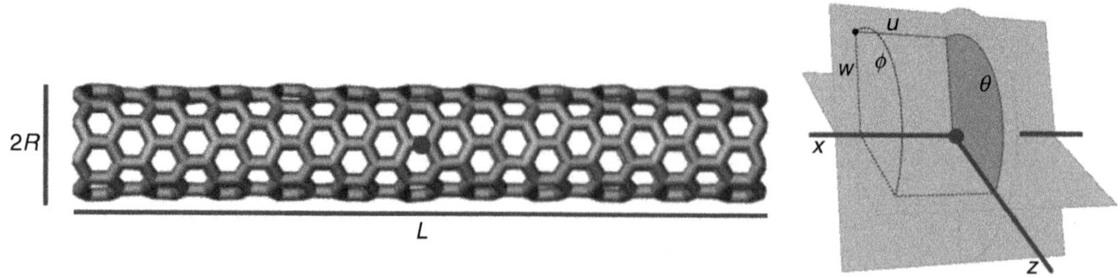

FIGURE 11.4 Coordinate system and displacement nomenclature.

the atoms located at $x = +L/2$. This type of kinematical imposition ensures full structural symmetry of the CNT atomic system and, for each increment, corresponds to $\Delta u = 0.05$ Å and $\Delta\phi = 1.0$ degree $= \pi/180$ rad between the end sections of the CNT.

Previous studies have shown that the buckling behavior of CNTs under pure shortening or pure twisting is greatly dependent on L and D [28,31]. To avoid such dependence as much as possible, CNTs with similar length-to-diameter aspect ratios, L/D =7.0, were selected. To study the buckling behavior of CNTs with different chirality under pure shortening and pure twisting, the following zigzag, armchair, and chiral CNT structures (Fig. 11.5) have been considered. The chiral vector (n, m) represents the way a graphene sheet can be wrapped to yield a given nanotube, with the pair of indices n and m denoting the number of unit vectors along two directions in the honeycomb crystal lattice of graphene. The

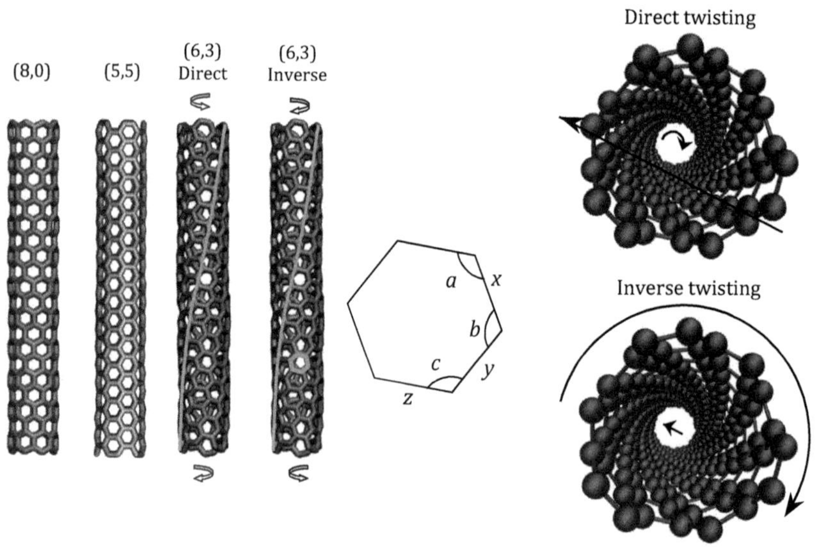

FIGURE 11.5 CNTs (8,0), (5,5), (6,3)-direct and (6,3)-inverse (direct and inverse twisting). Labeling of bonds and angles in a hexagonal cell of (6,3) [the *black line* (yellow in the web version) denotes the orientation of the zigzag line that spirals around the CNT in the initial nondeformed state].

wrapping of graphene sheets yields zig-zag CNTs if $m = 0$, arm-chair CNTs if $n = m$, and chiral CNTs if $m \neq 0$ and $m \neq n$:

- Zigzag (8,0) with $L = 46.3$ Å and $D = 6.2$ Å, having an aspect ratio $L/D = 7.4$ and 352 atoms.
- Armchair (5,5) with $L = 47.3$ Å and $D = 6.8$ Å, having an aspect ratio $L/D = 7.0$ and 380 atoms.
- Chiral (6,3) with $L = 41.0$ Å and $D = 6.2$ Å, having an aspect ratio $L/D = 6.6$ and 336 atoms.

Since the chiral CNT (6,3) displays an asymmetric lattice atomic structure, it has been analyzed under direct and inverse twisting rotation conditions, as seen in Fig. 11.5. Direct twisting corresponds to rotating both ends of the CNT in opposite directions following the rolling of the hexagonal lattice, while inverse twisting corresponds to rotating both CNT ends in opposite directions but opposing the rolling of the hexagonal lattice.

All simulations were performed at a temperature of 298.15 K using a Berendsen thermostat. The Newtonian equations of motion were integrated using the Verlet leapfrog predictor-corrector algorithm. A time step of 1 fs was used and, after each increment of Δu and $\Delta \phi$, the CNT was fully relaxed for a period of 30,000 time steps. The total energy, U, of the CNT calculated in each increment corresponds to the average value of the energy within the last 10,000 time steps. Ninety increments were performed in the simulations of the CNT under pure shortening and 180 increments were used for the analysis of CNT under pure twisting.

The three CNTs have been subjected to axial shortening and the results obtained from MD simulations are shown in Fig. 11.6. Fig. 11.6A shows the variation of the strain energy per atom, U, as a function of the axial strain $\varepsilon = u/L$. Fig. 11.6B displays the variation of the force per atom, F, with the axial shortening u (note that 1.0 eV/Å = 1.602 nN). From Fig. 11.6A, it is clear that the energy U grows quadratically with the strain ε and, at buckling, it suddenly drops. These drops are related to the buckling of CNTs (8,0), (5,5), and (6,3) for the critical strains $\varepsilon_c = 0.081$, $\varepsilon_c = 0.080$, and $\varepsilon_c = 0.094$, respectively. The zigzag

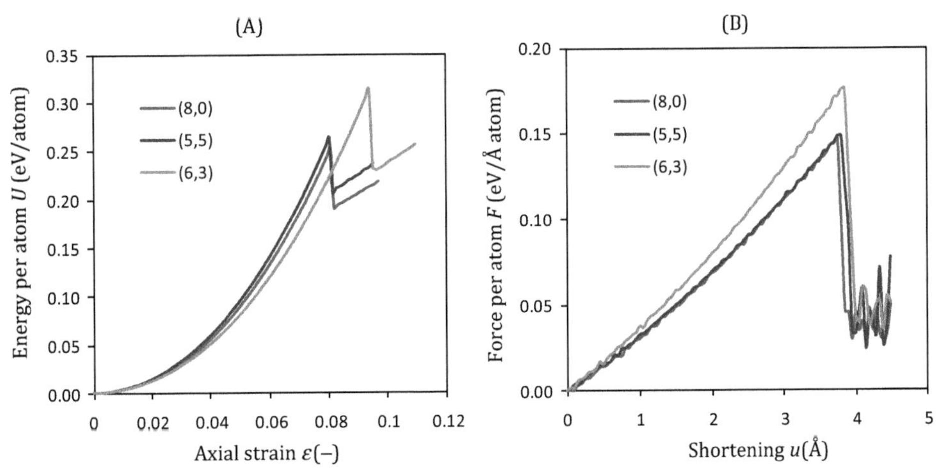

FIGURE 11.6 (A) Variation of energy U with axial strain ε and (B) variation of force F with shortening u.

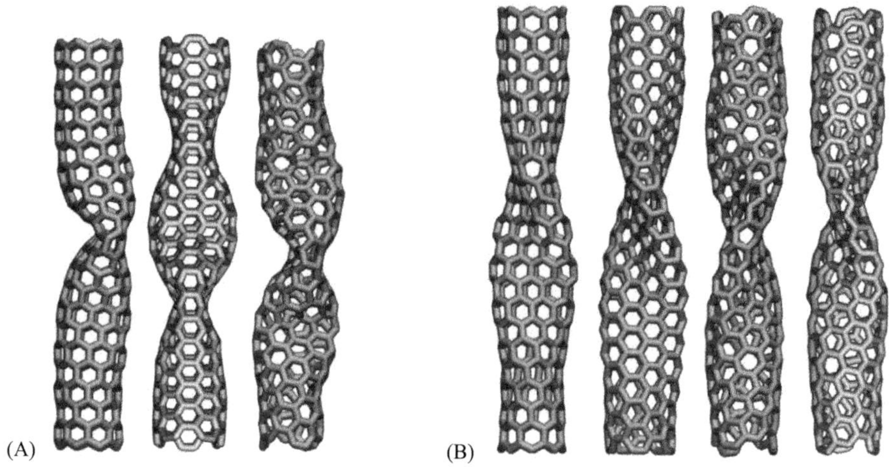

FIGURE 11.7 Critical modes shapes of CNTs (8,0), (5,5), (6,3) under (A) shortening and (B) twisting [third shape is for (6,3)-direct and fourth shape is for (6,3)-inverse].

and armchair CNTs have similar aspect ratios (L/D around 7) and exhibit similar critical strains. The very small difference between the ε_c values means that the critical strain does not appear to be closely dependent on these CNT structures. However, the chiral CNT has also a similar aspect ratio (L/D around 7) but it possesses a much higher critical strain (17.5% larger). It means that the critical strain of chiral CNTs depends on the atomic structure (CNT chirality). The critical buckling mode shapes of CNTs (8,0), (5,5), and (6,3) under pure shortening are shown in Fig. 11.7. It is seen that all CNTs buckle locally, similarly to a cylindrical thin shell. Nevertheless, buckling occurs for a different number of half-waves.

Based on the Sanders shell model, Silvestre et al. [32] proposed the following formula to calculate the critical strain ε_c for local and flexural buckling of CNTs:

$$\varepsilon_c = \begin{cases} \dfrac{h}{1.7R} \times Max\left(0.6;1 - \dfrac{L}{10R}\sqrt{\dfrac{h}{R}}\right) & \text{if } \dfrac{L}{D} \le 3.8\sqrt{\dfrac{R}{h}} \\ 2\left(\dfrac{\pi R}{L}\right)^2 & \text{if } \dfrac{L}{D} \ge 3.8\sqrt{\dfrac{R}{h}} \end{cases} \tag{11.41}$$

where h and R are the CNT thickness and radius, respectively. Comparing the aspect ratio L/D with its cross-section property $[3.8 \times (R/h)^{0.5}]$, a given CNT may be classified as short or long. The first expression corresponds to local buckling for short CNTs while the second one corresponds to flexural buckling of long CNTs. There has been much discussion concerning the graphene thickness value. Yakobson et al. [3] proposed h = 0.66 Å. However, studies [12,26] have shown that this value is slightly low for buckling analyses of CNTs. Huang et al. [12] proposed h = 0.734 Å while Wang et al. [26] suggested h = 0.76 Å. Following these investigations, h = 0.75 Å was adopted. Eq. (11.41) stipulates that these CNTs are short (i.e., they fail by local buckling), despite L/D = 7.0 being close to the transitional value

[$3.8 \times (R/h)^{0.5} = 8.0$]. From Eq. (11.41), $\varepsilon_c = 0.085$, $\varepsilon_c = 0.078$, and $\varepsilon_c = 0.085$ are obtained for the CNTs (8,0), (5,5), and (6,3), respectively. These values show good agreement with the MD results obtained for the zigzag and armchair CNTs. For the chiral CNT, the analytical value of ε_c is 10% lower than the MD result, meaning that continuum shell models are more accurate for zigzag and armchair CNTs than for chiral ones. This remark is in line with previous conclusions, showing the anisotropy of chiral CNTs. Fig. 11.6B shows that F increases linearly with u and the curves exhibit a huge drop when the CNTs buckle. It is interesting to note that zigzag (8,0) and armchair (5,5) CNTs possess the same precritical axial stiffness $\partial F/\partial u = 0.039$ eV/Å^2·atom, while the chiral (6,3) CNT has a higher precritical axial stiffness $\partial F/\partial u = 0.046$ eV/Å^2·atom. In this case, it is also possible to use continuum mechanics formulas to assess the CNT precritical axial stiffness, which is given by $\partial F/\partial u = Y(2\pi Rh)/L$. Using $Y = 1.06$ TPa and $h = 3.40$ Å (thick-walled CNT), one obtains $\partial F/\partial u = 0.043$ eV/Å^2·atom for zigzag (8,0) and armchair (5,5) CNTs, and $\partial F/\partial u = 0.051$ eV/Å^2·atom for the chiral (6,3) CNT. It is interesting to note that if $Y = 4.80$ TPa and $h = 0.75$ Å (thin-walled CNT) are used, exactly the same precritical axial stiffness values of the thick-walled CNT are obtained.

The three CNTs have also been subjected to pure twisting and the results obtained from MD simulations are shown in Fig. 11.8. Recall that the chiral CNT (6,3) has been analyzed twice, for direct and inverse twisting (Fig. 11.5B). Fig. 11.8A shows the variation of energy per atom, U, as a function of the angle of twist per unit of length, $\alpha = \phi/L$. Fig. 11.8B displays the variation of torque per atom, T, as a function of the twisting angle, ϕ. At each increment, the torque is calculated as the first derivative of the energy w.r.t. the twisting angle ϕ, that is, $T = \partial V/\partial \phi$ (note that 1.0 eV = 1.602 nN Å). Fig. 11.8A shows that all $U(\alpha)$ energy curves are quite similar. However, a more careful analysis shows that only the zigzag CNT (8,0) shows two small sudden drops in its $U(\alpha)$ curve at $\alpha_c = 0.056$ rad/Å and $\alpha_c = 0.065$ rad/Å, whereas the CNTs (5,5) and (6,3) show a continuous behavior in the

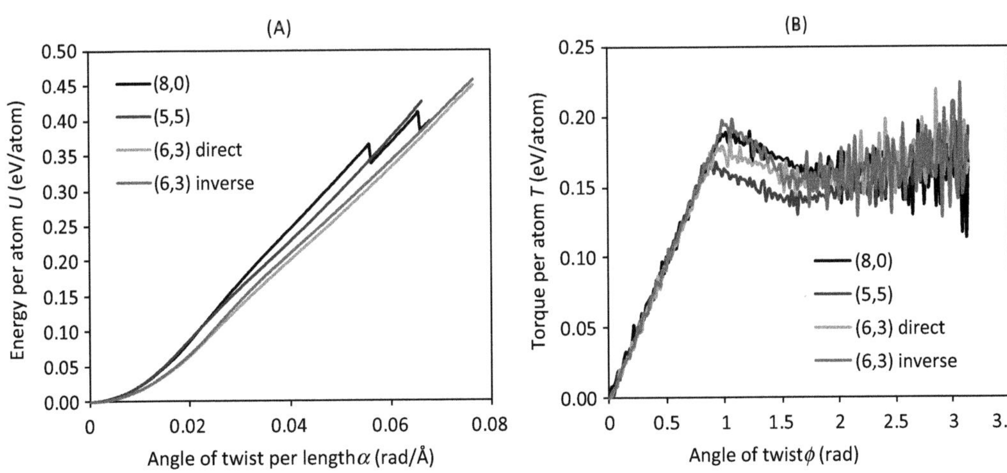

FIGURE 11.8 (A) Variation of energy U with α, and (B) variation of torque T with ϕ.

studied range of angles of twist. The behavior of the CNTs (5,5) and (6,3) $U(\alpha)$ curves contrasts with the previously discussed shortening $U(\varepsilon)$ curves, in which large sudden drops were always found. The main issue at this point is to answer the following two questions:

1. Why CNTs (5,5) and (6,3) do not exhibit sudden drops in their $U(\alpha)$ curves?
2. Why the $U(\alpha)$ curve of zigzag CNT (8,0) shows two small sudden drops?

To answer the first question, let us look at the trends of the $U(\alpha)$ curves. They change quadratically with α, for α smaller than 0.020–0.025 rad/Å. After that threshold value it seems that all $U(\alpha)$ curves start varying linearly with α. These facts are better explained if Fig. 11.8B is considered. The figure shows that for each CNT the torque T increases linearly with the twisting angle ϕ, until it reaches a certain limit and then starts decreasing [causing the "peaks" in the $T(\phi)$ curves]. These limit values correspond precisely to the transition between the "quadratic" and "linear" trends of the $U(\alpha)$ curves—($\alpha \sim 0.022$ rad/Å) = ($\phi \sim 1$ rad)/($L \sim 45$ Å)—which cannot be determined directly from the inspection of the $U(\alpha)$ curves in Fig. 11.8A. Indeed, all CNTs [including the zigzag CNT (8,0)] buckle when this limit is achieved, but the buckling phenomena take place softly, without a dynamic jump and abrupt loss of strain energy. Similar evidence was identified by Zhang and Shen [33] for armchair and zigzag CNTs under twisting. From the observation of Fig. 11.8B, it is concluded that all CNTs have similar precritical stiffness around $\partial T/\partial\phi = 0.19$–$0.20$ eV/atom. It is also possible to use continuum mechanics formulas to assess the CNT precritical torsional stiffness ($G = Y/2(1 + \nu)$, $\nu = 0.19$), which is given by $\partial T/\partial\phi = G(\pi/2)\{[R+(h/2)]^4 - [R-(h/2)]^4\}/L$. Using $Y = 1.06$ TPa and $h = 3.40$ Å (thick-walled CNT), one obtains $\partial T/\partial\phi = 0.226$ eV/atom for the zigzag CNT (8,0), $\partial T/\partial\phi = 0.260$ eV/atom for the armchair CNT (5,5), and $\partial T/\partial\phi = 0.268$ eV/atom for the chiral CNT (6,3). Considering the thick-walled CNTs, the estimated torsional stiffness of the zigzag CNT (8,0) is only 10% higher than the MD value, but the torsional stiffness predictions of the other CNTs are 30% higher than the MD values. On the other hand, if the thin-walled CNT properties ($Y = 4.80$ TPa and $h = 0.75$ Å) are used, one obtains $\partial T/\partial\phi = 0.176$ eV/atom for the zigzag (8,0) CNT, $\partial T/\partial\phi = 0.210$ eV/atom for the armchair CNT (5,5), and $\partial T/\partial\phi = 0.209$ eV/atom for the chiral CNT (6,3). These predictions, using thin-walled CNT properties, are much more accurate (5% maximum difference) than the thick-walled CNT counterparts.

From the observation of Fig. 11.8B, it can also be concluded that the CNTs buckle for slightly different critical angles of twist: $\phi_c = 1.01$ rad for CNT (8,0), $\phi_c = 0.84$ rad for CNT (5,5), $\phi_c = 0.94$ rad for direct twisting of CNT (6,3), and $\phi_c = 1.05$ rad for inverse twisting of CNT (6,3). From these results, it can be anticipated that armchair CNTs are more susceptible to buckling than the other configurations. Furthermore, it can be concluded that chiral CNTs are more sensitive to buckling when they are submitted to direct twisting (Fig. 11.5A) rather than inverse twisting (Fig. 11.5B). The critical buckling mode shapes of CNTs (8,0), (5,5), and (6,3) under pure twisting are shown in Fig. 11.7B. It is seen that all of them show a nonuniform helix deformed shape, consisting of an oval cross-section that spirals around the longitudinal axis between its ends. Moreover, the chirality of the CNT (6,3) must be correlated with distinct buckling behavior under inverse and direct twisting;

the direct twisting of CNT (6,3) implies an increased "curling" of the CNT helix structure, which means that one will reach ϕ_c sooner. In the case of inverse twisting, one is "straightening" the CNT helix structure and "delaying" the occurrence of buckling, which means that one will reach ϕ_c later.

Based on the Sanders shell model, Silvestre [34] proposed the following formula to calculate the critical angle of twist ϕ_c (in rad) for local and flexural buckling of CNTs:

$$\phi_c = \begin{cases} 1.7\left(\dfrac{h}{R}\right)^{1.25}\left(\dfrac{L}{R}\right)^{0.5} & \text{if } \dfrac{L}{D} \leq 4.36\sqrt{\dfrac{R}{h}} \quad (1) \\ \dfrac{2}{3}\left(\dfrac{h}{R}\right)^{1.5}\left(\dfrac{L}{R}\right)^{1.0} & \text{if } \dfrac{L}{D} \geq 4.36\sqrt{\dfrac{R}{h}} \quad (2) \end{cases} \tag{11.42}$$

Unlike in the shortening case, twisted CNTs always buckle locally, mostly with an oval buckle. There is also a limit aspect ratio [$4.36 \times (R/h)^{0.5}$] in twisting, but this corresponds to the transition between ranges where the boundary conditions influence or not the critical angle of twist. If the actual aspect ratio L/D is lower than that limit, the CNT is classified as short, the boundary conditions influence the ϕ_c values, and Eq. (11.42.1) applies. Conversely, if the actual aspect ratio L/D is higher than that limit, the CNT is classified as long, the boundary conditions (end caps) do not influence the ϕ_c values, and Eq. (11.42.2) applies. Using $h = 0.75$ Å (thin-walled CNT), a transitional aspect ratio of 9.0 is determined and the CNTs are classified as short (recall that $L/D = 7.0$). From the use of Eq. (11.42.1), the following approximated values of the critical angle of twist are obtained: $\phi_c = 1.07$ rad for CNT (8,0), $\phi_c = 0.95$ rad for CNT (5,5), and $\phi_c = 1.12$ rad for CNT (6,3). These approximate values agree fairly well with the MD results, showing errors around 10%. However, Eq. (11.42) does not depend on the direction of twisting and the approximate result for the chiral CNT (6,3) is closer to the MD value for inverse twisting. Using $h = 3.40$ Å (thick-walled), a transitional aspect ratio of 4.3 is determined and the CNTs are classified as long (recall that $L/D = 7.0$). From the use of Eq. (11.42.2), the following approximated values of the critical angle of twist are obtained: $\phi_c = 11.4$ rad for CNT (8,0), $\phi_c = 9.2$ rad for CNT (5,5), and $\phi_c = 10.1$ rad for CNT (6,3). These values are much higher (about 10 times!) than the MD results. Additionally, the CNTs buckle in local modes, as seen in Fig. 11.7B, rather than in flexural modes. It is concluded that CNTs can be modeled as thin-walled shells [35–37] but the degree of accuracy of the shell model depends on its complexity. Quoting Henry Louis Mencken, "*... for every complex problem there is an answer that is clear, simple, and wrong.*"

11.3.2 Influence of Combined Shortening–Twisting

Most of the existing MD simulations on the buckling behavior of CNTs have been performed for pure loading (either compression or bending or torsion). The buckling behavior of CNTs under combined loadings has only been performed in the context of continuum shell models [38–40] and frame models [41]. In practice, CNT-based spring elements in single-paddle torsional oscillator are often subjected to combined torsion and axial loading [42,43]. In this section, a brief account on the behavior of CNTs under combined shortening–twisting is

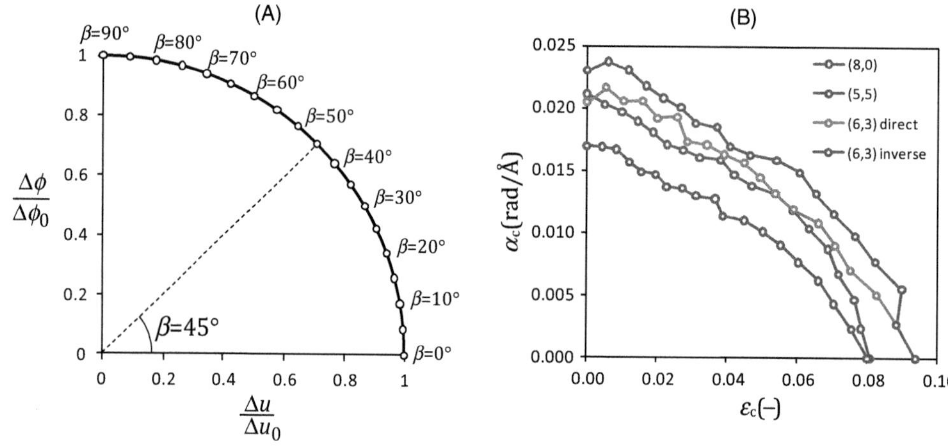

FIGURE 11.9 (A) Shortening–twisting combinations and (B) buckling interaction curves ε_c–α_c.

addressed [44]. To impose simultaneous shortening and twisting deformations to the CNT, the following relationship is adopted between the combined Δu and $\Delta\phi$:

$$\left(\frac{\Delta u}{\Delta u_0}\right)^2 + \left(\frac{\Delta\phi}{\Delta\phi_0}\right)^2 = 1 \qquad \Delta u = \Delta u_0 \cos\beta, \qquad \Delta\phi = \Delta\phi_0 \sin\beta \tag{11.43}$$

where an auxiliary parameter β (in arc degrees) is used. To have a full description of the plane (Δu, $\Delta\phi$), β was varied from 0 to 90 degree with increments of 5 degree. Thus, a total of 19 β values are considered, and depicted in Fig. 11.9A. For $\beta = 0$ degree, it is obtained $\Delta u = \Delta u_0$ and $\Delta\phi = 0$, that is, pure shortening behavior (without twisting). For $\beta = 90$ degree, it is obtained $\Delta\phi = \Delta\phi_0$ and $\Delta u = 0$, that is, pure twisting behavior (without shortening).

Let us turn our attention to the energy curves $U(\gamma)$ of CNTs under combined shortening–twisting conditions with the whole range of β parameter values as depicted in Fig. 11.5 for CNTs (8,0) and (5,5). The auxiliary parameter $\gamma = \sqrt{\varepsilon^2 + \alpha^2}$ is used to allow the simultaneous representation of all energy curves, including those corresponding to $\beta = 0$ degree (pure shortening) and $\beta = 90$ degree (pure twisting). Each $U(\gamma)$ curve corresponds to a given β value and the corresponding twisting–shortening rate, $\Delta\phi/\Delta u$, and the onset of buckling in each curve is denoted by a white circle. Independently of the CNT, it is clear that dynamic jumps happen more frequently and with higher amplitudes for lower β values. It means that decreasing the amount of twisting and increasing amount of shortening leads to an increasing tendency of CNTs to undergo dynamic jumps accompanied by sudden losses of strain energy. The evidence of a high drop of energy in CNTs under pure compression has been shown by several authors [27] and will not be detailed herein. Moreover, the existence of small sudden drops in the $U(g)$ curves is due to either dynamic jumps of CNTs to adjacent equilibrium configurations or, for very large twisting rotation (outside the scope of this investigation) to C—C bond breaks.

Fig. 11.9B shows the interaction diagram between the critical axial strain ε_c and the critical angle of twist per unit of length α_c. It exhibits nonlinear behavior. Despite the similar aspect ratio, L/D, of all studied CNTs, the interaction curves display different values for distinct CNT chirality. As the twisting–shortening rate $\Delta\phi/\Delta u$ decreases, the interaction curves ε_c–α_c of CNTs (8,0) and (5,5) become closer and coincide for $\beta = 0$ degree ($\Delta\phi/\Delta u \rightarrow 0$). As for the CNTs (6,3)-direct and (6,3)-inverse, it is seen that the curve corresponding to (6,3)-inverse is always exterior to the (6,3)-direct one, but to a low extent. It should be mentioned that Zhang and Shen [33] also found qualitatively similar curves for the F_c–T_c interaction of CNTs (8,8) and (14,0). If a nonlinear interactive curve is used to fit the MD data (minimizing the error using LSM or least squares method), the following expression is obtained:

$$\left(\frac{\varepsilon_c}{\varepsilon_c^0}\right)^{1.5} + \left(\frac{\alpha_c}{\alpha_c^0}\right)^{1.5} = 1 \quad (11.44)$$

where ε_c^0 is the critical strain for pure shortening and α_c^0 is the critical twisting angle per unit of length for pure twisting (circles lying in the vertical and horizontal axes of Fig. 11.8A). The curves corresponding to Eq. (11.44) follow the MD data fairly well. For a given value of the twisting–shortening rate, $\Delta\phi/\Delta u$, the expression $\alpha_c/\varepsilon_c = \Delta\phi/\Delta u$ should be used with Eq. (11.44) to determine the values of ε_c and α_c. To solve Eq. (11.44), only the values of ε_c^0 (pure shortening), α_c^0 (pure twisting), and the applied twisting–shortening rate $\Delta\phi/\Delta u$ are needed. These values can be determined approximately from available continuum mechanics formulae [28]. In Eq. (11.44), and regardless of the CNT chirality, the interaction coefficient for buckling of CNTs under combined shortening–twisting is equal to 1.5. To the authors' knowledge, this evidence (interaction factor of 1.5) has never been achieved by other investigations in the field of CNTs. This fact means that CNTs possess a shortening–twisting interaction curve between the linear interaction (coefficient 1.0) and quadratic interaction (coefficient 2.0). It should be mentioned that circular tubes at macroscale display a shortening–twisting interaction coefficient close to 2.0 (quadratic interaction). Due to the fact that the interaction coefficient of CNTs is much lower (25%) than that of macroscale tubes, it is concluded that CNTs are much more sensitive to buckling under shortening–twisting interaction than macroscale tubes.

11.3.3 Chirality and Anisotropic Effects

In this section, a brief account on the anisotropic behavior of (6,3) chiral CNT under combined tension–twisting is presented and discussed [45]. MD simulations were performed using the large-scale atomic/molecular massively parallel simulator code [46]. The C–C interaction parameters follow the latest version of the AIREBO potential proposed by Stuart et al. [47] and Brenner et al. [48]. The AIREBO potential is an improved version of Brenner's well-known second-generation REBO potential but includes a Lennard-Jones potential form to describe van der Waals long-range interactions and a torsional term for σ-bond torsions. In this work, the chiral CNT ($n = 6$, $m = 3$) with radius $R = 3.1$ Å, chiral angle $\theta = 21.2°$, length $L = 41.0$ Å, aspect ratio $L/D = 6.6$, and 336 C-atoms was considered. The explanation on the incremental loading procedure was given previously. The combined tension–twisting loading is imposed using a similar procedure to that described before for shortening–twisting. The results are presented and discussed next. First, the following two combinations (Δu vs. $\Delta\phi$) are adopted:

$$\text{Combination A:}\quad \Delta u = 0.03536\ \text{Å},\quad \Delta\phi = 0.01234\ \text{rad} \tag{11.45}$$

$$\text{Combination B:}\quad \Delta u = 0.02500\ \text{Å},\quad \Delta\phi = 0.01511\ \text{rad} \tag{11.46}$$

In combination A, the twist-to-tension ratio $\Delta\phi/\Delta u = 0.349$ rad/Å is low, which means that the CNT twists 0.349 rad for each angstrom that it is stretched. For combination B, the twist-to-tension ratio $\Delta\phi/\Delta u = 0.605$ rad/Å is high, which means that the CNT twists 0.605 rad for each angstrom that it is stretched. Fig. 11.10 shows a set of four snapshots of the CNT under several twist-to-tension ratios: (1) pure tension (first snapshot), (2) combination A (second snapshot), (3) combination B (third snapshot), and (4) pure twisting (fourth snapshot). Fig. 11.10A presents the initial (unloaded) configuration of the CNT while Fig. 11.11B–E presents the deformed shapes of the CNTs after 100, 200, 300, and 400 increments, respectively.

Fig. 11.11B1–E1 involve direct twisting of the CNT while the figures on the right side (Fig. 11.11B2–E2) involve inverse twisting of the CNT. From these figures, it is observed that the CNT degradation and collapse evolution is different if the CNT is under direct or inverse twisting. For combinations A (0.349 rad/Å, second snapshot) and B (0.605 rad/Å, third snapshot), the CNT also behaves differently for direct and inverse twisting—for instance, see the second snapshot in Fig. 11.11D1 and D2. The tensioned CNT under direct twisting (combination A) tends to buckle into a helix shape at early stages (200 increments; Fig. 11.11C1) without rupture. A clear rupture is visible only at 400 increments

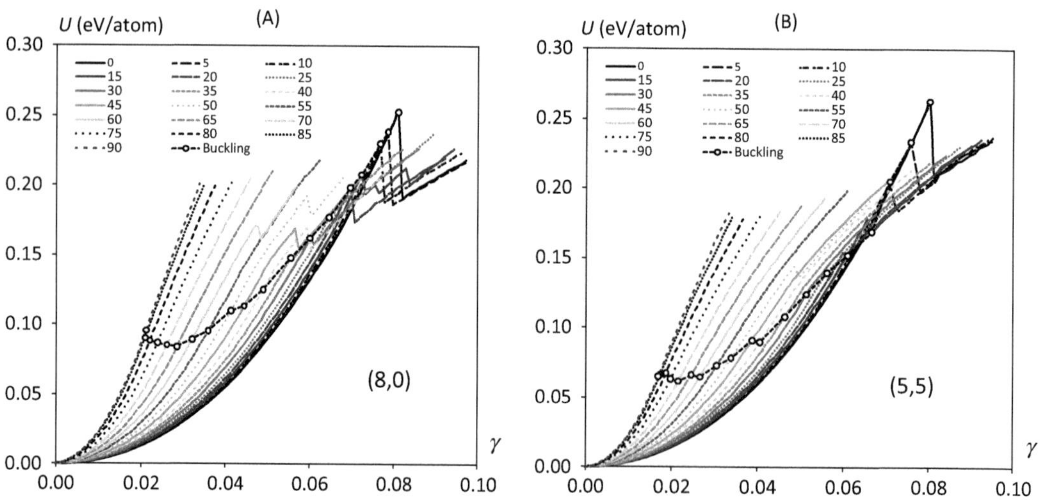

FIGURE 11.10 Variation of energy U with parameters γ and β for a given CNT: (A) (8,0) and (B) (5,5).

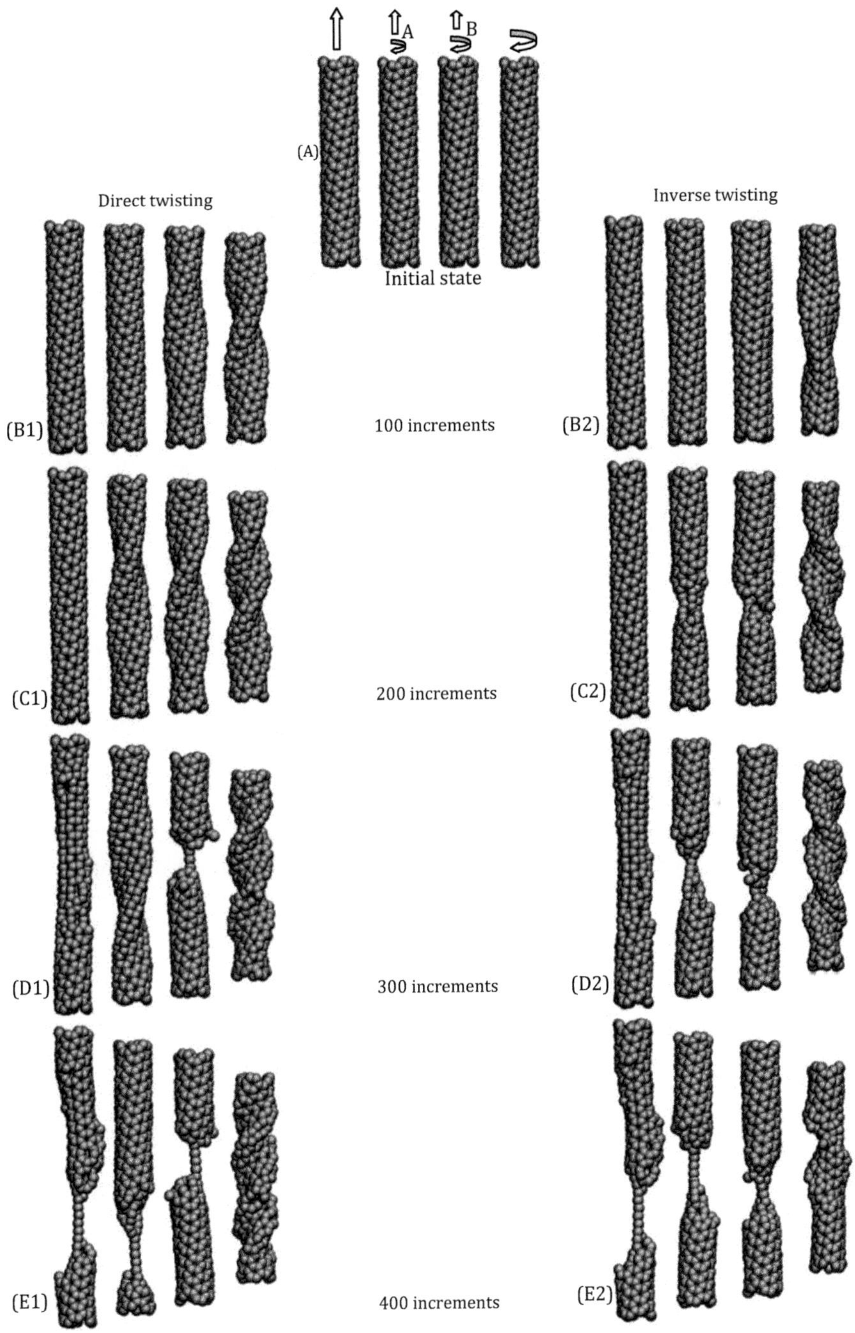

FIGURE 11.11 Deformed shapes of the CNT under tension to twist rates: (A) initial (unloaded), (B) after 100 increments, (C) after 200 increments, (D) after 300 increments, and (E) final (after 400 increments). (B1), (C1), (D1), (E1) for direct twisting and (B2), (C2), (D2), (E2) for inverse twisting.

(Fig. 11.11E1). The tensioned CNT under inverse twisting (combination A) does not tend to buckle into a helix shape and shows a localized rupture at an early stage (200 increments; Fig. 11.11C2). For combination B, a similar behavior is observed, but the CNT collapse occurs earlier for direct and inverse twisting.

To understand the chiral CNT anisotropic behavior and to explain its causes, further results are now presented at the C—C bond level. A kinematic analysis of the hexagon lattice (Fig. 11.5) is now required to understand how the deformation of the hexagon cell evolves and how it affects the CNT failure.

Consider two kinematical measures calculated at the onset of *first bond break*: (1) the C—C bond extension given by $\varepsilon_b = (L_b - L_0)/L_0$, where L_0 is the equilibrium (initial) length ($L_0 = 1.42$ Å) and L_b is the bond length at the breaking instant, and (2) the C—C angle distortion given by $\gamma_b = (A_b - A_0)/A_0$, where A_0 is the equilibrium (initial) angle ($A_0 = 120$ degree) and A is the angle at the breaking instant. The length $L_0 = 1.42$ Å represents the equilibrium distance between two C—C atoms in planar graphene while the angle $A_0 = 120$ degree represents the internal angle of a C—C hexagonal cell in planar graphene. In the case of rolled graphene (CNT), the interatomic equilibrium distance is slightly different and depends on the CNT radius. A brief study on the evolution of the first bond break with β is described next. For direct twisting, Fig. 11.12A and B shows the variation of ε_b and γ_b with β, respectively. Fig. 11.13 is similar to Fig. 11.12, but for inverse twisting. From the observation of Fig. 11.12 (direct twisting) and Fig. 11.13 (inverse twisting), the following remarks are possible:

- For $5 \leq \beta \leq 45$ degree (low-to-moderate twist-to-tension rate), the extension of *x*- and *y*-bonds of CNT under direct twisting (black and gray lines in Fig. 11.12A) remain unaltered with β while only *x*-bond extension of CNT under inverse twisting (black line in Fig. 11.13A) remains nearly constant. In the range $5 \leq \beta \leq 45$ degree, the extensions are always higher than 20%, denoting that the first bond break of the chiral

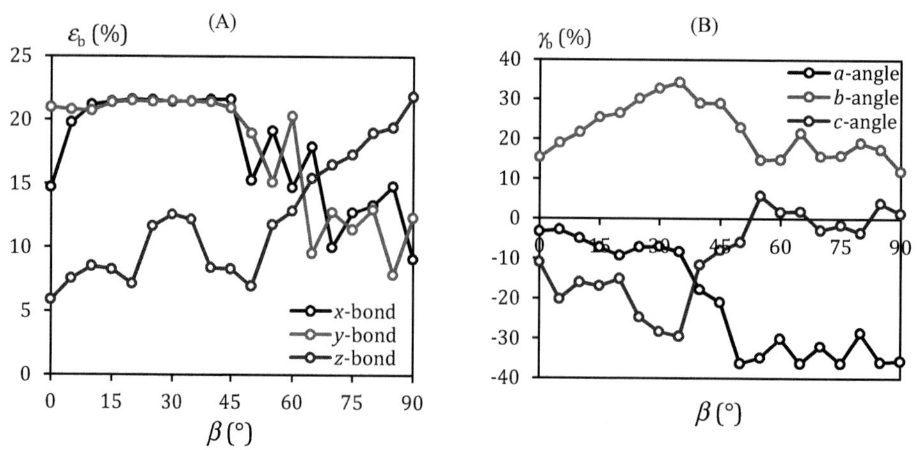

FIGURE 11.12 Direct twisting: variation, with β, of (A) bond extension ε_b and (B) angle distortion γ_b. The interested reader is referred to the electronic version of the chapter.

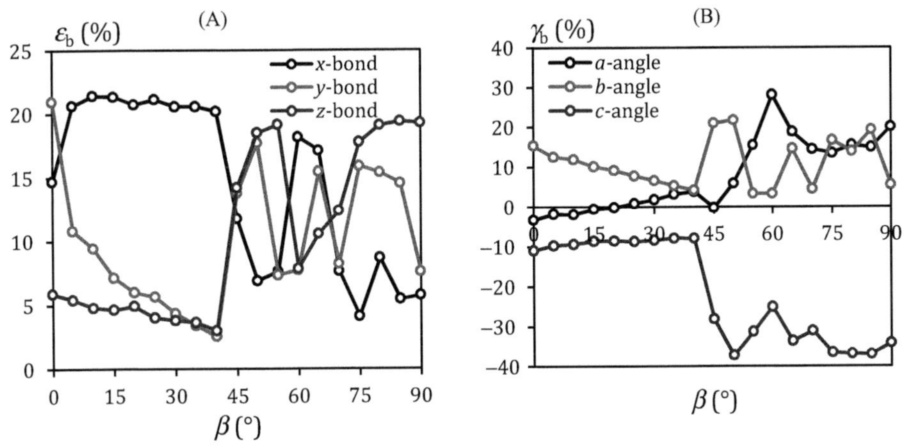

FIGURE 11.13 Inverse twisting: variation, with β, of (A) bond extension ε_b and (B) angle distortion γ_b. The interested reader is referred to the electronic version of the chapter.

CNT under low-to-moderate twist-to-tension rates occurs because the *x*- or *y*-bond breaks (direct) or *x*-bond breaks (inverse). Fig. 11.12B (direct twisting) shows that the *a*-angle distortion is always small ($\gamma_b \leq 10\%$) but the *b*- and *c*-angle distortions grow with β and achieve, for $\beta = 35$ degree, maximum values of $\gamma_b = 35\%$ and $\gamma_b = -30\%$, respectively. This scenario changes dramatically for $\beta = 40$ degree, regarding both bond extensions and angle distortions. Fig. 11.13B (inverse twisting) shows that angle distortions display a small distortion and limited variation ($-10.9\% \leq \gamma_b \leq 15.4\%$).

- For $45 \leq \beta \leq 90$ degree (moderate-to-high twist-to-tension rate), Figs. 11.12A and 11.13A show that bond extensions display high scatter due to the oval helix-shape of the chiral CNT, which is dynamically unstable. In this β range, the extensions are always lower than 20%, suggesting that the first bond break of the chiral CNT under moderate-to-high twist-to-tension rates is not likely to occur due to bond extension—the exception is the *z*-bond extension (dark gray line in Figs. 11.12A and 11.13A), which reaches extensions close to 20%. In the range $45 \leq \beta \leq 90$ degree, the angle distortions are higher than in the range $0 \leq \beta \leq 40$ degree. In particular, the distortions of *a*-angle (black line in Fig. 11.12B) and *c*-angle (dark gray line in Fig. 11.12B) achieve maximum values between 25% and 37%, denoting that the failure of chiral CNTs should be related with the limit value achieved by these angles.

Finally, Fig. 11.14 shows the failure interaction diagrams for a (6,3) chiral CNT under direct twisting (gray curve) and inverse twisting (black curve). Each point (u_f, ϕ_f) in these diagrams denote the relation between the axial stretching displacement (u_f) and twist rotation (ϕ_f) at the final failure (collapse) of CNT. Note that the points located at the vertical and horizontal axes correspond, respectively, to pure tension ($\beta = 0$ degree and $\phi_f = 0$) and pure twist ($\beta = 90$ degree and $u_f = 0$). The other points correspond to different values of the twist-to-tension rate ($0 < \beta < 90$ degree). The figure clearly shows the following:

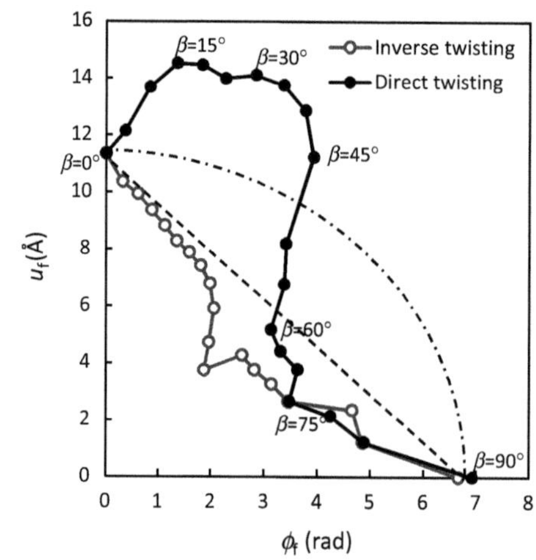

FIGURE 11.14 Tension–twisting interaction diagrams for (6,3) chiral CNT.

- The interaction diagrams for direct and inverse twisting are very dissimilar, showing that CNT fracture toughness is distinct in these cases. They are also very different from (1) the linear diagram denoted by the dashed line and (2) the quadratic diagram denoted by the dashed-pointed curve. The linear interaction is considered the most conservative (or safe) approach for the interactive failure of macroscale structures. The quadratic interaction is the usual approach to assess the ultimate strength of isotropic elastic–plastic circular tubes at macroscale.
- The interaction diagram for inverse twisting (gray curve) appears to be linear, but it is not. The interaction diagram for inverse twisting is inside the linear diagram, which means that the coupling between tension and inverse twisting is even worse for the CNT fracture toughness than the linear interaction. Thus, it can be concluded that inverse twisting influences very negatively the collapse behavior of chiral CNTs under tension. Even for very small increments of the twisting component (low ϕ values), the axial displacement at collapse u_f decreases immensely. The opposite also occurs, as tension harmfully influences the fracture toughness of chiral CNTs under inverse twisting; for small increments of the axial displacement component (low u values), the twisting rotation at collapse ϕ_f decreases a lot.
- The interaction diagram for direct twisting (black curve) is highly nonlinear. For low-to-moderate β values ($0 \leq \beta \leq 45$ degree), the interaction diagram is broadly outside the quadratic diagram, which means that the coupling between tension and direct twisting is highly beneficial for the CNT fracture toughness. Even for a very small increase of the twisting component (low ϕ values), the axial displacement at collapse u_f increases above the value for pure tension ($u_f = 11.35$ Å). Thus, it can be concluded that low-to-moderate direct twisting influences positively the collapse behavior of chiral CNTs under tension. For moderate-to-high β values ($60 \leq \beta \leq 90$ degree), the interaction diagram is inside the linear diagram, which means that the coupling

between tension and direct twisting becomes highly detrimental for the CNT collapse. In this case, the interaction diagrams for inverse and direct twisting become closer, showing similar kinematical quantities at CNT collapse—remember that the static variables at the onset of CNT collapse, such as torque T, are different for direct and inverse twisting due to the distinct CNT stiffness in each case.

The behavior of the chiral CNT under combined tension–twisting is strongly dependent on (1) the twist direction and (2) the twist-to-tension rate, thus evidencing its anisotropic nature. For the chiral CNT under low twist-to-tension ratios, direct twisting has no impact on the failure strain (it equals the pure tensile strain) while inverse twisting has a great influence on it (it is half the pure tensile strain). For low twist-to-tension ratios, it is shown that the chiral CNT accumulates much more energy when twisted directly than inversely. For direct twisting, the stored energy is even higher than that for pure tension. The addition of small-to-moderate direct twist to a tensioned CNT will increase its energy beyond that of pure tension! This fact is due to the stiffening effect arising from the direct twist action on the chiral CNT hexagonal lattice.

However, for high twist-to-tension ratios, the energy becomes lower than that of pure tension. For a low twist-to-tension ratio, both initial and postbuckling stiffnesses of the chiral CNT are not influenced by the twist direction. For high twist-to-tension ratios, the direction of twisting does not influence too much the initial stiffness of the CNT, but affects its postbuckling stiffness. The ductile (large deformation) behavior shown by the chiral CNT under pure twisting is lost when small-to-moderate tension is applied to the CNT. However, the postfailure mechanism of CNTs under combined tension–inverse-twisting is more ductile (a smoother decrease of forces) while that of CNTs under combined tension–direct-twisting is more brittle (a more sudden decrease of forces).

11.4 Final Remarks and Future Developments

In this chapter, the atomistic simulation of carbon nanotubes was presented and their behavior (stiffness, strength, and toughness) under local deformation was investigated, taking into account the contributions of the authors to the current state of the art. Different methods are usually applied to analyze the local instability of thin-walled tubes at macro- and nanoscale mostly because of the distinct nature of the structure, being continuum at macroscale and discrete at nanoscale. Despite this inherent difference, the phenomenon itself is rather similar and some characteristics are common at both scales. Bridging these two extreme scales are a few other scales (e.g., microscale and mesoscale). In the last decade, intensive research has been undertaken to develop methodologies and techniques to merge adjacent scales, and this led to the well-known multiscale methods. In all scales, the occurrence of local buckling always decreases the strength and stiffness of the thin-walled structure and, mostly, circumvents its targeted behavior of the structural member, composite material, or even nanodevice. Therefore, it is desired to avoid any kind of local instability from macroscale to nanoscale.

At nanoscale, this chapter focused on the CNT as a "thin-walled nanostructure." Because of the carbon sp^2 hybridization, no other material has displayed the combination of superlative mechanical properties attributed to CNTs, except graphene sheets (GSs), which are a hexagonal lattice of carbon atoms with 2D plane shape and have mechanical properties similar to CNTs. Because CNTs were discovered in 1991 [49] and GSs were first isolated 13 years later, the number of investigations is naturally much higher for CNTs than in the case of GSs. Regarding the developments for future research, they will certainly include both nanoparticles (CNTs and GSs). One of the ongoing application fields is the strengthening of nanocomposites (NCs) using CNTs and GSs. Because of its planar configuration, GSs are prone to local instabilities and their interaction at nanoscale with other molecules of the surrounding medium should be a topic of research.

Currently, the field of NCs is one of the most popular areas of research in all scientific disciplines. Regarding NCs for structural components, their highest requirements are (1) stiffness and strength, to resist several types of mechanical actions, and (2) damping and fatigue resistance, to dissipate energy from dynamic actions. These NC properties are due to the "symbiosis" between the matrix and nanoparticles (the former surrounds and bounds the latter, while the latter strengthens the former). Therefore, the development of multiscale methods (e.g., based on the finite element method) for NCs is an objective of research.

Acknowledgments

The authors gratefully acknowledge the financial support provided by FCT (Portuguese Foundation for Science and Technology), not only through several research projects but also due to the doctoral scholarships granted to its students. Finally, acknowledgments are also given to IDMEC (Instituto de Engenharia Mecânica) and CQE (Centro de Quimica Estrutural) for having provided all the necessary means to achieve the results reported in this document.

References

[1] A.C. Eringen, On differential equations of nonlocal elasticity and solutions of screw dislocation and surface waves, J. Appl. Phys. 54 (1983) 4703–4710.

[2] A.C. Eringen, Nonlocal Continuum Field Theories, Springer, New York, NY, (2002).

[3] B.I. Yakobson, C.J. Brabec, J. Bernholc, Nanomechanics of carbon tubes: instabilities beyond linear response, Phys. Rev. Lett. 76 (1996) 2511–2514.

[4] J.P. Lu, Elastic properties of carbon nanotubes and nanoropes, Phys. Rev. Lett. 79 (1997) 1297–1300.

[5] X. Zhou, J.J. Zhou, Z.C. Ou-Yang, Strain energy and Young's modulus of single-wall carbon nanotubes calculated from electronic energy-band theory, Phys. Rev. 62 (2000) 13692.

[6] K.N. Kudin, G.E. Scuseria, B.I. Yakobson, C_2F, BN, and C nanoshell elasticity from ab initio computations, Phys. Rev. 64 (2001) 235406.

[7] G.M. Odegard, T.S. Gates, L.M. Nicholson, K.E. Wise, Equivalent-continuum modeling of nano-structured materials, Comp. Sci. Technol. 62 (2002) 1869–1880.

[8] Y. Jin, F.G. Yuan, Simulation of elastic properties of single-walled carbon nanotubes, Comp. Sci. Technol. 63 (2003) 1507–1515.

[9] T. Vodenitcharova, L.C. Zhang, Effective wall thickness of a single-walled carbon nanotube, Phys. Rev. 68 (2003) 165401.

[10] A. Pantano, D.M. Parks, M.C. Boyce, Nonlinear structural mechanics based modeling of carbon nanotube deformation, Phys. Rev. Lett. 91 (2004) 145504.

[11] L. Wang, Q. Zheng, J.Z. Liu, Q. Jiang, Size dependence of the thin-shell model for carbon nanotubes, Phys. Rev. Lett. 95 (2005) 105501.

[12] Y. Huang, J. Wu, K.C. Hwang, Thickness of graphene and single-wall carbon nanotubes, Phys. Rev. 74 (2006) 245413.

[13] J. Peng, J. Wu, K.C. Hwang, J. Song, Y. Huang, Can a single-wall carbon nanotube be modeled as a thin shell?, J. Mech. Phys. Solids 56 (2008) 2213–2224.

[14] O.A. Shenderova, V.V. Zhirnov, Brenner DW, Carbon nanostructures, Crit. Rev. Solid State Mater. Sci. 27 (2002) 227–356.

[15] P. Lu, P.Q. Zhang, H.P. Lee, C.M. Wang, J.N. Reddy, Non-local elastic plate theories, Proc. R. Soc. 463 (2007) 3225–3240.

[16] J.P. Crocombette, F. Willaime, Ab initio electronic structure calculations for nuclear materials, in: R.J.M. Konings (Ed.), Comprehensive Nuclear Materials—Volume 1 Basic Aspects of Radiation Effects in Solids/Basic Aspects of Multi-Scale Modeling, Elsevier, Amsterdam, 2012, pp. 223–248 Chapter 1.08.

[17] D. Marx, J. Hutter, Ab initio molecular dynamics: theory and implementation, in: J. Grotendorst (Ed.), Modern Methods and Algorithms of Quantum Chemistry, volume 1 of NIC Series, John von Neumann Institute for Computing, 2000, pp. 301.

[18] W. Cai, J. Li, S. Yip, Molecular dynamics, in: R.J.M. Konings (Ed.), Comprehensive Nuclear Materials—Volume 1 Basic Aspects of Radiation Effects in Solids/Basic Aspects of Multi-Scale Modeling, Elsevier, Amsterdam, 2012, pp. 249–265 Chapter 1.09.

[19] M.P. Allen, D.J. Tildesley, Computer Simulation of Liquids, Oxford University Press, New York, (1987).

[20] V. Arnold, Mathematical Methods of Classical Mechanics, Springer, New York, (1978).

[21] D.W. Brenner, Empirical potential for hydrocarbons for use in simulating the chemical vapor deposition of diamond films, Phys. Rev. 42 (1990) 9458–9471.

[22] J.M. Sanz-Serna, M.P. Calvo, Numerical Hamiltonian Problems, Chapman & Hall, London, (1994).

[23] M.E. Tuckerman, Y. Liu, G. Ciccotti, G.J. Martyna, Non-Hamiltonian molecular dynamics: generalizing Hamiltonian phase space principles to non Hamiltonian systems, J. Chem. Phys. 115 (2001) 1678–1702.

[24] M. Griebel, A. Caglar, S. Knapek, G. Zumbusch, Numerische Simulation in der Molek uldynamik. Numerik, Algorithmen, Parallelisierung, Anwendungen, Springer, Berlin, Heidelberg, (2003).

[25] D. Frenkel, B. Smit, Understanding Molecular Simulation: From Algorithms to Applications, Academic Press, New York, (1996).

[26] Q. Wang, K.M. Liew, W.H. Duan, Modeling of the mechanical instability of carbon nanotubes, Carbon 46 (2008) 285–290.

[27] C.M. Wang, Y.Y. Zhang, Y. Xiang, J.N. Reddy, Recent studies on buckling of carbon nanotubes, Appl. Mech. Rev. 63 (2010) 030804.

[28] N. Silvestre, Length dependence of critical measures in single-walled carbon nanotubes, Int. J. Solids Struct. 45 (18–19) (2008) 4902–4920.

[29] N. Silvestre, B. Faria, J.N. Canongia Lopes, A molecular dynamics study on the thickness and post-critical strength of carbon nanotubes, Compos. Struct. 94 (4) (2012) 1352–1358.

[30] W. Smith, T.R. Forester, I.T. Todorov, The DL-Poly 2 User Manual, Version 2.21, STFC Daresbury Laboratory, Cheshire, UK, (2010).

[31] A.R. Ranjbartoreh, G. Wang, Molecular dynamic investigation of length dependency of single-walled carbon nanotube, Phys. E 43 (2010) 202–206.

[32] N. Silvestre, C.M. Wang, Y.Y. Zhang, Y. Xiang, Sanders shell model for buckling of single-walled carbon nanotubes with small aspect ratio, Compos. Struct. 93 (7) (2011) 1683–1691.

[33] C.L. Zhang, H.S. Shen, Buckling and postbuckling analysis of single-walled carbon nanotubes in thermal environments via molecular dynamics simulation, Carbon 44 (2006) 2608–2616.

[34] N. Silvestre, On the accuracy of shell models for torsional buckling of carbon nanotubes, Eur. J. Mech. A: Solids 32 (2012) 103–108.

[35] M. Arroyo, T. Belytschko, An atomistic-based finite deformation membrane for single layer crystalline films, J. Mech. Phys. Solids 50 (2002) 1941–1977.

[36] J. Wu, K.C. Hwang, Y. Huang, An atomistic-based finite-deformation shell theory for single-wall carbon nanotubes, J. Mech. Phys. Solids 56 (2008) 279–292.

[37] C.M. Wang, Z.Y. Tay, A.N.R. Chowdhury, W.H. Duan, Y.Y. Zhang, N. Silvestre, Examination of cylindrical shell theories for buckling of carbon nanotubes, Int. J. Struct. Stab. Dyn. 11 (6) (2011) 1035–1058.

[38] A.G. Arani, R. Rahmani, A. Arefmanesh, S. Golabi, Buckling analysis of multi-walled carbon nanotubes under combined loading considering the effect of small length scale, J. Mech. Sci. Technol. 22 (2008) 429–439.

[39] C.Q. Sun, K.X. Liu, Torsional buckling of multi-walled carbon nanotubes under combined axial and radial loadings, J. Phys. 41 (2008) 205404.

[40] X.Y. Wang, X. Wang, X.H. Xia, Eccentric compression stability of multi-walled carbon nanotubes embedded in an elastic matrix, Compos. Sci. Technol. 67 (2007) 1406–1414.

[41] C. Li, T.W. Chou, Modeling of elastic buckling of carbon nanotubes by molecular structural mechanics approach, Mech. Mater. 36 (2004) 1047–1055.

[42] P.A. Williams, S.J. Papadakis, A.M. Patel, M.R. Falvo, S. Washburn, R. Superfine, Torsional response and stiffening of individual multiwalled carbon nanotubes, Phys. Rev. Lett. 89 (2002) 255502.

[43] A.M. Fennimore, T.D. Yuzvinsky, W.Q. Han, M.S. Fuhrer, J. Cumings, A. Zettl, Rotational actuators based on carbon nanotubes, Nature 424 (2003) 408–410.

[44] B. Faria, N. Silvestre, J.N. Canongia Lopes, Interaction diagrams for carbon nanotubes under combined shortening-twisting, Compos. Sci. Technol. 71 (16) (2011) 1811–1818.

[45] B. Faria, N. Silvestre, J.N. Canongia Lopes, Induced anisotropy of chiral carbon nanotubes under combined tension-twisting, Mech. Mater. 58 (2013) 97–109.

[46] S. Plimpton, Fast parallel algorithms for short-range molecular dynamics, J. Comput. Phys. 117 (1995) 1–19.

[47] S.J. Stuart, A.B. Tutein, J.A. Harrison, A reactive potential for hydrocarbons with intermolecular interactions, J. Chem. Phys. 112 (2000) 6472–6486.

[48] D.W. Brenner, O.A. Shenderova, J.A. Harrison, S.J. Stuart, B. Ni, S.B. Sinnott, A second-generation reactive empirical bond order (REBO) potential energy expression for hydrocarbons, J. Phys.: Condens. Matter 14 (2002) 783–802.

[49] S. Iijima, Helical microtubules of graphitic carbon, Nature 354 (6348) (1991) 56–58.

12

Finite Element Modeling of Nanotubes

Androniki S. Tsiamaki, Georgios I. Giannopoulos, Stylianos K. Georgantzinos, Nick K. Anifantis

UNIVERSITY OF PATRAS, PATRAS, GREECE

CHAPTER OUTLINE

12.1 Introduction

The development and exploitation of novel materials can lead to the growth of new fields of research as well as new solutions to technological problems that have not been resolved up to now. One of the latest breakthrough sections in science and technology promising novel applications in different aspects of life is nanomaterials [1]. The most famous nanomaterials are the allotropes of carbon, the structure of which is usually made by one-atom-thick planar or curved sheets of sp^2-bonded carbon atoms densely packed in a honeycomb crystal lattice. The best known carbon allotropes are graphite, carbon nanotubes (CNTs), and fullerenes having graphene as the basic structural

Carbon Nanotube–Reinforced Polymers. http://dx.doi.org/10.1016/B978-0-323-48221-9.00012-1

element. A graphene monolayer is most easily visualized as an atomic-scale chicken wire made of carbon atoms and their bonds. The hexagonal lattice and the high strength of the carbon–carbon bond observed in the aforementioned materials lead to exceptional mechanical properties (200 times stronger than steel) [2,3]. Moreover, there are other similar inorganic materials with graphene-like geometric characteristics, such as boron nitride allotropes. Among the various nanomaterials, nanotubes have attracted most of the scientific interest owing to their excellent physicochemical, electrical, and mechanical properties. Single-walled carbon nanotubes (SWCNTs) and single-walled boron nitride nanotubes (SWBNNTs) may be generated by rolling the monolayers of graphene and hexagonal boron nitride into tubes, respectively. Multiwalled nanotubes (MWNTs) may be produced by rolling two or more parallel monolayers, which are kept connected due to van der Waals forces. Beyond their mechanical properties, nanotubes have remarkable electronic, optical, and thermal properties making them candidate materials for various applications. The combination of their properties, their small size, and low density also makes them very promising materials for applications in which the natural frequencies are key properties, such as molecular gas detectors and nanomechanical resonators. SWCNTs, for example, are the stiffest and strongest known fibers, also having remarkable electronic and conductive properties and many other unique characteristics [4]. However, these properties are obviously of limited value in individual tubes. The development of nanomaterial-based elastomeric composites could demonstrate the excellent energy absorption characteristics of the rubber component as well as the advanced structural properties of the nanomaterials. Due to the dramatically increasing number of applications of those materials in modern science and technology in which precise performance is required, the explicit relationships between their structure and mechanical properties should be revealed.

To characterize material properties even in nanoscale, experimental methods are always effective and reliable. However, given the high cost, the requirement of complex experimental equipment, and time-consuming measurements, theoretical methods seem to be more attractive for predicting the properties and behavior of nanomaterials. Several simulation approaches and numerical methods have been proposed for the computation of the mechanical and other properties of nanomaterials. From a theoretical point of view, the development of accurate and computationally effective numerical models for the evaluation of the elastic, fracture, and vibrational characteristics of nanomaterials is an important task especially for reasons of nanodesign. A number of efforts have already been presented in the open literature. They are mainly limited to continuum mechanics (CM) approaches varying from nonlinear continuum models to nonlocal continuum techniques. Regarding the mechanical vibrations of nanomaterials, corresponding theoretical studies have been carried out as well. Various techniques for analyzing the vibration of nanomaterials have been proposed in the literature, including nonlocal elasticity and density functional theories, lattice structure, and finite difference methods, such as molecular dynamics (MD) and molecular mechanics (MM) approaches. However, it must be mentioned that MD approaches are restricted to small length and time scale problems due to their computational cost, while continuum mechanics approaches fail to incorporate the atomistic structure of the nanomaterials and relevant nanostructural effects.

In this chapter, a numerical method based on molecular mechanics theory is proposed that can deal with any kind of problem regarding the elastic, plastic, and vibrational behavior of nanotubes. According to the proposed method, the nanotubes are modeled as three-dimensional (3D) frames of appropriate spring-like and, if required, mass-like finite elements in accordance with their exact nanostructural geometric characteristics. More specifically, the interatomic interactions and the potential energy functions that describe them are approximated by the use of 3D, two-noded, spring elements interconnecting atom pairs, while nanoscale inertia effects are included in the vibration analysis through the introduction of 3D mass elements representing the nucleus masses of the corresponding atoms. Furthermore, linear and nonlinear considerations can be made during modeling depending on the requirements and the limitations of the problem. The proposed technique permits efficient and accurate computations regarding the elastic, plastic, vibrational, and buckling properties of single-walled nanotubes (SWNTs) and MWNTs made of different atoms, having different chirality, size, and shape, under various boundary conditions. In any case, some representative results are included here as well, just to illustrate the good agreement between the proposed numerical scheme and other methods found in the literature.

12.2 Atomistic Geometry of Nanotubes

Numerous works report the main geometrical characteristics of nanotubes. Here, a brief discussion about nanotube geometry will be made just for reasons of completeness. A nanotube can be generated by rolling a sheet of the same atom consistency. For example, a graphene sheet that consists of hexagonal lattices of carbon atoms is ideally cut to make a cylinder representing a CNT. Two vectors, $\mathbf{a}_1$ and $\mathbf{a}_2$, are used to define the honeycomb lattice [5]. The chiral vector $\mathbf{C}_h$ is defined as

$$\mathbf{C}_h = n\mathbf{a}_1 + m\mathbf{a}_2 \tag{12.1}$$

where $\mathbf{a}_1$ and $\mathbf{a}_2$ are the basis lattice vectors and the integers n, m are the number of steps along the zigzag bonds, which are used to name a nanotube while characterizing their chirality. An (n,n) nanotube is usually called armchair while an $(n,0)$ one is called zigzag. In practical terms, the index n expresses how many hexagonal lattices correspond to a diametric ring of a nanotube, and the index m influences the slope of the arrangement of the hexagonal lattice along the longitudinal axis of the specific nanotube. The nanotube chiral angle ψ and diameter D are expressed, respectively by the following equations:

$$\tan\psi = \frac{\sqrt{3}m}{2n+m} \tag{12.2}$$

$$D = \frac{r_0\sqrt{3(n^2+nm+m^2)}}{\pi} \tag{12.3}$$

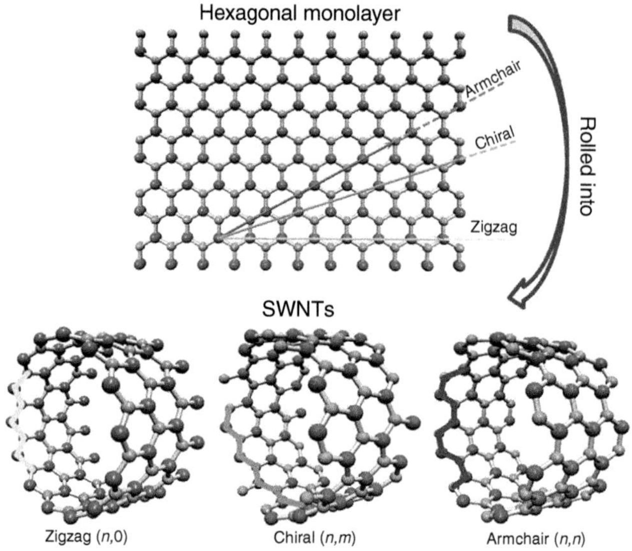

FIGURE 12.1 Single-walled carbon nanotubes *(SWNTs)* formed by rolling a graphene-like monolayer.

where r_0 is the distance between two neighboring bonded atoms, which, for example, is approximately equal to 0.1421 nm for a CNT and 0.145 nm for a boron nitride nanotube (BNNT). To define the geometry of an SWCNT completely, it is sufficient to specify only three geometric parameters of the tube. These are the values of the n and m indices that characterize the type of the tube and the length L of the nanotube. Thus, by defining these global design parameters, automatic geometric development of a specific tube can be easily programmed while a regenerative model may be developed for conducting various parametric analyses by using the aforementioned approach.

Most graphene-like nanotubes can be easily produced by utilizing the geometry of a honeycomb lattice sheet and then rolling it to a tubular shape as Fig. 12.1 illustrates. An SWBNNT may be produced by rolling an appropriately cut hexagonal boron nitride one-atom-thick sheet into a hollow cylindrical nanostructure.

The MWNTs consist of multiple rolled layers of hexagonal monolayers, such as graphene. MWNTs may be seen as concentric SWNTs having a constant interlayer distance. The interlayer distance is influenced by van der Waals forces and for multiwalled carbon nanotubes (MWCNTs) it is close to the distance between graphene layers in graphite, approximately equal to 0.34 nm.

12.3 Potential Energy Description

The potential energies due to interatomic interactions within a nanomaterial may be expressed by mathematical functions that are dependent on the distance and angle changes between interacting atoms. For example, molecular theory states that the potential energy between two atoms is a nonlinear function of their interdistance and that

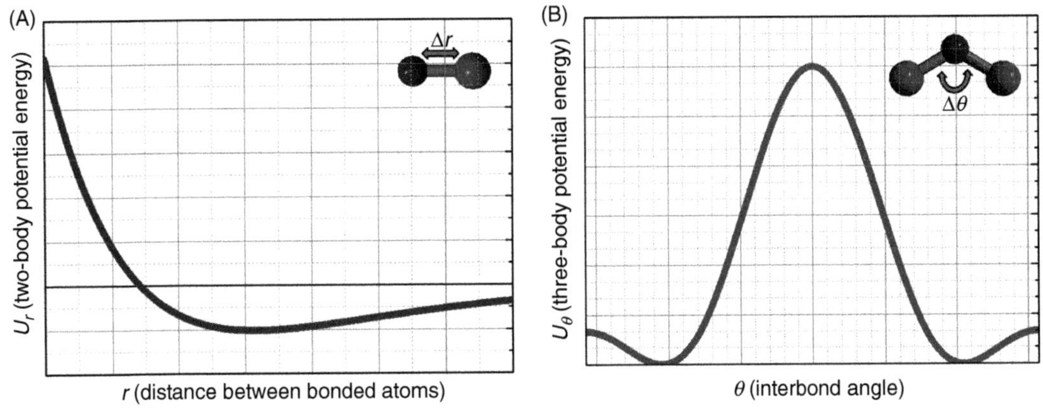

FIGURE 12.2 Potential energy variation for (A) the two-body interatomic interactions and (B) the three-body interatomic interactions.

the potential energy between two linked bonds is a nonlinear function of their interbond angle as Fig. 12.2A–B illustrate, respectively.

Generally, the total potential energy in a nanosystem, such as a nanotube may be expressed as [6]

$$U = \sum U_r + \sum U_\theta + \sum U_\phi + \sum U_\omega + \sum U_{\mathrm{vdw}} \quad (12.4)$$

where U_r, U_θ, U_ϕ, and U_ω represent the energies due to bond length, bond angle bending, dihedral angle torsion, and out-of-plane torsion changes. In addition, U_{vdw} is the energy due to the nonbonded van der Waals interactions.

It is computationally convenient to combine potential terms U_ϕ and U_ω in the single term U_τ, which describes the energy due to the total bond angle torsion change. For small strains, where linear analysis is appropriate, the simplest harmonic forms of the potential energies may be adopted, which reduce the computational cost:

$$U_r = \frac{1}{2}k_r(\Delta r)^2; \quad U_\theta = \frac{1}{2}k_\theta(\Delta\theta)^2; \quad U_\tau = U_\omega + U_\phi = \frac{1}{2}k_\tau(\Delta\phi)^2 \quad (12.5)$$

where the stiffness terms k_r, k_θ, and k_τ are the bond length, angle bending, and angle torsion resistance force constants, respectively, which at room temperature may be taken, respectively, equal to 652 nN/nm, 0.876 nN nm/rad^2, and 0.272 nN nm/rad^2 for CNTs [7], or equal to 486.5 nN/nm, 0.6952 nN nm/rad^2, and 0.6255 nN nm/rad^2 [8] for BNNTs.

When the fracture and large deformation problems of nanotubes are to be investigated, suitable nonlinear potential functions should be adopted. One common option is to follow the modified Morse potential, which is generally expressed by the following functions:

$$U_r = D_e[(1 - e^{-\beta\Delta r})^2 - 1]; \quad U_\theta = \frac{1}{2}k_\theta(\Delta\theta)^2[1 + k_{\mathrm{sextic}}(\Delta\theta)^4] \quad (12.6)$$

where the constants D_e, β, k_θ, and k_{sextic} for the CNTs may be taken equal to 6.03105×10^{-19} Nm, 2.625×10^{10} m^{-1}, 0.9×10^{-18} Nm/rad^2, and 0.754 rad^{-4}, respectively [9].

Evidently, there are many alternatives regarding the type of nonlinear potential energy functions that can be utilized in the proposed finite element formulation. Another effective representation of the potential field may be achieved via the Stillinger–Weber potential field model[10]:

$$U_r = A\left[B\left(\frac{s}{r_0+\Delta r}\right)^4 - 1\right]e^{[s/(r_0+\Delta r-b\sigma)]};\ U_\theta = 2\lambda e^{[\gamma s/(r_0-bs)]}\left[\cos(\theta_0+\Delta r)+\kappa\right]^2 \quad (12.7)$$

In the aforementioned equations A, B, s, δ, b, λ, γ, and κ are parameters that, for BNNTs, may be taken equal to 1.7448 nN nm, 0.8229, 0.1023 nm, 0.59, 2.3, 3.3678 nN nm, 0.79, and 0.79, respectively, while θ_0 is the equilibrium interbond distance that, for hexagonal nanostructures, is equal to $2\pi/3$.

The van der Waals forces may be neglected when an SWNT is to be modeled due to their weak effect. However, when an MWNT is analyzed, their effect between concentric layers is significant and may be incorporated by adopting the following Lenard-Jones potential [11]:

$$U_{\text{vdW}} = 4\varepsilon\left[\left(\frac{\sigma}{R}\right)^{12} - \left(\frac{\sigma}{R}\right)^6\right] \quad (12.8)$$

where R is the distance between two arbitrary carbon atoms and ε and σ are the Lenard-Jones parameters. Concerning a carbon atom pair, these parameters are $\varepsilon = 3.8655 \times 10^{-13}$ N nm and $\sigma = 0.34$ nm [11].

12.4 Modeling of Nanotube Interatomic Interactions

Since the second derivatives of the aforementioned potential energy functions represent stiffness constants or nonlinear stiffness variations, it seems numerically attractive to represent the force field within a nanotube with linear or nonlinear spring-like finite elements, respectively. When linear problems regarding the mechanical response of nanotubes are to be treated, it is convenient to follow a more simplified modeling technique due to the linearity of the relevant equations and small deformation theory.

12.4.1 Linear Analysis

For linear nanotube problems the line spring-like finite elements depicted in Fig. 12.3 are utilized for the representation of the nanomaterial. The elements are two-noded with three degrees of freedom per node, that is, three translations. Two types of elements denoted as el = 2b, 3b are utilized for the simulation of the interatomic interactions. The 2b elements simulate bond length change interatomic interaction, interconnect two bonded atoms, and are analyzed in the 3D local coordinate system $(\bar{x},\bar{y},\bar{z})$, which is shown in Fig. 12.3. Note that the $\bar{x}$-axis coincides with the line that connects the two bonded atoms. On the other hand, the 3b elements represent bond angle bending and torsion interatomic interactions, interconnect the two opposing atoms of two linked bonds, and

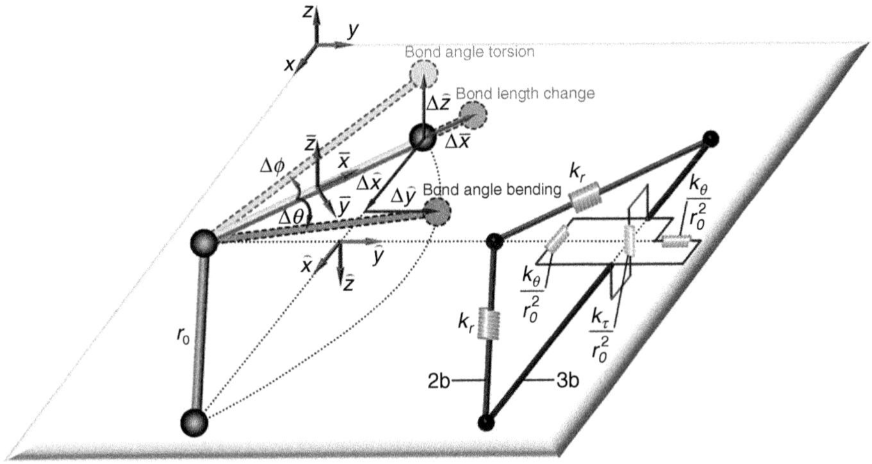

FIGURE 12.3 Finite element modeling of interatomic interactions for small strains.

are reanalyzed in a different local coordinate system defined as $(\hat{x},\hat{y},\hat{z})$. This coordinate system is positioned such that the $\hat{x}$-axis is aligned with the line that connects the two opposing nonbonded atoms, while $\hat{z}$ is vertical to the plane defined by the two linked bonds.

The elemental stiffness matrix of the spring elements e1 = 2b, 3b may be written as

$$\mathbf{K}^{\mathrm{el}} = \begin{bmatrix} \mathbf{k}^{\mathrm{el}} & -\mathbf{k}^{\mathrm{el}} \\ -\mathbf{k}^{\mathrm{el}} & \mathbf{k}^{\mathrm{el}} \end{bmatrix} \tag{12.9}$$

For small strains and given the applicably positioned local coordinate systems $(\bar{x},\bar{y},\bar{z})$ and $(\hat{x},\hat{y},\hat{z})$, it is easy to prove that

$$\Delta r = \Delta\bar{x};\ \Delta\theta = \frac{\sqrt{(\Delta\hat{x})^2 + (\Delta\hat{y})^2}}{r_0};\ \Delta\phi = \frac{\Delta\hat{z}}{r_0} \tag{12.10}$$

By substituting Eq. (12.10) into Eq. (12.5) and then differentiating twice, the following stiffness coefficients $\mathbf{k}^{\mathrm{e1}}$ with e1 = 2b, 3b may arise, respectively:

$$\mathbf{k}^{2\mathrm{b}} = \begin{bmatrix} k_{\bar{x}}^{2\mathrm{b}} & 0 & 0 \\ 0 & k_{\bar{y}}^{2\mathrm{b}} & 0 \\ 0 & 0 & k_{\bar{z}}^{2\mathrm{b}} \end{bmatrix} = \begin{bmatrix} \frac{\mathrm{d}^2 U_r}{\mathrm{d}(\Delta\bar{x})^2} & 0 & 0 \\ 0 & 0 & 0 \\ 0 & 0 & 0 \end{bmatrix} = \begin{bmatrix} k_r & 0 & 0 \\ 0 & 0 & 0 \\ 0 & 0 & 0 \end{bmatrix} \tag{12.11}$$

$$\mathbf{k}^{3\mathrm{b}} = \begin{bmatrix} k_{\hat{x}}^{3\mathrm{b}} & 0 & 0 \\ 0 & k_{\hat{y}}^{3\mathrm{b}} & 0 \\ 0 & 0 & k_{\hat{z}}^{3\mathrm{h}} \end{bmatrix} = \begin{bmatrix} \frac{\partial^2 U_\theta}{\partial(\Delta\hat{x})^2} & 0 & 0 \\ 0 & \frac{\partial^2 U_\theta}{\partial(\Delta\hat{y})^2} & 0 \\ 0 & 0 & \frac{\mathrm{d}^2 U_\tau}{\mathrm{d}(\Delta\hat{z})^2} \end{bmatrix} = \begin{bmatrix} \frac{k_\theta}{r_0^2} & 0 & 0 \\ 0 & \frac{k_\theta}{r_0^2} & 0 \\ 0 & 0 & \frac{k_\tau}{r_0^2} \end{bmatrix} \tag{12.12}$$

Note that in MWCNT analysis the van der Waals forces may not be omitted. Such forces may be incorporated into the formulation by the introduction of a straight spring element similar with the 2b one. According to Eq. (12.8), the specific element that interconnects two arbitrary atoms has the following stiffness along its axial direction:

$$\frac{d^2U_{vdw}}{dR^2} = 24\varepsilon\left(26\frac{\sigma^{12}}{R^{14}} - 7\frac{\sigma^6}{R^8}\right) \tag{12.13}$$

When the distance R is larger than 2.5σ, the effect of the van der Waals forces may be neglected.

12.4.2 Nonlinear Analysis

In the case of nanotube problems that are associated with large displacements, a more complex numerical scheme should be followed. Here, two types of two-noded spring-like finite elements with six degrees of freedom per node, that is, three translations and three rotations, are utilized as Fig. 12.4 illustrates. First, line spring-like finite elements, hereafter called 2B and expressed in $(\bar{x},\bar{y},\bar{z})$, are reutilized for the simulation of the force variation due to the bond length change, Δr, as well as the torsional moment due to the bond angle

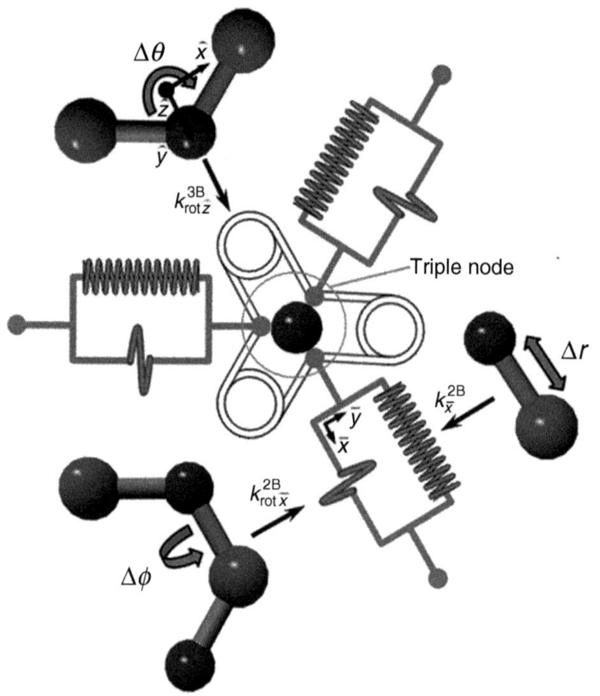

FIGURE 12.4 Finite element modeling of interatomic interactions for large strains.

torsion, $\Delta\phi$. Second, point spring-like finite elements, hereafter called 3B and expressed in $(\hat{x},\hat{y},\hat{z})$, are utilized for the simulation of the bending moment variation due to the inter-bond angle change, $\Delta\theta$.

The element stiffness matrix may be generally written for both elements as

$$\mathbf{K}^{\mathrm{EL}} = \begin{bmatrix} \mathbf{k}^{\mathrm{EL}} & -\mathbf{k}^{\mathrm{EL}} \\ -\mathbf{k}^{\mathrm{EL}} & \mathbf{k}^{\mathrm{EL}} \end{bmatrix} \tag{12.14}$$

where EL = 2B, 3B.

Evidently, according to the utilized local coordinate systems, there is

$$\Delta r = \Delta\bar{x}; \quad \Delta\phi = \Delta\mathrm{rot}\bar{x}; \quad \Delta\theta = \Delta\mathrm{rot}\hat{z} \tag{12.15}$$

Given these aforementioned geometric equations and by using the modified Morse or Stillinger–Weber potential nonlinear functions U_r and U_θ given by Eq. (12.6) or Eq. (12.7), respectively, the following stiffness coefficient matrices arise:

$$\mathbf{k}^{2\mathrm{B}} = \begin{bmatrix} k_{\bar{x}}^{2\mathrm{B}} & 0 & 0 & 0 & 0 & 0 \\ 0 & k_{\bar{y}}^{2\mathrm{B}} & 0 & 0 & 0 & 0 \\ 0 & 0 & k_{\bar{z}}^{2\mathrm{B}} & 0 & 0 & 0 \\ 0 & 0 & 0 & k_{\mathrm{rot}\bar{x}}^{2\mathrm{B}} & 0 & 0 \\ 0 & 0 & 0 & 0 & k_{\mathrm{rot}\bar{y}}^{2\mathrm{B}} & 0 \\ 0 & 0 & 0 & 0 & 0 & k_{\mathrm{rot}\bar{z}}^{2\mathrm{B}} \end{bmatrix} = \begin{bmatrix} \dfrac{\mathrm{d}^2U_r}{\mathrm{d}(\Delta\bar{x})^2} & 0 & 0 & 0 & 0 & 0 \\ 0 & \infty & 0 & 0 & 0 & 0 \\ 0 & 0 & \infty & 0 & 0 & 0 \\ 0 & 0 & 0 & \dfrac{\mathrm{d}^2U_\tau}{\mathrm{d}(\Delta\mathrm{rot}\bar{x})^2} & 0 & 0 \\ 0 & 0 & 0 & 0 & \infty & 0 \\ 0 & 0 & 0 & 0 & 0 & \infty \end{bmatrix} \tag{12.16}$$

$$\mathbf{k}^{3\mathrm{B}} = \begin{bmatrix} k_{\hat{x}}^{3\mathrm{B}} & 0 & 0 & 0 & 0 & 0 \\ 0 & k_{\hat{y}}^{3\mathrm{B}} & 0 & 0 & 0 & 0 \\ 0 & 0 & k_{\hat{z}}^{3\mathrm{B}} & 0 & 0 & 0 \\ 0 & 0 & 0 & k_{\mathrm{rot}\hat{x}}^{3\mathrm{B}} & 0 & 0 \\ 0 & 0 & 0 & 0 & k_{\mathrm{rot}\hat{y}}^{3\mathrm{B}} & 0 \\ 0 & 0 & 0 & 0 & 0 & k_{\mathrm{rot}\hat{z}}^{3\mathrm{B}} \end{bmatrix} = \begin{bmatrix} \infty & 0 & 0 & 0 & 0 & 0 \\ 0 & \infty & 0 & 0 & 0 & 0 \\ 0 & 0 & \infty & 0 & 0 & 0 \\ 0 & 0 & 0 & 0 & 0 & 0 \\ 0 & 0 & 0 & 0 & 0 & 0 \\ 0 & 0 & 0 & 0 & 0 & \dfrac{\mathrm{d}^2U_\theta}{\mathrm{d}(\Delta\mathrm{rot}\hat{z})^2} \end{bmatrix} \tag{12.17}$$

The infinity symbol used in the matrices is used to define a very high stiffness, which simultaneously allows convergence of the solution. Furthermore, a simple approximation can be made for U_τ, such as the one given in Eq. (12.5), since the second derivative of the specific potential shows adequate stability even for higher strains. It should be noted that the high stiffness coefficients in Eq. (12.16) are required to keep the bonds straight during loading and thus prevent their bending, which, from the molecular point of view, is not allowed. Furthermore, the very high coefficients of stiffness in Eq. (12.17) are utilized to assure that all triple nodes within the nanotube remains rigid at all times (Fig. 12.4).

12.5 Governing Equations

The equations that are used to solve each case problem and predict the mechanical response of the nanotubes will be presented next. Their numerical treatment will be discussed as well in detail. First, the equations regarding the linear elastic mechanical behavior of the nanotubes are presented. Second, the buckling as well as free vibration equations are demonstrated. Finally, the nonlinear mechanical nanotube problem is explained.

12.5.1 Linear Mechanical Behavior

By applying the conventional displacement formulation, the element equilibrium equation of both kinds of e1 = 2b, 3b elements may be expressed as

$$\mathbf{K}^{\mathrm{el}}\mathbf{U}^{\mathrm{el}} = \mathbf{F}^{\mathrm{el}} \tag{12.18}$$

where the element stiffness matrix $\mathbf{K}^{\mathrm{el}}$ is given by Eqs. (12.9), (12.11), and (12.12), $\mathbf{U}^{\mathrm{el}}$ is the element displacement vector and $\mathbf{F}^{\mathrm{el}}$ is the element force vector.

Considering that i and j are the two nodes of the specific spring-like finite elements, their generalized displacement and force vectors, respectively, may be written as

$$\begin{gathered}\mathbf{U}^{2\mathrm{b}} = \begin{bmatrix} u_{\overline{xi}}^{2\mathrm{b}} & u_{\overline{yi}}^{2\mathrm{b}} & u_{\overline{zi}}^{2\mathrm{b}} & u_{\overline{xj}}^{2\mathrm{b}} & u_{\overline{yj}}^{2\mathrm{b}} & u_{\overline{yj}}^{2\mathrm{b}} \end{bmatrix}^{\mathrm{T}}; \\ \mathbf{U}^{3\mathrm{b}} = \begin{bmatrix} u_{\overline{xi}}^{3\mathrm{b}} & u_{\overline{yi}}^{3\mathrm{b}} & u_{\overline{zi}}^{3\mathrm{b}} & u_{\overline{xj}}^{3\mathrm{b}} & u_{\overline{yj}}^{3\mathrm{b}} & u_{\overline{zj}}^{3\mathrm{b}} \end{bmatrix}^{\mathrm{T}}\end{gathered} \tag{12.19}$$

$$\begin{gathered}\mathbf{F}^{2\mathrm{b}} = \begin{bmatrix} f_{\overline{xi}}^{2\mathrm{b}} & f_{\overline{yi}}^{2\mathrm{b}} & f_{\overline{zi}}^{2\mathrm{b}} & f_{\overline{xj}}^{2\mathrm{b}} & f_{\overline{yj}}^{2\mathrm{b}} & f_{\overline{yj}}^{2\mathrm{b}} \end{bmatrix}^{\mathrm{T}}; \\ \mathbf{F}^{3\mathrm{b}} = \begin{bmatrix} f_{\overline{xi}}^{3\mathrm{b}} & f_{\overline{yi}}^{3\mathrm{b}} & f_{\overline{zi}}^{3\mathrm{b}} & f_{\overline{xj}}^{3\mathrm{b}} & f_{\overline{yj}}^{3\mathrm{b}} & f_{\overline{zj}}^{3\mathrm{b}} \end{bmatrix}^{\mathrm{T}}\end{gathered} \tag{12.20}$$

To predict the elastic mechanical properties of the nanotubes, a linear static analysis should be carried out. First, all the elemental equilibrium equations for every element of the nanotube are transformed to the global coordinate system (x, y, z). By assembling them according to the requirements of nodal equilibrium, the following system of linear equations is derived:

$$\mathbf{KU} = \mathbf{F} \tag{12.21}$$

where **K**, **U**, and **F** are the assembled stiffness matrix, displacement vector, and force vector, respectively, of the nanomaterial. Eq. (12.21) can be solved by taking into consideration all the boundary conditions and utilizing standard numerical techniques, such as Gauss elimination.

12.5.2 Linear Buckling Behavior

The problem of linear buckling (stability) is solved by applying a reference level of loading $\mathbf{F}_{\mathrm{ref}}$ to the nanotube. A standard linear static analysis is then carried out to obtain loads, which are needed to form the geometric stiffness matrix $\mathbf{K}_G$. The stability loads are then calculated [12] by solving the following eigenvalue problem:

$$[\mathbf{K}-\lambda\mathbf{K}_G]\mathbf{U}=0 \tag{12.22}$$

where λ is the multiplier to the reference load. The vector **U** is the eigenvector corresponding to the eigenvalue. In the present analysis, the eigenvalue problem is solved by using the Lanczos method. Only a small number of the lowest eigenvalues are normally calculated for stability analysis. The critical load $\mathbf{F}_{cr}$ that causes instability to the nanotube is usually of higher interest. It is associated with the lowest eigenvalue λ_{cr} and may be calculated from the equation

$$\mathbf{F}_{cr}=\lambda_{cr}\mathbf{F}_{ref} \tag{12.23}$$

12.5.3 Free Vibrational Behavior

During the vibration analysis of a nanotube, to simulate inertial effects, a 3D point mass-like finite element should be introduced at every atom position. The specific element, denoted hereafter as 1b, has three degrees of freedom per node, while its mass matrix according to the global coordinate system (x, y, z) is

$$\mathbf{M}^{1b}=\begin{bmatrix}\mathbf{m}^{1b} & 0\\ 0 & 0\end{bmatrix};\quad \mathbf{m}^{1b}=\begin{bmatrix}m & 0 & 0\\ 0 & m & 0\\ 0 & 0 & m\end{bmatrix} \tag{12.24}$$

where m is the nucleus mass of the simulated atom. The mass parameter m is equal to 1.9943×10^{-26} kg, 1.79479×10^{-26} kg, and 2.3254×10^{-26} kg for the carbon, boron, and nitrogen atoms, respectively.

The equilibrium equation of the adopted mass-like elements with respect to the global coordinate system (x, y, z) may be expressed as

$$\mathbf{M}^{1b}\ddot{\mathbf{U}}^{1b}=\mathbf{F}^{1b} \tag{12.25}$$

where $\mathbf{M}^{1b}$, $\ddot{\mathbf{U}}^{1b}$, and $\mathbf{F}^{1b}$ are the elemental mass matrix, acceleration, and force vector, respectively [13]. After the assembly of the global stiffness **M** and global mass **K** matrices from the elemental matrices just presented, and considering undamped free vibration for the nanotubes, the equation of motion becomes

$$\mathbf{M}\ddot{\mathbf{U}}+\mathbf{K}\mathbf{U}=0 \tag{12.26}$$

where the dot denotes the derivative with respect to time.

By assuming a harmonic motion of angular frequency ω_m of the nanotube, the last equation may be rewritten as

$$[\mathbf{K}-\omega_m^2\mathbf{M}]\mathbf{U}=0 \tag{12.27}$$

Applying the support conditions of the nanotube, the eigenvalue problem can be solved using common finite element procedures. The solution of the eigenvalue problem reveals the natural frequency, that is, $f_m = \omega_m / 2\pi$, and the corresponding mode shape for the mth mode of vibration.

12.5.4 Nonlinear Mechanical Behavior

As previously mentioned, the study of the mechanical plastic behavior of a nanotube requires the adoption of the EL = 2B, 3B nonlinear finite elements which, evidently, have the following equilibrium equation:

$$\mathbf{K}^{EL}\mathbf{U}^{EL} = \mathbf{F}^{EL} \tag{12.28}$$

The nonlinear element stiffness matrix $\mathbf{K}^{EL}$ may now be defined via Eqs. (12.14), (12.16), and (12.17). Furthermore, assuming that I and J are the two nodes of these nonlinear finite elements, their elemental stiffness matrix, displacement $\mathbf{U}^{EL}$ vector, and force vector $\mathbf{F}^{EL}$, respectively, may be written as

$$\begin{aligned} \mathbf{U}^{2B} &= \left[\, u^{2B}_{\bar{x}I} \; u^{2B}_{\bar{y}I} \; u^{2B}_{\bar{z}I} \; u^{2B}_{\text{rot}\bar{x}I} \; u^{2B}_{\text{rot}\bar{y}I} \; u^{2B}_{\text{rot}\bar{z}I} \; u^{2B}_{\bar{x}J} \; u^{2B}_{\bar{y}J} \; u^{2B}_{\bar{z}J} \; u^{2B}_{\text{rot}\bar{x}J} \; u^{2B}_{\text{rot}\bar{y}J} \; u^{2B}_{\text{rot}\bar{z}J} \,\right]^T; \\ \mathbf{U}^{3B} &= \left[\, u^{3B}_{\hat{x}I} \; u^{3B}_{\hat{y}I} \; u^{3B}_{\hat{z}I} \; u^{3B}_{\text{rot}\hat{x}I} \; u^{3B}_{\text{rot}\hat{y}I} \; u^{3B}_{\text{rot}\hat{z}I} \; u^{3B}_{\hat{x}J} \; u^{3B}_{\hat{y}J} \; u^{3B}_{\hat{z}J} \; u^{3B}_{\text{rot}\hat{x}J} \; u^{3B}_{\text{rot}\hat{y}J} \; u^{3B}_{\text{rot}\hat{z}J} \,\right]^T \end{aligned} \tag{12.29}$$

$$\begin{aligned} \mathbf{F}^{2B} &= \left[\, f^{2B}_{\bar{x}I} \; f^{2B}_{\bar{y}I} \; f^{2B}_{\bar{z}I} \; f^{2B}_{\text{rot}\bar{x}I} \; f^{2B}_{\text{rot}\bar{y}I} \; f^{2B}_{\text{rot}\bar{z}I} \; f^{2B}_{\bar{x}J} \; f^{2B}_{\bar{y}J} \; f^{2B}_{\bar{z}J} \; f^{2B}_{\text{rot}\bar{x}J} \; f^{2B}_{\text{rot}\bar{y}J} \; f^{2B}_{\text{rot}\bar{z}J} \,\right]^T; \\ \mathbf{F}^{3B} &= \left[\, f^{3B}_{\hat{x}I} \; f^{3B}_{\hat{y}I} \; f^{3B}_{\hat{z}I} \; f^{3B}_{\text{rot}\hat{x}I} \; f^{3B}_{\text{rot}\hat{y}I} \; f^{3B}_{\text{rot}\hat{z}I} \; f^{3B}_{\hat{x}J} \; f^{3B}_{\hat{y}J} \; f^{3B}_{\hat{z}J} \; f^{3B}_{\text{rot}\hat{x}J} \; f^{3B}_{\text{rot}\hat{y}J} \; f^{3B}_{\text{rot}\hat{z}J} \,\right]^T \end{aligned} \tag{12.30}$$

After discretizing the given nanotube geometry, the elemental stiffness equations for every 2B and 3B finite elements are transformed to the global coordinate system. Then the final system of equations may be constructed according to the requirements of nodal equilibrium as follows:

$$\mathbf{K}(\mathbf{U})\mathbf{U} = \mathbf{F} \tag{12.31}$$

where $\mathbf{K}(\mathbf{U})$ is the assembled deformation-dependent stiffness matrix.

By applying the known boundary conditions of the problem, that is, supports and loadings, the last equation may be solved via standard Newton–Raphson iterative procedures, which are required to treat the inherent nonlinearity of the global stiffness matrix $\mathbf{K}(\mathbf{U})$ due to the terms $d^2U_r / d(\Delta\bar{x})^2$ of Eq. (12.16) and $d^2U_\theta / d(\Delta\text{rot}\hat{z})^2$ of Eq. (12.17) arising from the selected potential field described by Eq. (12.6) or Eq. (12.7).

12.6 Results

First, five armchair and five zigzag SWCNTs as well as SWBNNTs of a length $L \approx 5D$ are tested under a small tensile static load, which causes only elastic deformations. The geometry and the finite element mesh utilized for an armchair and a zigzag SWCNT is illustrated in Fig. 12.5A–B, respectively.

The effective Young's modulus E of the SWCNTs and SWBNNTs as a function of their diameter is plotted in Fig. 12.6A–B, respectively. The computations correspond to an SWCNT

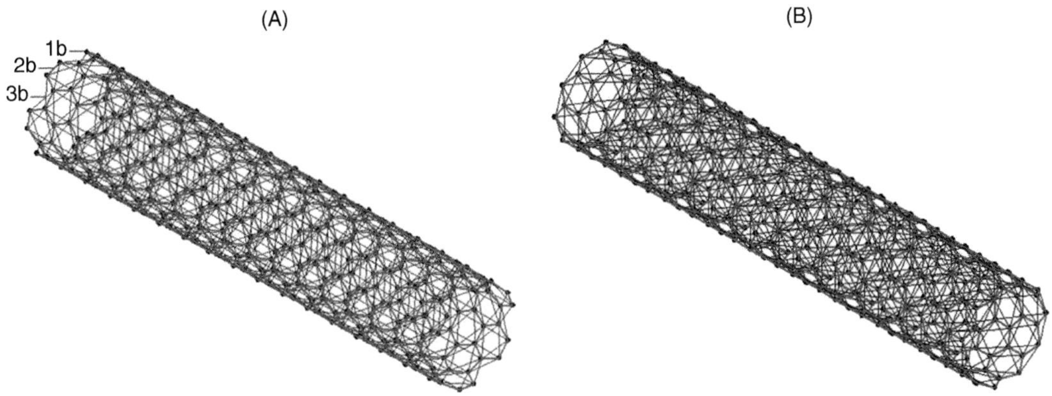

FIGURE 12.5 Utilized finite element model for (A) the (6,6) armchair and (B) the (10,0) zigzag SWNTs with an aspect ratio of 5.

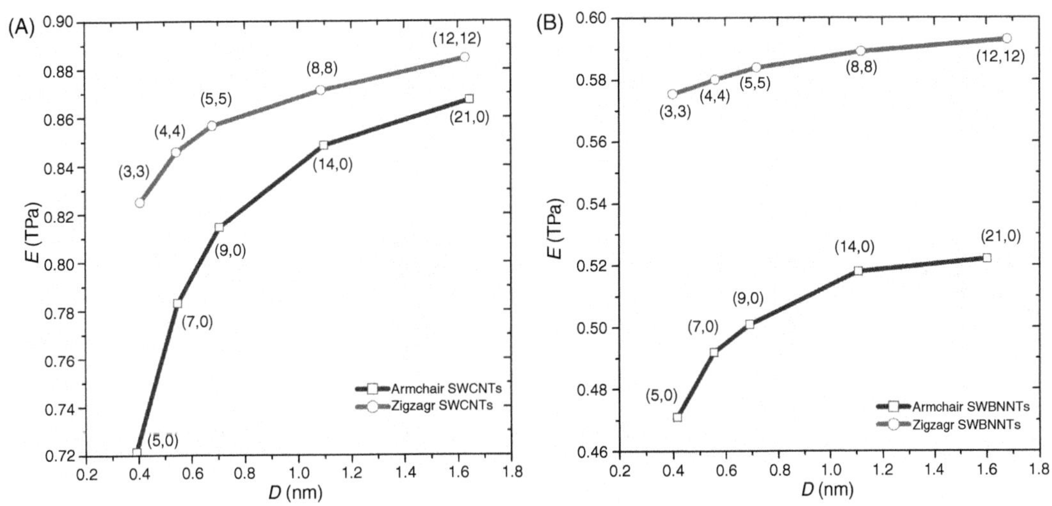

FIGURE 12.6 Effective Young's modulus of (A) SWCNTs and (B) SWBNNTs with respect to their diameter.

thickness of 0.34 nm and an SWBNNT thickness of $t = 0.333$ nm [14]. Armchair nanotubes seem to have a slightly higher Young's modulus in comparison to the zigzag ones of the same diameter. Moreover, obviously, the SWCNTs are stiffer than SWBNNTs regarding the axial direction.

Table 12.1 presents some average predictions of the proposed formulation concerning effective Young's modulus in contrast with relevant average results arising via different theoretical and experimental methods [15–20]. Good performance of the proposed method may be observed.

To validate the proposed method for the prediction of the critical compressive loadings before the buckling of an armchair and a zigzag SWCNT, Fig. 12.7A–B are presented, respectively. In this figure, the results from the present method are compared with other

Table 12.1 Comparisons Between the Present Method and Other Approaches Regarding the Effective Stiffness of Nanotube

Method	Type of Nanotube	*E* (TPa)
Experimental [15]	SWCNT	1.2000
MM (beam based) [16]	SWCNT	1.0300
MM (beam based) [17]	SWCNT	1.0293
Present	SWCNT	0.831
Experimental [18]	MWBNNT	0.550
Density functional theory [19]	SWBNNT	0.788
MM (beam based) [20]	SWBNNT	0.768
Present	SWBNNT	0.546

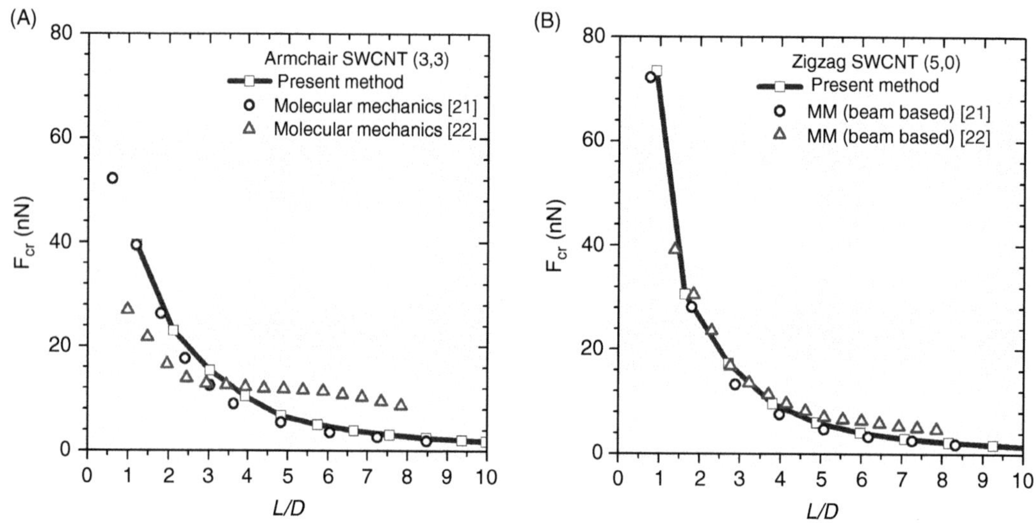

FIGURE 12.7 Compressive critical instability force versus aspect ratio for (A) the (3,3) armchair and (B) the (5,0) zigzag SWCNT.

available results from the literature, which have been computed by using beam-based molecular mechanics methods [21,22]. Good agreement between the three different numerical methods may be observed.

Next, the performance of the proposed numerical technique is investigated for the problem of the free vibration of nanotubes. Fig. 12.8A–B illustrate the nanotube diameter dependency of the three first natural frequencies of doubly clamped SWCNTs and SWBNNTs of an aspect ratio almost equal to 10. As shown, especially for smaller diameters, the armchair nanotubes present slightly higher natural frequencies that reduce as their diameter is increased. Finally, the SWCNTs present a higher fundamental frequency than SWBNNTs.

Fig. 12.9 illustrates the first three mode shapes of armchair and zigzag SWCNTs. Bending as well as axial modes of vibration are mainly observed. The first mode of

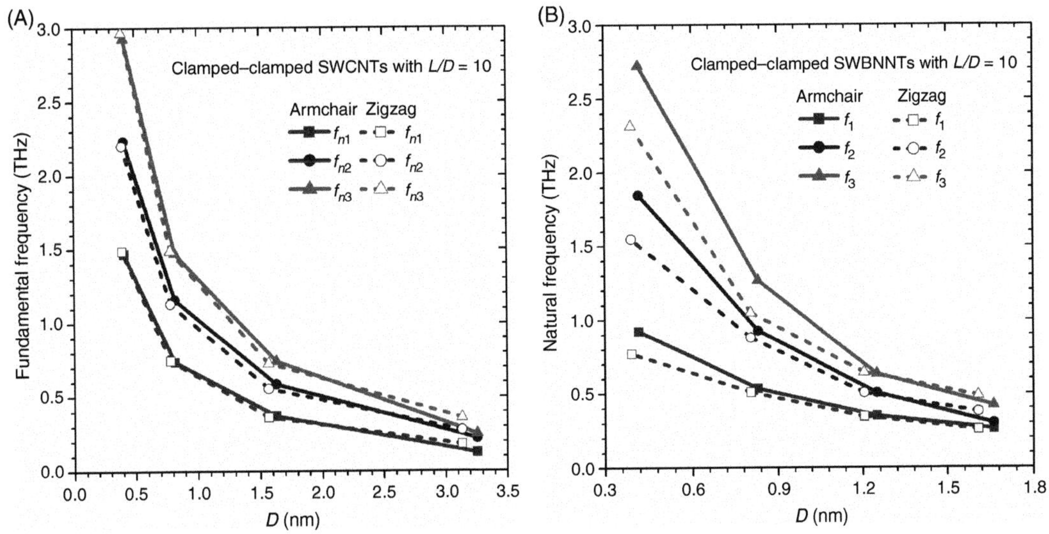

FIGURE 12.8 First three natural frequencies of doubly clamped (A) SWCNTs and (B) SWBNNTs with respect to their diameter.

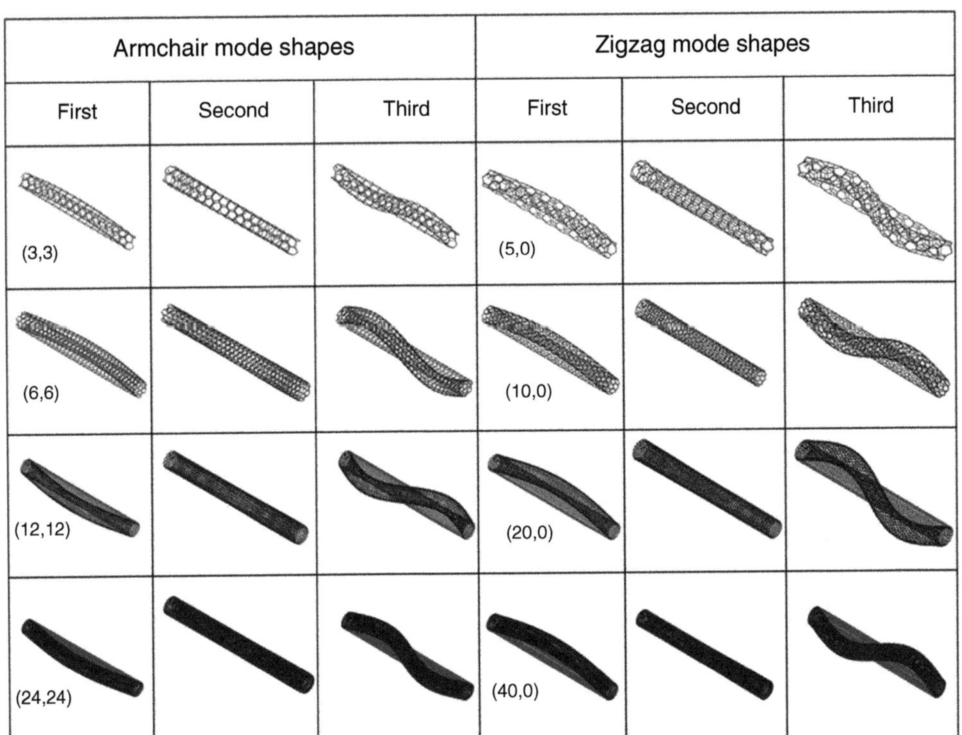

FIGURE 12.9 First three mode shapes of the SWCNTs for an aspect ratio of 10 and for the clamped–clamped edge support condition.

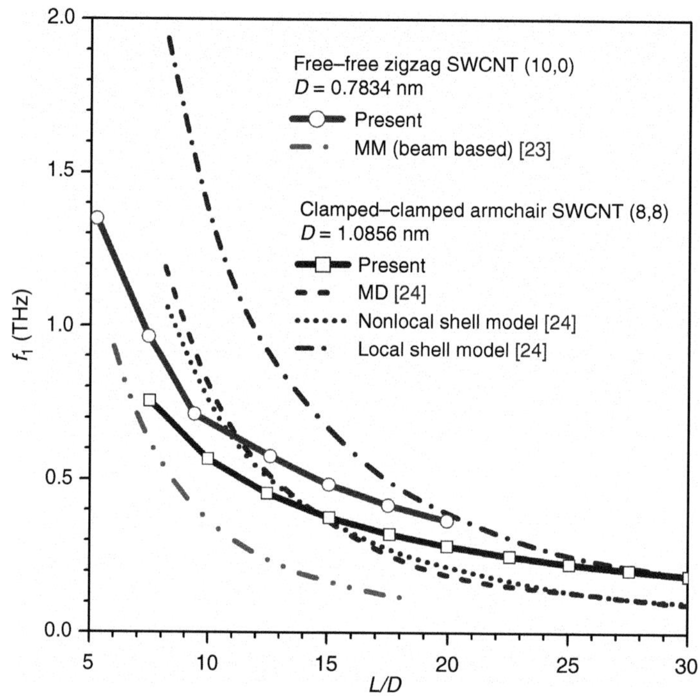

FIGURE 12.10 Fundamental natural frequency versus aspect ratio of SWCNTs calculated by the present MM method in contrast with other corresponding numerical predictions.

vibration is usually of the bending type while the second mode of vibration is of the axial type.

Fig. 12.10 is presented to illustrate the accuracy of the proposed finite element MM method. The specific figure illustrates the variation of the fundamental natural frequency with respect to the aspect ratio of two different SWCNTs, in terms of chirality and diameter, for two boundary condition cases.

In addition, to validate the proposed method concerning the vibrational behavior of SWBNNTs, Table 12.2 is presented, which includes a variety of comparisons concerning theoretical estimations. Good agreement between the present method and other theoretical numerical techniques may be observed.

Finally, the nonlinear version of the proposed formulation is tested by modeling the zigzag (20,0) SWCNT according to the modified Morse potential [Eq. (12.6)] and the zigzag (20,0) SWBNNT by utilizing, however, the Stillinger–Weber potential functions [Eq. (12.7)]. The nonlinear finite element model of the SWNTs is presented in Fig. 12.11. The thickness of the SWCNT and SWBNNT is typically taken equal to 0.34 and 0.333 nm, respectively. For both nanotubes an aspect ratio of about five is selected.

Table 12.2 Comparison of the Present Method With Other Predictions Regarding the Fundamental Frequency of SWBNNTs

Method	SWBNNT	*D* (nm)	*L/D* ≈	Support	f_1 (THz)
CM (analytical) [23]	—	0.800	10	Fixed-free	0.263
CM (FEM) [23]	—	0.800	10	Fixed-free	0.257
Present	(6,6)	0.831	10	Fixed-free	0.268
Present	(10,0)	0.799	10	Fixed-free	0.242
CM (analytical) [24]	(9,9)	1.246	10	Fixed-free	0.225
			15		0.101
			20		0.056
MM (beam based) [24]	(9,9)	1.246	10	Fixed-free	0.247
			15		0.093
			20		0.050
Present	(9,9)	1.246	10	Fixed-free	0.170
			15		0.116
			20		0.087
MD (400K temp.) [25]	(10,10)	1.385	5	Fixed-fixed	0.500
Present	(10,10)	1.381	5	Fixed-fixed	0.546
MD [26]	(5,5)	0.692	5	Fixed-fixed	1.735
Present	(5,5)	0.692	5	Fixed-fixed	1.269

CM, Continuum mechanics; MD, molecular dynamics; MM, molecular mechanics.

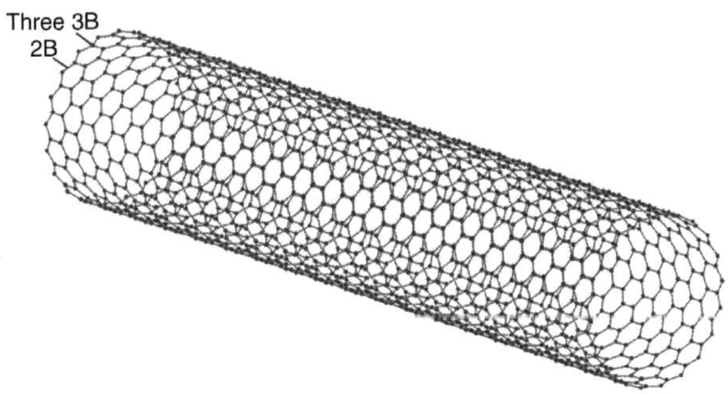

FIGURE 12.11 Finite element nonlinear model for the (20,0) SWNT.

The computed stress–strain behaviors of SWCNT and SWBNNT are illustrated in Fig. 12.12. The Young's modulus of the (20, 0) SWCNT is found to be equal to 936 GPa, its fracture occurs at a tensile stress of 98.4 GPa, and at a failure strain of 13.2%. On the other hand, the SWBNNT seems to be less brittle since it has a higher failure strain of about 21.5%. However, its strength as well as effective Young's modulus are found to be lower than those observed for the same-shaped SWCNT. The figure includes comparisons associated with the SWCNT, which prove the good agreement of the computed results with other theoretical solutions available in the literature.

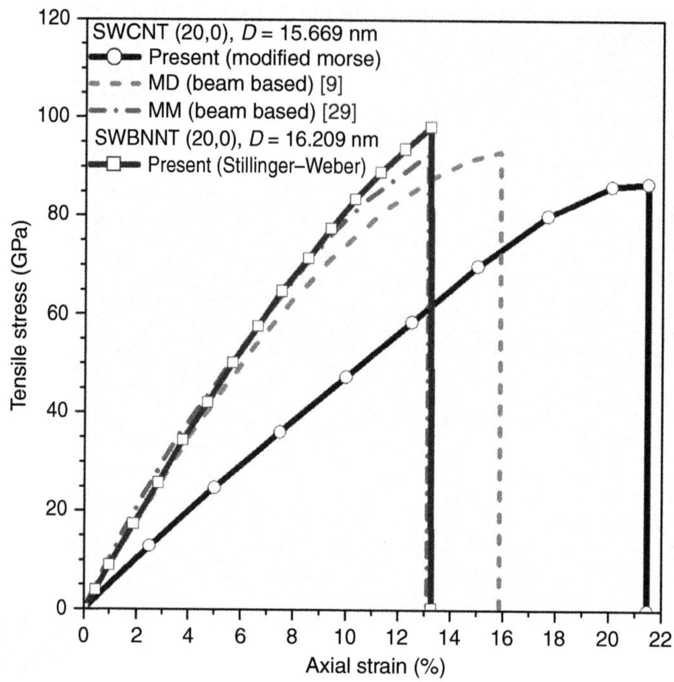

FIGURE 12.12 Tensile stress versus axial strain of the (20,0) SWCNT as well as the (20,0) SWBNNT.

12.7 Conclusions

In this chapter, a comprehensive 3D finite element method based on the use of linear or nonlinear spring-like and mass-like elements has been formulated for the investigation of the elastic and dynamic characteristics of nanotubes. A more efficient finite element modeling of a lower computational cost has been proposed for the linear elastic problems. The good accuracy of the adopted numerical scheme has been demonstrated for the linear and nonlinear problems via indicative comparisons with available data in the literature. The developed computational technique can be applied for the accurate simulation of the elastic as well as plastic mechanical response of various types of nanotubes. Properties, such as Young's modulus, tensile strength, failure strain, and critical and natural frequencies for a variety of boundary conditions and nanotube shapes may be estimated efficiently via the proposed numerical tool. The methodology presented may incorporate a variety of potential field functions without increasing the computational complexity.

References

[1] S. Iijima, Helical microtubules of graphitic carbon, Nature 354 (1991) 56–58.

[2] C. Lee, X. Wei, J.W. Kysar, J. Hone, Measurement of the elastic properties and intrinsic strength of monolayer graphene, Science 321 (5887) (2008) 385–388.

[3] K.S. Novoselov, D. Jiang, F. Schedin, T.J. Booth, V.V. Khotkevich, S.V. Morozov, A.K. Geim, Two-dimensional atomic crystals, Proc. Natl. Acad. Sci. USA 102 (30) (2005) 10451–10453.

[4] E.T. Thostenson, Z. Ren, T.W. Chou, Advances in the science and technology of carbon nanotubes and their composites: a review, Compos. Sci. Technol. 61 (2001) 1899–1912.

[5] M.S. Dresselhaus, G. Dresselhaus, R. Saito, Physics of carbon nanotubes, Carbon 33 (1995) 883–891.

[6] A.K. Rappe, C.J. Casewit, K.S. Colwell, W.A. Goddard, W.M. Skiff, UFF, a full periodic table force field for molecular mechanics and molecular dynamics simulations, J. Am. Chem. Soc. 114 (1992) 10024–10035.

[7] W.D. Cornell, P. Cieplak, C.I. Bayly, I.R. Gould, K.M. Merz, D.M. Ferguson, D.C. Spellmeyer, T. Fox, J.W. Caldwell, P.A. Kollman, A second generation force field for the simulation of proteins, nucleic acids, and organic molecules, J. Am. Chem. Soc. 117 (19) (1995) 5179–5197.

[8] L. Boldrin, F. Scarpa, R. Chowdhury, S. Adhikari, Effective mechanical properties of hexagonal boron nitride nanosheets, Nanotechnology 22 (50) (2011) 1–7 505702.

[9] T. Belytschko, S.P. Xiao, G.C. Schatz, R.S. Ruoff, Atomistic simulations of nanotube fracture, Phys. Rev. B 65 (2002) 235430.

[10] W.H. Moon, M.S. Son, H.J. Hwang, Theoretical study on structure of boron nitride fullerenes, Appl. Surf. Sci. 253 (2007) 7078–7081.

[11] C. Li, T.-W. Chou, Vibrational behaviors of multiwalled-carbon-nanotube-based nanomechanical resonators, Appl. Phys. Lett. 84 (1) (2004) 121–123.

[12] G.I. Giannopoulos, A.P. Tsiros, S.K. Georgantzinos, Prediction of elastic mechanical behavior and stability of single-walled carbon nanotubes using bar elements, Mech. Adv. Mater. Struct. 20 (9) (2013) 730–741.

[13] S.K. Georgantzinos, N.K. Anifantis, Carbon nanotube-based resonant nanomechanical sensors: a computational investigation of their behavior, Physica E 42 (2010) 1795–1801.

[14] O. Hod, Graphite and hexagonal boron-nitride have the same interlayer distance. Why?, J. Chem. Theory Comput. 8 (4) (2012) 1360–1369.

[15] W. Tombler, C. Zhou, J. Kong, H. Dai, L. Liu, C.S. Jayanthi, M. Tang, S.Y. Wu, Reversible electromechanical characteristics of carbon nanotubes under local probe manipulation, Nature 405 (2000) 769–772.

[16] C.W.S. To, Bending and shear moduli of single-walled carbon nanotubes, Finite Elem. Anal. Des. 42 (2006) 404–413.

[17] K.I. Tsepres, P. Papanikos, Finite element modeling of single-walled carbon nanotubes, Compos. B 36 (2005) 468–477.

[18] D. Golberg, P.M.F.J. Costa, O. Lourie, M. Mitome, X. Bai, K. Kurashima, C. Zhi, C. Tang, Y. Bando, Direct force measurements and kinking under elastic deformation of individual multiwalled boron nitride nanotubes, Nano Lett. 7 (7) (2007) 2146–2151.

[19] B. Baumeier, P. Krüger, J. Pollmann, Structural, elastic, and electronic properties of SiC, BN, and BeO nanotubes, Phys. Rev. B: Condens. Matter Mater. Phys. 76 (8) (2007) 1–10 085407.

[20] L. Jiang, W. Guo, A molecular mechanics study on size-dependent elastic properties of single-walled boron nitride nanotubes, J. Mech. Phys. Solids 59 (6) (2011) 1204–1213.

[21] N. Hu, K. Nunoya, D. Pan, T. Okabe, H. Fukunaga, Prediction of buckling characteristics of carbon nanotubes, Int. J. Solids Struct. 44 (2007) 6535–6550.

[22] C. Li, T.-W. Chou, Modeling of elastic buckling of carbon nanotubes by molecular structural mechanics approach, Mech. Mater. 36 (2004) 1047–1055.

[23] M.B. Panchal, S.H. Upadhyay, S.P. Harsha, Vibration analysis of single walled boron nitride nanotube based nanoresonators, J. Nanotechnol. Eng. Med. 3 (3) (2012) 1–5 031004.

[24] M.B. Panchal, S.H. Upadhyay, S.P. Harsha, An efficient finite element model for analysis of single walled boron nitride nanotube-based resonant nanomechanical sensors, Nano 8 (1) (2013) 1–16 1350011.

[25] A. Chandra, P.K. Patra, B. Bhattacharya, Thermal vibration characteristics of armchair boron-nitride nanotubes, J. Appl. Phys. 118 (23) (2015) 234503.

[26] R. Ansari, S. Ajori, A molecular dynamics study on the vibration of carbon and boron nitride double-walled hybrid nanotubes, Appl. Phys. A: Mater. Sci. Process. 120 (4) (2015) 1399–1406.

Further Reading

R. Ansari, H. Rouhi, R. Rajabiehfard, Free vibration analysis of single-walled carbon nanotubes using semi-analytical finite element, Int. J. Comput. Methods Eng. Sci. Mech. 13 (2012) 202–209.

R. Chowdhury, S. Adhikari, C.Y. Wang, F. Scarpa, A molecular mechanics approach for the vibration of single-walled carbon nanotubes, Comput. Mater. Sci. 48 (4) (2010) 730–735.

K.I. Tserpes, P. Papanikos, S.A. Tsirkas, A progressive fracture model for carbon nanotubes, Composites, Part B 37 (7–8) (2006) 662–669.

13

Multiscale Simulation of Impact Response of Carbon Nanotube/ Polymer Nanocomposites

Asimina K. Manta, Konstantinos I. Tserpes
UNIVERSITY OF PATRAS, PATRAS, GREECE

CHAPTER OUTLINE

Carbon Nanotube–Reinforced Polymers. http://dx.doi.org/10.1016/B978-0-323-48221-9.00013-3

13.1 Introduction

Nowadays, carbon fiber reinforced plastics (CFRPs) are widely used in more and more structures that need to be lightweight and enduring, especially in aeronautics–aerospace engineering and automotive applications. For instance, Airbus has achieved the manufacture of structures with the total incorporation of CFRP reaching 52%. One of the most important characteristics of these composites is their failure mechanisms, especially under dynamic loading. The main failure modes could be summarized to the cracking of the matrix, the carbon fiber debonding, and the delamination of the CFRP plies, finally leading to severe material degradation and catastrophic failure. It can be easily noted that these damage modes are mainly controlled by the material behavior of the polymeric matrix. To enhance the mechanical response of the polymer under dynamic loading, intense research efforts have been made in the last decade. One of the most promising methodologies is the reinforcement of the matrix with nanoparticles, such as carbon nanotubes (CNTs).

The findings on the enhancement of the static and fatigue properties of polymeric materials owing to CNT addition are already very promising [1,2]. However, a more detailed work is still to be done for the impact properties, which are closely related to fracture toughness characteristics.

Experimental works have reported a considerable enhancement of the impact behavior of nanocomposites and modified fiber composites and also the effect of several geometrical and material parameters, such as CNT content, CNT properties (functionalized, nonfunctionalized, etc.), dispersion, and interface properties [3–9]. The study on the detailed effects of these parameters, which is the first step toward optimization of CNT-reinforced materials, could be performed either by experimental iterations, which is a very cost- and time-consuming process, or by simulation.

Only a few works on the simulation of the impact behavior of CNT/polymer nanocomposites have been published [10–14]. Rafiee and Moghadam [10] employed a multiscale finite element (FE) model consisting of a single-walled CNT, the nonbonded interphase region, and the surrounding matrix to study the behaviors of nanocomposites under high strain rate loading in axial and transverse directions. The results revealed that even a small portion of CNTs improve the dynamic behaviors of the polymer against impact loading in the presence of nonbonded interactions in the interphase region. Khalili and Haghbin [11] have investigated the design parameters of nanocomposites under dynamic loads. The influence of diameter, chirality, and volume fraction of CNTs on representative volume elements (RVEs) of nanocomposites subjected to a tensile pulse have been considered. The results in the form of strain energy density and axial strain oscillations demonstrated a great improvement in the impact behavior of nanocomposites. It was concluded that the design procedure of nanocomposites subject to dynamic loads should be concentrated on the size and the volume fraction of CNTs instead of their chirality. Jama and Kiani [12] studied analytically the response of a functionally graded CNT-reinforced composite beam subjected to the action of an impacting mass. They performed parametric studies to examine the influences of thermal environment, the volume fraction of the CNTs, distribution of CNTs, the initial velocity of the impactor, and the impactor mass. Numerical results indicated that as the volume fraction of CNTs increased in the beam, peak contact force increased and the contact time decreased. Furthermore, it was found that temperature rise resulted in higher contact time duration and lower peak contact force. On the same modeling approach of a beam system, Papadopoulos and Tavlaki [13] illustrated the combined effect of interfacial shear strength and CNT weight fraction on the behavior of a CNT/polymer cantilever beam through a three-level multiscale technique. The results of the analyses showed that increasing CNT content could improve the damping properties of the composite structure, under the condition that the interfacial shear strength was low enough. Finally, Pandya and Naik [14] presented an analytical formulation for the prediction of ballistic impact behavior of multiwalled CNT (MWCNT) dispersed epoxy resin. The authors compared their analytical results with experimental ones, which were in accordance with the typical observations regarding the ballistic limit velocity and the energy absorbed by various mechanisms.

The aforementioned experimental and theoretical studies indicated that the presence of CNTs could improve the impact behavior of polymers, although there are many geometrical and material parameters whose effects have not been fully investigated and might trigger further enhancement of the polymers. Therefore, there is a need for reliable models capable of characterizing the effect of such parameters. In this chapter, a two-scale numerical model is developed for performing a detailed study on the effect of geometrical and material parameters on the low-velocity impact response of CNT/polymer nanocomposites. First, a unit cell including the carbon nanotube, the matrix, and the interphase developed between the carbon nanotube and the matrix was created to study the effect of the carbon nanotube volume fractions and geometrical characteristics, as well as the properties of the interphase. Then, under the assumption that the load–displacement curve obtained could be analyzed in accordance with the nanoindentation test theory, the material response of the unit cell was transferred to the scale of the specimen in terms of bilinear elastic–plastic properties. Finally, the specimen was loaded as stated in the ASTM D5628-10 [15] and the full energy dissipation and the reaction loads were calculated.

13.2 The Multiscale Approach

In the classical multiscale approach, simulations are performed at low scales (nano and micro) using either RVEs or representative unit cells to derive properties governed by material characteristics at these scales. The derived properties are assigned to regions of a meso or macro model with the use of which the behavior of the coupon or the structural element is finally simulated. For example, in the works of Tserpes and Chanteli [16,17], to derive the Young's modulus of a CNT/polymer nanocomposite, the elastic modulus of an RVE was derived as a function of geometrical and material parameters and then properly assigned into regions of the tension specimen. In the impact problem, the parameters needed as input in the macroscale cannot be easily derived from the low-scale simulation on the RVE. Consequently, in the present work, to study the impact response of the nanocomposite, the nanoindentation test was modeled because, for the specific simulation, the input parameters needed may be derived, based on the assumption of bilinear response of the polymeric matrix, from the load–deflection curve of the RVE and the use of nanoindentation theory.

13.3 RVEs

13.3.1 Geometry

The geometry of the RVE is that of a prismatic body with a rectangular cross-section. The RVE is comprised of the polymeric matrix, the MWCNT, and the interphase between them. The width of the RVE equals to

$$W_{RVE} = \sqrt{\frac{L_{CNT} \cdot A_{CNT}}{V_f \cdot L_{RVE}}} \tag{13.1}$$

where L_{CNT} is the CNT's length, A_{CNT} is the CNT's cross-section, L_{RVE} is the RVE's length, and V_f is the CNT's volume fraction. The CNT is seven-walled with a diameter of 50 nm. The thickness of each wall is 0.34 nm. The CNT is internal and did not span throughout the RVE's length. The CNT and the interphase were modeled as hollow cylinders. The thickness of the interphase is a parameter of the study. For impacting the RVE, a spherical impactor with a diameter of 25 nm was used. Fig. 13.1 illustrates the geometry of the RVE.

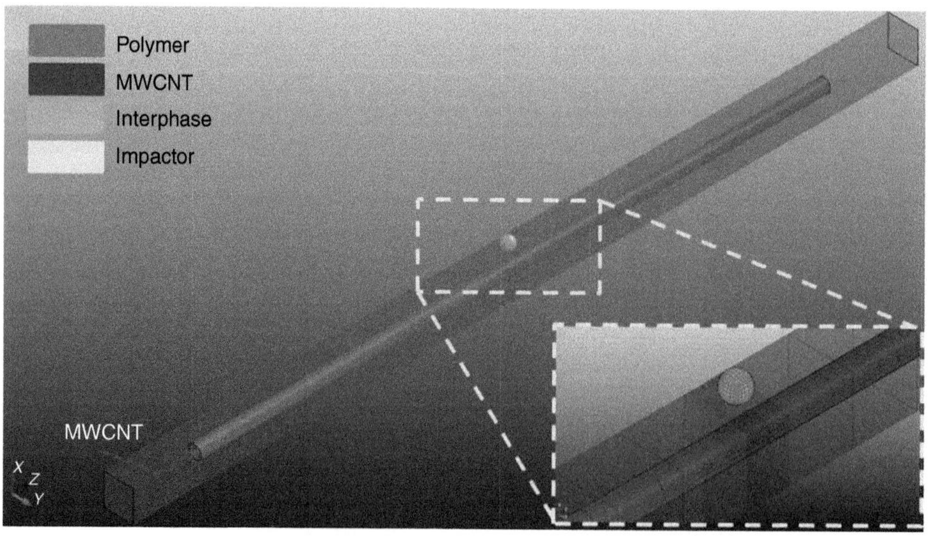

FIGURE 13.1 Schematic representation of the RVE. *MWCNT,* Multiwalled carbon nanotubes.

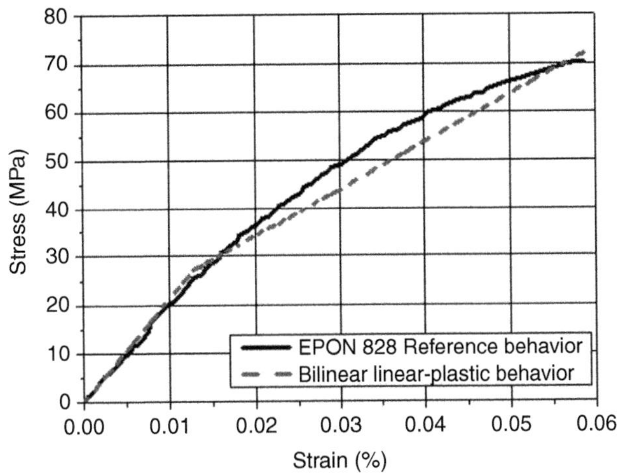

FIGURE 13.2 Reference and approximate stress–strain curves of the polymer.

13.3.2 Materials

13.3.2.1 Matrix

For the matrix, the EPON 828 resin has been used. This is a thermosetting material with a nonlinear tensile behavior, which has been simulated by the bilinear linear-plastic material model presented in Fig. 13.2. Polymeric materials exhibit viscoelastic–viscoplastic behavior as a reaction of the polymer chains to retain their mobility. However, knowing that the viscoplastic properties are strongly dependent on impact testing velocity, temperature, testing procedure, as well as type of applied load (tension, compression, bending, torsion, etc.), in literature, there were no available sufficient experimental data to characterize the viscoplastic behavior of the used resin. As a consequence, it was decided that the viscoplastic behavior of the matrix be neglected based on the assumption that at low impact velocities the effect of viscous behavior could be considered minor. Having assumed that the polymer possesses an elastic–plastic behavior, the most reliable material model to account for this behavior is the bilinear one since it is well established in FE codes and shows fewer complications on the solution of explicit codes. The material parameters of the bilinear linear-plastic material model used are listed in Table 13.1.

Table 13.1 Material Properties of the EPON 828 Resin

Young's modulus (MPa)	2122
Yield stress (MPa)	27.22
Poisson's ratio (—)	0.38
Tangent modulus of plastic region (MPa)	977
Plastic strain at fracture (—)	0.04589
Density (g/cm^3)	1.16

Table 13.2 Isotropic and Orthotropic Materials Properties of the MWCNT

Density	2.6
Isotropic properties	
Young's modulus (GPa)	1000
Poisson's ratio (—)	0.068
Orthotropic properties	
Axial Young's modulus (GPa)	1000
Transverse Young's modulus (GPa)	30
Shear moduli (GPa)	30
Poisson's ratio (—)	0.068

13.3.2.2 MWCNT

For the MWCNT, orthotropic material properties were used. Fracture of the MWCNT was not considered in the simulation since for the applied impact velocities the developed stresses are well beyond the threshold values. The material properties of the MWCNT are listed in Table 13.2.

13.3.2.3 Interphase

For the CNT/polymer interphase, a linear elastic behavior was adopted. The interphase properties were assumed to vary exponentially through the interphase thickness as described in Fig. 13.3. By assuming a continuity of the properties between the phases of the nanocomposite, at the polymer/interphase boundary the interphase properties are equal to the properties of the polymer, while at the interphase/CNT boundary they are equal to the CNT properties (Table 13.3). Between the two boundaries, the interphase property y follows the exponential distribution described by

$$y(t)=Y_{\mathrm{f}}\exp\left[\frac{\ln(Y_{\mathrm{m}})-\ln(Y_{\mathrm{f}})}{t_{\mathrm{int}}}t\right] \quad (13.2)$$

where Y_{f} is the property of the CNT, Y_{m} is the property of the matrix, $y(t)$ is the through thickness variable, and t_{int} is the interphase thickness. For the sake of computational

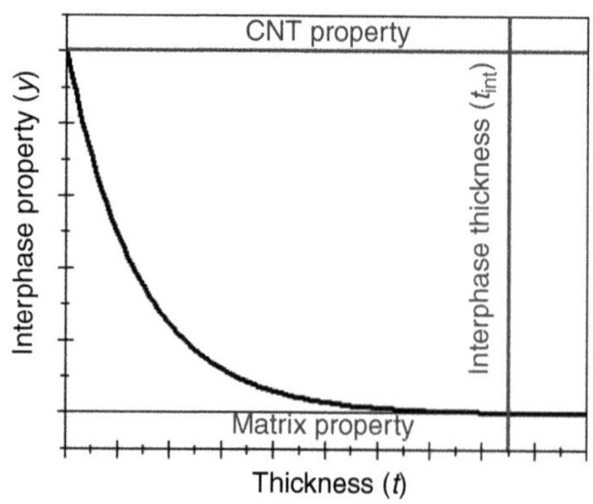

FIGURE 13.3 Exponential variation of interphase properties. *CNT,* Carbon nanotube.

Table 13.3 Material Properties of the Impactor

Young's modulus	E (GPa)	207
Poisson's ratio	ν (—)	0.3
Density	ρ (g/cm³)	7.83

simplicity, the equivalent interphase properties Y_{eq}, which are equal to the area under the curve y–t, were finally used in the RVE. The area under the curves is derived as

$$Y_{eq}t_{int} = \int_0^{t_{int}} y\,dt \tag{13.3}$$

Combining Eqs. (13.2) and (13.3), we get

$$Y_{eq}t_{int} = \int_0^{t_{int}} Y_f \exp\left[\frac{\ln(Y_m)-\ln(Y_f)}{t_{int}}t\right]dt \tag{13.4}$$

from which Y_{eq} is derived as

$$Y_{eq} = \frac{Y_m - Y_f}{\ln(Y_m)-\ln(Y_f)} \tag{13.5}$$

The equivalent interphase properties derived using Eq. (13.5) are listed in Table 13.4. Aiming to take into account the effect of the imperfect interphase, the interphase properties were reduced by a factor of Q, which takes the values of 0.25, 0.5, and 0.75.

Table 13.4 Summary Table of Elements' Mechanical Properties Representing MWCNT-Reinforced Polymer Regions

Interphase Thickness (nm)	Interphase Quality (%)	V_f	Impact Velocity (m/s)	Density ρ (g/cm³)	Elastic Modulus Under Impact E^* (MPa)	Poisson's Ratio	Yield Stress Under Impact (MPa)	Tangent Modulus in Plastic Region ETAN (MPa)
0.68	25	0.005	1	1.1672	444	0.37844	1.317	410
			5		1186		1.337	317
			10		538.8		1.354	155
		0.01	1	1.1744	532	0.37688	1.909	155
			5		1878		1.324	1348
			10		394		0.31099	258.85
		0.015	1	1.1816	500.44	0.37532	1.767	244.72
			5		1767		1.115	1222
			10		785.46		0.2307	217
	75	0.005	1	1.1672	643.9	0.37844	1.09428	222.5
			5		1096		1.173	782
			10		466.5		0.509	65.45
		0.01	1	1.1744	493	0.37688	1.682	139.5
			5		1768		1.079	1254
			10		432.6		0.308	258
		0.015	1	1.1816	487.46	0.37532	1.663	224.24
			5		1719		1.091	1211
			10		599		0.219	229.98

13.3.3 FE Models and Boundary Conditions

The FE model of the RVE was developed using the commercial explicit LS-DYNA FE code [18]. The RVE was modeled using the hexahedron SOLID ELFORM = 1 element. The specific element is characterized by an underintegrated constant stress, efficiency, accuracy, and usability under severe deformation, which makes it appropriate for impact problems, but needs hourglass stabilization. To stabilize the hourglass in the present analyses, the energy control (*CONTROL_HOURGLASS) was activated by setting the hourglass coefficient equal to 0.1. The contact between the RVEs and the impactor was modeled using the automatic surface-to-surface contact with the pinball segment-based contact being activated (SOFT = 2). The soft constraint is necessary when the difference between Young's moduli of the materials in contact is large. Fig. 13.4 shows a typical FE mesh of the RVE.

To comply with the standards of impact testing of plastics (ASTM D5628 [15]), all faces of the RVE, except for the impacted face and its opposite, were fully restrained. Three impact velocities, all of them being in the low-velocity range, were considered, namely, 1, 5, and 10 m/s.

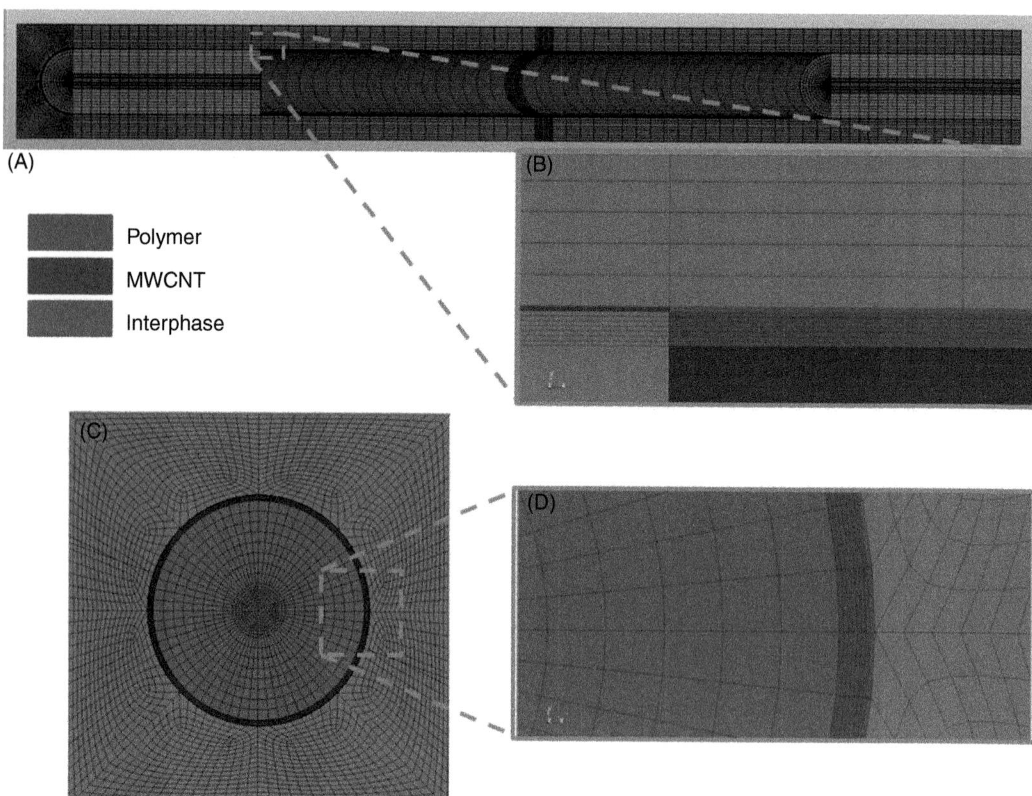

FIGURE 13.4 Typical FE mesh of the RVE and distinction between its constituents. (A) Top-view of half of the RVE. (B) Close-view of the CNT/polymer interphase's edge. (C) Side-view of the RVE. (D) Close-view of the side of the RVE at the interphase area.

13.4 Modeling of Nanoindentation Test

Taking into consideration the loading and boundary conditions in the RVE, it would be possible to simulate it with a nanoindentation test with impact loading rate and the use of a spherical nanoindenter. It is possible then to calculate the material mechanical properties under impact loading utilizing the mathematical analysis of this experimental procedure and thereafter to introduce these properties to a macroscale specimen, especially the one proposed by the ASTM D5628 [15] experimental standard.

First, it will be assumed that the total mechanical response of the RVE can be approached by a bilinear elastic–plastic stress–strain curve. To serve numerically the material model, the elastic modulus under impact, the Poisson's ratio, the critical stress under which plastic imprint occurs, the tangential modulus of the plastic region of the curve, as well as the plastic strain under which failure occurs should be calculated.

13.4.1 Specimen's Geometry, FE Model, and Boundary Conditions

In accordance with the ASTM D5628 standard [15], the specimen region which is under loading, is circular with diameter $d = 38.1$ mm and thickness $t = 1$ mm, while the boundary conditions require the specimen circumference to be fully fixed (all the degrees of freedom to be zero). As a consequence, an FE model for a circular plate was created. The circular region was meshed by 200 elements in the circumference of the plate, while shell elements with Belytschko–Tray formation and five integration points through the element thickness were used. The plate was loaded centrally by a spherical steel indenter of radius $r = 12.7$ mm at velocities of 1, 5, and 10 m/s. The indenter was simulated with finite elements of constant stress.

The nanocomposite architecture was approached by the random selection of 100, 200, and 300 elements for $V_f = 0.005$, 0.01, and 0.015, correspondingly, which were meant to represent the mechanical behavior of the CNT-reinforced material regions. The remaining elements modeled the neat polymeric volumes of the material. The mechanical properties of the CNT-embedded elements were approached by a bilinear elastic–plastic response, with material characteristics being derived by the methodology presented in Section 13.4.2 in detail. For this scale transition study, the material behavior obtained by the analysis of the CNT/interphase/matrix RVE were used for the cases of $L_{CNT}/L_{RVE} = 0.8$, $t_{int} = 0.68$ nm, interphase quality of 25 and 75%, as well as all studied V_f and impact velocities. The elements representing the polymeric matrix exhibited the material behavior presented in the Section 13.3.2.1, while the indenter was modeled as a rigid body with the mechanical properties of steel ($E = 200$ GPa, $v = 0.3$, $\rho = 7.83$ kg/cm^3). The FE model created in addition to the boundary and loading conditions is presented in Fig. 13.5.

13.4.2 Material Properties

13.4.2.1 Definition of Elastic Modulus Under Impact

The elastic modulus under impact E^* is defined in the specimen's unloading curve during the reverse movement of the indenter bounding from the specimen's surface. Especially,

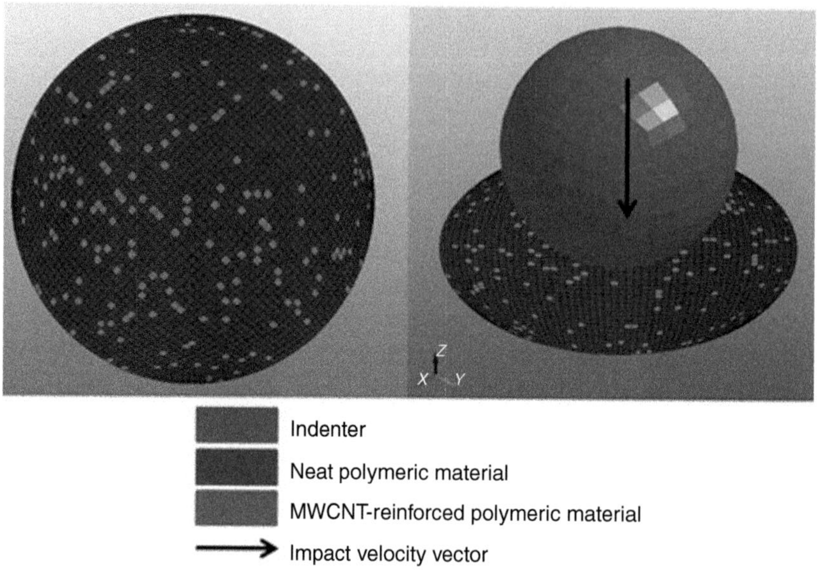

FIGURE 13.5 FE mesh of the circular specimen under impact loading in accordance with the ASTM D5628 standard [15].

this modulus is calculated by defining the slope of the force–displacement curve in the part of the unloading and the position of maximum force, which is reduced to the residual plastic imprint in the damaged surface as indicated in Eq. (13.6), where A is the area of the remaining plastic imprint, P is the impact load, and h is the deflection:

$$E^{*} = \frac{1}{2}\frac{\sqrt{\pi}}{\sqrt{A}}\frac{dP}{dh} \tag{13.6}$$

A derivation procedure example is given in the following lines for the case of $L_{\mathrm{CNT}}/L_{\mathrm{RVE}} = 0.8$, $t_{\mathrm{int}} = 0.68$ nm, $Q = 0.25$, $V_{\mathrm{f}} = 0.005$, and $u = 1$ m/s:

1. The unloading part of the load–deflection curve for this case is shown in Fig. 13.6. The curve was fitted by a second-order polynomial equation also shown in the figure.
2. The first derivative of the fitting equation is $dP/dh = -17.37x + 21.99$. The slope of the curve at the maximum load at $x = 0.55$ nm is $dP/dh = -12.44$ nN/nm.
3. By performing image analysis on the predicted contour of the remaining plastic deformation, shown in Fig. 13.7, the area of the imprint is calculated to be equal to 661.61 nm^2. Using Eq. (13.6), E^* is derived to be equal to 444 MPa.

13.4.2.2 Definition of Yield Stress Under Impact

The yield stress under impact σ_I^F equals to the ratio of the load, which causes the first plastic imprint, to the area of the remaining plastic footprint derived previously:

$$\sigma_F^I = \frac{P_{pl}}{A} = \frac{0.81 \text{ MPa}}{661.61 \text{ nm}^2} = 1.32 \text{ MPa} \tag{13.7}$$

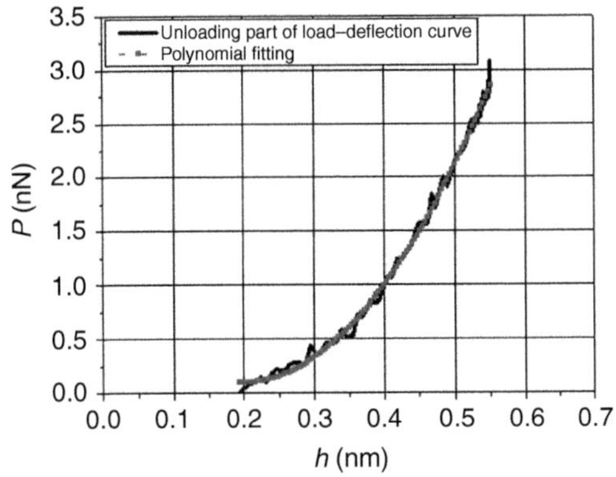

FIGURE 13.6 Example of calculating the elastic modulus under impact with the use of the uploading curve for $L_{CNT}/L_{RVE} = 0.8$, $t_{int} = 0.68$ nm, interphase quality 25%, $V_f = 0.005$, and impact velocity of 1 m/s.

FIGURE 13.7 Remaining plastic imprint for the studied case of $L_{CNT}/L_{RVE} = 0.8$, $t_{int} = 0.68$ nm, interphase quality 25%, $V_f = 0.005$, and impact velocity of 1 m/s.

13.4.2.3 Definition of Tangent Modulus of the Plastic Region

The tangent modulus of the plastic region, E_{tan}, is derived by applying the derivation process of E^* on the part of the load–deflection curve from the development of the first plastic imprint until the maximum load. The specific part of the curve is depicted in Fig. 13.8. The curve was fitted by a second-order polynomial equation also shown in the figure. The calculation of the tangent modulus for the case of $L_{CNT}/L_{RVE} = 0.8$, $t_{int} = 0.68$ nm, $Q = 0.25$, $V_f = 0.005$, and $u = 1$ m/s is as follows. The first derivative of the fitting equation is equal to $dP/dh = -75.17x - 29.92$. The slope of the curve at the maximum load at $x = 0.55$ nm is $dP/dh = -11.42$ nN/nm. Using the area of the remaining plastic imprint in Eq. (13.6), E_{tan} is derived to be equal to 410 MPa.

13.4.2.4 Approach of Poisson's Ratio and Plastic Strain to Fracture

Since fracture is developed only at the polymer, it is assumed that the plastic strain at fracture, $\varepsilon^F_{pl.}$, of the nanocomposite is the one of the polymer matrix, which equals to 0.0459. Finally, the Poisson's ratio is computed from the rule-of-mixtures: $\nu = \nu_{CNT}V_f + (1 - V_f)\nu_{pol.}$, where ν_{CNT} and ν_{Po} are the Poisson's ratios of the CNT and polymer, respectively.

Finally, the mechanical properties of elements representing MWCNT-reinforced polymer areas are summarized in Table 13.4.

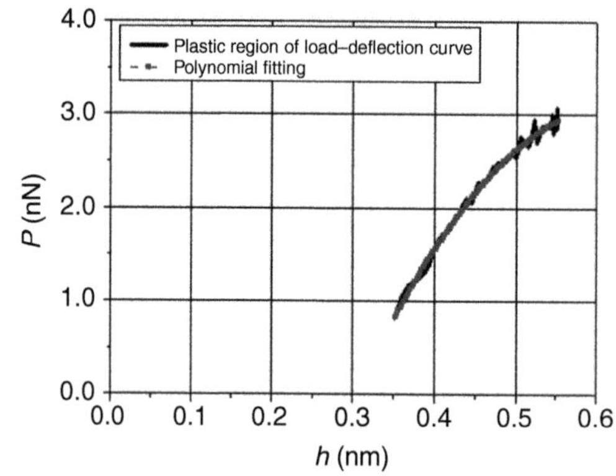

FIGURE 13.8 Example of calculating the tangent modulus of the plastic region under impact with the use of the force–displacement curve in the plastic region for $L_{CNT}/L_{RVE} = 0.8$, $t_{int} = 0.68$ nm, interphase quality 25%, $V_f = 0.005$, and impact velocity of 1 m/s.

13.5 Parametric Studies

Using the RVE, the effects of CNT's volume fraction (V_f), aspect ratio (AR), material properties (isotropic vs. orthotropic), and ratio of CNT's length to RVE's length (L_{CNT}/L_{RVE}) on the impact response of the nanocomposites have been studied concurrently with impact velocity. The range of parameters considered in the study is shown in Table 13.5.

13.5.1 Effect of the Presence of CNTs—Effect of V_f

Initially, the general effect of CNTs on the impact response of the polymer has been studied aiming to assess whether there is any reinforcement, by which mechanisms the reinforcement takes place, and how reinforcement might be affected by V_f. To this end, simulations have been performed for the case of AR = 10, $L_{CNT}/L_{RVE} = 0.8$, and isotropic CNT behavior. As the impact response is influenced by the specimen's dimensions and boundary conditions, in the comparison, for each V_f, RVEs of the same dimensions have been used for both the CNT/polymer and the neat polymer.

In Fig. 13.9 the force–displacement curves of the CNT/polymer of different V_f are compared against the respective curves of the neat polymer for the impact velocities of 1, 5, and 10 m/s. As can be seen, the presence of CNTs significantly enhances the impact response of the polymer by increasing the impact stiffness and the energy-absorbing capacity of the

Table 13.5 Range of Parameters Considered in the Present Study

Volume fraction	V_f	0.005, 0.01, 0.015
Aspect ratio	AR	10, 20, 30, 40, 50, 60
CNT's length to RVE's length ratio	L_{CNT}/L_{RVE}	0.4, 0.6, 0.8
CNT properties	—	Isotropic, orthotropic

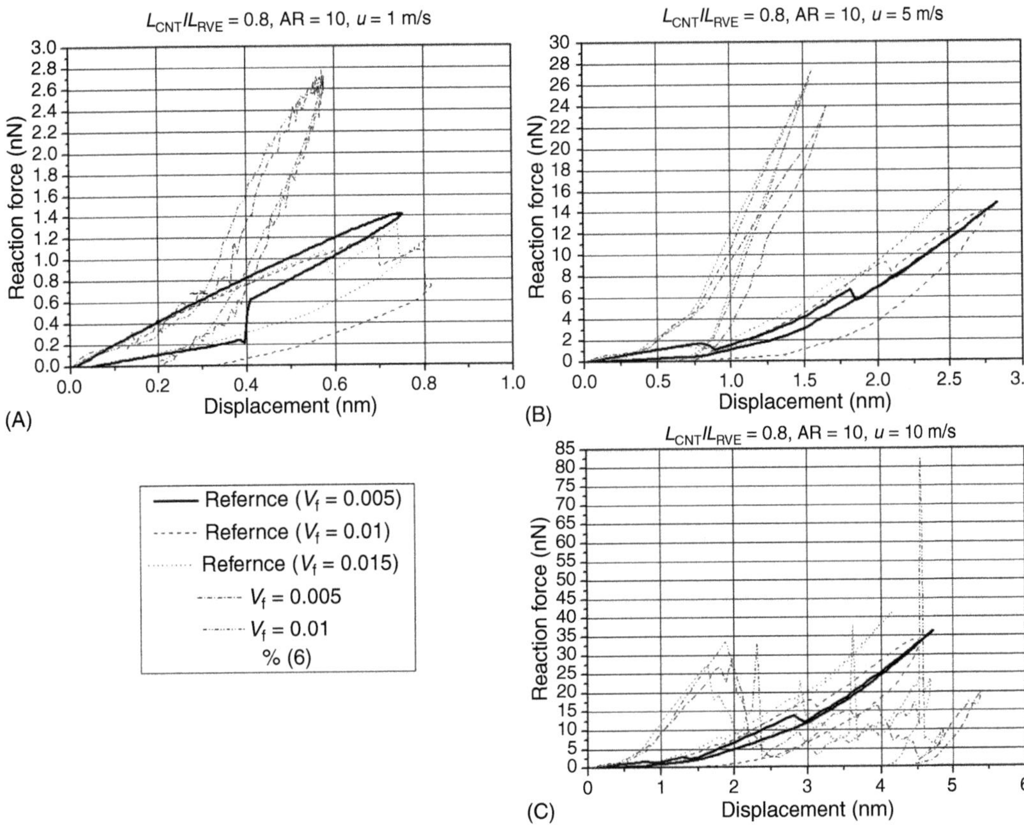

FIGURE 13.9 Variation of reaction force with regard to displacement in the polymer and the nanocomposite for different V_f values. (A) u = 1 m/s, (B) u = 5 m/s, and (C) u = 10 m/s.

material through the introduction of new energy distribution mechanisms (Fig. 13.10). For u = 1 m/s (Fig. 13.9A) and u = 5 m/s (Fig. 13.9B), by increasing the V_f, the maximum reaction force and impact stiffness increase. The increase is larger for u = 5 m/s. For u = 10 m/s (Fig. 13.9C), a totally different response, characterized by subsequent peaks and drops in the curve, has been computed for the CNT/polymer due to fracture of the polymer caused by the impact. In this case, the effect of V_f is not clear. The development of impact damage in the CNT/polymer at large impact velocities indicates that the CNT makes the polymer more susceptible to fracture since no damage has been predicted in the neat polymer for the same impact velocity. This finding is in agreement with experimental findings [6].

The area under the force–displacement curve corresponds to the strain energy stored in the material. In Fig. 13.11, the variation of strain energy with time is plotted for the CNT/polymer. For u = 1 m/s (Fig. 13.11A), the strain energy is similar for V_f = 0.01 and 0.015, while for V_f = 0.005 the absorbed strain energy is maximum. On the contrary, for u = 5 m/s (Fig. 13.11B), the V_f = 0.005 has the maximum absorbed strain energy, whereas for V_f = 0.01 the energy is minimized. Finally, for u = 10 m/s, the strain energy decreases

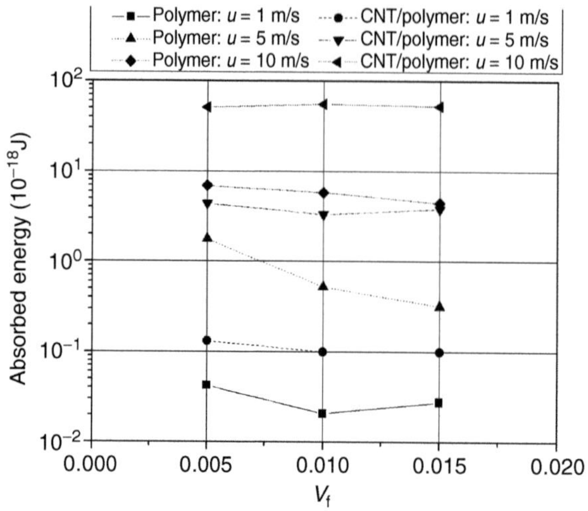

FIGURE 13.10 Variation of absorbed energy with regard to V_f in the polymer and nanocomposite for different impact velocities.

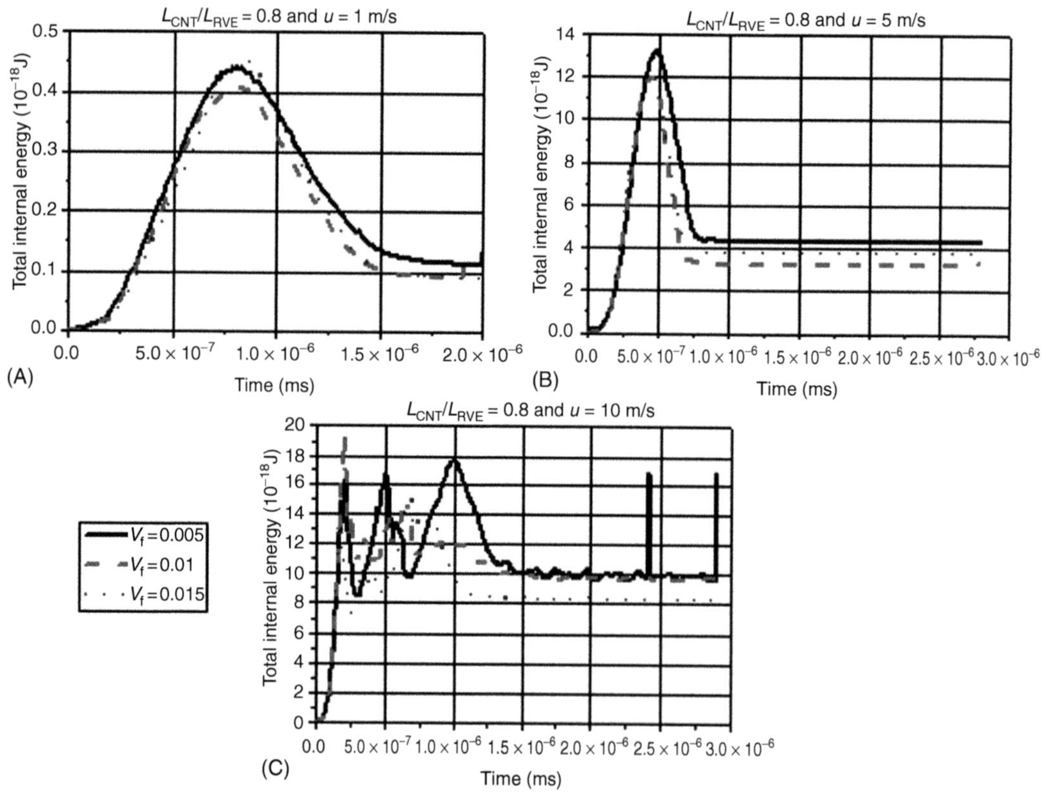

FIGURE 13.11 Variation of total energy with regard to time in the nanocomposite for different V_f values. (A) $u = 1$ m/s, (B) $u = 5$ m/s, and (C) $u = 10$ m/s.

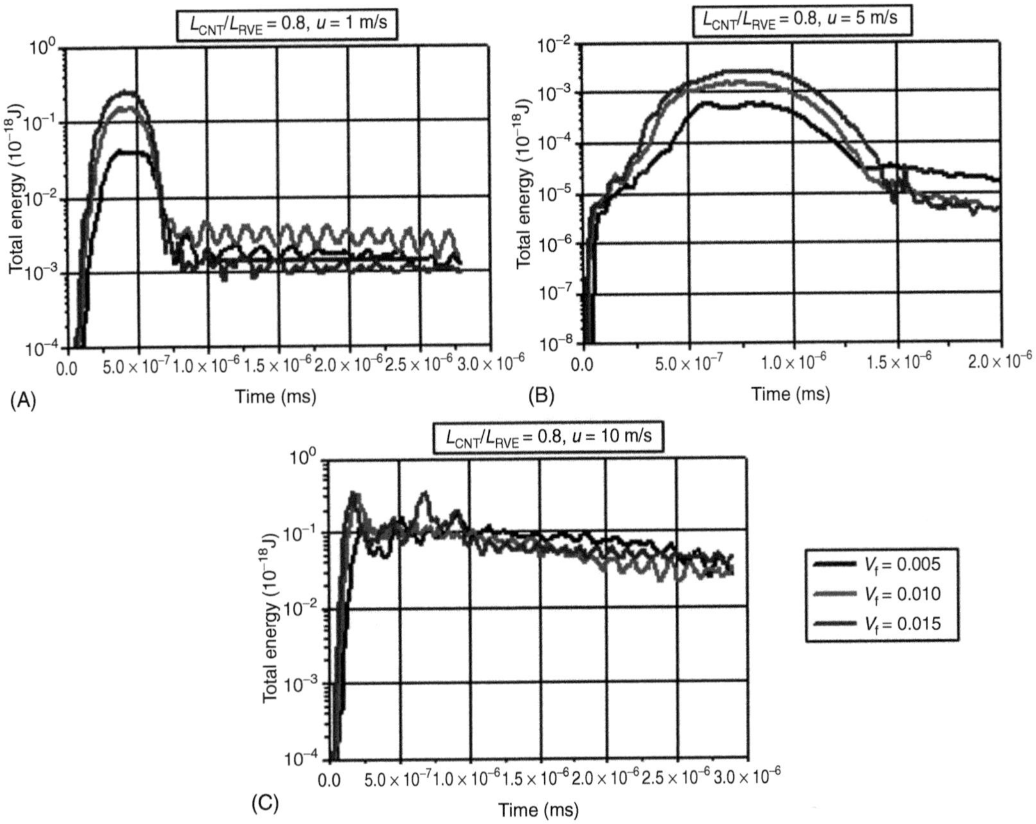

FIGURE 13.12 Variation of total energy with regard to time in the CNT for different V_f values. (A) $u = 1$ m/s, (B) $u = 5$ m/s, and (C) $u = 10$ m/s.

with increasing V_f. It is therefore concluded that the strain energy varies conversely with V_f. In Figs. 13.12 and 13.13, the variation of energy stored in the CNT and the polymer is plotted, respectively. In the polymer, the variation of strain energy is similar in shape and value to the one of the CNT/polymer presented in Fig. 13.12. This is attributed to the fact that the polymer remains the main energy storage despite the presence of CNT. The fluctuations in the curve of Fig. 13.12C are due to the development of fracture in the polymer. In the CNT, at the permanent condition, the variation of stored energy is sinusoidal, which indicates wave energy propagation. For all impact velocities, the maximum strain energy increases with increasing V_f. For $u = 1$ m/s, the remaining energy in the CNT is minimum for $V_f = 0.015$ and maximum for $V_f = 0.005$, while for $u = 5$ and 10 m/s the remaining energy is the same for $V_f = 0.01$ and 0.015. These findings suggest that the CNT stores part of the elastic energy in wave form and at the same time dissipates the strain front. The dissipation of the strain front is enhanced by increasing the V_f.

The time-dependent variation of energy consumed to fracture on the impact surface is shown in Fig. 13.14 for different V_f values for the case of u = 10 m/s. The maximum

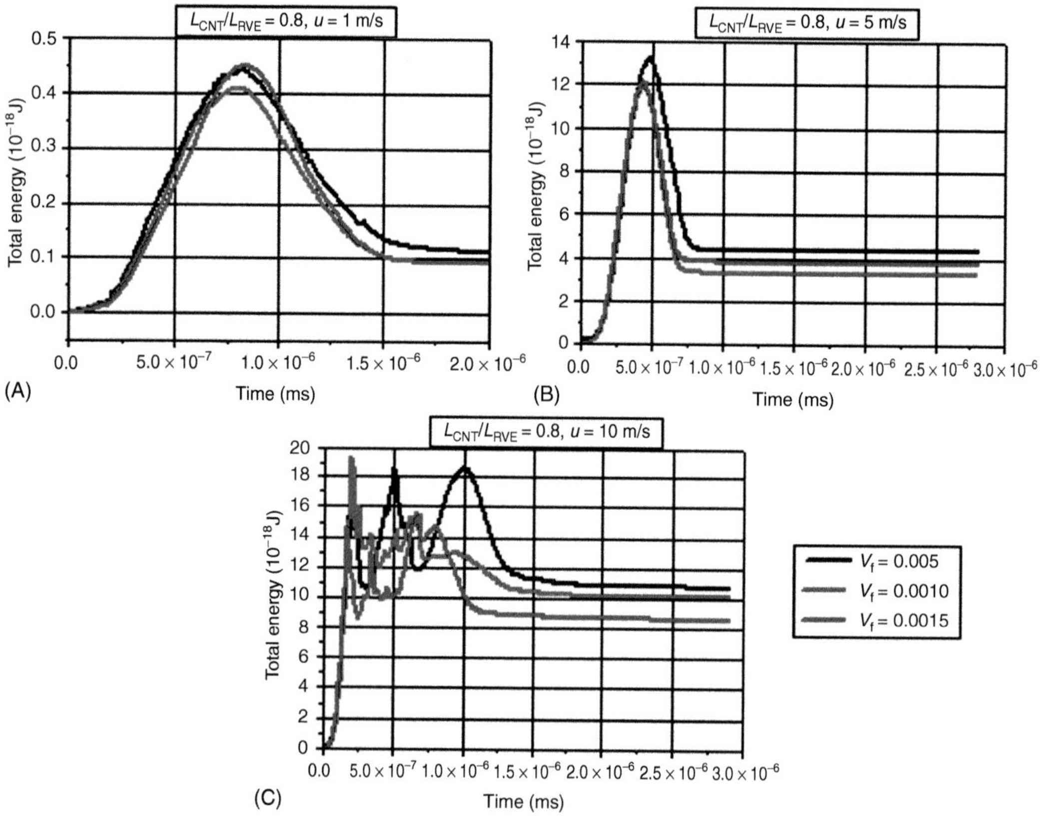

FIGURE 13.13 Variation of total energy with regard to time in the polymer for different V_f values. (A) $u = 1$ m/s, (B) $u = 5$ m/s, and (C) $u = 10$ m/s.

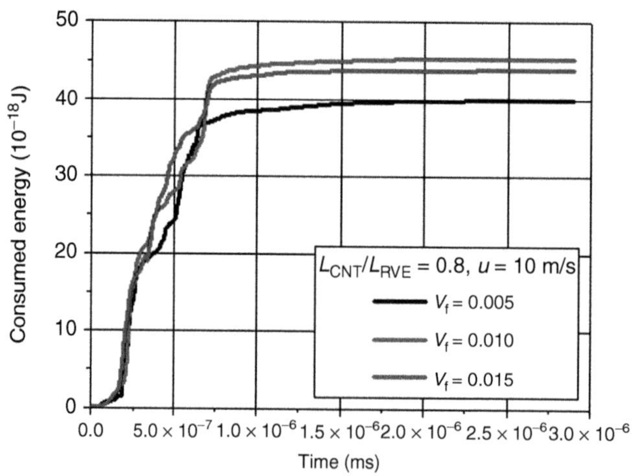

FIGURE 13.14 Variation of consumed energy with regard to time for different V_f values.

consumed energy was found for $V_f = 0.01$. This is in agreement with the findings for the absorbed energy presented in Fig. 13.10 and it is attributed to the interaction of the dissipated strain front with the applied strain front from which either a retraction or a reinforcement of the strain field occurs. This phenomenon is illustrated in the contour of Fig. 13.15A. As shown, the reflected strain field around the CNT has created a volume at the top of which the applied strain field gets milder and less intense than from the impacted surface. Although the plastic strains developed on the impacted surface tend to broaden to the width of the elastic field, the reflection of the previous elastic strain screenshots has managed to eliminate this effect. Fig. 13.15B shows the contour of the plastic strain for the same case.

13.5.2 Effect of Aspect Ratio

The effect of AR (CNT length to diameter ratio) has been studied based on the case $L_{CNT}/L_{RVE} = 0.8$, $V_f = 0.015$, and $u = 10$ m/s isotropic CNT properties. For the study, the maximum impact velocity ($u = 10$ m/s) has been selected to take into account all impact phenomena. From Fig. 13.16A and B it may be concluded that the AR has a small effect on the absorbed energy and reaction force. On the other hand, the increase of AR leads to the increase in total internal energy (Fig. 13.16C) and decrease in consumed energy of the nanocomposites (Fig. 13.16D). In general, the rather small effect of AR on the impact phenomenon is mainly attributed to the fact that the impact load is transverse to the CNT axis.

13.5.3 Effect of CNT Properties

In most numerical models of nanocomposites, the isotropic behavior is assumed for CNT for simplicity reasons. However, this is a rough estimation, since the transverse properties of the CNT are much smaller than the axial ones. In the present work, the effect of CNT behavior on the impact response of the nanocomposite has been studied by comparing the isotropic and orthotropic CNT behaviors.

Fig. 13.17 plots the variation of the absorbed energy of the nanocomposite with regard to V_f for the two material behaviors for different impact velocities. A large difference in the predicted absorbed energy between the nanocomposites with isotropic and orthotropic CNTs is observed for $u = 1$ m/s while for $u = 5$ and 10 m/s the difference is small. The large difference in the impact response of isotropic and orthotropic CNT behaviors at small impact velocities is due to the different strain energy stored in the CNT than in the polymer as already shown in Figs. 13.12 and 13.13. Specifically, because the propagation direction of the strain front is transverse to the CNT axis, the isotropic CNT (large Young's modulus in the transverse direction, 1 TPa) acts as a strain reflector, thus making the elastic energy be mainly stored in the polymer, whereas the orthotropic CNT (small Young's modulus in the transverse direction, 30 GPa) results in a smaller boundary modulus mismatch in the transverse direction, thus allowing more elastic energy to be stored in the CNT.

(A)

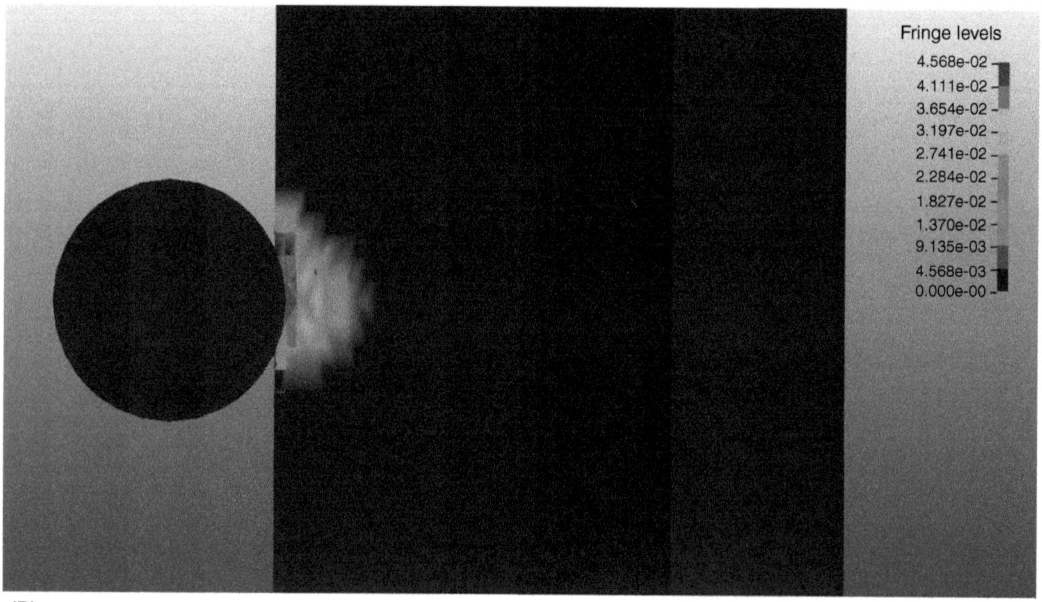

(B)

FIGURE 13.15 Contour plot of (A) elastic–plastic longitudinal strain and (B) plastic longitudinal strain in the lateral cross-section of the RVE ($V_f = 0.01$, AR = 50, $L_{RVE}/L_{CNT} = 0.8$, and $u = 10$ m/s).

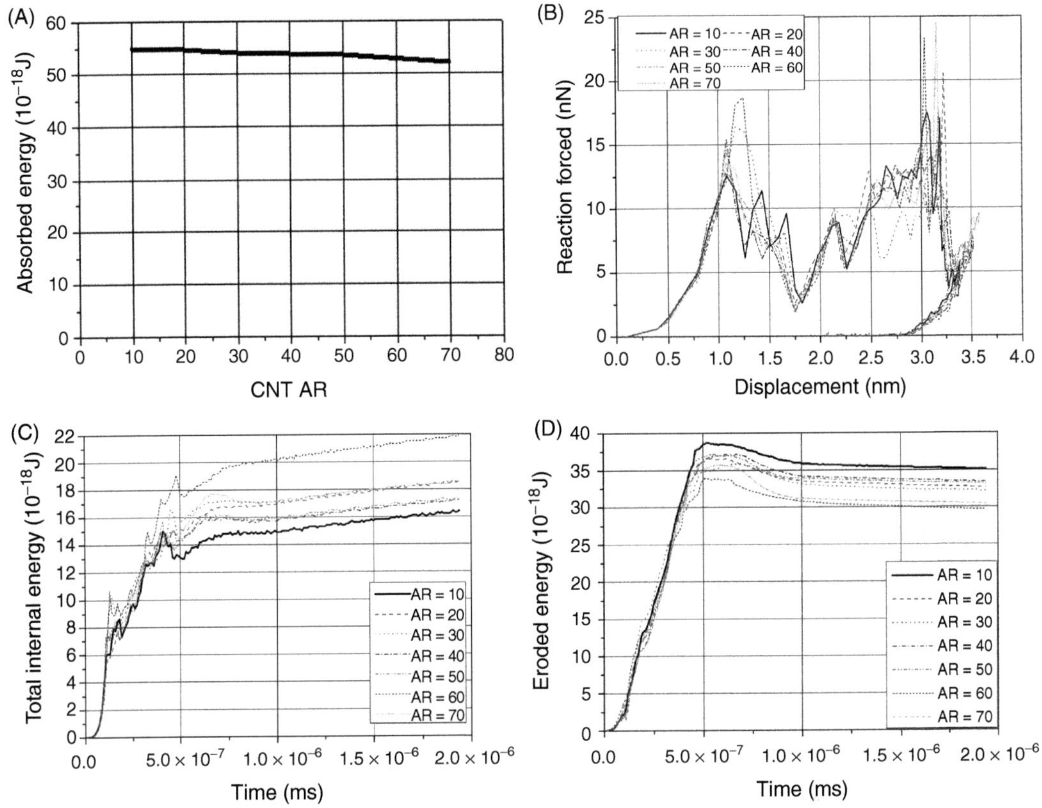

FIGURE 13.16 (A) Variation of absorbed energy with regard to Aspect ratio (AR). (B) Variation of reaction force with regard to displacement for different ARs. (C) Variation of total internal energy with regard to time for different ARs. (D) Variation of consumed energy with regard to time for different ARs.

13.5.4 Effect of L_{RVE}/L_{CNT} Ratio

By treating CNT as an internal reinforcement, its length-to-RVE-length ratio is introduced as an additional modeling parameter. In Fig. 13.18, the variation of the absorbed energy with V_f is plotted for different L_{CNT}/L_{RVE} values and impact velocities. For $u = 1$ m/s (Fig. 13.18A), the absorbed energy is the same for $L_{CNT}/L_{RVE} = 0.6$ and 0.8 while it is much larger for $L_{CNT}/L_{RVE} = 0.4$. This is because the smaller the L_{CNT}/L_{RVE} is, the thinner the polymer surrounding the CNT is. This configuration, at small impact velocities, leads to the increase of the amount of strain energy stored in the CNT. For $u = 5$ m/s (Fig. 13.18B), the variation of the absorbed energy with L_{CNT}/L_{RVE} becomes more complicated due to the wavy form of the energy distribution. Nevertheless, there are still two different trends, one for $L_{CNT}/L_{RVE} = 0.4$ and one for $L_{CNT}/L_{RVE} = 0.6$ and 0.8. For $u = 10$ m/s (Fig. 13.18C), the values of the absorbed energy almost coincide for all three L_{CNT}/L_{RVE} values. From a more detailed analysis of the impact energy distribution, it is found that for $L_{CNT}/L_{RVE} = 0.4$ and 0.6 the fracture energy increases with increasing V_f, while for $L_{CNT}/L_{RVE} = 0.8$ the fracture

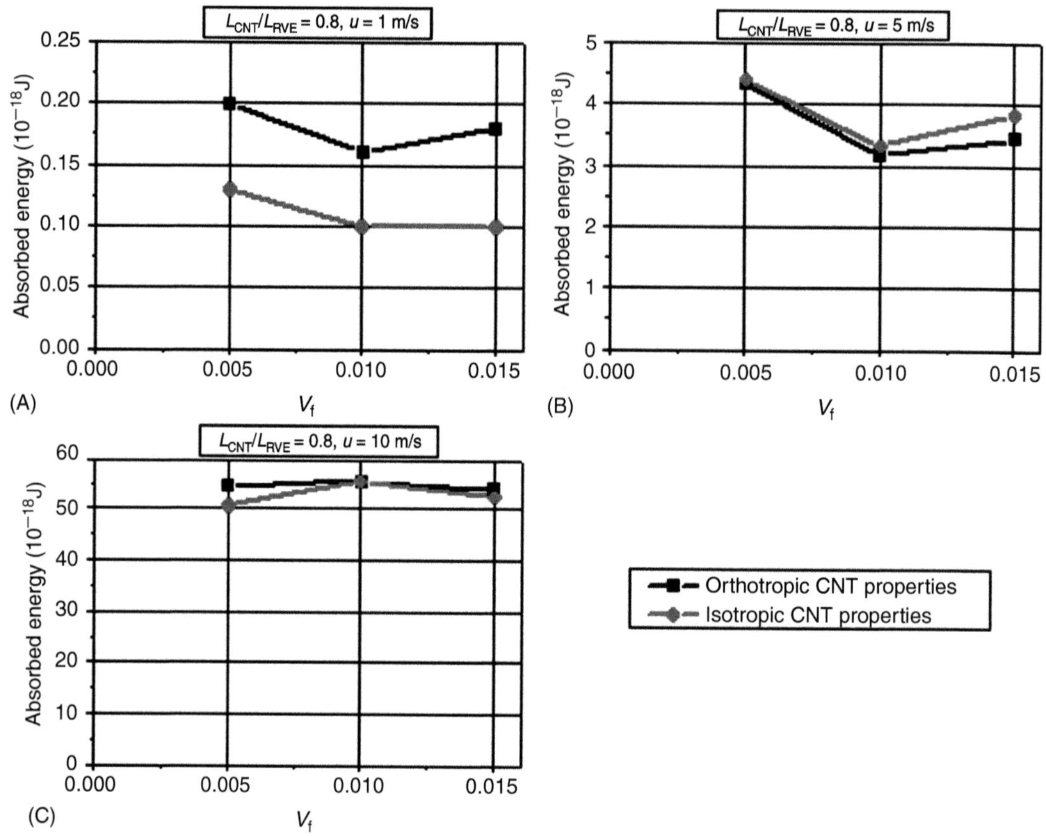

FIGURE 13.17 Variation of absorbed energy with regard to V_f for orthotropic and isotropic CNT behavior. (A) $u = 1$ m/s, (B) $u = 5$ m/s, and (C) $u = 10$ m/s.

energy decreases with increasing V_f. This effect is also related to the thickness of the polymer surrounding the CNT, as well as the applied impact velocity because the polymer thickness and the velocity of the developed strain field govern the ratio of the strain energy reaching the CNT and stored or reflected by the filler.

13.5.5 Mechanisms of Energy Absorption and Distribution

The mechanisms of energy absorption in the nanocomposite can be summarized into the following:

- elastic strain in CNT
- elastic strain in interphase
- elastic and plastic strains in polymer
- fracture of polymer for impact velocity $u = 10$ m/s

The mechanism of the impact energy distribution between the different components of the nanocomposite is schematically explained in Fig. 13.19. When the strain front

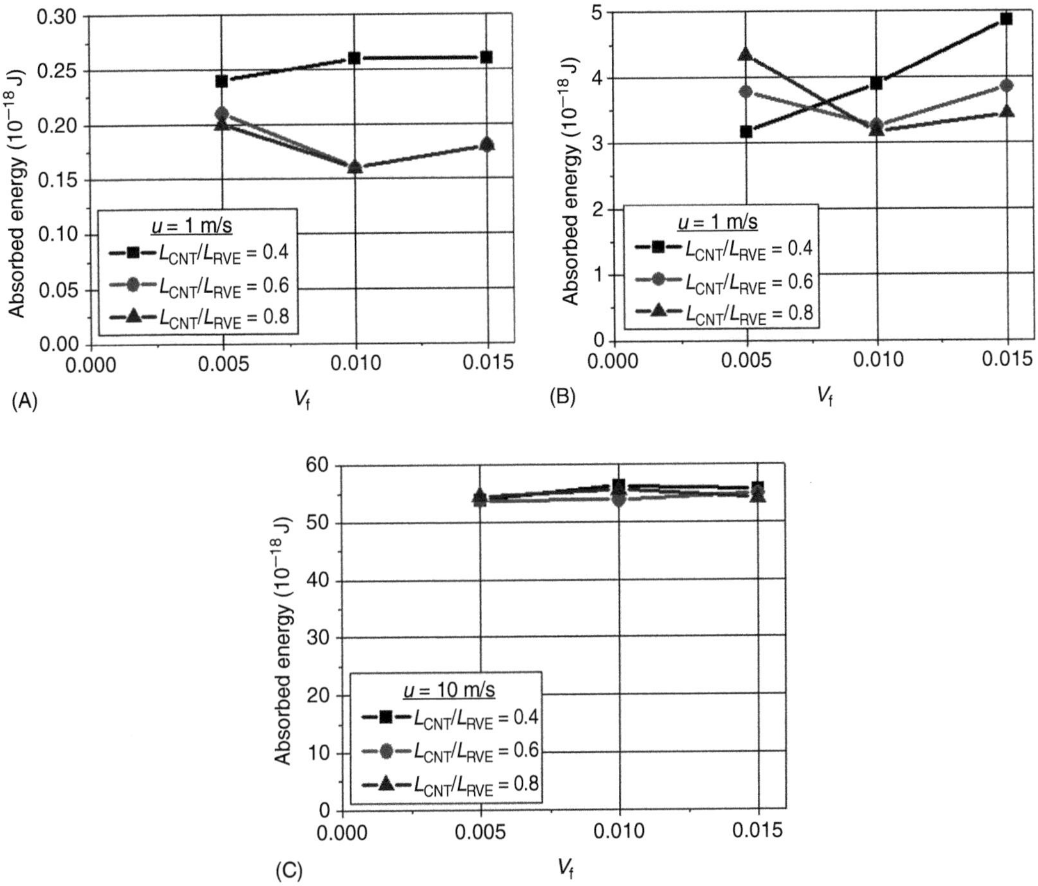

FIGURE 13.18 **Variation of absorbed energy with regard to V_f and L_{RVE}/L_{CNT}.** (A) u = 1 m/s, (B) u = 5 m/s, and (C) u = 10 m/s.

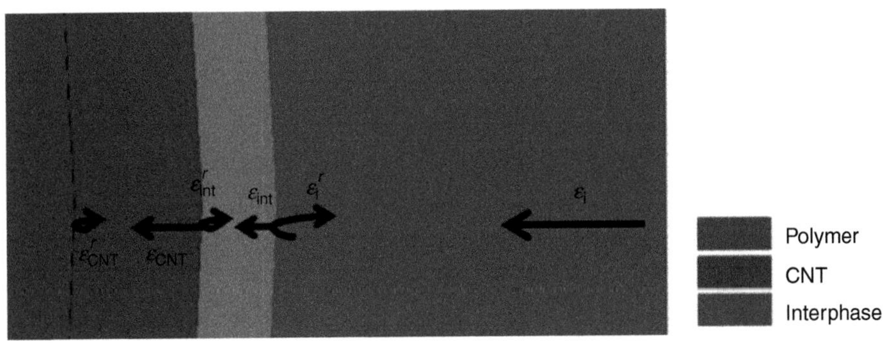

FIGURE 13.19 **Graphical representation of the strain fronts scattering on the nanocomposite's components.**

ε_i reaches the interphase, a part of it is reflected back to the polymer (ε_i^r) and the rest (ε_{int}) continues until it reaches the CNT. As the CNT is stiffer than the interphase, it reflects part of the strain front back to the interphase (ε_{int}^r) and absorbs the rest (ε_{CNT}). It stands

$$\begin{aligned} \varepsilon_i &= \varepsilon_i^r + \varepsilon_{int} && \text{(a)} \\ \varepsilon_{int} &= \varepsilon_{int}^r + \varepsilon_{CNT} && \text{(b)} \\ \varepsilon_{CNT} &= \varepsilon_{CNT}^r && \text{(c)} \end{aligned}$$

To better understand these mechanisms, the evolution of transverse and longitudinal strains in the direction of impact with regard to time are depicted in Figs. 13.20–13.25 for the case of the specimen with $V_f = 0.01$, interphase thickness $t_{int} = 0.68$ nm, and interphase quality of 100%. The case of $u = 10$ m/s is selected for this demonstration as it includes all energy absorption mechanisms. As can be seen, initially, the strain front progresses from the impact surface toward the interphase and the CNT. At the same time, a small plastic region develops at the impact surface with a lag with regard to the main strain front. When the strain front reaches the CNT it reflects radially through the length of the CNT. The reflected strain front moves in the opposite direction of initial strain causing a repeal between the two strain fronts. At $t = 0.8 \times 10^{-7}$ m/s, the reflected strain front reaches the impact surface and reflects back toward the interior of the RVE (Fig. 13.21). The new (reflected) strain front is broader than the initial one and when it reaches the CNT a part of it is reflected back and the rest moves circumferentially in the CNT until it reaches the edge of the RVE and reflected also (Fig. 13.22). This sequence is repeated during the impact phenomenon with the different strain components interacting with each other leading to weakening and enforcing. Also, during the impact phenomenon, the region of plastic strain continues to develop at the impact surface. This region acts as an undistorted body within the RVE and influences the amount of energy reflected at the impact surface. Finally, the repeated interaction of strain fronts within the material leads to the enhancement of strain at the impact surface, which causes fracture (Fig. 13.24).

13.5.6 Effect of Interphase Thickness

In Fig. 13.26, the variation of absorbed energy with regard to V_f and t_{int} is plotted for the different impact velocities and for interphase quality of 50 and 100%, respectively. The effect of t_{int} on the total absorbed energy is minor. However, the presence of the interphase affects the distribution of absorbed energy between the components of the nanocomposite. Specifically, the amount of energy absorbed by the polymer increases while the energy absorbed by the CNT decreases as can be derived by comparing Fig. 13.26 and Table 13.6. The maximum energy absorbed by the polymer is for the case of $u = 10$ m/s and $V_f = 0.005$. In this case, the increase of energy absorbed by the polymer leads to the decrease of energy consumed in the fracture of the polymer. This is an important aspect since the after-impact behavior of the nanocomposite is governed by the residual properties of the polymer, which degrade due to fracture. The interphase quality does not affect the absorbed energy.

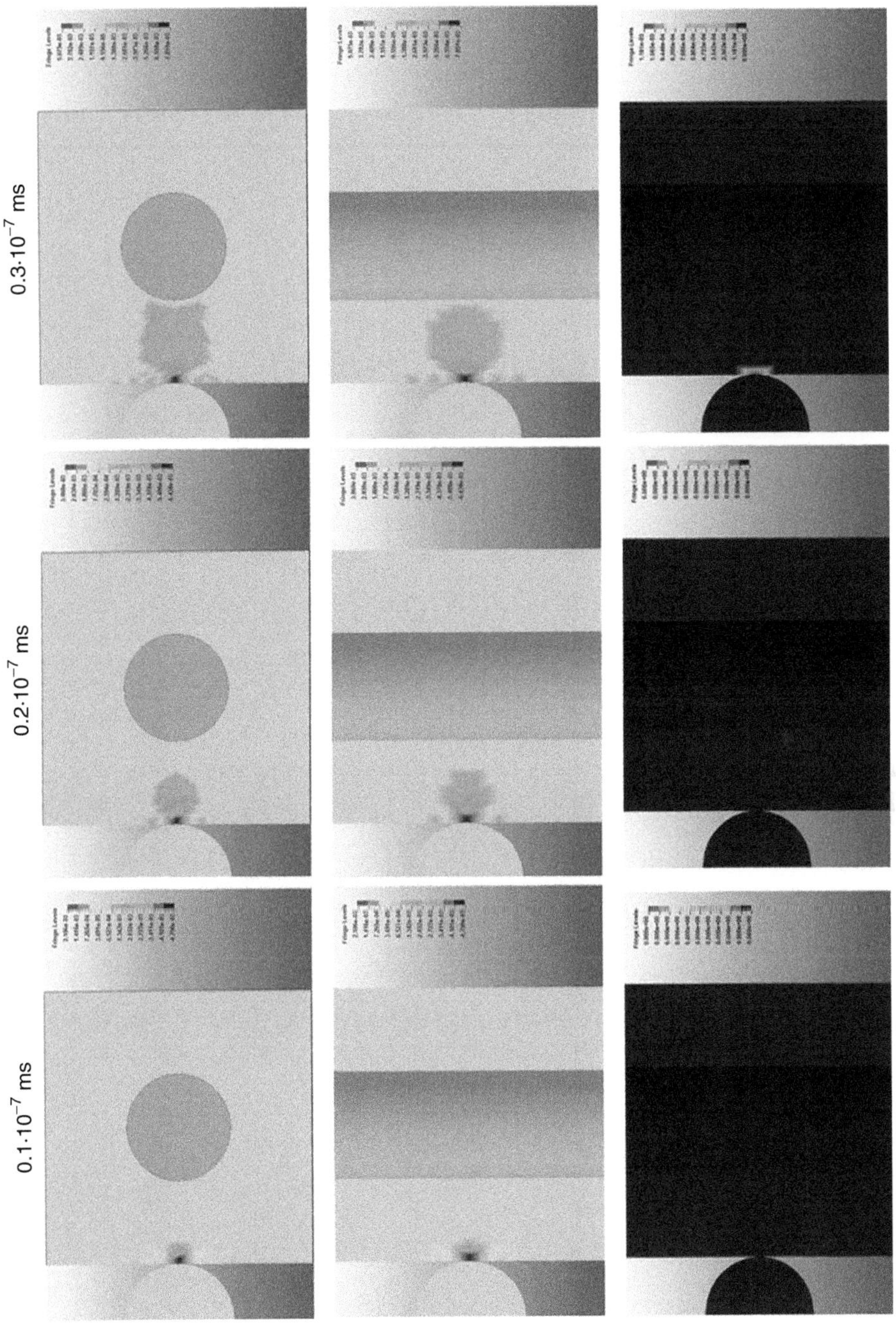

FIGURE 13.20 Development of transverse strain x for time $t = 0.1 \cdot 10^{-7}$–$0.3 \cdot 10^{-7}$ ms. First row, elastic strain in the transverse cross-section; second row, elastic strain x in the longitudinal cross-section; third row, plastic strain in the longitudinal cross-section.

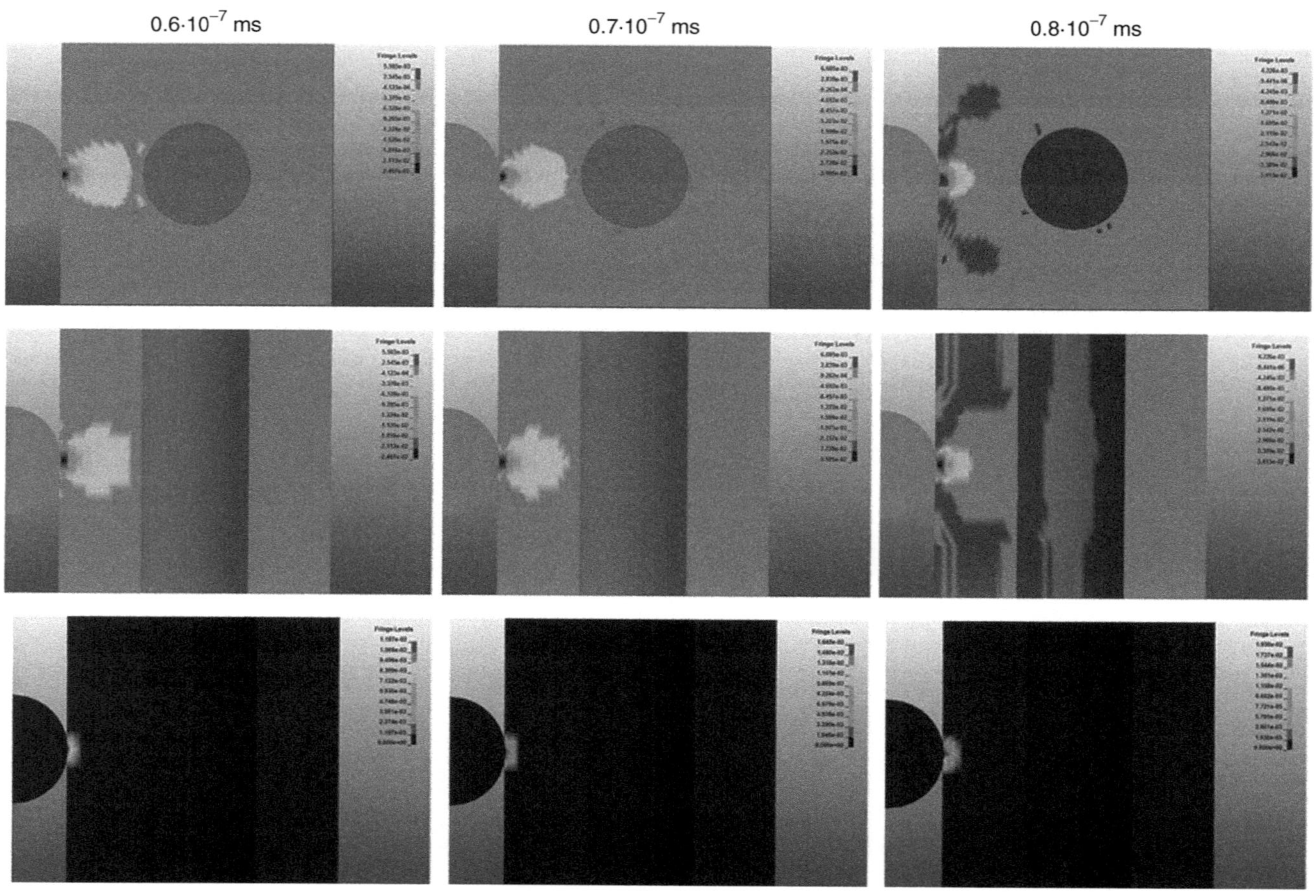

FIGURE 13.21 Development of transverse strain *x* for time $t = 0.6{\cdot}10^{-7}$–$0.8{\cdot}10^{-7}$ ms. First row, elastic strain in the transverse cross-section; second row, elastic strain *x* in the longitudinal cross-section; third row, plastic strain in the longitudinal cross-section.

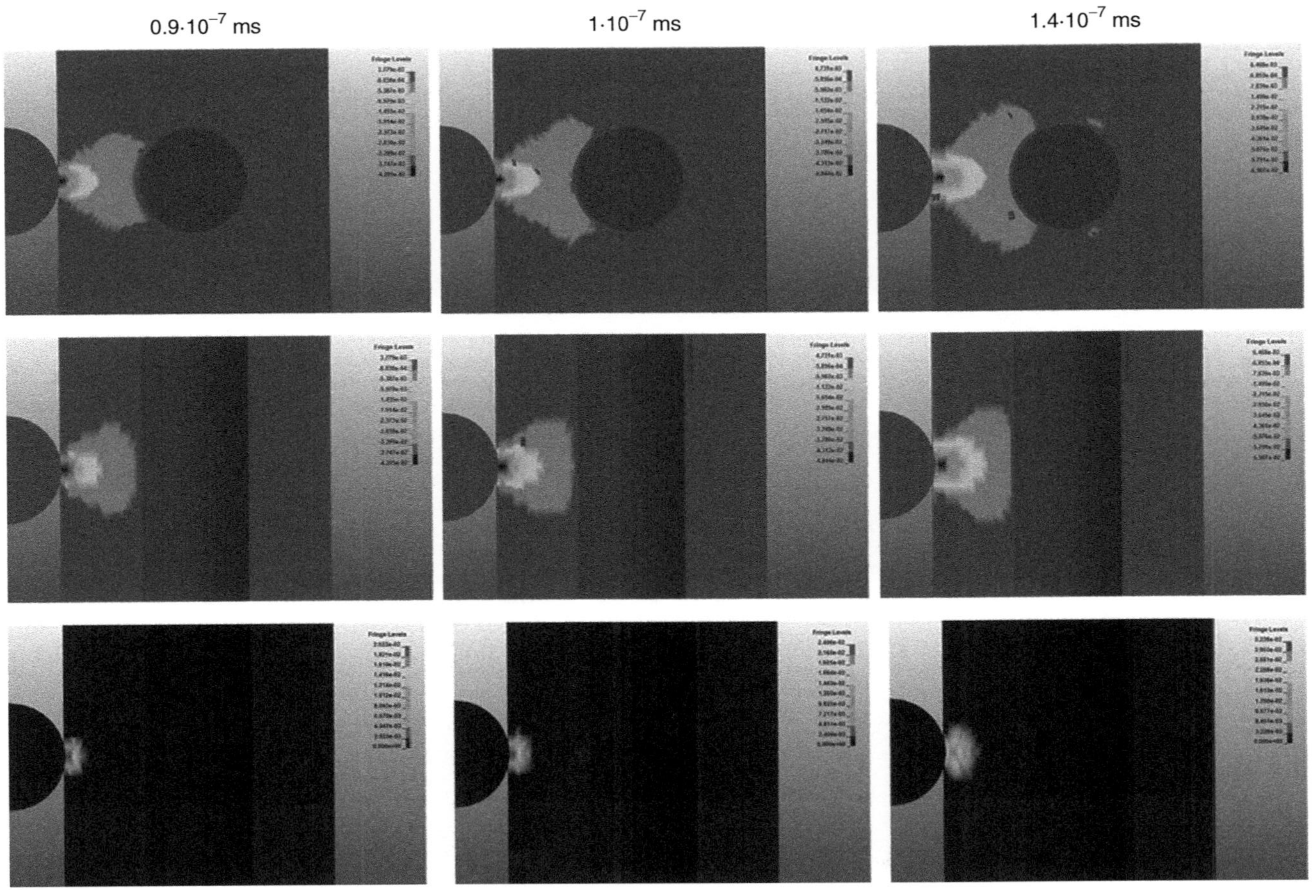

FIGURE 13.22 Development of transverse strain *x* for time $t = 0.9 \cdot 10^{-7}$–$1.4 \cdot 10^{-7}$ ms. First row, elastic strain in the transverse cross-section; second row, elastic strain *x* in the longitudinal cross-section; third row, plastic strain in the longitudinal cross-section.

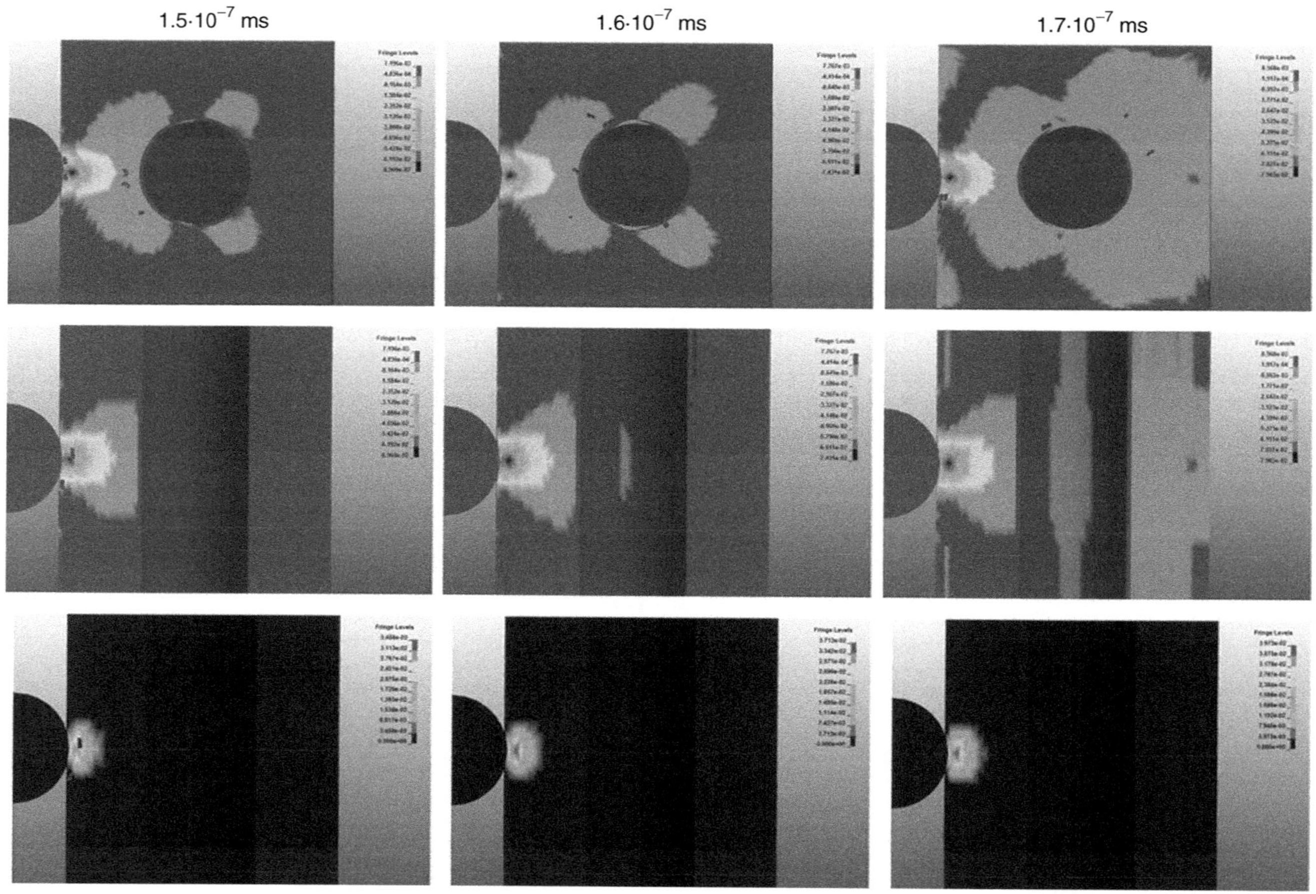

FIGURE 13.23 Development of transverse strain x for time $t = 1.5 \cdot 10^{-7}$–$1.7 \cdot 10^{-7}$ ms. First row, elastic strain in the transverse cross-section; second row, elastic strain x in the longitudinal cross-section; third row, plastic strain in the longitudinal cross-section.

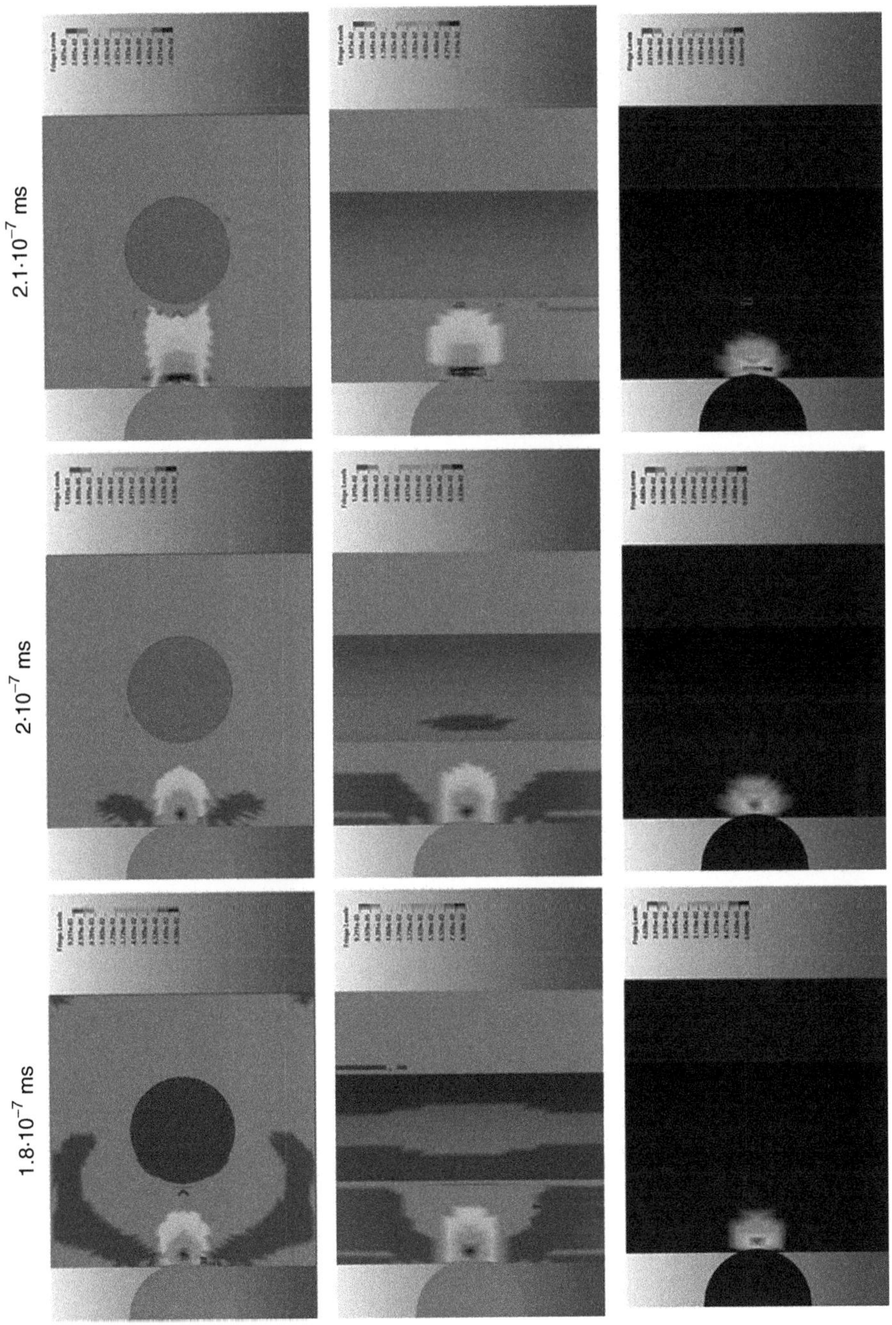

FIGURE 13.24 Development of transverse strain *x* for time $t = 1.8{\cdot}10^{-7}$–$2.1{\cdot}10^{-7}$ ms. First row, elastic strain in the transverse cross-section; second row, elastic strain *x* in the longitudinal cross-section; third row, plastic strain in the longitudinal cross-section.

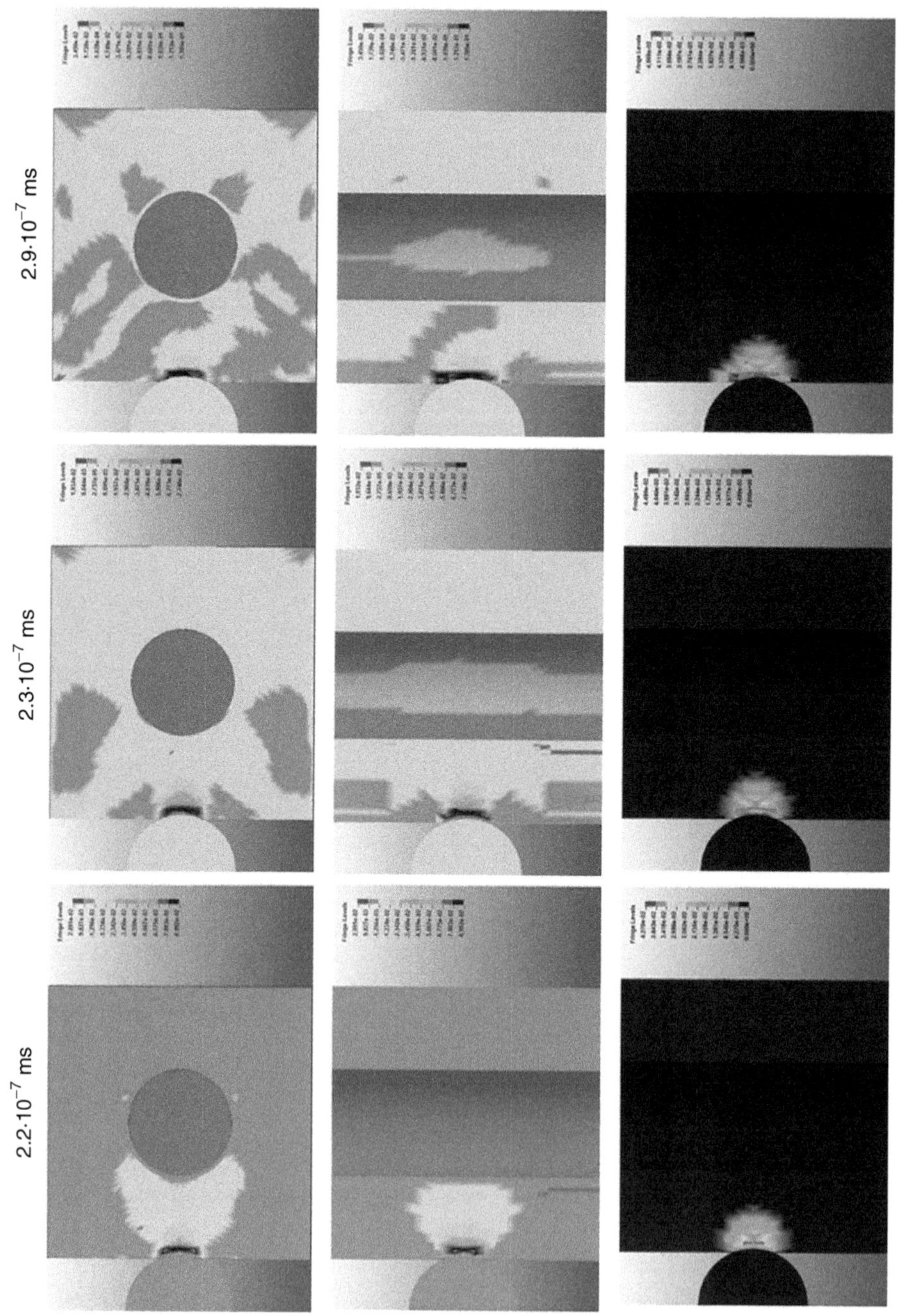

FIGURE 13.25 Development of transverse strain *x* for time *t* = $2.2{\cdot}10^{-7}$–$2.9{\cdot}10^{-7}$ ms. First row, elastic strain in the transverse cross-section; second row, elastic strain *x* in the longitudinal cross-section; third row, plastic strain in the longitudinal cross-section.

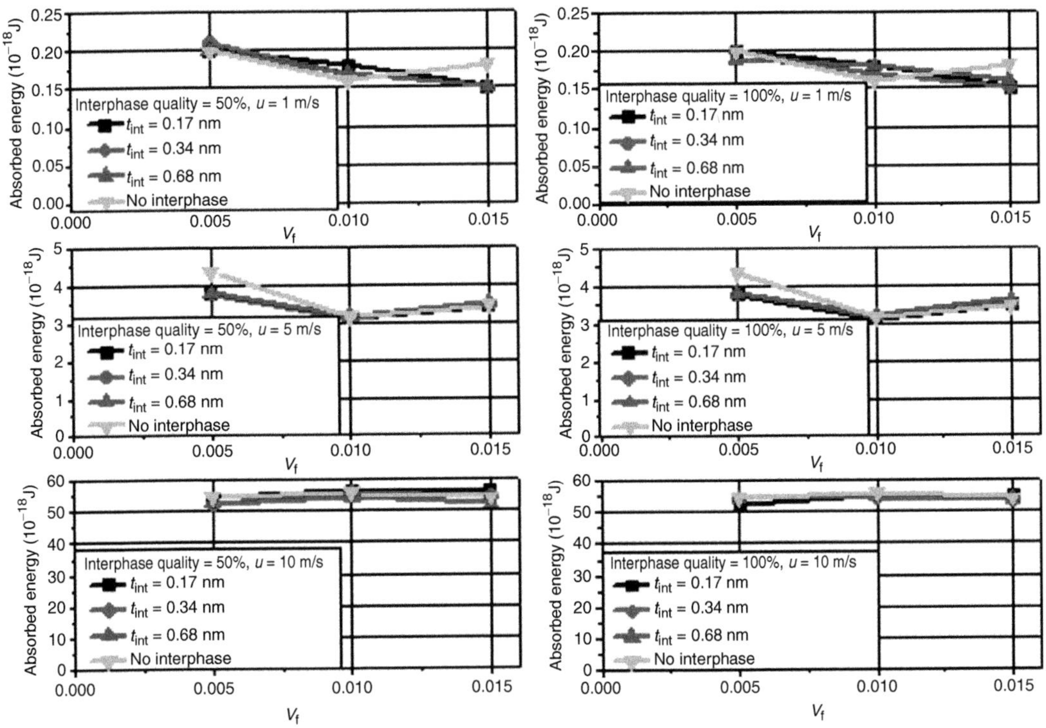

FIGURE 13.26 Effect of interphase thickness on the absorbed impact energy in the function of V_f for every case impact velocity and interphase quality of 50 and 100%.

By studying Table 13.6, some important observations can be derived relating to the variation of the polymer, MWCNT, and interphase absorbed energy. Generally, for every case of interphase thickness, the energy stored in the polymeric component is decreased with increasing volume fraction. On the other hand, the contribution of MWCNT on the absorbed energy is decreased with increasing interphase thickness, while the interphase contribution is increased for every impact velocity. Particularly, with increasing impact velocity, the decreasing contribution of MWCNT on the absorbed energy is more intense, while the corresponding interphase contribution increase is of the same order for every impact case. Finally, the interphase decreases the percentage of energy spent on the fracture phenomena on the impact surface, whereas the increasing interphase thickness led to decreasing ratio of energy consumed on fracture.

13.6 Simulation of Nanoindentation

Table 13.7 lists the results (absorbed energy and mechanisms of energy consumption) from the parametric study conducted on the nanoindentation specimen. Nanoindentation simulations were also performed on the specimen made from neat polymer for comparison reasons. It is observed that the energy absorption capability of the polymer is significantly

Table 13.6 Comparative Table of the Absorbed Energy Distribution in the Nanocomposites Components for Every Studied Case With Interphase Quality of 50 and 100%

Interphase Quality (%)	Interphase Thickness t_{int} (nm)	V_f	Velocity (m/s)	AAbsorbed Energy (10^{-18} J)	Ratio of Absorbed Energy to Impact Energy (%)	Percentage of Absorbed Energy in the Material Components (%)			
						Polymer	MWCNT	Interphase	Fracture
—	—	0.005	1	0.20	35	65	35	—	
			5	4.34	30	98	2	—	
			10	54.44	95	23	0.23	—	77.08
		0.01	1	0.16	28	59	41	—	
			5	3.17	22	97	3	—	
			10	55.59	97	19	0.226	—	80.55
		0.015	1	0.18	31	59	41	—	
			5	3.44	24	97	3	—	
			10	54.11	95	19	0.075	—	80.47
50	0.17	0.005	1	0.20	34.71	76.70	21.98	1.32	
			5	3.76	26.34	99.15	0.78	0.07	
			10	54.34	95.06	38.92	0.49	0.11	60.47
		0.01	1	0.18	31.03	71.29	26.27	1.51	
			5	3.12	21.83	98.59	0.54	0.09	
			10	56.41	98.67	19.79	0.17	0.07	79.97
		0.015	1	0.15	26.01	65.39	32.71	1.90	
			5	3.43	24.31	99.40	0.53	0.07	
			10	56.01	97.98	13.41	0.20	0.04	86.35
	0.34	0.005	1	0.21	36.52	77.52	20.72	1.76	
			5	3.78	37.09	99.19	0.72	0.09	
			10	52.34	91.56	35.09	0.38	0.14	64.40
		0.01	1	0.17	30.16	69.73	27.57	2.71	
			5	3.12	21.86	99.21	0.67	0.12	
			10	56.11	98.15	19.50	0.10	0.08	80.33
		0.015	1	0.15	27.09	64.885	32.105	3.010	
			5	3.46	24.20	99.361	0.531	0.108	
			10	54.27	94.93	18.497	0.198	0.061	81.24
	0.68	0.005	1	0.20	35.21	75.72	21.90	2.38	
			5	3.79	37.19	99.24	0.65	0.11	

			10	52.60	92.02	34.57	0.24	0.17	65.02
		0.01	1	0.17	29.98	69.56	27.43	3.01	
			5	3.16	31.06	99.17	0.69	0.14	
			10	54.45	95.25	20.50	0.07	0.12	79.32
		0.015	1	0.15	26.28	63.89	32.60	3.51	
			5	3.52	31.75	99.35	0.52	0.13	
			10	52.09	91.12	20.83	0.18	0.11	78.89
100	017	0005	1	020	3413	7397	2150	107	
			5	3.76	26.34	99.18	0.73	0.08	
			10	52.01	90.99	35.11	0.39	0.13	64.36
		0.01	1	0.18	31.03	72.41	26.33	1.26	
			5	3.12	21.83	99.25	0.66	0.09	
			10	55.19	96.54	20.32	0.10	0.08	79.50
		0.015	1	0.15	26.01	67.99	32.14	1.53	
			5	3.47	24.31	99.43	0.50	0.08	
			10	54.61	95.52	18.46	0.13	0.06	81.35
	0.34	0.005	1	0.20	34.98	77.56	21.00	1.44	
			5	3.79	26.53	99.29	0.63	0.08	
			10	54.22	94.84	32.20	0.24	0.17	67.40
		0.01	1	0.16	27.49	69.45	28.59	1.96	
			5	3.15	22.07	99.18	0.70	0.12	
			10	54.38	95.13	19.94	0.10	0.07	79.89
		0.015	1	0.15	26.27	66.251	31.559	2.190	
			5	3.51	24.59	99.264	0.623	0.113	
			10	53.83	94.15	18.512	0.081	0.055	81.35
	0.68	0.005	1	0.19	33.26	75.17	22.68	2.15	
			5	3.81	26.63	99.30	0.61	0.10	
			10	54.10	94.63	38.81	0.15	0.17	60.87
		0.01	1	0.18	31.52	74.11	23.68	2.21	
			5	3.20	22.39	99.15	0.71	0.14	
			10	54.72	95.72	18.79	0.03	0.07	81.10
		0.015	1	0.16	27.16	67.25	2.68	30.07	
			5	3.61	25.26	99.37	0.15	0.49	
			10	53.70	93.93	18.71	0.07	0.04	81.18

Table 13.7 Comparative Table of the Absorbed Energy Distribution in the Nanocomposites Components for Every Studied Case With Interphase Quality of 25 and 75%

Interphase Quality (%)	V_f	Impact Velocity u (m/s)	Absorbed Energy (mJ)	Ratio of Absorbed Energy to Impact Energy (%)	Percentage of Absorbed Energy in the Material Components (%)			
					Polymer	Reinforced Volume	Polymer Fracture	Reinforced Volume Fracture
—	0	1	8.80	11.74	100.00	—	—	—
		5	1263.71	67.45	57.02	—	42.98	—
		10	1456.94	19.44	56.32	—	43.68	—
25	0.005	1	9.64	12.86	96.77	3.23		
		5	457.39	24.41	59.75	1.72	35.64	2.89
		10	1257.47	16.78	55.67	0.00	42.80	1.53
	0.01	1	17.65	23.55	64.06	35.94	—	—
		5	1396.80	74.55	53.61	4.00	42.39	—
		10	1543.64	20.60	48.89	2.18	41.39	7.53
	0.015	1	19.51	26.04	52.28	47.72	—	—
		5	1442.91	77.01	51.05	6.36	42.59	—
		10	714.66	9.54	49.54	1.41	42.93	6.13
75	0.005	1	13.47	17.97	77.93	22.07	—	—
		5	1433.93	76.53	56.04	2.34	41.62	—
		10	1686.67	22.51	53.09	0.00	46.58	0.33
	0.01	1	17.77	23.71	64.80	35.20	—	—
		5	891.90	47.60	49.46	3.27	47.27	—
		10	1426.48	19.03	47.00	2.53	45.06	0.00
	0.015	1	8.15	10.88	4.94	95.06	—	—
		5	1268.63	67.71	50.27	5.64	44.09	—
		10	1203.94	16.06	49.62	0.67	43.42	6.29

enhanced by the presence of reinforcement. The enhancement is larger for larger V_f values as can be seen in Fig. 13.27. Fracture (hole) develops in the neat and reinforced polymers at $u = 5$ m/s. A detailed observation in the nanocomposite specimen reveals that the elements without reinforcement fail first.

It is important to state that the presence of reinforced regions in the material volume has led to an increase of the energy finally stored in the polymeric volume in the form of elastic–plastic strains, while a decrease was more intense with increasing volume fraction. The percentage of energy consumed on the fracture of the neat polymer varied with a similar trend as the stored energy in the polymer, as the obtained decrease ranged between 2% and 16.3% of the percentage of the absorbed energy consumed on fracture phenomena of the reference material. As far as the response of the reinforced regions under impact is concerned, the absorbed energy in the form of strains, as well as the consumption of energy on fracture, was increased with rising volume fraction.

Finally, the interphase quality has an important effect on the nanocomposite response under impact, as indicated in Table 13.7. Especially, the contribution of the polymer in the

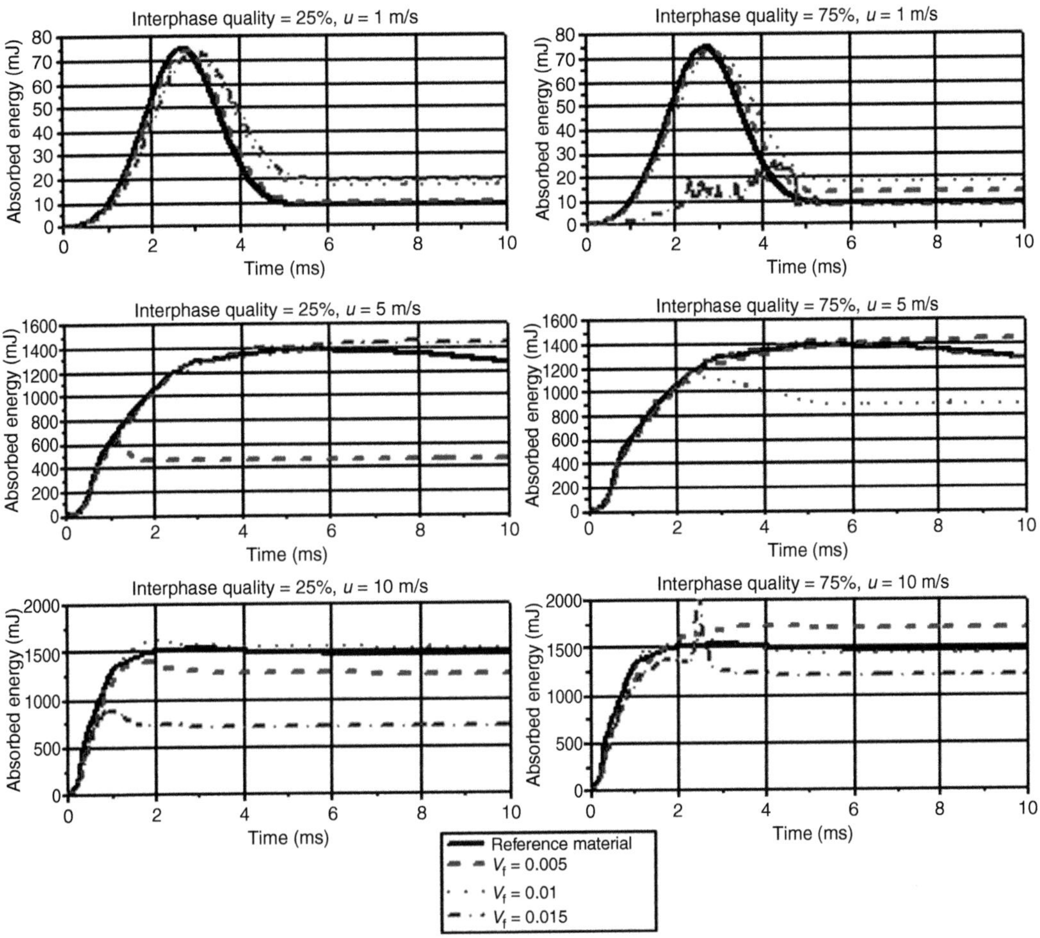

FIGURE 13.27 Comparative diagrams of the time variance of the absorbed energy for interphase quality of 25 and 75% for every case of volume fraction (V_f) and impact velocity (u), respectively.

stored strain energy decreased when the interphase quality rose, while the percentage of energy spent on fracture of the neat polymer region increased. In addition to this, the reinforced regions seem to store more strain energy with increasing interphase quality, while the corresponding percentage related to fracture is decreased.

13.7 Conclusions

This chapter describes a two-scale numerical model developed using the explicit FE code LS-DYNA for simulating the impact response of CNT/polymer nanocomposites. The model was initially used to conduct a study on the effect of geometrical and material parameters and then to simulate nanoindentation of the nanocomposite in an upper scale.

From the numerical simulations, the following conclusions are drawn:

- The presence of CNTs significantly enhances the impact response of the polymer by increasing the impact stiffness and the energy-absorbing capacity of the material. The enhancement is larger at larger impact velocities.
- With increasing V_f, the maximum reaction force and impact stiffness increase.
- The effect of the CNT's AR on the absorbed energy and reaction force is minor, while an increase of AR leads to an increase in the total internal energy and decrease in the consumed energy.
- The orthotropic behavior of CNTs gives a larger energy-absorbing capacity to the nanocomposites than the isotropic behavior at small impact velocities. At large impact velocities, the effect of CNT's behavior is small.
- The modeling parameter of L_{CNT}/L_{RVE} influences the numerical results as it determines the thickness of the polymer surrounding the CNT. The parameter can be determined through calibration with experiments or evaluation of the material's microstructure.
- The prediction of impact damage at large impact velocities indicates that the CNT makes the polymer more susceptible to fracture.
- The interphase thickness had a minor effect on the total absorbed energy, although its presence affected the energy-absorbed energy dissipations among the nanocomposite constituents. One of the most important contributions of the interphase on the impact response was the decrease of the energy consumed on the fracture phenomena on the impacted surface.
- Finally, the analysis of the macroscale specimen revealed that the energy absorption capability of the polymer was significantly enhanced by the presence of the reinforcement. Fracture took place at an even lower velocity than that noted at RVE level, while the unreinforced impacted areas were expected to fail first.

From the overall assessment of the present work, it is also concluded that the continuum modeling methods, such as the explicit FE analysis may be effectively used to simulate the performance of nanocomposites.

References

[1] A. Chanteli, K.I. Tserpes, Tensile behaviour of carbon nanotube/polypropylene composite material, Plast. Rubber Compos. 43 (2014) 330–336.

[2] A. Vavouliotis, P. Karapappas, T. Loutas, T. Voyatzi, A. Paipetis, V. Kostopoulos, Multistage fatigue life monitoring on carbon fibre reinforced polymers enhanced with multiwall carbon nanotubes, Plast. Rubber Compos. 38 (2009) 124–130.

[3] S. Laurenzi, R. Pastore, G. Giannini, M. Marchetti, Experimental study of impact resistance in multi-walled carbon nanotube reinforced epoxy, Compos. Struct. 99 (2013) 62–68.

[4] M. Tehrani, A. Boroujeni, T. Hartman, T. Haugh, S. Case, M. Al-Haik, Mechanical characterization and impact damage assessment of a woven carbon fiber reinforced carbon nanotube–epoxy composite, Compos. Sci. Technol. 75 (2013) 42–48.

[5] E. Solimana, M. Sheykab, M. Reda Taha, Low-velocity impact of thin woven carbon fabric composites incorporating multi-walled carbon nanotubes, Int. J. Impact Eng. 47 (2012) 39–47.

[6] M. Siegfried, C. Tola, M. Claes, S. Lomov, I. Verpoest, L. Gorbatikh, Impact and residual after impact properties of carbon fiber/epoxy composites modified with carbon nanotubes, Compos. Struct. 111 (2014) 488–496.

[7] W. da Silva, H. Ribeiro, J. Neves, A. Sousa, G. Silva, Improved impact strength of epoxy by the addition of functionalized multiwalled carbon nanotubes and reactive diluent, J. Appl. Polym. Sci. 132 (2015) 42587.

[8] G. Wu, L. Ma, L. Liu, Y. Wang, F. Xie, Z. Zhong, M. Zhao, B. Jiang, Y. Huang, Interface enhancement of carbon fiber reinforced methylphenylsilicone resin composites modified with silanized carbon nanotubes, Mater. Des. 89 (2016) 1343–1349.

[9] H. Ulus, T. Ustun, O. Sahin, S. Karabulut, V. Eskizeybek, A. Avci, Low-velocity impact behavior of carbon fiber/epoxy multiscale hybrid nanocomposites reinforced with multiwalled carbon nanotubes and boron nitride nanoplates, J. Compos. Mater. 24 (2015).

[10] R. Rafiee, R.M. Moghadam, Simulation of impact and post-impact behavior of carbon nanotube reinforced polymer using multi-scale finite element modeling, Comput. Mater. Sci. 63 (2012) 261–268.

[11] S.M. Khalili, A. Haghbin, Investigation on design parameters of single-walled carbon nanotube reinforced nanocomposites under impact loads, Compos. Struct. 98 (2013) 253–260.

[12] J. Jama, Y. Kiani, Low velocity impact response of functionally graded carbon nanotube reinforced composite beams in thermal environment, Compos. Struct. 132 (2015) 35–43.

[13] V. Papadopoulos, M. Tavlaki, The impact of interfacial properties on the macroscopic performance of carbon nanotube composites. A FE2-based multiscale study, Compos. Struct. 136 (2016) 582–592.

[14] K. Pandya, N. Naik, Analytical and experimental studies on ballistic impact behavior of carbon nanotube dispersed resin, Int. J. Impact Eng. 76 (2015) 49–59.

[15] ASTMD5628-10, Standard Test Method for Impact Resistance of Flat, Rigid Plastic Specimens by Means of a Falling Dart (Tup or Falling Mass), ASTM Int., 2010.

[16] K.I. Tserpes, A. Chanteli, Parametric numerical evaluation of the effective elastic properties of carbon nanotube-reinforced polymers, Compos. Struct. 99 (2013) 366–374.

[17] A. Chanteli, K.I. Tserpes, Finite element modeling of carbon nanotube agglomerates in polymers, Compos. Struct. 132 (2015) 1141–1148.

[18] Livermore Software Technology Corporation, 2014. LS-DYNA theory manual, CA, USA.

14

Theoretical Modeling of CNT–Polymer Interactions

Abbas Montazeri, Behzad Mehrafrooz

K.N. TOOSI UNIVERSITY OF TECHNOLOGY, TEHRAN, IRAN

CHAPTER OUTLINE

14.1 Introduction

In the last few decades, polymers have received a great deal of attention and are vastly utilized in advanced engineering applications such as car parts, electronic packaging, various optical components, semiconducting materials, insulating, liquid crystal display, and vibration damping. These various applications are due to their inherent characteristics such as being light weight, easy to change into complex shapes, and relatively inexpensive and low price. However, some drawbacks like low melting temperature, low stiffness and strength, and time-dependent liquid-like flow have limited their application. Therefore, ever-increasing incorporation of polymers into modern applications requires a deep understanding and quantitative prediction of their mechanical behavior. The discovery of carbon nanotubes (CNTs) has opened the door to enhancing the mechanical properties of polymer composites by adding them to the matrix materials. Previous experimental and numerical studies have confirmed that CNTs possess remarkable mechanical properties such as high stiffness and strength. These properties as well as their high aspect ratio and low density make CNTs ideal reinforcing elements in nanostructured composite materials. CNT-based polymer nanocomposites form a new class of lightweight super strong functional materials for structural applications, energy storage, molecular sensors, and

Carbon Nanotube–Reinforced Polymers. http://dx.doi.org/10.1016/B978-0-323-48221-9.00014-5

biomedical applications [1]. Recent experimental observations have demonstrated that substantial improvements in the elastic properties of a polymer can be obtained by using small volume fractions of CNTs as a reinforcing phase.

It has been well-established that the performance of CNT–polymer nanocomposites is greatly dependent on several parameters such as the alignment and dispersion of CNTs in the matrix. However, among all of parameters affecting the reinforcement phenomena, given CNT's large surface area to volume ratio, it seems that the interface between CNTs and the polymer is the most dominant factor. Since the load transferring occurs at the interface, the interfacial bonding characteristics play a crucial role in enhancing the stiffness of the host polymer and thus should be thoroughly studied [2]. Naturally, the interaction between CNT and its neighboring polymer molecules takes place through nonbonded van der Waals (vdW) interactions, which are relatively weak compared to covalent bonds. To overcome this drawback and improve the load transferring capability from the matrix to the CNT, various researches have been conducted to establish covalent cross-links in the interfacial region. This was accomplished by a chemical functionalization process in which elements (e.g., oxygen, fluorine, and nitrogen) or small organic groups such as hydroxyl and carbonyl have been introduced in the interfacial region to construct the covalent bond at this area [3]. On the other hand, several publications have appeared in recent decades reporting that this process also leads to the generation of some geometrical defects in the structure of CNTs. Consequently, from this point of view, applying covalent cross-links has potentially a negative impact on the mechanical behavior of the whole nanocomposite and thus should be considered carefully.

It is noted that in the case of strong interactions, since the nanotubes are on the same length scales as the polymer chains, the mobility of polymer segments near the CNT is significantly reduced compared to that of the chains existing in the bulk polymer. Subsequently, a new phase in the nanocomposite is formed, which is referred to as the *interphase* or *bounded polymer layer*. Thus, the interphase is a particular region of polymer chains in the neighborhood of CNTs having different properties from those of the matrix and nanotubes. There are several experimental and numerical data in the literature demonstrating the existence of the interphase layer. For example, the results obtained by differential scanning calorimetry indicated a second glass transition temperature (T_g) in the case of a CNT/epoxy nanocomposite attributed to the formation of this new layer [4]. Also, a substantial change in the viscoelastic behavior of the polymer matrix has been reported in some case studies due to the presence of the interphase region [5]. Moreover, the reduced movement of the polymer chains in this layer was detected by employing a solid-state nuclear magnetic resonance technique [6]. In addition, molecular dynamics (MD) simulations were accomplished by Seidel et al. [7] to provide supporting proofs for the existence of this region. Their findings depicted that attractive interaction between the CNT and the matrix makes an area having a polymer of higher density near the interface.

Aside from the formation of this new phase, it has been proven that the mechanical performance of nanotube-reinforced polymer composites is greatly affected by the interfacial characteristics. Therefore, before computing the overall behavior of this type of nanocomposites, particular attention should be devoted to analyzing this region. Regarding this, as will be discussed in the next section, some valuable experiments have been directed to address the interface's features. Despite these efforts, it should be emphasized that the interfacial behavior is chiefly dominated by atomistic scale interactions, which are difficult to control and measure experimentally. In addition, experimental surveys of nanocomposites demand the use of complicated processing methods and testing equipment that limit their applications in this case. To overcome these issues, numerical modeling has been successfully carried out to provide some crucial insights about the role of the interface. Generally, there are two main approaches implemented for modeling the mechanical behavior of nanostructured materials: continuum mechanics-based methods and atomistic methods. Although continuum-based approaches like finite element (FE) method and micromechanical analysis are rapid modeling techniques, these methods fail to consider interfacial interactions within the atomic level. It comes from the fact that the main hypothesis of these methods is the perfect bonding assumption between the CNT and the matrix [8]. Consequently, a higher enhancement modulus is calculated by these methods compared to the experimental results for CNT-reinforced nanocomposites [9,10]. In contrast, atomistic-based numerical methods such as MD simulations are applicable to obtain atomic-scale information about the interface. Therefore, atomistic-based numerical methods have gained substantial importance in probing atomistic interactions that happen at the interface of these nanocomposites. This chapter tries to do a systematic review of the current literature discussing the atomistic-based approaches implemented to model the CNT/polymer interactions together with the role of these interactions on the load transfer issues in this type of nanocomposites.

The remainder of the chapter is organized as follows. Section 14.2 outlines experimental investigations on the CNT–polymer nanocomposites. Next, Section 14.3 discusses numerical modeling techniques and examines current numerical techniques in the modeling of CNT–polymer interactions at different spatial and temporal scales. In Section 14.3, we focus on molecular mechanics (MM), molecular structural mechanics (MSM), all-atom (AA), and coarse-grained molecular dynamics (CGMD) simulation studies. Finally, Section 14.4 concludes the chapter.

14.2 Experimental Investigations

Recent advances in experimental techniques such as spectroscopy, wetting, optical tweezers, and atomic force microscopy (AFM) provided helpful analyses of nanoscale systems such as CNT–polymer nanocomposites. The current literature on experimental studies for characterizing the interaction between a CNT and its surrounding polymer can be broadly categorized into different classes and types. For example, considering the enhanced property of the composite, these studies can be classified based on the selective polymer properties enhanced due to the interfacial interactions. Among them, great efforts have been devoted to analyze the mechanical [11–13], electrical [14,15], thermal [16,17], and optical [18] properties of the host polymer. Additionally, from the perspective of the employed experimental procedure, the following main groups can be presented. Some researchers have applied spectroscopy techniques including Fourier transform infrared Raman spectroscopy and X-ray scattering, which are powerful techniques for material characterization. It is noteworthy that the spectroscopy techniques have been widely employed to characterize interfacial properties of CNTs and polymer matrices [9]. Different interaction types such as the nonbonded interaction between CNTs in a bundle [19,20] as well as hydrogen [21] and covalent bonds [22] between CNTs and the surrounding matrix can be characterized by these techniques. As an example, Ma et al. [23] evaluated the effects of adding amino and epoxide groups on the CNT–epoxy interactions based on the strain-sensitive Raman spectroscopy. In this technique, transferred mechanical loadings to CNTs can be captured by Raman spectroscopy. This happens through induced local strains to CNTs, which in turn caused a significant change in the C—C bond vibration.

AFM has evolved into a useful technique to investigate materials properties at the nanometer scale, which exhibits an extremely high accuracy in force–displacement measurement [24]. In the literature, AFM has been used in two different methods: (1) coating the AFM tip with a polymer and using a CNT substrate [25] and (2) attaching a CNT to the AFM tip and using a polymer substrate [26]. The latter group, in turn, can be divided into two main approaches: pullout [27] and peeling experiments [28]. Based on the nanopullout experiment, Tsuda et al. [29] proposed a simple and easy way to directly explore the interfacial shear strength (IFSS) of CNT-dispersed polymer composites. As shown in Fig. 14.1, in the introduced approach, AFM was utilized to manipulate the CNT in relation

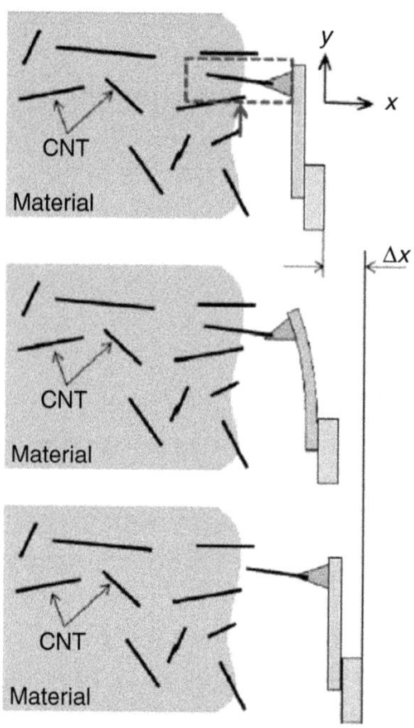

FIGURE 14.1 Schematic illustration of the nanopullout testing procedure [29].

to the polymer. Accordingly, the tensile load was imposed in the x direction using a piezoelectric actuator. Consequently, the pullout load was computed by determining the relative displacement of the AFM cantilever tip (Δx) after testing. It should be emphasized that in this experimental setup, (Δx) was evaluated from scanning electron microscopy images directly. The results implied that the adhesion has a strong correlation with physical and chemical phenomena at the interface such as the polymer properties near the nanotube and covalent bonding between the polymer and defect sites on the CNT.

Another efficient method in the literature is transmission electron microscopy, which can demonstrate the mechanics, fracture, and failure processes of the interface in real time. The aforementioned technique has been widely used to investigate interfacial adhesion between functionalized CNT–polymer nanocomposites [30,31]. Furthermore, at the macroscale, several studies have been conducted on the CNT–polymer bulk composites. The obtained results revealed improvement in mechanical properties such as Young's modulus and shear strength, which are correlated with continuum theories and modeling [20,32,33]. We refer the reader to other articles regarding a comprehensive review of the experimental approaches [34,35].

Totally, there are several reasons that have made experimental studies in the nanometer range a hard-to-reach task. Among them are the following: limitations imposed by the size of the specimen, lack of preciseness in obtaining data by indirect techniques, lack of useful methods to give nanometer-scale measurements in a direct way, the absence of techniques for the preparation of the specimen, and insufficient control over the distribution and adjustment of nanotubes. On the other hand, the aforementioned experimental techniques have gained unparalleled accuracy in both temporal and spatial scales. Given the challenges associated with directly probing materials at the nanoscale, it seems that it is difficult to explore the load transfer at the interface of these nanocomposites experimentally. Accordingly, exploiting theoretical methods to predict interface characteristics may reduce the need for costly and expensive experiments. Thus, numerical modeling as well as theoretical methods could be considered as predictive tools that complement experimental studies. More precisely, making new CNT–polymer nanocomposites would be easier if one could investigate their properties before actually producing them. Besides, the lack of a direct method for measuring interfacial bonding characteristics between the CNT and the surrounding polymer creates another restriction which, in turn, leads to errors in the results. Consequently, these issues demonstrate the importance and significance of the numerical modeling of materials across a very broad range of time and length scales and quite recently, considerable attention has been paid in this research scope. Hence, the next section is devoted to reviewing these techniques.

14.3 Numerical Modeling Techniques

So far, particular attention has been devoted to modeling the mechanical behavior of CNT–polymer nanocomposites at various temporal and spatial scales via different numerical techniques. It is noted that the macroscopic mechanical behavior of this type of composites is mainly controlled by the interplay of properties throughout these scales. Accordingly, to understand the interaction between the CNT and the polymeric matrix, it is crucial to elucidate the mechanism dominating the interactions at each scale. Fig. 14.2 demonstrates the classification of various methods employed for investigating CNT–polymer nanocomposites at each time and length scale. As can be seen, several approaches have been implemented for modeling CNT–polymer interactions, ranging from the highly detailed quantum mechanics (QM), through classical molecular modeling to process continuum mechanics. As depicted in Fig. 14.2, in the smallest length and time scale, QM is mostly used to explore the atomic structure. These studies involve the solution of Schrödinger's equation for each electron and can be exploited to any system of interest. In this scope, typical length scales are of the order of 0.1 nm. Although several researchers have employed the QM method to investigate the CNT–polymer interaction [36–38], this method is computationally exhaustive and thus is attributed to systems of small size. However, based on the QM calculations and Schrödinger's equation, it is possible to develop force fields that accurately describe atomic interactions. With the aforementioned force field in hand, molecular dynamics simulation can be used to solve Newton's equations.

At the next modeling scale, molecular modeling techniques that will be thoroughly discussed in this section are illustrated. These methods assume the presence of a separate molecular structure of material and can successfully be utilized to model larger systems having a higher number of atoms and molecules. In this level of description, the electronic degrees of freedom have been neglected. In the largest scale, continuum mechanics theories are applied assuming a continuous structure in the material. However, the multiscale methods, such as CGMD simulations, are used in the middle region as a new bridge to integrate nanomechanics to continuum mechanics. Despite the fact that MD simulations lead to many efficient and useful results in nanomaterials science, still many systems of interest require a huge cost of computation and CPU time. Hence, different coarse-grained (CG) models have been proposed in recent years. The CG model is an attempt to eliminate some degrees of freedom that are not important for the problem of interest and enables to extend the accessible time and length scales. It is noteworthy that multiscale modeling provides a connection between microscale and nanoscale. In addition, Fig. 14.2 presents an indication of what kind of properties can be computed at different scales from atomistic to continuum mechanics. This section reviews recent advances in modeling the interaction between CNTs and polymers via MM, MSM, MD, and CGMD.

As mentioned earlier, one of the most important issues about large-scale atomistic simulations is the role of computing resources. Nowadays, classical MD and Monte Carlo methods can be quite efficiently applied in modern supercomputers. Fig. 14.3 shows the effect of the historical development of computational power over several decades on atomistic modeling and predicting the emergence of extra large-scale atomistic simulation in the next few years.

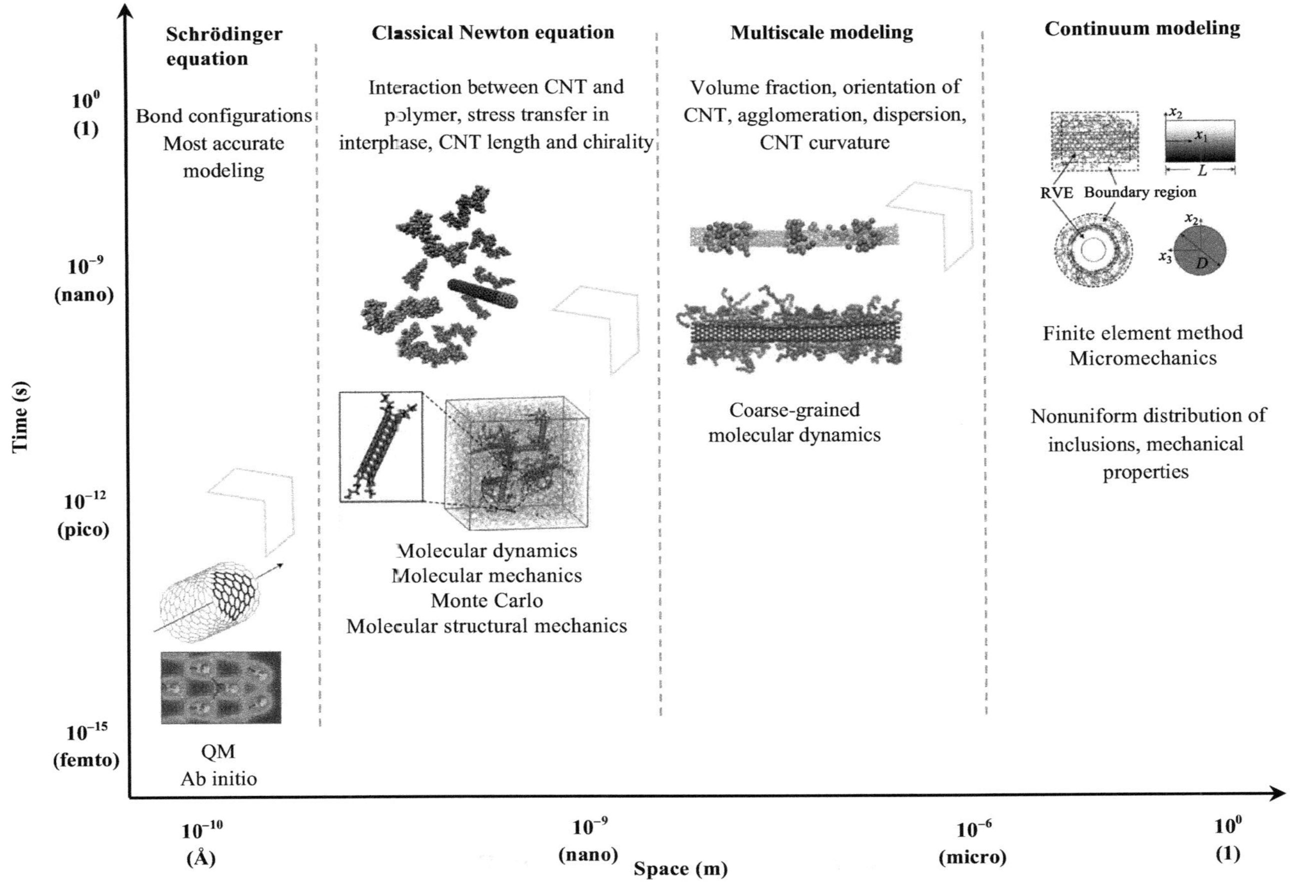

FIGURE 14.2 Various temporal and spatial scales of CNT–polymer nanocomposite modeling (QM, [37] and [39]; MD, [40] and [41]; CG, [42]; and continuum [43]).

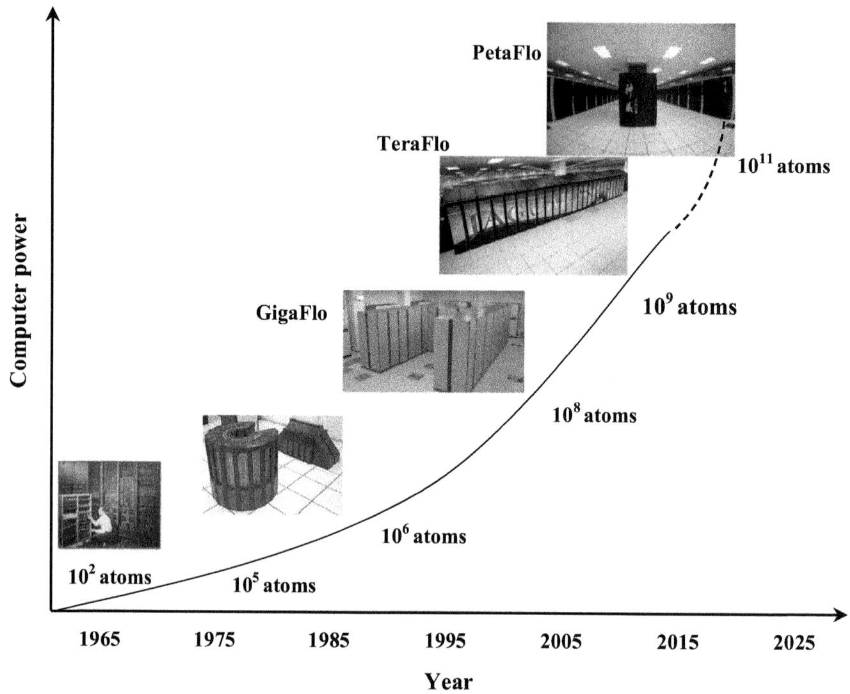

FIGURE 14.3 Development of computing power and atomistic simulation over the past decades.

14.3.1 Molecular Mechanics

The term MM refers to an approximation using Newtonian mechanics model interactions inside the molecule and intermolecular interactions. The potential energy of all systems in this approach is calculated using force fields. MM can be exploited to investigate molecular systems ranging in size and complexity from small to large material assemblies with many thousands to millions of atoms. In MM each atom is represented as an atom-centered interaction particle with an assigned radius, polarizability, and a constant net charge. In this method, nuclei and electrons are lumped into atom-like particles. Moreover, bonded interactions are treated as springs with an equilibrium distance equal to the experimental or calculated bond length.

In MM, the total energy of the CNT–polymer composite has the following form:

$$E_{tot} = E_{CNT} + E_{polymer} + E_{intermolecular} \quad (14.1)$$

where E_{total} is the total energy of the composite, E_{CNT} and $E_{polymer}$ denotes the potential energy of the CNT and the polymer matrix, respectively. Finally, $E_{intermolecular}$ indicates the energy of the interaction between the CNT and the polymer matrix. The potential energy of CNT can be determined usually by Brenner [44], AIREBO [45], and Tersoff [46] potentials. MM expresses the $E_{polymer}$ as a sum of Taylor series expansions for stretches of every pair of bonded atoms and adds additional potential energy terms coming from bond bending

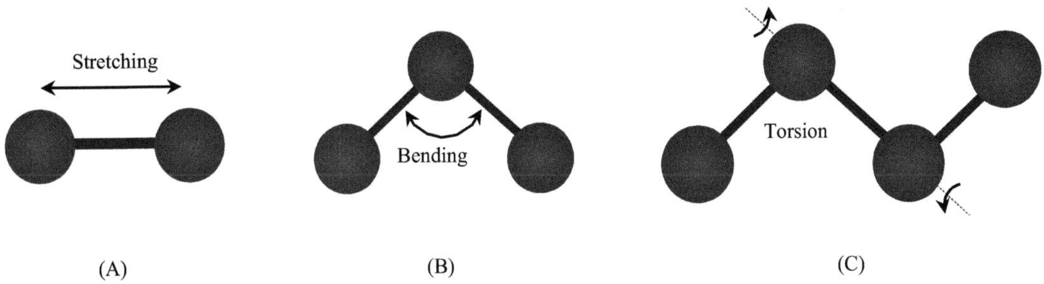

FIGURE 14.4 Bonded energy in MM. (A) Bond stretching, (B) bond bending, and (C) bond torsion.

and torsion, nonbonded vdW and electrostatics interactions, and also cross terms. Thus, the general expression for the $E_{polymer}$ is written as

$$E_{polymer} = E_{str} + E_{bend} + E_{tors} + E_{vdw} + E_{el} + E_{cross} \tag{14.2}$$

In Fig. 14.4 stretching, bending, and torsion terms are illustrated. In CNT–polymer composites, $E_{intermolecular}$, that is, the interactions between the two distinct phases, are usually the nonbonded or vdW forces. The most common potential used to model the vdW interactions is known as the Lennard-Jones (LJ) potential [47].

After calculating the energy of a molecule, MM adjusts the energy through changes in the bond lengths and angles to obtain the minimum energy structure. It is noteworthy that MM is different from MD. In MM the equilibrium condition of the system is determined by minimizing the potential energy, while MD solves the equation of motion based on certain time integration schemes.

MM and elasticity calculations have been used to study the interfacial properties of CNT–polystyrene composites by Liao and Li [48]. Their results showed that the CNT–polymer adhesion comes from the noncovalent intermolecular interactions between polymers and CNTs such as vdW and electrostatics interactions. Results of a CNT pullout simulation suggested that the interfacial shear stress of CNT/polystyrene (PS) is about 160 MPa. The energy variation of the CNT–PS composite against the displacement of the CNT from the matrix is shown in Fig. 14.4. The work required for fiber pullout can be calculated as the energy difference between the fully embedded tube and the complete pullout configuration. Accordingly, in the mentioned study, pullout energies for single- and double-walled tubes were found to be 158.7 and 154.4 kcal/mol, respectively. As shown in Fig. 14.5, the energy of a single-walled CNT (SWCNT) at pullout was 0 kcal/mol while that of the double-walled CNT (DWCNT) was approximately −100 kcal/mol. This negative energy of the DWCNT comes from the interaction between the outer and inner tubes.

Based on the MM simulation, Mokashi et al. [49] investigated the mechanical properties of crystalline polyethylene (PE) composites embedded with SWCNTs subjected to quasi-static tensile loading as illustrated in Fig. 14.6. Their observations revealed that the length of the CNT and the PE configuration play an important role in the tensile properties of nanocomposites. They also reported that a long CNT increased the modulus of crystalline while a short CNT caused a significant reduction in the modulus of amorphous

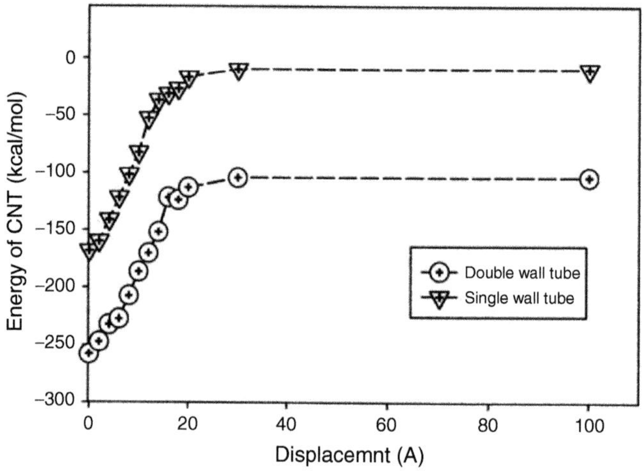

FIGURE 14.5 Pullout energies of SWCNT and DWCNT from the PS matrix [48].

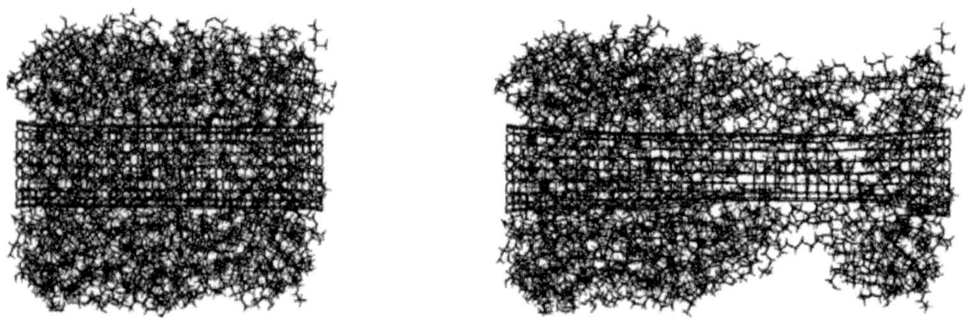

FIGURE 14.6 Tensile loading of nanotube–polyethylene (PE) (amorphous) composite [49]. (A) Equilibrium structure and (B) structure at nanotube fracture (35% strain).

PE. In Fig. 14.7, the comparison graphs for engineering stress versus strain for different volume fractions with short, embedded CNT are plotted. As Fig. 14.7 shows, the tensile strength of the composite was even less than pure amorphous PE, since the contact area of the CNT–PE interface was too small and therefore caused a poor load transfer at the interface. In this case, the short embedded CNT may act like a void, reducing the tensile strength of the matrix. Similar results were reported by Haghighatpanah and Bolton [50] by performing MD simulation on CNT–PE composites having short CNTs. Their findings revealed that short CNTs do not increase the Young's modulus of the system whereas long CNTs do.

Qian et al. [51] used MM and MD to investigate the nature of load transfer in an SWCNT bundle consisting of seven (10, 10) SWCNTs. Moreover, they also answered the question whether sufficient compressive stress from a surrounding polymer matrix on a bundle of SWCNTs could dramatically increase the load transfer between the SWCNTs in the bundle.

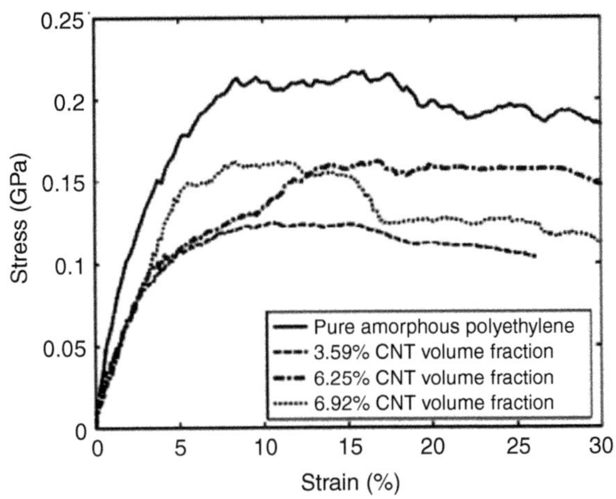

FIGURE 14.7 Engineering stress versus strain plots for amorphous PE and embedded nanotube composite systems [49].

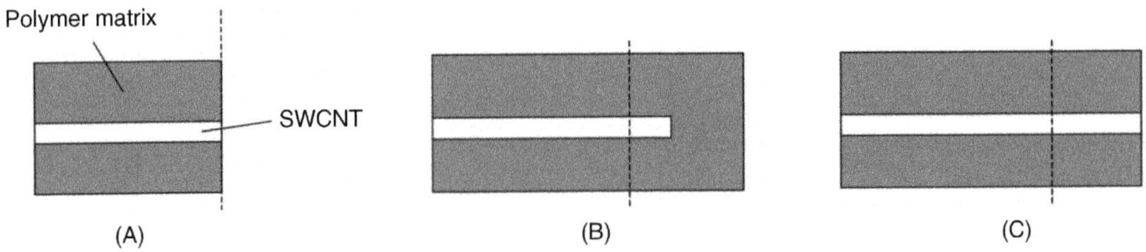

FIGURE 14.8 Simple analysis of uniaxial tension of a polymer reinforced by SWCNT [51]. (A) Initial system. (B) Assuming no bond between the tube and matrix, both will expand under increasing temperature. (C) Assuming bonding, the tube will stretch more and the matrix will stretch less compared to (B).

A simple analysis on the uniaxial thermal stretch (Fig. 14.8) in the polymer matrix reinforced by an SWCNT revealed a compressive stresses around the bundle accompanied by tensile strengths of epoxies and polyimides. It has also been reported that compressive stress in composites will not cause any significant increase in the load transfer between closest-packed SWCNTs.

Wong et al. [52] used MM simulations in combination with elasticity calculations to study the interfacial interaction and local fracture morphologies of CNT–PS rod and CNT/epoxy film composites. In the absence of chemical bonding at the interface, it was found that the nonbonded interactions result in CNT–polymer interfacial shear stress of about 138 and 186 MPa, respectively, for CNT/epoxy and CNT/PS. In this work, to study the effects of the local nonuniformity along the CNT including varying diameters, a system consisting of a CNT with diameter variation embedded in an array of linear PE was modeled. Then, the CNT was pulled through the polymer "brush" from the end with the smaller diameter. Fig. 14.9 demonstrates changes in the potential energy during the pull-out of CNT from the PS. As is clear from Fig. 14.9, when the portion of the CNT with larger diameter is drawn closer into the polymer, a jump in the potential energy of the system

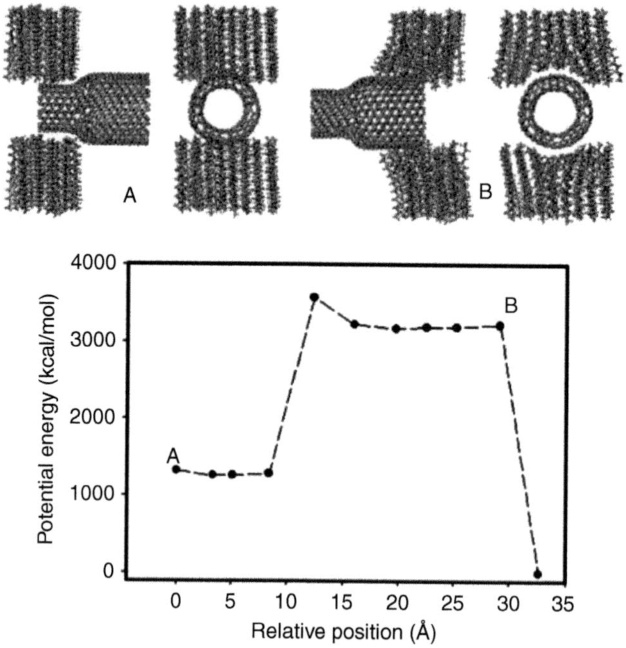

FIGURE 14.9 Molecular model of a CNT embedded in two layers of short linear PE array [52]. (A) Point of entry, and (B) near-pullout position.

takes place consequently causing deformation of the polymer. Based on the obtained results, it was found that the close contact of the CNT and the polymer and nonuniformity in the CNT structure can drastically enhance the interfacial interactions suggesting that mechanical interlocking can be an important contributor for CNT–polymer adhesion.

Lordi and Yao [53] performed energy minimization MM calculations to compute the binding energies and sliding frictional stresses between the nanotubes and different polymer matrices. This paper has clearly shown that binding energies and frictional forces play only a minor role in determining the strength of the interface, but helical polymer conformations are essential. They suggested that the strength of the nanotube–polymer interface may result from the molecular-level entanglement of the nanotube and the polymer matrix. They also predicted maximum frictional stresses from 18 to 135 MPa for sliding (10, 10) nanotubes within single polymer chains.

Based on MM and MD simulations, Zheng et al. [54] studied the influence of chemical functionalization on the interfacial bonding characteristics of SWNT-reinforced polymer composites, namely, PE and poly(methyl methacrylate) (PMMA) (Fig. 14.10). As indicated in Fig. 14.11, chemical functionalization of the SWNTs has been carried out by attaching functional groups to the surface of the CNTs through chemical covalent bonding and the functional groups were randomly end-grafted to the surface of the SWCNTs.

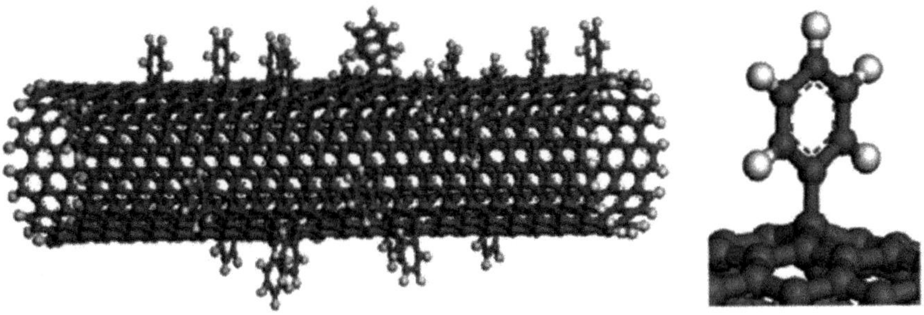

FIGURE 14.10 Illustration of a (10, 10) SWNT with phenyl groups randomly chemisorbed to 2.5% of the carbon atoms [54].

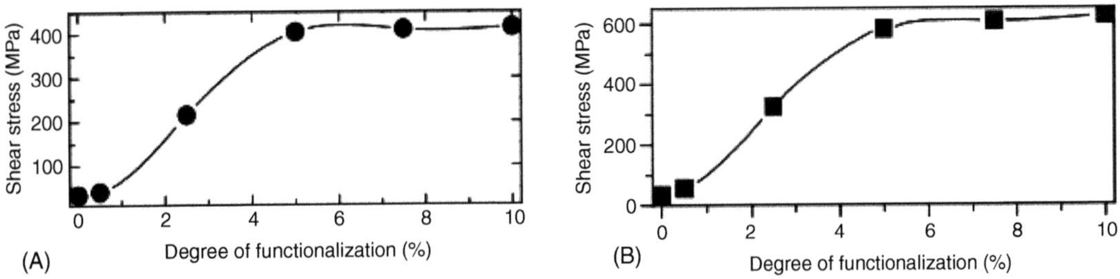

FIGURE 14.11 Influence of chemical functionalization on shear stress [54]. (A) SWNT–PMMA system. (B) SWNT–PE system.

In conclusion, it was shown that functionalization of nanotubes at low densities of functionalized carbon atoms drastically increases the interfacial bonding strength and shear stress between the nanotubes and the polymer matrix. Functionalization of as little as 5.0% of the nanotube carbon atoms caused an increase in the shear stress by about three orders of magnitude.

Totally, MM is a useful tool for modeling CNT–polymer interactions because it requires less computational resources compared to the quantum mechanical methods and thus, can model larger systems in a more effective way. It is also able to calculate properties such as optimized structures (minimum energy conformation) and enthalpy of formation. This method uses an appropriate algorithm such as steepest descent to find the molecular structure of a local energy minimum. However, it should be usually accompanied by another method such as MD, Monte Carlo, or continuum mechanics to find different aspects of the interfacial interactions.

14.3.2 Molecular Structural Mechanics

Due to the huge computational costs associated with atomistic modeling techniques such as MM and MD simulations, there has been a demand for developing approaches that analyze the mechanical response of nanostructures at the atomic scale but with proper length and time scales. Regarding this issue, Li and Chou [55] extended the theory of classical

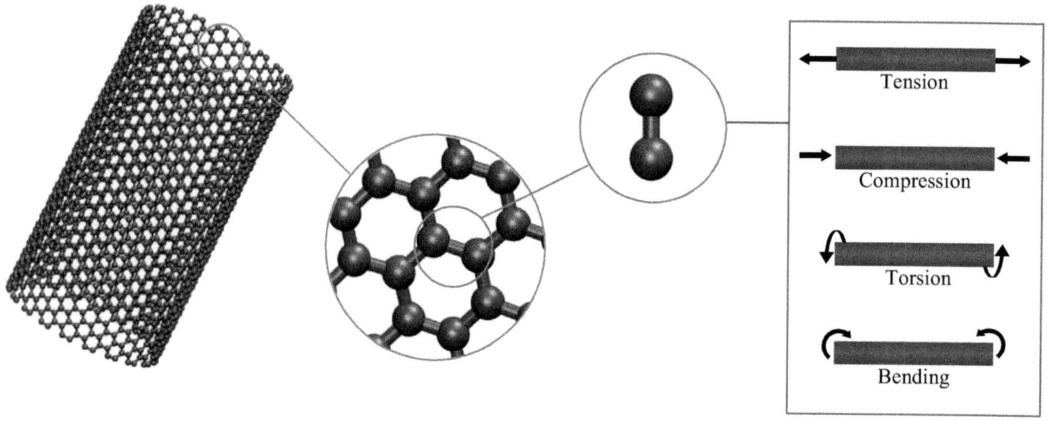

FIGURE 14.12 Schematic representation of the MSM model.

structural mechanics into the modeling of CNTs to introduce a new MSM method. In this method, an SWCNT is viewed as a space frame, where the covalent bonds are represented as connecting beams and the carbon atoms as joint nodes as illustrated in Fig. 14.12. The element used for the covalent bonds is a uniaxial element with tension, compression, torsion, and bending capabilities and has six degrees of freedom at each node: three translations in the x, y, and z directions and three rotations about the x, y, and z axes. This element is defined by cross-sectional area, moment of inertia, and material properties based on the energy equivalence between computational chemistry and structural mechanics. To this end, the force field constants of the covalent bonds are used as follows:

$$\frac{EA}{L} = K_r \tag{14.3a}$$

$$\frac{EI}{L} = K_\theta \tag{14.3b}$$

$$\frac{GJ}{L} = K_\varphi \tag{14.3c}$$

where the force field constants K_r, K_θ, and K_φ represent stretching, bending, and torsional stiffness of the covalent bonds, respectively. For C—C covalent bonds, the force field constants obtained experimentally in the context of chemistry by Cornell et al. [56] are used. These are well-known constants for modeling the carbon–carbon covalent bonds in CNTs and have been used successfully for modeling the static, dynamic, and thermal properties of CNTs and their composites. These values are listed in Table 14.1. Also, E and G denote moduli of elasticity and shear of the element, respectively. Moreover, A, I, J, and

Table 14.1 Force Field Constants of the Covalent Bonds of SWCNT [56]

K_r	6.52e-2 nN nm^{-1}
K_θ	8.76e-1 nN nm rad^{-2}
K_φ	2.78e-1 nN nm rad^{-2}

L are the cross-sectional area, the moment of inertia, the polar moment, and the length of the beam. The last parameter is assumed to be equal to the covalent bond distance of the carbon atoms (0.142 nm). Specific parameters of the element with a circular cross-section can be obtained from Eq. (14.3) as

$$d = 4\sqrt{\frac{K_\theta}{K_r}} \tag{14.4a}$$

$$E = \frac{K_r^2 L}{4\pi K_\theta} \tag{14.4b}$$

$$G = \frac{K_r^2 K_\varphi L}{8\pi K_\theta^2} \tag{14.4c}$$

where d is the cross-sectional diameter of the element.

With the beam element representing the lattice structure of CNT in hand, the developed MSM model can then be adapted into an FE model for the prediction of the mechanical properties of nanotube-reinforced composites. It is noted that from the traditional theoretical frame for evaluating the mechanical properties of composites, a possible approach for modeling CNT–polymer nanocomposites is to build up a representative volume element (RVE) constituted by a cubic or cylindrical body of the matrix with an embedded nanotube. In a similar manner, Li and Chou [57] were the first to develop a new MSM/FE concurrent multiscale method to study the mechanical behavior of SWCNT–polymer composites. To model the matrix, they considered the linear elastic behavior for polymer and analyzed the matrix deformation at the macroscopic scale by employing the continuum FE method.

As stated before, naturally, the nanotube and polymer matrix are assumed to be bonded by vdW interactions in the absence of any chemical functionalization. Therefore, in the proposed method, the bonding via vdW interactions is considered. These interatomic forces are defined based on the well-known LJ potential at the interface. For modeling these forces, nonlinear spring elements are utilized. To construct these elements, the distances between the nodes on the CNT wall and the nodes on the inner surface of the polymer matrix are computed. A spring element is then inserted between every two nodes with their distance smaller than the cut-off radius of the LJ potential (i.e., 0.85 nm) (Figs. 14.13 and 14.14). This

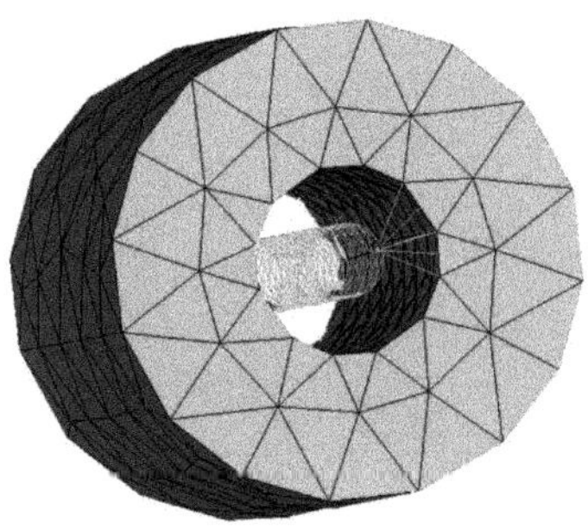

FIGURE 14.13 Spring elements implemented to model nonbonded interfacial interactions [58].

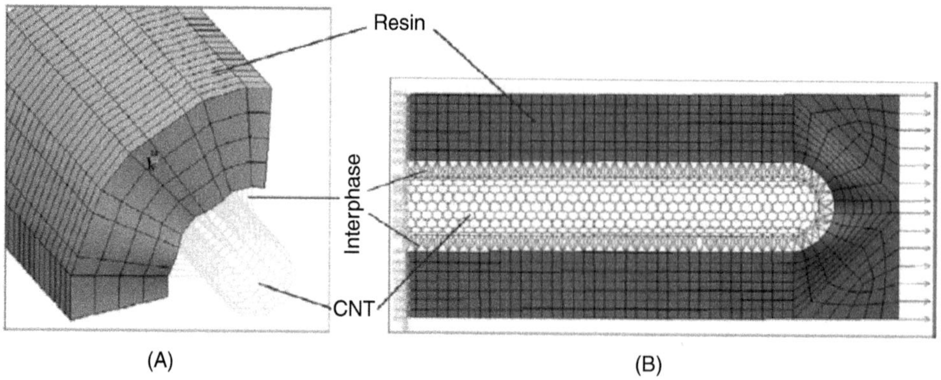

FIGURE 14.14 Spring elements representing the interfacial interactions in the case of composites reinforced with short CNT [59]. (A) Cut section of RVE and (B) side view of RVE with short CNT.

spring is a uniaxial tension–compression element with three degrees of freedom at each node: three translations in the x, y, and z directions. No bending or torsion is considered in this element. The spring stiffness of this element is determined by the second derivative of the LJ potential, as follows:

$$V_{LJ} = 4\varepsilon\left[\left(\frac{\sigma}{r}\right)^{12} - \left(\frac{\sigma}{r}\right)^{6}\right] \tag{14.5}$$

$$k = \frac{d^2V_{LJ}(r)}{dr^2} = \frac{624\varepsilon\sigma^{12}}{r^{14}} - \frac{168\varepsilon\sigma^{6}}{r^{8}} \tag{14.6}$$

where r is the interatomic distance and ε and σ denote the LJ parameters. For C–C van der Waals interactions, these parameters are 0.0566 kcal/mol and 0.34 nm, respectively [57]. Substitution of these values into Eq. (14.6) yields the spring stiffness of the interface elements as follows:

$$k = \frac{574.89188\times10^{-9}}{r^{14}} - \frac{1.00193\times10^{-4}}{r^{8}}\left[\frac{\text{nN}}{\text{nm}}\right] \tag{14.7}$$

For a better prediction of the interfacial interactions in this type of nanocomposites, Shokrieh and Rafiee [60] developed an adaptive vdW interaction (AVI) model. According to this model, as the FE analysis evolves, some vdW interactions are deactivated as their distances exceed 0.85 nm. Simultaneously, some new vdW interactions are created and activated based on the updated situation of the model. Employing this model, they examined the tensile behavior of an embedded CNT in a polymer matrix with a nonbonded interfacial region [59]. Additionally, conducting a systematic parametric study, they examined the effects of nanotube length on the reinforcement role of this nanostructure. Based on the nonbonded interface, it was found that the capped short CNTs enhance the elastic constants of polymeric matrices less efficiently than long ones. This was ascribed to the lower values for load transfer occurring in the interfacial region modeled via the AVI model [61]. Moreover, based on the adaptive vdW bonding method, they developed an equivalent long fiber for an embedded CNT in the polymer matrix. Comparing the elastic properties predicted by this method with that of the theoretical approaches such as the rule of mixture method, it was shown that the predicted data by this method are in more agreement with the experimental results [60]. Therefore, it seems that considering the interfacial interactions in CNT–polymer composites by nonbonded forces is a more realistic way than modeling them via perfect bonding as assumed in the simple rule of mixture formula. In addition to the reinforcement role of CNTs, other aspects of the mechanical behavior of this type of nanocomposites were also investigated using the AVI model by this group. For example, employing the introduced model, the influence of the nonbonded interface on the crack driving force was examined in the absence of the functionalization process [62]. It was concluded that replacing the lattice structure of CNT and its weak interfacial interactions by perfectly bonded solid fiber results in an overestimation in the prediction of the crack growth propagation in this case. The model was also successfully implemented to reveal the reinforcement role of CNTs in improving the dynamic behavior of polymers against impact loading [63]. In a similar way, the influence of CNT waviness on the reduction of stiffness in these nanocomposites was analyzed by the same group [64].

Referring to the nonbonded CNT/polymer interactions modeled by spring elements developed based on the MSM methodology, Montazeri and Naghdabadi investigated several aspects of CNT–polymer composites [58,65,66]. For example, in most of the simulation methods used to explore the mechanical behavior of these nanocomposites, linear elastic behavior has been assumed for the polymer matrix. Meanwhile, because nanotubes are on the same length scale as polymer chains, the polymer segments near the nanotubes are identified by a mobility that is different from that of the polymer chains in the bulk material. This reduced mobility results in significant differences between the viscoelastic behavior of the bulk polymer and that of the nanotube-reinforced polymer. Accordingly, using the MSM/FE multiscale model introduced here, the influence of considering a viscoelastic polymer instead of a linear elastic one on the mechanical behavior of CNT/polymer composites was investigated [58]. Also, due to the limited theoretical and experimental works available in the literature on the quantitative analysis of the effect of the interphase layer on the elastic constants of these nanocomposites, they proposed a new three-phase MSM/FE multiscale model [65]. Based on the modeling of the nonbonded interfacial interactions by the nonlinear spring elements introduced before, they employed a mathematical procedure to characterize the elastic properties of the interphase layer between the CNT and the polymer matrix in the RVE. Accordingly, the results obtained from two-phase and three-phase modeling (without and with considering the interphase layer) were compared with an AA MD analysis to find out the effect of the interphase on the mechanical properties of CNT-based nanocomposites. Finally, this group utilized the verified three-phase multiscale model to study the effect of CNT vacancy defects on the stability of SWCNT–polymer composites [66]. To this end, the buckling onset strain was calculated for perfect and defective CNTs in the polymer nanocomposites. There exist two different buckling regimes for thin-walled cylinders like CNTs under compressive axial loadings: (1) column (global) buckling and (2) shell (local) buckling, considered in this study. Additionally, the effect of interfacial mechanical characteristics on the stability behavior of defective nanocomposites was discussed.

It is worth mentioning that so far, several studies have been conducted to include the CNT functionalization in the proposed spring-based method for modeling the interface. Such a matter has been thoroughly examined by Rafiee and Pourazizi [67]. For this purpose, they studied two different cases resembling normal CNT and functionalized CNT. It was assumed that normal CNT interacts via nonlinear spring elements representing nonbonded vdW interactions. Meanwhile, for the other case, they introduced linear beam elements simulating covalent cross-links between the CNT and the polymer. The covalent bonds were established between those nodes in the CNT model incorporating defect and neighboring nodes on the polymer matrix.

In summary, molecular structural mechanics provides an effective route to investigate the interfacial interactions in these nanocomposites in a detailed atomistic way. However, as discussed in this section, its application is limited to the existence of force field constants implemented to define the mechanical characteristics of the elements representing the region. Although in the literature there are several cases utilizing the well-known LJ potential for modeling these interactions, it has been well-established that the potential works well only for nontreated CNTs. Meanwhile, various attempts have been directed in recent years to study the role of adding some functional groups at the interface to enhance the load transfer issue that cannot be followed via this method. Additionally, the harmonic potential hypothesis utilized to model the energy variations in the lattice structure of a nanotube is only valid for small deformation problems limiting the application of MSM. Furthermore, it should be emphasized that the method is basically unable to capture the trajectory of interfacial atoms resulting in a shift to other techniques such as MD simulation.

14.3.3 Molecular Dynamics

Ever since the MD method was proposed by Alder and Wainwright in the late 1950s and early 1960s [68], due to the ever-increasing power of computers and enormous advantage of parallel computing techniques, MD has been widely used to study the structure and dynamics of different materials, such as polymers or biological systems. MD simulation is a favorite technique for elucidating the atomistic mechanisms that solve each atom's equation of motion and follow the time evolution of the system. Such an evolution is called a trajectory. Solving Newton's classical second law for each atom requires the force acting on it be obtained from the summation of the spatial derivative of the atomic force field. More precisely, the interatomic interactions are described by a function of the atomic coordinates as follows:

$$U_{\text{tot}} = U(r_1, r_2, \cdots, r_N) \tag{14.8}$$

where N refers to the collection of all particles in the system. Given the interatomic force field, the force acting upon ith atom is obtained by the gradient with respect to the atomic displacements, as

$$F_j = m\frac{\mathrm{d}^2 r_j}{\mathrm{d}t^2} = -\nabla_{r_j} U(r_1, r_2, \cdots, r_N) = -\left(\frac{\partial U}{\partial x_i}, \frac{\partial U}{\partial y_i}, \frac{\partial U}{\partial z_i}\right) \tag{14.9}$$

Thus, the atomic force field model describes physical systems as collections of atoms kept together and can be cited as "the heart" of MD. Various force fields, such as DREIDING [69] and COMPASS [70], have been defined to model the polymer interaction, or AIREBO [45] for CNT modeling. However, some studies have been focused on developing or extending force fields to model CNT and polymer interactions. For example, Zhang et al. [71] adopted the UA (united atom) potential to simulate the interaction of CNT and polymers. In Ref. [72], AIREBO potential was developed to model the mechanical behavior of the CNT–polymer nanocomposites.

In modeling of the CNT and polymer interaction, it can be stated that MD is the most common method implemented among all of the other reviewed approaches. In other words, as MD performs modeling of the systems at the atomic level and is free of averaging errors, it can provide great detail about the understanding and improvement of interactions between the CNT and the polymer. However, large computational resources necessary to perform MD simulations serve as a drawback. Due to computational costs and limitations, MD simulations are often restricted to the time-scales that can be reached, limiting overall time spans to tens of nanoseconds. So far, several studies have been conducted on the MD modeling of the interaction of CNT with different polymers. Mainly two types of modeling, namely, introduction and pullout simulations, have been usually addressed in these simulations. In the first type, a CNT and several polymer chains are modeled to study their interaction, while in the latter one, a carbon nanotube is pulled out from polymer chains.

In the introduction simulation category, regarded as one of the pioneering studies in this research area, In het Panhuis et al. [73] investigated the interaction of a single carbon nanotube with poly(mphenylenevinylene-*co*-2,5-dioctyloxy-*p*-phenylenevinylene) (PmPV) polymer. The aforementioned modeling considered simple assumptions and assumed the SWCNT as a fixed and rigid body. Using MD simulation, Yang et al. [74] studied the interaction between an SWCNT and PPA [PS/poly(*p*-phenylenevinylene) (PPV)/poly(phenylacetylene)]. Their results indicated that the strength of interaction between the SWCNT and polymers is sensitive to the monomer structure. They reported that for the noncovalent binding of CNTs into composite structures, the polymers with aromatic rings serve as good choices. Zheng et al. [75] and Xie et al. [76] simulated the interaction energy between SWCNTs and polymers to study the influence of temperature, radius, and chirality on the interaction. Their results showed that the interactions between the SWCNTs and the studied polymers are negligibly influenced by chirality and temperature but strongly depend on the radius of SWCNTs.

In the scope of pullout simulations by applying a unidirectional force to the atoms of a CNT, Frankland et al. [77,78] studied the interfacial sliding during the pullout process of the nonfunctionalized and functionalized nanotube/PE composite systems. The main purpose of the paper was to draw attention to estimating the effective viscosity coefficient. Accordingly, they predicted that cross-links involving less than 1% of carbon atoms have a large influence on the properties of the nanotube–polyethylene interface. The effects of polymer matrix density, chemical cross-links in the interface, and geometrical

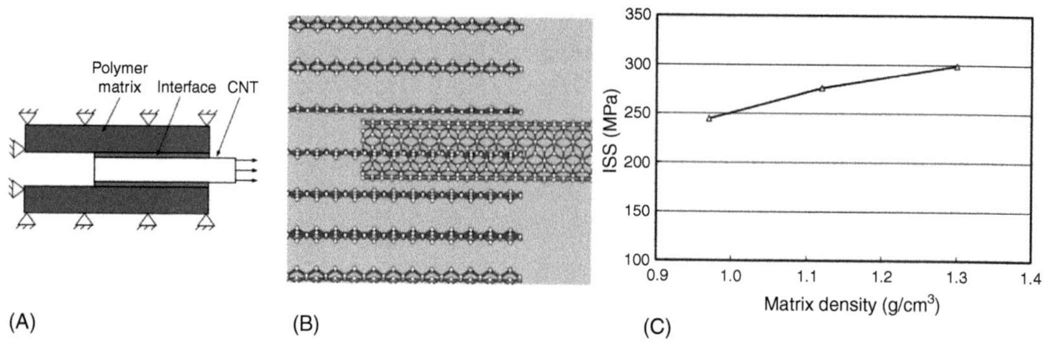

FIGURE 14.15 Computer simulation of CNT pullout from crystalline PE polymer [79]. (A) Schematic of partially embedded CNT pullout from polymer matrix. (B) Atomistic model of embedded CNT in polymer matrix without cross-links. (C) Variation in the IFSS with polymer matrix density.

defects in the CNTs on the pullout force have been investigated by Chowdhury and Okabe [79]. In this case, the studied system was composed of an SWCNT embedded in crystalline polyethylene as illustrated in Fig. 14.15A–B. Accordingly, the IFSS was estimated with variations of the total potential energy. As indicated in Fig. 14.15C, their results revealed that increasing the matrix density causes an increase in the interfacial shear strength. Their findings revealed that the IFSS increases from 310 to 1630 MPa for a functionalized (5, 5) CNT with only four chemical cross-links. However, they concluded that the presence of cross-links and their positions may increase or decrease the IFSS. For example, they illustrated a case where defects are present in CNTs leading to a reduction in the IFSS even in the presence of those links. Yu et al. [80] performed pullout simulation on PPS–SWCNT and PS–SWCNT and measured the interfacial shear stress through the following equation:

$$\tau = \frac{E_{\text{pullout}}}{\pi r L^2} \tag{14.10}$$

where r and L denote the radius and length of the SWCNT, respectively, and E_{pullout} represents the pullout energy defined as the difference between the interaction energy of the initial and final structures of the SWCNT and polymer. In the initial structure, CNT is embedded fully within the polymer, while at the final stage the SWCNT is completely pulled out from the host polymer. The interaction and IFSS are plotted in Fig. 14.16. The values of IFSS are consistent with the results of interfacial interaction energy and prove that orientation will induce interfacial enhancement.

Another useful and efficient method for exerting external forces to a molecular structure is steered MD (SMD) [81]. This method is based on the concept of adding a harmonic moving restraint to the center of mass of a group of atoms. Sindu and Sasmal [82] employed the SMD method on CNT and three epoxy resins, namely, diglycidyl ether of bisphenol A (DGEBA), diglycidyl ether of bisphenol F (DGEBF), and tetraglycidyl ether of diamino diphenyl methane (TGDDM). As depicted in Fig. 14.17, by applying constant velocity pulling, the force–displacement characteristics of the epoxy resins were studied and

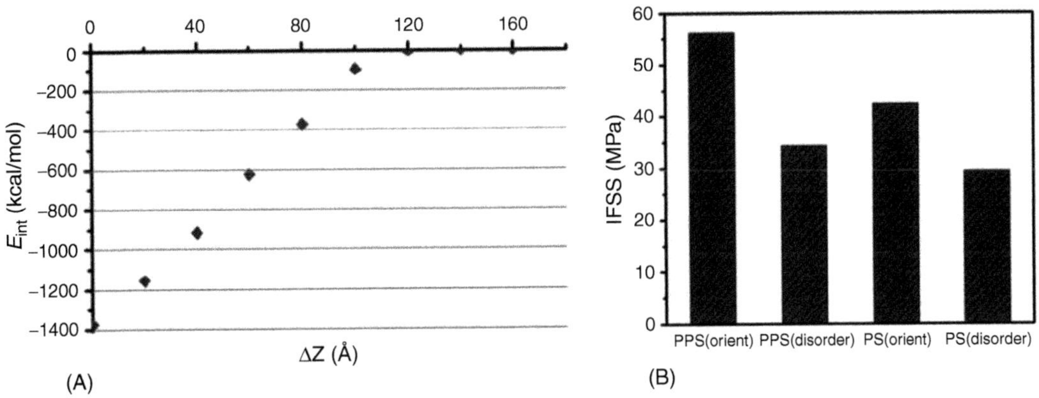

FIGURE 14.16 Energy evolution and IFSS during the pullout of the SWCNT from the PPS–SCWNT-oriented composite [80]. (A) Energy evolution and (B) value of IFSS for the four composites.

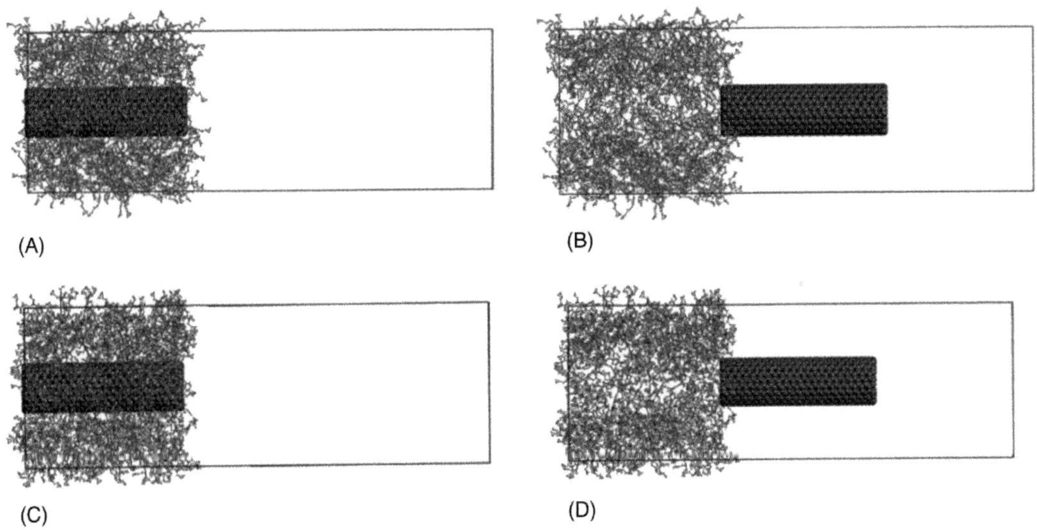

FIGURE 14.17 Two different states during pullout simulation of epoxy resins, DGEBA and TGDDM [82]. (A) Fully embedded (DGEBA–CNT), (B) fully outside (DGEBA–CNT), (C) fully embedded (TGDDM–CNT), and (D) fully outside (TGDDM–CNT).

the interaction of the epoxy resins and hardeners with CNT was investigated. In this paper, a parameter called affinity index has been proposed to quantify the interaction of CNT with an epoxy polymer. It is given as follows:

$$\text{Affinity index} = \left[\sum_{i=1}^{N}(r_i - C_{\mathrm{NT}})\right]_t - \left[\sum_{i=1}^{N}(r_i - C_{\mathrm{NT}})\right]_0 \tag{14.11}$$

where r_i is the coordinate of the ith atom of the polymer and C_{NT} is the geometric center of the CNT. The time evolution of the affinity index is shown in Fig. 14.18. As follows from

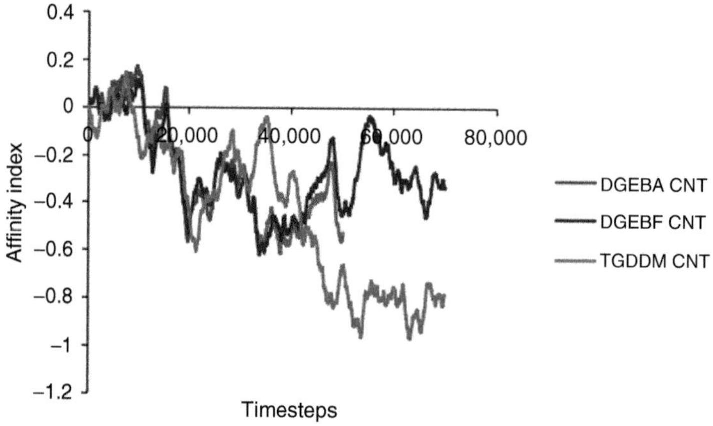

FIGURE 14.18 Time evolution of affinity index of the studied epoxy resins [82].

Fig. 14.18, it can be concluded that the TGDDM epoxy resin has the highest affinity with CNT compared to the other cases.

By applying a longitudinal tensile strain, Arash et al. [83] explored the fracture behavior of CNT-reinforced PMMA. The variation of the total potential energy of the CNT/PMMA composite under axial tensile strains is depicted in Fig. 14.19A. Additionally, a snapshot of the failure in the (5, 5) CNT/PMMA composite at the tensile strain of 1% is illustrated in Fig. 14.19B. They determined the yield strain equal to approximately 0.9%. This group also studied the effects of the aspect ratio of CNT reinforcements on the elastic properties, that is, Young's modulus and yield strength of the samples.

Generally, force fields are classified into two main classes: (1) reactive and (2) nonreactive force fields. Classical nonreactive force fields fail in modeling the rupture and formation of covalent bonds under large deformations. Therefore, the number of bonds during

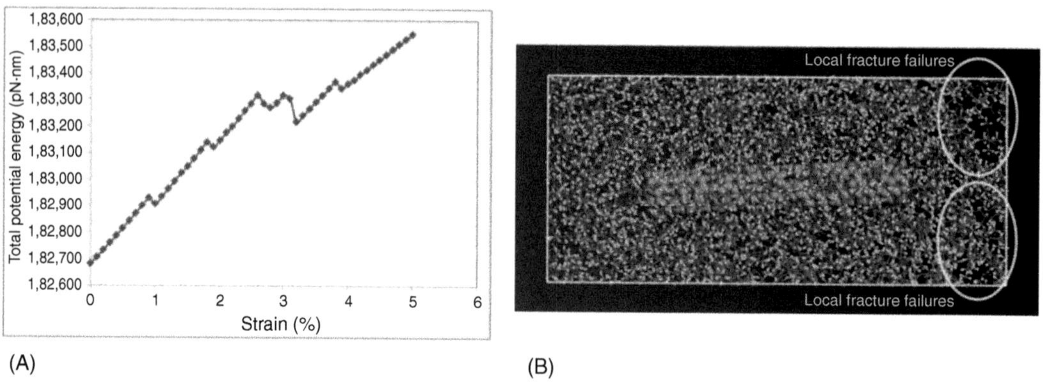

FIGURE 14.19 Mechanical behavior of the CNT/PMMA composite under tensile strain in the longitudinal direction [83]. (A) Variation of the total potential energy of the composite versus strain. (B) Snapshot of the composite at strain of 1% demonstrating local fracture failures in the composite due to the tensile loading.

the simulation is kept constant. Meanwhile, reactive force fields represent an important milestone in overcoming the mentioned limitations of nonreactive force fields and thus are able to simulate thousands of atoms with quantum mechanics accuracy. So far, several types of reactive force fields have been proposed in recent decades [44,45,84]. Zaminpayma and Mirabbaszadeh [85] were the first to utilize a reactive force field (ReaxFF [84]) to examine the interaction between SWCNTs and various polymers including poly(3-hexythiophene) (P3HT), poly(2-methoxy-5-(3-7-dimethyloctyloxy)-1,4-phenylenevinylene) (MDMO-PPV), and poly[[[(2-ethylhexyl)oxy]methoxy-1,4-phenylene]-1,2-ethenediyl] (MEH-PPV). In their work, the effect of temperature, chirality, and nanotube radius was investigated. This study shows that the interaction between the SWCNT and these polymers is not influenced by temperature, but strongly depend on the SWCNT radius and chirality. An overview of the MD modeling studies on the interaction between the CNT and the polymer is presented in Table 14.2. Table 14.2 categorizes the studies available in the literature based on the three different types, namely, pull out, introduction, and thermal properties investigation.

In summary, MD simulations have proven to be a useful tool for modeling the interaction between CNT and polymers, in particular by providing the possibility of investigating the mechanical properties and characterizing the interaction energy. Although at first sight, it might be tempting to obtain realistic characteristics of CNT–polymer nanocomposites in which all atoms are considered with detailed interactions, MD simulation is still computationally expensive. Therefore, this approach cannot be provided even by modern supercomputers and hence, only systems up to nanoscale can be modeled through it. Consequently, a further level of simplification is needed and thus other methods such as CGMD or multiscale simulations should be taken into account. These methods reduce the complexity of the simulated systems without losing all of the atomistic detail to perform fruitful simulations. In the next section, the CGMD method and its application on investigating CNT–polymer interactions will be reviewed.

14.3.4 Coarse-Grained Molecular Dynamics

CGMD is an elegant method aimed at modeling systems with few degrees of freedom absorbing some of the molecular details that are not important for the problem of interest. This usually means replacing the full atomistic representation of the system by a model in which small groups of atoms are treated as a bigger chain, namely, CG beads. These simplifications lead to removing spatial and temporal limits of simulations compared to detailed, for example, AA models and therefore significantly less required computational resources. Generally, in CG modeling, it will be impossible to provide accuracy of the AA model in every aspect. Hence, desired properties must be considered for the CG model to match with high accuracy. Because the CNT–polymer nanocomposites of micrometer size in detailed simulations are computationally prohibitively expensive, CGMD has been gaining more and more importance in recent years. The difference between AA and CGMD is presented next.

Table 14.2 Summary of Molecular Dynamics Studies in the Literature

Type	Reinforcement	Matrix	Box Size (Å)	TS[a] (fs) [TT[b] (ps)]	Conclusion	Improvement Property	Shortcoming/ Advantages	Force Fields	Platform	Year	References
Pullout	(10, 10) SWCNT	78 Chains of 42-mer PE	60 × 60 × 60	0.5 [50]	Effective viscosity coefficient of 0.2 cP for interfacial sliding		Developing a nanoscale friction law	REBO	DL-POLY	2002	[77]
Pullout	50.6-Å long (10, 10) SWCNT	Epon 862/ DETDA epoxy	82 × 82 × 52	0.2–2 [80]	Stronger SWNT–epoxy interaction for individual nanotubes	Molecular interaction and load transfer in SWCNT and epoxy resins	Investigating load transfer in the presence of SWCNT rope	MM	—	2005	[86]
Pull out	53-Å long (10, 10) SWCNT	178 Chains of 42-mer PE	53-Å long SWNT	—	10 times increase in shear strength with less than 1% cross-link between SWNT polymer	Shear strength		REBO LJ	DL-POLY	2002	[78]
Thermal test	20-Å long (10, 0) SWCNT	80 Chains of 10-mer PE	28 × 28 × 30	0.5 [300]	Addition of CNT increased glass transition temperature of the polymer	Glass transition temperature–thermal expansion and diffusion coefficients	Limited to small scale	Tersoff–Brenner and LJ and intrapolymer potential by Clarke et al. [87]	DL-POLY	2002	[88]
Thermal	190-Å long (10, 0) SWCNT	50 Chains of 100-mer PE	26 × 26 × 200	0.5 [—]	Shear strength of 46 MPa under tension	Shear strength	Clear analysis but not experimental validation	Amber LJ	DL-POLY	2006	[89]
Temperature	49.19-Å long (10, 10) SWCNT	1 Chain of 10-mer polyphenylacetylene	49.19 A CNT with single polymer chain	1 [200]	Strong effect of chirality but negligible effect of temperature on the interaction energy	Interfacial adhesion	Limited to small scale and not considering multiple polymer chains	COMPASS	Material Studio	2008	[90]

Introduction test	40.8-Å long (13, 6) SW-CNT	4-mer PmPV	Single polymer molecule	— [—]	Introduction test with vdW potential for SWNT–polymer interaction	Noncovalent binding energy	Limited to small scale and single molecule/experimental validation	COMPASS VdW	Cerius[2]	2003	[73]
Introduction test	97-Å long (10, 10) SWCNT	Epon 862/DETDA	Single polymer molecule	2 [40]	Polymer chains align wrap around SWNT	Wetting properties between SWNT and Epon 862 resin/DETDA agent	Not considering multiple polymer chains/separate investigation of Epon 862 and the curing agent did not simulate a cured nanocomposite	COMPASS	Material Studio	2004	[91]
Tensile test	(10, 10) SWCNT	PMMA and PmPV	—	1 [2000]	Increasing the strain elastic modulus especially in the tube longitudinal direction	Strain elastic modulus		DREIDING COMPASS	Cerius[2]	2007	[92]
Pullout	59.03-Å long (5, 5) SWCNT	Chain length of 20, 40, 60 $Å^3$ (PE)	59.03 Å SW-CNT long	0.5 [1600]	Dispersion angle of polymer wrapped around SWNTs dominates the load transfer	Interaction force	New aspect for analyzing by employing SMD simulations	AIREBO LJ bond angle torsional	LAMMPS	2013	[71]

(*Continued*)

Table 14.2 Summary of Molecular Dynamics Studies in the Literature (*cont.*)

Type	Reinforcement	Matrix	Box Size (Å)	TS[a] (fs) [TT[b] (ps)]	Conclusion	Improvement Property	Shortcoming/ Advantages	Force Fields	Platform	Year	References
Pullout	(10, 10) SWCNT	Epon 862 and EPI-CURE curing agent	46 × 46 × 63	1 [400]	The increased interface shear strength can either enhance or reduce the damping stability of the composite	Interface shear strength	Multiscale hierarchical modeling and large scale modeling	Tersoff LJ OPLS-UA	DL_POLY	2011	[93]
Thermal	(10, 10) SWCNT	PE, various chain lengths	25- to 75-nm long CNT	0.5 [1000]	The thermal resistance of CNT junctions can be significantly improved through modifying the molecular structure at the interface	Thermal resistance	Novel idea with AA simulations	EAM AIREBO LJ	LAMMPS	2009	[72]
Pull out	97-Å long (9, 9) SWCNT	Epoxy	15 nm CNT	—	Functionalization can improve the interfacial shear strength between CNT and polymer, while can reduce Young's modulus of the interphase region	Young's modulus	Continuum modeling	REBO stretching bending dihedral inversion LJ	LAMMPS	2016	[94]

Thermal	(3, 3)	Epoxy (DGEBA) and (4,4′-DDS)	—	1 [4000]	Functionalization of the CNTs facilitates the transfer of both mechanical load and thermal energy	Interfacial interactions	Independently control the functionalization not varied the aspect ratio of the CNTs	AMBER LJ	LAMMPS	2014	[40]
Tension simulation	62.8-Å long (15,0) SWCNT	Polypropylene	Varied	1 [3000]	Significant role of both weakened interface and polymer-densified interphase in the elastoplastic behavior	Elastic and plastic behavior	Multiscale approach/modeling local elastoplastic behavior of the interphase region	PCFF	LAMMPS	2013	[95]
Introduction test	73-Å long SWCNT with varied (m, n)	P3HT, MDMO-PPV, MEH-PPV	Varied	1 [100]	Weak influence of temperature but high influence for chirality and SWCNT radius on the interaction energy	Interaction energy	Novel idea for using a reactive force field	ReaxFF	LAMMPS	2012	[85]

[a]Time step.
[b]Total time.

Given an atomistic Hamiltonian for an AA MD modeling:

$$H^{AA}=\sum_{i=1}^{n}\frac{\boldsymbol{p}_i^2}{2m_i}+U^{AA}(\boldsymbol{r}_1,\ldots,\boldsymbol{r}_n) \tag{14.12}$$

where $\mathbf{r}_i$ and $\mathbf{p}_i$ are Cartesian coordinates and momenta of the ith atom and U^{AA} its potential energy. In the same way the Hamiltonian of a CG modeling takes the form

$$H^{CG}=\sum_{\alpha=1}^{N}\frac{\boldsymbol{P}_\alpha^2}{2M_\alpha}+U^{CG}(\boldsymbol{R}_1,\ldots,\boldsymbol{R}_N) \tag{14.13}$$

where $\mathbf{R}_\alpha$ and $\mathbf{P}_\alpha$ are Cartesian coordinates and momenta of the αth CG bead and U^{CG} is CG potential energy. The connection between AA and CG beads is formally given by application of the mapping operator, M:

$$\mathrm{M}(r^i)=\left\{\mathrm{M}_1(\boldsymbol{r}^i),\ldots,\mathrm{M}_N(\boldsymbol{r}^i)\right\} \tag{14.14}$$

The mapping describes how the CG beads are constructed from the underlying AA simulation. A commonly used mapping is to consider the center of mass of a subset of atoms to represent a CG bead. Similar to the AA MD, again the potential is the heart of the CG simulations and similarly, defining a proper potential for the CGMD plays a particularly vital role. Hence, proper interaction potentials are of central concern to any CG model and deserve the most hard-to-reach task. All the previous studies on deriving CG force fields have common objectives such as removing bounded degrees of freedom while minimizing deviation from the AA model of the system.

The literature on the proposed CG models for polymers shows a variety of approaches that can be classified as UA [96], tangent hard spheres (pearl-necklace) [97], ellipsoidal [98], and bead-spring [99]. Additionally, there are ready-to-use force fields, such as the MARTINI force fields, which are extended to modeling polymers [100,101]. For several years, great effort has been devoted to the development of CG models for polymer materials [102–104]. Most of these CG potentials are nonreactive; however, Deng and Shi [105] proposed a reactive CG model for polydisperse polymers. Several publications have appeared in recent years proposing the same CG models to simulate CNTs. Cranford and Buehler [106] developed various CGMD potentials to attain a representative mesoscopic buckypaper model of CNTs. Zhao et al. [107] parameterized the CG stretching, bending, and torsion potential terms of SWCNTs to study their static and dynamic behaviors. Although up to now, much research on CG models of CNT and polymers has been done, to the authors' knowledge, CNT-reinforced polymer composites have been scarcely investigated from the point of CG modeling and few publications have been conducted on this issue, which will be reviewed next.

A comprehensive CG model of CNT–polymer composites has been developed by Arash et al. [108] using the strain energy conservation between CG models and full atomistic systems. A significant feature of the proposed CG model was the ability to capture interactions between polymer chains and CNT. Then they utilized the model to investigate the elastic properties of short and long CNT–PMMA polymer composites. An AA and CG RVE

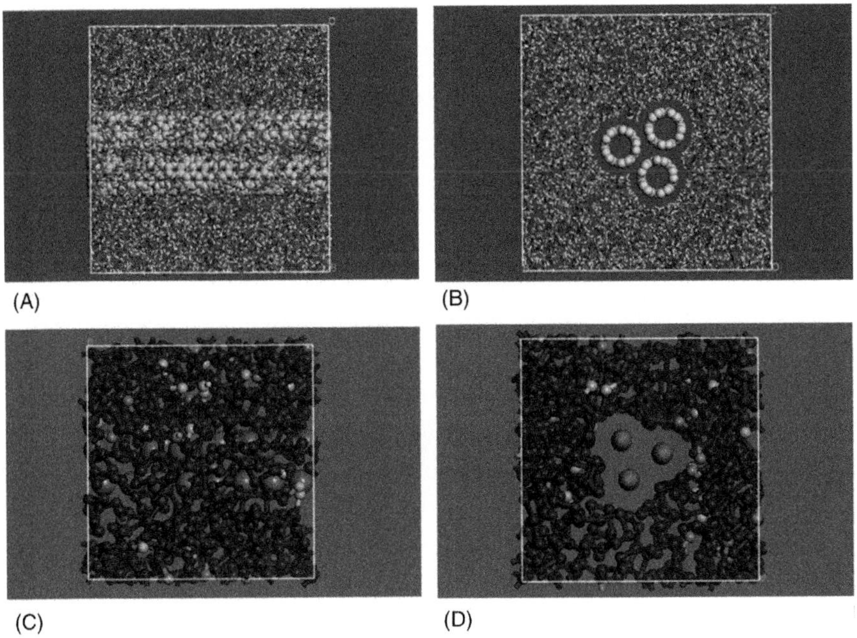

FIGURE 14.20 PMMA polymer matrix with a size of 5 × 5 × 5 nm^3 reinforced by a nanotube rope made of three (5, 5) SWCNTs [108]. (A) Side view of AA model, (B) top view of AA model, (C) side view of CG model, and (D) top view of CG model.

of the polymer matrix reinforced by a three-CNT bundle is illustrated in Fig. 14.20. The simulation results revealed that the CG model is able to estimate the mechanical properties of the nanocomposites with high accuracy and low computational cost as the number of atoms is decreased from 11,144 atoms to 684 beads and the time step is increased 10-fold from 1 to 10 fs. Moreover, the effect of the volume fraction of CNT reinforcements on the Young's modulus of the nanocomposites was evaluated by CG and AA methods as is seen in Table 14.3. The results show that the measured Young's moduli for the PMMA matrix composite reinforced by CNT bundles are broadly consistent with the values obtained via AA molecular simulation.

Table 14.3 Mechanical Properties of the PMMA Polymer Matrix With Different Volume Fraction of CNTs [108]

	Young's Modulus	
Volume Fraction of CNTs (%)	**CG model (GPa)**	**AA (GPa)**
10 (3 CNTs)	73.39	72.54
13.38 (4 CNTs)	97.13	96.05
16.73 (5 CNTs)	122.05	121.30
20 (6 CNTs)	145.49	145.34

Table 14.4 Effect of RVE Size on the Mechanical Properties of CNT–PMMA Nanocomposite [109]

RVE Size (nm^3)	Number of Beads (Number of Equivalent AA Atoms)	Young's Modulus (GPa)	Tensile Strength (MPa)
20 × 20 × 20	48,520 (734,780)	3.35	104.51
30 × 30 × 30	163,750 (2,479,882)	3.27	102.15
40 × 40 × 40	388,160 (5,878,240)	3.22	101.71
50 × 50 × 50	756,912 (11,462,568)	3.24	101.85

In a paper by Arash et al. [109], the mechanical behavior of a PMMA polymer matrix reinforced by short (5, 5) SWCNTs under tension in elastic and plastic regimes was reported using a CG model. In this study, they used the previously proposed and examined CG model [108]. The effect of the nanotube weight fraction on the mechanical properties, that is, the Young's modulus, yield strength, tensile strength, and critical strain of the CNT/polymer composites was studied in detail. Furthermore, to find a proper size for the RVE, an enlargement test was undertaken. Using the CG method, the size of the box enlarged from 20 to 50 nm and the polymer matrix was reinforced by randomly distributed (5, 5) CNT of 10 nm length at a weight fraction of about 10 wt.%. Table 14.4 presents the simulation results. As can be seen, employing the CG model leads to modeling of such a huge system with an equivalent size of 11 million atoms. Moreover, in this work, the proper size of the RVE was proposed to be equal to $50 \times 50 \times 50$ nm^3.

In another similar study, the same group [110] proposed an effective calculation method based on the CG simulations for predicting the J-integral of CNT/polymer composites. The CG model of the composite system contained 121,116 beads, which was equivalent to a full atomistic system with 2,065,020 atoms. Additionally, the dependence of the J-integral on the crack length and the RVE size was discussed. The simulation results showed that 10-nm-long (5, 5) CNT reinforcements with a weight fraction of 10 wt.% can significantly enhance the J-integral of the PMMA polymer matrix as much as three times. The proposed approach provided an effective methodology for predicting the J-integral of polymers and their composites without requiring knowledge about the crack tip position or the continuum fields around the tip.

Mousavi et al. [111] have also developed a CG model to investigate the CNT–polymer nanocomposite effectively. They employed the aforementioned model to investigate the elastic properties of a cross-linked short CNT–PMMA nanocomposite. In this study, 121,120 beads, equivalent to 2,065,091 atoms, were simulated in a $60 \times 60 \times 5$ nm^3 RVE box. Fig. 14.21 illustrates the RVE. The results suggested that functionalization of CNTs using methylene cross-links is a vital factor toward significantly increasing the elastic properties of randomly distributed short CNT–PMMA nanocomposites. The influences of Ethylene glycol dimethylacrylate (EGDMA) cross-links between polymer chains on the Young's moduli of the composites are showed in Table 14.5.

As aforementioned, studies in CG modeling of CNT–polymer nanocomposites exhibit high potential and has been gaining importance in recent years. It seems that more and

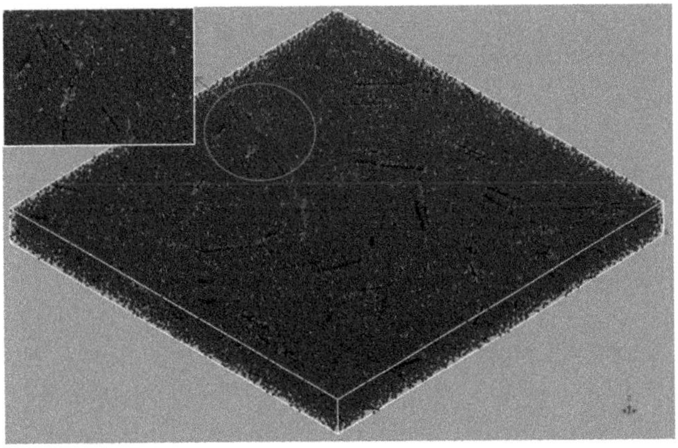

FIGURE 14.21 CNT (10 wt.%)–PMMA composite with EGDMA cross-links between polymer chains and 2CH cross-links between polymer chains and CNTs [111].

Table 14.5 Effect of Cross-Links Between PMMA Chains on the Young's Modulus of CNT [111]

Mole Fraction of the Polymers Cross-Link (%)	E_x (GPa)	E_y (GPa)	E_{av} (GPa)
0	3:98	4:01	4.00
2	3:99	4:12	4.10
6	4:25	4:15	4.20
8	4:29	4:20	4.25

more considerable attention will be paid to this issue soon. In addition to the mentioned advantages of this method, CG modeling has also some possible disadvantages. One question still unanswered is whether some atomistic details crucial for the phenomena of interest may absorb at larger length- and time-scales of CG modeling. Unfortunately, any CG method intrinsically consists of some loss of information compared to the AA modeling. Thus, a great deal of care must be taken into account to ensure that the CG model is capable of preserving the essential properties from the AA model.

14.4 Concluding Remarks

CNT–polymer nanocomposites are among the most interesting materials discovered during the last 20 years. These nanocomposites possess many unique features due to their excellent processability characteristics of the polymer and superior magnetic, electrical, and mechanical properties of the CNTs. From the experimental point of view, since the interfacial interaction is principally dominated by the atomistic scale phenomena, it has been proven difficult to obtain reliable experimental data on CNT–polymer interactions. Additionally, experimental studies usually suffer from several limitations such as the inability to measure load transfer at the interface, the nanometer size of the specimen, and the high cost of experiments. Therefore, numerical modeling techniques can provide helpful insights into interactions behavior that are not directly accessible to the experiment. In this chapter, a comprehensive review of the numerical modeling approaches implemented to study the CNT–polymer interactions was presented. The numerical techniques were divided into four main groups, namely, MM, MSM, MD, and CGMD. MM and MSM can well capture the interfacial interaction energy and optimized structures (minimum energy conformation). However, they fail in generating a trajectory of the system over a period of time. In MD, the interfacial interactions can be monitored in a detailed way within some timeframe. In some cases, CNT–polymer nanocomposites are too complicated to be efficiently modeled by classical molecular modeling techniques due to the large size of the system and also the large time-scale of specific interaction patterns. In such systems, CGMD provides successfully a useful and promising solution.

References

[1] J. Gou, K. Lau, Modeling and simulation of carbon nanotube/polymer composites, Handbook of Theoretical and Computational Nanotechnology, American Scientific Publishers, Valencia, 2005, pp. 20051–58883.

[2] S.A. Shandiz, A. Montazeri, A novel MD-based procedure to obtain the interphase Young's modulus in nanocomposites, Comput. Mater. Sci. 113 (2016) 104–111.

[3] K.A. Wepasnick, B.A. Smith, J.L. Bitter, D.H. Fairbrother, Chemical and structural characterization of carbon nanotube surfaces, Anal. Bioanal. Chem. 396 (2010) 1003–1014.

[4] Y.S. Lipatov, T.T. Alekseeva, Phase-Separated Interpenetrating Polymer Networks, Springer, Berlin, Heidelberg, (2007).

[5] H. Liu, L.C. Brinson, Reinforcing efficiency of nanoparticles: a simple comparison for polymer nanocomposites, Compos. Sci. Technol. 68 (2008) 1502–1512.

[6] A. Domján, P.T. Mezey, J. Varga, Behavior of the interphase region of an amphiphilic polymer conetwork swollen in polar and nonpolar solvent, Macromolecules 45 (2012) 1037–1040.

[7] G. Seidel, D. Lagoudas, S. Frankland, T. Gates, Modeling functionally graded interphase regions in carbon nanotube reinforced composites, 20th American Society for Composites Technical Conference, Drexel University, Philadelphia, PA, USA, 2006.

[8] Y. Li, G.D. Seidel, Multiscale modeling of the effects of nanoscale load transfer on the effective elastic properties of unfunctionalized carbon nanotube–polyethylene nanocomposites, Model. Simul. Mater. Sci. Eng. 22 (2014) 025023.

[9] X. Chen, Y. Liu, Square representative volume elements for evaluating the effective material properties of carbon nanotube-based composites, Comput. Mater. Sci. 29 (2004) 1–11.

[10] D.C. Hammerand, G.D. Seidel, D.C. Lagoudas, Computational micromechanics of clustering and interphase effects in carbon nanotube composites, Mech. Adv. Mater. Struct. 14 (2007) 277–294.

[11] R. Andrews, M. Weisenberger, Carbon nanotube polymer composites, Curr. Opin. Solid State Mater. Sci. 8 (2004) 31–37.

[12] A.K.-T. Lau, D. Hui, The revolutionary creation of new advanced materials—carbon nanotube composites, Compos. B 33 (2002) 263–277.

[13] Y. Ren, F. Li, H.-M. Cheng, K. Liao, Tension–tension fatigue behavior of unidirectional single-walled carbon nanotube reinforced epoxy composite, Carbon 41 (2003) 2177–2179.

[14] F. Du, R.C. Scogna, W. Zhou, S. Brand, J.E. Fischer, K.I. Winey, Nanotube networks in polymer nanocomposites: rheology and electrical conductivity, Macromolecules 37 (2004) 9048–9055.

[15] E. Kymakis, I. Alexandou, G. Amaratunga, Single-walled carbon nanotube–polymer composites: electrical, optical and structural investigation, Synth. Met. 127 (2002) 59–62.

[16] M. Biercuk, M.C. Llaguno, M. Radosavljevic, J. Hyun, A.T. Johnson, J.E. Fischer, Carbon nanotube composites for thermal management, Appl. Phys. Lett. 80 (2002) 2767–2769.

[17] T. Kashiwagi, E. Grulke, J. Hilding, K. Groth, R. Harris, K. Butler, et al. Thermal and flammability properties of polypropylene/carbon nanotube nanocomposites, Polymer 45 (2004) 4227–4239.

[18] E. Kymakis, G. Amaratunga, Single-wall carbon nanotube/conjugated polymer photovoltaic devices, Appl. Phys. Lett. 80 (2002) 112–114.

[19] S. Lefrant, J. Buisson, J. Schreiber, O. Chauvet, M. Baibarac, I. Baltog, Raman studies of carbon nanotubes and polymer nanotube composites, Mol. Cryst. Liq. Cryst. 415 (2004) 125–132.

[20] L. Schadler, S. Giannaris, P. Ajayan, Load transfer in carbon nanotube epoxy composites, Appl. Phys. Lett. 73 (1998) 3842–3844.

[21] A. Rasheed, H.G. Chae, S. Kumar, M.D. Dadmun, Polymer nanotube nanocomposites: correlating intermolecular interaction to ultimate properties, Polymer 47 (2006) 4734–4741.

[22] M. Baibarac, I. Baltog, S. Lefrant, Raman spectroscopic evidence for interfacial interactions in poly (bithiophene)/single-walled carbon nanotube composites, Carbon 47 (2009) 1389–1398.

[23] P.-C. Ma, Q.-B. Zheng, E. Mäder, J.-K. Kim, Behavior of load transfer in functionalized carbon nanotube/epoxy nanocomposites, Polymer 53 (2012) 6081–6088.

[24] F. Leite, L. Mattoso, O. Oliveira, P. Herrmann, The atomic force spectroscopy as a tool to investigate surface forces: basic principles and applications, Modern Research and Educational Topics in Microscopy, Formatex, Badajoz, Spain, 2007, pp. 747–757.

[25] X. Li, W. Chen, Q. Zhan, L. Dai, L. Sowards, M. Pender, et al. Direct measurements of interactions between polypeptides and carbon nanotubes, J. Phys. Chem. B 110 (2006) 12621–12625.

[26] C. Bernard, S. Marsaudon, R. Boisgard, J.-P. Aimé, Competition of elastic and adhesive properties of carbon nanotubes anchored to atomic force microscopy tips, Nanotechnology 19 (2007) 035709.

[27] A.H. Barber, S.R. Cohen, H.D. Wagner, Measurement of carbon nanotube–polymer interfacial strength, Appl. Phys. Lett. 82 (2003) 4140–4142.

[28] M. Strus, L. Zalamea, A. Raman, R. Pipes, C. Nguyen, E. Stach, Peeling force spectroscopy: exposing the adhesive nanomechanics of one-dimensional nanostructures, Nano Lett. 8 (2008) 544–550.

[29] T. Tsuda, T. Ogasawara, F. Deng, N. Takeda, Direct measurements of interfacial shear strength of multi-walled carbon nanotube/PEEK composite using a nano-pullout method, Compos. Sci. Technol. 71 (2011) 1295–1300.

[30] J.-M. Yuan, Z.-F. Fan, X.-H. Chen, X.-H. Chen, Z.-J. Wu, L.-P. He, Preparation of polystyrene–multi-walled carbon nanotube composites with individual-dispersed nanotubes and strong interfacial adhesion, Polymer 50 (2009) 3285–3291.

[31] C.-M. Chang, Y.-L. Liu, Functionalization of multi-walled carbon nanotubes with non-reactive polymers through an ozone-mediated process for the preparation of a wide range of high performance polymer/carbon nanotube composites, Carbon 48 (2010) 1289–1297.

[32] H.D. Wagner, Nanotube–polymer adhesion: a mechanics approach, Chem. Phys. Lett. 361 (2002) 57–61.

[33] E.T. Thostenson, T.-W. Chou, Aligned multi-walled carbon nanotube-reinforced composites: processing and mechanical characterization, J. Phys. D 35 (2002) L77.

[34] M. Rahmat, P. Hubert, Carbon nanotube–polymer interactions in nanocomposites: a review, Compos. Sci. Technol. 72 (2011) 72–84.

[35] A. Desai, M. Haque, Mechanics of the interface for carbon nanotube–polymer composites, Thin-Walled Struct. 43 (2005) 1787–1803.

[36] K. Mylvaganam, L. Zhang, Nanotube functionalization and polymer grafting: an ab initio study, J. Phys. Chem. B 108 (2004) 15009–15012.

[37] K. Mylvaganam, L. Zhang, Chemical bonding in polyethylene-nanotube composites: a quantum mechanics prediction, J. Phys. Chem. B 108 (2004) 5217–5220.

[38] M. Simeoni, C. De Luca, S. Picozzi, S. Santucci, B. Delley, Interaction between zigzag single-wall carbon nanotubes and polymers: a density-functional study, J. Chem. Phys. 122 (2005) 214710.

[39] J. Moon, S. Yang, M. Cho, Interfacial strengthening between graphene and polymer through Stone-Thrower-Wales defects: Ab initio and molecular dynamics simulations, Carbon 118 (2017) 66–77.

[40] K.S. Khare, F. Khabaz, R. Khare, Effect of carbon nanotube functionalization on mechanical and thermal properties of cross-linked epoxy–carbon nanotube nanocomposites: role of strengthening the interfacial interactions, ACS Appl. Mater. Interfaces 6 (2014) 6098–6110.

[41] X. Peng, S.A. Meguid, Molecular dynamics simulations of the buckling behavior of defective carbon nanotubes embedded in epoxy nanocomposites, Eur. Polym. J. 93 (2017) 246–258.

[42] H. Lee, Molecular modeling of PEGylated peptides, dendrimers, and single-walled carbon nanotubes for biomedical applications, Polymers 6 (2014) 776–798.

[43] G. Odegard, T. Gates, K. Wise, C. Park, E. Siochi, Constitutive modeling of nanotube–reinforced polymer composites, Compos. Sci. Technol. 63 (2003) 1671–1687.

[44] D.W. Brenner, O.A. Shenderova, J.A. Harrison, S.J. Stuart, B. Ni, S.B. Sinnott, A second-generation reactive empirical bond order (REBO) potential energy expression for hydrocarbons, J. Phys. Condens. Matter 14 (2002) 783.

[45] S.J. Stuart, A.B. Tutein, J.A. Harrison, A reactive potential for hydrocarbons with intermolecular interactions, J. Chem. Phys. 112 (2000) 6472–6486.

[46] J. Tersoff, New empirical approach for the structure and energy of covalent systems, Phys. Rev. B 37 (1988) 6991.

[47] J.E. Lennard-Jones, The electronic structure of some diatomic molecules, Trans. Faraday Soc. 25 (1929) 668–686.

[48] K. Liao, S. Li, Interfacial characteristics of a carbon nanotube–polystyrene composite system, Appl. Phys. Lett. 79 (2001) 4225–4227.

[49] V.V. Mokashi, D. Qian, Y. Liu, A study on the tensile response and fracture in carbon nanotube-based composites using molecular mechanics, Compos. Sci. Technol. 67 (2007) 530–540.

[50] S. Haghighatpanah, K. Bolton, Molecular-level computational studies of single wall carbon nanotube–polyethylene composites, Comput. Mater. Sci. 69 (2013) 443–454.

[51] D. Qian, W.K. Liu, R.S. Ruoff, Load transfer mechanism in carbon nanotube ropes, Compos. Sci. Technol. 63 (2003) 1561–1569.

[52] M. Wong, M. Paramsothy, X. Xu, Y. Ren, S. Li, K. Liao, Physical interactions at carbon nanotube–polymer interface, Polymer 44 (2003) 7757–7764.

[53] V. Lordi, N. Yao, Molecular mechanics of binding in carbon-nanotube–polymer composites, J. Mater. Res. 15 (2000) 2770–2779.

[54] Q. Zheng, Q. Xue, K. Yan, X. Gao, Q. Li, L. Hao, Effect of chemisorption on the interfacial bonding characteristics of carbon nanotube–polymer composites, Polymer 49 (2008) 800–808.

[55] C. Li, T.-W. Chou, A structural mechanics approach for the analysis of carbon nanotubes, Int. J. Solids Struct. 40 (2003) 2487–2499.

[56] W.D. Cornell, P. Cieplak, C.I. Bayly, I.R. Gould, K.M. Merz, D.M. Ferguson, et al. A second generation force field for the simulation of proteins, nucleic acids, and organic molecules, J. Am. Chem. Soc. 117 (1995) 5179–5197.

[57] C. Li, T.-W. Chou, Multiscale modeling of compressive behavior of carbon nanotube/polymer composites, Compos. Sci. Technol. 66 (2006) 2409–2414.

[58] A. Montazeri, R. Naghdabadi, Study the effect of viscoelastic matrix model on the stability of CNT/polymer composites by multiscale modeling, Polym. Compos. 30 (2009) 1545–1551.

[59] M.M. Shokrieh, R. Rafiee, On the tensile behavior of an embedded carbon nanotube in polymer matrix with non-bonded interphase region, Compos. Struct. 92 (2010) 647–652.

[60] M.M. Shokrieh, R. Rafiee, Prediction of mechanical properties of an embedded carbon nanotube in polymer matrix based on developing an equivalent long fiber, Mech. Res. Commun. 37 (2010) 235–240.

[61] M.M. Shokrieh, R. Rafiee, Investigation of nanotube length effect on the reinforcement efficiency in carbon nanotube based composites, Compos. Struct. 92 (2010) 2415–2420.

[62] R. Rafiee, A. Fereidoon, M. Heidarhaei, Influence of non-bonded interphase on crack driving force in carbon nanotube reinforced polymer, Comput. Mater. Sci. 56 (2012) 25–28.

[63] R. Rafiee, R.M. Moghadam, Simulation of impact and post-impact behavior of carbon nanotube reinforced polymer using multi-scale finite element modeling, Comput. Mater. Sci. 63 (2012) 261–268.

[64] R. Rafiee, Influence of carbon nanotube waviness on the stiffness reduction of CNT/polymer composites, Compos. Struct. 97 (2013) 304–309.

[65] A. Montazeri, R. Naghdabadi, Investigation of the interphase effects on the mechanical behavior of carbon nanotube polymer composites by multiscale modeling, J. Appl. Polym. Sci. 117 (2010) 361–367.

[66] A. Montazeri, R. Naghdabadi, Investigating the effect of carbon nanotube defects on the column and shell buckling of carbon nanotube–polymer composites using multiscale modeling, Int. J. Multiscale Comput. Eng. 7 (2009) 431–444.

[67] R. Rafiee, R. Pourazizi, Influence of CNT functionalization on the interphase region between CNT and polymer, Comput. Mater. Sci. 96 (2015) 573–578.

[68] B. Alder, T. Wainwright, Phase transition for a hard sphere system, J. Chem. Phys. 27 (1957) 1208.

[69] S.L. Mayo, B.D. Olafson, W.A. Goddard, DREIDING: a generic force field for molecular simulations, J. Phys. Chem. 94 (1990) 8897–8909.

[70] H. Sun, COMPASS: an ab initio force-field optimized for condensed-phase applications overview with details on alkane and benzene compounds, J. Phys. Chem. B 102 (1998) 7338–7364.

[71] Y. Zhang, J. Zhao, N. Wei, J. Jiang, Y. Gong, T. Rabczuk, Effects of the dispersion of polymer wrapped two neighbouring single walled carbon nanotubes (SWNTs) on nanoengineering load transfer, Compos. B 45 (2013) 1714–1721.

[72] Z. Xu, M.J. Buehler, Nanoengineering heat transfer performance at carbon nanotube interfaces, ACS Nano 3 (2009) 2767–2775.

[73] M. In het Panhuis, A. Maiti, A.B. Dalton, A. van den Noort, J.N. Coleman, B. McCarthy, et al. Selective interaction in a polymer-single-wall carbon nanotube composite, J. Phys. Chem. B 107 (2003) 478–482.

[74] M. Yang, V. Koutsos, M. Zaiser, Interactions between polymers and carbon nanotubes: a molecular dynamics study, J. Phys. Chem. B 109 (2005) 10009–10014.

[75] Q. Zheng, Q. Xue, K. Yan, L. Hao, Q. Li, X. Gao, Investigation of molecular interactions between SWNT and polyethylene/polypropylene/polystyrene/polyaniline molecules, J. Phys. Chem. C 111 (2007) 4628–4635.

[76] J. Xie, Q. Xue, H. Chen, A. Keller, M. Dong, Different factors' effect on the SWNT-fluorocarbon resin interaction: a MD simulation study, Comput. Mater. Sci. 49 (2010) 148–157.

[77] S. Frankland, V.M. Harik, Analysis of carbon nanotube pull-out from a polymer matrix, Surf. Sci. 525 (2003) L103–L108.

[78] S. Frankland, A. Caglar, D. Brenner, M. Griebel, Molecular simulation of the influence of chemical cross-links on the shear strength of carbon nanotube–polymer interfaces, J. Phys. Chem. B 106 (2002) 3046–3048.

[79] S. Chowdhury, T. Okabe, Computer simulation of carbon nanotube pull-out from polymer by the molecular dynamics method, Compos. A 38 (2007) 747–754.

[80] B. Yu, S. Fu, Z. Wu, H. Bai, N. Ning, Q. Fu, Molecular dynamics simulations of orientation induced interfacial enhancement between single walled carbon nanotube and aromatic polymers chains, Compos. A 73 (2015) 155–165.

[81] S. Izrailev, S. Stepaniants, B. Isralewitz, D. Kosztin, H. Lu, F. Molnar, et al. Steered molecular dynamics, in: P. Deuflhard, J. Hermans, B. Leimkuhler, A.E. Mark, S. Reich, R.D. Skeel (Eds.), Computational Molecular Dynamics: Challenges, Methods, Ideas: Lecture Notes in Computational Science and Engineering, vol. 4, Springer, Berlin, Heidelberg, 1999, pp. 39–65.

[82] B. Sindu, S. Sasmal, Evaluation of mechanical characteristics of nano modified epoxy based polymers using molecular dynamics, Comput. Mater. Sci. 96 (2015) 146–158.

[83] B. Arash, Q. Wang, V. Varadan, Mechanical properties of carbon nanotube/polymer composites, Sci. Rep. 4 (2014) 1–8.

[84] A.C. Van Duin, S. Dasgupta, F. Lorant, W.A. Goddard, ReaxFF: a reactive force field for hydrocarbons, J. Phys. Chem. A 105 (2001) 9396–9409.

[85] E. Zaminpayma, K. Mirabbaszadeh, Interaction between single-walled carbon nanotubes and polymers: a molecular dynamics simulation study with reactive force field, Comput. Mater. Sci. 58 (2012) 7–11.

[86] J. Gou, Z. Liang, C. Zhang, B. Wang, Computational analysis of effect of single-walled carbon nanotube rope on molecular interaction and load transfer of nanocomposites, Compos. B 36 (2005) 524–533.

[87] D. Brown, J.H. Clarke, M. Okuda, T. Yamazaki, The preparation of polymer melt samples for computer simulation studies, J. Chem. Phys. 100 (1994) 6011–6018.

[88] C. Wei, D. Srivastava, K. Cho, Thermal expansion and diffusion coefficients of carbon nanotube-polymer composites, Nano Lett. 2 (2002) 647–650.

[89] C. Wei, Adhesion and reinforcement in carbon nanotube polymer composite, Appl. Phys. Lett. 88 (2006) 093108.

[90] H. Chen, Q. Xue, Q. Zheng, J. Xie, K. Yan, Influence of nanotube chirality, temperature, and chemical modification on the interfacial bonding between carbon nanotubes and polyphenylacetylene, J. Phys. Chem. C 112 (2008) 16514–16520.

[91] Z. Liang, J.J. Gou, C. Zhang, B. Wang, L. Kramer, Investigation of molecular interactions between (10, 10) single-walled nanotube and Epon 862 resin/DETDA curing agent molecules, Mater. Sci. Eng. A 365 (2004) 228–234.

[92] Y. Han, J. Elliott, Molecular dynamics simulations of the elastic properties of polymer/carbon nanotube composites, Comput. Mater. Sci. 39 (2007) 315–323.

[93] A. Liu, K. Wang, C.E. Bakis, Effect of functionalization of single-wall carbon nanotubes (SWNTs) on the damping characteristics of SWNT-based epoxy composites via multiscale analysis, Compos. A 42 (2011) 1748–1755.

[94] R. Rafiee, M. Mahdavi, Characterizing nanotube–polymer interaction using molecular dynamics simulation, Comput. Mater. Sci. 112 (2016) 356–363.

[95] S. Yang, S. Yu, J. Ryu, J.-M. Cho, W. Kyoung, D.-S. Han, et al. Nonlinear multiscale modeling approach to characterize elastoplastic behavior of CNT/polymer nanocomposites considering the interphase and interfacial imperfection, Int. J. Plast. 41 (2013) 124–146.

[96] J.-P. Ryckaert, A. Bellemans, Molecular dynamics of liquid n-butane near its boiling point, Chem. Phys. Lett. 30 (1975) 123–125.

[97] D. Rapaport, Molecular dynamics study of a polymer chain in solution, J. Chem. Phys. 71 (1979) 3299–3303.

[98] K.M. Zimmer, D.W. Heermann, A new model with non-spherical interactions for dense polymer systems, J. Comput. Aided Mater. Des. 2 (1995) 1–8.

[99] M. Bishop, M. Kalos, H. Frisch, Molecular dynamics of polymeric systems, J. Chem. Phys. 70 (1979) 1299–1304.

[100] G. Rossi, L. Monticelli, S.R. Puisto, I. Vattulainen, T. Ala-Nissila, Coarse-graining polymers with the MARTINI force-field: polystyrene as a benchmark case, Soft Matter 7 (2011) 698–708.

[101] E. Panizon, D. Bochicchio, L. Monticelli, G. Rossi, MARTINI coarse-grained models of polyethylene and polypropylene, J. Phys. Chem. B 119 (2015) 8209–8216.

[102] D. Reith, M. Pütz, F. Müller-Plathe, Deriving effective mesoscale potentials from atomistic simulations, J. Comput. Chem. 24 (2003) 1624–1636.

[103] D. Fritz, V.A. Harmandaris, K. Kremer, N.F. van der Vegt, Coarse-grained polymer melts based on isolated atomistic chains: simulation of polystyrene of different tacticities, Macromolecules 42 (2009) 7579–7588.

[104] D.D. Hsu, W. Xia, S.G. Arturo, S. Keten, Systematic method for thermomechanically consistent coarse-graining: a universal model for methacrylate-based polymers, J. Chem. Theory Comput. 10 (2014) 2514–2527.

[105] B. Deng, Y. Shi, A reactive coarse-grained model for polydisperse polymers, Polymer 98 (2016) 88–99.

[106] S.W. Cranford, M.J. Buehler, In silico assembly and nanomechanical characterization of carbon nanotube buckypaper, Nanotechnology 21 (2010) 265706.

[107] J. Zhao, J.-W. Jiang, L. Wang, W. Guo, T. Rabczuk, Coarse-grained potentials of single-walled carbon nanotubes, J. Mech. Phys. Solids 71 (2014) 197–218.

[108] B. Arash, H.S. Park, T. Rabczuk, Mechanical properties of carbon nanotube reinforced polymer nanocomposites: a coarse-grained model, Compos. B 80 (2015) 92–100.

[109] B. Arash, H.S. Park, T. Rabczuk, Tensile fracture behavior of short carbon nanotube reinforced polymer composites: a coarse-grained model, Compos. Struct. 134 (2015) 981–988.

[110] B. Arash, H.S. Park, T. Rabczuk, Coarse-grained model of the J-integral of carbon nanotube reinforced polymer composites, Carbon 96 (2016) 1084–1092.

[111] A.A. Mousavi, B. Arash, X. Zhuang, T. Rabczuk, A coarse-grained model for the elastic properties of cross linked short carbon nanotube/polymer composites, Compos. B 95 (2016) 404–411.

15

Continuum/Finite Element Modeling of Carbon Nanotube–Reinforced Polymers

Nam Vu-Bac*, Timon Rabczuk*, Xiaoying Zhuang**,†

*INSTITUTE OF STRUCTURAL MECHANICS, BAUHAUS UNIVERSITY WEIMAR, WEIMAR, GERMANY; **TONGJI UNIVERSITY, SHANGHAI, CHINA; †INSTITUTE OF CONTINUUM MECHANICS, LEIBNIZ UNIVERSITY HANNOVER, HANNOVER, GERMANY

CHAPTER OUTLINE

15.1 Introduction

Due to superior mechanical, thermal, and electrical properties [1–4], carbon nanotubes (CNTs) are considered as an ideal reinforcement in a polymer matrix (CNRPs). Although substantial experiments have been performed [5,6], computational modeling is a strong and complementary alternative. The behavior of CNRPs is complex and influenced by many factors, such as temperature, deformation state, and the interface/interphase behavior between the CNT reinforcement and polymer matrix. Models at one length scale

Carbon Nanotube–Reinforced Polymers. http://dx.doi.org/10.1016/B978-0-323-48221-9.00015-7

cannot properly predict a variety of macroscopic material parameters [7]. Therefore, multiscale models have been developed. They can be categorized into concurrent, semiconcurrent, and hierarchical multiscale models. The bridging scale method [8], bridging domain method [9], and quasicontinuum method [10] are among popular concurrent multiscale methods. However, studying the molecular behavior of CNPRs is anywhere from difficult to unfeasible due to the high computational costs [11–13]. A concurrent or semiconcurrent multiscale approach for polymeric nanocomposites (PNCs) has been proposed in Refs. [14,15]. However, a continuum approach has been utilized at the fine scale. Furthermore, most of the multiscale methods deal with intact materials, and there are comparatively few contributions on multiscale methods for fracture as proposed in Refs. [16–20]. Hierarchical multiscale approaches seem computationally more feasible. An interesting hierarchical multiscale model was proposed by Shokrieh and Rafiee [21], bridging models over four length scale, from the nanoscale to the macroscale. This approach was exploited by Ghasemi et al. [22] to optimize the fine-scale structure of CNRPs for several applications.

Quantum mechanics seems unfeasible to model CNRPs. Popular approaches at the nanoscale are based on molecular dynamics (MD) or united atom (UA-MD); for example, the works by Odegard et al. [23], Vu-Bac et al. [24,25], and Zhao et al. [26]. However, these models are restricted to extremely small-length scales. Recently, coarse-grained models have been developed that are based on a quasicontinuum approach [27–30]. They allow modeling much larger length scales and successfully predict mechanical properties [27], as well as fracture-related material parameters [28,29]. They were extended to account for cross-links between the polymer chains and covalent interactions with the CNTs. There are also several analytical models accounting for the interface behavior among CNRPs that have the potential of being used in cohesive zone models [31–35]; see also the overviews given by Ben et al. [36] and Rafiee et al. [37].

Numerous “micromodels” can be found in the literature that were employed to predict the material behavior of CNPRs. Odegard et al. [23] indicated that continuum micromechanical models cannot be used directly to obtain the material behavior of the CNRPs, as the lattice structure of the CNT is not considered. Furthermore, these models do not account for the interphase between the CNT and the polymer matrix adequately. To capture the interphase, that is governed by van der Waals (vdWs) interactions, realistically various studies of atomistic-based continuum models have been developed; see, for example, the works by Wernik and Meguid [11], Shokrieh and Rafiee [21], and Giannopoulos et al. [38]. Studies based on phase-field models aiming to predict fracture-related material parameters were done [39]. Mesoscale models accounting for micromechanical behavior were described by Sheng et al. [40], Spencer et al. [41], and Silani et al. [42]. In those models, finite element method (FEM) was used to obtain the macroscopic properties of the material. Later, the effect of the morphological factors was quantified via sensitivity analysis by Vu-Bac et al. [43].

To adequately predict the mechanical properties of PNCs, a vast number of approaches have been proposed. However, those model predictions often deviate to results obtained from experiments. Spanos et al. [44] pointed out that this inconsistency can be alleviated by a stochastic–mechanical model using the Monte–Carlo finite element (FE) approaches.

Nevertheless, their model considered only the volume fraction as an uncertain parameter, whereas many others, such as the length, orientation, concentration, curvature, and distribution of fibers, were given as deterministic parameters. An interesting model accounting for all aforementioned effects to predict the mechanical properties of CNRPs was proposed by Shokrieh and Rafiee [21]. In Ref. [45], the correlation between input parameters, such as the length and the curvature of the equivalent fiber, was quantified within an uncertainty quantification framework. Assessing the predictive power of the underlying models and determining the influence of uncertain input parameter with respect to an output of interest is indeed crucial for computational material design of CNRPs. Several approaches have been developed accounting for uncertainties in the material parameters [46–48]. They have even been extended to optimize the material (the fine-scale structure) with respect to a certain application [22,49–51] in the context of probabilistic optimization. In Ref. [52], a Bayesian method was employed to compare different models and their ability to predict certain homogenized properties. However, this approach was based on experimental data and not numerical predictions.

This chapter gives a short overview of CNRP models at four different length scales, that is, the nano-, micro-, meso-, and macroscale. The focus is on (*continuum*) models to predict the mechanical behavior, though there exist numerous contributions on predicting other aspects, such a thermal properties; see, for example, the work by Mortazvi et al. [53–57]. Instead of providing an extensive literature overview, we focus on some "benchmark" models, which we will explain briefly.

15.2 Models at the Nanoscale

15.2.1 Equivalent-Continuum Model

An equivalent-continuum model for the PNRPs was proposed by Odegard et al. [23] in which an effective fiber was developed from MD simulations. In this model, the discrete nature of the atomic and the interfacial interactions at an atomistic level between the CNT and the polymer matrix are naturally simulated. Based on the energy balance between the total potential energy of the atomistic model and the equivalent-continuum (continuum) models via an intermediate truss model, the properties of an effective fiber can be obtained and employed in micromechanical models. The modeling procedure can be summarized as follows:

- At first, a representative volume element (RVE), including the CNT and the polymer molecules, is constructed at a molecular level. This step is performed within the MD framework.
- MD simulations are then used to obtain the mechanical properties on which an equivalent-continuum model is developed. A homogenization technique that equates the strain energies in the molecular models with those in the equivalent-continuum models under identically applied loads is employed to obtain the effective fiber. It should be noted that an equivalent-truss model of the RVE where the mechanical

properties of the truss element are linked to the force constants in a molecular model can be used as an intermediate step.

- The effective fiber is subsequently inserted into micromechanical models to determine the macroscopic behavior of the PNCs.

This approach was extended by several researchers [11,58–61]. For instance, stress–strain responses of the PNCs were studied using MD simulations by Frankland et al. [58]. Later, Wernik and Meguid [11] presented an atomistic-based continuum model to study fracture mechanism of the CNT under uniaxial tension and the phenomena of torsional buckling. Subsequently, the key ingredients of the model in Ref. [23] are summarized.

15.2.1.1 Molecular Potential Energy

The total potential energy of the molecular model, contributed from the covalent and noncovalent interactions, is written in terms of the sum of the individual energy as follows:

$$\Pi^m = \sum_a K_a^\rho (\rho_a - P_a)^2 + \sum_a (K_a^\theta - \Theta_a)^2 + \sum_a D_a^{IJ}\left[\frac{1}{2}\left(\frac{\rho^{IJ}}{\rho_a}\right)^{12} - \left(\frac{\rho^{IJ}}{\rho_a}\right)^6\right], \tag{15.1}$$

where P_a and ρ_a are the equilibrium and current (deformed) length of bond a, respectively; Θ_a and θ_a denote the respective equilibrium and current angle of bond-angle number a; K_a^ρ and K_a^θ refer to the force constants associated with the bond stretching and bond-angle bending number a, respectively. The well depth and vdWs distance for interaction a can be expressed as in Ref. [62]

$$\begin{aligned} D_a^{IJ} &= \sqrt{D_a^I \cdot D_a^J} \\ \rho_a^{IJ} &= \sqrt{\rho_a^I \cdot \rho_a^I} \end{aligned} \tag{15.2}$$

where I and J denote the two atoms belonging to a vdWs interaction. Values of the parameters involved in this molecular model can be found in Ref. [23].

15.2.1.2 Truss Model

Odegard et al. [63] used a pin-jointed truss model to simulate either a covalent or noncovalent interaction between atoms. The deformed covalent and noncovalent interactions in the molecular models are replaced by the axially deformed truss elements. Consequently, the atomic lattice of the CNT is modeled by a truss model based on the energy balance between the truss model and the molecular model. Note that there is no bending term in the truss model and hence, the total strain energy Π^t is given by

$$\Pi^t = \sum_b \sum_a \frac{A_a^b E_a^b}{2R_a^b}(r_a^b - R_a^b)^2, \tag{15.3}$$

with A_a^b and E_a^b being the cross-sectional area and Young's modulus of the truss element a of the hexagon b (Fig. 15.1); R_a^b and r_a^b denote the current and initial lengths of the truss

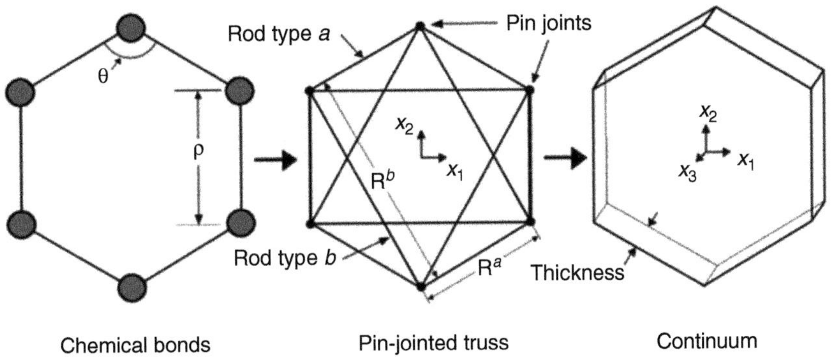

FIGURE 15.1 Equivalent-truss modeling at molecular level [63].

elements, respectively. For small deformations, the Young's moduli of the truss elements type α (bond-stretching element) and type β (bond-angle-bending element) can be determined by

$$\begin{aligned} E_a^{\alpha} &= \frac{2K_a^{\alpha} R_a^{\alpha}}{A_a^{\alpha}}, \\ E_a^{\beta} &= \frac{32K_a^{\theta} R_a^{\alpha}}{A_a^{\alpha} R_a^{\beta}}\left(\sin\frac{\Theta_a}{2}\right)^2, \end{aligned} \tag{15.4}$$

K_a^{ρ}, K_a^{θ}, and Θ_a are aforementioned parameters in Eq. (15.1).

The vdWs strain energy in Eq. (15.1) is then equated to the one given in Eq. (15.3), yielding the Young's modulus:

$$E_a^{vdw} = \frac{2\rho_a^{IJ} D_a^{IJ}}{A_a^{\delta}\left(\rho_a - \rho_a^{IJ}\right)^2}\left[\frac{1}{2}\left(\frac{\rho_a^{IJ}}{\rho^{IJ}}\right)^{12} - \left(\frac{\rho_a^{IJ}}{\rho^{IJ}}\right)^{6}\right] \tag{15.5}$$

Note that the Young's modulus value is assigned discretely according to the range of interatomic distance for each pair of atoms to reduce the computational costs. Details of the implementation can be found in Ref. [23].

15.2.1.3 Continuum Model

Based on the established equivalent-truss model, the equivalent-continuum model was built as shown in Fig. 15.2. By equating the total strain energy of the equivalent-truss to the equivalent-continuum model under the same boundary conditions (BCs), the mechanical properties of the effective solid fiber are obtained. For a transversely isotropic fiber, five linear-elastic material parameters are required. Each of these parameters were determined by five different independent simulations corresponding to the five BCs. The obtained mechanical properties of the effective fiber were then used in subsequent micromechanical analysis, see Ref. [23] for details.

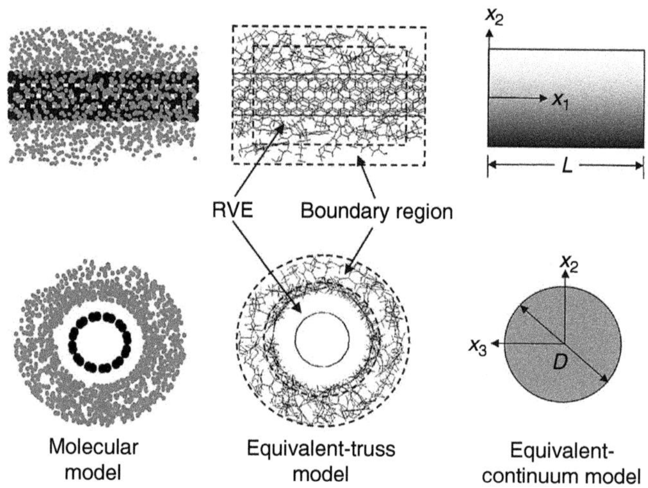

FIGURE 15.2 Equivalent-continuum model for a representative volume element (RVE) of the polymeric nanocomposites (PNCs) [23].

15.2.1.4 Effective-Fiber Constitutive Model

For the transversely isotropic effective fiber, the stress–strain relationship is expressed as

$$\sigma_{ij} = C^{f}_{ijkl}\varepsilon_{kl}, \tag{15.6}$$

where σ_{ij} and ε_{kl} ($i, j, k, l = 1, 2, 3$) are the components of the stress and strain tensor, respectively; C^{f}_{ijkl} is the tangent modulus of the effective fiber. The Dirichlet and Neumann BCs are prescribed as:

$$u_i(B) = \varepsilon_{ij}x_j, \tag{15.7}$$

$$T_i(B) = \sigma_{ij}n_j, \tag{15.8}$$

where B denotes the boundary surface, x_j is shown in Fig. 15.2, and n_j denotes the normal vector to the boundary surface B. The strain energy is given by

$$\Pi^{f} = \frac{V}{2}\sigma_{ij}\varepsilon ij = \frac{\pi D^2 L}{8}\sigma_{ij}\varepsilon ij, \tag{15.9}$$

where V, D, and L denote the respective volume, diameter, and length of the effective fiber illustrated in Fig. 15.2. To determine the material parameters of the transversely isotropic effective fiber, independent simulations must be performed and the parameters are obtained from the energy equivalence between the equivalent-truss and the equivalent-continuum models. The longitudinal Young's modulus and the Poisson's ratio were determined from uniaxially tensile simulations in x_1-direction characterized by Eq. (15.10). The longitudinal shear modulus can be computed by prescribing a pure shear strain in the $x_1 - x_2$ direction characterized by Eq. (15.11). For the transverse shear modulus, the BCs are applied as presented in Eq. (15.12). The transverse bulk modulus can be determined by applying isotropic expansion in x_2 and x_3 described by Eq. (15.13).

$$u_1(B)=ex_1,\quad u_2(B)=0,\quad u_3(B)=0, \tag{15.10}$$

$$u_1(B)=\frac{e}{2}x_2,\quad u_2(B)=\frac{e}{2}x_1,\quad u_3(B)=0, \tag{15.11}$$

$$u_1(B)=0,\quad u_2(B)=\frac{e}{2}x_3,\quad u_3(B)=ex_3, \tag{15.12}$$

$$u_1(B)=0,\quad u_2(B)=ex_2,\quad u_3(B)=\frac{e}{2}x_2, \tag{15.13}$$

where e is selected according to a specific BC, see Ref. [23].

15.2.2 Interfacial Shear Stress Characteristic

As mentioned earlier, the properties of CNRPs strongly depend on the interface/interphase behavior between the polymer matrix and the CNT [64]. Therefore, the load transfer mechanism through the interfacial region is crucial in enhancing the mechanical and thermal properties of the PNCs. MD simulations of pulling out single fibers of the polymer matrix are commonly used to study the interfacial shear strength of conventional fiber–reinforced polymer composites [65–69].

15.2.2.1 Potential Functions

To study interafacial shear stress (ISS) characteristic and influence of parameters, such as CNT radius, temperature, and pull-out velocity on the ISS, Vu-Bac et al. [25] performed pull-out tests with MD simulations. The intra-CNT interactions are described by the adaptive intermolecular reactive empirical bond order (AIREBO) potential, as suggested by Brenner et al. [70]. In this potential (E_{airebo}) intermolecular interactions between nonbonded atoms (E^{IJ}) and torsional interactions of four consecutive bonded atoms ($E^{TORSION}$) are added to the REBO potential (E^{REBO}).

$$E_{airebo}=\frac{1}{2}\sum_{i}\sum_{j\neq i}\left[E_{ij}^{REBO}+E_{ij}^{LJ}+\sum_{k\neq i,j}\sum_{l\neq i,j,k}E_{kijl}^{TORSION}\right] \tag{15.15}$$

where i, j, k, and l denote the atom's index; see Ref. [71] for more details. The matrix is made of polyethylene (PE). A UA model is employed where CH_2 atoms are grouped into a monomer (UA) to represent the PE molecules. The intrapolymer potential [72] is formulated by

$$E_{total}=E_{bond}(r)+E_{angle}(\theta)+E_{torsion}(\phi)+E_{vdW}(r). \tag{15.16}$$

where the functional forms of bond stretching, bond-angle bending, and bond torsion are given by

$$E_{bond}=\frac{1}{2}c_b\left(r-r_{eq}\right)^2; \tag{15.17}$$

$$E_{angle} = \frac{1}{2} c_\theta \left(\cos(\theta) - \cos\left(\theta_{eq}\right)\right)^2; \tag{15.18}$$

$$E_{torsion} = \frac{1}{2} \sum_{i=1}^{4} c_n \cos^i(\phi); \tag{15.19}$$

with c_b and c_θ being the constants associated with the bond length and bond angle potentials, respectively; r_{eq} and θ_{eq} are the respective bond length and bond angle at the equilibrium state; and the coefficients for the dihedral multiharmonic are denoted by c_i, $i = 1, \ldots, 4$. The vdWs interactions are formulated by the Lennard–Jones potential as follows

$$E_{vdW} = \begin{cases} 4\varepsilon\left[\left(\frac{\sigma}{\varepsilon}\right)^{12} - \left(\frac{\sigma}{\varepsilon}\right)^{6}\right] & r \le r_c \\ 0 & r > r_c \end{cases}, \tag{15.20}$$

where r is the distance between the two monomers, σ and ε are the distance at zero energy and the energy well depth, respectively; $r_c = 10.5$ Å was chosen as the cutoff radius. The interactions between monomers are modeled by a force field, whose parameters are briefly provided in Table 15.1. Furthermore, the interaction between the CNT and the PE matrix is described by the truncated Lennard–Jones (LJ) 6–12 with $\varepsilon = 0.1102$ kcal/mol and $\sigma = 3.65$ Å (Fig. 15.3) [25].

15.2.2.2 Pull-Out Simulations

The pull-out process is shown in Fig. 15.4. During the pull-out process, the potential mean force (PMF)—the pull-out energy—and the total potential energy are computed and plotted, as shown in Fig. 15.5. The total potential energy and the PMF of the system increases with increasing displacement of the CNT and levels off after it has been completely pulled out of the PE matrix. These results are in good agreement to previous results reported by Liao and Li [65] and Gou et al. [67].

Based on the PMF, the ISS is then obtained by Eq. (15.22).

Table 15.1 United Atom (UA) Model and Parameters of the Force Field

Parameters	Values
c_b	350 kcal/mol
r_{eq}	1.53 Å
c_θ	60 kcal/mol/rad^2
θ_{eq}	1.911 rad
c_1	1.736 kcal/mol
c_2	−4.490 kcal/mol
c_3	0.776 kcal/mol
c_4	6.990 kcal/mol
σ	4.01 Å
ε	0.112 kcal/mol

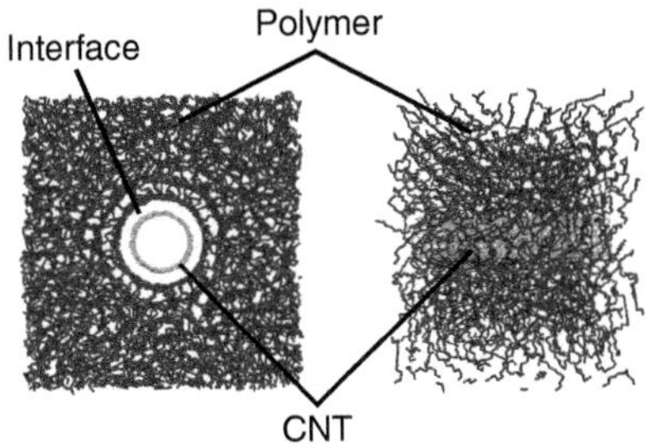

FIGURE 15.3 Cross-section and longitudinal side of the carbon nanotube (CNT) reinforced polymer composites.

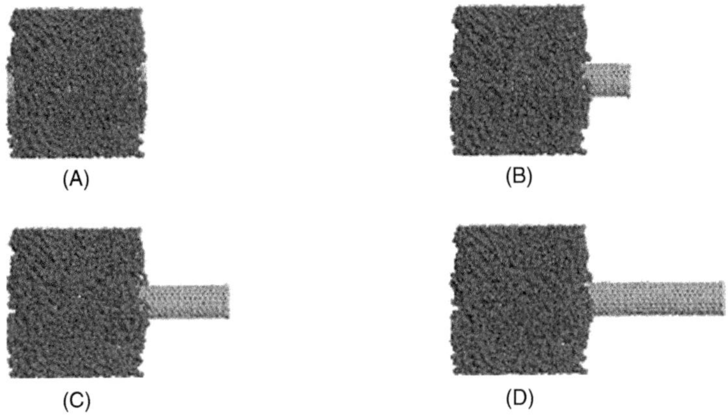

FIGURE 15.4 Snapshots of pull-out simulations. The CNT is pulled out from polyethylene matrix at (A) $x = 0$ nm, (B) $x = 2$ nm, (C) $x = 4$ nm, (D) $x = 6.4$ nm [25].

$$E_{pullout} = \int_0^L 2\pi r(L - x)\tau_i \, dx = \pi r \tau_i L^2 \tag{15.21}$$

$$\tau_i = \frac{E_{pullout}}{\pi r L^2} \tag{15.22}$$

where r and L denote the respective radius and length of the CNT inside the PE matrix, x is the displacement of the CNT, and τ_i is the ISS.

15.3 Models at the Microscale

As the computations using MD simulations are prohibitively expensive, even one cubic micron cannot be simulated [11]. Reaching reasonable time scales is another issue, and strain rates in MD simulations are commonly out of classical engineering applications. To reduce the computational cost, atomistic-based continuum models have been employed. Within an atomistic-based continuum model, the CNT and the polymer matrix is commonly replaced by an equivalent fiber, where the covalent interactions in the CNT are modeled by structural elements. Particularly, the *C—C* bonds are described by 3D beam

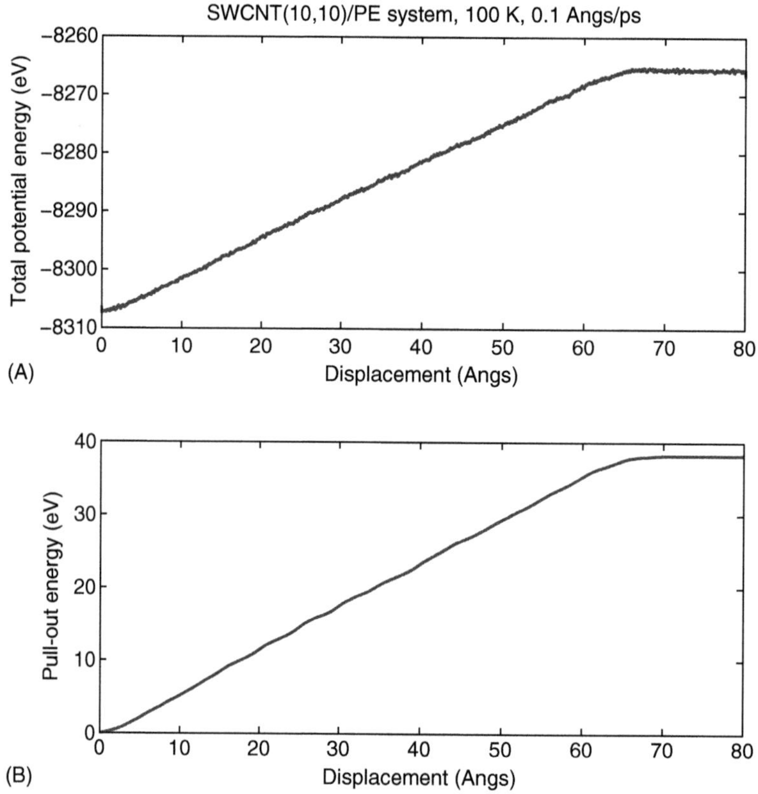

FIGURE 15.5 Energy evolution during the pull-out process. (A) Total potential energy versus displacement and (B) Pull-out energy (potential mean force-PMF) versus displacement [25].

elements. Joint nodes connecting the beams are used to simulate the C atoms as illustrated in Fig. 15.6. The total energy is computed as the sum of the energies contributed by the covalent and noncovalent bonds provided by Rappé et al. [73], as follows:

$$E_{total} = \sum E_{bond} + \sum E_{angle} + \sum E_{torsion} + \sum E_{vdW}. \tag{15.23}$$

with E_{bond}, E_{angle}, and $E_{torsion}$ being energies caused by the bond stretching, bond-angle bending, and bond torsion, respectively; V_{vdW} refers to the energy contributed by the vdWs interactions. The strain energy of the system is given by Li and Chou [74]

$$E_{bond} = \frac{1}{2}k_r\left(r - r_{eq}\right)^2 = \frac{1}{2}k_r(\Delta r)^2, \tag{15.24a}$$

$$E_{angle} = \frac{1}{2}k_\theta\left(\theta - \theta_{eq}\right)^2 = \frac{1}{2}k_\theta(\Delta\theta)^2, \tag{15.24b}$$

$$E_{torsion} = \frac{1}{2}k_\phi(\Delta\phi)^2, \tag{15.24c}$$

with k_r, k_θ, and k_ϕ being force constants associated with the respective bond stretching, bond-angle bending, and bond torsion; Δr, $\Delta\theta$, and $\Delta\phi$ refer to deformations due to the

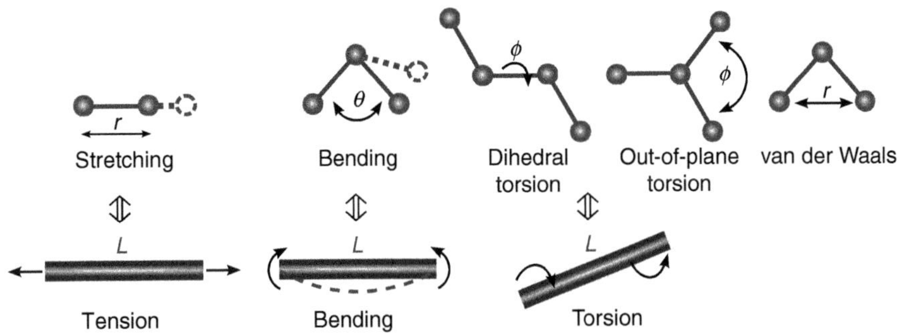

FIGURE 15.6 Modeling of bond stretching, bond angle bending, bond dihedral torison, out-of-plane torsion energies, and van der Waals (vdWs) interaction by the molecular structural mechanics approach. Molecular mechanics model (top row) and structural mechanics model (bottom row) [45].

respective bond stretching, the bond-angle bending, and the bond torsion. Linking molecular mechanics to structural mechanics based on the energy equivalence, the structural properties (*EA*, *EI*, and *GJ*) of the beam can be obtained from the molecular constants k_r, k_θ, and k_ϕ

$$\frac{EA}{L} = k_r, \quad \frac{EI}{L} = k_\theta, \quad \frac{GJ}{L} = k_\phi, \tag{15.25}$$

Due to the high density compared to the CNT structure, the polymer matrix is commonly modeled as a continuum. This model will be briefly described in the next section. The interphase between the CNT and the polymer matrix is described by a truncated Lennard–Jones (LJ) 6–12 force field provided by:

$$F_{vdw} = 24\frac{\varepsilon}{\sigma}\left[2\left(\frac{\sigma}{r}\right)^{13} - \left(\frac{\sigma}{r}\right)^{7}\right] \tag{15.26}$$

Another microscale model for CNRPs was presented by Wernik and Meguid [11], where the modified Morse potential is used for the CNT. As this potential is capable of accounting for the broken C—C bonds, the atomistic-based continuum model can be exploited to simulate fracture in the CNT (Fig. 15.7). The bond stretching and bond-angle bending are formulated by

$$E_{total} = E_{bond}(r) + E_{angle}(\theta), \tag{15.27}$$

$$E_{bond} = D_e\left(1 - \exp^{-\beta\left(r - r_{eq}\right)} - 1\right), \tag{15.28}$$

$$E_{angle} = \frac{1}{2}k_\theta\left(\theta - \theta_{eq}\right)^2\left[1 + k_{sextic}\left(\theta - \theta_{eq}\right)\right], \tag{15.29}$$

with r and θ being the current bond length and the current bond angle, respectively; D_e is the separation energy; and β denotes a constant controlling the "width" of the potential. The potential parameters shown in Table 15.2 are provided by Belytschko et al. [75].

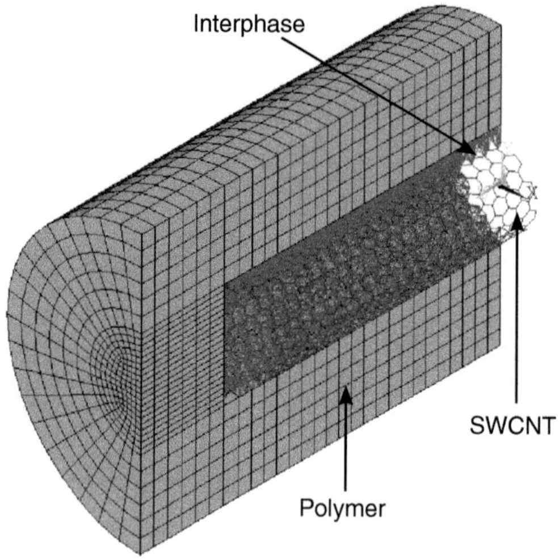

FIGURE 15.7 Finite element (FE) discretization of half a unit cell of the PNCs [45].

Table 15.2 Force Field Parameters for Modified Morse Potential

Parameters	Values
r_{eq}	1.421×10^{-10} m
D_e	6.03105×10^{-19} N·m
B	2.625×10^{10} m^{-1}
θ_{eq}	2.094 rad
k_θ	0.876×10^{-18} N·m/rad^{-2}
k_{sextic}	0.754 rad^{-4}

Wenik and Meguid [11] used beam elements to describe the bond stretching, while nonlinear rotational spring elements coupling translational and rotational degrees of freedom were utilized in modeling the bond-angle bending. The equivalent structural properties of the beams can be obtained based on the link between structural mechanics and molecular mechanics, as outlined earlier. The uniaxial force in C—C bond is derived from the stretching energy in Eq. (15.28) as follows:

$$F = 2\beta D_e\left(1 - \exp^{-\beta\varepsilon r}\right)\exp^{-\beta\varepsilon r} \tag{15.30}$$

where $\varepsilon = r - r_{eq}/r$. The moment resultant in the springs is determined by taking derivative of the potential associated with bond-angle bending in Eq. (15.29), as follows:

$$M = k_\theta \Delta\theta\left[1 + 3k_{sextic}(\Delta\theta)^4\right]. \tag{15.31}$$

Similar to Fig. 15.8, 3D solid elements are used to model the polymer matrix. The vdWs interactions are simulated by the truss rod elements, which connect a node on the CNT

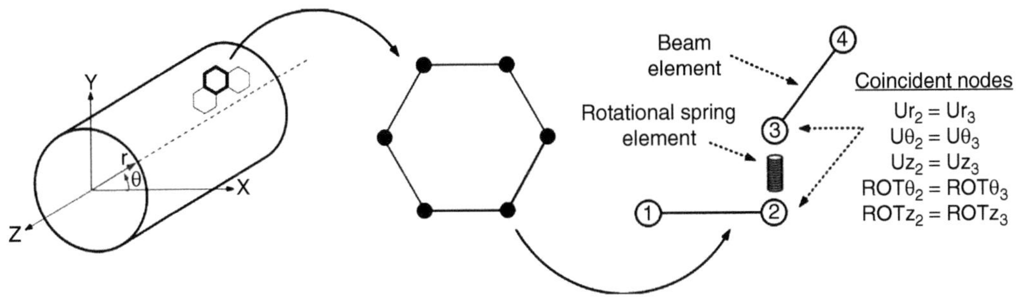

FIGURE 15.8 The space frame structure of the CNT and its structural components [11].

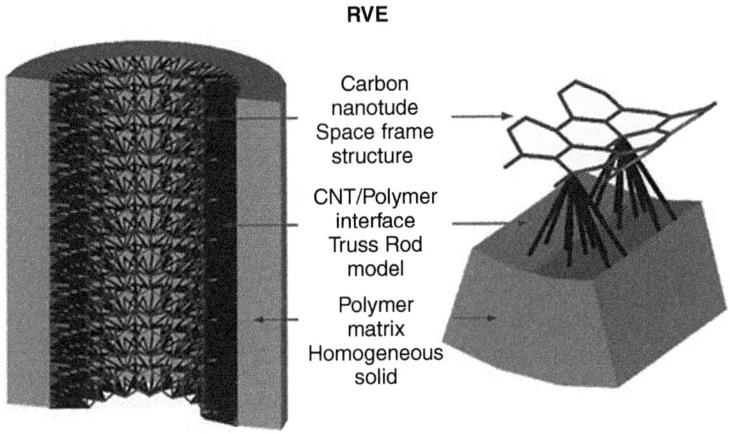

FIGURE 15.9 A 3D unit cell of the PNCs [11].

to a node of the solid element (Fig. 15.9). The Lennard–Jones potential is described by the truncated vdWs interactions, as shown in Eq. (15.26).

Once a unit cell containing the CNT and the surrounding polymer matrix has been established, the five independent material parameters of the transversely isotropic equivalent fiber are determined in the same manner used in Section 15.2.2. Those properties of the equivalent fiber are then used in the micromechanics equations to obtain macroscopic behavior of the PNCs.

15.4 Models at the Mesoscale

As presented in the previous section, effective fibers were developed for use in micromechanical models of the continuum-based composite. As observed, the properties of the PNCs depend strongly on the second-phase fibers Besides that, the morphology of the PNCs, such as the shape of the fibers, their volume fraction, distribution in the polymer matrix, and the ratio of mechanical properties between the fiber and the polymer matrix,

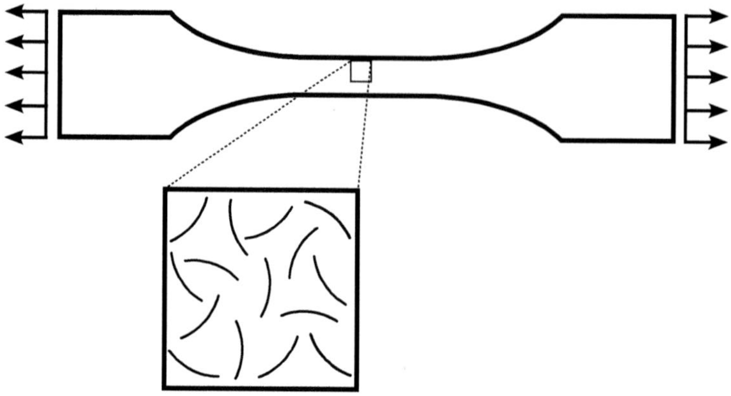

FIGURE 15.10 Length scales of carbon nanotube–reinforced polymer (CNRPs) from macro- to mesoscales.

plays an important role in the material performance. The morphology of the CNRPs has agglomerations in local regions where the CNT highly concentrates within the polymer matrix (Fig. 15.10). In this section, the fully random distributed CNT within the polymer matrix will be considered. Though this is not mandatory, models at the mesoscale are often based on RVEs. Therefore, they will be discussed in this content.

15.4.1 RVE Definition

According to Hill [13], an RVE is "structurally entirely typical of the whole mixture on average." An RVE must include enough information to represent a model heterogeneous material. By using the RVE approach, the heterogeneous material is substituted by an equivalent homogeneous material and the original properties is homogenized (averaged) by different methods. In this way, we avoid simulating an infinite medium, but still obtain the average properties of the heterogeneous material.

To ensure the characteristic of the RVE from its definition, an adequate RVE size must be determined based on the so-called "sample enlargement" approach shown in Fig. 15.11. First, we randomly generate a statistical volume element (SVE) on which sensitivity of the overall properties to the size of the SVE (and spatially random components within the SVE) is then estimated. If the sensitivity is insignificant, the SVE can be considered as an RVE, and the obtained properties over the SVE is representative for the material properties. As the SVE is randomly generated, the overall properties of an ensemble can be determined as shown in Ref. [43]

$$\langle R \rangle = \frac{1}{M} \sum_{k=1}^{M} R^{(k)} \tag{15.32}$$

where $R^{(k)}$ is the properties obtained from the k-th RVE, $k = (1, 2, \ldots, M)$, with M being the number of realizations in the ensemble.

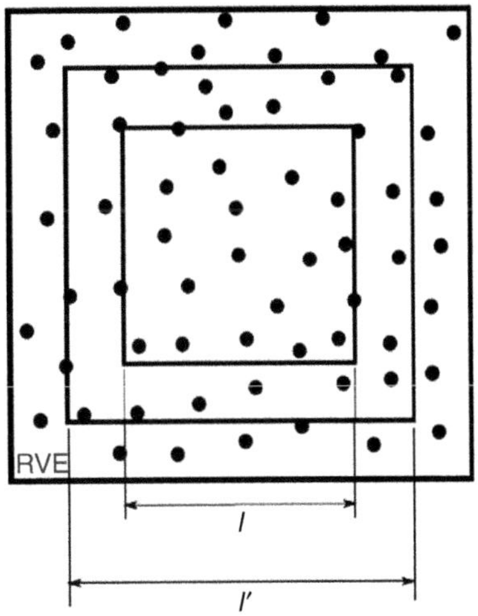

FIGURE 15.11 Sample enlargement technique [42].

15.4.2 Boundary Conditions

In simulations, three types of BCs, such as uniform displacement, uniform traction, and mixed-type BCs consisting of PBCs, can be applied on the RVE. For the sake of computational efficiency, the PBCs are commonly used to deform the RVE, while the uniform displacement and uniform traction usually provide computational bounds on the properties in case the RVE is infinitively large [42]. It is well known that a fast convergence of the effective properties can be obtained over the RVE size [76–78], when PBCs, as shown in Fig. 15.12, are employed [43].

The CNRPs are commonly assumed to be isotropic material. Two independent parameters, that is, the bulk modulus K and the shear modulus μ, are needed to obtain the effective Young's moduli of the CNRPs. PBCs will be applied to determine these two elastic parameters. By prescribing a horizontal strain δ as shown in Fig. 15.12, the corresponding stress and strain are $\sigma_{22} = 0$ and $\varepsilon_{11} = \delta$, while ε_{22} and σ_{11} can be computed from the homogenization approach. Consequently, the effective Young's modulus of the PNCs E can be determined.

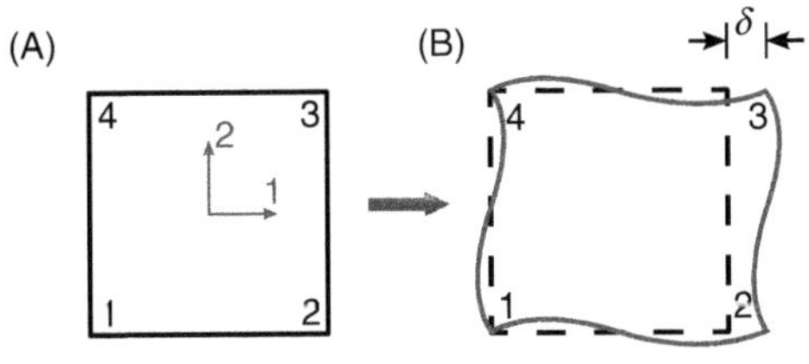

FIGURE 15.12 Boundary conditions (BC) imposed on the RVE. (A) An undeformed RVE and (B) a deformed RVE [43].

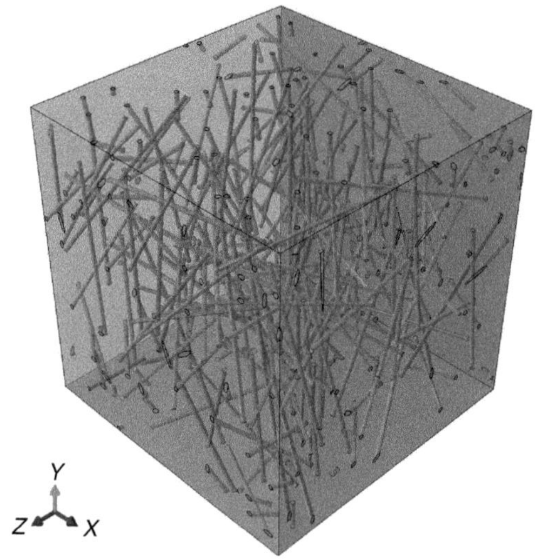

FIGURE 15.13 A 3D simulation box of the PNCs [53].

The procedure can be summarized by the following steps:

- Determine the number of fibers based on the RVE size, the aspect ratio of the fiber, and the volume fraction of the fiber within the matrix.
- Randomly generate the fibers inside the RVE based on a specific probability density function of distribution of the number of fibers, as illustrated in Fig. 15.13. The random sequential addition algorithm is an efficient tool to generate the mesoscale structure of the CNRP [79–82]. PBCs should also be considered. In other words: the missing part of the fiber that is cut on one side has to appear on the opposite side of the RVE.
- Discretize the RVE and the fibers. A typical FE mesh of an RVE is shown in Fig. 15.14.
- Formulate a set of equations for the PBCs that will be solved to obtain the desired material response via computational homogenization.

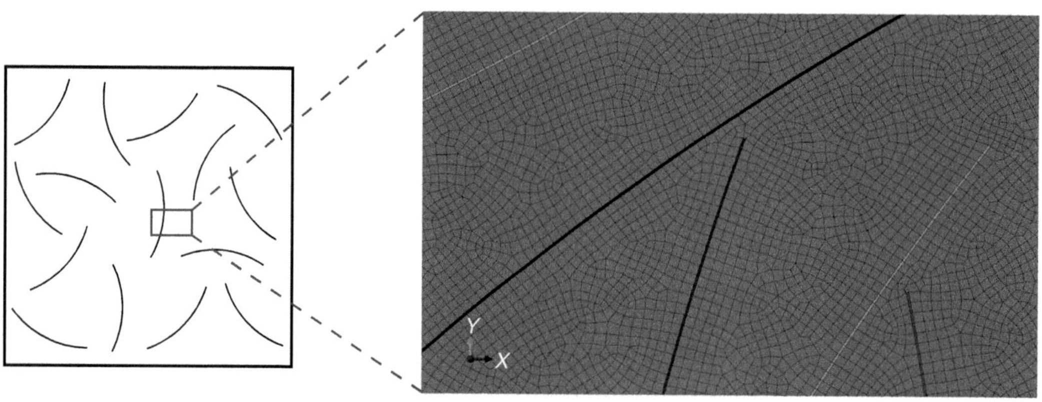

FIGURE 15.14 A zoom-out of detailed view of an FE mesh of the PNC model.

15.4.3 RVE Size

At first, a number of realizations (ensemble of RVEs) of the material model whose size is l are generated randomly. The same volume fraction of fiber is used for these realizations. The mechanical properties (e.g., Young's modulus) of each realization are then evaluated. These properties over realizations show statistical scatter values when their sizes l are still small. Therefore, averaging on a set of realizations of the same size will be required to obtain the ensemble average of the properties of the RVE. The same steps are repeated for a set of realizations whose size is $l' > l$. By gradually increasing the size of realizations until the following criteria shown by Eq. (15.33) is satisfied, l' is considered as a sufficiently large size of the RVE.

$$\frac{\left|\langle E\rangle_{l'}^{(1)} - \langle E\rangle_{l}^{(1)}\right|}{(l'-l)\cdot\langle E\rangle_{l'}^{(1)}} < \mathrm{Tol}^{a} \tag{15.33}$$

where $\langle E\rangle_{l}^{(1)}$ and $\langle E\rangle_{l'}^{(1)}$ refer to the first realization of the ensemble of RVEs, whose sizes are l and l', respectively. Tol^{a} designates the stopping criteria. For example, the resultant of the elastic modulus in Fig. 15.15 is converged when the REV size is beyond 15,000 nm.

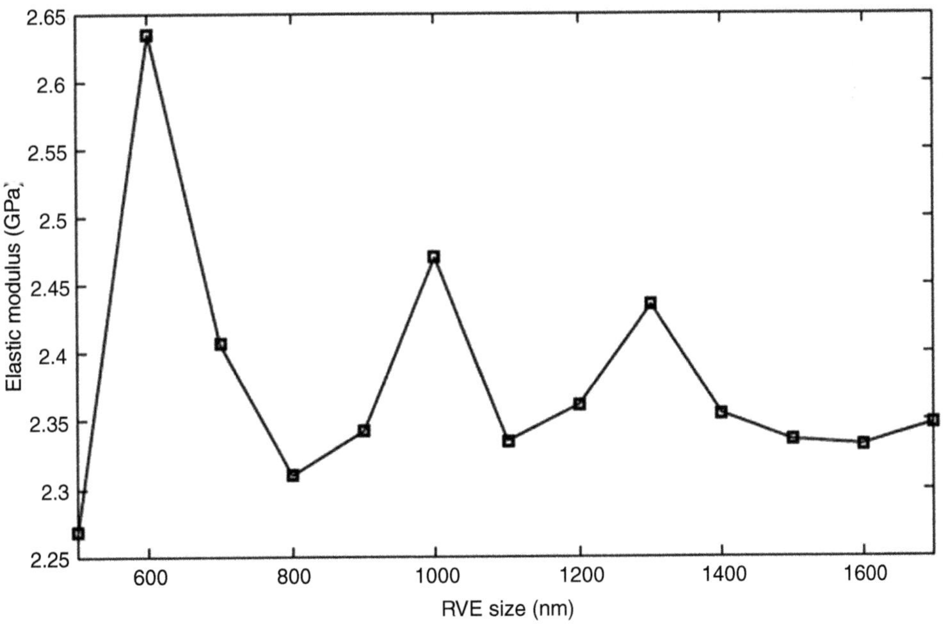

FIGURE 15.15 Plot of the elastic modulus as a function of RVE size [42].

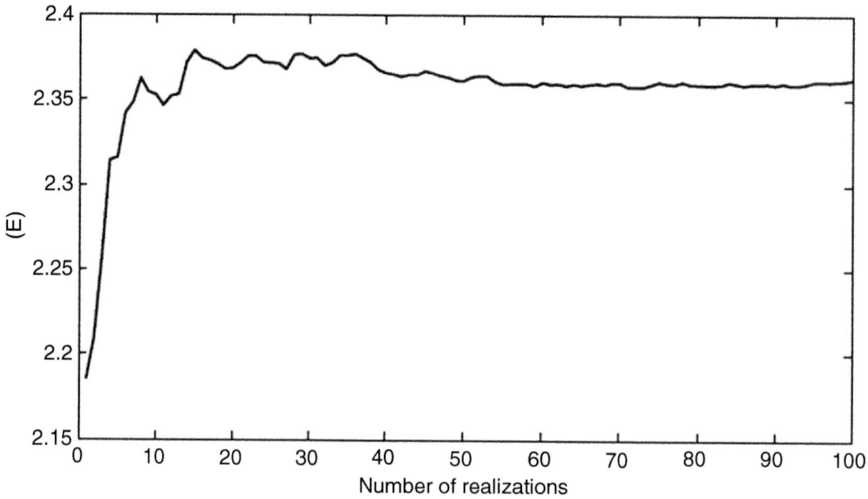

FIGURE 15.16 Convergence of the average elastic modulus over the number of realizations [42].

15.4.4 Ensemble Averaging

As the fibers are randomly generated within the RVE, it is necessary that the ensemble of realizations be a good statistical representation of the PNC. To determine a proper number of realizations, the ensemble averages must be evaluated as:

$$\left|\frac{\langle E\rangle^{(2J)} - \langle E\rangle_l^{(J)}}{\langle E\rangle^{(2J)}}\right| < \mathrm{Tol}^b \tag{15.34}$$

where $\langle \bullet \rangle^{(J)}$ and $\langle \bullet \rangle^{(2J)}$ refer to the properties obtained from an ensemble average using the j realization and the $2j$ realizations, respectively. For example, the average elastic over the number of realizations was depicted in Ref. [42] for PNCs (Fig. 15.16). It can be seen that the convergence is obtained after 100 realizations.

15.5 Models at the Macroscale

At the macroscale, the CNRP is commonly assumed as a continuous homogenized medium. Various models have been proposed to predict the macroscopic material properties. The constitutive model of the CNRPs can be written in terms of an averaged form as follows:

$$\langle \sigma \rangle = \mathbf{C}\langle \varepsilon \rangle. \tag{15.35}$$

where the averaged tangent modulus **C** is given in Refs. [83,84] as

$$C = \left(V_m C_m + V_f \boldsymbol{C}_f \boldsymbol{A}\right)\left(V_m \boldsymbol{I} + V_f \langle \boldsymbol{A} \rangle\right)^{-1} \tag{15.36}$$

where **I** refers to the fourth-order identity tensor; $\mathbf{C}_f$ and $\mathbf{C}_m$ are the respective tangent moduli of the fiber and the polymer matrix; the volume fractions of the fibers and the polymer matrix are denoted by V_f and V_m, respectively. The average strain $\boldsymbol{\varepsilon}_f$ can be described by

$$\epsilon_f = A\epsilon_m, \tag{15.37}$$

where the strain-concentration tensor **A** is determined by

$$A = \left[I + S(C_m)^{-1}(C_f - C_m)\right]^{-1}. \tag{15.38}$$

with **S** being the Eshelby tensor. Given a transversely isotropic CNRP reinforced with straight fibers along the x direction, the components of **S** are given by Ref. [85]

$$\begin{aligned}
&S_{1111} = S_{3333} = \frac{5-4v_m}{8(1-v_m)}, \quad S_{1122} = S_{3322} = \frac{v_m}{2(1-v_m)},\\
&S_{1133} = S_{3311} = \frac{4v_m - 1}{8(1-v_m)}, \quad S_{2323} = S_{1212} = \frac{1}{4}\\
&S_{1313} = \frac{3-4v_m}{8(1-v_m)},
\end{aligned} \tag{15.39}$$

where v_m is the Poisson's ratio of the isotropic elastic polymer matrix. Inserting S into Eq. (15.38), the strain concentration tensor **A** can be determined as

$$\begin{aligned}
&A_{1111} = A_{3333} = -\frac{E_m(1-v_m)\{E_m + 2k_r(1+v_m)[m_r(3-4v_m) + k_r(2-4v_m)]\}}{(-1+2v_m)[E_m + 2k_r(1+v_m)][E_m + 2m_r(3 - v_m - 4v_m^2)]},\\
&A_{1133} = A_{3311} = \frac{E_m(1-v_m)\{E_m(1-4v_m) + 2(v_m+1)[m_r(3-4v_m) + k_r(2-4v_m)]\}}{(-1+2v_m)[E_m + 2k_r(1+v_m)][E_m + 2m_r(3 - v_m - 4v_m^2)]},\\
&A_{1122} = A_{3322} = \frac{l_r(1 - v_m - 2v_m^2) - E_m V_m}{(-1+2v_m)[E_m + 2k_r(1+v_m)]}, \quad A_{2222} = 1,\\
&A_{2323} = A_{1212} = \frac{E_m}{E_m + 2pr(1+v_m)}, \quad A_{1313} = \frac{2E_m(1-v_m)}{E_m + 2m_r(3 - v_m - 4v_m^2)}
\end{aligned} \tag{15.40}$$

where k_r, l_r, m_r, n_r, and p_r denote Hill's elastic moduli for the fiber. Subsequently, by replacing the obtained tensor **A** into Eq. (15.36), we can obtain the effective tangent modulus **C** of the CNRP. The effective modulus of the composite is then given by

$$E_{\parallel} = n = \frac{l^2}{k}, \quad E_{\perp} = \frac{4m(kn - l^2)}{kn - l^2 + mn} \tag{15.41}$$

in which the Hill's elastic moduli are provided by Ref. [84]

$$
\begin{aligned}
k &= \frac{E_m\left\{E_m c_m + 2k_r\left(1+v_m\left[1+c_r\left(1-2v_m\right)\right]\right)\right\}}{2(1+v_m)\left[E_m\left(1+c_r-2v_m\right)+2c_m k_r\left(1-v_m-2v_m^2\right)\right]}, \\
l &= \frac{E_m\left\{c_m v_m\left[E_m+2k_r\left(1+v_m\right)\right]+2c_r l_r\left(1-v_m^2\right)\right\}}{2(1+v_m)\left[E_m\left(1+c_r-2v_m\right)+2c_m k_r\left(1-v_m-2v_m^2\right)\right]}, \\
n &= \frac{E_m^2 c_m\left(1+c_r-c_m v_m\right)+2c_m c_r\left(k_r n_r-l_r^2\right)\left(1+v_m\right)^2\left(1-2v_m\right)}{2(1+v_m)\left[E_m\left(1+c_r-2v_m\right)+2c_m k_r\left(1-v_m-2v_m^2\right)\right]} \\
&\quad + \frac{E_m\left[2c_m^2 k_r\left(1-v_m\right)+c_r n_r\left(1-2v_m+c_r\right)-4c_m l_r v_m\right]}{E_m\left(1+c_r-2v_m\right)+2c_m k_r\left(1-v_m-2v_m^2\right)}, \\
p &= \frac{E_m\left[E_m C_m+2\left(1+c_r\right)p_r\left(1+v_m\right)\right]}{2(1+v_m)\left[E_m\left(1+c_r\right)+2c_m p_r\left(1+v_m\right)\right]}, \\
m &= \frac{E_m\left[E_m c_m+2m_r\left(1+v_m\right)\left(3+c_r-4v_m\right)\right]}{2(1+v_m)\left\{E_m\left[c_m+4c_r\left(1-v_m\right)\right]+2c_m m_r\left(3-v_m-4v_m^2\right)\right\}},
\end{aligned}
\tag{15.42}
$$

It should be noted that these equations are valid for CNRPs reinforced with aligned, straight fibers. For CNRPs reinforced by randomly wavy-oriented fibers, extensions have been provided by Shokrieh and Rafiee [21], Vu-Bac et al. [43,45], and Shi et al. [84].

15.6 Conclusions

Nanostructured materials have been developed for many decades. As they are lightweight and cost effective, they are widely used in the industry. However, the challenge remains of coupling physical description at the molecular level to component design at the continuum level. Hence, a predictive physics-based model is essential for a comprehensive understanding of material behavior.

In this chapter, various models from nano- to macroscales have been summarized. The macroscopic properties of the CNRPs are obtained from continuum models whose parameters are known from molecular level. Atomic details of nanostructured materials are hence incorporated into continuum constitutive models by using the homogenization technique. These models can be extensively applied to a wide range of materials whose systems are so complex that multiscale modeling is needed to predict their behavior. It is believed that these models will provide an efficient and effective tool for the design of PNCs.

References

[1] M.S.P. Shaffer, A.H. Windle, Fabrication and characterization of carbon nanotube/poly(vinyl alcohol) composites, Adv. Mater. 11 (1999) 937–941.

[2] P.M. Ajayan, L.S. Schadler, C. Giannaris, A. Rubio, Single-walled carbon nanotube polymer composites: strength and weakness, Adv. Mater. 12 (2000) 750–753.

[3] H. Dai, Carbon nanotubes: opportunities and challenges, Surf. Sci. 500 (2002) 218–241.

[4] K.T. Lau, C. Gu, D. Hui, A critical review on nanotube and nanotube/nanoclay related polymer composite materials, Compos. B 37 (2006) 425–436.

[5] E.T. Thostenson, C. Li, T.-W. Chou, Nanocomposites in context, Compos. Sci. Technol. 65 (2005) 491–516.

[6] Q.H. Zeng, A.B. Yu, G.Q. Lu, Multiscale modeling and simulation of polymer nanocomposites, Prog. Polym. Sci. 33 (2008) 191–269.

[7] J.L. Bouvard, D.K. Ward, D. Hossain, S. Nouranian, E.B. Marin, M.F. Horstemeyer, Review of hierarchical multiscale modeling to describe the mechanical behavior of amorphous polymers, J. Eng. Mater. Technol. 131 (2009) 041206.

[8] G.J. Wagner, W.K. Liu, Coupling of atomistic and continuum simulations using a bridging scale decomposition, J. Comput. Phys. 190 (2003) 249–274.

[9] S.P. Xiao, T. Belytschko, A bridging domain method for coupling continua with molecular dynamics, Comp. Meth. Appl. Mech. Eng. 193 (2004) 1645–1669.

[10] E.B. Tadmor, M. Ortiz, R. Phillips, Quasicontinuum analysis of defects in solids, Philos. Mag. A 73 (1996) 1529–1563.

[11] J.M. Wernik, S.A. Meguid, Multiscale modeling of the nonlinear response of nano-reinforced polymers, Acta Mech. 1 (2011) 1–16.

[12] Y. Li, S. Tang, B.C. Abberton, M. Krger, C. Burkhart, B. Jiang, G.J. Papakonstantopoulos, M. Poldneff, W.K. Liu, A predictive multiscale computational framework for viscoelastic properties of linear polymers, Polymer 53 (2012) 5935–5952.

[13] R. Hill, Elastic properties of reinforced solids: some theoretical principles, J. Mech. Phys. Solids 11 (1963) 357–372.

[14] M. Silani, H. Talebi, A.M. Hamouda, T. Rabczuk, Nonlocal damage modelling in clay/epoxy nanocomposites using a multiscale approach, J. Comp. Sci. 15 (2016) 18–23.

[15] M. Silani, S. Ziaei-Rad, H. Talebi, T. Rabczuk, A semi-concurrent multiscale approach for modeling damage in nanocomposites, Theor. Appl. Fract. Mech. 74 (2014) 30–38.

[16] H. Talebi, M. Silani, S.P. Bordas, P. Kerfriden, T. Rabczuk, A computational library for multiscale modeling of material failure, Comp. Mech. 53 (5) (2014) 1047–1071.

[17] H. Talebi, M. Silani, T. Rabczuk, Concurrent multiscale modeling of three-dimensional crack and dislocation propagation, Adv. Eng. Softw. 80 (2015) 82–92.

[18] H. Talebi, M. Silani, S.P. Bordas, P. Kerfriden, T. Rabczuk, Molecular dynamics/XFEM coupling by a three-dimensional extended bridging domain with applications to dynamic brittle fracture, Int. J. Multisc. Comp. Eng. 11 (2013) 527–541.

[19] P.R. Budarapu, R. Gracie, S.W. Yang, X. Zhuang, T. Rabczuk, Efficient coarse graining in multiscale modeling of fracture, Theor. Appl. Fract. Mech. 69 (2014) 126–143.

[20] P. Budarapu, R. Gracie, S. Bordas, T. Rabczuk, An adaptive multiscale method for quasi-static crack growth, Comp. Mech. 53 (2014) 1129–1148.

[21] M.M. Shokrieh, R. Rafiee, Stochastic multi-scale modeling of CNT/polymer composites, Comp. Mater. Sci. 50 (2010) 437–446.

[22] H. Ghasemi, R. Rafiee, X. Zhuang, J. Muthu, T. Rabczuk, Uncertainties propagation in metamodel-based probabilistic optimization of CNT/polymer composite structure using stochastic multiscale modeling, Comp. Mater. Sci. 85 (2014) 295–305.

[23] G.M. Odegard, T.S. Gates, K.E. Wise, C. Park, E.J. Siochi, Constitutive modeling of nanotube-reinforced polymer composites, Compos. Sci. Technol. 63 (2003) 1671–1687.

[24] N. Vu-Bac, T. Lahmer, H. Keitel, J. Zhao, X. Zhuang, T. Rabczuk, Stochastic predictions of bulk properties of amorphous polyethylene based on molecular dynamics simulations, Mech. Mater. 68 (2014) 70–84.

[25] N. Vu-Bac, T. Lahmer, Y. Zhang, X. Zhuang, T. Rabczuk, Stochastic predictions of interfacial characteristic of polymeric nanocomposites (PNCs), Compos. B 59 (2014) 80–95.

[26] J. Zhao, Z. Zhang, W. Guo, T. Rabczuk, Size-dependent elastic properties of crystalline polymers via a molecular mechanics model, Appl. Phys. Lett. 99 (2011) 241902.

[27] B. Arash, H. Park, T. Rabczuk, Mechanical properties of carbon nanotube bundle reinforced polymer nanocomposites: a coarse-grained approach, Compos. B 80 (2015) 92–100.

[28] B. Arash, H.S. Park, T. Rabczuk, Tensile fracture behavior of short carbon nanotube reinforced polymer composites: a coarse-grained model, Compos. Struct. 134 (2015) 981–988.

[29] B. Arash, H.S. Park, T. Rabczuk, Coarse-grained model of the J-integral fracture toughness in short carbon nanotube reinforced polymer composites, Carbon 96 (2016) 1084–1092.

[30] A.A. Mousavi, B. Arash, X. Zhuang, T. Rabczuk, A coarse-grained model for the elastic properties of cross-linked short carbon nanotube/polymer composites, Compos. B 95 (2016) 404–411.

[31] J. Zhao, Y. Jia, N. Wei, T. Rabczuk, Binding energy and mechanical stability of two parallel and crossing carbon nanotubes, Proc. Royal Soc. Lond. A 471 (2015) 20150229.

[32] J. Zhao, L. Lu, Z. Zhang, W. Guo, T. Rabczuk, Continuum modeling of the cohesive energy for the interfaces between films, spheres, coats and substrates, Comp. Mater. Sci. 96 (2015) 432–438.

[33] J. Zhao, J.W. Jiang, L. Wang, W. Guo, T. Rabczuk, Coarse-grained potentials of single-walled carbon nanotubes, J. Mech. Phys. Solid. 71 (2014) 197–218.

[34] Y. Zhang, J. Zhao, Y. Jia, T. Mabrouki, Y. Gong, N. Wei, T. Rabczuk, An analytical solution on interface debonding for large diameter carbon nanotube-reinforced composite with functionally graded variation interphase, Compos. Struct. 104 (2013) 261–269.

[35] Y. Zhang, J. Zhao, N. Wei, J.W. Jiang, T. Rabczuk, Effects of the dispersion of polymer wrapped two neighbouring single walled carbon nanotubes (SWNTs) on nanoengineering load transfer, Compos. B 45 (2013) 1714–1721.

[36] S. Ben, J. Zhao, Y. Zhang, T. Rabczuk, The interface strength and debonding for composite structures: review and recent developments, Compos. Struct. 129 (2015) 8–26.

[37] R. Rafiee, T. Rabczuk, R. Pourazizi, J. Zhao, Y. Zhang, Challenges of the modeling methods for investigating the interaction between the CNT and the surrounding polymer, Adv. Mater. Sci. Eng. 2013 (2013) 183026.

[38] G.I. Giannopoulos, S.K. Georgantzinos, N.K. Anifantis, A semi-continuum finite element approach to evaluate the Young's modulus of single-walled carbon nanotube reinforced composites, Compos. B 41 (8) (2010) 594–601.

[39] M.A. Msekh, M. Silani, M. Jamshidian, P. Areias, X. Zhuang, G. Zi, P. He, T. Rabczuk, Predictions of J integral and tensile strength of clay/epoxy nanocomposites material using phase field model, Compos. B 93 (2016) 97–114.

[40] N. Sheng, M.C. Boyce, D.M. Parks, G.C. Rutledge, J.I. Abes, R.E. Cohen, Multiscale micromechanical modeling of polymer/clay nanocomposites and the effective clay particle, Polymer 45 (2004) 487–506.

[41] P.E. Spencer, J. Sweeney, Modeling of polymer clay nanocomposites for a multiscale approach, Nano- and Micromechanics of Polymer Blends and Composites, Carl Hanser Verlag GmbH & Co. KG, München, (2009) [pp. 545–578].

[42] M. Silani, H. Talebi, S. Ziaei-Rad, P. Kerfriden, S.P. Bordas, T. Rabczuk, Stochastic modeling of clay/epoxy nanocomposites, Compos. Struct. 118 (2014) 241–249.

[43] N. Vu-Bac, M. Silani, T. Lahmer, X. Zhuang, T. Rabczuk, A unified framework for stochastic predictions of mechanical properties of polymeric nanocomposites, Comp. Mater. Sci. 96 (2015) 520–535.

[44] P.D. Spanos, A.A. Kontsos, A multiscale Monte-Carlo finite element method for determining mechanical properties of polymer nanocomposites, J. Probabil. Eng. Mech. 23 (2008) 456–470.

[45] N. Vu-Bac, R. Rafiee, X. Zhuang, T. Lahmer, T. Rabczuk, Uncertainty quantification for multiscale modeling of polymer nanocomposites with correlated parameters, Compos. B 68 (2015) 446–464.

[46] A. Almasi, M. Silani, H. Talebi, T. Rabczuk, Stochastic analysis of the interphase effects on the mechanical properties of clay/epoxy nano-composites, Compos. Struct. 133 (2015) 1302–1312.

[47] K.M. Hamdia, M.A. Msekh, M. Silani, N. Vu-Bac, X. Zhuang, T. Nguyen-Thoi, T. Rabczuk, Uncertainty quantification of the fracture properties of polymeric nanocomposites based on phase field modeling, Compos. Struct. 133 (2015) 1177–1190.

[48] K.M. Hamdia, T. Lahmer, T. Nguyen-Thoi, T. Rabczuk, Predicting the fracture toughness of PNCs: a stochastic approach based on ANN and ANFIS, Comp. Mater. Sci. 102 (2015) 304–313.

[49] H. Ghasemi, R. Brighenti, X. Zhuang, J. Muthu, T. Rabczuk, Optimal fiber content and distribution in fiber-reinforced solids using a reliability and NURBS based sequential optimization approach, Struct. Multidiscip. Optim. 51 (2015) 99–112.

[50] H. Ghasemi, P. Kerfriden, S.P. Bordas, J. Muthu, G. Zi, T. Rabczuk, Interfacial shear stress optimization in sandwich beams with polymeric core using non-uniform distribution of reinforcing ingredients, Compos. Struct. 120 (2015) 221–230.

[51] H. Ghasemi, R. Brighenti, X. Zhuang, J. Muthu, T. Rabczuk, Optimization of fiber distribution in fiber-reinforced composite by using NURBS functions, Comp. Mater. Sci. 83 (2014) 463–473.

[52] K.M. Hamdia, X. Zhuang, P. He, T. Rabczuk, Fracture toughness of polymeric particle nanocomposites: evaluation of Models performance using Bayesian method, Compos. Sci. Technol. 126 (2016) 122–129.

[53] B. He, B. Mortazavi, X. Zhuang, T. Rabczuk, Modeling Kapitza resistance of two-phase composite material, Compos. Struct. 152 (2016) 939–946.

[54] B. Mortazavi, J. Bardon, J.A.S. Bomfim, S. Ahzi, A statistical approach for the evaluation of mechanical properties of silica/epoxy nanocomposite: verification by experiments, Comp. Mater. Sci. 59 (2012) 108–113.

[55] B. Mortazavi, F. Hassouna, A. Laachachi, A. Rajabpour, S. Ahzi, D. Chapron, V. Toniazzo, D. Ruch, Experimental and multiscale modeling of thermal conductivity and elastic properties of PLA/expanded graphite polymer nanocomposites, Thermochim. Acta 552 (2013) 106–113.

[56] B. Mortazavi, O. Benzerara, H. Meyer, J. Bardon, S. Ahzi, Combined molecular dynamics-finite element multiscale modeling of thermal conduction in graphene epoxy nanocomposite, Carbon 60 (2013) 356–365.

[57] N. Burger, A. Laachachi, B. Mortazavi, M. Ferriol, M. Lutz, V. Toniazzo, D. Ruch, Alignments and network of graphite fillers to improve thermal conductivity of epoxy-based composites, Int. J. Heat Mass Trans. 89 (2015) 505–513.

[58] S.J.V. Frankland, V.M. Harik, G.M. Odegard, D.W. Brenner, T.S. Gates, The stress strain behavior of poly-mernanotube composites from molecular dynamics simulation, Compos. Sci. Technol. 63 (2003) 1655–1661.

[59] P. Papanikos, D.D. Nikolopoulos, K.I. Tserpes, Equivalent beams for carbon nanotubes, Comp. Mater. Sci. 43 (2008) 345–352.

[60] S.A. Meguid, J.M. Wernik, Z.Q. Cheng, Atomistic-based continuum representation of the effective properties of nano-reinforced epoxies, Int. J. Solid. Struct. 47 (2010) 1723–1736.

[61] J.M. Wernik, S.A. Meguid, Atomistic-based continuum modeling of the nonlinear behavior of carbon nanotubes, Acta Mech. 212 (2010) 167–179.

[62] A.K. Rappé, C.J. Casewit, Molecular Mechanics across Chemistry, University Science Books, Sausalito, CA, (1997).

[63] G.M. Odegard, T.S. Gates, L.M. Nicholson, K.E. Wise, Equivalent-continuum modeling of nano-structured materials, Compos. Sci. Technol. 62 (2002) 1869–1880.

[64] A.V. Desai, M.A. Haque, Mechanics of the interface for carbon nanotube polymer composites, Thin Wall. Struct. 43 (2011) 1854–1860.

[65] K. Liao, S. Li, Interfacial characteristics of a carbon nanotube-polystyrene composite system, Appl. Phys. Lett. 79 (2001) 4225–4227.

[66] S.J.V. Frankland, V.M. Harik, Analysis of carbon nanotube pull-out from a polymer matrix, Surf. Sci. 525 (2003) L103–L108.

[67] J. Gou, B. Minaie, B. Wang, Z. Liang, C. Zhang, Computational and experimental study of interfacial bonding of single-walled nanotube reinforced composites, Comp. Mater. Sci. 31 (2004) 225–236.

[68] S.C. Chowdhury, T. Okabe, Computer simulation of carbon nanotube pull-out from polymer by the molecular dynamics method, Compos. A 38 (2007) 747–754.

[69] Q. Zheng, D. Xia, Q. Xue, K. Yan, X. Gao, Q. Li, Computational analysis of effect of modification on the interfacial characteristics of a carbon nanotube-polyethylene composites system, Appl. Surf. Sci. 255 (2009) 3524–3543.

[70] D.W. Brenner, O.A. Shenderova, J.A. Harrison, S.J. Stuart, B. Ni, S.B. Sinnott, A second-generation reactive empirical bond order (REBO) potential energy expression for hydrocarbons, J. Phys. Condens. Mat. 14 (2002) 783–802.

[71] S.J. Stuart, A.B. Tutein, J.A. Harrison, A reactive potential for hydrocarbons with intermolecular interactions, J. Chem. Phys. 112 (2000) 6472–6486.

[72] N. Vu-Bac, P.M.A. Areias, T. Rabczuk, A multiscale multisurface constitutive model for the thermoplastic behavior of polyethylene, Polymer 105 (2016) 327–338.

[73] A.K. Rappé, C.J. Casewit, K.S. Colwell, A full periodic-table force-field for molecular mechanics and molecular dynamics simulations, J. Amer. Chem. Soc. 114 (1992) 10024–10035.

[74] C.Y. Li, T.-W. Chou, Structural mechanics approach for the analysis of carbon nanotubes, Int. J. Solids Struct. 40 (2003) 2487–2499.

[75] T. Belytschko, S.P. Xiao, G.C. Schatz, R.S. Ruoff, Atomistic simulations of nanotube fracture, Phys. Rev. B 65 (2002) 235430.

[76] R. Hill, The essential structure of constitutive laws for metal composites and polycrystals, J. Mech. Phys. Solids 15 (1967) 79–95.

[77] Z. Hashin, Analysis of composite materials, J. Appl. Mech. 50 (1983) 481–505.

[78] E.W.C. Coenen, V.G. Kouznetsova, M.G.D. Geers, Novel boundary conditions for strain localization analyses in microstructural volume elements, Int. J. Numer. Meth. Eng. 90 (2012) 1–21.

[79] X. Zhuang, Q. Wang, H. Zhu, A 3D computational homogenization model for porous material and parameters identification, Comp. Mater. Sci. 96 (2015) 536–548.

[80] X. Zhuang, Q. Wang, H. Zhu, Multiscale modelling of hydro-mechanical couplings in quasi-brittle materials, Int. J. Fract. (2016) 1–27.

[81] H. Zhu, Q. Wang, X. Zhuang, A nonlinear semi-concurrent multiscale method for fractures, Int. J. Impact Eng. 87 (2016) 65–82.

[82] X. Zhuang, Q. Wang, H. Zhu, Effective properties of composites with periodic random packing of ellipsoids, Materials 10 (2017) 112.

[83] Y. Benveniste, A new approach to the application of Mori-Tanaka's theory in composite materials, Mech. Mater. 6 (1987) 147–157.

[84] D.L. Shi, X.Q. Feng, Y.Y. Huang, K.C. Hwang, H. Gao, The Effect of nanotube waviness and agglomeration on the elastic property of carbon nanotube-reinforced composites, J. Eng. Mater. Technol. 126 (2004) 250–257.

[85] J.D. Eshelby, The determination of the elastic field of an ellipsoidal inclusion, and related problems, Proc. Royal Soc. Lond. A 241 (1957) 376–396.

Further Reading

T. Mori, K. Tanaka, Average stress in matrix and average elastic energy of materials with misfitting inclusions, Acta Metall. 21 (1973) 571–574.

L.S. Schadler, S.C. Giannaris, P.M. Ajayan, Load transfer in carbon nanotube epoxy composites, Appl. Phys. Lett. 73 (1998) 3842–3844.

M. Silani, H. Talebi, S. Ziaei-Rad, P. Kerfriden, S.P. Bordas, T. Rabczuk, Stochastic modelling of clay/epoxy nanocomposites, Compos. Struct. 118 (2014) 241–249.

V. Varshney, S.S. Patnaik, A.K. Roy, B.L. Farmer, A molecular dynamics study of epoxy-based networks: cross-linking procedure and prediction of molecular and material properties, Macromolecules 41 (2008) 6837–6842.

N. Vu-Bac, M.A. Bessa, T. Rabczuk, W.K. Liu, A multiscale model for the quasi-static thermo-plastic behavior of highly cross-linked glassy polymers, Macromolecules 48 (2015) 6713–6723.

16

Multiscale Continuum Modeling of Carbon Nanotube–Reinforced Polymers

Konstantinos I. Tserpes, Aggeliki Chanteli
UNIVERSITY OF PATRAS, PATRAS, GREECE

CHAPTER OUTLINE

Carbon Nanotube–Reinforced Polymers. http://dx.doi.org/10.1016/B978-0-323-48221-9.00016-9

16.1 Introduction

From the extensive study conducted in the past 15 years on the properties (mechanical, electrical, and thermal) of carbon nanotubes (CNTs), it became apparent that CNTs may act as reinforcing agents in polymers. Aircraft, automotive, and even sport manufacturing industries are incorporating CNT-reinforced polymers (nanocomposites) into their products. Due to the increasing demand for quantity and quality of nanocomposites, further development in the manufacturing process is expected. In this context, the manufacturing-related parameters affecting the reinforcement need to be recognized and their effects need to be fully characterized. Some of these parameters are the wavy shape and high aspect ratio of CNTs, the CNT quality (defects arising during production and abrasion due to functionalization), CNT dispersion (formation of agglomerates), and the adhesion of CNTs with the polymer (interphase strength). The detailed understanding of the effects of the aforementioned parameters on the macroscopic properties of nanocomposites requires a thorough experimental program combining mechanical, electrical, and thermal tests with microscopy investigations, which, however, could be disburdened by using reliable simulation models.

The reported simulation models, developed for evaluating the effective elastic properties of nanocomposites, may be categorized into molecular dynamics simulations [1–5], analytical approaches [6–9], and continuum mechanics approaches [10,11]. While molecular dynamics simulations offer accurate predictions, they are focused on small time calculations (such as picoseconds or nanoseconds) and are limited at the atomistic length analysis scale. Analytical modeling approaches are used for the micro/mesoscale estimation of the elastic properties of nanocomposites having limited versatility to incorporate multiple important factors and complex geometries. On the other hand, continuum mechanics methods, mainly in the form of finite element analysis (FEA), offer the potential to be used for the parametric evaluation of the effective elastic properties of nanocomposites by efficiently incorporating multiple geometrical and material parameters and facilitating the transition of analysis from the nanoscale to the macroscale, based on the presumption that they incorporate the necessary information from the atomic scale regarding the properties of chemical bonds and the properties of the interphase. The majority of the reported FEA models are based on the development of a representative volume element (RVE) or a representative (or repeating) unit cell (RUC). The differences among the proposed FEA models mainly refer to the RVE and RUC geometries. It is important to note that several of the developed FE models have not been validated either against experiments or against other validated models.

In the present chapter, a two-scale FE-based model is developed for predicting the effective Young's moduli and yield strengths of nanocomposites by accounting the characteristics of CNT waviness, CNT agglomeration, and CNT/polymer interphase. The model is initially used to perform a parametric study and then, to predict the Young's modulus and yield strength of multiwalled CNT–reinforced polypropylene (MWCNT/PP). The characteristics of the agglomerates have been drawn from scanning electron microscopy

(SEM) images and those of the interphase from atomic force microscopy (AFM) images and nanoindentation tests. The final predictions of the model have been validated against tension tests conducted for this purpose.

16.2 The Method of Continuum Multiscale Modeling

Multiscale modeling approaches are common for materials whose constituents lie at different length scales, such as nanocomposites. Herein, the basic requirement is to transfer the individual and combined effects of geometrical and material parameters, defined at the nanoscale, to the effective material behavior measured in the macroscale.

Numerical analyses at lower scales, for example, nano- to microscale, are usually accompanied by a homogenization technique based on prescribed displacement boundary conditions (periodic boundary conditions). This homogenization technique provides predictions for the effective material properties of the nanocomposite, which are applied to the next scale analysis level (specimen level). Both the multiscale methodology and the homogenization technique are strongly attached to the concepts of RVE and RUC. These two concepts, although being distinguished, in the literature often coincide. The concept of RVE characterizes heterogeneous materials with macroscopically or statistically homogeneous microstructures at an appropriate scale, and an RUC characterizes periodic heterogeneous materials [12]. While the determination of RVE's geometry and size is a specific independent field [13], the majority of the reported models employ approximate RVE models to investigate specific effects [11,14–16]. In the present work, an approximate RVE of a single wavy CNT and an approximate RUC of a CNT agglomerate have been used.

16.3 Models of the RVE and RUC

16.3.1 RVE of a Single Wavy CNT

16.3.1.1 Geometry

The RVE represents a single wavy MWCNT embedded into the polymer and the MWCNT/polymer interphase. A typical geometry of the RVE is shown in Fig. 16.1. The RVE is prismatic with a square cross-section. The MWCNT-to-RVE length ratio L_{CNT}/L_{RVE} is 0.6. The MWCNT and the MWCNT/polymer interphase have been modeled as hollow tubes. The interphase shares the same wavy shape with the MWCNT; it has a constant thickness of 0.68 nm and closed ends. The MWCNT's length is determined by the value of the MWCNT's aspect ratio, that is, $L_{CNT}/D_{CNT}=20$. The MWCNT's waviness is described by Fisher et al. [17].

$$y = a\cos\left(\frac{2\pi z}{\lambda}\right) \tag{16.1}$$

where $\lambda/D_{CNT} = L_{CNT}/D_{CNT} = 20$ and a = 12. Four types of waviness have been considered (w = 0.02, 0.04, 0.06, and 0.08).

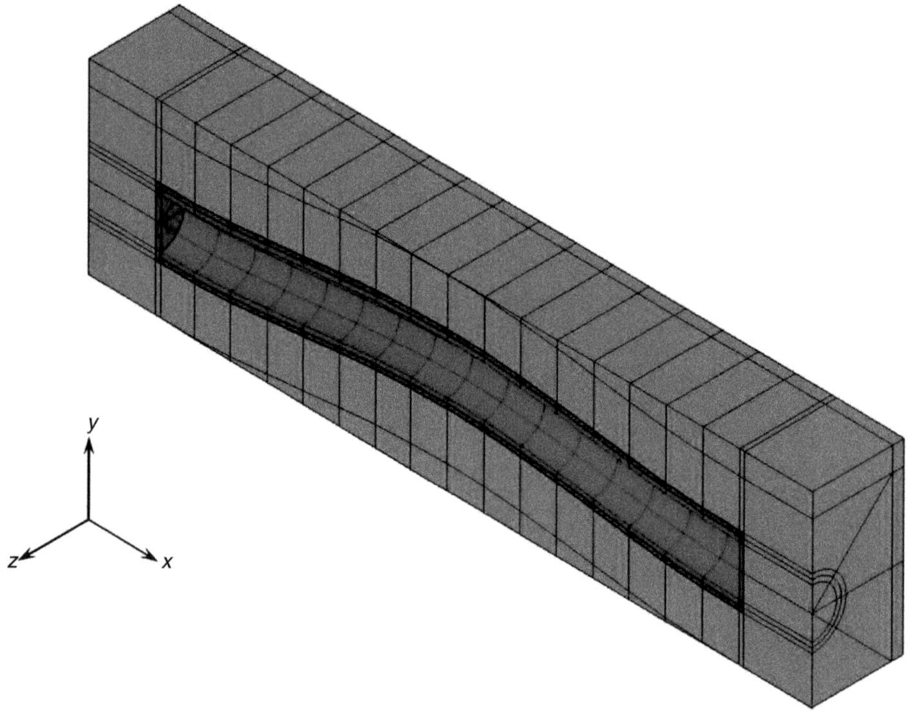

FIGURE 16.1 Typical view of the representative volume element (RVE) model comprising a wavy carbon nanotube (CNT) and the polymer/CNT interphase [area around tube (*red* in the web version)].

16.3.1.2 Materials

The polymer matrix used in the RVE is the M21 epoxy for which an isotropic linear elastic behavior has been assumed, as the load applied in the RVE is small. The Young's modulus and Poisson's ratio of the M21 epoxy are 3.5 GPa and 0.3, respectively. The Young's modulus of the interphase is 1.75 GPa.

16.3.1.3 FE Model of the RVE

Analyses have been conducted by means of a 3D FE model of the developed RVE using the ANSYS commercial FE code [18]. All model constituents have been represented by the SOLID185 element [18]. Views of a typical FE mesh of the RVE are shown in Fig. 16.2. The adopted FE mesh has been selected from a convergence study. To derive the effective elastic moduli properties of the nanocomposite, the RVE has been loaded in axial tension in the three directions and in shear in the three planes by applying small displacements. The elastic and shear moduli have been simply derived using the Hooke's law and the Poisson's ratio has been taken as the negative ratio between longitudinal and transverse strains. In addition, periodic boundary conditions (constraint equations in opposite faces) have been applied in the model for the homogenized representation of the macroscopic

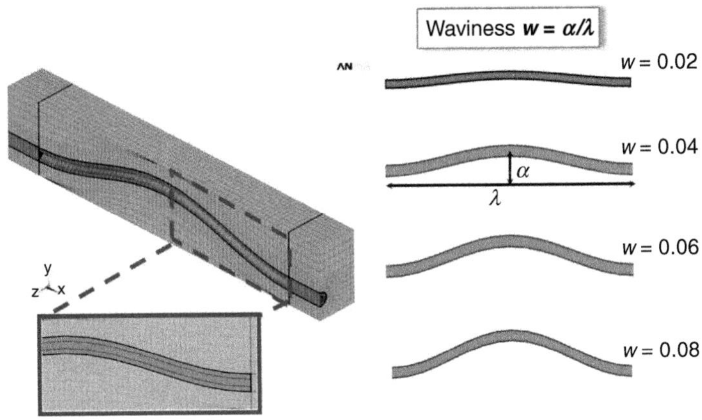

FIGURE 16.2 Views of a typical finite element (FE) mesh and the different CNT waviness cases.

behavior of the material. These boundary conditions should be applied to the RVE in such a way that, except the strains in the direction, in which the effective coefficients have to be calculated, all other mechanical strains are zero and, as a load, uniform unidirectional displacement (prescribed displacement condition) is applied [19].

16.3.2 RUC of a CNT Agglomerate

Poor dispersion of CNTs leads to the formation of agglomerates, which have been well denoted from experiments and microscope observations and which can act as defects causing, in the worst scenario, a decrease in the mechanical properties of the polymer nanocomposite [20–22] and in the best scenario, a smaller than anticipated increase in the properties [23–30]. Although the negative effect of the CNT agglomerates presence is recognized, the investigation about the role of agglomerates' characteristics is still evolving.

Several numerical and analytical models of CNT agglomerates in nanocomposites have been reported in the literature [15,16,31–35]. The analytical models [31–33,36] treat CNT agglomerates as inclusions and employ the Mori–Tanaka micromechanics theory for the predictions of the effective material properties connected to the morphology of CNT-reinforced polymers. The parameters considered in these models are the CNT volume fraction and the degree of agglomeration. On the other hand, most reported numerical models use RUCs at which the CNTs are stacked into simple geometrical configurations, which deviate from the actual configuration of CNT agglomerates [32,33]. The works that have attempted to model the exact structure of CNT agglomerates are very few [33,34]. Therefore, there is still a need for developing models that will represent the exact structure of CNT agglomerates and will be used for conducting studies on the effects of agglomerate's geometrical characteristics on the effective mechanical properties of nanocomposites.

16.3.2.1 Geometry

The RUC of the MWCNT agglomerate is cubic; the width of its side is determined according to the MWCNTs volume fraction from

$$W_{RUC} = \sqrt[3]{\frac{k\dfrac{\pi d_{CNT}^2 L_{CNT}}{4}}{V_{CNT}}} \tag{16.2}$$

where k is the number of MWCNTs in the agglomerate, d_{CNT} and L_{CNT} are the MWCNT's diameter and length, respectively, and V_{CNT} is the MWCNT's volume fraction. The main challenge during the development of the RUC model is the realistic representation of the agglomerate's geometry. To surpass this challenge, the open source TexGen software [37] has been used for designing the CNTs' wavy form. TexGen software [37] has advanced features in the geometrical representation of textile materials, offering also a direct link to FE codes. The CNTs have a solid circular cross-sectional area due to TexGen design limitations. The CNT's diameter is 29 nm and the length is equal to 210 nm. Parameters as the CNT's waviness, multitude, and topology have been encountered in the agglomerate based on the restrictions of the modeling capacity. Eventually, three different configurations of the RUC have been considered; namely 3, 5, and 10 CNTs, representing a minimum, a medium, and an extreme density case. In Fig. 16.3, typical geometries of the three

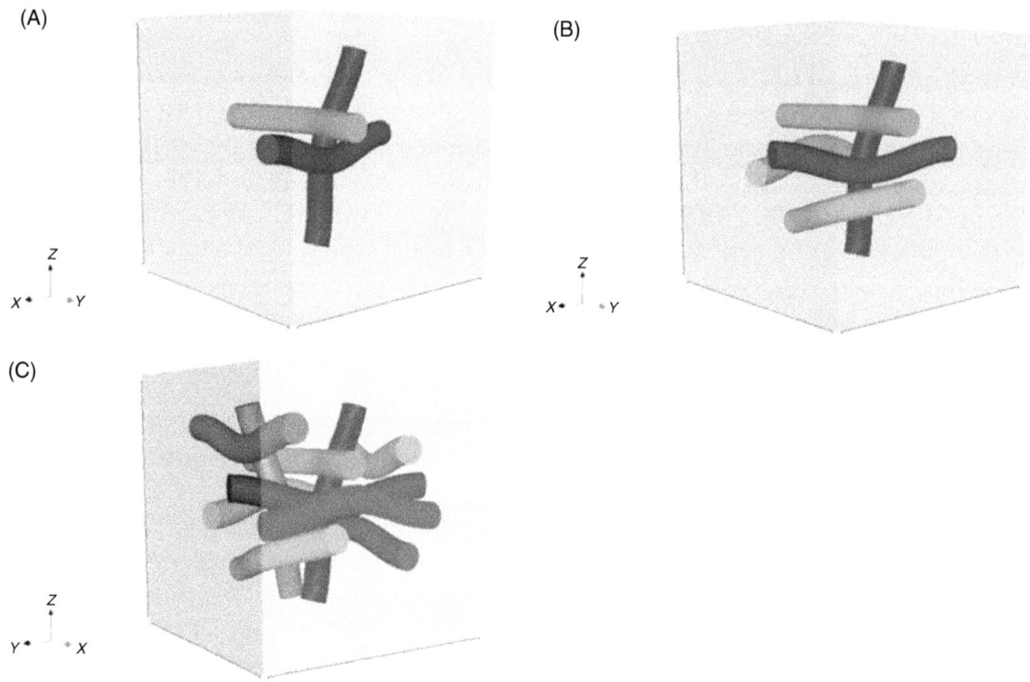

FIGURE 16.3 Representative (or repeating) unit cell (RUC) geometries. (A) 3-CNT agglomerates, (B) 5-CNT agglomerates, and (C) 10-CNT agglomerates.

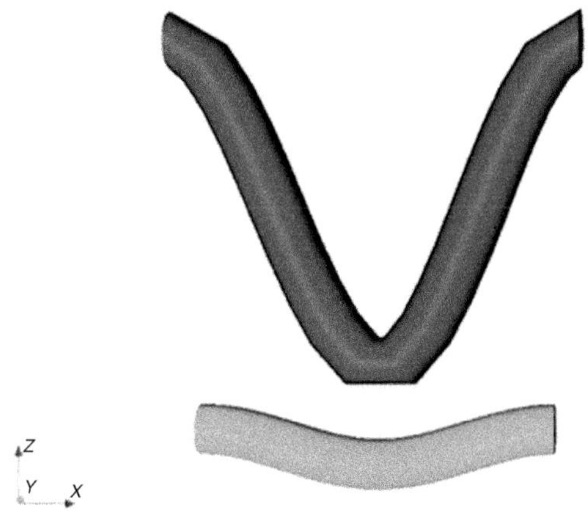

FIGURE 16.4 Schematic representation of the two CNT waviness cases: extreme waviness (upper image) and moderate waviness (lower image).

RUCs are shown. The CNT wavy shape is also described by the Eq. (16.1). To obtain a more realistic shape of the agglomerates in all the previously mentioned cases, an extra tilt of 15 degrees has been assigned to each CNT. Two waviness cases have been considered in the RUC model: a moderate waviness (a = 10.5) and an extreme waviness (a = 82.26). In Fig. 16.4, schematic representations of the two waviness cases are depicted. The effect of CNT topology has been studied by means of the RUC with 5 CNTs by considering the cases of 1 CNT, 2 CNTs, and 3 CNTs aligned to the loading axis. In the RUCs with 3 CNTs, a uniform arrangement has been applied (1 CNT per axis) in all cases and in the RUC with 10 CNTs, a close to uniform arrangement.

16.3.2.2 Materials

For the matrix of the RUC model, the polypropylene (PP) thermoplastic material with a Young's modulus of 1.64 GPa and a Poisson's ratio of 0.3 has been used. The CNTs have been also modeled as isotropic materials with a Young's modulus of 1 TPa and a Poisson's ratio of 0.3.

16.3.2.3 FE Models of the RUC

Two different FE meshes of the RUCs have been generated using the ANSYS Workbench [18]: a voxel-based mesh used for performing the linear analyses and a mesh with smooth surfaces used for performing the nonlinear analyses.

The voxel-based mesh has been solved using the ANSYS FE code [18]. In this case, no interphase has been considered. The matrix and the CNTs have been represented using the 3D ANSYS SOLID185 element [18]. Typical voxel-based FE meshes of the RUC and the CNT agglomerates are shown in Fig. 16.5. The local coordinate system of the cubic elements has been rotated properly according to the orientation of the elements with regard to the global coordinate system of the RUC. The effective elastic properties of the CNT

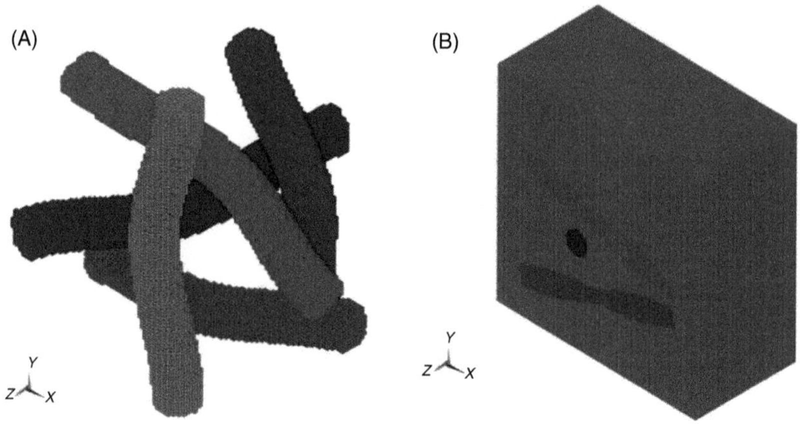

FIGURE 16.5 Views of the voxel FE mesh. (A) Details of the 5-CNT agglomerate and (B) view of the half RUC model.

agglomerate have been derived using the same procedure that has been applied for the RVE (see Section 16.3.1.3).

The mesh with the smooth surfaces has been solved using the LS-DYNA [38]. All parts of the RUC have been modeled using ELFORM 1 solid elements. In this case, interphase has been also considered. The FE mesh of the MWCNTs surrounded by the interphase is shown in Fig. 16.6A while the half FE mesh of an MWCNT surrounded by the interphase is shown in Fig. 16.6B. For the elements of the MWCNTs and the interphase, an elastic

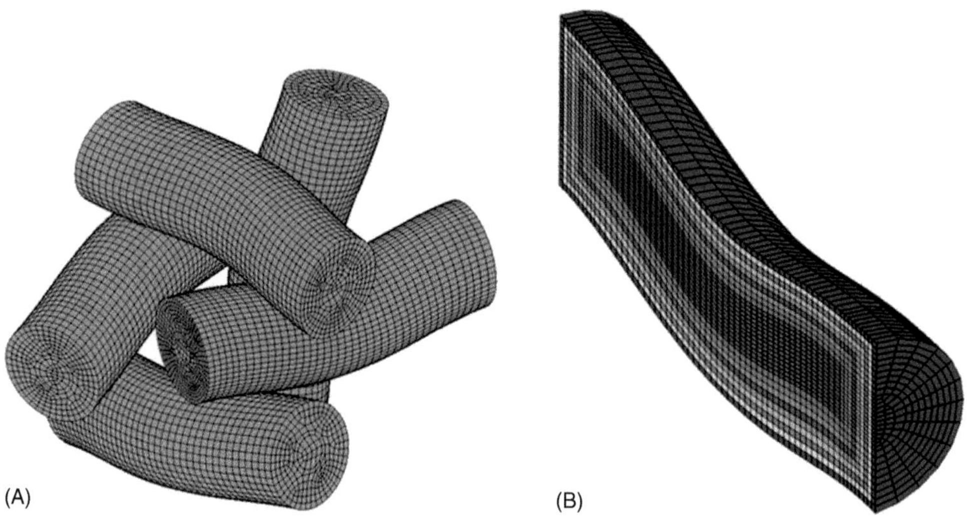

FIGURE 16.6 Views of the FE mesh. (A) Details of the multiwalled carbon nanotubes (MWCNTs) FE mesh surrounded by the interphase [grid-like elements (*blue* in the web version)] and (B) details of the half FE mesh of an MWCNT surrounded by nine layers of interphase elements.

behavior has been simulated using the elastic material model MAT_001. For the polymer matrix, an elastoplastic behavior has been simulated using the material model MAT_089. This is an elastoplastic material model with an arbitrary stress versus strain curve and arbitrary strain rate dependency. It is intended for applications where the elastic and plastic sections of the response are not clearly distinguishable. The input to the material model is the true stress–true strain curve and the true strain to failure (maximum strain) as defined from the uniaxial tension test. MAT_089 includes failure and damage. The strain used for failure and damage calculation, ε_{pm}, is based on an approximation of the greatest value of maximum principal strain encountered during the analysis:

$$\varepsilon_{\text{pm}} = \max_{i \le n}\left(\varepsilon_{\text{H}}^{i} + \varepsilon_{\text{vm}}^{i}\right) \tag{16.3}$$

where n is the current time step; $\max_{i \le n}(...)$ is the maximum strain value attained by the argument during the calculation; ε_x, ε_y, and ε_z are the cumulative strain in the local x-, y-, or z-direction $\left(\varepsilon_x + \varepsilon_y + \varepsilon_z\right)/3$; ε_{vm} is $\sqrt{(2/3) \times tr\left(\varepsilon'^{T}, \varepsilon'\right)}$, the equivalent uniaxial strain; and ε' is deviatoric strain tensor, where each ε_x, ε_y, and ε_z is cumulative.

By considering failure and damage, MAT_089 is capable of simulating the softening part of the stress–strain curve and predicting the maximum stress, which is regarded as the yield strength.

16.4 Parametric Studies

Using the RVE and RUC, a parametric study on the effect of various parameters on the effective elastic properties of the nanocomposite material has been conducted. The ranges of the parameters considered in the study are listed in Table 16.1.

16.4.1 RVE: Effect of CNT Waviness

In the diagrams of Fig. 16.7 the variations of the effective elastic and shear moduli, as well as of the Poisson's ratio with respect to the ratio $L_{\text{CNT}}/L_{\text{RVE}}$ and for different cases of CNT waviness w, are presented. The CNT waviness, $w = 0$ corresponds to the straight CNT and varies from 0.02 to 0.08. As can be seen, by increasing the w, the effective axial modulus E_X decreases (10% reduction occurs for $w = 0.08$). On the contrary, the shear modulus G_{XY} and the Poisson's ratio ν_{XY} increases. The rest of the effective elastic properties are unaffected by the CNT waviness.

Table 16.1 List of Parameters and Their Considerations

Parameter	Range
CNT volume fraction	1%–5%
CNT waviness	Extreme, moderate
CNT multitude	3, 5, and 10
CNT topology	1–3 CNTs in the *X*-direction

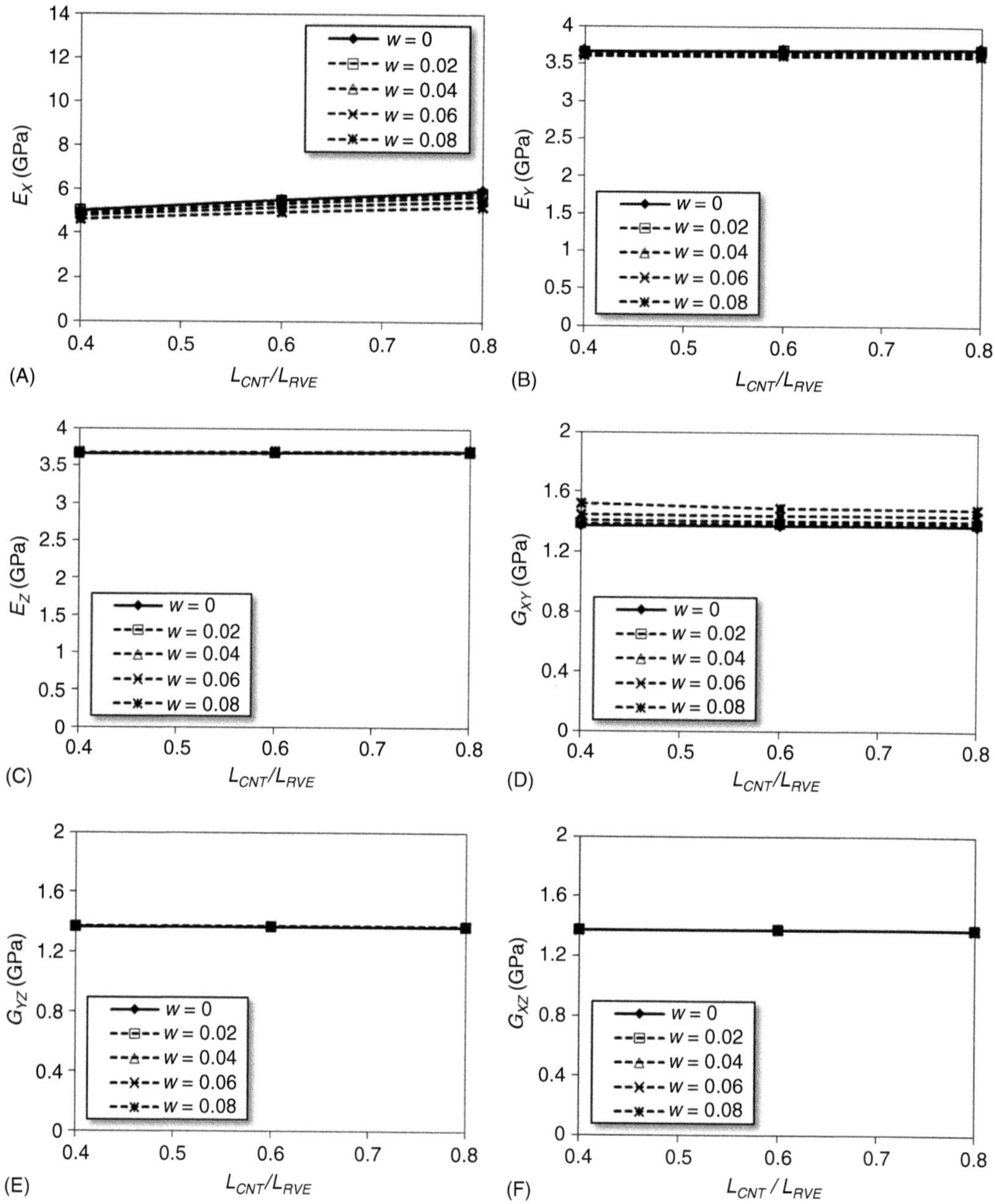

FIGURE 16.7 Predicted variations of (A) E_X, (B) E_Y, (C) E_Z, (D) G_{XY}, (E) G_{YZ}, (F) G_{XZ}, and (G) ν_{XY} with respect to the L_{CNT}/L_{RVE} ratio and the waviness *w*.

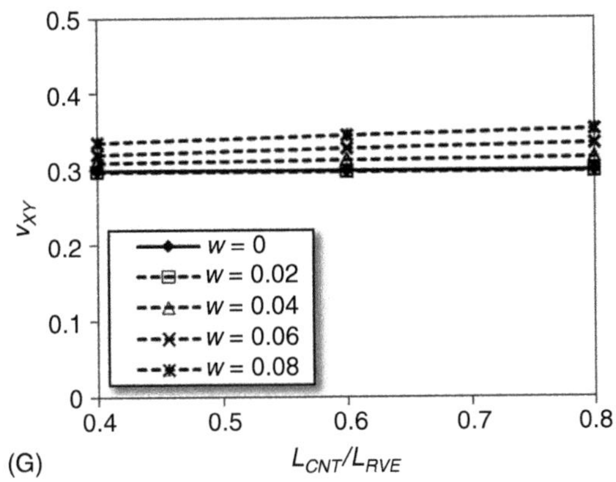

FIGURE 16.7 *(Cont.)*

16.4.2 RUC: Effect of CNT Volume Fraction and Multitude

The effect of CNT volume fraction is not trivial for CNT agglomerates as it is for single straight CNTs due to the complex architecture of the agglomerates. On the other hand, it is important to ascertain how the multitude of CNTs in the agglomerate may affect the degree of reinforcement through the formation of different load paths between the CNTs. The diagrams in Fig. 16.8 plot the variations of E_X, E_Y, E_Z, G_{XY}, G_{YZ}, G_{XZ}, and v_{XY} with regard to CNT volume fraction V_{CNT} for agglomerates with different multitude. For $V_{CNT} = 1\%$, a small increase in all elastic properties has been predicted. With an increase in the V_{CNT}, E_X and G_{XY} increase while the variation of v_{XY} depends on the multitude of CNTs; it increases for 3 CNTs and decreases for 5 and 10 CNTs. The predicted variations of E_Y and E_Z are similar with E_X, the variations of G_{XZ} and G_{YZ} are similar with G_{XY}, and the variations of v_{XZ} and v_{YZ} are similar to v_{XY}. The increase in E_X is inversely proportional to the multitude of CNTs. This finding is explained by the fact that the largest load is solely transferred by the aligned CNTs toward the loading direction.

16.4.3 RUC: Effect of CNT Waviness

In Fig. 16.9, the predicted values of E_X for the single straight CNT, the single waved CNT of moderate waviness, and the agglomerates of moderate and extreme waviness are compared. The comparison between the single CNTs (with straight and wavy shapes) shows clearly that the waviness reduces the degree of reinforcement offered by the CNT in the polymer. The reduction increases with an increase in the wave amplitude (from moderate to extreme waviness). The opposite trend is observed for the agglomerates: the agglomerate of extreme waviness has a larger Young's modulus than the agglomerate of moderate waviness. The reason for this unexpected finding is purely geometrical: while in the case of moderate waviness 1 out of 3 CNTs is aligned to the loading direction, in the case of extreme waviness the larger curling of CNTs results to the alignment in the loading direction of the other CNT parts, thus increasing the load-carrying capability of the agglomerate.

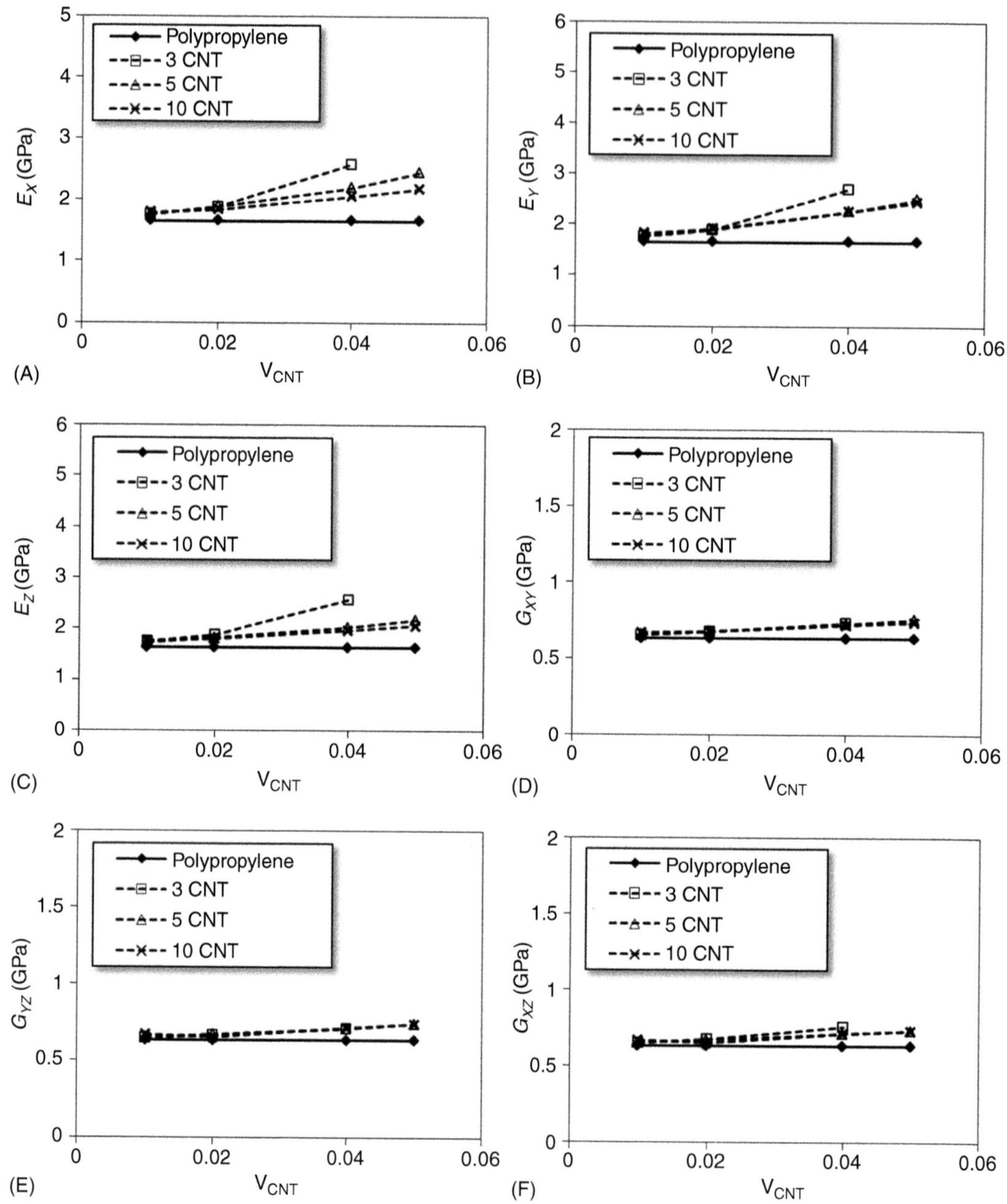

FIGURE 16.8 Predicted variations of (A) E_X, (B) E_Y, (C) E_Z, (D) G_{XY}, (E) G_{YZ}, (F) G_{XZ}, and (G) ν_{XY} with respect to the CNT volume fraction V_{CNT} and the CNT's multitude. *PP*, Polypropylene.

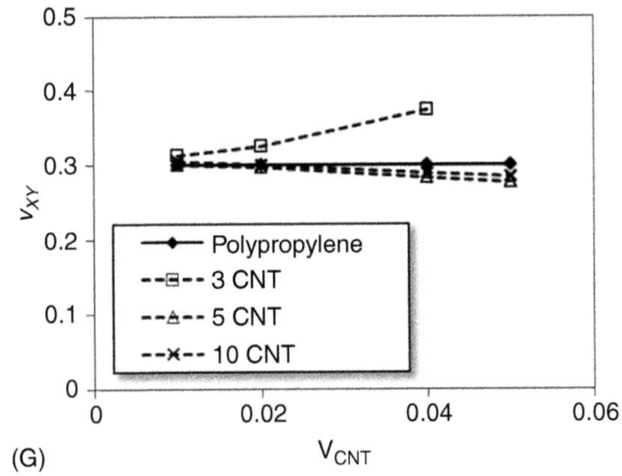

FIGURE 16.8 *(Cont.)*

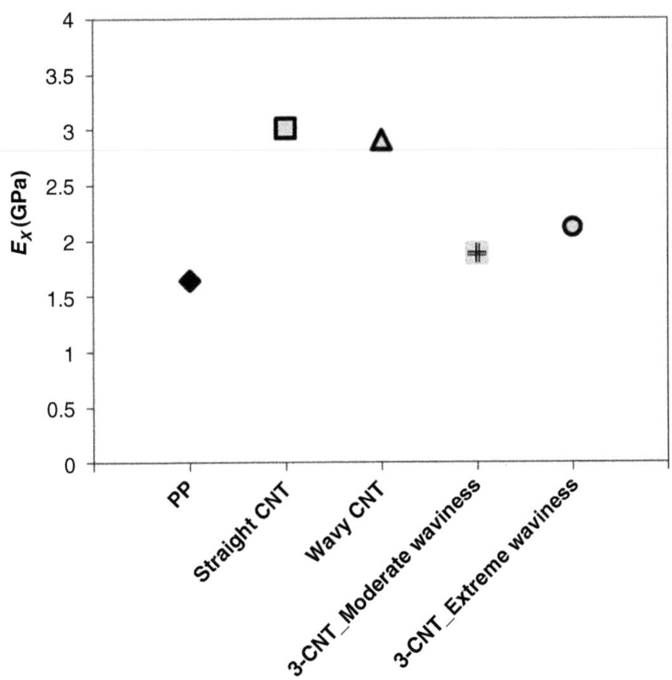

FIGURE 16.9 Effect of CNT waviness on longitudinal modulus, E_X. The agglomerate has three uniformly oriented CNTs and the CNT volume fraction is 2%.

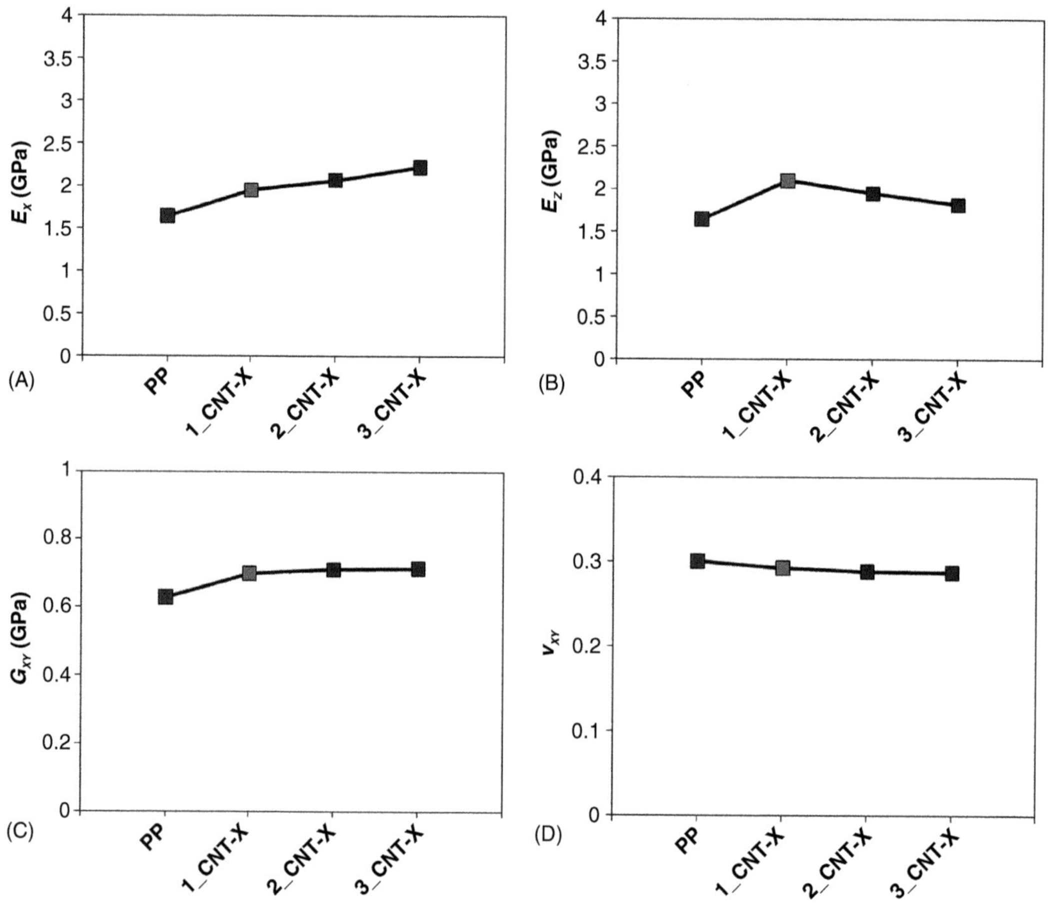

FIGURE 16.10 Predicted variation of (A) E_X, (B) E_Z, (C) G_{XY}, and (D) v_{XY}, with respect to the CNT's topology.

16.4.4 RUC: Effect of CNT Topology

The effect of CNT topology on the homogenized elastic properties of the RUC is illustrated in Fig. 16.10. From the figure it is concluded that the effect on the elastic moduli is balanced between the different directions. The alignment of more CNTs in the *X*-direction leads to the increase of E_X and to the decrease of E_Y and E_Z. On the other hand, the effect of CNTs topology on the shear moduli and Poisson's ratio is minor. It is mentioned that these findings are for CNTs of moderate waviness. In the case of extreme waviness the trend, especially for the elastic moduli, is expected to be different due to the explanation given in the previous section.

16.5 Experiments

Following the parametric study, the model has been applied to predict the Young's modulus and yield strength of the MWCNT/PP nanocomposite in the contents of 2 and 5 wt.%. In this section, the tensile tests, the SEM and AFM morphological characterizations, and the nanoindentation tests conducted on the MWCNT/PP nanocomposite material for obtaining input data for the model and for validating the numerical predictions are described.

16.5.1 Tension Tests

Tension tests on the PP and MWCNT/PP specimens of 2 and 5 wt.% have been conducted according to the ASTM D638 standard [39] at room temperature using an electromechanical testing machine (Tinius Olsen H5KT), which incorporated a tension/compression transducer (load cell of 5 kN) operated at a crosshead speed of 1 mm/min. Six specimens have been tested for each material. Strains in the specimens have been measured using strain gauges and the Digital Image Correlation system ARAMIS 5M [40].

The recorded tensile stress–strain curves of the PP and MWCNT/PP materials are shown in Fig. 16.11. The curves of the PP material show a small scatter. On the contrary, the curves of the MWCNT/PP material show a large scatter, which is indicative of the varying reinforcement quality. The comparison between the curves of the two materials reveals a

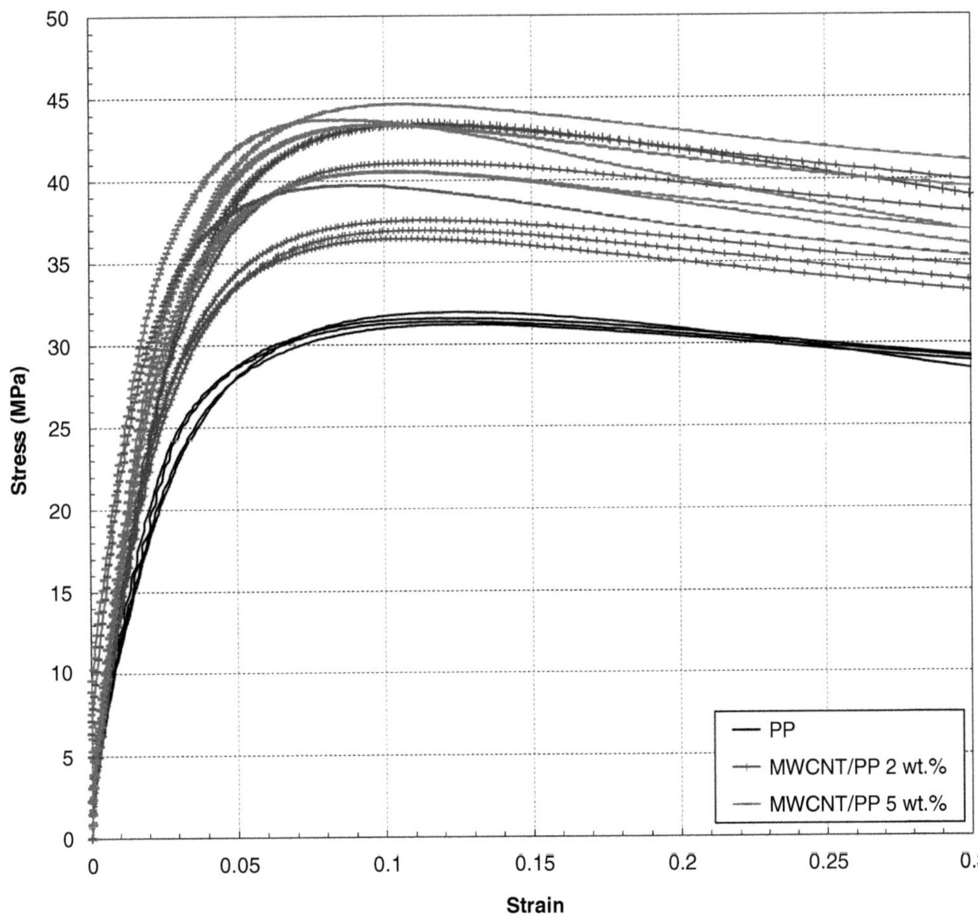

FIGURE 16.11 Stress–strain curves measured by ARAMIS system for all tests.

clear enhancement of Young's modulus and yield stress, taken as the maximum stress, of the PP material by the addition of MWCNTs.

16.5.2 Scanning Electron Microscopy Tests

SEM tests have been conducted to characterize the morphology of the MWCNT/PP and derive data for numerical modeling. The inspection has been performed using a LaB6 electron gun integrated with an e-beam lithography system. The technical data of the microscope system are: 30-kV accelerating voltage, 3-nm resolution, 300.000 magnification, 1 pA to 1 μA beam current, five-axis *X-Y-Z-R-T* stage, and low-vacuum mode for nonconductive sample observation. SEM images reveal that the MWCNT/PP specimens can be divided with regard to the reinforcement quality in two categories: one with good dispersion and wetting of MWCNTs, which led to the minimal formation of agglomerates, and one with poor dispersion and wetting of MWCNTs, which led to the formation of many large agglomerates. The categorization of specimens is verified by the results of tension tests through the division of tensile stress–strain curves into two groups, an upper and a lower, which correspond to the categories of good and poor dispersion, respectively.

16.5.3 Nanoindentation Tests

To estimate the Young's modulus of the interphase between PP and MWCNTs, nanoindentation tests have been conducted using a nanomechanical test instrument with a load range of 1–30,000 μN. Displacement has been recorded as a function of applied load in the rate of 1 nN/0.04 nm. The test instrument is equipped with a scanning probe microscope in which a sharp probe tip moves in a raster scan pattern across the sample's surface using a three-axis piezopositioner. Tests have been conducted in a clean area environment under conditions of 45% humidity and 23°C temperature. To enable operation under a closed loop displacement control, the feedback control option has been used. The tests have been conducted using a Berkovich diamond indenter (three-side pyramid) with an average radius of 100 nm. The loading and unloading segment times have been 40 s and the holding time has been 3 s (to avoid residual viscoelasticity) at various depths ranging from 20 to 2500 nm and at fixed depth of 300 nm. Prior to indentation, the area function of the indenter's tip has been calibrated in fused silica. The depth profile is described by trapezoidal loading–unloading curves.

Representative AFM images of the MWCNT/PP 2 wt.%_03 specimen are shown in Fig. 16.12. The images show the formation of agglomerates with a diameter of approximately 10 μm, which is in accordance with the measurements taken from SEM images. Overall, the good wetting of the MWCNTs can be acknowledged. The specific specimen belongs to the upper group of the stress–strain curves, and it has been expected to present many single dispersed MWCNTs, thus facilitating the observation of the interphase. Due to the high level of roughness of the tested area, there have been difficulties on the focus of isolated MWCNTs, and the observation of the interphase has been impossible.

The Young's modulus of the MWCNT/PP material, E, has been derived from the experimental nanoindentation load–displacement curves using the Oliver–Pharr method [41].

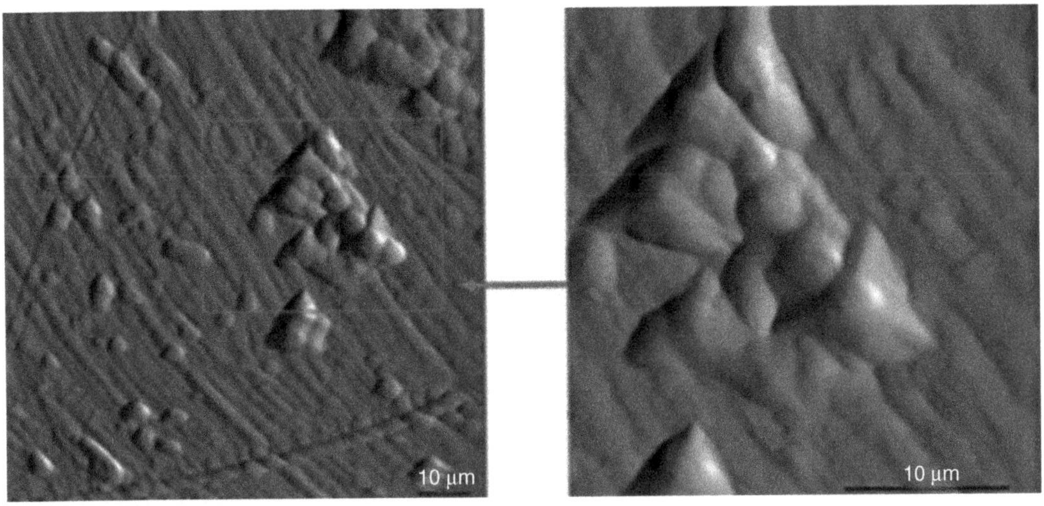

FIGURE 16.12 Atomic force microscopy (AFM) images of a selected area in MWCNT/PP 2 wt.%_03 specimen with CNT agglomerate of approximately 10 -μm diameter.

The variation of E with regard to displacement is plotted in Fig. 16.13. For displacements between 0 and 500 nm, the variation of E shows a critical decrease possible due to local stress concentrations appearing when the indenter moves from concentrated CNTs to areas full of CNTs agglomerates [42,43]; while for displacements above 500 nm, E varies within a smaller range, which is reported also in another study [42]. Mapping of the Young's modulus measured at the depth of 300 nm within a square grid of size of $70 \times 70\ \mu m^2$ is

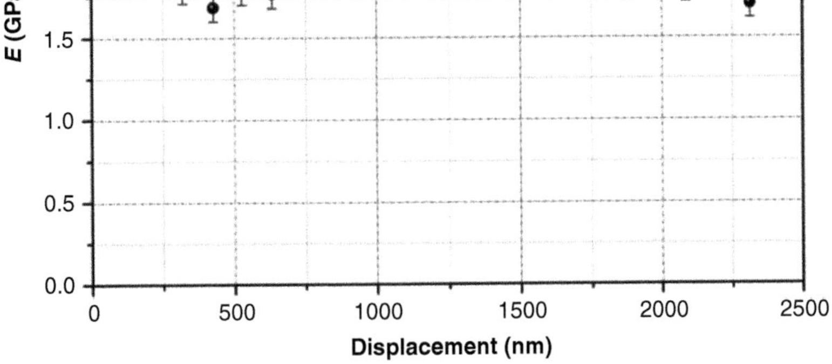

FIGURE 16.13 Variation of Young's modulus *E* with respect to the displacement.

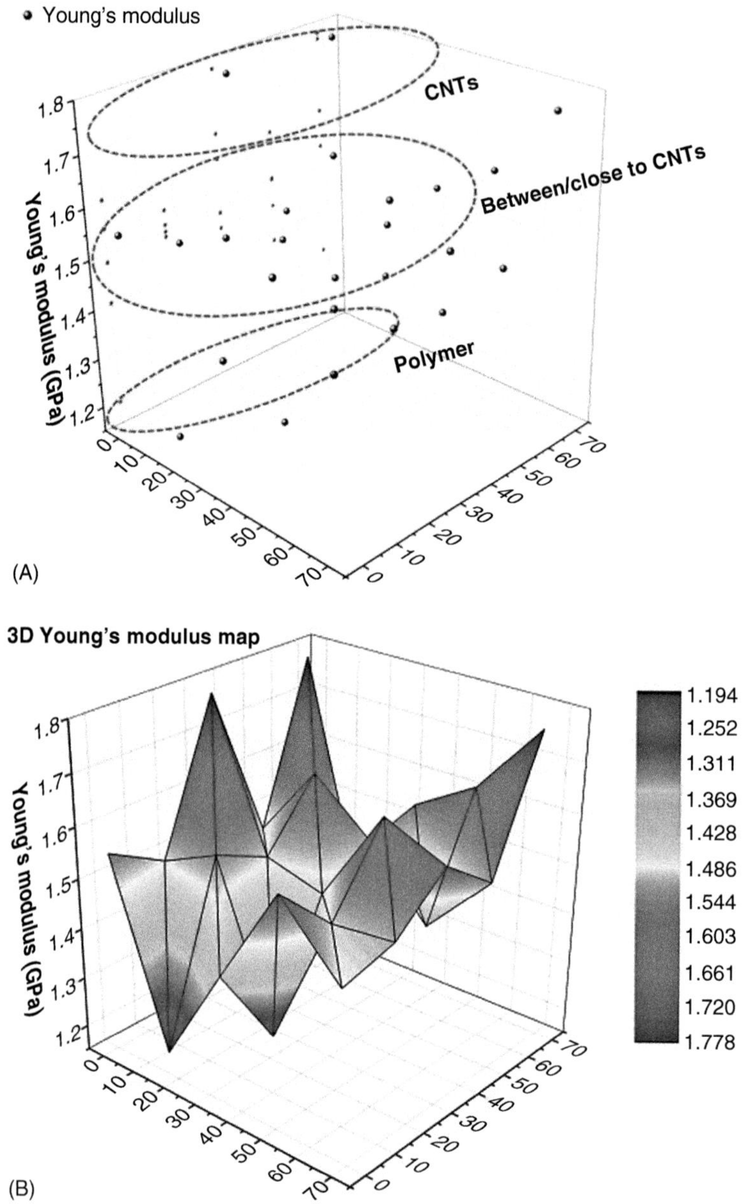

FIGURE 16.14 (A) Variation of Young's modulus *E* with respect to the displacement and (B) 3D distribution diagram of Young's modulus, *E*, values at the depth of 300 nm within a square grid of size of 70 × 70 μm².

shown in Fig. 16.14A, while the spatial distribution of *E* is shown in Fig. 16.14B. By projecting these values and taking into account the range of Young's moduli of the neat PP material as measured from the tension tests, the material subjected to nanoindentation may be divided into three regions with regard to Young's modulus: the region of polymer matrix with $E = 1.51–1.76$ GPa, the region of interphase with $E = 1.20–1.50$ GPa, and the region between/close to the MWCNTs (MWCNT agglomerate) with $E > 1.50$ GPa.

16.6 Modeling of the MWCNT/PP Tension Specimen

16.6.1 Image Processing

The characteristics of the RUCs, which have been used to model the MWCNT/PP specimens, have been determined from processing and analysis of SEM images performed using the ImageJ software [44]. Image analysis has been based on the fact that the areas filled with MWCNTs are in general brighter in SEM image than the unfilled areas (pure resin). The output from ImageJ [44] is a map of black-and-white areas in the form of a matrix. The original SEM images have been reconstructed using a MATLAB routine developed for this purpose. In Fig. 16.15, a reconstructed image is compared with the original SEM image. The comparison indicates that the reconstructed images represent sufficiently the agglomeration state in the specimens. In Fig. 16.15B, the uniformly dispersed small white spots represent the well-dispersed MWCNT agglomerates of small density while the few large white spots represent the few agglomerates of large density. The analysis indicates that the agglomerates of large density occupy the 1.72% of Fig. 16.15B image. Complementary to the information drawn from the reconstructed images, the diameters of the agglomerates have been measured from SEM images using the ImageJ software [44]. As it is not computationally feasible to model the actual CNT agglomerates, the analogy of diameters between small and large agglomerates has been used for selecting the appropriate RUCs More details about this process can be found in Ref. [45].

In addition to the information about the CNT agglomerates occupation percentages drawn from reconstructed images, the diameters of the agglomerates have been also measured from the original SEM images using the ImageJ software [44]. The direct representation of the actual size of the CNT agglomerates has been computationally limited and therefore an analogy has been introduce at this point between small and large agglomerates for the selection of the appropriate RUC models. The analogy has been based on the diameter ratios taken from the SEM measurements and the RUC width ratios. The selected RUCs for the 2-wt.% specimen are the RUCs of 3 and 5 CNTs and for the 5-wt.% specimen the RUCs of 5 and 10 CNTs.

16.6.2 FE Models

16.6.2.1 Linear Analysis

A 3D FE model of the tensile specimen has been developed using the ANSYS commercial FE code [18]. The CNT/PP material has been represented using the 3D ANSYS SOLID185 element [18]. To simulate the tensile loading, boundary conditions have been applied at the ends of the specimen captured by the tabs. One side has been fully restrained and at the other side an axial displacement has been applied. For the elements of the specimen's

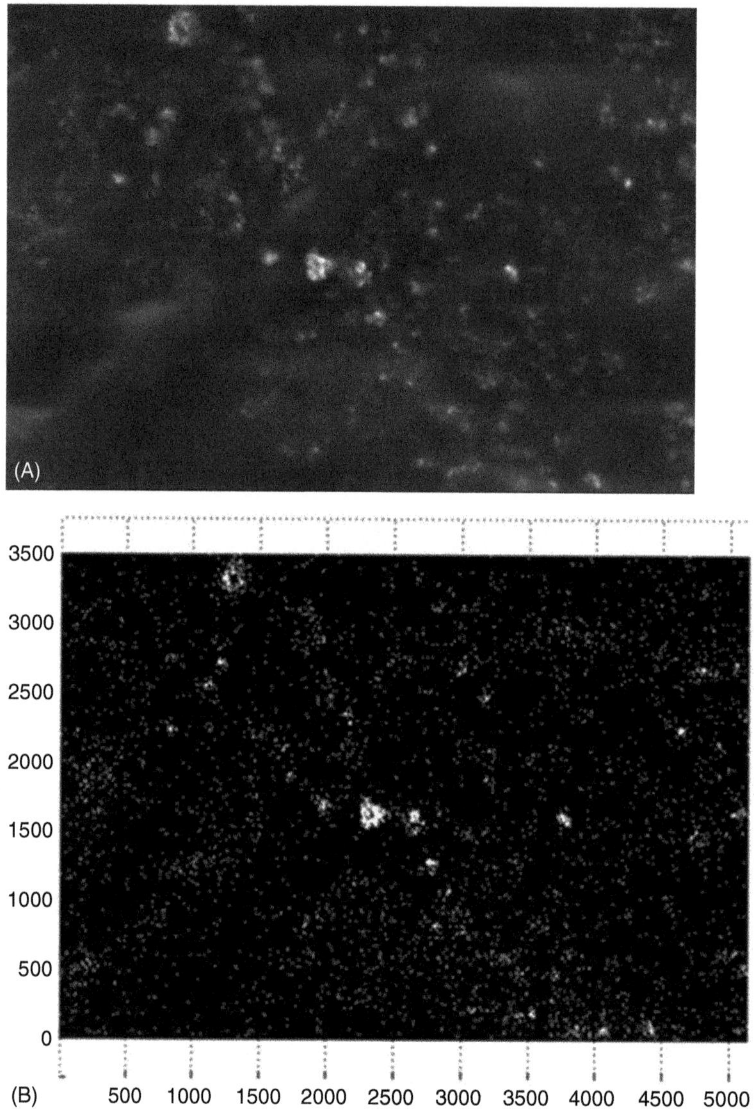

FIGURE 16.15 (A) Original scanning electron microscopy (SEM) image from an MWCNT/PP specimen of 2 wt.% and (B) reconstructed image from MATLAB.

FE model, an isotropic material behavior has been assumed and the Young's modulus of the elements has been computed from the RUC models of the CNT agglomerates. The 2-wt.% specimen has been modeled using the RUCs of 5 and 10 CNTs in the analogy of 96.44 and 3.56%, respectively. This choice has been based on the assumption that the reinforcement characteristics identified from the SEM images apply to the entire specimen. The elements that had properties of the denser CNT agglomerates RUCs (5 CNTs for the

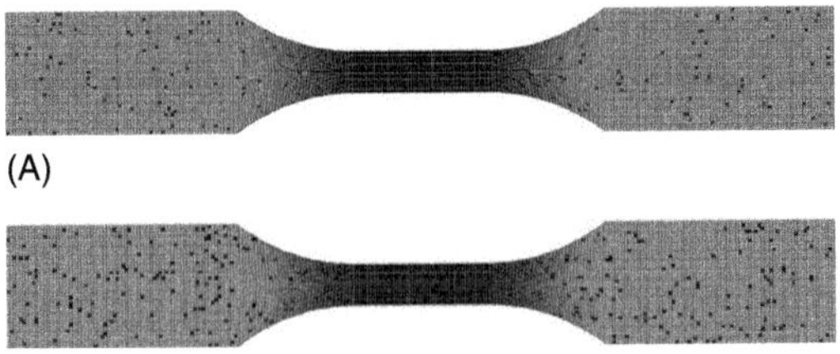

FIGURE 16.16 (A) Assignment of properties of the RUC with 5 CNTs at the elements of the tensile specimen (*blue* in the web version) and (B) assignment of properties of the RUC with 10 CNTs at the elements of the tensile specimen (*red* in the web version).

2-wt.% specimen and 10 CNTs for the 5-wt.% specimen) have been selected randomly (Fig. 16.16).

A very important issue arising in the modeling of the RUCs is the conversion of MWCNT weight ratio to volume fraction. During the conversion, the basic parameters are the number of CNT walls and the density of CNTs. The Plasticyl PP2001 nanocomposite material [46] contains the type NC7000 of MWCNTs for which the manufacturer reports seven walls, an outer diameter of 9.5 nm, and a density range from 1.3 to 2.0 g/cm^3. However, the measurements from the SEM images showed an average CNT diameter of 29 nm. This finding indicates that MWCNTs of more than the reported seven walls have been eventually produced. Given the nanotube density is a function of diameter [47–50], in the present work the density ρ has been derived from the following equation, taken from Ref. [50]

$$\rho_{\mathrm{CNT}} = \frac{4000}{1315}\left[\frac{n}{d_{\mathrm{out}}} - \left(2d_{\mathrm{s-s}}\sum_{i=0}^{n-1} i\right)\Big/ d_{\mathrm{out}}^2\right] \tag{16.4}$$

where d_{out} is the outer diameter of MWCNTs (=29 nm), $d_{\mathrm{s-s}}$ is the interlayer distance set to 0.34 nm, and n (=16) is the number of walls derived using the d_{out} and n. The calculated result from Eq. (16.4) gives a density equal to 1.38 g/cm^3. Eventually, the weight ratio of 2% converts to a volume fraction of 1.3%, and the weight ratio of 5% to a volume fraction of 3.3%. These MWCNT volume fractions have been applied to the respective RUCs. In the RUCs, the Young's modulus of the PP matrix is set equal to the average experimental value of 1.64 GPa.

16.6.2.2 Nonlinear Analysis

In the nonlinear analysis, performed to predict the yield strength of the MWCNT/PP material, the interphase between the two materials has been also considered, as it is expected to influence load transfer at high loads. The interphase has been modeled as a hollow cupped cylinder surrounding the MWCNT. Due to the poor conditions of material roughness in AFM images, it wasn't feasible to obtain information for the interphase thickness. Thus, the value of 30 nm, measured in Ref. [16] for the same nanocomposite material, has been used in the model. Based on the findings of nanoindentation tests, the Young's modulus

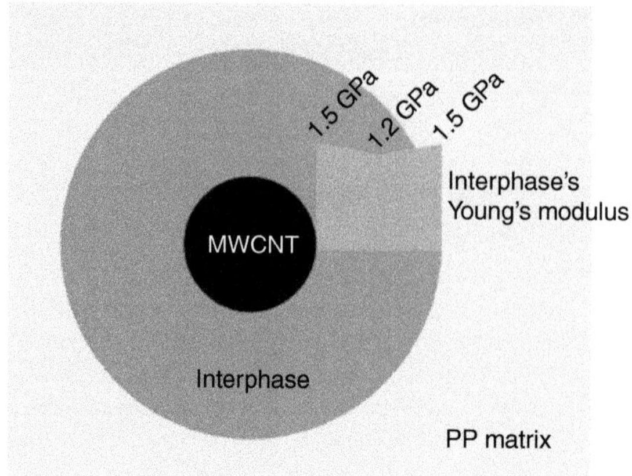

FIGURE 16.17 Schematic representation of the variation of interphase's Young's modulus.

of the interphase has been modeled as shown in Fig. 16.17. It has been assumed that at the boundaries with the matrix and the MWCNT, the interphase Young's modulus gets the maximum measured value (1.50 GPa) and decreases linearly toward half thickness of the interphase where it gets the minimum measured value (1.20 GPa)

FE models of the RUCs have been developed in ANSYS Workbench [18]. The geometries of the agglomerates have been determined after the morphological analysis of SEM images of the 2-wt.% MWCNT/PP material. The material properties of the interphase have been determined from AFM images and nanoindentation tests conducted on the MWCNT/PP specimens as described in Section 16.5. The FE models of the RUCs have been solved using the LS-DYNA explicit FE code [38]. The nonlinear behavior of the PP material has been modeled using an elastic–plastic material model for which input data have been obtained from the tension tests. The simulated behavior of the RUCs has been assigned to the elements of the 3D FE model of a tension specimen, which has been used to predict the yield strength of the MWCNT/PP material that is regarded as the yield strength according to the ASTM D638 specification [39].

16.7 Numerical Results

16.7.1 Young's Modulus of the MWCNT/PP Material

The predicted values of Young's modulus are compared with the average experimental values for the two nanocomposites and the neat PP in Fig. 16.18. In the diagrams, the standard deviations of the experimental values and the reference material values are also included. Although for the 5-wt.% material the model results agree very well with the test (Fig. 16.18A), for the case of 2-wt.% there is a significant deviation, which still lies within the standard deviation of the mean experimental value (Fig. 16.18B). This finding suggests that while the model is capable of accurately predicting the Young's moduli of materials at which the agglomeration effect is prominent, that is, the 5-wt.% material, it underestimates

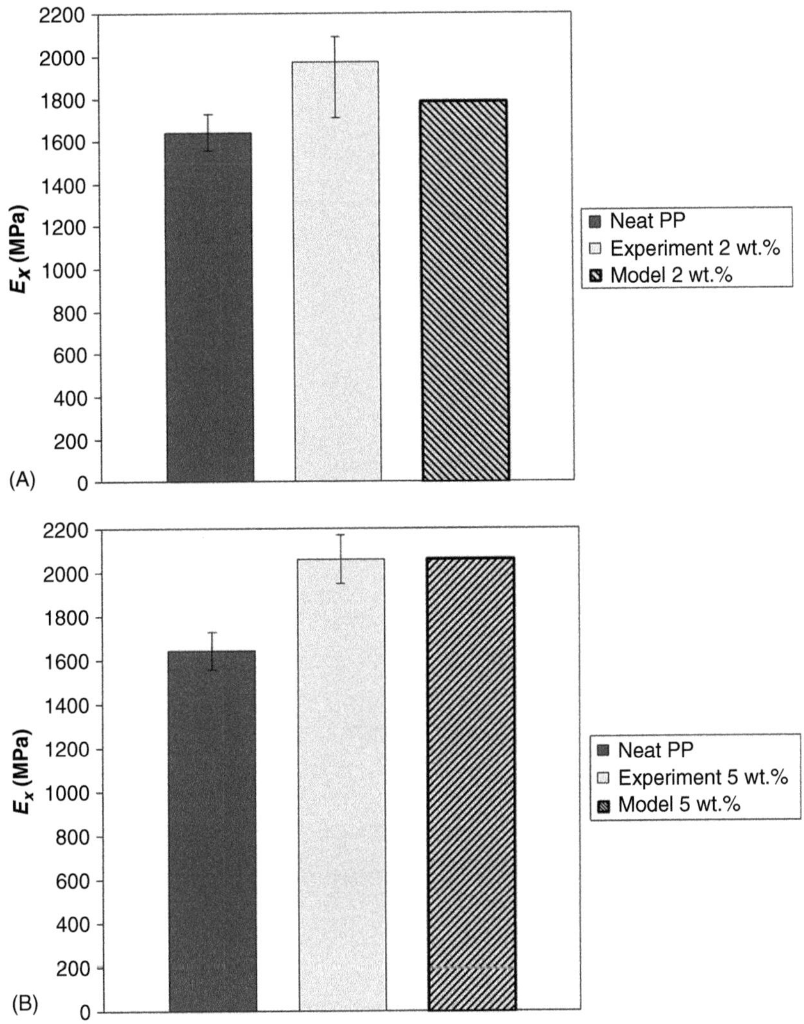

FIGURE 16.18 Comparison between predicted and experimental Young's moduli values for (A) the case of 2-wt.% MWCNT/PP specimens and (B) the case of 5-wt.% MWCNT/PP specimens. The reference values of the PP specimen and the standard deviations of the average experimental values are also included.

the prediction for the Young's moduli of materials at which the formation of agglomeration is minor, that is, the 2-wt.% material. Therefore, for modeling materials with characteristics of a good CNT dispersion, a more realistic approach could be succeeded by the use of RVEs with single wavy randomly oriented CNTs, such as the RVE model.

16.7.2 Yield Strength of the MWCNT/PP Material

Fig. 16.19 compares the experimental tensile stress–strain curves of the MWCNT/PP 2-wt.% specimens with the predicted curve (t_{int} = 30 nm). At small strains, the numerical

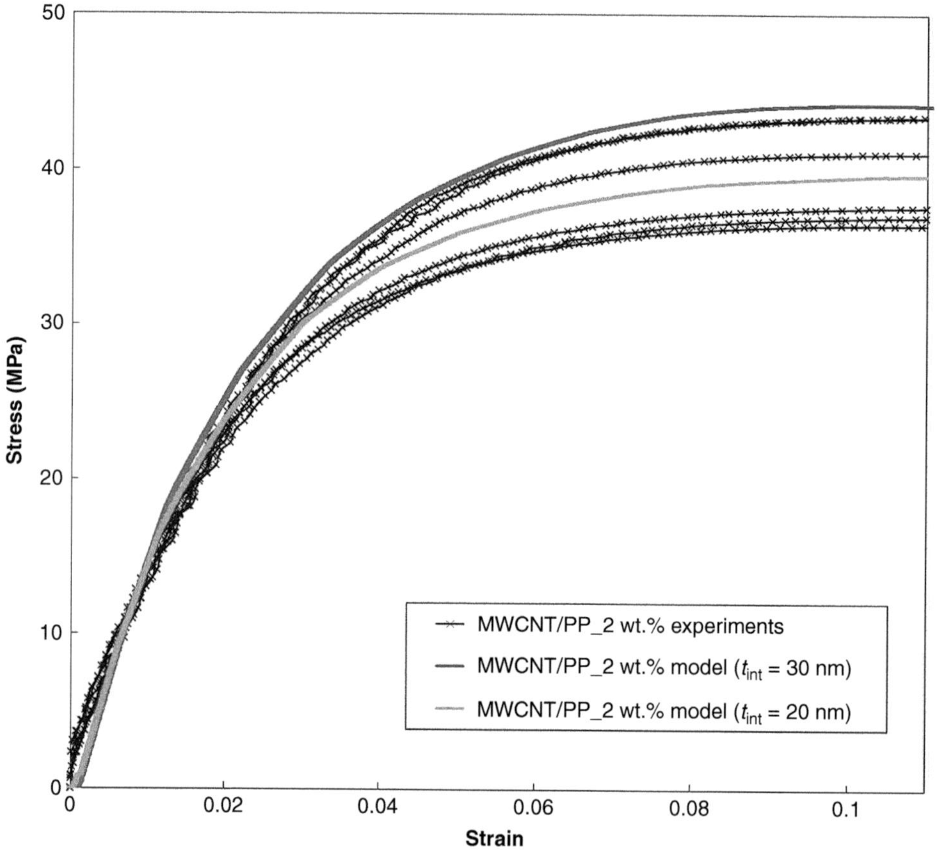

FIGURE 16.19 Stress–strain experimental curves for the 2-wt.% specimen compared with the numerical stress–strain curves [*solid curved lines* (*blue* and *green* in the web version)] for the cases of interphase thickness values, t_{int} = 30 and 20 nm, respectively.

curve coincides with the experimental curves. At strains larger than 0.02, the numerical curve starts to deviate and follow the upper experimental curve, thus overestimating the average experimental yield strength of the MWCNT/PP material as shown in the bar chart of Fig. 16.20. The difference in the correlation between numerical and experimental curves is attributed to the imperfect modeling of the interphase, which at small strains has no major effect on the tensile behavior of the specimens, but at large strains dominates the tensile behavior of the nanocomposite by dominating the load transfer between the MWCNTs and the polymer. Imperfect modeling of the interphase lies in the use of an approximate value for the thickness and/or the assumption of linear elastic behavior. As no data could be found for the variation of Young's modulus of the interphase with regard to the applied load, it has been decided to perform an additional analysis using a different interphase thickness (20 nm). The results from the additional analysis show a significant influence of interphase thickness on the predicted yield strength of the MWCNT/PP material: with

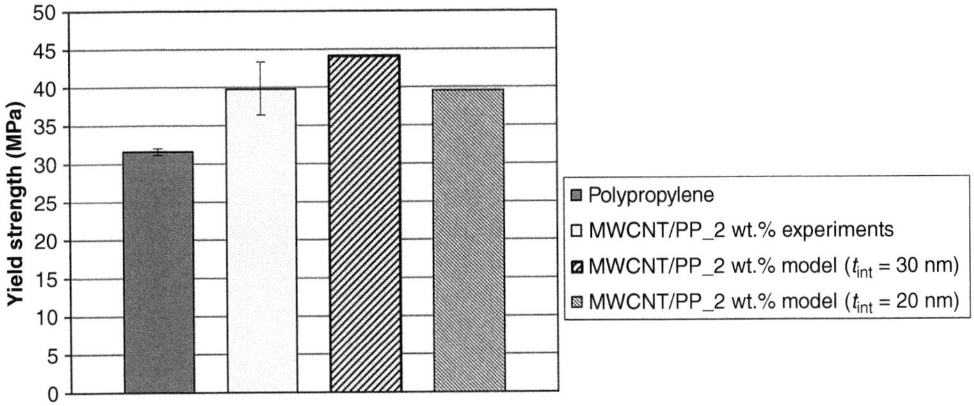

FIGURE 16.20 Comparison of the predicted yield strength values by the models with the respective experimental value.

decreasing the interphase thickness, the predicted yield strength decreases. This finding is closely related to the consideration of a linear elastic behavior for the interphase in contrast to the nonlinear behavior of the PP material. In Fig. 16.19, the curve predicted by the model using t_{int} = 20 nm is also included. As may be seen, this curve falls well within the band of experimental curves, and for this case the predicted yield strength, also included in the bar chart of Fig. 16.20, coincides with the average experimental yield strength.

16.8 Conclusions

In this chapter, a two-scale continuum model developed for predicting the effective mechanical properties of CNT/polymer nanocomposites by accounting for CNT waviness, agglomeration, and CNT/polymer interphase is described. The model is based on the FE method and has been applied to perform a parametric study and predict the Young's modulus and yield strength of an MWCNT/PP nanocomposite. In this context, tension tests, SEM, and AFM investigations, as well as nanoindentation tests have been conducted to obtain input data for the model and validate the final model predictions.

The parametric studies conducted using the RVE and RUC show the following:

- CNT's wavy shape mostly affects the effective axial elastic modulus.
- With increasing the multitude of CNTs in the agglomerate, the elastic properties decrease.
- With increasing the waviness of CNTs, the elastic properties increase, as curling of CNTs leads to the alignment of larger parts of the CNTs.
- The effect of CNT topology is balanced between the different directions; the largest elastic modulus appears in the direction with the more aligned CNTs, and the shear moduli and Poisson's ratios are rather independent of CNT topology.

The comparison between model predictions on the Young's modulus of the MWCNT/PP material and experimental values reveals that the model is capable of accurately predicting the Young's moduli of materials at which the agglomeration effect is prominent; however, it tends to underestimate the Young's moduli of materials at which agglomeration is minor.

The model predictions of the yield strength of the MWCNT/PP reveal the significant effect of interphase's characteristics (thickness and behavior) on the simulated behavior of the nanocomposite at large strains, and consequently on the predicted yield strength. For linear elastic behavior of the interphase and a thickness of 30 nm, the model overestimates the average experimental yield strength, while for the thickness of 20 nm the model prediction coincides with the experimental value.

Acknowledgment

The authors would like to thank Prof. C. Charitidis and his colleagues for conducting the Nanoindentation tests at the R-NanoLab situated at the School of Chemical Engineering (Department of Materials Science and Engineering) of National Technical University of Athens (NTUA)—Greece and for providing a thorough analysis of the experimental results

References

[1] Y. Han, J. Elliott, Molecular dynamics simulations of the elastic properties of polymer/carbon nanotube composites, Comput. Mater. Sci. 39 (2) (2007) 315–323.

[2] S. Yang, S. Yu, W. Kyoung, D-S. Han, M. Cho, Multiscale modeling of size-dependent elastic properties of carbon nanotube/polymer nanocomposites with interfacial imperfections, Polymer 53 (2) (2012) 623–633.

[3] V.V. Mokashi, D. Qian, Y. Liu, A study on the tensile response and fracture in carbon nanotube-based composites using molecular mechanics, Compos. Sci. Technol. 67 (3-4) (2007) 530–540.

[4] Q. Dewei, J. Hinkley, G. He, Molecular dynamics simulation of thermal and mechanical properties of polyimide–carbon-nanotube composites, Modelling Simul. Mater. Sci. Eng. 13 (2005) 493–507.

[5] R. Zhu, E. Pan, A.K. Roy, Molecular dynamics study of the stress–strain behavior of carbon-nanotube reinforced Epon 862 composites, Mater. Sci. Eng. A 447 (2007) 51–57.

[6] K. Li, S. Saigal, Micromechanical modeling of stress transfer in carbon nanotube reinforced polymer composites, Mater. Sci. Eng. A 457 (1-2) (2007) 44–57.

[7] C. Hammerand Daniel, D. Seidel Gary, C. Lagoudas Dimitris, Computational micromechanics of clustering and interphase effects in carbon nanotube composites, Mech. Adv. Mater. Struct. 14 (2007) 277–294.

[8] Young Seok Song, Jae Ryoun Youn, Modeling of effective elastic properties for polymer based carbon nanotube composites, Polymer 47 (5) (2006) 1741–1748.

[9] A.V. Desai, M.A. Haque, Mechanics of the interface for carbon nanotube–polymer composites, Thin-Wall Struct. 43 (11) (2005) 1787–1803.

[10] M.R. Ayatollahi, S. Shadlou, M.M. Shokrieh, Multiscale modeling for mechanical properties of carbon nanotube reinforced nanocomposites subjected to different types of loading, Compos. Struct. 93 (9) (2011) 2250–2259.

[11] X.L. Chen, Y.J. Liu, Square representative volume elements for evaluating the effective material properties of carbon nanotube-based composites, Comput. Mater. Sci. 29 (1) (2004) 1–11.

[12] A. Drago, M.J. Pindera, Micro-macromechanical analysis of heterogeneous materials: macroscopically homogeneous vs periodic microstructures, Compos. Sci. Technol. 67 (2007) 1243–1263.

[13] D. Savvas, G. Stefanou, V. Papadopoulos, M. Papadrakakis, Effect of waviness and orientation of carbon nanotubes on random apparent material properties and RVE size of CNT-reinforced composites, Compos. Struct. 152 (2016) 870–882.

[14] Y.J. Liu, X.L. Chen, Evaluations of the effective material properties of carbon nanotube-based composites using a nanoscale representative volume element, Mechan. Mater. 35 (2003) 69–81.

[15] M.A. Bhuiyan, R.V. Pucha, J. Worthy, M. Karevan, K. Kalaitzidou, Defining the lower and upper limit of the effective modulus of CNT/polypropylene composites through integration of modeling and experiments, Compos. Struct. 95 (2013) 80–87.

[16] M.A. Bhuiyan, R.V. Pucha, M. Karevan, K. Kalaitzidou, Tensile modulus of carbon nanotube/polypropylene composites: a computational study based on experimental characterization, Comp. Mater. Sci. 50 (2011) 2347–2353.

[17] F.T. Fisher, R.D. Bradshaw, L.C. Brinson, Fiber waviness in nanotube-reinforced polymer composites—I: modulus predictions using effective nanotube properties, Compos. Sci. Technol. 63 (2003) 1689–1703.

[18] ANSYS 11.0 Documentation, ANSYS Inc.

[19] S. Kurukuri, A comprehensive study: boundary conditions for representative volume elements (RVE) of composites, Homogenization of Damaged Concrete Meso-Structures Using Representative Volume Elements—Implementation and Application to SLang, MSc Thesis, Institute of Structural Mechanics (SLA), Rheinisch-Westfälische Technische Hochschule (RWTH) Aachen University, Aachen, Germany, 2005.

[20] D. Bikiaris, Microstructure and properties of polypropylene/carbon nanotube nanocomposites, Materials 3 (2010) 2884–2946.

[21] A.G. Osorio, L.A. Dos Santos, C.P. Bergmann, Evaluation of the mechanical properties and microstructure of hydroxyapatite reinforced with carbon nanotubes, Rev. Adv. Mater. Sci. 27 (2011) 58–63.

[22] M. Zaidi, S. Joshi, M. Kumar, D. Sharma, A. Kumar, S. Alam, et al. Modifications of mechanical, thermal, and electrical characteristics of epoxy through dispersion of multi-walled carbon nanotubes in supercritical carbon dioxide, Carbon Lett. 14 (4) (2013) 218–227.

[23] K. Prashantha, J. Soulestin, M.F. Lacrampe, P. Krawczak, G. Dupin, M. Claes, Masterbatch-based multi-walled carbon nanotube filled polypropylene nanocomposites: assessment of rheological and mechanical properties, Compos. Sci. Technol. 69 (2009) 1756–1763.

[24] S.H. Park, P.R. Bandaru, Improved mechanical properties of carbon nanotube/polymer composites through the use of carboxyl-epoxide functional group linkages, Polymer 51 (22) (2010) 1–7.

[25] S. Pegel, P. Pötschke, G. Petzold, I. Alig, S.M. Dudkin, D. Lellinger, Dispersion, agglomeration, and network formation of multiwalled carbon nanotubes in polycarbonate melts, Polymer 49 (2008) 974–984.

[26] S.R. Bakshi, R.G. Batista, A. Agarwal, Quantification of carbon nanotube distribution and property correlation in nanocomposites, Compos. A 40 (2009) 1311–1318.

[27] M.D. Haslam, B. Raeymaekers, A composite index to quantify dispersion of carbon nanotubes in polymer-based composite materials, Compos. B 55 (2013) 16–21.

[28] A. Chanteli, K.I. Tserpes, Tensile behavior of carbon nanotube/polypropylene composite material, Plast. Rubber Compos. 43 (2014) 330–336.

[29] M.M. Rahman, S. Zainuddinb, M.V. Hosurb, J.E. Malone, M.B.A. Salam, A. Kumar, et al. Improvements in mechanical and thermo-mechanical properties of e-glass/epoxy composites using amino functionalized MWCNTs, Compos. Struct. 94 (2012) 2397–2406.

[30] A. Montazeri, J. Javadpour, A. Khavandi, A. Tcharkhtchi, A. Mohajeri, Mechanical properties of multi-walled carbon nanotube/epoxy composites, Mater. Des. 31 (9) (2010) 4202–4208.

[31] Q. Yang, X. He, X. Liu, F. Leng, Y.W. Mai, The effective properties and local aggregation effect of CNT/SMP composites, Compos. B 43 (1) (2012) 33–38.

[32] C.S. Jarali, S.R. Basavaraddi, B. Kiefer, S.C. Pilli, Y.C. Lu, Modeling of the effective elastic properties of multifunctional carbon nanocomposites due to agglomeration of straight circular carbon nanotubes in a polymer matrix, J. Appl. Mech. 81 (2) (2013) 1–11.

[33] R. Moradi-Dastjerdi, A. Pourasghar, M. Foroutan, The effects of carbon nanotube orientation and aggregation on vibrational behavior of functionally graded nanocomposite cylinders by a mesh-free method, Acta Mech. 224 (11) (2013) 2817–2832.

[34] V.S. Romanov, S.V. Lomov, I. Verpoest, L. Gorbatikh, Modelling evidence of stress concentration mitigation at the micro-scale in polymer composites by the addition of carbon nanotubes, Carbon 82 (2015) 184–194.

[35] N. Montinaro, A. Pantano, Parameters influencing the stiffness of composites reinforced by carbon nanotubes: a numerical–analytical approach, Compos. Struct. 109 (2014) 246–252.

[36] D.L. Shi, X.Q. Feng, Y. Huang, K.C. Hwang, H. Gao, The effect of nanotube waviness and agglomeration on the elastic property of carbon nanotube-reinforced composites, J. Eng. Mater. Technol. 126 (3) (2004) 250–257.

[37] TexGen. Available from: http://texgen.sourceforge.net/index.php/Main_Page.

[38] LS-DYNA Theory Manual, Livermore Software Technology Corporation, Livermore, CA, 2014.

[39] ASTM D638-14, Standard Test Method for Tensile Properties of Plastics, ASTM International, West Conshohocken, PA, 2014. Available from: www.astm.org.

[40] ARAMIS 5M Digital Image Correlation System, Optical Measuring Techniques, Braunschweig, Germany, 2008.

[41] W.C. Oliver, G.M. Pharr, An improved technique for determining hardness and elastic modulus using load and displacement sensing indentation experiments, J. Mater. Res. 7 (1992) 1564–1583.

[42] D. Penumadu, A. Dutta, G.M. Pharr, B.S. Files, Mechanical properties of blended single-wall carbon nanotube composites, J. Mater. Res. 18 (2003) 1849–1853.

[43] L. Hyukjae, M. Shankar, H. Peng, S. Donglu, N. Suhasini, Y.H. Yun, S. Vesselin, J.S. Mark, Characterization of carbon nanotube/nanofiber-reinforced polymer composites using an instrumented indentation technique, Compos. B 38 (2007) 58–65.

[44] C.A. Schneider, W.S. Rasband, K.W. Eliceiri, NIH Image to ImageJ: 25 years of image analysis, Nat. Methods 9 (2012) 671–675.

[45] A. Chanteli, K.I. Tserpes, Finite element modeling of carbon nanotube agglomerates in polymers, Compos. Struct. 132 (2015) 1141–1148.

[46] PLASTICYL TM PP2001 Technical Data Sheet, NANOCYL, 2009. Available from: http://www.nanocyl.com/product/plasticyl-pp2001/.

[47] Y. Lin, A. Dichiara, D. He, P. Haghi-Ashtiani, J. Bai, Uniform diameter multi-walled carbon nanotubes with a controlled wall number obtained by a simple chemical vapour deposition method, Chem. Phys. Lett. 554 (2012) 137–142.

[48] E.G. Ordoñez-Casanova, M. Román-Aguirre, A. Aguilar-Elguezabal, F. Espinosa- Magaña, Synthesis of carbon nanotubes of few walls using aliphatic alcohols as a carbon source, Materials 6 (2013) 2534–2542.

[49] R. Xie, G. Zhong, C. Zhang, B. Chen, C.S. Esconjauregui, J. Robertson, Diameter and wall number control of carbon nanotubes by chemical vapour deposition, J. Appl. Phys. 114 (2013) 244–302.

[50] C. Laurent, E. Flahaut, A. Peigney, The weight and density of carbon nanotubes versus the number of walls and diameter, Carbon 48 (2010) 2994–2996.

17

Nonlinear Multiscale Modeling of CNT/Polymer Nanocomposites

David Weidt*, Łukasz Figiel**,†

*UNIVERSITY OF LIMERICK, LIMERICK, IRELAND; **INTERNATIONAL INSTITUTE FOR NANOCOMPOSITES MANUFACTURING, WMG, UNIVERSITY OF WARWICK, COVENTRY, UNITED KINGDOM; †CENTRE OF MOLECULAR AND MACROMOLECULAR STUDIES, POLISH ACADEMY OF SCIENCES, ŁÓDŹ, POLAND

CHAPTER OUTLINE

17.1 Introduction

In the aerospace and wind-turbine industries, epoxy resin-based carbon fiber-reinforced plastic (CFRP) laminates promise to meet current environmental demands for the reduction of fuel consumption, efficient energy generation, and long-term structural and environmental durability. When compared to metals, they are very attractive due to their good specific stiffness/strength, improved fatigue life, and corrosion resistance.

However, one of the remaining issues that still compromise the use of epoxy-based laminates in high-performance applications is their relatively brittle behavior when exposed to in-service impact events such as those caused by foreign objects. Therefore, the

Carbon Nanotube–Reinforced Polymers. http://dx.doi.org/10.1016/B978-0-323-48221-9.00017-0

improvement of the energy absorption characteristics of these materials is an important research goal for material scientists and engineers working with composite materials.

Applying a surface coating composed of carbon nanotubes (CNTs) and epoxy or incorporating CNTs into the epoxy matrix of CFRP laminates offers a means to improve the energy absorption of these laminates at low CNT concentrations. Enhancement in the energy absorption characteristics is believed to be caused by two deformation mechanisms: (1) improved nonlinear finite deformation of epoxy in compression due to the presence of CNTs and (2) CNT crack bridging mechanisms accompanied by CNT pullout and interface debonding in tension.

Nevertheless, material composition (e.g., CNT volume fraction and aspect ratio) and the nanocomposite morphology induced by the manufacturing method (e.g., CNT distribution, orientation, and functionalization) have a significant impact on the macroscopic properties and thus on the energy absorption of CNT/epoxy nanocomposites. Theoretically, CNTs aligned and optimally distributed within the epoxy matrix promise to provide the maximum enhancement in stiffness in the direction of CNT alignment. However, obtaining homogenous CNT dispersions and a controlled degree of CNT alignment within polymeric matrices still remains a significant research challenge.

Hence, understanding the relationship between process-induced morphology and the resulting macroscopic behavior is an important engineering task. Advanced material modeling can complement experimental work and assist experimentalists in exploring this relationship systematically.

A systematic investigation of morphological parameters such as CNT aspect ratio or CNT distribution/orientation solely by experiments is time-consuming and imposes significant challenges on nanocomposite preparation. For example, an increase in CNT volume fraction is often observed to lead to changes in the agglomeration behavior and thus to a different final CNT distribution when applied dispersion/functionalization methods do not work to satisfaction [1]. Hence, the effect of CNT volume fraction on the macroscopic behavior can be a combined effect composed of CNT volume fraction and CNT distribution. Experimental studies are prone to the interaction of two or more factors and thus they are frequently unable to apportion the effect of a single parameter on the mechanical response. On the other hand, advanced computational material modeling allows for the study of morphology–property relationships in a more systematic manner, for example, by changing one factor at a time. Hence, advanced computational material modeling is therefore a complementary approach to experimental work in exploring the morphology–property relationships of nanocomposites. In addition, it carries an economic advantage over costly and time-consuming experimental work.

However, the development of a robust and reliable nanocomposite computational modeling approach poses other challenges. The mechanical response of CNT/epoxy nanocomposites is complex due to its hierarchical structure, which results from several length- and timescales involved, spanning the nanometer scale and the macroscale. Further elucidation in the following paragraphs provides a short excursion into the different length scales of the hierarchical structure of nanocomposites.

At the atomistic scale, CNTs consist of a small number of carbon atoms arranged in hexagons with a given chirality and possible defects. Variations in those structural characteristics can alter the mechanical properties of isolated CNTs. The same applies to the interface, where the interaction between the CNT and the polymer is affected by intra- and interatomic interactions like interatomic bonding, discrete distribution of bonding, as well as bond breaking and rebonding mechanisms. The corresponding timescale for atoms and molecules in motion is the femtosecond scale [2].

In comparison, the microscale of the nanocomposite can be characterized by CNT agglomerations, that is, local differences between CNT concentrations that may result in zones of different stiffness compared to the surrounding nanocomposite. Concerning the timescales, structural relaxation or motion in polymer chains may take several seconds.

Hence, robust macroscopic predictions need to account for those aspects at different scales. Holistic multiscale computational modeling techniques must be developed and applied to cope with that range of time and length scales to predict the macroscopic response of nanocomposites.

Clearly, for the model to complement and then replace experimental tests with simulations, it requires the model prediction to be an accurate representation of the real material behavior. Thus, robust multiscale modeling demands experimental validation to confirm its degree of accuracy. The comparison between experimental outcome and model predictions allows to draw conclusions about the robustness of the applied multiscale modeling approach and answers the question whether sufficient detail was included within the multiscale framework. Validation can be conducted under specific conditions, for example, for certain nanocomposite morphologies or in a smaller range of strain rates, to save on costs of the experiments, and then used to evaluate the material behavior in the full range of the operating regime. However, predictions outside the validated regime have to be handled with care since the accuracy might decrease when approaching extremes of the operating regime.

This work follows a combined experimental–computational approach to develop a reliable multiscale model for the prediction of nonlinear stress–strain characteristics of CNT/epoxy nanocomposites. The focus herein is put on the nonlinear compressive behavior of those materials. Despite its practical importance, only few experimental studies have focused on the nonlinear compressive response of CNT/epoxy nanocomposites [3]. To this date, the authors are aware of only few experimental research contributions investigating the nonlinear compressive behavior of CNT/epoxy nanocomposites [4]. Similarly, the compressive response of CNT/polymer nanocomposites has only been sparsely researched using computational models. Nanocomposite behavior under compressive loading (e.g., CNT instability and damping behavior) was predicted using molecular dynamics and multiscale finite element (FE) modeling [5–7]. The proposed models did not, however, consider sufficiently large deformations (important as an energy-absorbing mechanism) and nanocomposite morphological effects (e.g., CNT orientation and nanoparticle interactions) for the prediction of the nanocomposite mechanical response.

Therefore, the main goal of this work was to develop a nonlinear multiscale computational model for the prediction of the nonlinear mechanical response of CNT/epoxy nanocomposites under compressive loads [8,9]. The following length scales were used and combined in a sequential manner to develop the model. In particular, CNT and interface properties were predicted from molecular mechanics/dynamics and used in a mesoscale model that was formulated within the representative volume element (RVE) concept, nonlinear FE framework, and employed first-order nonlinear homogenization. Moreover, the model was also complemented by a physically based constitutive "glass-rubber" model [10] able to capture the nonlinear strain-rate-dependent behavior of the epoxy matrix. In addition, the effective fiber (EF) concept based on molecular mechanics calculations of CNTs [11,12] was used to represent the continuum mechanical response of single-walled CNTs (SWCNTs) and multiwalled CNTs (MWCNTs).

The robustness of the multiscale modeling approach was evaluated by comparing the predictions with the authors' experimental results. The validation focused on randomly oriented and distributed CNT/epoxy nanocomposites subjected to uniaxial compression under quasistatic strain rates.

17.2 Experimental Part

17.2.1 Material Preparation

Nonfunctionalized MWCNTs (NANOCYL NC3100), produced via a catalytic chemical vapor deposition process, were obtained from Nanocyl S.A. (Belgium). The average outer diameter of the nanotubes is about 9.5 nm and the average length is 1.5 μm. The purity is above 95 mass%. The number of sidewalls of 18 MWCNTs, which were randomly picked and investigated via transmission electron microscopy, is between 6 and 19. Amino-functionalized MWCNTs (HDPlas), also produced via catalytic chemical vapor deposition process, were obtained from Cheap Tubes (USA) as functionalized using a split plasma process. The purity is above 99 mass%. The outer diameter and length range between 13–18 nm and 3–30 μm, respectively. The content of amino-groups (NH_2) is 7 ± 1.5 mass%. Other functionalities are N—H and O=C—N—H_2.

The resin was a standard diglycidyl ether of bisphenol A (Epon 828). The monomer with a molecular weight of 377 was obtained from Polyscience Europe GmbH (Germany). Glycolitic polypropyleneoxide triamine (Jeffamine T403 hardener) with a molecular weight of 440 was provided by Alfa Chemicals Ltd. (United Kingdom).

Nanocomposite and pure epoxy samples (Fig. 17.1), which will serve as reference material, were produced using two production routes: (1) a nonsolvent method and (2) a solvent method using chloroform. Functionalized CNTs were used together with the solvent chloroform, because they are more stable in organic solvents such as chloroform than their nonfunctionalized counterparts. A relatively time-consuming production route was used for the nonsolvent method. Preliminary tests showed that adding the whole CNT mass fraction in one step results in a relatively high viscosity mixture, which could not be dispersed successfully. Using the solvent chloroform as dispersion medium enabled the incorporation and dispersion of the whole CNT mass fraction in

FIGURE 17.1 (A) Pure epoxy sample versus nanocomposite sample. Porosity of nanocomposite samples (CNT mass fraction of 0.3% of nonfunctionalized CNTs) produced via (B) nonsolvent method and (C) solvent method; pores are visible as *white* and *black spots*.

one step. Hence, the production time was reduced significantly. Specifics of the samples of the batches are listed in Table 17.1 (batch notation, functionalization, solvent, and CNT mass fraction).

Nanocomposites produced using the nonsolvent method exhibit mass fractions of 0.1, 0.2, and 0.3% of nonfunctionalized MWCNTs. A CNT concentration of 0.1 mass% was added first to the epoxy monomer and the mixture was stirred manually for 10 min. This was followed by mild water bath sonication for 10 h (45 kHz, 150 W). CNT mass fractions of 0.1 and 0.2% were obtained stepwise by adding 0.1 mass% to the dispersed CNT/epoxy mixture followed by a repetition of the sonication process. Between those steps the mixture was left to rest overnight. In the case of 0.2 and 0.3 mass%, at the start and the middle of the second and the third sonication steps, the mixture was stirred manually for 15 min to break off larger agglomerates at the beaker wall. After obtaining the final CNT mass fraction, the hardener was poured into the beaker containing the CNT/epoxy mixture

Table 17.1 Batch Notations, Preparation Route, and CNT Type/Content

Batch	Name	Solvent	CNT Functionalization	CNT Mass Fraction (%)
1	Epoxy1 (NS)	O	O	0.00
2	Epoxy2 (NS)	O	O	0.00
3	Nanocomposite (NS, NF) MF = 0.1%	O	O	0.10
4	Nanocomposite (NS, NF) MF = 0.2%	O	O	0.20
5	Nanocomposite (NS, NF) MF = 0.3%	O	O	0.30
6	Epoxy1 (S)	X	O	0.00
7	Nanocomposite (S, F) MF = 0.3%	X	X	0.30
8	Nanocomposite (S, F) MF = 0.45%	X	X	0.45
9	Nanocomposite (S, F) MF = 0.6%	X	X	0.60
10	Nanocomposite (S, NF) MF = 0.3%	X	O	0.30

F, Functionalized CNTs; NF, nonfunctionalized CNTs; NS, nonsolvent preparation method; O, without solvent/CNT functionalization; S, solvent preparation method; X, with solvent/CNT functionalization.

(resin/hardener ratio of 100:42). The final mixture was then stirred manually for 25 min followed by two short sonication steps of 15 min to degas the mixture. The warmed water was changed in between those sonication steps. Then, the mixture was poured into tubular metal molds. These molds consisted of two pieces glued together. The tubular part is open at both ends and a cylindrical metal bar is slid into it. The open molds were placed in an oven and the material was cured at 75°C for 2 h at ambient pressure. A postcure of 2.5 h at 125°C was applied.

Nanocomposites produced using the solvent method exhibit mass fractions of 0.3, 0.45, and 0.6% of functionalized MWCNTs and 0.3% of nonfunctionalized MWCNTs. Requested final mass fractions of CNTs were first added to chloroform (CNT concentration in chloroform: 2 mg/mL). The mixture, placed in air-sealed vials, was predispersed for 2 h using mild water bath sonication (45 kHz, 150 W). Then it was added to the epoxy monomer and manually stirred for 3 min. The open beaker was placed in a water bath at 60°C and sonicated for 4 h to degas the chloroform (45 kHz, 150 W). This was followed by automatic stirring on a hot plate at 65°C and 90 rpms overnight for 12 h. Trapped air and possible rest of the solvents were degassed in a vacuum oven by applying four cycles of vacuum (one cycle: 20 min at 850 mbar, 10 min at 1000 mbar) at 80°C. The mixture was then sonicated (water bath) at 65°C for 2 h to counteract possible reagglomeration that might have happened overnight or while applying the vacuum. After the addition of the hardener (resin/hardener ratio 100:42) and manual agitation for 25 min, the material was poured into the metal molds and placed in a vacuum oven for four cycles (one cycle: 20 min at 850 mbar, 10 min at 1000 mbar) at 35°C. In such a way, possible bubbles incorporated previously by mixing the filled epoxy with the hardener were degassed. The open molds were placed in an oven and the material was cured at 75°C for 2 h at ambient pressure. A postcure of 2.5 h at 125°C was also applied here.

Pure epoxy reference samples were produced using the same production route without stirring/sonicating the pure epoxy monomer. Released samples were ground down and polished wet to obtain parallel surfaces and their final sample length. The cylindrical samples have a length of 20 mm and a diameter of 12.7 mm (Fig. 17.1A).

By using the aforementioned manufacturing route, nanocomposite samples could be prepared without major difficulties. However, samples with a mass fraction of 0.3% of nonfunctionalized CNTs exhibited bubbles, which could not be removed via sonication or the application of a vacuum. All other samples are free of visible bubbles. Fig. 17.1B–C shows the grade of porosity for those samples.

17.2.2 Mechanical Behavior

Compression tests were performed using a DARTEC 100 kN testing facility, a 25 mm linear variable differential transformer displacement transducer, and the Stress Analysis Data System 6000. A linear variable differential transformer displacement transducer and the load-cell of the DARTEC machine were connected to the System 6000, and calibrated for the force–displacement recording. The parallel sample surfaces were covered by a thin layer of petroleum jelly and the samples were placed between polished metal plates that were attached to the main grip bodies. The DARTEC machine was programmed to test at a constant speed of 0.2 mm/s. The displacement of the upper sample surface was measured indirectly at an angle iron mounted to the main grip body. The load–displacement behavior was recorded for the subsequent data analysis. Batches 1 and 2 (Table 17.1) of pure epoxy were produced (nonsolvent method) and tested separately to evaluate the reproducibility of the production process and experimental testing. The average stress–strain behavior of the two batches was found to be nearly identical, which herein was assumed to confirm their reproducibility. The batches 1, 2–5, and 6–10 were tested on three different days and the samples of these batches were tested in a random order. The true-stress–true-strain curves for relevant batch realizations were averaged, as shown in Fig. 17.2.

In general, relatively small enhancements of the Young's modulus were obtained by incorporating CNTs into the epoxy matrix (both for solvent and nonsolvent methods). The Young's modulus was evaluated as a slope between two data points of 10 and 30 MPa true stress (not shown in Fig. 17.2). Nonfunctionalized CNTs showed a negligible reinforcement effect (increase in Young's modulus smaller than 1%) for CNT concentrations of 0.1 and 0.2 mass%. A CNT mass fraction of 0.3% resulted in a significant drop of the nanocomposite's Young's modulus for both preparation methods, which was attributed to poor

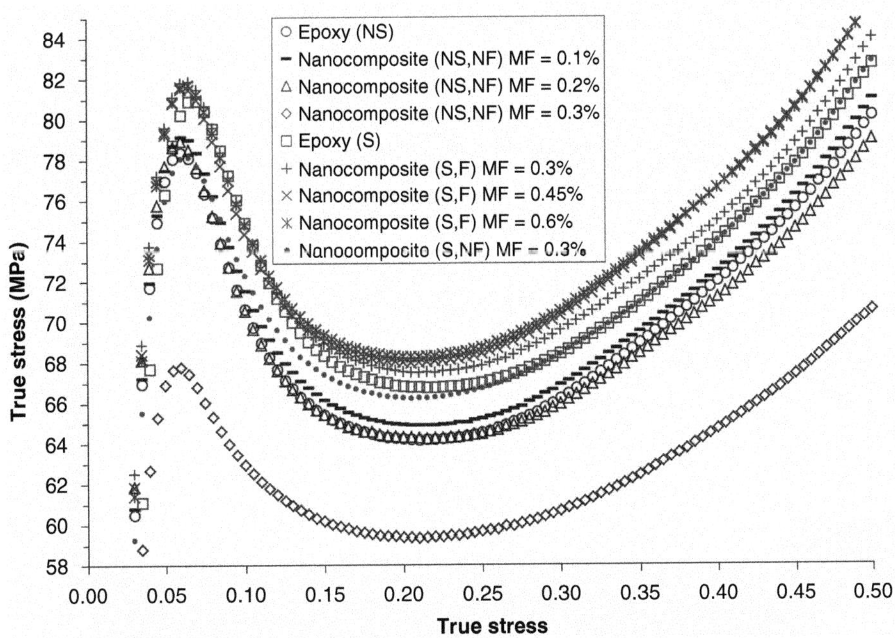

FIGURE 17.2 Portion of average experimental stress–strain curves of epoxy and nanocomposites of various CNT mass fractions (MF) prepared using functionalized/nonfunctionalized (F/NF) CNTs and a solvent/nonsolvent preparation method (S/NS).

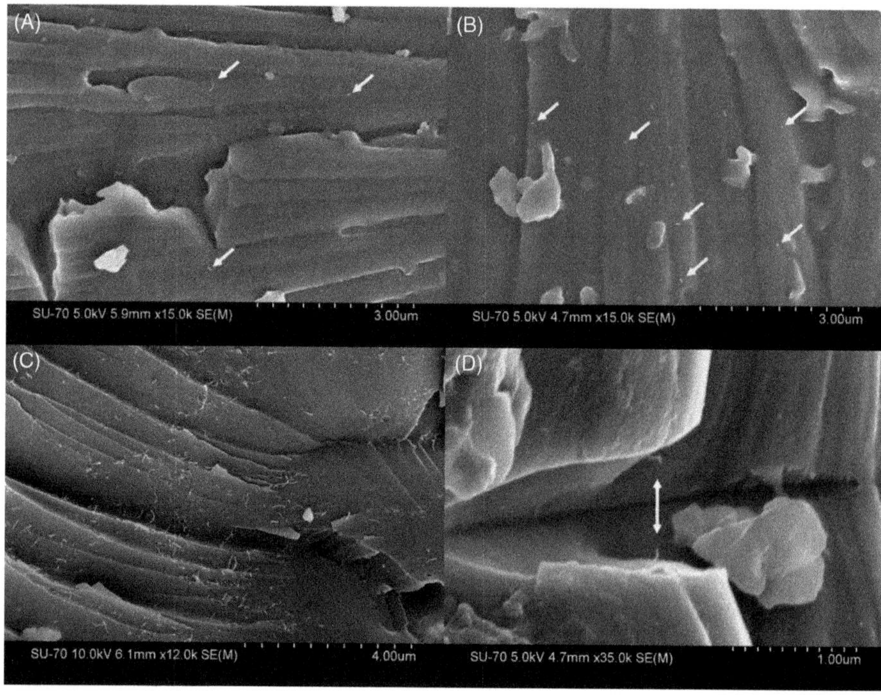

FIGURE 17.3 SEM micrographs of nanocomposites of various CNT MF with nonfunctionalized CNTs and a nonsolvent preparation method. (A) MF = 0.1% and (B) MF = 0.2%, isolated CNTs *(white arrows)*; (C) MF = 0.3%, CNT-rich regions/agglomerations; (D) MF = 0.2%, fractured CNT after crack bridging mechanism *(white arrow)*; white particles are a result of the postprocessing using a Dremel.

dispersion (Fig. 17.3). However, the samples produced with the solvent method exhibited a larger Young's modulus than their nonsolvent counterparts indicating a better dispersion state. Functionalized CNTs show a better reinforcement capability than their nonfunctionalized counterparts. An increase in Young's modulus of around 2.3 and 1.1% was obtained for samples composed of 0.3 and 0.6 mass% of functionalized CNTs. In contrast, the intermediate CNT concentration of 0.45 mass% shows a negligible reinforcement effect below 0.5%. Again, the decrease of the nanocomposite's Young's modulus at a higher CNT content (i.e., above 0.45 mass%) is possibly caused by an insufficient state of dispersion, which was confirmed by morphology studies (Fig. 17.4).

Qualitatively similar results were obtained for the yield peak stress. However, the effect of the presence of CNTs on yield peak stress is less significant than on Young's modulus. A CNT concentration of 0.1 mass% of nonfunctionalized CNTs resulted in a 1% increase in yield peak stress. Further increase in CNT concentration decreases the yield peak stresses when compared to the previous concentration. Likewise, for the case of functionalized CNTs, an enhancement of 1.1% in yield peak stress was obtained for 0.3 mass%. CNT concentrations of 0.45 and 0.6 mass% show negligible enhancements (smaller than 1%) of yield peak stresses.

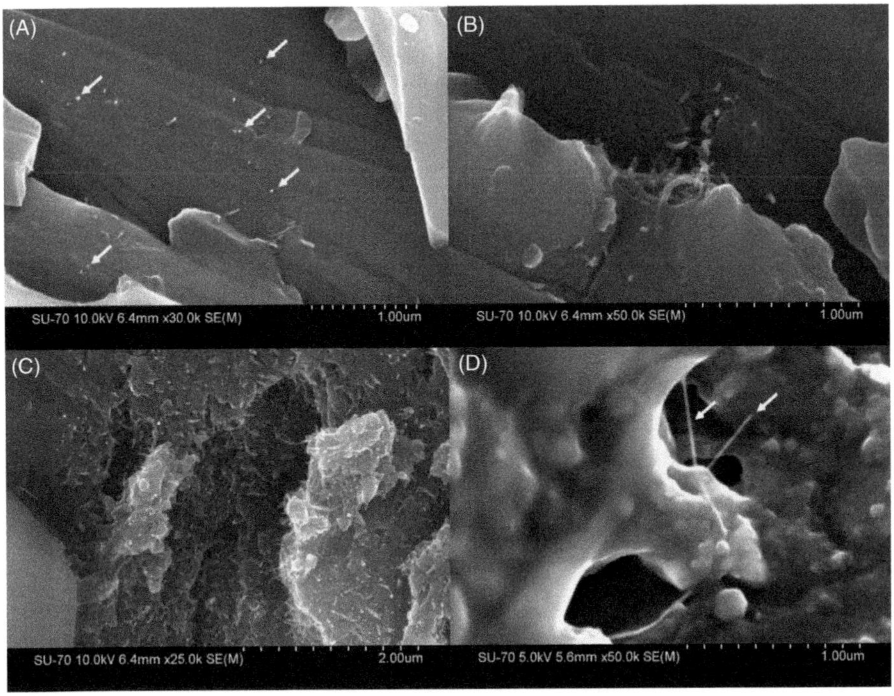

FIGURE 17.4 SEM micrographs of fracture surfaces of nanocomposite samples of various CNT MF of functionalized CNTs. (A) MF = 0.3%, good state of dispersion with CNTs broken close to the fracture surface *(white arrows)*; (B) MF = 0.45%, small CNT agglomeration with incomplete impregnation; (C) MF = 0.45%, large CNT agglomeration, similar agglomeration sizes obtained for the case MF = 0.6%; (D) MF = 0.3%, CNT crack bridging *(white arrows)*.

The nanocomposite samples tend to break at earlier stages of strain than the samples made of pure epoxy. The general trend is that the larger the CNT concentration the smaller is the true strain at break. This was observed for both types of CNTs, functionalized and nonfunctionalized. However, a smaller true strain at break was obtained for a CNT concentration of 0.1 mass% than for 0.2 mass% of nonfunctionalized CNTs. A possible reason can be the small difference in the dispersion state of these batches, which cannot be visualized by scanning electron microscopy (SEM), and can have a significant effect on the true strain at break due to stress concentrations through agglomerates. Second, it is possible that the doubled sonication time for the case of 0.2 mass% shortened the CNTs, compared to 0.1 mass%, and thus entangled them more effectively, or decreased their tendency to form agglomerates by reducing their specific surface area. Furthermore, when considering the incorporated CNT as preexistent crack, the CNT length reduction can be linked to a larger stress intensity factor and thus to a better toughness.

The ultimate compressive strength was associated with the stress at break. The strength of the nanocomposite samples was found to be smaller than for pure epoxy, which was decreasing with an increase in CNT content—this trend was a consequence of the trends observed for the strain at break of nanocomposites.

The largest stress enhancements within the postyield regime at 30% of true strain were achieved for 0.45 and 0.6 mass% of functionalized nanotubes. In comparison to pure epoxy using the same solvent-assisted preparation method, approximately 3% larger stress values were achieved here. The nanocomposite composed of 0.3 mass% of functionalized CNTs, which exhibited the best result in terms of Young's modulus and yield peak stress, shows a less significant stress enhancement in the postyield regime. An improved behavior (when compared to pure epoxy) for nonfunctionalized CNTs was only obtained for a mass fraction of 0.1%. Concentrations of 0.2 and 0.3 mass% exhibited smaller stresses within the postyield regime.

Altogether, the reinforcement effect of CNTs on the finite strain compressive properties was shown to be small. The reasons that limit their reinforcement efficiency are expected to be caused by CNT curvature, CNT defects, CNT misalignment, poor interfacial bonding between CNTs and epoxy, and poor dispersions/agglomerations for larger CNT mass fractions.

Thus, nanocomposite samples composed of 0.3 mass% of functionalized CNTs and prepared using a solvent method were chosen for the computational validation. Those samples showed the largest effect of CNT incorporation on the stress–strain response for the highest CNT mass fraction that did not result in significant CNT agglomerations at fracture surfaces—agglomerations are the features that cannot be captured by the current computational model.

17.3 Nonlinear Multiscale Nanocomposite Model

17.3.1 Modeling Approach

The applied multiscale modeling approach involves four distinct length scales (after Weidt and Figiel [8]):

1. The nanoscale, where mechanical properties of SWCNTs and MWCNTs are derived from molecular mechanics following the work of Shen and Li [11,12], and properties for CNT/epoxy interfaces are based on molecular dynamics simulations [13].
2. The microscale, where the discrete CNT is represented by a homogenized solid cylinder using the EF concept and where phenomena related to the polymer network are captured by a nonlinear continuum constitutive model for the epoxy matrix [10].
3. The mesoscale, where the nanocomposite morphology is reconstructed using the RVE concept under assumptions of global periodicity.
4. The macroscale, where the nanocomposite stress–strain response is predicted using numerical homogenization of the RVE response.

Moreover, practical implementation of the modeling approach requires an efficient route for automated model generation. The model generation involves several aspects (Fig. 17.5): (1) the reconstruction of the nanocomposite morphology, (2) the definition of the material properties, (3) the controlled meshing of the RVE, and (4) the formulation of

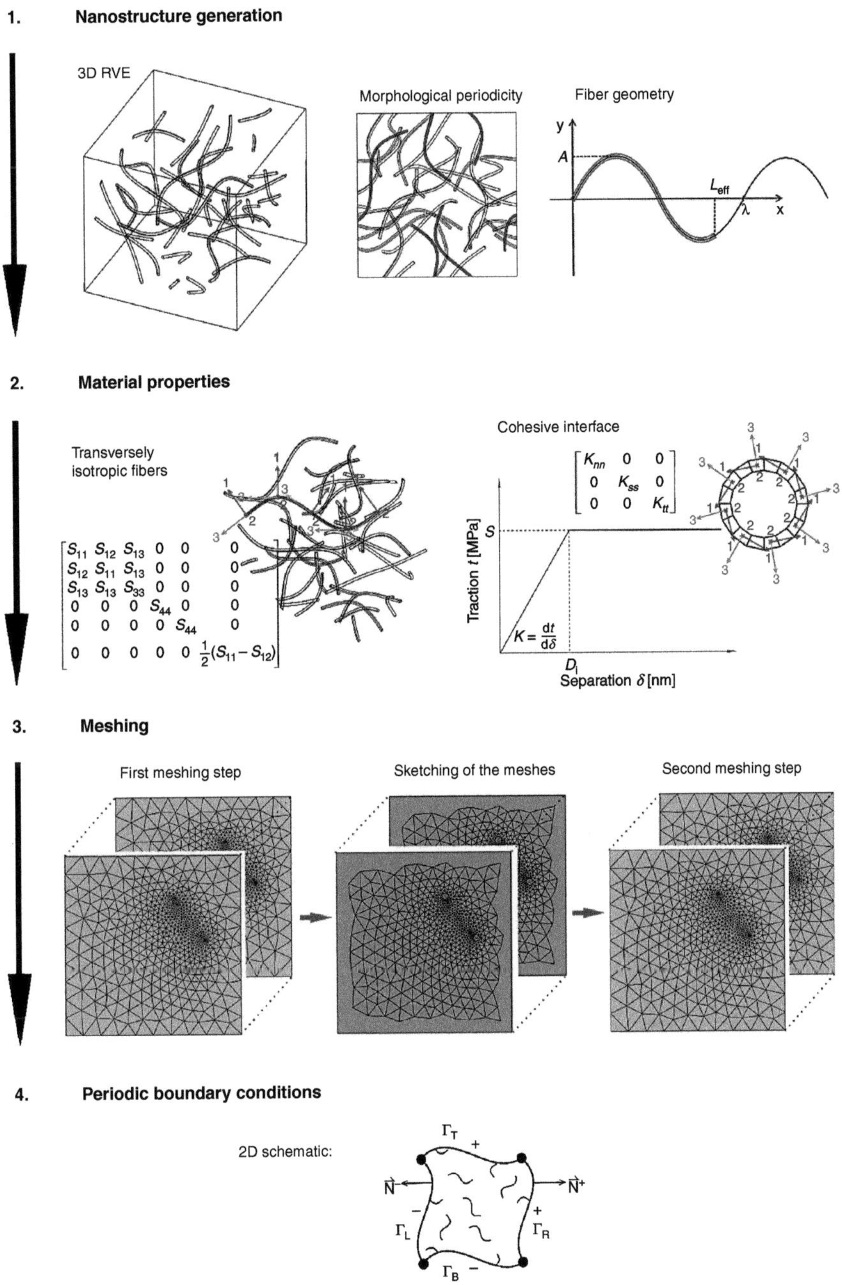

FIGURE 17.5 Schematic of main generation steps for the nanocomposite model. 3D morphology generation to account for periodicity of wavy fiber geometries; definition of the material properties, for example, transversely isotropic fibers with evolving local coordinate systems along the fiber axis and cohesive interface between the fiber and the matrix based on vdW interaction; meshing of the RVE using a two-step meshing procedure to obtain identical meshes on opposing RVE faces; formulation of periodic boundary conditions.

the periodic boundary conditions. In detail, after the definition of the RVE input parameters, the complete procedure up to the writing of the INPUT-file was programmed allowing an automated model generation solely within the ABAQUS environment.

17.3.2 CNT Representation

Each CNT was modeled as an effective continuum fiber with transversely isotropic properties. A diameter of 15.5 nm was chosen for the effective fiber representation. The inner diameter of 4 nm in combination with the interlayer distance of 0.34 nm provides a wall number of 18, which was used for the calculation of the transversely isotropic material properties. The amplitude and wavelength were determined from SEM micrographs of CNTs in powdered form. Those values were assumed not to change significantly when incorporating the CNTs into the epoxy matrix. However, it must be mentioned that the determination of amplitude and wavelength is a delicate task because CNT shape in reality is arbitrary and does not follow a sinusoidal curve. A considerable number of 74 CNTs were used to determine the means of the amplitude and wavelength. The length used for the calculation of the transversely isotropic properties and the length of the effective fiber was set to 911.87 nm. This value is substantially lower than the values provided by CNT manufacturers, but this is justified by the fact that it actually reflects more accurately the effective CNT length contributing to stress transfer [9].

The transversely isotropic elastic properties of the effective CNT fiber were calculated by using the equations of Shen and Li [11,12]. For that purpose, each wall of the MWCNT was assumed to exhibit an armchair structure. The properties of the effective fiber were calculated assuming perfectly straight CNTs. The CNT wavelength was used as input for the CNT length. Then the properties were applied within local coordinate systems along the wavy CNT axis. The results are shown in Table 17.2 (after Weidt and Figiel [9]).

17.3.3 CNT/Polymer Interface

To account for different CNT surface treatments (as mentioned in Section 17.2), the model contained a cohesive zone interface (with the thickness equal to 2 Å) between the lateral surface of the EF and the epoxy matrix. Two extreme cases of surface treatment were considered: (1) imperfect interfacial bonding and (2) perfect bonding, to

Table 17.2 Molecular Mechanics Results for HDPlas (After Weidt and Figiel [9])

	Effective Fiber (Wavy)	Effective Fiber (Straight)
Longitudinal modulus $E_{33}^{tension}/E_{33}^{compression}$ (GPa)	687.21/1051.1	713.18/1051.1
Major Poisson's ratio $\nu_{31}^{tension}/\nu_{31}^{compression}$	0.12807/0.13686	0.12900/0.13686
Longitudinal shear modulus G_{31} (GPa)	204.71	208.77
In-plane bulk modulus K_{12} (GPa)	112.11	112.11
In-plane shear modulus G_{12} (GPa)	2.0256	2.0256

Table 17.3 Cohesive Zone Parameters (After Weidt and Figiel [9])

Parameter	Perfect Bonding	Imperfect Bonding
Mode 1 stiffness K_{nn} (Pa/Å)	1.23761e + 09	1.23761e + 09
Mode 2/3 stiffness K_{ss}, K_{tt} (Pa/Å)	4.3500e + 08	1.20e + 06
Strength S (all modes) (Pa)	1.0e + 12	2.7e + 06
Damage evolution displacement D_e (nm)[a]	9119	9119

[a]Damage evolution displacement D_e corresponds to the failure displacement D_f subtracted by the damage initiation displacement D_i.

compare the model with experimental data. In the first case, the cohesive zone parameters were chosen to capture secondary van der Waals (vdW) bonding with a sliding-like behavior. For the second case, perfect bonding was modeled using the cohesive zone concept with its parameters associated with the linear elastic properties of the epoxy matrix. Interfacial damage was prevented throughout the simulation by setting interfacial strength artificially high. Such a formulation of the cohesive zone was shown to provide for an RVE the same macroscopic stress–strain response as an RVE modeled without the cohesive zone. The complete set of cohesive zone parameters is shown in Table 17.3.

17.3.4 Constitutive Model for Polymer Matrix

The nonlinear and rate-dependent epoxy behavior was represented using a physically based constitutive model proposed and validated by Buckley et al. [10] for thermosetting resins. The model accounts for the strain softening and adiabatic heating effect (active upon applied impact loads), and enabled capturing the basic phenomena of epoxy matrices subjected to varying strain rates. The model assumes two sources of energy:

1. Perturbation of interatomic potentials (relaxed by thermally activated flow events).
2. Entropy-elastic perturbation of molecular conformations (arising from preferred molecular alignment).

The model was described within the framework of continuum, and its kinematics is driven by the deformation gradient tensor $\boldsymbol{F}$. The tensor was decomposed multiplicatively into its volumetric $\boldsymbol{F}_{\text{vol}}$ and isochoric (deviatoric) $\bar{\boldsymbol{F}}$ parts:

$$\boldsymbol{F} = \boldsymbol{F}_{\text{vol}}\bar{\boldsymbol{F}} \quad \text{with} \quad \bar{\boldsymbol{F}} = J^{-1/3}\boldsymbol{F} \quad \text{and} \quad J = \det \boldsymbol{F}, \tag{17.1}$$

where J is the volume ratio.

The nonlinear response is contained within the isochoric part of the deformation gradient ($\bar{\boldsymbol{F}}$). Thus, additional measures of deviatoric deformation such as left the Cauchy–Green tensor $\bar{\boldsymbol{B}}$, velocity gradient $\bar{\boldsymbol{L}}$, rate of deformation $\bar{\boldsymbol{D}}$, and continuum spin $\boldsymbol{W}$ are defined based on $\bar{\boldsymbol{F}}$:

$$\bar{\boldsymbol{B}} = \bar{\boldsymbol{F}}\bar{\boldsymbol{F}}^{\mathrm{T}}, \quad \bar{\boldsymbol{L}} = \dot{\bar{\boldsymbol{F}}}\bar{\boldsymbol{F}}^{-1}, \quad \bar{\boldsymbol{D}} = \frac{1}{2}(\bar{\boldsymbol{L}} + \bar{\boldsymbol{L}}^{\mathrm{T}}), \quad \boldsymbol{W} = \frac{1}{2}(\bar{\boldsymbol{L}} - \bar{\boldsymbol{L}}^{\mathrm{T}}), \tag{17.2}$$

where the symbol • denotes the material time derivative. Furthermore, additive decomposition of the isochoric rate of deformation $\bar{D}$ into elastic $\bar{D}^e$ and viscous $\bar{D}^v$ parts was invoked:

$$\bar{D} = \bar{D}^e + \bar{D}^v. \tag{17.3}$$

There are two deviatoric stress contributions: (1) bond-stretching stress $\bar{S}^b$ (b-part of the model) and (2) conformational stress $\bar{S}^c$ (described in terms of isotropic hyperelasticity) (c-part of the model) as a result of the two energy sources just stated.

The total Cauchy stress σ is the sum of these two tensor contributions and the volumetric stress tensor:

$$\sigma = \bar{S}^b + \bar{S}^c + \sigma_m \mathbf{I} \quad \text{with} \quad \sigma_m = K_b \ln J, \tag{17.4}$$

where σ_m, $\mathbf{I}$, and K_b are the mean (hydrostatic) stress, second-order identity tensor, and bulk modulus, respectively.

Concerning the b-part of the model, the objective rate of deviatoric bond-stretch stress is represented as a single relaxation time approximation of the real multimode response [10]:

$$\overset{\circ}{\bar{S}}{}^b = 2G^b \bar{D} - \frac{\bar{S}^b}{\tau}, \tag{17.5}$$

and G^b is the shear modulus associated with bond stretching, and τ is the relaxation time. Taking the objective rate of deviatoric bond-stretch stress $\overset{\circ}{\bar{S}}{}^b$ as the Jaumann rate, the deviatoric Cauchy b-stress $\bar{S}^b$ is then determined from the following differential equation for the rate of deviatoric bond-stretch stress $\overset{\bullet}{\bar{S}}{}^b$ [10]:

$$\overset{\bullet}{\bar{S}}{}^b = \overset{\circ}{\bar{S}}{}^b + W\bar{S}^b - \bar{S}^b W. \tag{17.6}$$

The c-part of the model applies the physical theory of rubber elasticity due to Ref. [14] neglecting, however, entanglement effects due to the high crosslink density of the epoxy resins.

The deviatoric Cauchy c-stress $\bar{S}^c$ is obtained by differentiating the Edwards–Vilgis strain-energy function A_c with respect to elastic network principal stretches $\bar{\lambda}_i$ [10]:

$$\bar{S}_i^c = \bar{\lambda}_i \frac{\partial A_c}{\partial \bar{\lambda}_i} + p(i\,\text{not summed}), \quad \bar{S}^c = \bar{S}_i^c \boldsymbol{q}_i \otimes \boldsymbol{q}_i, \tag{17.7}$$

where $\bar{\lambda}_i^2$ and $\boldsymbol{q}_i$ are the eigenvalues and eigenvectors of the isochoric left Cauchy–Green tensor $\bar{\boldsymbol{B}}$, and pressure $p = -(1/3)\text{tr}(\sigma)$.

The model accounts for several features such as strain rate dependence, structural rejuvenation, and adiabatic heating. Strain rate dependence and structural evolution are both driven by the viscous part of the deformation rate tensor, in particular by the relaxation time τ [10]:

$$\tau = a_\sigma \tau_0, \tag{17.8}$$

where a_σ is the time-stress shift factor and τ_0 is the relaxation time in the linear viscoelastic limit. To account for the strain rate-dependent behavior, the stress-shift factor a_σ is based on a three-dimensional (3D) generalization of the Eyring rate kinetics [10]:

$$a_\sigma = \frac{V_s \tau_{\text{oct}}^{\text{b}}}{2RT} \frac{\exp(-V_p \sigma_m / RT)}{\sinh(V_s \tau_{\text{oct}}^{\text{b}} / 2RT)}, \tag{17.9}$$

where $\tau_{\text{oct}}^{\text{b}}$ is the octahedral shear component of the bond-stretch stress, V_p and V_s are pressure and shear activation volumes, R is the universal gas constant and T is the temperature.

Using Tool's fictive temperature concept [15], structural rejuvenation, which manifests itself as a softening regime in the constitutive model, is incorporated in the τ_0 parameter given by Ref. [10]

$$\tau_0 = \tau_0^* \exp\left(\frac{\Delta H_0}{RT} - \frac{\Delta H_0}{RT^*} + \frac{C}{T_{f\sigma} - T_\infty} - \frac{C}{T_{f\sigma}^* - T_\infty} \right), \tag{17.10}$$

and the viscoelastic relaxation time τ_0^* at some reference temperature T^*, and reference structure $T_{f\sigma}^*$, the Vogel equation constant C, the Vogel temperature T_∞, the activation enthalpy ΔH_0 associated with structural change, and the fictive temperature $T_{f\sigma}$ as a measure for the structural state/rejuvenation of the epoxy resin. The latter was assumed to be driven by the sum of the invariants of the viscous strain increments $\bar{\varepsilon}^v$ [10]:

$$T_{f\sigma} = T_{f\sigma 0} + (T_{f\sigma\infty} - T_{f\sigma 0}) \left[1 - \exp\left(-\frac{\bar{\varepsilon}^v}{\varepsilon_0^v} \right) \right], \tag{17.11}$$

where the final fictive temperature $T_{f\sigma\infty}$ and the strain range over which rejuvenation occurs, ε_0^v, were taken as material constants.

The thermomechanical coupling is based on the following differential equation for the rate of adiabatic heating [10]:

$$\dot{T} = \frac{1}{\rho c} \left[\bar{\sigma} : \bar{D} - \bar{\sigma}^{\text{b}} : \bar{D}^{\text{e}} \right] - \frac{\varphi \Delta c \dot{T}_{f\sigma}}{c}, \tag{17.12}$$

where $\varphi \Delta c$ is the effective specific heat difference, c is the specific heat, ρ is the density, and $\bar{\sigma}$ and $\bar{\sigma}^{\text{b}}$ are the deviatoric Cauchy stress and deviatoric bond-stretching Cauchy stress, respectively.

The original model parameters obtained for the standard bisphenol A resin epoxy resin reported in Ref. [10] did not provide a satisfactorily agreement with the current experimental data for Epon 828/Jeffamine T403 epoxy using the solvent method. Therefore, some model parameters were modified according to the current experimental data and they are listed in Table 17.4. Values of the remaining model parameters were used as in the work of Buckley et al. [10].

Table 17.4 Constitutive Model Parameters Obtained by Comparison to Experimental Stress–Strain Data of Pure Epoxy Prepared Using the Solvent Method (After Weidt and Figiel [9])

Parameter	Epon 828/Jeffamine T403
Shear modulus G (Pa)	0.87e + 9
Bulk modulus K (Pa)	3.078e + 9
Glass transition temperature T_g (K)	363[a]
Density ρ (kg/m^3)	1140[a]
Number density of crosslinks N_c (m^{-3})	5.7e + 27
Inextensibility factor of network α	0.297
Final fictive temperature $T_{f\sigma\infty}$ (K)	372
Linear viscoelastic relaxation time τ_0 (s)	2.02e + 6
Rejuvenation strain range ε_0^v	0.339

[a]Parameters obtained by inference from data with similar materials.

17.4 Results and Discussion

The comparison between experimental data and the model predictions is addressed in this section. The computational model was validated under quasistatic strain rates and then used to elucidate the effect of CNT mass fraction and the effect of epoxy–CNT interface bonding on the finite strain compressive response of the nanocomposite. Premature termination of the simulation (due to singular behavior of the epoxy model at large strains) restricted the computational study to the evaluation of the macroscopic behavior up to the yield peak stress in the case of imperfect bonding, and prior to the onset of strain-softening–strain-hardening for the case of perfect bonding. To give an insight into the effect of CNT waviness, RVEs comprising straight CNTs were investigated additionally. Both cases, imperfect bonding based on vdW interaction and perfect bonding, were considered for those straight CNTs.

17.4.1 Nonlinear Stress–Strain Curves

As an example, Fig. 17.6 shows the average computational predictions and average experimental stress–strain response for the pure epoxy and nanocomposites (with the perfectly bonded CNT/epoxy interface). The predicted Young's modulus and yield peak stress are relatively close to the experimental data as shown previously. However, the stresses within the softening regime (beyond 10% of true strain) were overpredicted. The possible reasons are the assumption of perfect bonding applied to the computational model and the inaccurate prediction of the softening by the epoxy model. Perfect bonding keeps the stress transfer relatively high throughout the full strain range. In reality, however, one can expect that bonds between the CNT and epoxy resin can change/break at larger strains and the stress transfer can reduce, and the interactions between the epoxy and CNTs can change

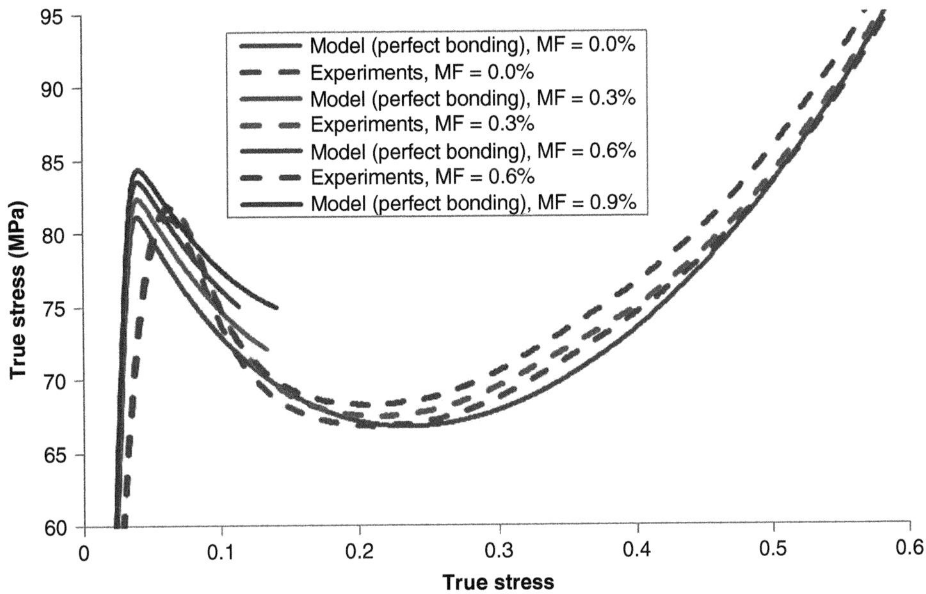

FIGURE 17.6 Portion of the average nonlinear stress–strain response, model predictions for wavy CNTs, and experimental results.

to the sliding-like behavior (with friction and vdW interactions). Then, for a smaller stress-transfer efficiency due to vdW interaction, a real sample is then believed to change its stress–strain response at larger applied strains toward the behavior of pure epoxy. Furthermore, one can see that the true strain value, at which the yield peak occurs, is captured inaccurately by the computational model. This may be attributed to the single relaxation time approximation for the real multimode response in the constitutive model for pure epoxy. The implementation of a relaxation spectrum is believed to provide a more accurate fit to the preyield part of experimental stress–strain curves. Additionally, it may improve the predictions in the softening regime because the structural evolution/rejuvenation is incorporated in the relaxation time parameter. However, that multimode approach is computationally more expensive.

Generally, an increase in CNT mass fraction leads to an increase in the nanocomposite's Young's modulus for perfectly bonded CNTs and CNTs interacting with the polymer through vdW forces. In both cases, and within the investigated range of CNT mass fraction, an approximately linear relation between CNT mass fraction and Young's modulus was obtained (see Ref. [9] for more details). The lack of perfect linearity resulted from statistical scatter obtained for reconstructed RVE morphologies. Hence, a small preferred orientation of CNTs within an RVE can slightly disturb the output as was found for the perfectly bonded model with 0.6 mass% of CNTs. To quantify if the preferred orientation exists for a given mass fraction, volume averaging over all CNT orientations was carried out. The study of the preferred average orientation of CNTs requires in general a 3Dl consideration.

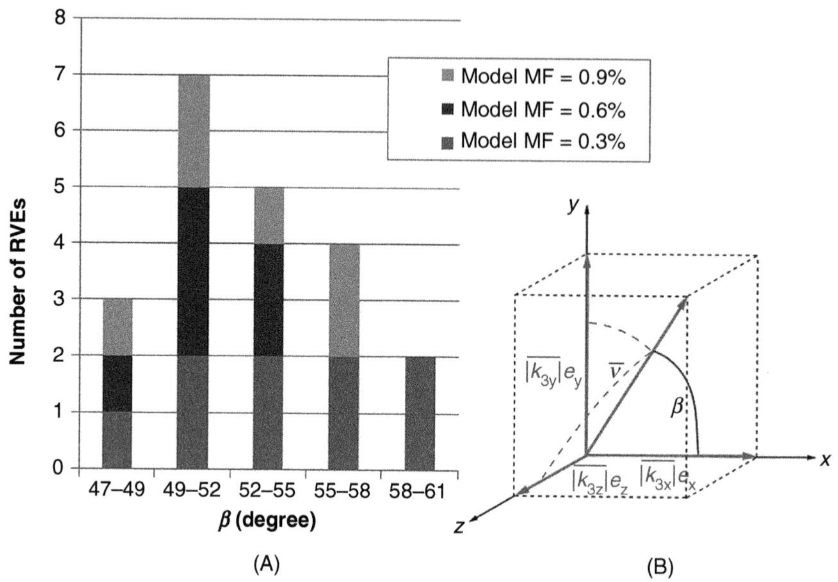

FIGURE 17.7 (A) Distribution of β obtained for RVEs of various mass fractions for perfectly bonded CNTs. The length of each bar shows the number of RVEs (hence the total number of RVEs is 9, 6, and 6 for MF = 0.3, 0.6, and 0.9%, respectively). The arithmetic mean of β was computed to be 54.12, 51.13, and 53.26 degree for 0.3, 0.6, and 0.9% CNT mass fractions, respectively. (B) Location of angle β.

However, following the work of Fu and Lauke [16] (here the CNTs instead of short fibers), only a single angle was used to relate the average CNT orientation within the RVEs to the average Young's modulus (instead of the strength as investigated by Fu and Lauke [16]) in the loading direction. For that, the angle β between the loading direction x and the average vector $\bar{\boldsymbol{v}}$ of the CNT phase (Fig. 17.7) was calculated. The latter was obtained by volume-averaging of the vector $\boldsymbol{v}$, which is associated with a CNT axis and defined in terms of direction cosines k_{3x}, k_{3y}, and k_{3z}—these direction cosines relate the local orientation of a CNT in its axial direction (3-3) and axes x, y, and z of the global coordinate system. The direction cosines k_{3x}, k_{3y}, and k_{3z} are the components (row 3, columns 1–3) of the direction cosine matrix (3 × 3) available in ABAQUS using the member *localCoordSystem* of a field value object that is stored as tensor with respect to a local coordinate system. The complete set of equations involved in the calculation of angle β is given as

$$\beta = \arccos\left(\frac{\overline{|k_{3x}|}}{|\bar{\boldsymbol{v}}|}\right) \quad \text{with} \quad \bar{\boldsymbol{v}} = \begin{pmatrix} \overline{|k_{3x}|} \\ \overline{|k_{3y}|} \\ \overline{|k_{3z}|} \end{pmatrix} = \frac{1}{V_{\text{CNT}}} \int_{V_{\text{CNT}}} \boldsymbol{v}\, \mathrm{d}V_{\text{CNT}} \quad \text{and} \quad \boldsymbol{v} = \begin{pmatrix} |k_{3x}| \\ |k_{3y}| \\ |k_{3z}| \end{pmatrix}, \qquad (17.13)$$

where $k_{3x} = \boldsymbol{e}_3 \cdot \boldsymbol{e}_x = |\boldsymbol{e}_3||\boldsymbol{e}_x|\cos(\beta_i^{3x})$ (and analogically for k_{3y} and k_{3z}), while $\boldsymbol{e}_3$ and $\boldsymbol{e}_x$ are the unit vectors associated with 3-axis and x-axis (Fig. 17.7), respectively, and β_i^{3x} is the angle between axis 3 and x at the i-th integration point within a given CNT. Then, V_{CNT} in Eq. (17.13) denotes the volume of the CNT phase. Absolute values (|.|) of the components of the local orientation vectors were used here to consider the average orientation in the range from 0 to $\pi/2$. In the case of random CNT orientation and for RVE sizes (and thus CNT numbers) tending to infinity, one can assume that $\overline{|k_{3x}|}=\overline{|k_{3y}|}=\overline{|k_{3z}|}$ holds, which in turn gives $\beta=\arccos\left(1/\sqrt{3}\right)\approx 54.74$ degree.

In Fig. 17.7, one can see the distribution of β obtained by evaluating Eq. (17.13), from which the arithmetic mean of β was found to be equal to β_m = 54.12, 51.13, and 53.26 degree for 0.3, 0.6, and 0.9% CNT mass fractions, respectively. The reconstructed RVEs contained CNTs with a preferred orientation for the intermediate mass fraction and thus affected slightly the linear relationship between the Young's modulus and mass fractions (Fig. 17.7). As one can see from Fig. 17.7, there is an increased fraction of CNTs aligned toward the loading direction for the case of 0.6 mass% CNTs when compared to the other CNT mass fractions as indicated by the concentration on the left side of the distribution.

As a result, the CNT phase carried higher stresses (as shown in Ref. [9]). In general, an increase in CNT mass fraction resulted in a decrease in stress within the matrix phase. In contrast, when neglecting the previously mentioned case with a slightly preferred orientation toward the loading axis, the average stresses within the CNT phase experienced hardly any changes. It was found that vdW interactions significantly reduce the stress transfer efficiency and therefore reduced the average stress within the CNT phase by half. The average matrix stresses exceed the stresses of pure epoxy only in the softening regime.

VdW interaction limits the reinforcement efficiency considerably as shown in terms of normalized Young's modulus (see Ref. [9]). In general, the simulation results suggest that a three-fold increase in Young's modulus can be achieved through modification of the interface properties from imperfect (vdW bonding) to perfect bonding. The computational models predicted an enhancement of the Young's modulus of around 9% for 0.9 mass% of CNTs, perfectly bonded to the epoxy matrix. In comparison, the computational study predicted a slight increase (i.e., below 3%) in the case of imperfect bonding.

Young's modulus predictions by the model with perfect bonding is closer to the experimental results of CNT mass fraction of 0.3% (functionalized CNTs and solvent preparation), as compared with the corresponding case of imperfect bonding. It suggests that the actual interfacial shear strength of the CNT–polymer bond is larger than assumed by the imperfect bonding case based on vdW interaction. The model with imperfect bonding underpredicted the experimental results. The models predicted an increase of 3% (perfect bonding) and an increase of 1% (vdW interactions), as compared with 2.3% increase found experimentally.

The computational study predicted that an increased CNT mass fraction leads to increased nanocomposite yield peak. However, when compared to the increase in Young's modulus, the effect of CNT mass fraction on yield peak stress has only half the magnitude ($\sim$1.015 versus $\sim$1.03 for perfect bonding and 0.3 mass% of CNTs). Additionally, an interesting effect was observed—the relation between CNT mass fraction and normalized yield peak stress is nonlinear with an approximate linear behavior up to MF = 0.6%; further increase in CNT mass fractions resulted in a less pronounced increase in nanocomposite yield peak stress, especially for the case of vdW interactions. It is believed that the reason

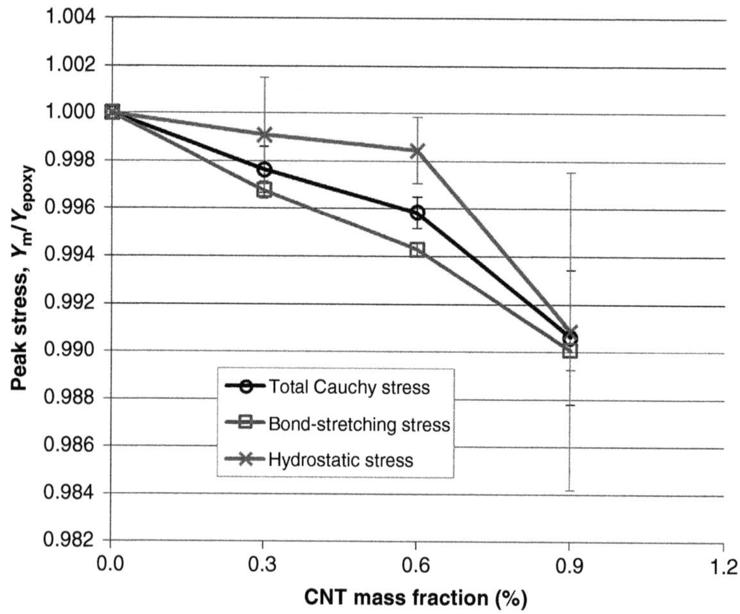

FIGURE 17.8 Effect of CNT mass fraction on the normalized (yield) peak stresses in the matrix. The average matrix Cauchy stress, average matrix bond-stretching stress, and average matrix hydrostatic stress are shown along with results obtained from RVEs comprising wavy and imperfectly bonded CNTs based on vdW interaction. Error bars show standard error of the mean.

for this is the average stress reduction in the matrix phase, caused by a drop in the average hydrostatic stress of the matrix phase with increasing CNT mass fraction. This reduction of hydrostatic stress in the matrix counteracted the increase in stress related to the increase in the CNT mass fraction. While the average matrix peak stresses of the bond-stretching part of the model decreased approximately linearly with increasing CNT mass fraction, the average hydrostatic peak stress was found to decrease nonlinearly with the CNT loading (Fig. 17.8). This led to a nonlinear decrease of the average peak stress (total Cauchy stress) within the matrix phase, and hence it is believed that this finally resulted in the nonlinear trend of the nanocomposite yield peak stress with CNT loading, as shown in Ref. [9]. This effect can be attributed to the waviness of CNTs—that is, for RVEs containing randomly oriented and distributed straight CNTs, the yield peak stress increases linearly with increasing CNT volume fraction and the hydrostatic stress of the matrix phase is increased when compared to pure epoxy. Similarly, straight CNTs led to an increase in average matrix hydrostatic stress (as shown in Ref. [9]).

Material expansion in the lateral directions (as a result of macroscopic compression) can effectively be reduced by a network of straight CNTs but not for wavy CNTs. Straight CNTs oriented in the lateral direction will carry a higher load due to an efficient shear stress transfer and will counteract the lateral expansion, which finally leads to an enhanced stress state within the matrix phase. Wavy CNTs, however, experience a lower stress transfer, and they need to be straightened before their axial modulus can be exploited fully. Prior to this straightening, the local bending deformation of the CNT occurs under small load, which increases the effective CNT length, that is, the length of the end-to-end segment. This is believed to facilitate the lateral expansion that leads to a reduction of the average hydrostatic stress and to a reduced stress state within the matrix phase.

As in the case of Young's modulus, changing the assumption from perfect bonding to imperfect bonding based on vdW interactions led to a considerable property reduction for the yield peak stress. The computational models predicted an enhancement of the yield peak stress around 3% for 0.6 mass% of CNTs perfectly bonded to the epoxy matrix. In contrast, the computational study predicted a relative increase of around 1% for the case of imperfect bonding at the same mass fraction. The experimental result for the normalized nanocomposite yield peak fell between model predictions for perfect and imperfect bonding. Models predicted an increase of around 0.5% (imperfect bonding) and 1.5% (perfect bonding) of the nanocomposite yield peak stress, in comparison with 1% increase found by experiments.

For the nanocomposite's Young's modulus and yield peak stress, it was found that the experimentally obtained increase of 2.3% in Young's modulus and 1.1% in yield peak stress could be approximately predicted by using the cohesive zone parameters in Table 17.3 with an interfacial shear strength of 30 MPa. RVEs with such an interfacial shear strength value predicted a 2% increase in Young's modulus and a 1% increase in yield peak stress.

17.4.2 Effects of CNT Waviness and van der Waals Interactions

Besides agglomeration, waviness and imperfect epoxy–CNT bonding are considered as main factors limiting the reinforcement capabilities of CNTs within nanocomposites. Thus, the stress–strain curves were determined for different interface interactions (perfect bonding and vdW interactions) and degrees of waviness (straight and sinusoidally wavy). An example of the average stress–strain curves showing the effect of vdW interactions for sinusoidally wavy CNTs is shown in Fig. 17.9.

The effects of waviness and vdW interaction on the nanocomposite's Young's modulus and yield peak stress were studied for a CNT mass fraction of 0.6%, and compared with the case of straight and perfectly bonded CNTs, respectively. The effects of vdW interaction on the nanocomposite's Young's modulus and yield peak stress were normalized with respect to corresponding quantities for perfectly bonded CNTs. The effect of imperfect bonding based on vdW interaction has the greater magnitude when compared to the effect of CNT waviness, which is discussed later. Of the two factors limiting the reinforcement efficiency of CNTs (i.e., imperfect interface bonding and waviness), vdW interaction was found to have the most detrimental effect on the nanocomposite's Young's modulus. For RVEs containing wavy CNTs, the Young's modulus was reduced by 4.5% and for the case of straight CNTs by 7.5% due to vdW interaction. Interestingly, vdW interaction did not affect the yield peak by the same magnitude. The nanocomposite yield peak stress

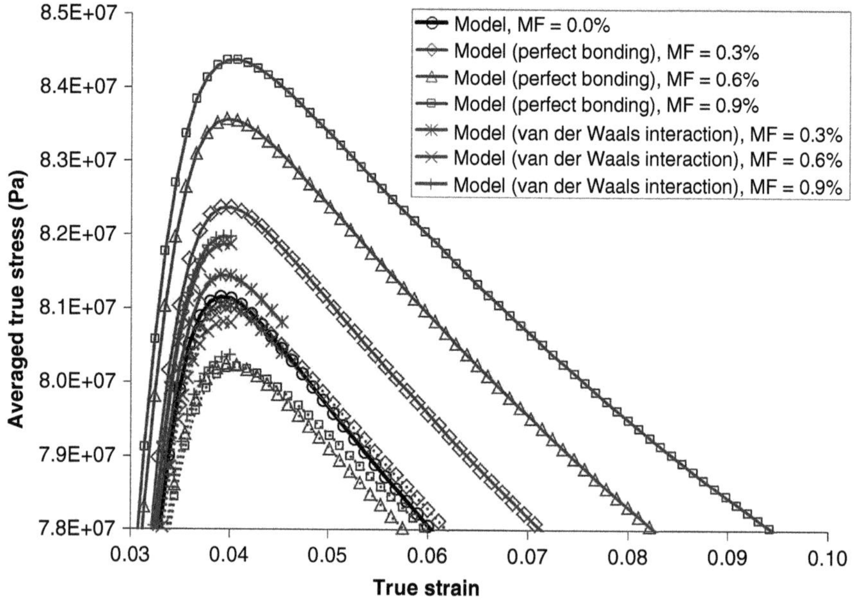

FIGURE 17.9 Average stress–strain response of the matrix phase (dotted lines) and nanocomposite (continuous lines) obtained from RVEs comprising wavy CNTs.

was reduced by 2% for wavy CNTs and by 5% for straight CNTs—a possible reason for that may be the polymer matrix response governing primarily the yield peak behavior of the nanocomposite, and hence it is less affected by the quality of interface bonding. Therefore, vdW interaction primarily limited the stress transfer efficiency.

CNT waviness (normalized with respect to relevant quantities obtained for straight CNTs) reduced the nanocomposite's Young's modulus by 1.5% for imperfectly bonded CNTs and by 4% for perfectly bonded CNTs. It is noteworthy to mention that the same reductions were obtained when studying the effect of CNT waviness on the nanocomposite yield peak stress. Waviness reduced the shear stress transfer, and hence, significantly lowered stresses within the CNT phase (see Ref. [9]). Moreover, the average stress within the CNT phase leveled off prior to the nanocomposite yield strain for wavy CNTs. In contrast, RVEs containing straight CNTs exhibited average stresses within the CNT phase that increased beyond the nanocomposite yield peak. In that case, the efficient stress transfer from the matrix across the perfectly bonded interface ensures that CNTs carry the load efficiently with increasing deformation. A major effect of CNT waviness was on the total stress state of the matrix. Compared to pure epoxy, straight CNTs increased the average matrix stress, while wavy CNTs decreased the average matrix stress. Hence, besides the reduced stress transfer efficiency, an additional contribution to the lowered yield peak stresses arose from the reduced total stress state of the matrix. This led finally to an equalized effect on yield peak stress when compared to the effect on Young's modulus.

17.5 Conclusions

A holistic 3D multiscale nonlinear computational model for CNT/epoxy nanocomposites was developed in this work. The model accounts for different degrees of bonding at the epoxy/CNT interface by considering two limiting cases: vdW and perfect bonding. It also takes CNT waviness into consideration. The model was validated with the results of experimental programs carried out for a range of CNT/epoxy nanocomposites subjected to quasistatic uniaxial compression. Generally, the model predicted well the nanocomposite response with functionalized CNTs in terms of experimental Young's modulus and yield peak stress. It was found that the cohesive zone parameters based on vdW interaction provide a relatively accurate prediction of Young's modulus and yield peak stress, when the interfacial shear strength is increased to 30 MPa.

After validation with experiments, the model was used to predict the effects of CNT mass fraction, waviness, and imperfect bonding on the compressive response of the nanocomposite. The computational model predicted a linear increase of the nanocomposite's Young's modulus with increasing CNT mass fraction. The yield peak stress, however, exhibited a nonlinear behavior and was found to saturate for the case of vdW bonding at higher mass fractions. This nonlinear behavior is believed to originate from the CNT waviness, which was shown to have a direct effect on the average stress within the matrix phase. When compared to the behavior of pure epoxy, at the nanocomposite yield peak, those average stresses decreased nonlinearly with increasing CNT mass fraction. This counteracted the increase in stress related to the increase in the CNT mass fraction, which finally results in the nonlinear behavior of the yield peak stress. Under a relatively small load, wavy CNTs oriented in the lateral direction will experience local bending deformation, which can then lead to CNT straightening, and thus a significant increase in the effective length of the CNT. This is believed to facilitate the lateral expansion of the RVE (and thus of the nanocomposite) that leads to a reduction of the hydrostatic stress and therefore to a reduced total stress state within the matrix phase. An enhanced stress state within the matrix was obtained at later stages of strain. In contrast, straight CNTs resulted in an increased average stress within the matrix phase, right from the beginning of the deformation, possibly due to an increase in hydrostatic stress of the matrix phase. This coincided with the findings for straight CNTs, where additionally a linear relation between yield peak stress and CNT volume fraction was found. Hence, straight CNTs counteracted the lateral expansion, which was manifested in an increased average hydrostatic stress and Cauchy stress within the matrix phase.

At larger strains and within the postyield regime, the model with perfectly bonded CNT/epoxy interface and experimental results deviated increasingly, which may be caused by bond breaking at the CNT/epoxy interface, or by the epoxy constitutive model due to the single relaxation time approximation. Premature termination of the analysis with models with imperfectly bonded CNT/epoxy interface did not allow studying the postyield behavior of the nanocomposite; however, it is believed that the predictions of the model with imperfectly bonded interface would shift toward the experimental results within this regime.

Altogether, a relatively weak effect of CNT incorporation and CNT mass fraction on the mechanical properties was found by experiments. The computational model captured well quantitatively and qualitatively that weak reinforcement effect of CNTs. This was possible by accounting for the CNT waviness and by varying the bonding degree of the CNT/epoxy interface. Both factors were shown to cause a significant reduction in the nanocomposite's Young's modulus and yield peak stress. vdW interaction was found to have the most detrimental effect on the nanocomposite's Young's modulus.

In terms of energy absorption capabilities, the modeling results suggest using defect-free and straight CNTs, well-bonded to the epoxy matrix—this combination is expected to lead to a linear relation between CNT mass fraction and yield peak stress. Additionally, straight CNTs are expected to lead to an enhanced hydrostatic stress state within the matrix material and thus are believed to lead to an early nanocomposite hardening. Furthermore, those CNTs can effectively increase their stress-load after nanocomposite yield. It is, however, not clear whether those defect-free CNTs remain straight when embedded in the polymer.

References

[1] C.H. Tseng, C.C. Wang, C.Y. Chen, Functionalizing carbon nanotubes by plasma modification for the preparation of covalent-integrated epoxy composites, Chem. Mater. 19 (2) (2007) 308–315.

[2] A.H. Zewail, Femtochemistry: atomic-scale dynamics of the chemical bond, J. Phys. Chem. A 104 (24) (2000) 5660–5694.

[3] K.G. Dassios, S. Musso, C. Galiotis, Compressive behavior of MWCNT/epoxy composite mats, Compos. Sci. Technol. 72 (9) (2012) 1027–1033.

[4] T. Gómez-Del Río, J. Rodríguez, R.A. Pearson, Compressive properties of nanoparticle modified epoxy resin at different strain rates, Compos. B 57 (2014) 173–179.

[5] S. Namilae, N. Chandra, Role of atomic scale interfaces in the compressive behavior of carbon nanotubes in composites, Compos. Sci. Technol. 66 (13) (2006) 2030–2038.

[6] R. Rafiee, R.M. Moghadam, Simulation of impact and post-impact behavior of carbon nanotube reinforced polymer using multi-scale finite element modeling, Comput. Mater. Sci. 63 (2012) 261–268.

[7] C. Li, T.W. Chou, Multiscale modeling of compressive behavior of carbon nanotube/polymer composites, Compos. Sci. Technol. 66 (14) (2006) 2409–2414.

[8] D. Weidt, Ł. Figiel, Finite strain compressive behaviour of CNT/epoxy nanocomposites: 2D versus 3D RVE-based modelling, Comput. Mater. Sci. 82 (2014) 298–309.

[9] D. Weidt, Ł. Figiel, Effect of CNT waviness and van der Waals interaction on the nonlinear compressive behaviour of epoxy/CNT nanocomposites, Compos. Sci. Technol. 115 (2015) 52–59.

[10] C.P. Buckley, P.J. Dooling, J. Harding, C. Ruiz, Deformation of thermosetting resins at impact rates of strain. Part 2: constitutive model with rejuvenation, J. Mech. Phys. Solids 52 (10) (2004) 2355–2377.

[11] L. Shen, J. Li, Transversely isotropic elastic properties of single-walled carbon nanotubes, Phys. Rev. B 69 (4) (2004) 45414.

[12] L. Shen, J. Li, Transversely isotropic elastic properties of multiwalled carbon nanotubes, Phys. Rev. B 71 (3) (2005) 35412.

[13] S.J.V. Frankland, A. Caglar, D.W. Brenner, M. Griebel, Molecular simulation of the influence of chemical cross-links on the shear strength of carbon nanotube–polymer interfaces, J. Phys. Chem. B 106 (12) (2002) 3046–3048.

[14] S.F. Edwards, T. Vilgis, The effect of entanglements in rubber elasticity, Polymer 27 (4) (1986) 483–492.

[15] A.Q. Tool, Relation between inelastic deformability and thermal expansion of glass in its annealing range, J. Am. Chem. Soc. 29 (9) (1946) 240–253.

[16] S.Y. Fu, B. Lauke, Effects of fiber length and fiber orientation distributions on the tensile strength of short-fiber-reinforced polymers, Compos. Sci. Technol. 56 (1996) 1179–1190.

Further Reading

C.P. Buckley, J. Harding, J.P. Hou, C. Ruiz, A. Trojanowski, Deformation of thermosetting resins at impact rates of strain. Part I: experimental study, J. Mech. Phys. Solids 49 (7) (2001) 1517–1538.

18

Computational Multiscale Modeling of Carbon Nanotube–Reinforced Polymers

Mohammad Silani*, Timon Rabczuk**, Xiaoying Zhuang†,‡

*ISFAHAN UNIVERSITY OF TECHNOLOGY, ISFAHAN, IRAN; **INSTITUTE OF STRUCTURAL MECHANICS, BAUHAUS UNIVERSITY WEIMAR WEIMAR, GERMANY; †TONGJI UNIVERSITY, SHANGHAI, CHINA; ‡INSTITUTE OF CONTINUUM MECHANICS, LEIBNIZ UNIVERSITY HANNOVER, HANNOVER, GERMANY

CHAPTER OUTLINE

18.1 Introduction

The macroscopic behavior of carbon nanotube–reinforced polymers (CNRPs) are governed by several phenomena that usually occur at the scales below the macroscale. The interaction between carbon nanotubes and the polymeric matrix and the presence of interface and interphase regions in CNRPs make it very difficult to predict their material behavior. Multiscale methods are a class of computational methods that can explicate these scale-dependent behaviors. Multiscale methods are powerful tools that can predict the material behaviors based on the fine-scale details (Fig. 18.1).

Generally, all multiscale methods can be divided to hierarchical or sequential multiscale methods, semiconcurrent multiscale methods, and concurrent multiscale methods (Fig. 18.2). Hierarchical multiscale methods pass the information from the fine scale to the coarse scale but not vice versa. To scale up the results in continuum domains, computational homogenization is usually adopted. For atomistic domains, Cauchy–Born rule can

Carbon Nanotube–Reinforced Polymers. http://dx.doi.org/10.1016/B978-0-323-48221-9.00018-2

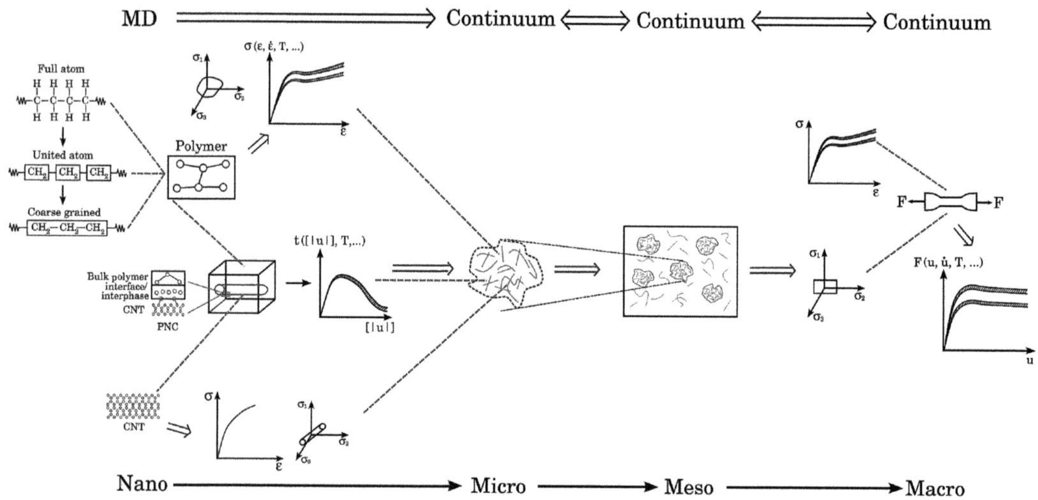

FIGURE 18.1 Computational multiscale modeling framework for carbon nanotube–reinforced polymers (CNRPs). ***CNT, Carbon nanotubes; PNC, polymer nanocomposites; MD, molecular dynamics.***

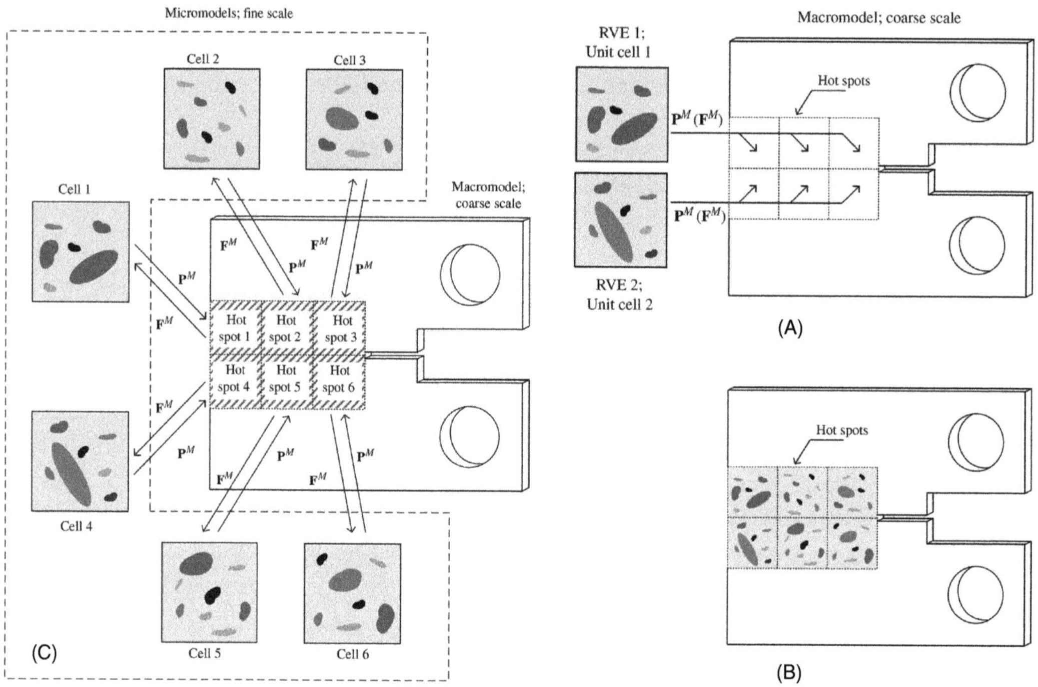

FIGURE 18.2 Schematic representation of (A) an hierarchical or sequential, (B) concurrent, and (C) semiconcurrent multiscale methods [7]. *RVE*, Representative volume element.

be used instead. There are several research studies to predict the mechanical properties of intact polymeric nanocomposites using the hierarchical multiscale method [1–6].

The existence of disparate length scales is the basic requirement for the application of homogenization theories [8]. In case of CNRPs, it states that the representative volume element (RVE) size should be much less than the specimen sample and the CNRP's feature sizes should be much smaller than the RVE size. For problems involving fracture, the crack is also considered as the CNRP's feature and when the crack propagates, this requirement is violated. Another problem arises when the periodic boundary conditions (PBC) at the fine scale is used. In this case, when a crack touches the boundary, the resulting displacement jump violates the PBC.

Semiconcurrent multiscale methods use a two-way information passing technique. A classical semiconcurrent multiscale method coupling two continuum models is the FE^2 method proposed by Feyel and Chaboche [9]. This method was developed for intact materials, and then, Kouznetsova [10] extended this method for material failure [11,12]. Also the Cauchy–Born rule [13] and the virtual atom cluster proposed in Refs. [14–16] can be also considered as semiconcurrent multiscale approaches coupling an atomistic model with a continuum model; refer to the contributions from Silani et al. [17,18], who coupled different commercial packages with such an approach. However, while the Cauchy–Born rule is applicable for crystalline structures, it is not well suited to model amorphous polymer chains at the nanoscale. Semiconcurrent multiscale methods seem particularly suitable for nonlinear problems, as hierarchical models require simulations for all deformation states to extract the coarse-scale material behavior, while in the semiconcurrent approach, only the deformation states that actually occur in the simulation need to be considered. Nonetheless, there are some contributions employing sequential multiscale approaches to extract nonlinear material constitutive behavior for polymeric materials [5,6,19].

In concurrent multiscale methods, both scales are considered simultaneously and the fine scale is directly embedded into the coarse scale. Several concurrent multiscale methods, such as the Arlequin method [20], bridging domain method (BDM) [21,22], and quasicontinuum method [13], have been proposed that can be used to predict intact CNRP's behaviors. These methods can be classified into interface coupling methods and handshake coupling methods. It is believed that handshake coupling methods are more suitable for dynamic applications, as they can avoid artificial wave reflection from coupling region [23]. However, Miller and Tadmore [24] compared 14 different multiscale approaches for 2 quasistatic benchmark problems and showed that interface coupling methods are more accurate. Concurrent multiscale methods can be used to couple continuum–continuum [25], as well as continuum–atomistic domains [22,24,26]. These methods are mainly developed for intact materials. Far less work has been devoted to concurrent multiscale methods for fracture [14,23,27,28]. Using concurrent multiscale methods for fracture avoids the transfer of length scale around the crack tip and adaptively adjusts the fine-scale domain when cracks propagate in the fine-scale domain [15]. An efficient approach to detect cracks and crack tips in the fine-scale domain including an associated adaptive refinement and coarsening scheme in 2D and 3D has been proposed in Ref. [15].

As fracture is stochastic by nature and the prediction of macroscopic material parameters are influenced by many factors at the fine scale, it is of utmost importance to obtain upper and lower bounds of the predictions. In other words, it is important to quantitatively judge the reliability of the underlying model to predict certain macroscopic properties. Recently, uncertainty quantification of the fracture properties of polymeric nanocomposites has been done by Hamdia et al. [29] based on phase field modeling, and by Vu-Bac et al. in a multiscale framework [30]. They also presented a software framework for probabilistic sensitivity analysis of polymeric nanocomposites [31,32]. The effect of different stochastic inputs on the macroscopic properties of polymeric nanocomposites have also been investigated by Silani and coworkers [33,34].

Besides uncertainty quantification, optimization methods are required to design the "best" material with regard to a specific application. Optimization of a CNRP specimen considering uncertainties needs effective and robust optimization methods, taking into account the uncertain nature of all input variables (over multiple scales). These optimization methods seek to maximize the fracture toughness, residual strength, and ductility of the CNRPs. At the same time, the weight, cost, and probability of structural level resonance should be minimized.

It should be noted that not every optimized geometric design can be manufactured in practice. The key ingredients of the optimization are the definition of the objective functions, model equations, physical constraints, and design variables. In case of conflicting objectives, the final solution should be a trade-off between them that leads to a multiobjective problem. For example, Ghasemi et al. optimized the fiber distribution of a polymeric composite using NURBS functions [35,36] and metamodel-based probabilistic optimization [37] for CNRPs over four length scales.

18.2 Hierarchical Multiscale Methods for CNRPs

CNRPs are often considered homogeneous at the macroscopic level. To obtain macroscopic or "homogenized" or "effective" material properties for CNRPs with a fine-scale microscopic structure, the homogenization theory can be utilized. The aim of the homogenization theory is to study the thermomechanical behavior of heterogeneous materials, such as CNRPs. The behavior of CNRPs are usually the outcome of phenomena that occur at one to several length scales below the macroscale. Practically, it is not feasible to consider all length scales in a unified framework (a few researches have recently proposed nested scale approaches but they are usually limited to linear elastic case) and hence, a two-scale framework is considered here.

In homogenization theory and for arbitrary additive quantities, the volume averaging operator $\langle \bullet \rangle$ can be defined as proposed by Nemat-Nasser and Hori [8]:

$$\langle \bullet \rangle = \frac{1}{|\Omega|}\int_{\Omega} \bullet \, dV. \tag{18.1}$$

where $|\Omega|$ is the microvolume. For CNRPs, the microvolume is usually the RVE volume. Using Eq. (18.1), one can calculate the average stress (Σ) and average strain (E) over the

RVE. In linear elasticity, the effective constitutive relation is basically relationships between averages, and hence, the extracted properties are the effective elastic constants of the CNRPs. Using appropriate boundary conditions, the boundary value problem can be solved at the RVE level and the effective mechanical properties D^{eff} can be calculated as:

$$\Sigma = \boldsymbol{D}^{\text{eff}} : \boldsymbol{E} \tag{18.2}$$

18.3 Semiconcurrent Multiscale Methods for CNRPs

Semiconcurrent multiscale methods use a reciprocal information passing technique between the fine scale and the coarse scale. At the coarse scale, strain measures (such as deformation gradient) are known and stress measures must be calculated. Therefore, and in most cases, the deformation gradient of the coarse scale is passed to the fine scale and the stress tensor is sent back using homogenization theory [10].

The numerical scheme of semiconcurrent multiscale method for CNRPs can be described as follows: in every integration point of the macro model of CNRPs, an RVE (fine-scale model) is attached. The RVE contains the details of carbon nanotubes, as well as interfaces and interphases (when needed). There are different approaches for applying boundary conditions to the aforementioned RVE. The most common types are PBC, uniform strains, and uniform stresses [38,39]. From an implementation point of view, it is more common to use uniform strain boundary conditions. The uniform strain boundary condition can be expressed as:

$$u = \mathbb{E} \cdot x \text{ for every } x \text{ on } \partial\Omega \tag{18.3}$$

where u is the displacement vector of each point with microscale coordinate x,$\mathbb{E}$ is the strain measure, which comes from the coarse scale, and $\partial\Omega$ denotes the boundary of the RVE. Using the higher level (macroscale) deformation gradient, F, the displacement vector of the boundary nodes of the RVE can be calculated as:

$$\mathbf{u} = \mathbf{x} - \mathbf{X} = \mathbf{F}\mathbf{X} - \mathbf{X} = (\mathbf{F} - \mathbf{I})\mathbf{X} \tag{18.4}$$

where X and x is the fine-scale coordinate in the initial and current configuration, respectively.

Inside the CNRP's RVE, any kind of micromechanical constitutive behavior can be considered. Depending on the phenomena under consideration, any configurational complexity, such as defects and voids, can be considered inside RVE. Using the finite element method and homogenization techniques, the stress tensor for the next higher level, S^{cs}, can be calculated as:

$$\langle \boldsymbol{S}^{cs} \rangle \overset{def}{=} \frac{1}{\Omega_{fs}} \int_{\Omega_{fs}} \boldsymbol{S}^{fs} \, d\Omega \tag{18.5}$$

in which the superscript fs denotes the fine-scale domain. This homogenized stress tensor will be sent back to the corresponding integration point of the coarse-scale model. This

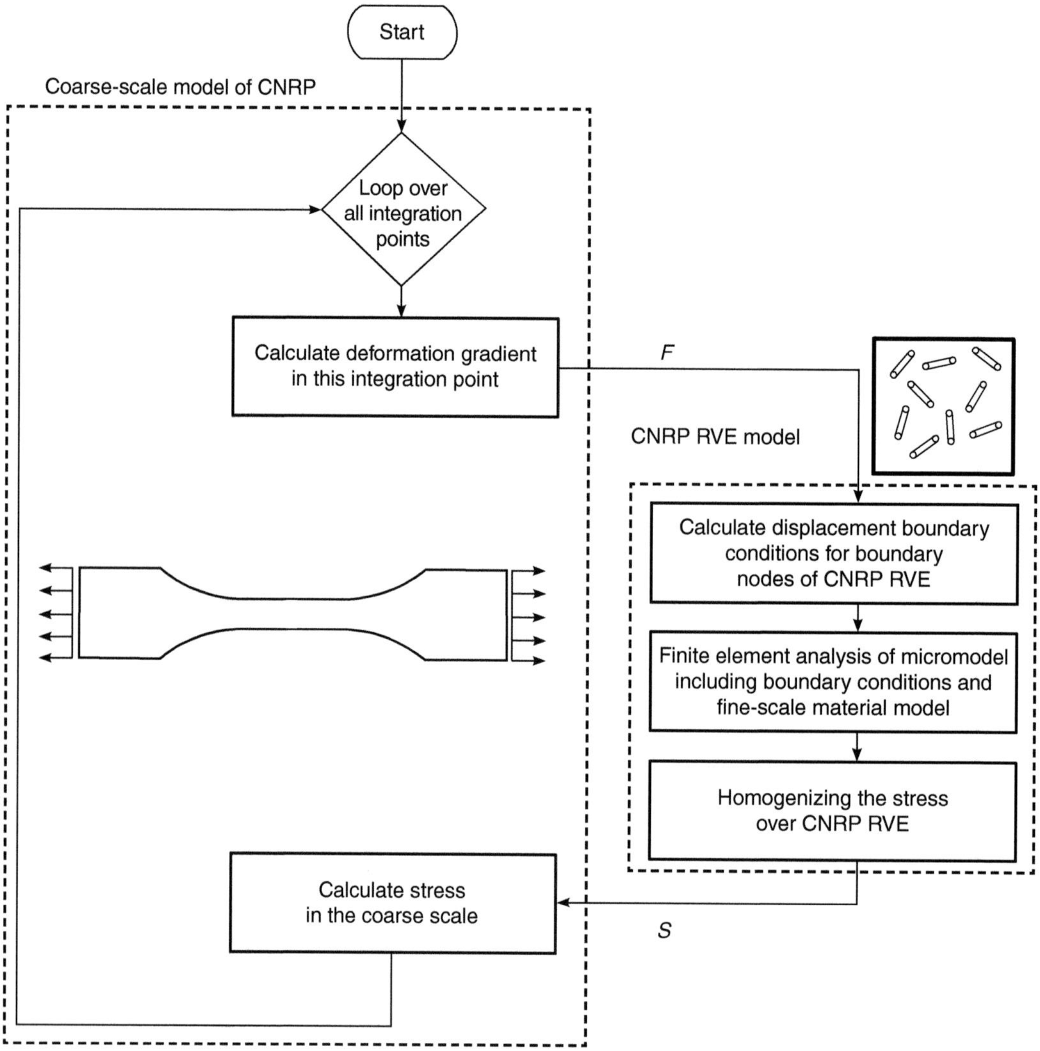

FIGURE 18.3 Schematic representation of semiconcurrent multiscale method for CNRPs.

process should be done for all integration points and in all loading increments of coarse scale [18]. Fig. 18.3 schematically shows the semiconcurrent multiscale method for CNRPs.

18.4 Concurrent Multiscale Methods for CNRPs

As already mentioned in the introduction, numerous concurrent multiscale methods for CNRPs have been developed and some of them have already been extended to fracture problems. Here the Arlequin method is presented, coupling two continuum domains [27]. The concept of this method is very similar to the BDM, which is originally presented to

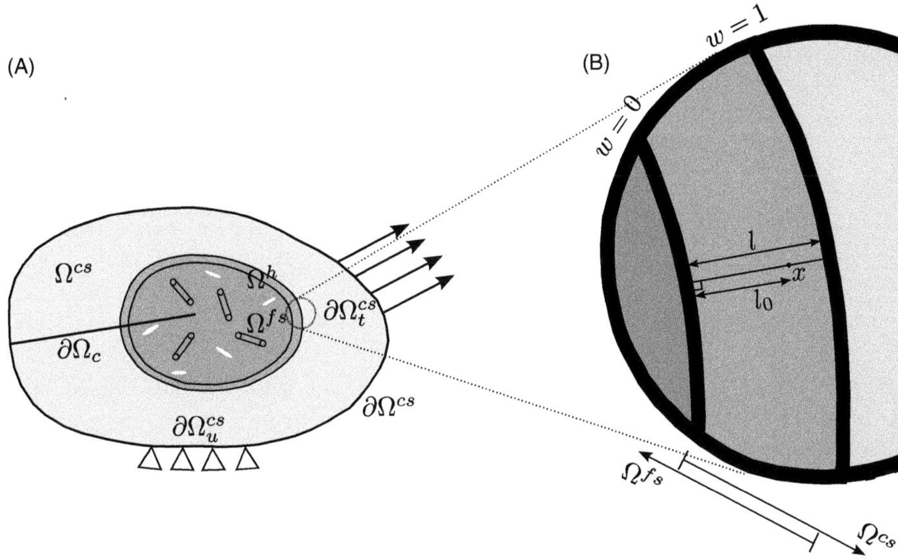

FIGURE 18.4 (A) The relation between fine- and coarse-scale areas and (B) the geometry of handshake domain, from Ref. [41].

couple an atomistic domain to a continuum one. Both methods can handle cracks in the coupling area, which is a big advantage of concurrent multiscale methods as compared to others [40].

In concurrent multiscale methods, both scales are considered at the same time. The problem domain can be considered as $\Omega = \Omega^{fs} \cup \Omega^{cs}$, with the outer boundary of $\partial\Omega^{cs}$ in which $\partial\Omega^{cs} = \partial\Omega_t^{cs} \cup \partial\Omega_u^{cs} \cup \partial\Omega_c^{cs}$ and $\partial\Omega_t^{cs} \cap \partial\Omega_u^{cs} = \theta$, $\partial\Omega_c^{cs} \cap \partial\Omega_u^{cs} = \theta$, $\partial\Omega_t^{cs} \cap \partial\Omega_c^{cs} = \theta$; here the subscripts u, t, and c refers to "displacement-," "traction-," and "crack," respectively (in case of fracture modeling). In concurrent multiscale methods, the part of the model that is of particular interest (e.g., crack tip in fracture simulations) is modeled by a finer scale. This fine-scale domain usually contains features of the micro/nanostructure of the CNRPs (Fig. 18.4).

As mentioned earlier, there are generally two types of coupling in concurrent multiscale methods: interface coupling and handshake coupling methods. In interface coupling methods, there is a sharp interface between fine scales and coarse scales, which results in spurious wave reflection at the coupling interfaces. Interface coupling method can be used for static cases without any problem [42]. Unlike interface coupling methods, in handshake coupling methods the coarse scale is continuously blended into a fine scale in a finite band, the so-called handshake domain: $\Omega^h = \Omega^{cs} \cap \Omega^{fs}$. Both Arlequin and BDMs use handshake coupling to connect the fine scale to the coarse scale. To avoid multiple energy calculation over the handshake region, the Arlequin and BDMs use a linear weighting of the fine-scale energy (W^{fs}) and coarse-scale energy (W^{cs}) inside handshake region:

$$W(\boldsymbol{u}) = w(X)W^{cs} + (1 - w(X))W^{fs} + W^h \tag{18.6}$$

where W^h denotes the energy due to the coupling of the two scales and $w(\mathbf{X})$ is the weighting function that is required to fulfill the following conditions:

$$w(X)=\begin{cases} 1 & \forall X\in\Omega^{cs}\setminus\Omega^{fs} \\ [0,1] & \forall X\in\Omega^{h} \\ 0 & \forall X\in\Omega^{fs}\setminus\Omega^{cs} \end{cases} \tag{18.7}$$

One can use a normalized distance function as the weighting function over the handshake region:

$$w(X)=\frac{l(X)}{l_0}. \tag{18.8}$$

As shown in Fig. 18.4, $l(\mathbf{X})$ is the orthogonal projection of any points with the coordinates of X on the coarse-scale interior boundary and l_0 is the length of this orthogonal projection to the fine-scale boundary [41].

Considering the energy contribution from the enforcement of compatibility between the fine and coarse displacements in the handshake domain, the variational formulation can be expressed: Find $\boldsymbol{u}^i\in\mathcal{U}\forall\delta\boldsymbol{u}^i\in\mathcal{U}_0\ (i=fs,cs)$ such that:

$$\delta W=w(X)\delta W^{cs}+(1-w(X))\delta W^{fs}+\delta W^{h}=0 \tag{18.9}$$

using approximation spaces of admissible trial u and test functions δu

$$\begin{aligned} \mathcal{U} &= \{\boldsymbol{u}^i\,|\,\boldsymbol{u}^i\in\mathcal{H}^1,\boldsymbol{u}^i=\bar{\boldsymbol{u}}^i \text{ on } \partial\Omega_u, \mathrm{u}^i \text{ discontinuous on } \partial\Omega_c\} \\ \mathcal{U}_0 &= \{\delta\boldsymbol{u}^i\,|\,\delta\boldsymbol{u}^i\in\mathcal{H}^1,\delta\boldsymbol{u}^i=0 \text{ on } \partial\Omega_u, \delta\boldsymbol{u}^i \text{ discontinuous on } \partial\Omega_c\} \end{aligned} \tag{18.10}$$

The variation of energy in coarse and fine scales can be written as:

$$\delta W^{cs}=\delta W^{cs}_{int}(\boldsymbol{u}^{cs},\boldsymbol{u}^{fs})-\delta W^{cs}_{ext}(\boldsymbol{u}^{cs})-\delta W^{cs}_{coh}(\boldsymbol{u}^{cs}) \tag{18.11}$$

$$\delta W^{fs}=\delta W^{fs}_{int}(\boldsymbol{u}^{cs},\boldsymbol{u}^{fs})-\delta W^{fs}_{ext}(\boldsymbol{u}^{fs})-\delta W^{fs}_{coh}(\boldsymbol{u}^{fs}) \tag{18.12}$$

with

$$\begin{aligned} \delta W^{cs}_{int}(\boldsymbol{u}^{cs},\boldsymbol{u}^{fs}) &= \int_{\Omega^{cs}}\sigma(\boldsymbol{u}^{cs},\boldsymbol{u}^{fs}):\mathrm{grad}^{\mathrm{sym}}(\delta\boldsymbol{u}^{cs})\,\mathrm{d}\Omega \\ \delta W^{cs}_{ext}(\boldsymbol{u}^{cs}) &= \int_{\Omega^{cs}}\boldsymbol{f}.\delta\boldsymbol{u}^{cs}\,\mathrm{d}\Omega+\int_{\partial\Omega^{cs}_t}\boldsymbol{t}.\delta\boldsymbol{u}^{cs}\,\mathrm{d}\partial\Omega \\ \delta W^{cs}_{coh}(\boldsymbol{u}^{cs}) &= \int_{\partial\Omega^{cs}_{disc}}\boldsymbol{t}^c.[\![\delta\boldsymbol{u}^{cs}]\!]\,\mathrm{d}\Gamma \end{aligned} \tag{18.13}$$

$$\begin{aligned} \delta W^{fs}_{int}(\boldsymbol{u}^{cs},\boldsymbol{u}^{fs}) &= \int_{\Omega^{fs}}\sigma(\boldsymbol{u}^{cs},\boldsymbol{u}^{fs}):\mathrm{grad}^{\mathrm{sym}}(\delta\boldsymbol{u}^{fs})\,\mathrm{d}\Omega \\ \delta W^{fs}_{ext}(\boldsymbol{u}^{fs}) &= \int_{\Omega^{fs}}\boldsymbol{f}.\delta\boldsymbol{u}^{fs}\,\mathrm{d}\Omega \\ \delta W^{fs}_{coh}(\boldsymbol{u}^{fs}) &= \int_{\partial\Omega^{fs}_{disc}}\boldsymbol{t}^c.[\![\delta\boldsymbol{u}^{fs}]\!]\,\mathrm{d}\Gamma \end{aligned} \tag{18.14}$$

where f denotes the body force, t are the traction forces, and t_c are the cohesive forces in case of cohesive cracks. The term δW^h refers to the variation of the energy due to the coupling of two scales. It is common to use Lagrange multiplies (λ) to couple coarse scales and fine scales in the case of Arlequin and BDM methods. Using Lagrange multiplier, the variation of the coupling energy can be written as:

$$\delta W^h = \int_{\Omega^h} \delta \lambda_I^T g + \int_{\Omega^h} \lambda^T \delta g \tag{18.15}$$

where g is the gap of the coarse and fine scales and is equal to $g = u^{cs} - u^{fs}$. The Lagrange multiplier field λ (and δλ) can be discretized as the displacement field.

Instead of the L_2 coupling method described earlier, the H^1 coupling method proposed in [43,44] can also be used :

$$\delta W^h = \int_{\Omega^h} \delta \lambda_I^T g + \int_{\Omega^h} \lambda^T \delta g + \int_{\Omega_h} l^2 (\delta \Lambda^T G + \Lambda \delta G)\, d\Omega \tag{18.16}$$

with $\Lambda = \nabla \otimes \lambda$ and $G = \nabla \otimes u^{cs} - \nabla \otimes u^{fs}$. They tested the H^1 coupling but the results did not improve compared to the L^2 coupling. Considering the higher effort, it seems advisable to take advantage of the L^2 coupling. It is also important to study the size of the bridging domain Ω^h.

18.4.1 Concurrent Multiscale Methods for Fracture in CNRPs

Cracks can initiate and propagate at different scales. There are different methods to model fracture in the atomistic and continuum domains. In case of continuum domains, the displacement discretization should be accounted for displacement jump over the crack faces. One can use the eXtended Finite Element Method (XFEM) [45] to model cracks on both fine and coarse scales. In this method, the displacement approximation can be decomposed into a continuous part u^{cont} and a discontinuous part $u^{discont}$ [46]:

$$u^h(X) = \underbrace{\sum_{I \in \mathcal{N}} N_I(X) u_I}_{u^{cont}} + \underbrace{\sum_{I \in \mathcal{N}_b} N_I(X) \bar{H}(f(X)) a_I}_{u^{discont}} \tag{18.17}$$

where $\mathcal{N}$ is the set of all nodes in the domain and $\mathcal{N}_b$ is the set of nodes that belong to all elements that are completely cut by the crack. The nodal parameters u_I and a_I, are usual and enriched degrees of freedom and H or $\bar{H}$ is the enrichment function or shifted enrichment function, respectively:

$$H(f(X)) = \begin{cases} 1 & \text{if } f(X) > 0 \\ 0 & \text{if } f(X) < 0 \end{cases} \tag{18.18}$$

with

$$f(X) = sign[n \cdot (X_I - X)] \min_{X_I \in \partial\Omega_c} \|X_I - X\| \tag{18.19}$$

and $\bar{H}(f(X)) = H(f(X)) - H(f(X_I))$ where n is the outward normal to the crack surface. The shifting avoids difficulties due to blending. The cracks usually cut through the entire element and hence, do not require a tip enrichment [47].

In case of fracture, the Lagrange multiplier field λ (and δλ) can be discretized as the displacement field:

$$\lambda^h(X) = \sum_{I\in\mathcal{N}} N_I(X)\Lambda_I + \sum_{I\in\mathcal{N}_b} N_I(X)\bar{H}(f_I(X))\Lambda_I \tag{18.20}$$

$$\delta\lambda^h(X) = \sum_{I\in\mathcal{N}} N_I(X)\delta\Lambda_I + \sum_{I\in\mathcal{N}_b} N_I(X)\bar{H}(f_I(X))\delta\Lambda_I \tag{18.21}$$

Gracie and Belytschko [48] noted in the context of coupling an atomistic model to a continuum model through the Arlequin method that by omitting the enriched part for the discretization of the Lagrange multiplier field, that is, the second term on the RHS of Eqs. (18.20) and (18.21), more accurate results are obtained (i.e., when comparing the results of the coupled model to results of full fine-scale simulations). They argue that the missing surface energy in the continuum coupling justifies a less accurate coupling.

There are several contributions on the fracture modeling using concurrent multiscale methods. Talebi et al. presented a computational library for multiscale modeling of material failure [40]. They extended their method to model 3D cracks and dislocations propagation [23]. They also presented a molecular dynamics/eXtended Finite Element Method coupling by a 3D extended bridging domain [26]. This method can be used easily for dynamic brittle fracture. There are also some extensions of concurrent multiscale method for coarse graining and adaptivity [14,15]. These methods can adaptively refine the fine scale while the crack grows and coarse grain the regions far from the crack tip.

18.5 Challenges and Concluding Remarks

The thermomechanical properties of CNRPs come from phenomena at lower scales. Therefore, it is very important to adopt efficient and appropriate multiscale techniques for modeling polymeric nanocomposites. These methods should account for all important mechanisms, such as defects, cracks, and voids. Different multiscale methods have been developed that can be divided into hierarchical, semiconcurrent, and concurrent multiscale methods.

It is believed that semiconcurrent multiscale methods are more suitable for nonlinear cases, as they account only for deformation states that really occur in a simulation. It is also more convenient to couple different packages specially for different scales using semiconcurrent methods. For example, semiconcurrent methods can use molecular dynamic packages at the fine scale and finite element packages at the coarse scale. Parallelization is also simple. However, the transfer of length scale between hierarchical multiscale methods are the most versatile and efficient tools for modeling CNRPs. These methods are mainly based on computational homogenization and are well developed for CNRPs. Nonetheless, the application of these methods for modeling damage and fracture is still challenging. The existence of disparate length scale and application of PBC when the cracks touches the boundaries of RVE should be investigated in more detail. Different scales are still challenging.

Concurrent multiscale methods are more suitable for modeling fracture phenomena in CNRPs. They can easily couple a molecular domain directly to a continuum one using BDM, though numerous challenges in dynamic applications remain: bridging the disparate time scales and avoiding spurious wave reflections are just two of them. Coarse graining and refinement are also applicable in these methods. To date, there is not a commercial package that can handle handshake coupling methods with cracks, refinement, and coarse-graining capabilities.

References

[1] B. Arash, H.S. Park, T. Rabczuk, Mechanical properties of carbon nanotube reinforced polymer nanocomposites: a coarse-grained model, Compos. B 80 (2015) 92–100.

[2] B. Arash, H.S. Park, T. Rabczuk, Tensile fracture behavior of short carbon nanotube reinforced polymer composites: a coarse-grained model, Compos. Struct. 134 (2015) 981–988.

[3] B. Arash, H.S. Park, T. Rabczuk, Coarse-grained model of the *j*-integral of carbon nanotube reinforced polymer composites, Carbon 96 (2016) 1084–1092.

[4] A.A. Mousavi, B. Arash, X. Zhuang, T. Rabczuk, A coarse-grained model for the elastic properties of cross linked short carbon nanotube/polymer composites, Compos. B 95 (2016) 404–411.

[5] N. Vu-Bac, P.M.A. Areias, T. Rabczuk, A multiscale multisurface constitutive model for the thermoplastic behavior of polyethylene, Polymer 105 (2016) 327–338.

[6] N. Vu-Bac, M.A. Bessa, T. Rabczuk, W.K. Liu, A multiscale model for the quasi-static thermo-plastic behavior of highly cross-linked glassy polymers, Macromolecules 48 (18) (2015) 6713–6723.

[7] T. Belytschko, J.H. Song, Coarse-graining of multiscale crack propagation, Int. J. Numer. Meth. Eng. 81 (5) (2010) 537–563.

[8] S. Nemat-Nasser, M. Hori, Micromechanics: Overall Properties of Heterogeneous Materials, Elsevier, Amsterdam, (1993).

[9] F. Feyel, J.-L. Chaboche, FE^2 multiscale approach for modeling the elastoviscoplastic behavior of long fiber SiC/Ti composite materials, Comp. Meth. Appl. Mech. Eng. 183 (2000) 309–330.

[10] V. Kouznetsova, Computational Homogenization for the Multi-Scale Analysis of Multi-Phase Materials, PhD Thesis, Netherlands Institute for Metals Research, The Netherlands, 2002.

[11] V.P. Nguyen, O. Lloberas-Valls, M. Stroeven, L.J. Sluys, Homogenization-based multiscale crack modelling: from micro diffusive damage to macro cracks, Comp. Meth. Appl. Mech. Eng. 200 (9–12) (2010) 1220–1236.

[12] T. Belytschko, S. Loehnert, J.-H. Song, Multiscale aggregating discontinuities: a method for circumventing loss of material stability, Int. J. Numer. Meth. Eng. 73 (2007) 869–894.

[13] E.B. Tadmor, M. Ortiz, R. Phillips, Quasicontinuum analysis of defects in solids, Philos. Mag. A 73 (6) (1996) 1529–1563.

[14] P.R. Budarapu, R. Gracie, S.P.A. Bordas, T. Rabczuk, An adaptive multiscale method for quasi-static crack growth, Comput. Mech. 53 (6) (2014) 1129–1148.

[15] P.R. Budarapu, R. Gracie, S.-W. Yang, X. Zhuang, T. Rabczuk, Efficient coarse graining in multiscale modeling of fracture, Theoret. Appl. Fract. Mech. 69 (2014) 126–143.

[16] H.S. Park, E. Karpov, P.A. Klein, W.K. Liu, The bridging scale for two-dimensional atomistic/continuum coupling, Philos. Mag. A 85 (1) (2005) 79–113.

[17] M. Silani, H. Talebi, D. Arnold, S. Ziaei-Rad, T. Rabczuk, On the coupling of a commercial finite element package with LAMMPS for multiscale modeling of materials, Steel Res. Int. (2012) 1371–1374.

[18] M. Silani, S. Ziaei-Rad, H. Talebi, T. Rabczuk, A semi-concurrent multiscale approach for modeling damage in nanocomposites, Theor. Appl. Fract. Mech. 74 (2014) 30–38.

[19] P. Areias, N. Vu-Bac, T. Rabczuk, A multisurface constitutive model for highly cross-linked polymers with yield data obtained from molecular dynamics simulations, Int. J. Mech. Mater. Des. 2016 (2016) 1–16.

[20] H.B. Dhia, The Arlequin method: a partition of models for concurrent multiscale analyses, Proceedings of the Challenges in Computational Mechanics Workshop, 2006 (10, p. 12).

[21] S. Tang, T.Y. Hou, W.K. Liu, A mathematical framework of the bridging scale method, Int. J. Numer. Meth. Eng. 65 (10) (2006) 1688–1713.

[22] S.P. Xiao, T. Belytschko, A bridging domain method for coupling continua with molecular dynamics, Comp. Meth. Appl. Mech. Eng. 193 (2004) 1645–1669.

[23] H. Talebi, M. Silani, T. Rabczuk, Concurrent multiscale modeling of three-dimensional crack and dislocation propagation, Adv. Eng. Softw. 80 (2015) 82–92.

[24] R.E. Miller, E.B. Tadmor, A unified framework and performance benchmark of fourteen multiscale atomistic/continuum coupling methods, Model Simul. Mater. Sc. 17 (2009) 053001.

[25] M. Silani, H. Talebi, S. Ziaei-Rad, A. Magid Hamouda, G. Zi, T. Rabczuk, A three dimensional extended Arlequin method for dynamic fracture, Comput. Mater. Sci. 96 (2015) 425–431.

[26] H. Talebi, M. Silani, S.P.A. Bordas, P. Kerfriden, T. Rabczuk, Molecular dynamics/XFEM coupling by a three-dimensional extended bridging domain with applications to dynamic brittle fracture, Int. J. Multiscale Comput. Eng. 11 (6) (2013) 527–541.

[27] M. Silani, H. Talebi, A.M. Hamouda, T. Rabczuk, Nonlocal damage modelling in clay/epoxy nanocomposites using a multiscale approach, J. Comput. Sci. 15 (2016) 18–23.

[28] S.-W. Yang, P.R. Budarapu, D.R. Mahapatra, S.P.A. Bordas, G. Zi, T. Rabczuk, A meshless adaptive multiscale method for fracture, Comput. Mater. Sci. 96 (2015) 382–395.

[29] K.M. Hamdia, M.A. Msekh, M. Silani, N. Vu-Bac, X. Zhuang, T. Nguyen-Thoi, T. Rabczuk, Uncertainty quantification of the fracture properties of polymeric nanocomposites based on phase field modeling, Compos. Struct. 133 (2015) 1177–1190.

[30] N. Vu-Bac, R. Rafiee, X. Zhuang, T. Lahmer, T. Rabczuk, Uncertainty quantification for multiscale modeling of polymer nanocomposites with correlated parameters, Compos. B 68 (2015) 446–464.

[31] N. Vu-Bac, T. Lahmer, X. Zhuang, T. Nguyen-Thoi, T. Rabczuk, A software framework for probabilistic sensitivity analysis for computationally expensive models, Adv. Eng. Softw. 100 (2016) 19–31.

[32] N. Vu-Bac, M. Silani, T. Lahmer, X. Zhuang, T. Rabczuk, A unified framework for stochastic predictions of mechanical properties of polymeric nanocomposites, Comput. Mater. Sci. 96 (2015) 520–535.

[33] A. Almasi, M. Silani, H. Talebi, T. Rabczuk, Stochastic analysis of the interphase effects on the mechanical properties of clay/epoxy nanocomposites, Compos. Struct. 133 (2015) 1302–1312.

[34] M. Silani, H. Talebi, S. Ziaei-Rad, P. Kerfriden, S.P.A. Bordas, Rabczuk S T., Stochastic modelling of clay/epoxy nanocomposites, Compos. Struct. 118 (2014) 241–249.

[35] H. Ghasemi, R. Brighenti, X. Zhuang, J. Muthu, T. Rabczuk, Optimization of fiber distribution in fiber reinforced composite by using NURBS functions, Comput. Mater. Sci. 83 (2014) 463–473.

[36] H. Ghasemi, R. Brighenti, X. Zhuang, J. Muthu, T. Rabczuk, Optimal fiber content and distribution in fiber-reinforced solids using a reliability and NURBS based sequential optimization approach, Struct. Multidiscip. Optim. 51 (1) (2015) 99–112.

[37] H. Ghasemi, R. Rafiee, X. Zhuang, J. Muthu, T. Rabczuk, Uncertainties propagation in metamodel-based probabilistic optimization of CNT/polymer composite structure using stochastic multi-scale modeling, Comput. Mater. Sci. 85 (2014) 295–305.

[38] Z. Hashin, Analysis of composite materials, J. Appl. Mech 50 (2) (1983) 481–505.

[39] R. Hill, The essential structure of constitutive laws for metal composites and polycrystals, J. Mech. Phys. Solids 15 (1967) 79–95.

[40] H. Talebi, M. Silani, S.P.A. Bordas, P. Kerfriden, T. Rabczuk, A computational library for multiscale modeling of material failure, Comput. Mech. 53 (5) (2014) 1047–1071.

[41] S.P. Xiao, T. Belytschko, A bridging domain method for coupling continua with molecular dynamics, Comp. Meth. Appl. Mech. Eng. 193 (17–20) (2004) 1645–1669.

[42] T. Belytschko, S.P. Xiao, Coupling methods for continuum model with molecular model, Int. J. Multisc. Comput. Eng. 1 (1) (2003).

[43] P.A. Guidault, T. Belytschko, On the l2 and the h1 couplings for an overlapping domain decomposition method using Lagrange multipliers, Int. J. Numer. Meth. Eng. 70 (3) (2007) 322–350.

[44] P.A. Guidault, T. Belytschko, Bridging domain methods for coupled atomistic-continuum models with l2 or h1 couplings, Int. J. Numer. Meth. Eng. 77 (11) (2009) 1566–1592.

[45] N. Moës, J. Dolbow, T. Belytschko, A finite element method for crack growth without remeshing, Int. J. Numer. Meth. Eng. 46 (1999) 131–150.

[46] T. Belytschko, T. Black, Elastic crack growth in finite elements with minimal remeshing, Int. J. Numer. Meth. Eng. 45 (5) (1999) 601–620.

[47] T. Rabczuk, G. Zi, A. Gerstenberger, W.A. Wall, A new crack tip element for the phantom node method with arbitrary cohesive cracks, Int. J. Numer. Meth. Eng. 75 (2008) 577–599.

[48] R. Gracie, T. Belytschko, Concurrently coupled atomistic and xfem models for dislocations and cracks, Int. J. Numer. Meth. Eng. 78 (3) (2009) 354–378.

19

Macroscopic Elastic Properties of Nonbonded Wavy Carbon Nanotube Composites

Saeed Herasati*, Liangchi Zhang**, Majid Elyasi†

*YASOUJ UNIVERSITY, YASOUJ, IRAN; **THE UNIVERSITY OF NEW SOUTH WALES, SYDNEY, NSW, AUSTRALIA; †BABOL NOSHIRVANI UNIVERSITY OF TECHNOLOGY, BABOL, IRAN

CHAPTER OUTLINE

19.1 Introduction

In recent years, many nanomaterials, such as graphene, nanofibers, nanoparticles, and carbon nanotubes (CNTs) have been used as the reinforcement in a wide range of nanocomposites. It is expected that only a small weight percentage of such nanoadditives can lead to a significant improvement in the mechanical properties of nanocomposites [1–3]. Among these nanomaterials, CNTs are the most popular due to their high strength (greater than steel), small density (lighter than aluminum), and high conductivity (greater than copper) [4]. The exceptional mechanical properties of CNTs have been shown by many experimental and theoretical studies [5]. Hence, CNTs have been considered as a promising reinforcing additive to make extraordinary engineering nanocomposites. The overall mechanical properties of CNT composites are affected by many factors, such as CNT waviness [6], dispersion [7], functionalization [8], van der Waals (vdW) interaction [9],

Carbon Nanotube–Reinforced Polymers. http://dx.doi.org/10.1016/B978-0-323-48221-9.00019-4

agglomeration [10], and CNT–matrix interphase [11]. The discussion in this chapter will focus on the effects of the CNT–matrix interphase and CNT waviness.

The CNT–matrix interphase is a transient region between the external surface of a CNT and the bulk matrix material of a CNT composite, which is usually called the interphase. In comparison to the small diameter of a CNT, the thickness of a CNT–matrix interphase is considerable. The interphase links the CNT to the matrix, and thus plays an important role in determining the macroscopic properties of a CNT composite [12,13]. There are two distinct interphase regions for a nonbonded CNT composite. One is the vdW gap between the external surface atoms of a CNT and the internal surface of the matrix surrounding the CNT [14]. The other is the adsorption layer, which often has a greater density than that of the bulk matrix material itself [15,16]. For convenience, this layer is often called a dense interphase. On the other hand, due to the small flexural rigidity of CNTs, most manufactured CNT composites include highly wavy CNTs. Hence, the interphase property and waviness can significantly influence the macroscopic properties of a CNT composite. However, due to the molecular-sized CNTs and CNT–matrix interphases, experimental investigations have been very difficult and costly. Theoretical characterization has thus a practical and useful method to explore the macroscopic properties of nanocomposites. Moreover, examination and analysis by transmission electron microscopy [17] and macroscopic mechanical testing [18], if not in situ, can only evaluate the performance of a CNT composite after its fabrication and testing. A theoretical analysis can be used to reveal the mechanisms of reinforcement or to provide a way to optimize the mechanical properties beforehand.

Over the past decades, various theoretical methods aiming to understand and predict the mechanical properties of CNT composites have been developed [19–21]. Similar to the classical methods used to investigate the microfiber-reinforced composites, a representative volume element (RVE) containing the matrix and the reinforcement can also be applied to the study of a CNT composite. However, due to the nanoscale dimensions of CNTs, the nanoscaled RVE (NRVE) cannot reflect all the influential factors, such as the waviness and agglomeration of CNTs. A multiscale approach is therefore necessary. In the literature, a variety of methods have been proposed. For example, the CNT composite RVE have been analyzed by continuum micromechanics [22], finite element (FE) method [23], atomistic simulations [15,24–26], and semicontinuum FE approach [27,28]. Nevertheless, it has been found that the continuum and semicontinuum methods are unable to accommodate the effects of the details of the geometries and properties of nanoinclusions and the interphase regions. For example, a semicontinuum FE approach treats the vdW interaction as spring elements to connect a continuum matrix to the CNT atoms and constructs the CNT interatomic bonds as beam elements. This kind of treatment is convenient as it can integrate all simulations in a single finite element code to study regular structures, such as multiwalled CNTs (MWCNT) [29], but the spring element that simply connects the internal surface of the matrix cannot easily adopt the fact that the vdW interaction is nonlocal and that each atom of the CNT interacts with several atoms of the matrix and the CNT within the cutting radius. Moreover, compared to the atomistic methods, the former is unable to predict the real equilibrated location of atoms of a CNT and its adjacent matrix.

The detailed reinforcing mechanisms of CNT composites have been determined frequently by atomistic methods. For example, utilizing the atomistic methods, the effects of the CNT–matrix interphase [24], the bonding and functionalization [30,31], and the stress–strain properties of nanoscale molecular mechanisms [26,32] have been studied. Although atomistic simulations can characterize the physical parameters of a nanocomposite RVE, due to the high computational cost, an atomistic model can only deal with a few nanometers in dimension and a timescale of a few nanoseconds. Therefore, atomistic models have mainly been used to investigate nanoscale issues around the nanoinclusions, such as in determining the geometry and mechanical properties of interphase regions of an RVE, which need less computational cost. A continuum analysis is still used for a greater length scale, though it lacks the capacity to reflect nanoscale effects as aforementioned.

This chapter will introduce an effective multiscale method to characterize the nanocomposites reinforced by CNTs and to explore the effect of the interphase and waviness on their overall elastic properties. The discussion in this chapter is arranged as follows. Section 19.2 will describe the general background for the discussion of the multiscale modeling of CNT composites including NRVE and microscale RVE (MRVE). Section 19.3 will discuss the simplifications of NRVE (three-phase model) including a discussion about the equivalent solid fiber (ESF), dense interphase, and bulk matrix. A case study on the effect of the interphase on the overall elastic properties will then be given in Section 19.4. Finally, Section 19.5 will provide and summarize key issues on the multiscale modeling of CNT composites.

19.2 Multiscale Modeling of CNT Composites

Although numerous traditional models based on continuum micromechanics have been utilized, they cannot address the detailed atomistic mechanisms between CNTs and their surrounding matrix. Therefore, appropriated atomistic simulation tools, such as classical molecular dynamics (MD) or molecular mechanics (MM) are necessary to resolve the nanoscopic problems. However, time and length limitations of atomistic simulations confine them to very small models to include a small portion of a CNT and the matrix adjacent to the CNT. Thus, an appropriate solution is multiscale modeling to bridge the nano-, meso-, and macroscales hierarchically. It is expected that a proper multiscale model will enable us to understand the properties of CNT composites at a desired level of approximation, and eventually lead to appropriate design of the materials by making full use of the strength of the individual phases in the composites.

A strategy of the multiscale modeling is illustrated in Fig. 19.1. The basic element includes a straight CNT embedded in a matrix [amorphous polyvinyl chloride (PVC) matrix in the present case]. Here, a nanoscale atomic representative volume element (A-NRVE) includes a CNT and sufficient quantity of adjacent matrix consisting of an interphase region and related nanoscopic regions. The mechanical properties of the NRVE will be extracted as explained in the following section. The A-NRVE has transversely isotropic

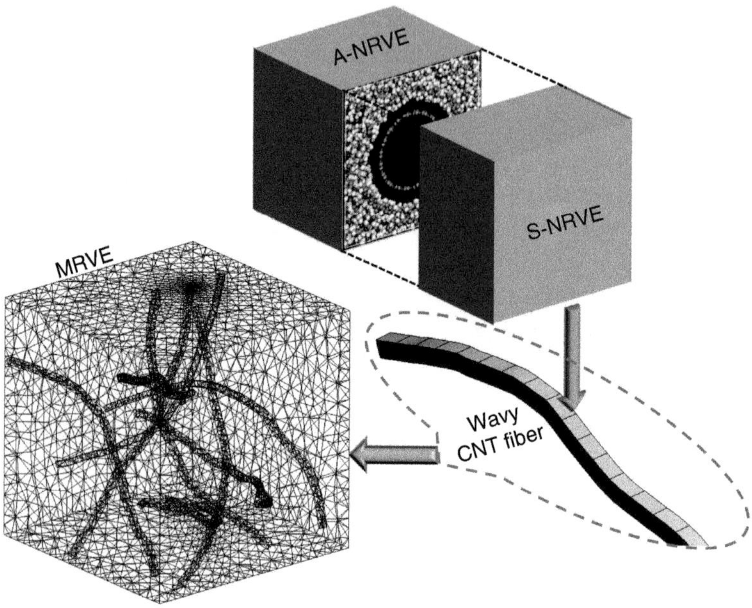

FIGURE 19.1 Schematic diagram of the multiscale modeling strategy for characterizing carbon nanotubes (CNT) composites from the atomistic scale nanoscale representative volume element (NRVE) to the microscale RVE (MRVE).

properties, as will be discussed in Section 19.2.1.2, and can be treated as a solid NRVE (S-NRVE) with the same mechanical properties of A-NRVE through numerical homogenization. This solid element, which has taken the key nanoscopic effects into account, can be used as a bridge element for FE modeling to significantly reduce the computational cost in comparison with an atomistic simulation.

A series of S-NRVEs can then be assembled to build up a randomly wavy CNT fiber, as illustrated in Fig. 19.1, which will be called a CNT fiber hereafter. Wavy CNT fibers are assembled in an MRVE with the surrounding PVC matrix. Such an MRVE is ready to be used for an efficient FE analysis but with the atomistic and interphase effects included.

19.2.1 Nanorepresentative Volume Element

In a CNT composite, the volume fraction of CNTs is always low. If the CNTs are dispersed well in the composite, it is rational to assume that each CNT is surrounded fully by the matrix, and that the distance between individual CNTs is much larger than the thickness of an individual CNT–matrix interphase. Hence, the contact between CNTs and that between CNT–matrix interphases do not need to be considered in the modeling. As a result, the A-NRVE under a periodic boundary condition can be big enough to avoid any artificial effect of interphase or CNT contacts.

19.2.1.1 Modeling

The mechanical properties of A-NRVE were extracted with the aid of MM. The initial model was equilibrated by means of MD. The mechanical properties were then extracted with the aid of MM. The atomistic simulation was conducted using the commercial package Material Studio Forcite modules [33]. The COMPASS force field, an optimized molecular potential, was used to model the atomic interactions. It is one of the most suitable force fields for investigating organic molecules, inorganic small molecules and polymers, and polymeric composites [15,25,29,34–37], originally developed and parameterized by Sun [38] using ab initio simulations and a wide range of experimental observations.

PVC reinforced by single-walled CNTs (SWCNT) is used as a typical polymer matrix CNT composite. Since PVC is an amorphous polymer, the atomistic RVE model contains an SWCNT at its center surrounded by random chains of PVC molecules. Two constituents of the nanocomposite, SWCNT and PVC, were first simulated separately to obtain their individual properties and to verify the COMPASS force field and the simulation approach. An armchair SWCNT with different diameters was then investigated. To avoid the effect of free surfaces, the periodic boundary conditions (PBCs) were enforced. Thus, in detail, four schemes of the atomistic simulation were generated:

- *Scheme 1*: With the aid of the Material Studio visualizer, a (10, 10) armchair SWCNT (Fig. 19.2A) was generated. The SWCNT is infinitely long and aligned with the *z*-axis. To do so, PBCs were enforced along all the three directions. Thus, a two-dimensional (2D) array of SWCNTs with infinite length was simulated. To avoid the side effect

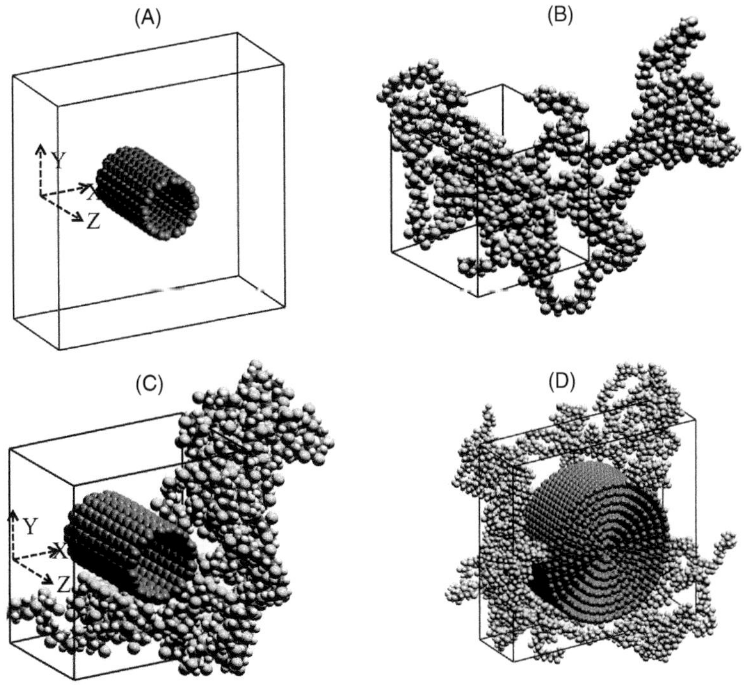

FIGURE 19.2 Four distinct atomistic schemes. (A) Scheme 1, a single single-walled carbon nanaotubes (SWCNT); (B) Scheme 2, the pure amorphous polyvinyl chloride (PVC); (C) Scheme 3, the SWCNT–PVC composite; and (D) Scheme 4, the multiwalled carbon nanaotubes (MWCNT)–PVC composite, where the dark gray balls represent carbon atoms, the big light gray ones indicate chlorine atoms, and the white small ones show hydrogen atoms [16].

from neighboring CNTs, the lateral size of the cell was made greater than the vdW cutting radius. Here, the initial dimensions of the simulation cell were taken as $100 \times 100 \times 24.6$ Å. To compare the longitudinal elastic constant of SWCNT with those published in the literature, a uniaxial strain was applied along the z-direction and the elastic constants were calculated.

- *Scheme 2*: The Accelyrs Amorphous Cell package, which is based on the method proposed by Theodorou and Suter [39] and Meirovitch [40], was used to generate a random-walk polymer chain. A cube of amorphous PVC with the chemical formula of $(CH_2{=}CHCl)_n$, where n is the degree of polymerization, was generated as a reference. The cell contains a long chain of PVC with $n = 400$ (Fig. 19.2B). Previous studies reported that shorter chains would show liquid-like behavior. The PVC model was initially generated with a low density of 0.2 g/cm^3 and then cooled down and densified in a stepwise manner with the aid of MD simulation. This process can help to avoid any artificial void in the PVC cube; otherwise, erroneous results for density and mechanical properties may appear. In this process, the cell dimensions were reduced by 2% for every relaxation of 50 ps at 650 K, until the density reached that of the bulk PVC at room temperature. Using MD simulation, under the NPT ensemble (fixed Number of atoms, Pressure and Temperature), the atomic system was then equilibrated for 1 ns at a temperature of 650 K and pressure of 1 atm. Finally, the atomic system was cooled down to room temperature with a step of 50 K for every relaxation of 0.5 ns under the same NPT ensemble. In the MD simulation, the time step was 1 fs and the Brendsen algorithm [41] was employed to keep the temperature constant with a decay constant of 0.005 ps, which showed as the minimum quantity before any dynamic instability in the atomistic system. Also, the cell pressure was controlled by the Parrinello and Rahman [42] algorithm, which is able to change the cell dimensions and angles independently to keep a homogeneous internal stress. Finally, the cell dimensions were averaged for the last 1 ns of the equilibrium process at room temperature and then equilibrated one more nanosecond under the NVT ensemble (fixed Number of atoms, Volume and Temperature). Averaging the cell dimensions gives a bulk PVC without any residual stress, which makes it ready for MM stress analysis.
- *Scheme 3*: This scheme includes a composite A-NRVE of an SWCNT surrounded by a PVC chain, as shown in Fig. 19.2C. To obtain a systematic result verifying the theoretical model, several A-NRVEs including a range of diameters of SWCNTs from 6.76 Å of the SWCNT (5, 5) to 26.18 Å of the SWCNT (20, 20) was used (a typical range for SWCNTs). Since the aspect ratio of a typical SWCNT is big, usually around 1000 or more, it is reasonable to assume that the SWCNT is infinitely long and can be represented by the PBCs in the z-direction. It is therefore rational to assume that the longitudinal strains in the SWCNT and the PVC are the same. The PVC in this scheme is a random-walk chain with $n = 400$. The equilibrium process is the same as that described for Scheme 2.
- *Scheme 4*: To study the effect of the internal walls of an MWCNT on the geometry of the interphase, an armchair CNT (5, 5)–(40, 40) of eight walls was embedded in a PVC matrix of four random-walk chain molecules with $n = 400$, as shown in Fig. 19.2D, following the same method explained in Scheme 3.

19.2.1.2 Stress Analysis and Elastic Constants

The elastic constants of A-NRVE can be obtained with the following procedure, which is the same as that in continuum mechanics, except the method for atomistic stress calculation. The generalized Hooke's law for the stress–strain relation can be written as

$$\sigma_{lm} = C_{lmnk}\varepsilon_{nk}, \tag{19.1}$$

where C_{lmnk} is the stiffness tensor and $\sigma_{lm}, \varepsilon_{nk}$ are the stress and strain tensors, respectively. Since the stress and strain tensors are symmetric, the Voigt notation (i.e., 1 = *xx*, 2 = *yy*, 3 = *zz*, 4 = *xz*, 5 = *yz*, 6 = *xy*) is convenient, which gives rise to

$$\sigma_i = C_{ij}\varepsilon_j \qquad i,j = 1:6, \tag{19.2}$$

where C_{ij} are the stiffness matrix components. Here, the static method of stress–strain analysis based on the work of Theodorou and Suter was implemented [39]. They proved that for glassy atactic polypropylene, the contribution of atomic configuration changes and the vibrational part of the Helmholtz free energy is negligible. With these conditions, the changes of the free energy can be replaced by minimum potential energy changes under small deformation as

$$C_{ij} = \frac{1}{V}\frac{\partial A}{\partial \varepsilon_{ij}}\bigg|_T = \frac{1}{V}\frac{\partial^2 U_{\min}^{\text{pot}}}{\partial \varepsilon_{ij}^2}\bigg|_T \tag{19.3}$$

in which A is the free Helmholtz energy and $U_{\min}^{\text{pot}}$ is the local minimum point of total potential energy. It is noted that this relation is valid only under very small deformation; otherwise, the contribution changes of the system configuration cannot be ignored. In reality, a numeric second-order derivation of Eq. (19.3) has numerical errors. Thus, it is preferable to use the linear stress–strain relation of Eq. (19.2). To evaluate the stiffness matrix by Eq. (19.2), a prescribed small strain was applied to the cell and the total potential energy based on the MM relations was minimized (i.e., $U_{\min}^{\text{pot}}$). The average *lm* component of stress was then calculated using the static part of the Virial equation [43], that is,

$$\sigma_{lm} = \frac{1}{\bar{V}}\sum_{\alpha}\sum_{\beta} F_l^{\alpha\beta} \cdot r_m^{\alpha\beta}, \tag{19.4}$$

where $\bar{V}$ is the total volume of the A-RVE, and $F_l^{\alpha\beta}$ and $r_m^{\alpha\beta}$ are, respectively, the interatomic force and distance between atoms α and β.

To calculate the stiffness matrix components, in an MM analysis, only one strain component (ε_1) was applied while the others were set to zero. Six stress components were then calculated. With the linear relation of stress–strain Eq. (19.2) six components of a single column of the stiffness matrix could then be determined. The process is repeated with the remaining five components of strain to evaluate all components of the stiffness matrix. In each cycle of the process, the strain of 0.2% was applied stepwise with an increment of 0.02%, during which the potential energy of the system was minimized. The stresses were then calculated using Eq. (19.4). More details can be found in Ref. [16].

For the stress analysis of SWCNT (Scheme 1), only a uniaxial tensile test in the z-direction was performed. For the SWCNT (10, 10), the longitudinal Young's modulus is 0.98 TPa if the CNT wall thickness was set as 3.4 Å [44–47]. It should be noted that Wang and Zhang [48] thoroughly studied the different choices of CNT thickness in literatures [48–50] and concluded that the most appropriate CNT thickness is 6.17 Å [49]. However, for the sake of comparison with relevant works, here the CNT wall thickness was assumed to be 3.4 Å. The Poisson's ratio obtained is 0.28, which is the same as that obtained by Lu [47] from empirical potentials and is close to the result from tight-binding methods [51].

For the equilibrated pure PVC, the final density obtained was 1.42 g/cm^3, which is within the range of those in the literature PVC density of 1.3–1.45 g/cm^3 [52]. The Young's moduli in three directions thus obtained are E_{xx} = 3.58 GPa, E_{yy} = 3.54 GPa, and E_{zz} = 3.62 GPa, which indicate the isotropy owing to the randomly distributed polymer chains. The average Young's modulus and Poisson's ratio are therefore 3.55 GPa and 0.34, respectively, which are in good agreement with the experimental results available [53]. These results indicate that the COMPASS force field and the MM simulations with the relaxed atomic models can render good results for isolated PVC and CNT and literally can be used for analyzing the PVC–SWCNT composite.

The resulting stiffness matrix of the composite exhibits an anisotropic behavior, which is expected because of the engagement of the SWCNT. However, the numerical results indicate that the stiffness matrix of the composite, C, is approximately transversely isotropic [16] and hence the number of elastic constants can be reduced to five.

19.2.2 Microscale Representative Volume Element

At the second stage of the multiscale modeling, it is necessary to assemble the wavy CNTs in an MRVE. As the CNTs are at a nanoscale, the MRVE includes numerous CNTs and statistically it is big enough to be a representative of a macroscale nanocomposite.

The micrograph images of the CNT composites show that manufactured CNTs are highly wavy. Thus, it is necessary to find a way to generate the most appropriate shape to represent the CNT waviness. In the literature, some models have been developed to study the effect of waviness. Some of them treated a CNT as a uniform fiber with a sinusoidal wavy shape [54,55], or simplified them to be of a bow shape [56,57]. Obviously, such assumptions are inconsistent with the actual geometries of wavy CNTs in a CNT composite [58,59].

Extensive investigations have been carried out to characterize the waviness morphology and to understand the waviness effect of the CNTs and microfibers on the elastic properties of composites, both experimentally [60] and analytically, with the aid of micromechanics [56,61]; numerically with the FE analysis [62]; or a combination of the FE–micromechanics methods [6,63]. Most of the micromechanical investigations on the waviness of CNTs have been based on the Mori–Tanaka method [64]. Since this method was developed for straight inclusions, some attempts have been made to modify the Mori–Tanaka method to accommodate the CNT waviness. Obtaining the concentration tensor for the inclusions of different geometries is the most challenging issue of this method. For example, Bradshaw et al. [63] used the FE technique to determine the dilute stress concentration tensor of a wavy CNT in an RVE. This concentration tensor was then used in the Mori–Tanaka method to describe the behavior of the wavy CNTs. It was found that the modified Mori–Tanaka method gave acceptable results only for microscale inclusions of a simple geometry. It is therefore necessary to develop a proper characterization method that can be integrated with the finite element technique [16] for more reliable results. To this end, Clarke et al. [60] used some 2D cross-sectional images of fibers in a composite to figure out the three-dimensional (3D) structure of the microfibers. Bhuiyan et al. [65], using a similar technique of image analysis, characterized the probability distribution functions of the diameter, orientation, dispersion, and waviness of the CNTs. Misiego and Pipes [66] tried to extract the morphologies of CNTs by examining the CNT network in the subsurface of a composite up to the depth of approximately 250 nm with the aid of the voltage-contrast scanning electron microscopy.

Based on the work of Herasati and Zhang [67], we will introduce a method to characterize the waviness of CNTs, which can be used to construct the wavy CNT composites and to study the effects of waviness and alignment of CNTs on the elastic properties of their composites. The CNT waviness is characterized with the aid of micrograph images of CNTs in composites. An algorithm will be developed to construct an RVE containing a wavy CNT. Then an MRVE model will be established to carry out the engineering-scale (i.e., macroscopic scale) analysis of the elastic properties of CNT composites.

19.2.2.1 Modeling the Waviness of CNTs

The NRVE characterized in Section 19.2.1 is the same as a cut segment of a squared cross-sectional fiber surrounding the CNT and adjacent matrix. Thus, each wavy CNT is generated as a cord including a sequence of 3D small straight segments as shown schematically in Fig. 19.3. The total number of straight segments (N_s) is a function of the CNT volume fraction and the external diameter of CNTs. For each wavy CNT, a set of points (P_i) on their neutral line is generated. The initial point (P_0) of the first segment is selected randomly on one of the boundary planes of the cubic MRVE. The direction of the first segment is selected randomly. Each new point is generated on the base of a right circular cone. The apex of the cone is located at the end point of the last segment (P_{i-1}) and the axis of the cone is coincident with the orientation of that segment. The length of the generatrix line is equal to the walking step (d_s). The deflection angle (θ_i) and the rotational angle (ψ_i) are assigned randomly. To avoid close contact of the CNTs, the distance of each new generated point from the previously generated points is calculated to ensure that it is more than the minimum predefined distance (d_{min}). Otherwise, the algorithm will keep searching for a path throughout the chains until such condition is reached. If this path cannot be identified in certain iterations (J_{Cmax}), the current chain is discarded and a new chain will be established. To reduce the shear lag effects, a minimum chain length is defined (L_{Cmin}) in the algorithm as a constraint.

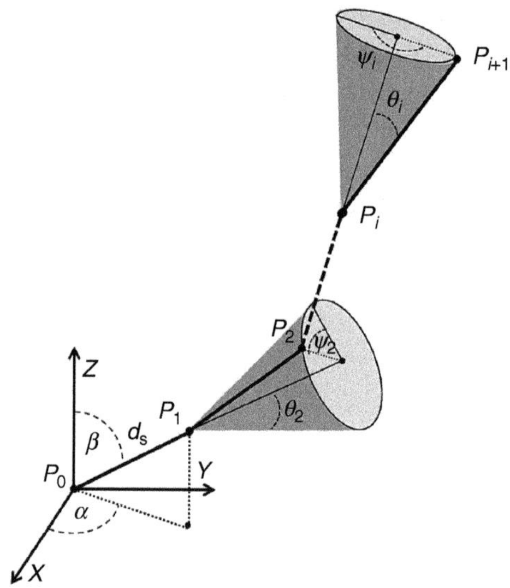

FIGURE 19.3 Schematic diagram of the random walking chains [67].

The detailed algorithm is described in more detail in Ref. [67]. The algorithm was implemented using a MATLAB code to generate the end points of the segments of a wavy CNT. This provides the input data for a standard FE code, such as ANSYS, to make the model and perform the required stress analysis.

The deflection angle (θ_i) is the most important parameter of the algorithm, which needs a deeper study to illuminate its nature and distribution. The best sample for the characterization of θ_i is a thin film sample with randomly oriented in-plane CNTs in which the segments have no out-of-plane component, because in this case the difference of the slope angle of a CNT on the available 2D image gives the deflection angle of the *i*th segment. The experimental data of Qian et al. [58] have been utilized as a reference, in which they dispersed MWCNTs homogeneously through a polystyrene matrix with a weight percentage of 1% (equivalent to a volume fraction of 0.487%). They made thin film samples with randomly oriented in-plane CNTs, which is appropriate as a lamina with in-plane (2D) distribution of the CNTs. A transmission electron microscopy (TEM) image of the sample of such a thin film is shown in Fig. 19.4A.

To analyze the waviness in a more consistent manner, the CNTs with approximately the same diameter of 30 nm were chosen, which is equal to the average diameter of the CNTs of the nanocomposite sample used in the experiment of Qian et al. [58]. The end points of segments (P_i) have been marked and the coordinate of each point was collected with the aid of the Plot Digitizer software as shown in Fig. 19.4A. The investigations showed that the short walking distance (d_s) can increase the acquisition error, while the long walking distance cannot capture sharp curvatures [67]. It was found that the walking distance equal to the CNT diameter is an appropriated quantity. Therefore, $d_s = 30$ nm was selected to characterize the waviness of CNTs. Since the CNTs are assumed to have a 2D (in-plane) distribution,

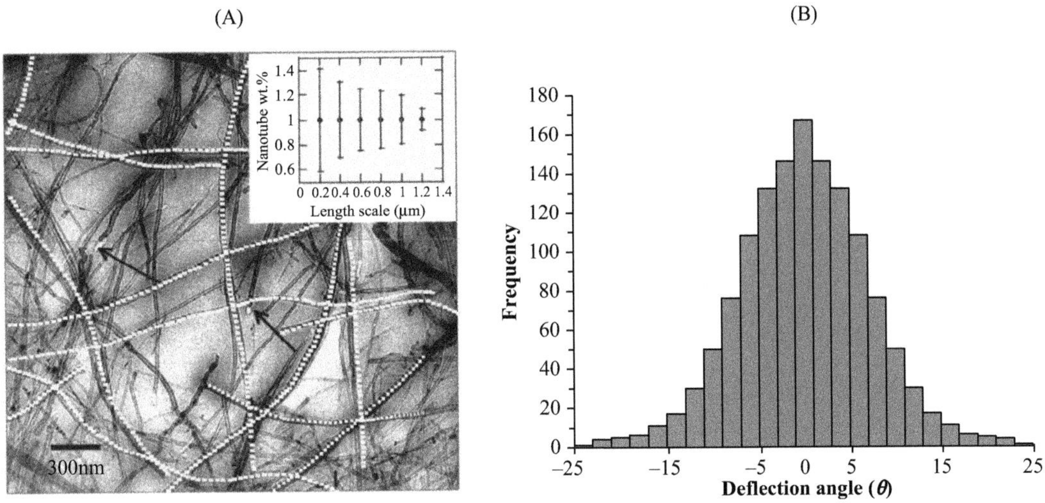

FIGURE 19.4 (A) TEM image of the sample of thin film for characterizing the waviness of the CNTs with a diameter of 30 nm, and (B) histogram of the deflection angle from the scanned data of the TEM image [67].

the deflection angle of each segment (θ_i) is equal to the angle between the orientation of two sequential segments. Using the coordinates of P_i, the deflection angles of all the segments were analyzed with a MATLAB code. Because the rotational angle (ψ) around the cone axis is in the range of (0, 2π), the sign of the deflection angle (θ) is not important. Thus the absolute value of θ for each segment can be calculated and mirrored to the negative side, giving rise to the symmetric histogram shown in Fig. 19.4B. The histogram demonstrates that the deflection angle (θ) has a normal Gaussian (normal) distribution; thus in the algorithm for the generation of random wavy CNTs, the deflection angles (θ_i) are assigned randomly with a Gaussian distribution. Fig. 19.4B shows that the maximum deflection angle or waviness angle of the CNTs in the examined sample composite is $\theta_{max} = 25$ degrees. Fig. 19.3B is based on 740 data points of Fig. 19.4A. An error analysis of this sample showed that 300 points can provide sufficiently accurate results, indicating that an analysis on a few CNTs would provide an acceptable representation of the CNT waviness in a composite.

To perform a stress analysis, the RVE of the composite can now be generated using the values of θ_{max} obtained earlier. To understand the effects of waviness and alignment of CNTs, it is necessary to construct different CNT distribution schemes of aligned (one-dimensional or 1D schemes), in-plane (2D schemes), and spatial (3D schemes) with a variety of waviness angles, as shown in Fig. 19.5.

To construct 1D schemes, for each CNT, a local narrow cube is defined, which is constrained in the x and y directions and has the same length of RVE in the z-direction. The lateral dimensions of the narrow cube ($\Delta x = \Delta y$), depend on the average maximum amplitude of the aligned CNTs, which could be measured from the micrograph images [68]. In the present study $\Delta x = \Delta y = 200$ nm were used, because they are close to the maximum amplitude of the wavy CNTs in the experimental characterization of Qian et al. [58].

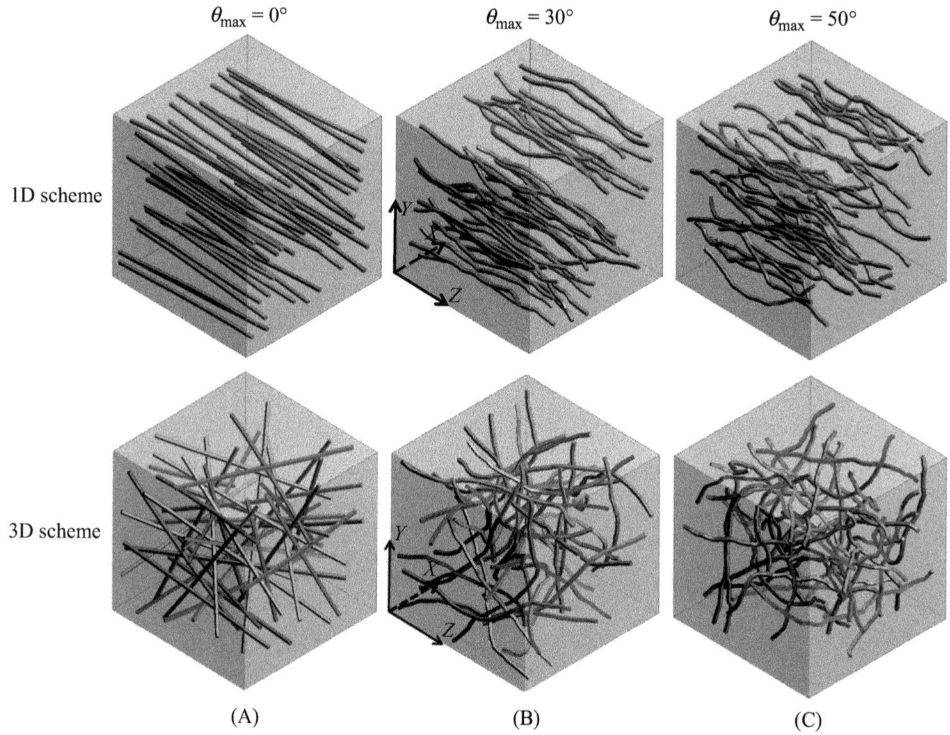

FIGURE 19.5 Sample composite schemes with a CNT volume fraction of 0.5% and waviness angles for 1D schemes and 3D schemes. (A) $\theta_{max} = 0$ degrees, (B) $\theta_{max} = 30$ degrees, and (C) $\theta_{max} = 50$ degrees [67].

19.2.2.2 Finite Element Analysis of MRVE

In this section, commercially available FE code, ANSYS, is utilized to carry out an efficient 3D stress analysis, which allows a convenient characterization of the macroscopic elastic properties of a CNT composite. The ANSYS Workbench was utilized to generate the initial FE models including mesh generation. The models were then transferred to ANSYS APDL to gain the script file for the stress analysis. In the present case study, the MWCNTs had a diameter of 30 nm; thus the effect of the CNT–matrix interphase on the macroscopic Young's modulus of the CNT composite could be ignored without causing any noticeable error according to Wang et al. [69].

For the simplicity of the FE meshing but without losing generality, it is assumed that a CNT can be regarded as a solid fiber of a square cross-section with the same volume fraction, which is called a CNT fiber hereafter for convenience.

The mechanical properties and the cross-sectional area of the CNT fiber is equivalent to that of MWCNTs. The wavy CNT fibers could then be easily meshed using the 20-node brick element (Solid 186) in ANSYS, as shown in Fig. 19.6A. The surrounding matrix, however, was meshed with the 10-node tetrahedron elements (Solid 187), as demonstrated in

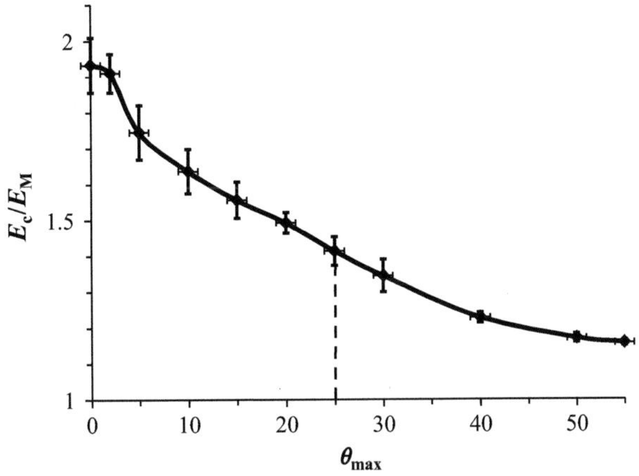

FIGURE 19.6 Relative Young's moduli of the composite to the matrix (E_c/E_M) with different waviness angles (θ_{max}) for the 2D scheme and a volume fraction of 0.5%.

Fig. 19.6B. Each CNT fiber could be meshed as a sequence of brick elements. In the majority of the previous studies in the literature, CNTs have been treated as an isotropic material [70,71]. Nevertheless, CNTs have a hollow structure and overall, as an additive material in a nanocomposite, shows a transversely isotropic mechanical property [72,73] as we do in this study. Thus, the coordinate system of a CNT fiber must follow the direction of a CNT in each point. A script file was developed to adjust the principal axis of the local coordinate system of each element in parallel to the neutral axis of the CNT fiber.

The same procedure explained in Section 19.2.1.2 for NRVEs was taken into account to determine the stiffness matrix of the MRVE, except for the stress analysis in which the total normal force of each face was calculated from the nodes involved. The average normal stresses were then calculated by dividing the total normal reaction force to the related area. The elastic moduli were subsequently determined from the stiffness matrix [16].

For the FE stress analysis, the elastic properties of the blank polystyrene matrix from the work by Qian et al. [58], that is, E_m = 1190 MPa with the Poisson's ratio of ν_m = 0.38 were used. Assuming that the length of the nanotube is bigger than 1 μm, the transversely isotropic properties of an MWCNT of diameter 30 nm was obtained with the method proposed by Shen and Li [72], which gave rise to the elastic properties of E_{zz} = 1.1 TPa, $E_{xx} = E_{yy}$ = 14 GPa, $G_{zx} = G_{zy}$ = 190 GPa, $\nu_{zx} = \nu_{zx}$ = 0.13, and ν_{xy} = 0.91.

19.2.2.3 Validation of Method

To validate the developed method, for MRVEs including wavy CNTs, in-plane distribution of CNTs with a volume fraction of 0.5 were made, which is the same as the experimental work of Qian et al. [58]. Although the waviness angle of the experimental work was quantified as 25 degrees, for a more comprehensive discussion, several models with the waviness angles between 0 and 55 degrees were made and analyzed with the procedure explained earlier. The in-plane Young's modulus ($E_{xx} = E_{yy}$) is then calculated and normalized to that of the matrix as shown in Fig. 19.6. The graph shows a relative Young's modulus of 1.41 at the waviness angle of 25 degrees as shown on the dashed line, which is remarkably the same as the results of the experimental work of Qian et al. [58]. The graph also shows that increasing the waviness angle can reduce the Young's modulus dramatically.

19.3 Simplifications on NRVE (Three-Phase Model)

In Section 19.2.1, the NRVE has been characterized with the aid of MD and MM simulations. The atomistic simulation is a time-consuming process. Thus, probably a better solution is a continuum-based model to evaluate the transversely isotropic elastic properties of the NRVE, which takes into account all the atomistic interactions. To do so, a priority is to identify probable distinct regions of the NRVE and to characterize them.

Different regions have been addressed in the literature. For example, Shokrieh and Rafiee [28] proposed an equivalent solid fiber (ESF) model in which a CNT and the equilibrium distance between the CNT wall and inner surface of the matrix, denoted as the "vdW gap" interphase, were considered to be a solid cylinder. In some works, the vdW gap interphase was assumed as the only interphase region [28,74,75] with certain thickness and mechanical properties. However, considering that the interphase matrix has a significant effect on the mechanical properties of nanocomposites [13,69], these models are insufficient for resolving all the details. Most of the works aforementioned revealed that the density of the interphase matrix of the CNT–polymer composites is bigger than the density of the bulk matrix, which can be considered as a criterion for characterizing this region.

For the SWCNT–PVC composite subjected to the present investigation, one can immediately identify the four regions, that is, the SWCNT, the vdW gap interphase, the interphase PVC matrix region, and the bulk PVC matrix, which play different roles. We consider the volume inside the PVC as a single phase, which contains the SWCNT under the vdW forces and is the same as the ESF discussed earlier. This volume will be treated as a solid fiber in the continuum model and thus named the equivalent solid fiber. Adjacent to the ESF, the PVC layer of a certain thickness is named the dense interphase. The outmost region is the PVC matrix with the identical mechanical properties as characterized using the model of pure PVC (Scheme 2). With these considerations, the three-phase model is illustrated in Fig. 19.7. These regions are characterized in the following subsections.

19.3.1 Phase 1: Equivalent Solid Fiber (ESF)

A nonbonded SWCNT surrounded by PVC matrix can be treated as a solid fiber. However, due to the vdW interaction between the CNT and the matrix, its mechanical properties may show a big discrepancy with the results of the free-standing one. In other studies, the weak interface has been approximated by a low-stiffness layer [15] or nonlinear springs [28]. However, there is a nonlocal effect of the vdW interaction wherein each CNT atom interacts with several atoms of the adjacent PVC and the CNT ones as schematically shown in

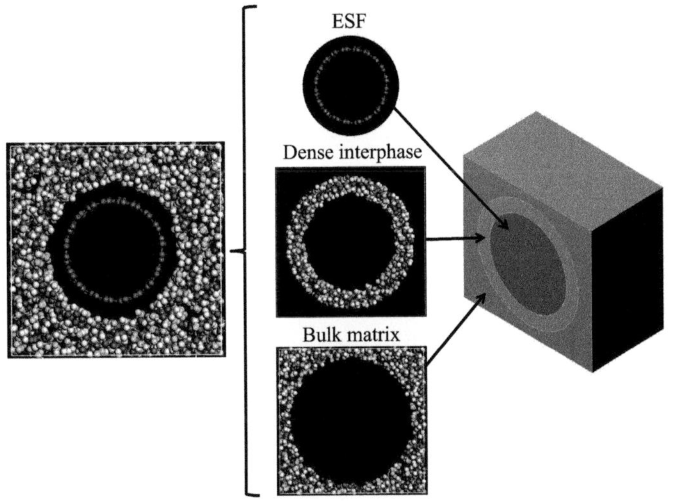

FIGURE 19.7 Three-phase RVE model: Phase 1, ESF; Phase 2, dense interphase; and Phase 3, bulk matrix [16].

Fig. 19.8A, which is very difficult to consider by the continuum models explained earlier. A more rational solution is to characterize the ESF inside the composite directly. Since the effect of the weak interface between the SWCNT and the matrix has been involved in the characterization of the ESF, it can now be assumed with a perfect bonding in a continuum model.

To characterize the ESF, the same procedure as described in Section 19.2.1.2 is carried out. However, six small strains should now be applied on the boundaries of the ESF. To do so, a prescribed strain field is first applied to every atom in the composite, as shown in Fig. 19.8. The PVC chain is then kept fixed while the SWCNT atoms are allowed to rearrange themselves to minimize the total potential energy. This means that a constant strain is applied only on the ESF boundary through the matrix. The resultant stress components of the SWCNT are calculated based on Eq. (19.4), in which $\overline{V}$ represents the volume of the ESF not the entire RVE volume. Besides the method explained earlier, a simplified CNT

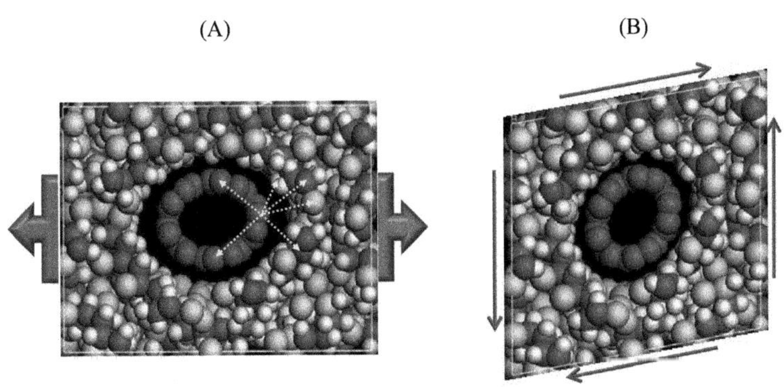

FIGURE 19.8 Sample strain state to characterize ESF. (A) Tensile, and (B) pure shear strain [16].

cluster approach has also been developed by Herasati et al. [16], which can help to characterize ESF with a simple method and much less computational cost.

19.3.2 Phase 2: Dense Interphase

Although there is no clear boundary between the dense interphase region and the bulk matrix, the variation of density can be used as a reasonable criterion to determine the thickness of this region as used in some works [13]. As a criterion, the dense interphase region ends when the local density of the material in this region reaches that of the bulk matrix. In this work, the local density at different radius R is determined by dividing the atomic mass of a cylindrical shell of thickness $t_c = 0.3$ Å with the corresponding volume as shown in Fig. 19.9A. When such a density is scanned radially from the center outward, the inner boundary of the matrix of radius R_0 (the radius of ESF) is determined when the PVC density becomes nonzero. The thickness of the vdW gap is then $h = R_0 - R_{CNT}$. The density variation of the PVC is shown in Fig. 19.5B, where the horizontal axis is $T_m = R - R_0$.

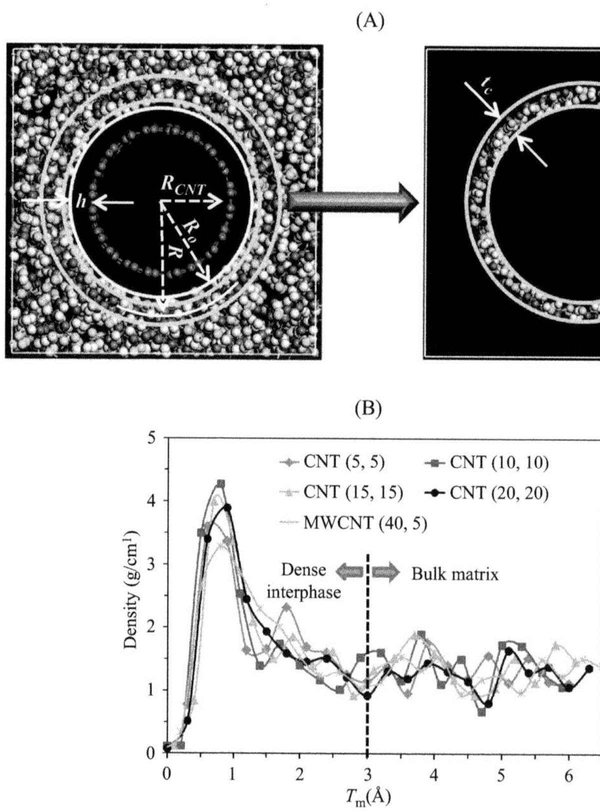

FIGURE 19.9 (A) Concentric cylinders to explore the local density of the matrix, and (B) the variation of the local density in the matrix with the radial thickness of the matrix [16].

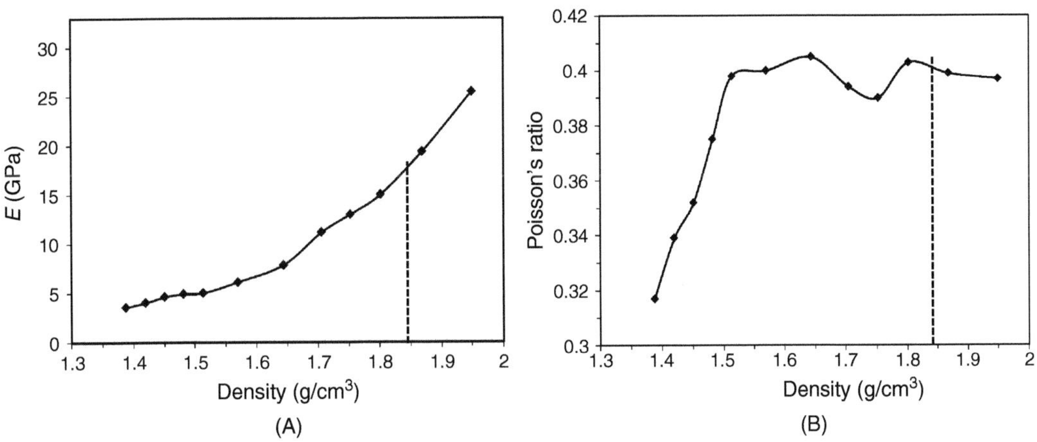

FIGURE 19.10 Elastic constants of PVC against density. (A) Young's modulus, and (B) Poisson's ratio [16].

Fig. 19.5B shows that the variations of the local density for MWCNTs and different diameters of SWCNTs are almost similar showing that the diameter of a CNT and the number of walls does not alter the interphase thickness. The hump of the density variation ends at around $T_m = 1.5$ Å. The density then fluctuates and reduces slowly until about $T_m = 3$ Å. Beyond this, the average density approaches the bulk PVC density of 1.42 g/cm^3. This result indicates that the thickness of the PVC interphase can be treated as a constant for different CNTs.

Since density seems to be the only variable parameter of the matrix in the interphase region, in this work, it is used as the only parameter to determine the mechanical properties of this dense PVC region. To do so, the cell dimensions of the PVC model (Scheme 2) are reduced stepwise and equilibrated by MD simulation for 0.1 ns at each step. The mechanical properties of PVC at different densities are then calculated by using the method described in Section 2.2. The variation of the Young's modulus and Poisson's ratio of the PVC with the density are shown in Fig. 19.10A and B, respectively. The average density of the dense interphase region of Fig. 19.9 is quantified as 1.84 g/cm^3. Thus, the Young's modulus and Poisson's ratio obtained from Fig. 19.10 are 17 GPa and 0.4, respectively.

19.3.3 Phase 3: Bulk Matrix

The bulk matrix region surrounds the dense interphase and has the same density and mechanical properties as the pure PVC. The mechanical properties can be obtained from literature [52] or by atomistic simulations. In this work, the atomistic results for pure PVC were used.

19.4 A Case Study on the Effect of Interphase

Although there is no doubt about the interphase regions of CNT composites, their effects on the overall elastic properties of nanocomposites need to be considered. Thus, in this section, the three-phase model established in Section 19.3 is utilized to characterize

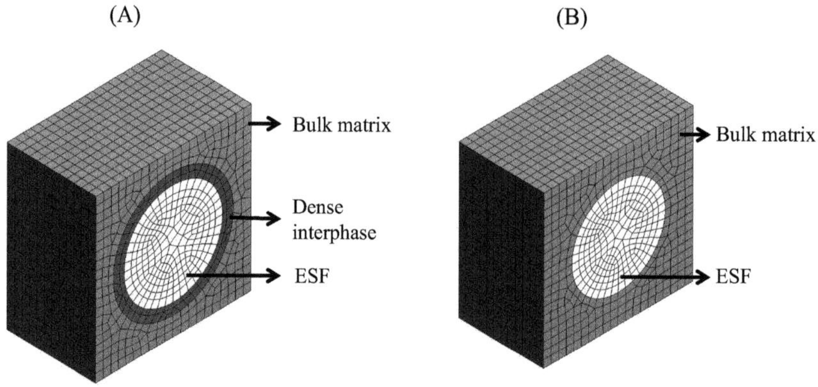

FIGURE 19.11 Two FE models. (A) Three-phase model, and (B) two-phase model.

S-NRVEs with and without interphase regions as shown in Fig. 19.11. A series of S-NRVEs can then be assembled to build randomly wavy SWCNT fibers and to make MRVEs ready for stress–strain analysis by FE as explained before.

To explore the influence of the dense interphase, a two-phase model was generated in which the dense interphase layer was eliminated as shown in Fig. 19.11B. Furthermore, to explore the influence of the vdW gap interphase, the three-phase and two-phase models are reexamined by substituting the ESF with the equivalent free-standing SWCNTs. The elastic properties of S-NRVE were then extracted with the method explained in Section 19.2.2.2.

In this study, several configurations of solid fibers with a variety of waviness angles and diameters with a fixed volume fraction of 0.35% have been generated and assembled in MRVEs. The mechanical properties of each MRVE with the following distinct interphase cases are obtained: (A) all phases included, (B) without any vdW gap, (C) no dense interphase, and (D) both interphases are ignored. To study the effect of waviness and diameter of CNTs, a 3D scheme of SWCNT (10, 10) with a variety of waviness angles are initially considered, then with a fixed waviness angle (i.e., $\theta_{max} = 30$ degrees here). A variety of diameters from SWCNT (5, 5) to (15, 15) are modeled and characterized.

Fig. 19.12A shows the variation of the relative Young's modulus (E_c/E_M) of the CNT composite using the 3D scheme of SWCNT (10, 10). The results show that ignoring the effect of vdW gap will overestimate the Young's modulus and that ignoring the dense interphase will underestimate it. When the vdW gap interphase and dense interphases are ignored, the results approach those including all phases. This indicates that with SWCNT (10, 10), the overall influence of the interphases as a whole is insignificant. Fig. 19.12B shows that SWCNT diameter has a big influence on the composite properties. When the diameter is greater than SWCNT (10, 10), the interphase effect becomes greater. A closer look at Fig. 19.12B shows that the upper and lower bonds, which are cases (B) and (C), respectively, approach each other with increasing SWCNT diameter. They should converge to case (A), after a big enough diameter of CNTs. This is consistent with the findings of Wang et al. [69] and Odegard et al. [13].

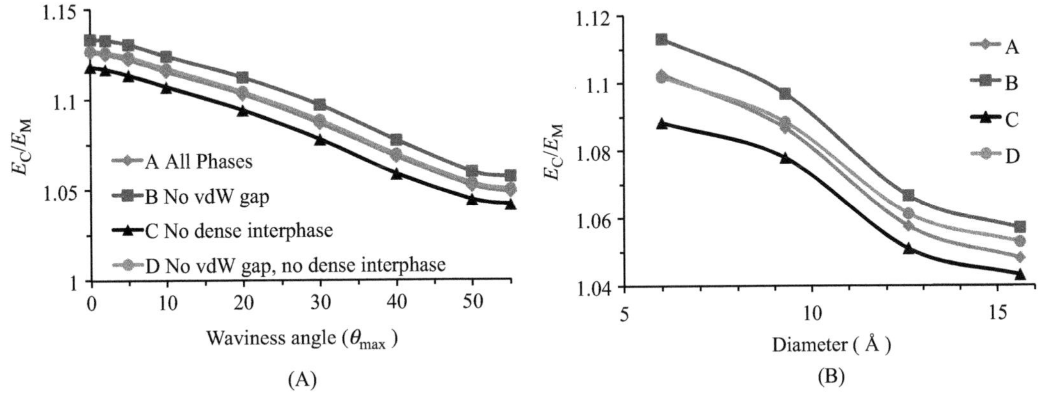

FIGURE 19.12 Relative Young's modulus of the CNT composite with 3D schemes. (A) SWCNT (10, 10) and different waviness angles, and (B) with a variety of diameters from SWCNT (5, 5) to (15, 15) under a given $\theta_{max} = 30$ degrees [76].

19.5 Conclusions

This chapter has presented a comprehensive multiscale technique for the characterization of the elastic moduli of nonbonded CNT composites. The effects of SWCNT–matrix interphases and SWCNT waviness on the macroscopic elastic properties of SWCNT composites had been studied. The RVE of the nanocomposite system has been systematically modeled and characterized by atomistic simulation. A method to characterize and generate wavy CNTs has also been introduced. A series of NRVEs has been assembled to build randomly wavy CNT fibers surrounded by polymer matrix in MRVEs. The method has been verified by the experimental results. To reduce the calculation cost of the atomistic simulation of NRVEs, a continuum-based three-phase model has been developed, which is capable of mimicking the CNT under vdW forces, the dense interphase PVC with properly determined thickness and density, and the bulk matrix. The effect of the interphase on the overall elastic properties of nanocomposites has also been studied, addressing the influential factors, such as the vdW gap interphase, the dense interphase, the waviness of SWCNTs, and the nanotube diameter. The study in this chapter has concluded that the average density of the interphase can be used as a parameter to determine the mechanical properties of the dense PVC region, and that the CNT diameter and number of walls have a negligible effect on the equilibrium thickness of the vdW gap and on the density, thickness, and properties of the dense interphase. The results indicate that ignoring either the vdW interaction region or the interphase matrix layer can bring about ambiguous results and that in some circumstances the overall influence of the interphases as a whole is insignificant.

References

[1] F.H. Gojny, et al., Influence of different carbon nanotubes on the mechanical properties of epoxy matrix composites—a comparative study, Compos. Sci. Technol. 65 (15–16) (2005) 2300–2313.

[2] J.N. Coleman, et al., Small but strong: a review of the mechanical properties of carbon nanotube–polymer composites, Carbon 44 (9) (2006) 1624–1652.

[3] L.C. Zhang, I. Zarudi, K.Q. Xiao, Novel behaviour of friction and wear of epoxy composites reinforced by carbon nanotubes, Wear 261 (7–8) (2006) 806–811.

[4] M. Moniruzzaman, K.I. Winey, Polymer nanocomposites containing carbon nanotubes, Macromolecules 39 (16) (2006) 5194–5205.

[5] M.M. Shokrieh, R. Rafiee, A review of the mechanical properties of isolated carbon nanotubes and carbon nanotube composites, Mech. Compos. Mater. 46 (2) (2010) 155–172.

[6] F.T. Fisher, R.D. Bradshaw, L.C. Brinson, Fiber waviness in nanotube-reinforced polymer composites-I: modulus predictions using effective nanotube properties, Compos. Sci. Technol. 63 (11) (2003) 1689–1703.

[7] K. Xiao, L. Zhang, Effective separation and alignment of long entangled carbon nanotubes in epoxy, J. Mater. Sci. 40 (24) (2005) 6513–6516.

[8] K. Mylvaganam, L.C. Zhang, Chemical bonding in polyethylene–nanotube composites: a quantum mechanics prediction, J. Phys. Chem. 108 (17) (2004) 5217–5220.

[9] H. Tan, et al., The effect of van der Waals-based interface cohesive law on carbon nanotube-reinforced composite materials, Compos. Sci. Technol. 67 (14) (2007) 2941–2946.

[10] M.M. Shokrieh, R. Rafiee, Stochastic multi-scale modeling of CNT/polymer composites, Comput. Mater. Sci. 50 (2) (2010) 437–446.

[11] B. Paliwal, M. Cherkaoui, Estimation of anisotropic elastic properties of nanocomposites using atomistic-continuum interphase model, Int. J. Solids Struct. 49 (18) (2012) 2424–2438.

[12] I. Sevostianov, M. Kachanov, Effect of interphase layers on the overall elastic and conductive properties of matrix composites. Applications to nanosize inclusion, Int. J. Solids Struct. 44 (3–4) (2007) 1304–1315.

[13] G.M. Odegard, T.C. Clancy, T.S. Gates, Modeling of the mechanical properties of nanoparticle/polymer composites, Polymer 46 (2) (2005) 553–562.

[14] J.M. Wernik, B.J. Cornwell-Mott, S.A. Meguid, Determination of the interfacial properties of carbon nanotube reinforced polymer composites using atomistic-based continuum model, Int. J. Solids Struct. 49 (13) (2012) 1852–1863.

[15] S. Yang, et al., Multiscale modeling of size-dependent elastic properties of carbon nanotube/polymer nanocomposites with interfacial imperfections, Polymer 53 (2) (2012) 623–633.

[16] S. Herasati, L.C. Zhang, H.H. Ruan, A new method for characterizing the interphase regions of carbon nanotube composites, Int. J. Solids Struct. 51 (9) (2014) 1781–1791.

[17] I. Zarudi, L.C. Zhang, Structural evolution of carbon nanotubes in composites under contact sliding stresses, Appl. Phys. Lett. 88 (24) (2006) 243102–243103.

[18] K.Q. Xiao, L.C. Zhang, I. Zarudi, Mechanical and rheological properties of carbon nanotube-reinforced polyethylene composites, Compos. Sci. Technol. 67 (2) (2007) 177–182.

[19] M. Rahmat, P. Hubert, Carbon nanotube–polymer interactions in nanocomposites: a review, Compos. Sci. Technol. 72 (1) (2011) 72–84.

[20] J.M. Wernik, S.A. Meguid, Recent developments in multifunctional nanocomposites using carbon nanotubes, Appl. Mech. Rev. 63 (5) (2010) 050801–050840.

[21] H. Hu, et al., Characterizing and modeling mechanical properties of nanocomposites—review and evaluation, J. Miner. Mater. Charact. Eng. 9 (4) (2010) 275–319.

[22] X.L. Gao, K. Li, A shear-lag model for carbon nanotube-reinforced polymer composites, Int. J. Solids Struct. 42 (5–6) (2005) 1649–1667.

[23] U.A. Joshi, S.C. Sharma, S.P. Harsha, Effect of carbon nanotube orientation on the mechanical properties of nanocomposites, Composites Part B 43 (4) (2012) 2063–2071.

[24] G.M. Odegard, et al., Constitutive modeling of nanotube-reinforced polymer composites, Compos. Sci. Technol. 63 (11) (2003) 1671–1687.

[25] M. Grujicic, Y.P. Sun, K.L. Koudela, The effect of covalent functionalization of carbon nanotube reinforcements on the atomic-level mechanical properties of poly-vinyl-ester-epoxy, Appl. Surf. Sci. 253 (6) (2007) 3009–3021.

[26] R. Zhu, E. Pan, A.K. Roy, Molecular dynamics study of the stress-strain behavior of carbon-nanotube reinforced Epon 862 composites, Mater. Sci. Eng. 447 (1–2) (2007) 51–57.

[27] G.I. Giannopoulos, S.K. Georgantzinos, N.K. Anifantis, A semi-continuum finite element approach to evaluate the Young's modulus of single-walled carbon nanotube reinforced composites, Composites Part B 41 (8) (2010) 594–601.

[28] M.M. Shokrieh, R. Rafiee, Prediction of mechanical properties of an embedded carbon nanotube in polymer matrix based on developing an equivalent long fiber, Mech. Res. Commun. 37 (2) (2010) 235–240.

[29] N. Hu, et al., Prediction of buckling characteristics of carbon nanotubes, Int. J. Solids Struct. 44 (20) (2007) 6535–6550.

[30] S.J.V. Frankland, et al., Molecular simulation of the influence of chemical cross-links on the shear strength of carbon nanotube–polymer interfaces, J. Phys. Chem. 106 (12) (2002) 3046–3048.

[31] K. Mylvaganam, L.C. Zhang, W.C.D. Cheong, The effect of interface chemical bonds on the behaviour of nanotube–polyethylene composites under nano-particle impacts, J. Comput. Theor. Nanosci. 4 (1) (2007) 122–126.

[32] M. Griebel, J. Hamaekers, Molecular dynamics simulations of the elastic moduli of polymer–carbon nanotube composites, Comput. Methods Appl. Mech. Eng. 193 (17–20) (2004) 1773–1788.

[33] Accelrys Inc., Materials Studio User's Manual.

[34] Q. Wang, et al., Modeling of fracture of carbon nanotubes with vacancy defect, Phys. Rev. 75 (20) (2007) 201405.

[35] Y.D. Kuang, X.Q. He, Young's moduli of functionalized single-wall carbon nanotubes under tensile loading, Compos. Sci. Technol. 69 (2) (2009) 169–175.

[36] S. Yu, S. Yang, M. Cho, Multi-scale modeling of cross-linked epoxy nanocomposites, Polymer 50 (3) (2009) 945–952.

[37] Y. Li, et al., Pull-out simulations on interfacial properties of carbon nanotube-reinforced polymer nanocomposites, Comput. Mater. Sci. 50 (6) (2011) 1854–1860.

[38] H. Sun, COMPASS: an ab initio force-field optimized for condensed-phase applications overview with details on alkane and benzene compounds, J. Phys. Chem. 102 (38) (1998) 7338–7364.

[39] D.N. Theodorou, U.W. Suter, Detailed molecular structure of a vinyl polymer glass, Macromolecules 18 (7) (1985) 1467–1478.

[40] H. Meirovitch, Computer simulation of self-avoiding walks: testing the scanning method, J. Chem. Phys. 79 (1) (1983) 502–508.

[41] H.J.C. Berendsen, et al., Molecular dynamics with coupling to an external bath, J. Chem. Phys. 81 (8) (1984) 3684–3690.

[42] M. Parrinello, A. Rahman, Polymorphic transitions in single crystals: a new molecular dynamics method, J. Appl. Phys. 52 (12) (1981) 7182–7190.

[43] D.H. Tsai, The virial theorem and stress calculation in molecular dynamics, J. Chem. Phys. 70 (3) (1979) 1375–1382.

[44] M.-F. Yu, et al., Tensile loading of ropes of single wall carbon nanotubes and their mechanical properties, Phys. Rev. Lett. 84 (24) (2000) 5552–5555.

[45] N. Yao, V. Lordi, Young's modulus of single-walled carbon nanotubes, J. Appl. Phys. 84 (4) (1998) 1939–1943.

[46] C. Li, T.W. Chou, A structural mechanics approach for the analysis of carbon nanotubes, Int. J. Solids Struct. 40 (10) (2003) 2487–2499.

[47] J.P. Lu, Elastic properties of carbon nanotubes and nanoropes, Phys. Rev. Lett. 79 (7) (1997) 1297–1300.

[48] C.Y. Wang, L.C. Zhang, A critical assessment of the elastic properties and effective wall thickness of single-walled carbon nanotubes, Nanotechnology 19 (7) (2008) 075705.

[49] T. Vodenitcharova, L.C. Zhang, Effective wall thickness of a single-walled carbon nanotube, Phys. Rev. 68 (16) (2003) 165401.

[50] Y. Huang, J. Wu, K.C. Hwang, Thickness of graphene and single-wall carbon nanotubes, Phys. Rev. 74 (24) (2006) 245413.

[51] E. Hernández, et al., Elastic properties of C and $BxCyNz$ composite nanotubes, Phys. Rev. Lett. 80 (20) (1998) 4502–4505.

[52] W.V. Titow, PVC Technology, Springer, Netherlands, (1984).

[53] J. Dyment, H. Ziebland, The tensile properties of some plastics at low temperatures, J. Appl. Chem. 8 (4) (1958) 203–206.

[54] G. Karami, M. Garnich, Micromechanical study of thermoelastic behavior of composites with periodic fiber waviness, Composites Part B 36 (3) (2005) 241–248.

[55] C.H. Tsai, et al., The effect of inclusion waviness and waviness distribution on elastic properties of fiber-reinforced composites, Composites Part B 42 (1) (2011) 62–70.

[56] J. Nafar Dastgerdi, G. Marquis, M. Salimi, The effect of nanotubes waviness on mechanical properties of CNT/SMP composites, Compos. Sci. Technol. 86 (0) (2013) 164–169.

[57] L.H. Shao, et al., Prediction of effective moduli of carbon nanotube–reinforced composites with waviness and debonding, Compos. Struct. 87 (3) (2009) 274–281.

[58] D. Qian, et al., Load transfer and deformation mechanisms in carbon nanotube–polystyrene composites, Appl. Phys. Lett. 76 (20) (2000) 2868–2870.

[59] M.S.P. Shaffer, A.H. Windle, Fabrication and characterization of carbon nanotube/poly(vinyl alcohol) composites, Adv. Mater. 11 (11) (1999) 937–941.

[60] A.R. Clarke, G. Archenhold, N.C. Davidson, A novel technique for determining the 3D spatial distribution of glass fibres in polymer composites, Compos. Sci. Technol. 55 (1) (1995) 75–91.

[61] E. Shady, Y. Gowayed, Effect of nanotube geometry on the elastic properties of nanocomposites, Compos. Sci. Technol. 70 (10) (2010) 1476–1481.

[62] M. Motamedi, M. Eskandari, M. Yeganeh, Effect of straight and wavy carbon nanotube on the reinforcement modulus in nonlinear elastic matrix nanocomposites, Mater. Des. 34 (0) (2012) 603–608.

[63] R.D. Bradshaw, F.T. Fisher, L.C. Brinson, Fiber waviness in nanotube-reinforced polymer composites. II: Modeling via numerical approximation of the dilute strain concentration tensor, Compos. Sci. Technol. 63 (11) (2003) 1705–1722.

[64] T. Mori, K. Tanaka, Average stress in matrix and average elastic energy of materials with misfitting inclusions, Acta Metall. 21 (5) (1973) 571–574.

[65] M.A. Bhuiyan, et al., Understanding the effect of CNT characteristics on the tensile modulus of CNT reinforced polypropylene using finite element analysis, Comput. Mater. Sci. 79 (2013) 368–376.

[66] C.R. Misiego, R.B. Pipes, Dispersion and its relation to carbon nanotube concentration in polyimide nanocomposites, Compos. Sci. Technol. 85 (2013) 43–49.

[67] S. Herasati, L. Zhang, A new method for characterizing and modeling the waviness and alignment of carbon nanotubes in composites, Compos. Sci. Technol. 100 (2014) 136–142.

[68] N.J. Ginga, W. Chen, S.K. Sitaraman, Waviness reduces effective modulus of carbon nanotube forests by several orders of magnitude, Carbon 66 (2013) 57–66.

[69] H.W. Wang, et al., Nanoreinforced polymer composites: 3D FEM modeling with effective interface concept, Compos. Sci. Technol. 71 (7) (2011) 980–988.

[70] F.T. Fisher, R.D. Bradshaw, L.C. Brinson, Effects of nanotube waviness on the modulus of nanotube-reinforced polymers, Appl. Phys. Lett. 80 (24) (2002) 4647–4649.

[71] C. Li, T.W. Chou, Failure of carbon nanotube/polymer composites and the effect of nanotube waviness, Composites, Part A 40 (10) (2009) 1580–1586.

[72] L. Shen, J. Li, Transversely isotropic elastic properties of multiwalled carbon nanotubes, Phys. Rev. 71 (3) (2005) 035412.

[73] S. Herasati, L.C. Zhang, Elastic properties of single-walled carbon nanotube clusters: dependence on hydrostatic pressure, Comput. Mater. Sci. 86 (0) (2014) 93–98.

[74] J.L. Tsai, S.H. Tzeng, Y.T. Chiu, Characterizing elastic properties of carbon nanotubes/polyimide nanocomposites using multi-scale simulation, Composites Part B 41 (1) (2010) 106–115.

[75] N. Hu, et al., Prediction of elastic properties of carbon nanotube reinforced composites, Proc. R. Soc. 461 (2058) (2005) 1685–1710.

[76] S. Herasati, L. Zhang, Interphase effect on the macroscopic elastic properties of non-bonded single-walled carbon nanotube composites, Composites Part B 77 (0) (2015) 52–58.

20

Stochastic Multiscale Modeling of CNT/Polymer

Roham Rafiee, Vahid Firouzbakht

UNIVERSITY OF TEHRAN, TEHRAN, IRAN

CHAPTER OUTLINE

20.1 Introduction

Carbon is one of the most abundant elements in the world, which forms a very large number of allotropes. Among them, carbon nanotube (CNT) possesses outstanding and exceptional mechanical, electrical, and thermal properties [1–4] accompanied with low density and high aspect ratio [5,6]. This exceptional nature has encouraged researchers to incorporate CNT as a new generation of reinforcing agent into polymeric matrices. Experimental studies revealed that the exploitation of a small amount of CNTs into the polymeric matrix leads to a significant growth in the mechanical properties of the polymer [5–7]. As a matter of fact, carbon nanotube–reinforced polymers (CNTRPs) have established a new era in the field of advanced materials. Therefore, predicting the mechanical properties of the CNT/polymer composites on the basis of their constituents' properties not only plays a leading role in their evolving process, but can also provide a clear insight into their future industrial applications.

Referring to a comprehensive literature review on typical theoretical studies performed to predict the Young's modulus of CNTRPs, it can be easily concluded that they suffer from a variety of drawbacks and/or invalid assumptions [8]. Different investigations have tried

Carbon Nanotube–Reinforced Polymers. http://dx.doi.org/10.1016/B978-0-323-48221-9.00020-0

to predict the mechanical properties of CNTRPs using available micromechanics rules originally developed and suitable for the case of fibrous composites. Although developed rules can estimate the mechanical properties of traditional composite materials with an acceptable level of accuracy, their direct implementation for the specific case of CNTRPs are not allowed. Micromechanics rules have been basically developed for those cases wherein the reinforcing agents appear at the scale of micro and thus they are not capable of distinguishing the difference between micro- and nanoscales. Since the starting scale of micromechanics rules is micro, they simply neglect molecular interactions of CNT occurring at nanoscale by simply replacing the lattice structure of CNT with a continuum medium. Moreover, the majority of these rules assume perfect bonding between the reinforcing agent and the surrounding polymer, which is not pertinent to the case of CNTRP. The load transferring mechanism from the matrix to the CNT, which has an important role in defining the efficiency of CNT's capability in reinforcing polymers at microscale, is accomplished through weakly nonbonded van der Waals (vdW) interactions. Those specific sets of micromechanics rules taking into account an intermediate phase between the reinforcing agent and the polymer cannot be used either, since no material regions exist between the CNT and the surrounding polymer, but instead molecular interactions accounts for the load transferring issue. CNTs are inherently intended to be concentrated in local aggregates in the form of bundled CNTs presenting a heterogeneous material region. Thus, the investigated material region cannot be treated as a homogenous medium because CNT dispersion does not follow a consistent and uniform pattern. Therefore, localized homogenous regions need to be distinguished at lower scales and then homogenization techniques can be used at the proper scale. In addition to CNT agglomeration, they experience nonstraight shapes in the matrix, which stems from their large aspect ratios and low bending stiffness. CNT agglomeration and CNT curvature cannot be captured through the utilization of micromechanics rules. Above all, CNT dispersion and curvature issues, which significantly diminish CNT effectiveness in reinforcing polymers, are also induced during the processing of CNTRPs. The random nature of these uncertainties necessitates a stochastic modeling procedure that accounts for arbitrary CNT curvature shapes and nonuniform dispersion patterns.

Consequently, a significant reported discrepancy in open literature between experimental observations and theoretical estimations for the mechanical properties of CNTRP is attributed to micromechanics overlooking the aforementioned phenomena [9–19].

In this chapter, a hierarchical multiscale modeling procedure is developed for predicting the mechanical properties of CNTRPs capturing all involved scales starting from the nanoscale and lasting to the macroscale passing in-between the micro- and mesoscales. Processing-induced inconsistencies are also treated as random parameters addressing associated uncertainties, and thus stochastic implementation of the developed modeling is performed. The involved scales and associated phenomena are schematically depicted in Fig. 20.1 for the specific case of CNTRPs wherein CNT appears at nanoscale as the reinforcing agent and the mechanical properties of CNTRP are intended to be extracted at bulk material level known as the scale of macro.

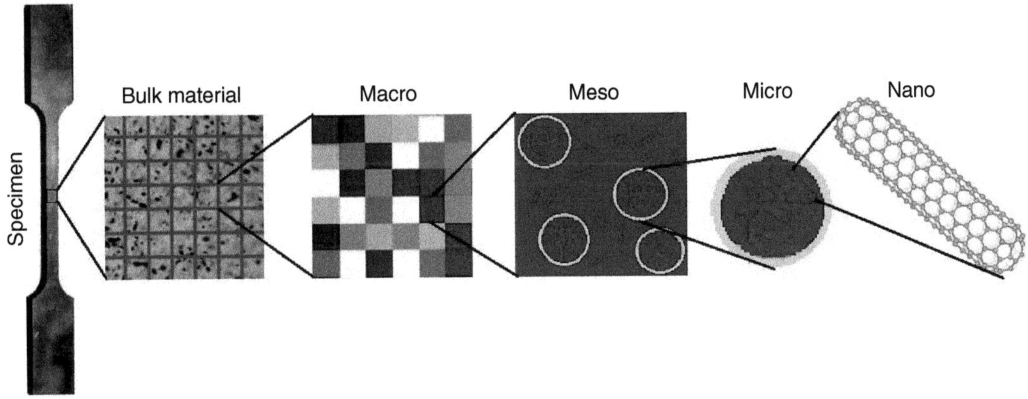

FIGURE 20.1 Length scales spectrum of carbon nanotube–reinforced polymers (CNTRP) for the multiscale modeling.

20.2 Definition of RVEs for Each Scale

Following the general routine of hierarchical modeling, the outputs of each scale are fed into the very next upper scale as input data. A proper multiscale modeling of CNTRP demands the definition of separate representative volume elements (RVE) at each scale to facilitate the recognition of corresponding effective parameters. Considering the introduced scales in Fig. 20.1, the effective parameters of each scale are presented in Table 20.1 in conjunction with the output of each level of simulation.

Prior to developing the multiscale modeling, the bulk material region is partitioned into regular constitutive blocks as the model preparation activity. Each block is assigned a random volume fraction in conformance with the overall Young's modulus of the investigated CNTRP. In other words, the average of all the assigned volume fractions is consistent with the total volume fraction of CNTRP. Consequently, inhomogeneity of the material region arising from the nonuniform dispersion of CNTs is captured. In the next stage, multiscale modeling starts through a bottom-up modeling procedure and the mechanical properties of each constitutive block is estimated.

Table 20.1 Effective Parameters of Each Scale in Multiscale Modeling

Scale	Effective Length	Parameters	Output
Nano	nm	Covalent carbon—carbon bond configurations	Young's modulus of isolated CNT
Micro	nm/μm	Interaction between CNT and polymer, stress transfer in interphase	Young's modulus of CNT and interphase
Meso	μm	CNT orientations, agglomeration, and curvature	Mechanical properties of constitutive blocks
Macro	mm	Nonuniform dispersion of CNTs	Mechanical properties of CNTRP

CNT, Carbon nanotubes; CNTRP, carbon nanotube–reinforced polymers.

20.3 Multiscale Modeling

The multiscale modeling developed in this chapter is a hierarchical one at meso- and macroscales, while specifically at nano- and microscales, concurrent multiscale modeling is conducted. The results of the nano/microscale modeling are fed into the upper scale of meso as input information. The main reason for performing concurrent modeling at nano/microscale is that for the purpose of modeling the stress transferring issue from the matrix to the CNT through simulating the interaction between the CNT and the surrounding polymer, it aims to keep the lattice structure of the CNT. The detailed procedures of modeling at each scale are outlined in the next sections.

20.3.1 Nano/Microscale Modeling

A concurrent multiscale finite element (FE) modeling is employed. The RVE consists of a CNT at nanoscale and the surrounding polymer at microscale while the CNT–polymer interaction is modeled simultaneously.

For the purpose of modeling CNT at nanoscale, nanoscale continuum technique is employed wherein instead of replacing the whole CNT with a continuum structure, each C—C bond is replaced with a virtual continuum element as a rational compromise in modeling. The mechanical properties of this equivalent continuum element are obtained by establishing a linkage between molecular space and structural mechanics; that is, molecular interactions of C—C bonds are captured by a structural member whose properties are dictated by governing interatomic potentials. Li and Chou [20] developed a molecular structural mechanics technique by establishing an analogy between the linear interatomic potential energies of the molecular structure and strain energies of a beam element resembling a C—C bond. The established analogy is schematically presented in Fig. 20.2.

Following the aforementioned analogy method, Shokrieh and Rafiee developed a closed-form solution to predict the Young's modulus of CNT [22]. They replaced each C—C bond with a beam element and subsequently the lattice molecular structure is substituted with an equivalent discrete frame structure [22]. A CNT with chiral index of (10, 10) is constructed in an ANSYS commercial FE package. One end of the model is fixed while the other end is subjected to axial displacement. The obtained Young's modulus is compared with the estimated results by analytical solution [22] and a very good agreement implies the proper modeling of CNT at nanoscale.

The efficiency of CNT in reinforcing polymers not only depends on the mechanical properties of isolated CNTs, but also on the load transferring mechanism from the matrix to the CNT at nano/microscale through CNT–polymer interactions, which play an important role in defining the degree to which mechanical properties are enhanced in CNTRP. Generally, interactions between CNT and the surrounding polymer takes place through weakly nonbonded vdW and electrostatics forces [23]. Electrostatic forces are negligible in comparison with vdW forces, so just vdW interactions are considered.

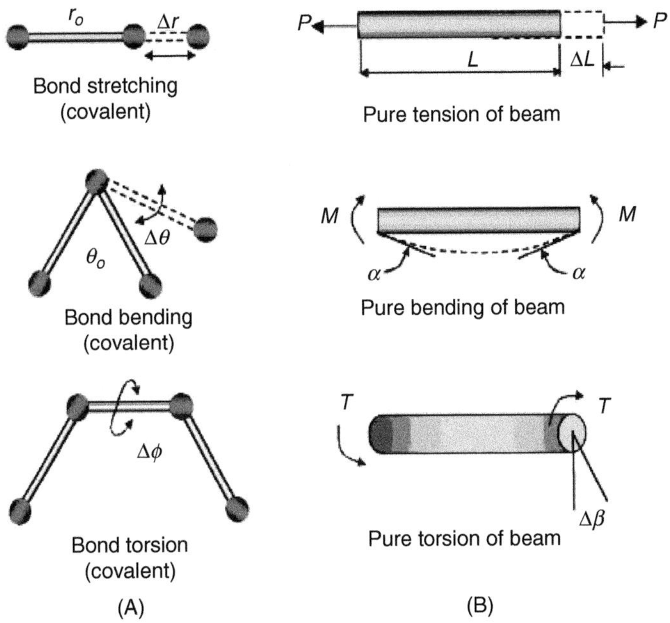

FIGURE 20.2 Energy equivalence between (A) molecular mechanics and (B) structural mechanics [21].

Since the volume fraction of CNT in CNTRP is usually less than 2%, the volume of the surrounding polymer is significantly higher than that of CNT; thus, as a reasonable compromise in modeling and for the sake of reducing computational efforts, continuum solid elements are used for modeling the polymer while the lattice structure of the CNT is kept following the explained procedure of nanoscale continuum mechanics modeling. The outer radius of the surrounding polymer is chosen in a manner that reflects 95% volume fraction for resin and 5% for embedded CNT. The size of the solid elements for the polymer is chosen to be as small as the length of the hexagonal rings of CNTs in the axial direction. The strategy of defining the size of the polymer elements can be justified by the fact that the molecular chain of the polymer cannot penetrate into CNTs, so the selection of a smaller size is not permissible. Considering the linearly isotropic behavior, the mechanical properties of the resin are chosen in a manner that they resemble the available epoxy resins in the market. Assuming a virtual thickness of 0.34 nm for CNT as the distance between adjacent walls in multiwalled CNTs, the center of the carbon atoms are assumed to be placed at the midsection of the tube thickness. The innermost layer of the resin is located at the same position of the outer surface of the CNT. The nodes of the CNTs are connected to the neighboring nodes placed at the inner surface of the resin elements wherein the distance is less than 0.85 nm as the cut-off distance for the vdW force obtained on the basis of the Lennard-Jones "6–12" potential. The vdW force is a highly nonlinear force described by two distinct regions of attraction and repulsion. Subsequently, a nonlinear spring element is employed and its parameters are adjusted appropriately to simulate the vdW force between the CNT and the surrounding polymer.

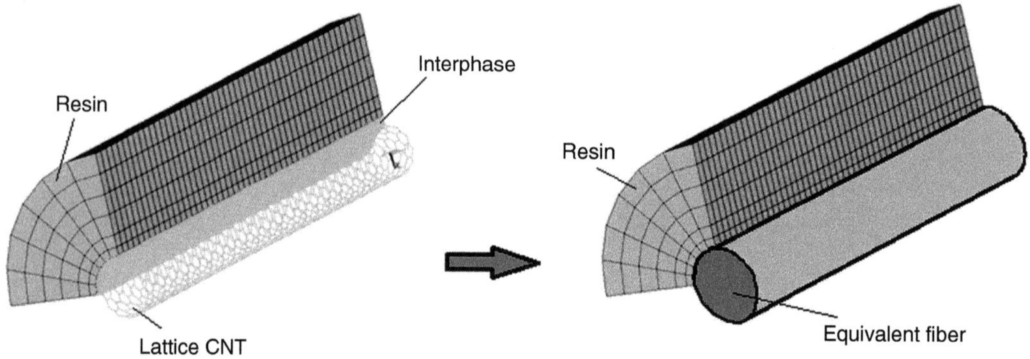

FIGURE 20.3 Obtaining mechanical properties of an equivalent fiber as a replacement of carbon nanotube (CNT) and interphase region.

The FE model constructed is subjected to different loadings and boundary conditions to obtain the mechanical properties of the investigated RVE. A detailed explanation of loadings can be found in the research by Rafiee and Shokrieh [24], and for the sake of brevity, applied loadings are briefly explained here. The RVE is subjected to axial strain at one end while the other end is totally fixed, and then the longitudinal modulus is obtained. Then, an internal pressure is applied to all the nodes of the embedded CNT in the matrix while just one node is fixed to avoid rigid body motion. The transverse modulus and Poisson's ratio of the RVE are obtained as the output of this analysis. Finally, one end of the RVE is totally fixed and the other end is exposed to torsional torque and thus in-plane shear modulus is calculated. It is noteworthy that in all cases nonlinear analysis is conducted due to the presence of nonlinear spring elements resembling vdW interactions. Obtaining the mechanical properties of the investigated RVE, the CNT and the interphase region are replaced with a virtual equivalent fiber, which is assumed to be perfectly bonded to the resin. Having in hand the mechanical properties of the simulated RVE, the mechanical properties of the developed equivalent fiber can be obtained considering another RVE containing an equivalent fiber and the same matrix. The strategy of converting CNT and the interphase to an equivalent fiber is presented in Fig. 20.3.

Following the mentioned procedure in Fig. 20.3, the obtained mechanical properties for the developed equivalent fiber are presented in Table 20.2. It should be mentioned that both long and short CNTs are investigated. For the case of long CNT, the length of the CNT and RVE are equal, while for the case of short CNT the length of the CNT is smaller than the RVE length. It has been observed that the length of the CNT has a significant

Table 20.2 Effective Properties of Developed Equivalent Fiber [24]

Longitudinal Modulus (GPa)	Transverse Modulus (GPa)	In-plane Shear Modulus (GPa)	Poisson's Ratio
58~642.4	11.27	5.13	0.284

effect on the longitudinal modulus of the RVE, since the higher CNT lengths contain more population of vdW links and other properties are not affected by CNT length at all. Therefore, the upper bound for the longitudinal Young's modulus reflected in Table 20.2, that is, 642.4 GPa, belongs to the case of long CNT. An extensive analysis of the influence of CNT length on the longitudinal modulus of an equivalent fiber have been performed by Rafiee and Shokrieh reporting effective longitudinal modulus of the equivalent fiber with respect to the CNT length varying between 50 and 1550 nm [24].

As the outcome of the performed analysis at this scale, the CNT and the interphase region are replaced with a virtual equivalent fiber behaving as a transversely isotropic material. The advantage of developing an equivalent fiber at this scale is twofold: (1) The influence of nonbonded interactions are taken into account in the mechanical properties of the equivalent fiber, and (2) the developed equivalent fiber is perfectly bonded to the matrix and thus micromechanics rules can be employed in the upper scales.

20.3.2 Mesoscale Modeling

As was explained in the previous section, as in the output of nano/microscale analysis the CNT and its interphase are replaced with an equivalent fiber, the reinforcing agent is the equivalent fiber at mesoscale. The RVE at mesoscale is a constitutive block wherein equivalent fibers are assumed to be either locally agglomerated in spherical regions or dispersed outside the spheres (Fig. 20.1). Moreover, an equivalent fiber is randomly oriented in a different direction while experiencing arbitrary nonstraight shapes.

As a novel method, irregular tessellation technique is used to create partitioned polygons inside the RVE at mesoscale, based on the Voronoi approach [25] as depicted in Fig. 20.4. Each constitutive polygon contains a single special point, called generator or seed. All the other points located inside the polygon and in the neighboring areas of a seed are at the lowest distance to the seed rather than the other points. All points placed on the edges of each polygon are at the same distance to the two neighbor seeds. Therefore, if it is assumed that each specific seed represents only a single aggregate, the locations of the aggregate can be taken into account by employing the explained strategy. Finally, each obtained polygon is mapped into a spherical element assuming its center is located at the same position of the corresponding polygon seed. Each constitutive block at mesoscale is converted to a series of adjacent spheres with different radii. The magnitude of the radius and also the surface area of each sphere vary from very low to high values depending on the size of the generating polygon. Consequently, each concentrated aggregation is considered as the center of a separate sphere while all the other points located in the neighboring areas in the same sphere have the shortest distance to it, rather than all the other center points placed inside other spheres. As a result, mesoscale RVE is altered to tessellated spherical elements, varying in diameter and surrounded with spherical surfaces of the matrix. The spheres are adjacent and tangent with each other; no interference happens between them. The explained procedure is schematically shown in Fig. 20.4.

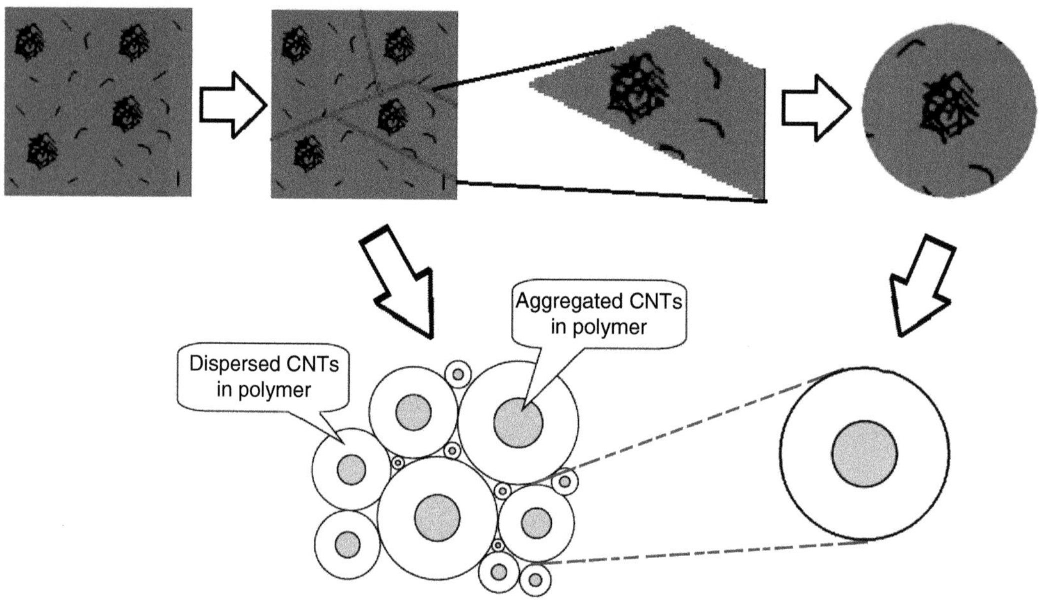

FIGURE 20.4 **Voronoi tessellation at mesoscale and obtained CSA model.**

In constant to regular tessellation technique, the locations of concentrated aggregates are also taken into account. Moreover, it is possible to treat the locations of aggregates as another random parameter in the modeling procedure. For the purpose of constructing polygons in the mesoscale RVE mentioned in Fig. 20.4, the coordinates of seeds have to be defined in the *x*–*y* plane, remembering the fact that the number of seeds is correlated to the number of polygons.

With the aid of statistical science, Bayes algorithm is used to determine the locations of the seeds. This algorithm demonstrates formation of random independent variables and their relationships. As a principle, there is a probability distribution for each entity, and by observing new data and reasoning about the probability distribution, optimal decisions can be taken. This is the fundamental of Bayes algorithm [26]. To take advantage of this algorithm, the coordinate of a seed point is taken as the mean value for normal distribution of other points located on the distribution curves centered on seeds with different distances from each other.

In fact, all of the points on the *x*–*y* plane are characterized with several devoted numbers; each one is related to a specific seed. Among them, the biggest value means that it is the closer point to the corresponding seed. This is true for all the points, so the biggest value devoted to each point can be easily chosen as its indicator and other numbers will be simply ignored. Thus, each point is attached to a single distribution curve. In comparison to other seeds, the points located on a distribution curve have the shortest distance to the curve peak (the coordination of its seed). Looking at all the curves together, they generate conic surfaces. Those specific points with the minimum distance from the seeds are all located at the foot of conic surfaces. If the spatial cones are projected to the *x–y* plane, Voronoi polygons are shaped. This procedure is mentioned for a specific case in Fig. 20.5.

Looking at conic surfaces from the top view in Fig. 20.5, the random distributions of aggregates are tangible. After obtaining irregularly tessellated RVE at mesoscale, they are converted to the composite sphere assemblage (CSA) model in accordance with the depicted procedure in Fig. 20.4.

Finally, the CSA method is employed to extract the mechanical properties of this spherical assemblage embedded in the tessellated structure [27,28]. Applying the CSA method, each of the polygons can be considered as a spherical element whose inner circle represents aggregated CNTs in the polymer surrounded by fully dispersed CNTs in the polymer as the outer circle. Also, each of the polygons has its own portion of CNT volume fraction, dedicated to constitutive blocks at mesoscale. The amount is not deterministic and defined randomly. The properties of these two distinct phases in the circular elements should be obtained to calculate the overall mechanical properties of each polygon. Since the size of each

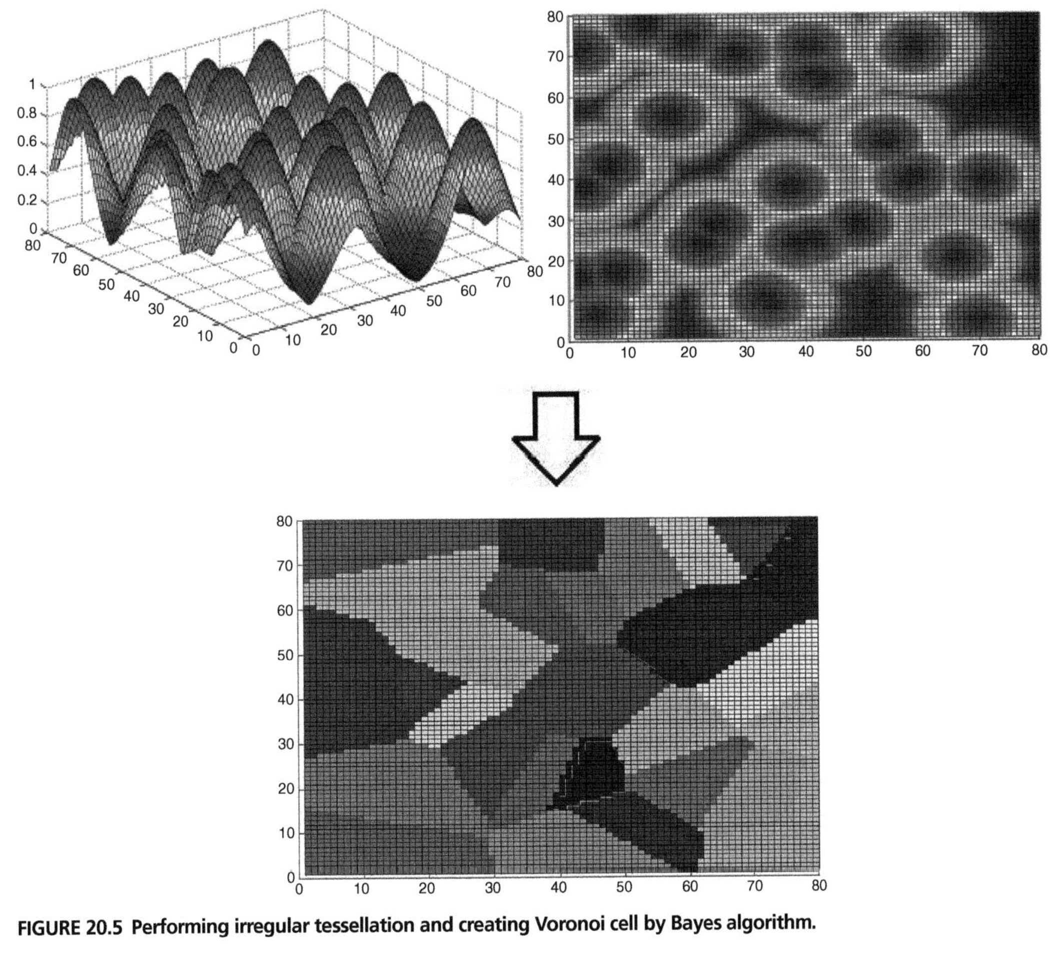

FIGURE 20.5 Performing irregular tessellation and creating Voronoi cell by Bayes algorithm.

spherical element and the density of inside aggregates are unique, a random logic should be employed to determine the local volume fractions for each phase. Therefore, random coefficients representing the degree of CNT concentration of each element is defined. In this case, the volume fraction assigned to each spherical element can be divided into two phases [29]:

$$V_{\mathrm{CNT}} = V_{\mathrm{CNT}}^{\mathrm{inclusion}} + V_{\mathrm{CNT}}^{m} \tag{20.1}$$

The random coefficients representing CNT concentration/agglomeration phenomena are defined as follows [29]:

$$\mu = \frac{V_{\mathrm{inclusion}}}{V}, \quad \kappa = \frac{V_{\mathrm{CNT}}^{\mathrm{inclusion}}}{V_{\mathrm{CNT}}} \tag{20.2}$$

where $V_{\mathrm{inclusion}}$ is the volume of spherical elements and V is the sum of all the sphere volumes. μ is defined as the ratio of total aggregated particles to the total volume of RVE. κ represents the ratio of CNT volume fraction inside the aggregates to the total CNT volume fraction assigned to spherical particles. The local volume fraction of CNTs allocated to each phase can be obtained from the following relations [29]:

$$\frac{V_{\mathrm{CNT}}^{\mathrm{inclusion}}}{V_{\mathrm{inclusion}}} = \frac{f_{\mathrm{CNT}}\kappa}{\mu}, \quad \frac{V_{\mathrm{CNT}}^{m}}{V - V_{\mathrm{inclusion}}} = \frac{f_{\mathrm{CNT}}(1-\kappa)}{1-\mu} \tag{20.3}$$

where f_{CNT} is defined as

$$f_{\mathrm{CNT}} = \frac{V_{\mathrm{CNT}}}{V} \tag{20.4}$$

Combining the mentioned relations with the micromechanics rule developed by Hashin [27] and Christensen [28], the mechanical properties of each phase can be extracted as follows:

$$K_{\mathrm{in}} = K_m + \frac{(f_{\mathrm{CNT}}\kappa/\mu)(K_r - K_m)}{1 + \dfrac{[1-(f_{\mathrm{CNT}}\kappa/\mu)](K_r - K_m)}{[K_m + (4/3)G_m]}} \tag{20.5}$$

$$K_{\mathrm{out}} = K_m + \frac{\{[f_{\mathrm{CNT}}(1-\kappa)]/(1-\mu)\}(K_r - K_m)}{1 + \dfrac{(1-\{[f_{\mathrm{CNT}}(1-\kappa)]/(1-\mu)\})(K_r - K_m)}{[K_m + (4/3)G_m]}} \tag{20.6}$$

$$G_{\mathrm{in}} = G_m + \frac{f_{\mathrm{CNT}}\kappa/\mu}{\dfrac{1}{G_r - G_m} + \dfrac{6(K_m + 2G_m)[1-(f_{\mathrm{CNT}}\kappa/\mu)]}{5G_m(3K_m + 4G_m)}} \tag{20.7}$$

$$G_{\mathrm{out}} = G_m + \frac{[f_{\mathrm{CNT}}(1-\kappa)]/(1-\mu)}{\dfrac{1}{G_r - G_m} + \dfrac{6(K_m + 2G_m)(1-\{[f_{\mathrm{CNT}}(1-\kappa)]/(1-\mu)\})}{5G_m(3K_m + 4G_m)}} \tag{20.8}$$

It should be noted that items subscripted by "in" and "out" are related to inner sphere or aggregate and outer sphere or fully dispersed CNT in the matrix, respectively. Moreover, the subscripts "m" and "r" denote matrix and reinforcing agent, respectively. As has already been mentioned, at mesoscale the reinforcing agent is the developed equivalent fiber whose properties are obtained and reported in Table 20.2. Wang et al. [30] have statistically reported that CNT lengths vary between 50 and 1700 nm. Rafiee and Shokrieh [24] have concluded that the mean length of CNTs can be used as a deterministic value instead of random selection of CNT lengths. Thus, a mean length of 500 nm is selected in this study, the corresponding longitudinal modulus of which was reported as 348 GPa.

Finally, the mechanical properties of each polygon element, which has been already converted to the two-phase spherical inclusion, can be extracted as follows:

$$G = G_{out} + \frac{\mu}{\dfrac{1}{G_{in} - G_{out}} + \dfrac{6(K_{out} + 2G_{out})(1-\mu)}{5G_{out}(3K_{out} + 4G_{out})}} \tag{20.9}$$

$$K = K_{out} + \frac{\mu(K_{in} - K_{out})}{1 + \dfrac{(1-\mu)(K_{in} - K_{out})}{K_{out} + (4/3)G_{out}}} \tag{20.10}$$

$$E = \frac{9KG}{3K + G} \tag{20.11}$$

It is noteworthy that the random orientation of equivalent fiber is intrinsically considered through Eqs. (20.9)–(20.11) wherein the investigated RVE is treated as an isotropic behavior.

For capturing CNT waviness, a phenomenological technique is used [31]. As the equivalent fiber behavior is categorized under the transversely isotropic class of materials, the module of a unit-cell containing an embedded nonstraight fiber in the matrix can get bounded between two extreme values calculated as the longitudinal and transverse modules of the unit-cell. Subsequently, the effective modulus of the nonstraight equivalent fiber is placed between bounding values of the longitudinal and transverse equivalent moduli of the straight one. To describe the bounding technique mathematically, Eqs. (20.12) and (20.13) are presented for longitudinal and transverse directions, respectively [31]:

$$E_T^{\text{Straight}} \le E_L^{\text{Wavy}} \le E_L^{\text{Straight}} \tag{20.12}$$

$$E_T^{\text{Straight}} \le E_T^{\text{Wavy}} \le E_L^{\text{Straight}} \tag{20.13}$$

where E_L^{Straight} and E_T^{Straight} are 348 and 11.27 GPa for 500 nm equivalent fiber, respectively. Therefore, any randomly selected values between the extreme bounds obtained for the straight equivalent fiber resemble any arbitrary nonstraight shapes of CNTs regardless of characterizing the geometry. Subsequently, the mechanical properties assigned to straight equivalent fiber presented in Eqs. (20.5)–(20.8) are replaced with randomly devoted values between bounding extremes.

20.3.3 Macroscale Modeling

At the uppermost scale of macro, the mechanical properties of each constitutive block have already been obtained at the previous scale of meso while random volume fractions have been assigned to each block at the beginning of the modeling procedure. As has been previously explained, the assigned random volume fractions to each constitutive block simulated the nonuniform dispersion of CNTs in the matrix resulting in inhomogeneous material. The mean value of the properties associated with each constitutive block at macroscale is calculated to obtain the overall mechanical properties of the investigated bulk material region.

20.4 Integrated Modeling Procedure

A comprehensive flowchart of the developed multiscale modeling in this research is presented in Fig. 20.6 as an integrated procedure focusing on meso- and macroscales. The output analysis at nano/micro is reflected as equivalent fiber in the flowchart. The MATLAB platform is chosen and a computer code is generated to execute the developed model.

The developed algorithm for creating Voronoi polygons at mesoscale is fully a stochastic process, so that none of the blocks follow the same tessellation pattern. For instance, two produced polygons at mesoscale are depicted in Fig. 20.7.

The random parameters involved can be summarized as follows:

- CNT agglomeration (the portion of CNTs in local concentrated areas)
- distribution of aggregates
- CNT curvature
- CNT orientation
- CNT dispersion (inhomogeneous material region)

The variety of incorporated random variables demands a convergence study to extract sufficient numbers of constitutive blocks at macroscale, the number of Voronoi polygons at mesoscale, and also the number of produced random samples, globally. A convergence criterion of 0.1% is considered as the coefficient of variation for the Young's modulus of the bulk material region at macroscale. It is observed that by producing 300 realizations of the stochastic method on a 30×30 region, consisting of seven seeds in each block, the requirement of convergence criterion is met. The convergence trend is also shown in Fig. 20.8.

Along with the modeling procedure, a case study is accomplished to verify that the developed simulation procedure is capable of addressing random positions of aggregates as an important novel feature of study. On the other hand, the sensitivity of the model to the positions of aggregated CNTs is also investigated. Thus, a series of modifications is applied to the computer code to generate some random patterns of polygons created by the irregular tessellation technique, while keeping the value of the assigned CNT volume fraction ($f = 1.5\%$) and agglomeration indexing parameters ($\mu = 0.23, \kappa = 0.56$) constant

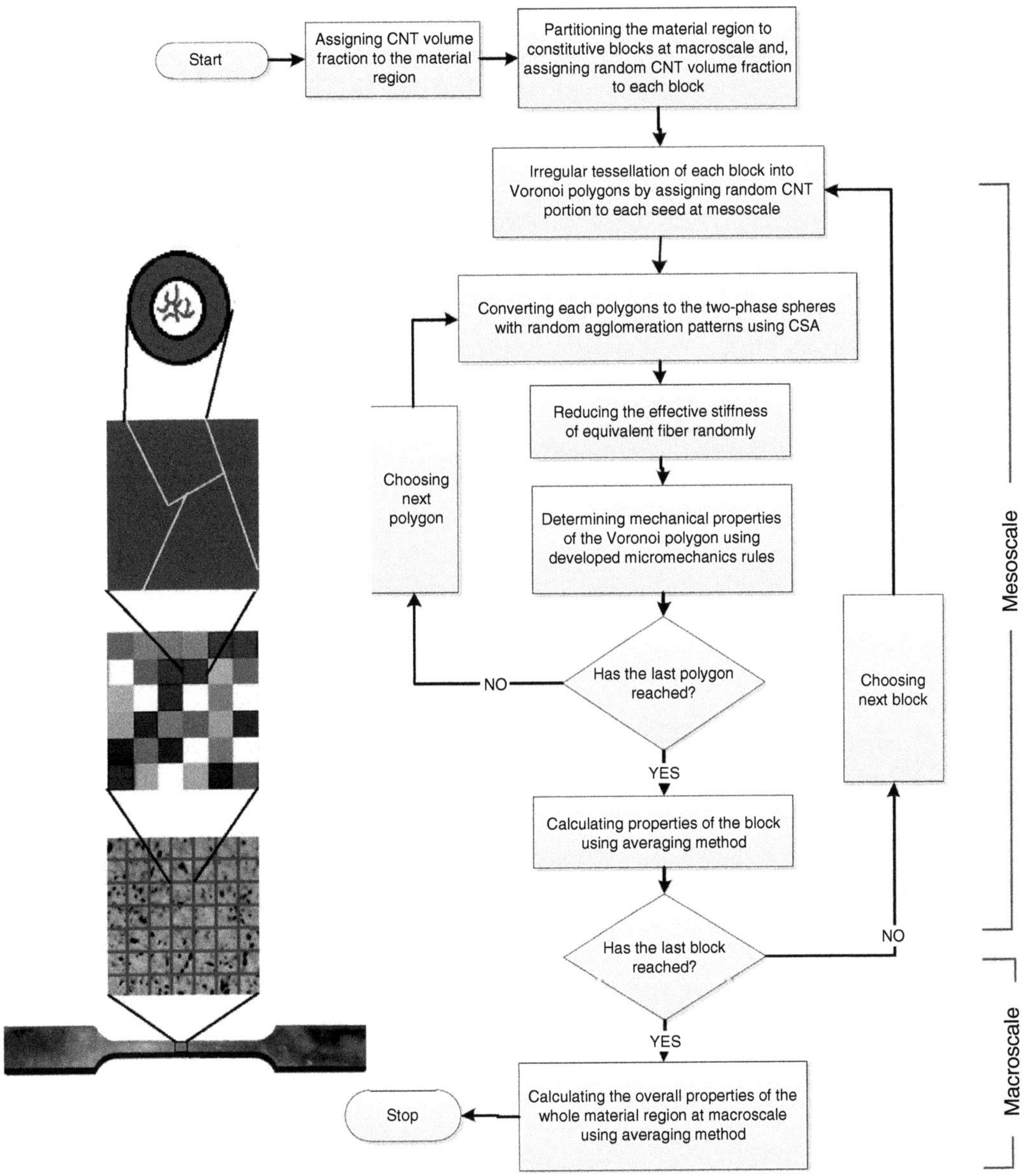

FIGURE 20.6 Flowchart of the developed stochastic multiscale modeling procedure.

for all constitutive blocks at mesoscale. The Young's modulus of the resin is considered as 1.5 GPa. As a result, the Young's modulus for the material region is obtained between 2.07 and 2.11 GPa for different patterns. Therefore, it can be easily concluded that the presented model is capable of sensing and capturing different positions of CNT aggregations. This significant capability originate from the irregular tessellation technique. The mentioned scenario is schematically depicted in Fig. 20.9.

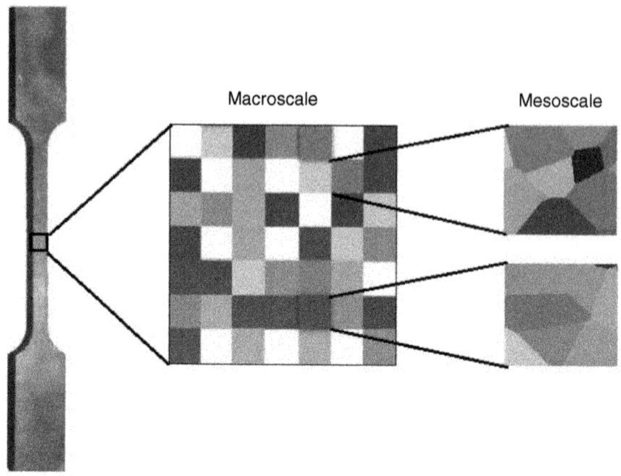

FIGURE 20.7 Random patterns for constitutive polygons.

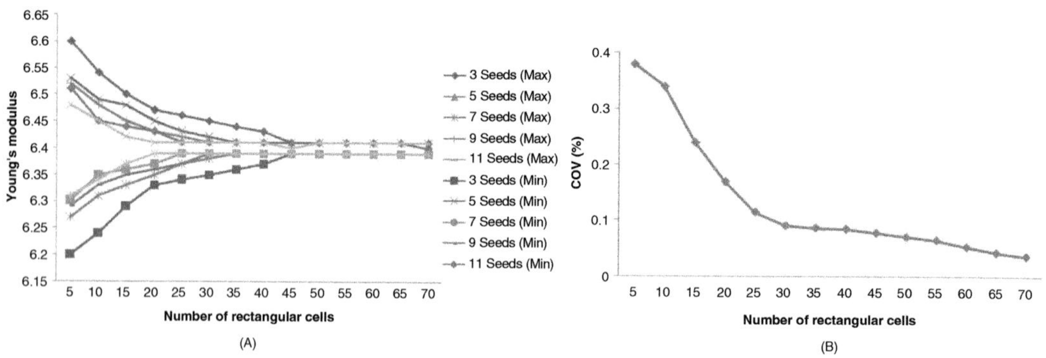

FIGURE 20.8 Convergence trend. Mesh density at (A) macroscale and (B) COV.

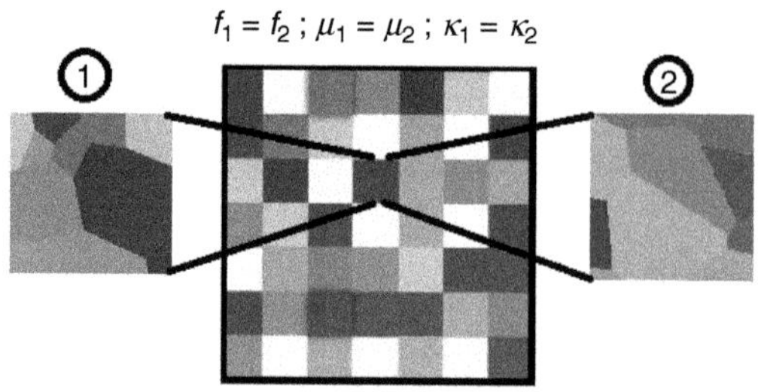

FIGURE 20.9 Random patterns for constitutive generated polygons.

20.5 Model Validation

The validity of a developed model is investigated by comparing the estimated results with the experimentally measured ones. For this purpose, two scenarios are employed. First, an experimental program is conducted to measure the properties of CNTRP and compare them with the results of theoretical modeling. Second, for establishing confidence toward the proper performance of the developed modeling procedure, the estimated results are also compared with available experimental observations in open literature.

Several CNT/polymer specimens were fabricated and subjected to tensile test in a laboratory. The incorporated CNTs are multiwalled with carbon purity of 98%. They have an average diameter of 60–80 nm and average length of 75 μm. Low-viscosity epoxy resin, commercially known as Crestapol 1212, was used with Perkador CH50 hardener. Tip sonication method was used to initially disperse the multiwalled carbon nanotubes into acetone solution. Having the epoxy resin added to the solution, the mixture bath was exposed to sonication for a reasonable period of time, between 30 and 120 min depending on the CNT contents. To avoid negative consequences of the overheating phenomenon, the power amplitude of the sonicator was adjusted to the level of 60% while the mixture was also surrounded by ice bath. Afterward, the solution was heated in a magnetic stirring tool and vacuum oven to make acetone evaporate from the mixture. Then, the final mixture was cast in a metal mold and cured for 5 h at a temperature of 80–100°C. Casting the tensile specimens, they were polished mechanically. Subsequently, a series of uniaxial tensile tests were carried out on the specimens at room temperature, while their dimensions were in accordance with ASTM D 638 (Type I). Five samples were fabricated for each specific CNT content and also five other samples from the neat resin for the sake of comparison. Theoretically predicted values are compared with the measured properties and presented in Table 20.3. It can be easily seen that the theoretical predictions are in a very good agreement with experimental data.

Table 20.3 Experimentally Measured and Theoretically Predicted Young's Modulus

	Measured Young's Modulus			
CNT (wt.%)	**Mean (GPa)**	**STD**	**Predicted Young's Modulus (GPa)**	**Error (%)**
0.00	2.71	0.094	2.71	0.0
0.25	2.86	0.047	2.782	−2.73
0.50	3.12	0.053	3.025	−2.79
1.00	3.18	0.042	3.089	−2.86
1.50	3.41	0.118	3.306	−3
2.00	3.65	0.051	3.663	0.3
3.00	3.88	0.057	3.866	−0.36
4.00	4.15	0.058	4.284	3.2
6.00	4.36	0.056	4.423	1.4
8.00	4.45	0.079	4.442	−0.1
10.00	4.79	0.057	4.591	−4.15

STD, Standard deviation.

Table 20.4 Comparison Between Predicted Results With Published Experimental Data

	Constituent Properties		CNTRP Modulus (GPa)		
Researcher(s)	**Matrix Modulus (GPa)**	**CNT Volume Fraction (%)**	**Experimentally Measured**	**Theoretically Estimated**	**Error (%)**
Villoria and Miravete [12]	2.875	0.12	2.909	2.917	−0.2
	2.587	0.47	2.659	2.651	0.3
Ogasawara et al. [32]	2.84	2.3	3.07	2.96	3.5
	2.84	5.4	3.28	3.19	2.7
	2.84	10.3	3.90	3.75	3.8
Kanagaraj et al. [33]	1.095	0.11	1.169	1.155	1.1
	1.095	0.22	1.228	1.232	−0.3
	1.095	0.33	1.287	1.286	0.07
	1.095	0.44	1.338	1.323	1.09
Xiao et al. [34]	0.235	1	0.261	0.252	3.4
	0.235	3	0.284	0.264	0.7
	0.235	5	0.386	0.359	6.9
	0.235	10	0.444	0.418	5.8

The developed multiscale modeling is also used to predict available experimental observations in open literature. The results are compared in Table 20.4.

A comparison between experimental results and the theoretically predicted ones reveals that the developed modeling procedure is capable of capturing involved parameters efficiently while insignificant overestimations are observed in rare cases.

20.6 Conclusions

A novel hierarchical multiscale approach is developed to simulate CNT-based composites. Covering a full-range of scales, the pioneering approach especially focused on the phenomena arising with the macroscale and less investigated mesoscales. A comprehensive study of lower scales of nano and micro is performed. It is illustrated clearly that a proper modeling should initially insist to keep the lattice structure of the CNT by simulating CNT as an equivalent discrete structure. Having the surrounding matrix simulated at microscale, the equivalent fiber technique is employed to consider the nonbonded interphase region and better demonstrate the CNT and surrounding polymer interactions at microscale. Using the Bayes classification algorithm combined with the Voronoi method, a novel technique is developed for the irregular tessellation of the constitutive block at mesoscale. Based on the irregular pattern and to capture aggregated CNTs, each constitutive block is tessellated to the heterogeneous polygons, so that every polygon just bears a single bundle of aggregated CNTs. Consequently, the locations of aggregates are stochastically taken into account. Inspiring composite cylinder assemblage model [27,28], polygons are converted to biphase spheres one by one, each of which contains an aggregated bundle of CNTs surrounded by dispersed CNT in the matrix. To have the agglomeration phenomenon taken into account, the proposed formulation by Hashin [27] and Christensen [28] are modified and the properties of each polygon are obtained subsequently. On the results of the equivalent fiber theory, the curvature and nonstraight shapes of CNTs are modeled using phenomenological bounding technique. Treating all random parameters including agglomeration, waviness pattern, and nonuniform dispersion of CNTs, a robust full stochastic multiscale modeling is implemented. Regular tessellation technique is used to tessellate the material region at macroscale to the rectangular constitutive blocks and is assigned with a random volume fraction of CNT to address nonuniform dispersion. Comparing the results with those of the experimental observation, very good and close consistency is observed.

References

[1] H. Dai, Carbon nanotubes: opportunities and challenges, Surf. Sci. 500 (1–3) (2002) 218–241.

[2] I. Kang, Y.Y. Heung, J.H. Kim, J.W. Lee, R. Gollapudi, S. Subramaniam, Introduction to carbon nanotube and nanofiber smart materials, Composites Part B 37 (6) (2006) 382–394.

[3] J.P. Salvetat-Delmotte, A. Rubio, Mechanical properties of carbon nanotubes: a fiber digest for beginners, Carbon 40 (10) (2002) 1729–1734.

[4] K.T. Lau, C. Gu, D. Hui, A critical review on nanotube and nanotube/nanoclay related polymer composite materials, Composites Part B 37 (6) (2006) 425–436.

[5] E.T. Thostenson, Z. Ren, T.-W. Chou, Advances in the science and technology of carbon nanotubes and their composites: a review, Compos. Sci. Technol. 61 (13) (2001) 1899–1912.

[6] E.T. Thostenson, C.Y. Li, T.-W. Chou, Nanocomposites in context, Compos. Sci. Technol. 65 (3–4) (2005) 491–516.

[7] D. Qian, E. Dickey, R. Andrews, T. Rantell, Load transfer and deformation mechanisms in carbon nanotube–polystyrene composites, Appl. Phys. Lett. 76 (20) (2000) 2868–2870.

[8] M.M. Shokrieh, R. Rafiee, A review of mechanical properties of isolated carbon nanotubes and carbon nanotube composites, J. Mech. Compos. Mater. 46 (2) (2010) 155–172.

[9] G.M. Odegard, T.S. Gates, K.E. Wise, C. Park, E.J. Siochi, Constitutive modeling of nanotube–reinforced polymer composites, Compos. Sci. Technol. 63 (11) (2003) 1671–1687.

[10] B. Ashrafi, P. Hubert, Modeling the elastic properties of carbon nanotube array/polymer composites, Compos. Sci. Technol. 66 (3–4) (2006) 387–396.

[11] Y. Han, J. Elliott, Molecular dynamics simulations of the elastic properties of polymer/carbon nanotube composites, Comput. Mater. Sci. 39 (2) (2007) 315–323.

[12] R.G. Villoria, A. Miravete, Mechanical model to evaluate the effect of the dispersion in nanocomposites, Acta Mater. 55 (9) (2007) 3025–3031.

[13] K.I. Tserpes, P. Panikos, G. Labeas, Sp.G. Panterlakis, Multi-scale modeling of tensile behavior of carbon nanotube-reinforced composites, Theor. Appl. Fract. Mech. 49 (1) (2008) 51–60.

[14] S.J.V. Frankland, V.M. Harik, G.M. Odegard, D.W. Brenner, T.S. Gates, The stress–strain behavior of polymer–nanotube composites from molecular dynamics simulation, Compos. Sci. Technol. 63 (11) (2003) 1655–1661.

[15] V.V. Mokashi, D. Qian, Y. Liu, A study on the tensile response and fracture in carbon nanotube-based composites using molecular mechanics, Compos. Sci. Technol. 67 (3–4) (2007) 530–540.

[16] M.M. Shokrieh, R. Rafiee, On the tensile behavior of an embedded carbon nanotube in polymer matrix with non-bonded interphase region, Compos. Struct. 92 (3) (2010) 647–652.

[17] A. Selmi, C. Friebel, I. Doghri, H. Hassis, Prediction of the elastic properties of single walled carbon nanotube reinforced polymers: a comparative study of several micromechanical models, Compos. Sci. Technol. 67 (10) (2007) 2071–2084.

[18] G.D. Seidel, D.C. Lagoudas, Micromechanical analysis of the effective elastic properties of carbon nanotube reinforced composites, Mech. Mater. 38 (8–10) (2006) 884–907.

[19] Y. Liu, N. Nishimura, Y. Otani, Large-scale modeling of carbon-nanotube composites by a fast multipole boundary element method, Comput. Mater. Sci. 34 (2) (2005) 173–187.

[20] C. Li, T.W. Chou, A structural mechanics approach for the analysis of carbon nanotubes, Int. J. Solids Struct. 40 (10) (2003) 2487–2499.

[21] A.L. Kalamkarov, A.V. Georgiades, S.K. Rokkam, V.P. Veedu, M.N. Ghasemi-Nejhad, Analytical and numerical techniques to predict carbon nanotubes properties, Int. J. Solids Struct. 43 (22–23) (2006) 6832–6854.

[22] M.M. Shokrieh, R. Rafiee, Prediction of Young's modulus of graphene sheets and carbon nanotubes using nanoscale continuum mechanics approach, Mater. Des. 31 (2) (2010) 790–795.

[23] V. Lordi, N. Yao, Molecular mechanics of binding in carbon-nanotube–polymer composites, J. Mater. Res. 15 (12) (2000) 2770–2779.

[24] R. Rafiee, M.M. Shokrieh, Prediction of mechanical properties of CNT based composites using multiscale modeling and stochastic analysis, in: K.I. Tserpes, N. Silvestre (Eds.), Modeling of Carbon Nanotubes, Graphene and their Composites, Springer Series in Materials Science, Springer International Publishing, Switzerland, 2014, pp. 201–238.

[25] A. Okabe, B. Boots, K. Kokichi Sugihara, S.N. Chiu, Spatial Tessellations—Concepts and Applications of Voronoi Diagrams, second ed., John Wiley and Sons Ltd, Chichester, (2000).

[26] I. Androutsopoulos, J. Koutsias, K. Chandrinos, C. Spyropoulos. An experimental comparison of Naive Bayesian and keyword-based anti-spam filtering with encrypted personal e-mail messages. In: Proceedings of the 23rd ACM SIGIR Conference; 2000 July 24–28; Athens, Greece. pp. 160–167.

[27] Z. Hashin, The elastic moduli of heterogeneous materials, J. Appl. Mech. 29 (1) (1962) 143–150.

[28] R.M. Christensen, Mechanics of Composite Materials, Wiley-Interscience, New York, (1979).

[29] D.L. Shi, X.Q. Feng, Y.Y. Huang, K.C. Hwang, H. Gao, The effect of nanotube waviness and agglomeration on the elastic property of carbon nanotube-reinforced composite, J. Eng. Mater. Technol. 126 (3) (2004) 250–257.

[30] S. Wang, Z. Liang, B. Wang, C. Zhang, Statistical characterization of single-walled carbon nanotube length distribution, Nanotechnology 17 (2006) 634–639.

[31] R. Rafiee, Influence of carbon nanotube waviness on the stiffness reduction of CNT/polymer composites, Compos. Struct. 97 (2013) 304–309.

[32] T. Ogasawara, Y. Ishida, T. Ishikawa, R. Yokota, Characterization of multi-walled carbon nanotube/phenylethynyl terminated polyimide composites, Composites Part A 35 (1) (2004) 67–74.

[33] S. Kanagaraj, F.R. Varanda, T.V. Zhiltsova, M.S.A. Oliveira, J.A.O. Simoes, Mechanical properties of high density polyethylene/carbon nanotube composites, Compos. Sci. Technol. 67 (15–16) (2007) 3071–3077.

[34] S.P. Xiao, L.C. Zhang, I. Zarudi, Mechanical and rheological properties of carbon nanotube-reinforced polyethylene composites, Compos. Sci. Technol. 67 (2) (2007) 177–182.

21

Stochastic Modeling of CNT-Grown Fibers

Roham Rafiee, Amin Ghorbanhosseini
UNIVERSITY OF TEHRAN, TEHRAN, IRAN

CHAPTER OUTLINE

21.1 Introduction

Industries, particularly the aerospace industry, are looking for stronger and lighter materials for their new generation of products. Unique mechanical, electrical, and physical properties are among the reasons for developing polymeric nanocomposites. Another reason for the development of polymeric nanocomposites and the increased research in this field is the discovery of carbon nanotubes (CNTs) in 1991. By incorporating CNTs into polymers, extraordinary conductivity and strength are obtained, which makes CNTs a promising candidate for a wide range of applications. The new context of fuzzy fibers is an interesting field for the application of nanocomposites containing CNTs. In this novel material, instead of dispersing CNTs into the polymeric matrix, CNTs are grown on the surface of a core microfiber. Utilizing this novel

Carbon Nanotube–Reinforced Polymers. http://dx.doi.org/10.1016/B978-0-323-48221-9.00021-2

fiber instead of conventional fibers for reinforcing polymeric matrices, fuzzy fiber–reinforced polymers are achieved.

The main advantages of fuzzy fiber–reinforced laminates arising from the direct growth of CNTs on the surface of a core fiber are as follows:

- Improved mechanical properties particularly enhanced transverse Young's modulus in unidirectional (UD) laminates overcoming the weakness of UD composites along the matrix direction.
- Prevention of CNT agglomeration, which is the main drawback of CNT-reinforced polymers wherein the accumulation of carbon nanotubes cluster in the resin due to their distribution on the fibers.
- Increased quality of bonding between the main fiber and the surrounding matrix by increasing the stiffness of the interphase between the CNT and the polymer; this is attributed to the existence of CNTs.

Fuzzy fibers can appear in two general configurations depending on the pattern of CNT distribution around the core fiber. CNTs can be either radially aligned or randomly oriented on the surface of the core fiber [1].

The main objective of this chapter is to develop a hierarchical multiscale modeling to predict the mechanical properties of fuzzy fibers on the basis of the bottom-up modeling process. The developed multiscale modeling is also extended to the level of fuzzy fiber–reinforced laminates to compare the predicted results with available experimental data. Stochastic implementation of the developed multiscale modeling is conducted taking into account random patterns of CNT curvature and inhomogeneous CNT distributions along the length of the core fiber. The introduced uncertainties are induced during the processing of fuzzy fiber.

21.2 Modeling Framework

For the purpose of extracting the mechanical properties of fuzzy fibers, two distinct phases can be distinguished in the configuration of a single fuzzy fiber: (1) a core fiber and (2) the surrounding CNT-reinforced polymer. The individual constituents of fuzzy fiber–reinforced laminates are the matrix and the CNT-coated fiber known as the fuzzy fiber and matrix wherein CNT/polymer nanocomposites located around the core fiber play the role of an interphase between the fiber and the resin. Therefore, fuzzy fiber–reinforced laminates can be categorized as multireinforcement materials wherein the CNT and core fiber can be viewed as the reinforcing agent at nano- and microscales, respectively.

Reinforcing agents appear at nano- and microscales while the mechanical properties are intended to be predicted at the macroscale. This wide range of involved scales implies the necessity of developing a full-range multiscale modeling technique to capture all the

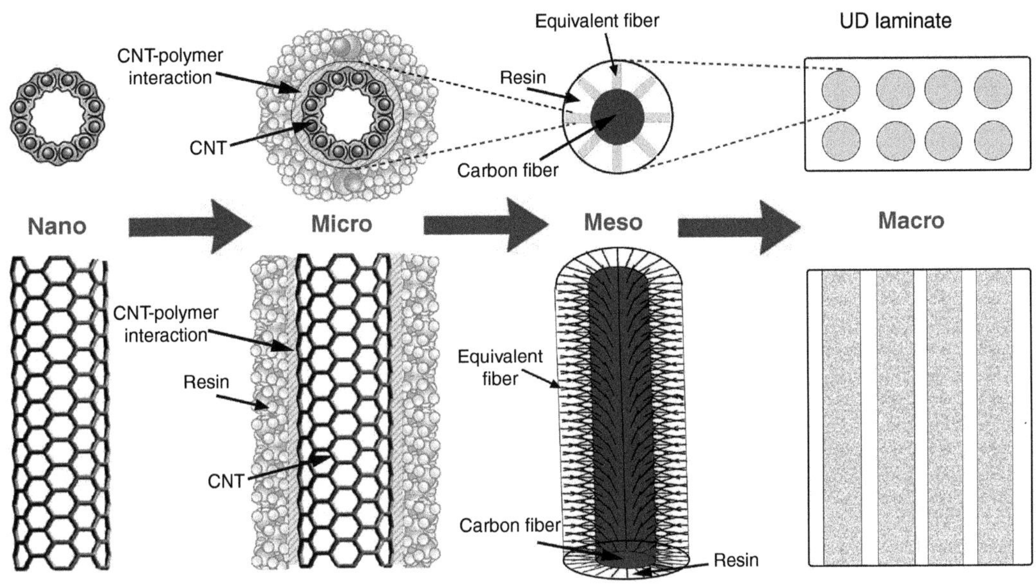

FIGURE 21.1 Scales involved in the simulation of fuzzy fiber–reinforced laminates.

effective parameters properly. A schematic representation of the involved scales of analysis is presented in Fig. 21.1.

As the first step of modeling, a top-down scanning is required to recognize the effective parameters at each scale separately. Then, bottom-up modeling is conducted starting from nanoscale and going up to macroscale wherein the output of modeling at each scale is fed into the upper scale as input data based on a hierarchical scheme. Finally, the modeling is performed stochastically accounting for random parameters associated with processing-induced uncertainties. Subsequently, the developed modeling procedure encompasses three various and sequential stages:

1. top-down scanning,
2. bottom-up modeling, and
3. stochastic simulation.

21.3 Top-Down Scanning

Appropriate representative volume elements (RVEs) are defined for each scale of analysis separately to address the corresponding effective parameters. Recalling from Fig. 21.1, all parameters defining the mechanical properties of fuzzy fiber–reinforced laminates are inserted in Table 21.1. The effectiveness of each parameter is also assigned to a specific scale of analysis.

Table 21.1 Effective Parameter for Each Scale Simulation

Effective Scale	Parameter(s)
Nano	CNT molecular interactions
Micro	Interactions between CNT and surrounding resin
Meso	CNT orientation (radially aligned or randomly oriented)
	CNT curvature, CNT volume fraction
	Mechanical properties of core fiber
Macro	Mechanical properties of fuzzy fiber
	Mechanical properties of matrix
	Matrix volume fraction

As can be understood from Table 21.1, the effective scale of all random issues (i.e., CNT curvature and CNT volume fraction) along the core fiber's length for the specific case of radially aligned CNT around the core fiber is mesoscale. In addition to CNT curvature and volume fraction, CNT orientation is also treated as a random parameter for the specific case of a fuzzy fiber with random orientations of CNTs.

21.4 Bottom-Up Modeling

As is evident from Fig. 21.1, modeling starts from the very first scale of nano and lasts in the final scale of macro. The mechanical properties of a single fuzzy fiber are characterized as the output of mesoscale analysis. However, modeling is extended to the macroscale and a comparison is performed between the predicted mechanical properties for a fuzzy fiber–reinforced laminate and published data for validation purposes. The detailed procedure of the modeling at each scale is outlined in the next sections.

21.4.1 Nanoscale Modeling

The first step of modeling deals with nanoscale modeling focusing on an isolated CNT as the reinforcing agent operating at the nanoscale. Over the past years, many studies have been conducted to find the mechanical properties of CNTs. At this scale, the study of carbon–carbon covalent bonds is performed in the atomic structure of CNTs. Studies on carbon nanotubes can be classified into two categories: (1) atomic modeling or molecular dynamics (MD) simulation and (2) continuum modeling as analytical/numerical [2]. Atomic modeling is basically conducted either by solving the Schrödinger equation directly or replacing the Schrödinger equation with Newton's second law to extract the positions of atoms. Owing to the complexity of the governing formulations as well as their intensive computational efforts, these methods are not only limited to systems consisting of a few atoms, but are also only applicable to very short timescales. Thus, the application of continuum modeling is preferred in the research community, especially when CNTs are modeled in conjunction with other media instead of isolated

CNTs. In this research, a CNT is replaced with a hollow CNT with a thickness of 0.34 nm and a Young's modulus of 1 TPa as a widely accepted and reported value in literature [2,3].

21.4.2 Microscale Modeling

The essential requirement of modeling the behavior of CNT-reinforced polymers at microscale is a clear understanding of its relationship with its surrounding resin. Actually, besides the mechanical properties of the CNT and resin as constituent materials, CNT–polymer interactions accounting for load transferring from the matrix to the CNT substantially influence the mechanical properties of CNT-reinforced polymers. In case of weakness in the intermediate phase between the CNT and the polymer, no load is transferred to the CNT and the mechanical properties of the nanocomposite will be even less or equal to the resin properties. At microscale, a proper RVE consists of a CNT, the surrounding polymer, and CNT–polymer interactions.

Naturally, the connection between the CNT and the surrounding resin is established through weakly nonbonded van der Waals (vdW) interaction bonds in the absence of chemical functionalization of CNTs. In some studies, CNT is perfectly bonded to the resin neglecting the important role of vdW interaction. In this category of studies, perfect bonding is considered between the CNT and the polymer and thus conventional micromechanics rules are employed. This compromise in modeling leads to an overestimation of the mechanical properties and is thus not recommended [4–7]. As an improvement to this method, a distinct continuum material region is considered between the CNT and the polymer, that is, a continuum interphase is assumed. The main drawback of this modeling technique can be found in defining the mechanical properties of the interphase region. The mechanical properties of this phase is unclear and it is usually chosen as having a Young's modulus between that of the matrix and the CNT without any experimental basis [8,9].

Subsequently, semicontinuum modeling was employed by some researchers to overcome the aforementioned difficulties [10–18]. In this approach, vdW interactions between the CNT and the polymer are simulated using continuum truss, beam, or spring elements. The behavior of these connecting elements is adapted from molecular space by establishing an analogy between molecular space and structural mechanics. Therefore, instead of assuming perfect bonding between the CNT and the polymer, vdW interactions between them are simulated using continuum elements. Since, the level of continuity is shifted to the molecular interaction instead of utilizing a continuum interphase between the CNT and the polymer, this category of modeling is referred to as "semicontinuum." Although semicontinuum techniques have a proficiency leaning toward continuum modeling in terms of accuracy and time for analysis, the process of defining the relationship between the CNT and the resin and using the appropriate elements and time-consuming analysis are considered major challenges in using these methods.

As an alternative approach in this study and a reasonable compromise in modeling procedure, the interaction between the CNT and the polymer is modeled using an interface instead of a continuum or semicontinuum interphase. The CNT and the polymer are modeled as continuum media while a zero-thickness interface is placed between them. The mechanical behavior of the interface is dictated by the cohesive zone model (CZM). CZM is one of the most modern evolutions in the field of fracture mechanics, and with the aid of commercial finite element (FE) packages, provides the possibility of separation between two different phases. Thus, CZM can be utilized to simulate the interactions between the CNT and the surrounding resin. CZM introduces no physical area, but it describes the adhesion forces that occur at the time of separation. In CZM, the mechanical behavior of the interface is described by establishing a correlation between the traction vector on the interface and the resultant interfacial separation. The mathematical representation of this correlation is given by a constitutive law between two phases in the form of a traction–separation curve. The separation energy can be calculated using Lennard-Jones "6–12" potential for vdW interactions [19].

Bilinear CZM is defined in the ABAQUS commercial FE package adjusting the parameters properly to resemble vdW interactions. Adjacent nodes on the inner surface of the polymer and the outer surface of the CNT are linked using a defined cohesive behavior. Therefore, cohesive behavior is employed instead of cohesive elements to reduce the run-time. The volume of resin is selected in a manner that reflects 95% volume fraction for resin and 5% volume fraction for CNT. The Young's modulus and Poisson's ratio of the resin is chosen as 5 GPa and 0.3, respectively.

The constructed FE model is subjected to different loading and boundary conditions to obtain the mechanical properties of the investigated microscale RVE. Due to the nonlinear behavior of CZM adapted from the nonlinear behavior of vdW interaction, nonlinear FE analysis is conducted. After obtaining the mechanical properties of the microscale RVE, the CNT and its interface are converted into a virtual equivalent fiber. The equivalent fiber is a continuum fiber perfectly bonded to the matrix. The strategy of converting CNT and CNT–polymer interaction into an equivalent fiber is presented in Fig. 21.2.

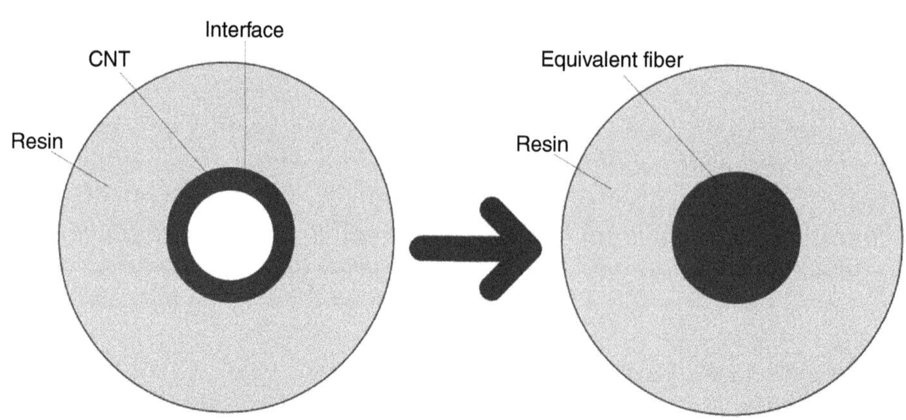

FIGURE 21.2 Replacing CNT and CNT–polymer interaction with a virtual equivalent fiber.

Table 21.2 Effective Properties of Equivalent Fiber

	E_L (GPa)	$E_T = E_Z$ (GPa)	G_{ZT} (GPa)	ν_{LT}
Semicontinuum modeling [18]	649.37	11.27	5.13	0.284
Present study	642.27	12.5	4.88	0.28

Odegard et al. were the first to develop the idea of equivalent fiber [4]. They calculated the total energy of a system consisting of a functionalized CNT and resin through MD simulation and replaced the investigated unit-cell with an RVE containing the equivalent fiber and resin. Equating the total molecular energy through MD simulations with the total strain energy of the RVE, mechanical properties of the equivalent fiber were extracted [4].

In this research, having in hand the mechanical properties of the investigated microscale RVE as the output of the FE analysis (Fig. 21.2, left part), the CNT and the interface are replaced with an equivalent fiber as a single inclusion, which is perfectly bonded to the resin (Fig. 21.2, right part). Thus, the mechanical properties of the developed equivalent fiber can be computed using micromechanics rules inversely, that is, the mechanical properties of the resin and the RVE containing the equivalent fiber are fed into appropriate micromechanics rules to extract the mechanical properties of the equivalent fiber. Implying a transversely isotropic behavior, the obtained mechanical properties of the developed equivalent fiber are presented in Table 21.2 and compared with the results obtained through semicontinuum modeling. A transversely isotropic behavior is observed for the developed equivalent fiber.

In line with the development of the equivalent fiber, the micromechanics rules can be used in the upper scales of analysis considering the fact that the reinforcing agent is the equivalent fiber instead of CNT as the output of the microscale analysis.

21.4.3 Mesoscale Modeling

At mesoscale, the mechanical properties of a single fuzzy fiber are characterized. The RVE of the mesoscale modeling contains a core fiber perfectly bonded to the surrounding CNT/polymer nanocomposites. The surrounding nanocomposites appear in two different configurations: (1) radially aligned CNTs and (2) randomly oriented CNTs. Both configurations are analyzed in this section. It should be emphasized that because the output of the performed analysis is at microscale, the reinforcing agent is viewed as the developed equivalent fiber. Therefore, the properties of CNT/polymer nanocomposites can be computed with any desired CNT volume fraction. After obtaining the mechanical properties of the surrounding CNT/polymer nanocomposites, the mechanical properties of the fuzzy fiber are computed using the mechanical properties of its constituents, namely, the core fiber and the surrounding CNT/polymer nanocomposites.

As reflected in Table 21.1, CNT orientation, volume fraction, and curvature are captured at this scale. The first two parameters are addressed in the next section; however, the last one is captured during the stochastic implementation of modeling due to its random behavior.

21.4.3.1 Radially Aligned CNT-Reinforced Polymer

To extract the properties of nanocomposites containing CNTs with a radial distribution pattern, the properties of the nanocomposites with aligned CNTs are first obtained and then converted to a radially aligned CNT by converting the Cartesian coordinates to the cylindrical one.

The main step in solving every elastic problem is to extract the relationship between the components of the strain tensor and the corresponding members of the stress tensor. This correlation is referred to as the constitutive law [20]:

$$\{\bar{\sigma}\}=[C]\{\bar{\varepsilon}\} \tag{21.1}$$

where C denotes the stiffness tensor and the overhead lines imply volumetric average. For composite materials, the stiffness matrix of composites can be obtained using the stiffness matrices of their constituents [21]:

$$C=C_{\mathrm{m}}+f_{\mathrm{r}}(C_{\mathrm{r}}-C_{\mathrm{m}})A_{\mathrm{r}} \tag{21.2}$$

where A is the strain concentration matrix and f stands for volume fraction. The m and r subscripts stand for matrix and reinforcing agent, respectively. Different approaches can be used to obtain the strain concentration matrix. In this research, the Mori–Tanaka method is employed to obtain the mechanical properties of aligned CNT-reinforced polymers [22]:

$$A_{\mathrm{r}}^{\mathrm{MT}}=A_{\mathrm{s}}^{dil}\left[f_{\mathrm{m}}I+f_{\mathrm{r}}A_{\mathrm{r}}^{dil}\right]^{-1} \tag{21.3}$$

where

$$A_{\mathrm{r}}^{dil}=\left[I+S_{\mathrm{r}}C_{\mathrm{m}}^{-1}(C_{\mathrm{r}}-C_{\mathrm{m}})\right]^{-1} \tag{21.4}$$

where I and S_{r} represent the identity matrix and Eshelby matrix, respectively. The Eshelby matrix for a cylindrical inclusion is just a function of the Poisson's ratio of the matrix [23].

The stiffness matrix of the developed equivalent fiber, as a transversely isotropic material, is obtained using the following equation:

$$C_{\mathrm{r}}=\begin{bmatrix} \frac{1}{E_{\mathrm{L}}} & -\frac{\nu_{\mathrm{LT}}}{E_{\mathrm{L}}} & -\frac{\nu_{\mathrm{LZ}}}{E_{\mathrm{L}}} & 0 & 0 & 0 \\ -\frac{\nu_{\mathrm{LT}}}{E_{\mathrm{L}}} & \frac{1}{E_{\mathrm{T}}} & -\frac{\nu_{\mathrm{TZ}}}{E_{\mathrm{T}}} & 0 & 0 & 0 \\ -\frac{\nu_{\mathrm{LZ}}}{E_{\mathrm{L}}} & -\frac{\nu_{\mathrm{TZ}}}{E_{\mathrm{T}}} & \frac{1}{E_{\mathrm{Z}}} & 0 & 0 & 0 \\ 0 & 0 & 0 & \frac{1}{G_{\mathrm{XZ}}} & 0 & 0 \\ 0 & 0 & 0 & 0 & \frac{1}{G_{\mathrm{TZ}}} & 0 \\ 0 & 0 & 0 & 0 & 0 & \frac{2(1+\nu_{\mathrm{LT}})}{E_{\mathrm{L}}} \end{bmatrix}^{-1} \tag{21.5}$$

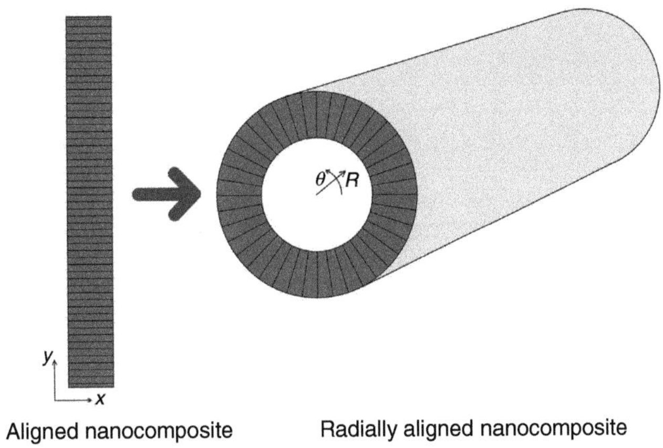

FIGURE 21.3 Converting aligned CNT-reinforced composites to the radially aligned one.

All reflected parameters in Eq. (21.5) are presented in Table 21.2, except for ν_{TZ}, which is taken as 0.14 [23]. The stiffness matrix of the resin material can be also obtained using a simplified form of Eq. (21.5), considering an isotropic behavior for the resin, that is, $E_L = E_T = E_Z$, $\nu_{TZ} = \nu_{TL} = \nu_{LZ}$, and $G_{TZ} = 2(1 + \nu_{TZ})/E_L$. The L, T, and Z subscripts represent the local coordinate systems of the equivalent fiber; L shows an axial direction while T and Z have transverse directions.

Consequently, having in hand the mechanical properties of the resin and the developed equivalent fiber and also their volume fractions, the mechanical properties of the aligned CNT-reinforced polymer are obtained. Then, the Cartesian coordinate is mapped into the cylindrical coordinate to obtain the mechanical properties of the radially aligned CNT-reinforced polymer. The conversion strategy is schematically depicted in Fig. 21.3.

The stiffness matrix in the cylindrical coordinates in terms of the stiffness matrix in Cartesian coordinates is expressed as follows [20]:

$$[C_{CYL}] = [T_\sigma][C_{X\text{-}Y}][T_\varepsilon]^{-1} \tag{21.6}$$

where C_{X-Y} is the stiffness matrix in Cartesian coordinates, $C_{R-\theta}$ is the stiffness matrix in cylindrical coordinates, and T_ε and T_σ are stress and strain transformation matrices, respectively, expressed as

$$[T_\sigma] = \begin{bmatrix} \cos^2(\theta) & \sin^2(\theta) & 2\sin(\theta)\cos(\theta) & 0 & 0 & 0 \\ \sin^2(\theta) & \cos^2(\theta) & -2\sin(\theta)\cos(\theta) & 0 & 0 & 0 \\ -\sin(\theta)\cos(\theta) & \sin(\theta)\cos(\theta) & \cos^2(\theta)-\sin^2(\theta) & 0 & 0 & 0 \\ 0 & 0 & 0 & \cos(\theta) & \sin(\theta) & 0 \\ 0 & 0 & 0 & -\sin(\theta) & \cos(\theta) & 0 \\ 0 & 0 & 0 & 0 & 0 & 1 \end{bmatrix} \tag{21.7}$$

$$[T_\varepsilon]=\begin{bmatrix} \cos^2(\theta) & \sin^2(\theta) & \sin(\theta)\cos(\theta) & 0 & 0 & 0 \\ \sin^2(\theta) & \cos^2(\theta) & -\sin(\theta)\cos(\theta) & 0 & 0 & 0 \\ -2\sin(\theta)\cos(\theta) & 2\sin(\theta)\cos(\theta) & \cos^2(\theta)-\sin^2(\theta) & 0 & 0 & 0 \\ 0 & 0 & 0 & \cos(\theta) & \sin(\theta) & 0 \\ 0 & 0 & 0 & -\sin(\theta) & \cos(\theta) & 0 \\ 0 & 0 & 0 & 0 & 0 & 1 \end{bmatrix} \tag{21.8}$$

To make the results independent of the rotation angle, the average properties based on the angle is calculated by the following equation [24]:

$$[C_{\mathrm{avg}}]_{ij}=\frac{1}{\pi(R^2-a^2)}\int_0^{2\pi}\int_a^R [C_{R\text{-}\theta}]_{ij}\,\mathrm{r}\,\mathrm{dr}\,\mathrm{d}\theta \tag{21.9}$$

where $\left[C_{\mathrm{avg}}\right]_{ij}$ represents the volume average of the CNT/polymer nanocomposite stiffness matrix in cylindrical coordinate system. R and a stand for outer and inner radius of the CNT/polymer nanocomposites around the core fiber, respectively.

The obtained properties for the radially aligned CNT-reinforced polymer implies a transversely isotropic behavior with axis of symmetry along the radial direction.

21.4.3.2 Randomly Oriented CNT-Reinforced Polymer

In contrast to the transversely isotropic behavior of radially aligned CNT-reinforced polymers, the oriented CNT/polymer nanocomposites are categorized under isotropic materials. For randomly oriented CNT-reinforced polymer, a suitable form of the Mori–Tanaka formulation is employed. For this purpose, below formulation is employed to calculate Hill's elastic moduli for the reinforcing agent [22]:

$$C_r=\begin{bmatrix} \frac{1}{E_L} & -\frac{\nu_{LT}}{E_L} & -\frac{\nu_{LZ}}{E_L} & 0 & 0 & 0 \\ -\frac{\nu_{LT}}{E_L} & \frac{1}{E_T} & -\frac{\nu_{TZ}}{E_T} & 0 & 0 & 0 \\ -\frac{\nu_{LZ}}{E_L} & -\frac{\nu_{TZ}}{E_T} & \frac{1}{E_Z} & 0 & 0 & 0 \\ 0 & 0 & 0 & \frac{1}{G_{LZ}} & 0 & 0 \\ 0 & 0 & 0 & 0 & \frac{1}{G_{TZ}} & 0 \\ 0 & 0 & 0 & 0 & 0 & \frac{2(1+\nu_{LT})}{E_L} \end{bmatrix}^{-1} = \begin{bmatrix} n_r & l_r & k_r-m_r & 0 & 0 & 0 \\ l_r & k_r+m_r & l_r & 0 & 0 & 0 \\ k_r-m_r & l_r & k_r+m_r & 0 & 0 & 0 \\ 0 & 0 & 0 & p_r & 0 & 0 \\ 0 & 0 & 0 & 0 & m_r & 0 \\ 0 & 0 & 0 & 0 & 0 & p_r \end{bmatrix} \tag{21.10}$$

After obtaining n, l, k, m, and p for the equivalent fiber, the following equations are employed [22]:

$$\alpha_r=\frac{3(K_m+G_m)+k_r-l_r}{3(G_m+K_m)} \tag{21.11}$$

$$\beta_r = \frac{1}{5}\left\{\frac{4G_m+2k_r+l_r}{3(G_m+k_r)} + \frac{4G_m}{G_m+p_r} + \frac{2\left[G_m(3K_m+G_m)+G_m(3K_m+7G_m)\right]}{G_m(3K_m+G_m)+m_r(3K_m+7G_m)}\right\} \tag{21.12}$$

$$\delta_r = \frac{1}{3}\left[n_r + 2l_r + \frac{(2k_r+l_r)(3K_m+2G_m-l_r)}{G_m+k_r}\right] \tag{21.13}$$

$$\eta_r = \frac{1}{5}\left[\frac{2}{3}(n_r-l_r) + \frac{8G_m p_r}{G_m+p_r} + \frac{8m_r G_m(3K_m+4G_m)}{3K_m(m_r+G_m)+G_m(7m_r+G_m)} + \frac{2(k_r-l_r)(2G_m+l_r)}{3(G_m+k_r)}\right] \tag{21.14}$$

Then, the bulk modulus (K) and the shear modulus of the CNT/polymer nanocomposite as an isotropic material are obtained as follows [25]:

$$K_{\text{CNT/Polymer}} = K_m + \frac{V_r(\delta_r - 3K_m\alpha_r)}{3(V_m + V_{\text{CNT}}\alpha_r)} \tag{21.15}$$

$$G_{\text{CNT/Polymer}} = G_m + \frac{V_r(\eta_r - 2G_m\beta_r)}{2(V_m + V_{\text{CNT}}\beta_r)} \tag{21.16}$$

Finally, having in hand the bulk modulus and shear modulus of the CNT/polymer nanocomposite around the core fiber, its Young's modulus and Poisson's ratio can be obtained.

21.4.3.3 Single Fuzzy Fiber

The mechanical properties of the CNT/polymer nanocomposite around the core fiber in both radially aligned and randomly oriented CNTs are obtained in previous sections. Now, the concentric cylinder method is employed to extract the mechanical properties of the single fuzzy fiber using the mechanical properties of the core fiber and the surrounding CNT/polymer nanocomposite. It is noteworthy that carbon fiber is considered as a transversely isotropic material with axis of symmetry along the longitudinal direction. The radially aligned CNT/polymer nanocomposite is also a transversely isotropic material with axis of symmetry along the radial direction, and the randomly oriented CNT/polymer nanocomposite is an isotropic material. Transverse bulk modulus, axial Young's modulus, axial Poisson's ratio, and in-planc shear modulus are obtained using the following equations [24]:

$$K^{\text{FuzzyFiber}} = \frac{k_{\text{CNTRP}}(k_{\text{CF}}+k_{\text{CNTRP}})V_{\text{CNTRP}} + k_{\text{CF}}\left(k_{\text{CNTRP}}+G_{xy}\big|_{\text{CNTRP}}\right)V_{\text{CF}}}{(k_{\text{CF}}+k_{\text{CNTRP}})V_{\text{CNTRP}} + \left(k_{\text{CNTRP}}+G_{xy}\big|_{\text{CNTRP}}\right)V_{\text{CF}}} \tag{21.17}$$

$$E_z\big|_{\text{FuzzyFiber}} = E_{\text{CNTRP}}V_{\text{CNTRP}} + E_{\text{CF}}V_{\text{CF}} + \frac{4\left(\nu_{xy}\big|_{\text{CF}} - \nu_{xy}\big|_{\text{CNTRP}}\right)}{3\left(\frac{\nu_{xy}|_{\text{CNTRP}}}{K_{\text{CF}}} + \frac{\nu_{xy}|_{\text{CF}}}{K_{\text{CNTRP}}} + \frac{1}{G_{xy}|_{\text{CNTRP}}}\right)} \tag{21.18}$$

$$\nu_{zx}\big|_{\text{FuzzyFiber}} = \nu_{zx}\big|_{\text{CNTRP}}V_{\text{CNTRP}} + \nu_{zx}\big|_{\text{CF}}V_{\text{CF}} + \frac{\left(\nu_{zx}\big|_{\text{CF}} - \nu_{zx}\big|_{\text{CNTRP}}\right)\left(\frac{\nu_{zx}|_{\text{CF}}}{K_{\text{CNTRP}}} - \frac{\nu_{zx}|_{\text{CNTRP}}}{K_{\text{CF}}}\right)\nu_{zx}\big|_{\text{CF}}\,\nu_{zx}\big|_{\text{CNTRP}}}{\left(\frac{\nu_{zx}|_{\text{CNTRP}}}{K_{\text{CF}}} + \frac{\nu_{zx}|_{\text{CF}}}{K_{\text{CNTRP}}} + \frac{1}{G_{zx}|_{\text{CNTRP}}}\right)} \tag{21.19}$$

$$G_{zx}\big|_{\text{FuzzyFiber}} = G_{zx}\big|_{\text{CNTRP}} \frac{G_{zx}\big|_{\text{CNTRP}} v_{zx}\big|_{\text{CNTRP}} + G_{zx}\big|_{\text{CNTRP}} (v_{zx}\big|_{\text{CF}} + 1)}{G_{zx}\big|_{\text{CF}} v_{zx}\big|_{\text{CNTRP}} G_{zx}\big|_{\text{CNTRP}} (v_{zx}\big|_{\text{CF}} + 1)} \tag{21.20}$$

It is worth mentioning that the "CNTRP" and "CF" subscripts stand for CNT/polymer nanocomposites and carbon fiber, respectively. x, y, and z denote the global coordinate system of a fuzzy fiber where z is placed along its axial direction. V represents volume fraction.

The transverse shear module is also obtained using the following equation [26]:

$$G_{xy}\big|_{\text{FuzzyFiber}} = \frac{G_{xy}\big|_{\text{CNTRP}} [K_{\text{CNTRP}}(G_{xy}\big|_{\text{CNTRP}} + G_{xy}\big|_{\text{CF}}) + 2G_{xy}\big|_{\text{CNTRP}} G_{xy}\big|_{\text{CF}} + K_{\text{CNTRP}}(G_{xy}\big|_{\text{CF}} - G_{xy}\big|_{\text{CNTRP}}) V_{CF}]}{K_{\text{CNTRP}}(G_{xy}\big|_{\text{CNTRP}} + G_{xy}\big|_{\text{CF}}) + 2G_{xy}\big|_{\text{CNTRP}} G_{xy}\big|_{\text{CF}} - (K_{\text{CNTRP}} + 2G_{xy}\big|_{\text{CNTRP}})(G_{xy}\big|_{\text{CF}} - G_{xy}\big|_{\text{CNTRP}}) V_{CF}} \tag{21.21}$$

Moreover, Poisson's ratio and the transverse modulus of single fuzzy fiber can be calculated using the developed relation by Sullivan and Hashin [27]:

$$m = 1 + \frac{4K^{\text{FuzzyFiber}} v_{zx}\big|^2_{\text{FuzzyFiber}}}{E_z\big|_{\text{FuzzyFiber}}} \tag{21.22}$$

$$v_{xy} = \frac{K^{\text{FuzzyFiber}} - mG_{xy}\big|_{\text{FuzzyFiber}}}{K^{\text{FuzzyFiber}} + mG_{xy}\big|_{\text{FuzzyFiber}}} \tag{21.23}$$

$$E_x\big|_{\text{FuzzyFiber}} = 2(1 + v_{xy}\big|_{\text{FuzzyFiber}}) G_{xy}\big|_{\text{FuzzyFiber}} \tag{21.24}$$

21.4.4 Macroscale Modeling

In this section, the determined properties in the preceding section and the Halpin–Tsai method are used for characterizing a unidirectional fuzzy fiber–reinforced polymer. The empirical micromechanics rules for obtaining the mechanical properties of long fiber–reinforced composites are expresses as follows [28]:

$$\begin{aligned} E_{\text{L}} &= E_{\text{m}} \frac{(1 + 2\zeta \eta_{\text{L}} V_{\text{f}})}{(1 - \eta_{\text{T}} V_{\text{f}})}, \eta_{\text{L}} = \frac{E_{\text{ffL}}/E_{\text{m}} - 1}{E_{\text{ffL}}/E_{\text{m}} + \zeta}, \zeta = 2l/D \\ E_{\text{T}} &= E_{\text{m}} \frac{(1 + 2\eta_{\text{T}} V_{\text{f}})}{(1 - \eta_{\text{T}} V_{\text{f}})}, \eta_{\text{T}} = \frac{E_{\text{ffT}}/E_{\text{m}} - 1}{E_{\text{ffT}}/E_{\text{m}} + 2} \end{aligned} \tag{21.25}$$

where E_{L} and E_{T} imply longitudinal and transverse Young's modulus of the UD fuzzy fiber–reinforced laminate, respectively. V_{F} denotes fuzzy fiber volume fraction. L and T stand for the length and diameter of the fuzzy fiber. E_{ffL} and E_{ffT} stand for the longitudinal and transverse modulus of the fuzzy fiber obtained in Section 21.4.3.3, respectively. For the case of short fuzzy fiber–reinforced laminate, the following formulation is employed [28]:

$$E = \frac{3}{8} E_{\text{L}} + \frac{5}{8} E_{\text{T}} \tag{21.26}$$

where E_{L} and E_{T} are obtained using Eq. (21.25).

21.5 Stochastic Modeling

In this section, stochastic modeling is conducted taking into account the CNT curvature and CNT volume fraction along the length of the fuzzy fiber as random variables. A phenomenological method is used to capture the influence of CNT curvature in the developed multiscale modeling procedure [29]. Considering the transversely isotropic behavior of the developed equivalent fiber (Table 21.2), Young's moduli of the equivalent fiber in both longitudinal and transverse directions are randomly chosen between two bounding values dictated by longitudinal and transverse modulus of the equivalent fiber, that is, 642.27 and 12.5 GPa. Therefore, regardless of the geometrical shape of nonstraight CNT, the influence of CNT curvature on the mechanical properties are addressed. This procedure is schematically illustrated in Fig. 21.4.

The Monte Carlo simulation technique is used to perform the developed multiscale modeling stochastically. The Monte Carlo technique is suitable for computer-based simulations. The main procedure for the calculation of the Monte Carlo simulation algorithm is the same for any simple or complex process. First, a random number is determined then the probability of an event is compared with the generated random number. This procedure can be repeated several times and a measurable output produced for each realization. In stochastic multiscale modeling of single fuzzy fiber or fuzzy fiber–reinforced laminate relying on the Monte Carlo technique, the whole developed modeling procedure is executed several times and in each realization the mechanical properties of the equivalent fiber are selected randomly. Moreover, CNT volume fraction is also generated randomly in each realization to capture the inhomogeneous dispersion of CNTs around the core fiber. The random selection of the CNT volume fraction is performed at each produced sample in a manner that the mean value is equal to the defined mean value by the user with 2% standard deviation. The overall flowchart of the developed stochastic modeling procedure is shown in Fig. 21.5. It should be pointed out that the developed modeling procedure results in mechanical properties of the single fuzzy fiber being

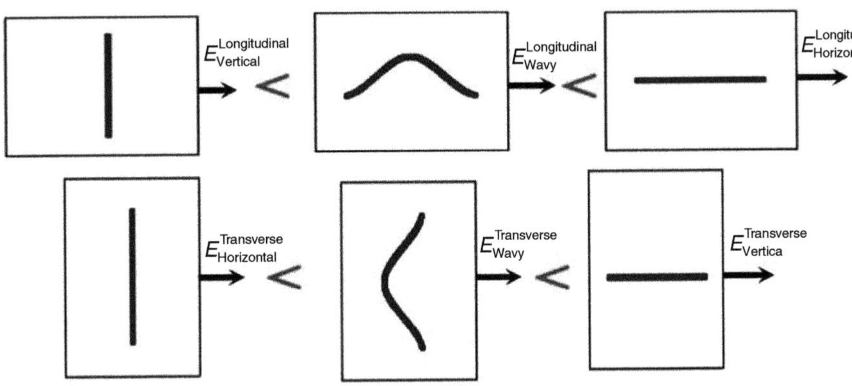

FIGURE 21.4 Simulating nonstraight CNT using random selection of properties [29].

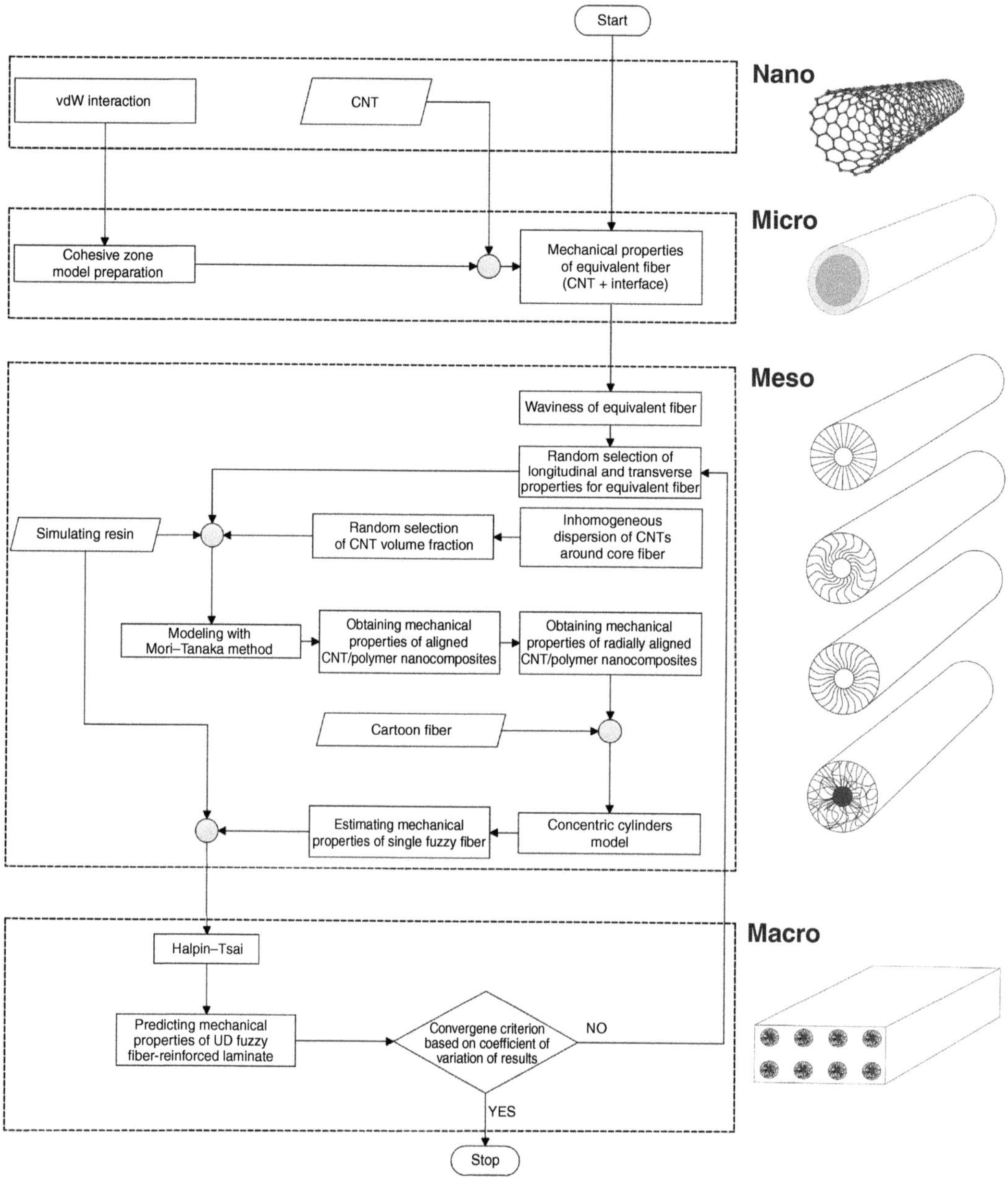

FIGURE 21.5 Flowchart of stochastic multiscale modeling.

obtained at mesoscale and the mechanical properties of fuzzy fiber–reinforced laminate being obtained at macroscale.

As a convergence criterion, the generated random samples are considered to be sufficient when the coefficient of variation for the investigated property falls below 1%.

21.6 Model Validation

The proper performance of the model is examined in this section comparing the estimated results with experimental observations available in literature.

Motivated by the experimental program conducted by Sager et al. [30], the modeling procedure is executed for the same investigated fuzzy fiber. Sager et al. [30] have successfully grown multiwalled CNTs on T650 carbon fibers using the chemical carbon vapor deposition method. Then, tensile strength and modulus are measured using the single-fiber tensile test on a fuzzy fiber with randomly oriented CNTs. The same mechanical properties for the carbon fiber and epoxy matrix reported by Sager et al. are fed into the modeling procedure and longitudinal Young's modulus of the fuzzy fiber is estimated. A comparison between the reported observation and the estimated value along with the number of generated samples is presented in Table 21.3.

Kulkarni et al. [25] have experimentally measured the transverse modulus of fuzzy fiber–reinforced laminate. In their experiment, CNTs were grown on IM7 carbon fiber and EPON862 is used as the matrix. The fabricated composite plates using a resin transfer molding process had a fiber volume fraction of 40% and a CNT volume fraction of 2%. They have observed that by adding just 2% of CNT, the transverse modulus of the fuzzy fiber–reinforced laminate has increased 100.36% in comparison with UD composites containing untreated carbon fiber [25].

A comparison between the results obtained through theoretical modeling and experimental data is presented in Table 21.3.

Shazed et al. [31] fabricated a composite with short randomly oriented fuzzy fibers. Propylene is used as the host matrix. CNT volume fraction is declared as 6.38×10^{-4}. The developed multiscale modeling is executed for this study and the comparison of results is presented in Table 21.3.

A very good agreement between theoretical results and experimental data is observed for the first two cases reflected in Table 21.3. This implies the proper performance of the

Table 21.3 Comparison Between Experimental Observations and Estimated Properties

Investigated Material	Measured Property	Experiment (GPa)	Theoretical Prediction [Error] (GPa)	Number of Realizations
Fuzzy fiber with radially aligned CNT	Longitudinal modulus of fuzzy fiber	175	180.88 ± 5.56 [3.36%↑]	1000
Long fuzzy fiber–reinforced laminate	Transverse modulus of laminate	10.2 ± 1.3	10.00 ± 1.16 [2%↓]	1500
Short fuzzy fiber–reinforced laminate	Young's modulus	3.52	5.29 ± 0.05 [50.28%↑]	1500

developed modeling procedure. However, the percentage of error increased in the last case. This can be attributed to the assumption of perfect bonding between the fuzzy fiber and the surrounding polymer, which is not pertinent to the specific experiment conducted in that research [31]. Moreover, it is also probable that the developed modeling procedure is not applicable to the very low contents of CNT.

21.7 Parametric Study

It is evident from the aforementioned experimental studies [30,31] that the longitudinal modulus of a fuzzy fiber is less than the untreated core carbon fiber. Although fuzzy fibers present less longitudinal modulus in comparison with untreated carbon fibers, the main advantage of CNT growth on carbon fiber can be traced to improved transverse modulus. Thus, the influence of CNT volume fraction on the transverse Young's modulus of fuzzy fiber–reinforced laminates is investigated. Considering a longitudinal and transverse moduli of the carbon fiber as 276 and 18.5 GPa, respectively, and the epoxy resin as having a Young's modulus of 2.7 GPa, the obtained trend is presented in Fig. 21.6.

The influence of the longitudinal modulus of the developed equivalent fiber on both longitudinal and transverse moduli of a single fuzzy fiber containing radially aligned CNTs is also studied. According to the explained procedure for capturing nonstraight shapes of CNT, this value is randomly selected to be between 12.5 and 642.2 GPa in the stochastic modeling procedure. Therefore, longitudinal modulus is varied in the aforementioned intervals and the degree to which it affects the overall properties of a fuzzy fiber is presented in Fig. 21.7.

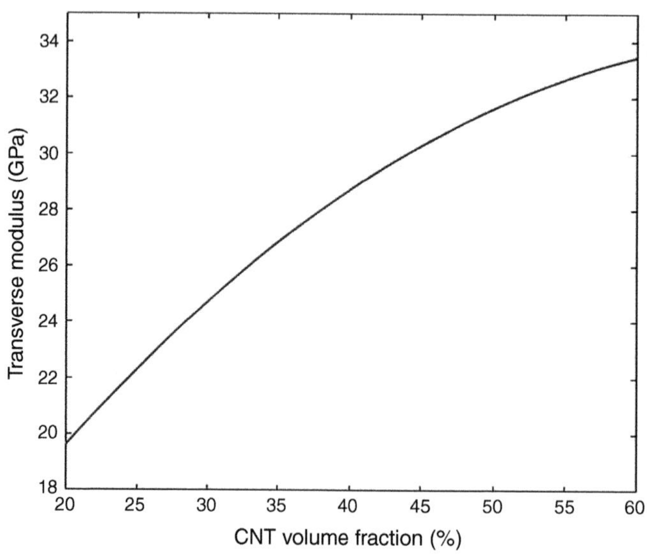

FIGURE 21.6 Transverse Young's modulus of fuzzy fiber (randomly oriented CNTs) against CNT volume fraction.

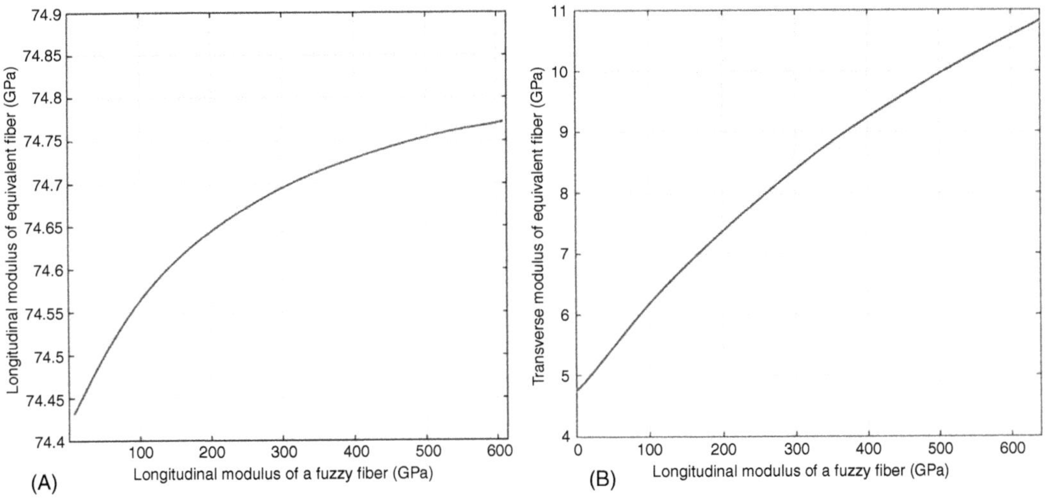

FIGURE 21.7 Effect of longitudinal modulus of equivalent fiber on longitudinal (A) and transverse (B) moduli of a fuzzy fiber.

Similarly, the influence of transverse modulus of the developed equivalent fiber on the longitudinal and transverse moduli of a fuzzy fiber is presented in Fig. 21.8.

A quantitative comparison between the reported trends implies the fact that the influence of longitudinal modulus of the equivalent fiber on the mechanical properties of a fuzzy fiber is more pronounced than the transverse modulus of the equivalent fiber. Therefore, it can be interpreted that the main contributors for enhancing the properties of a fuzzy fiber are the mechanical properties of the equivalent fiber highlighting the important role of CNT–polymer interactions on the results.

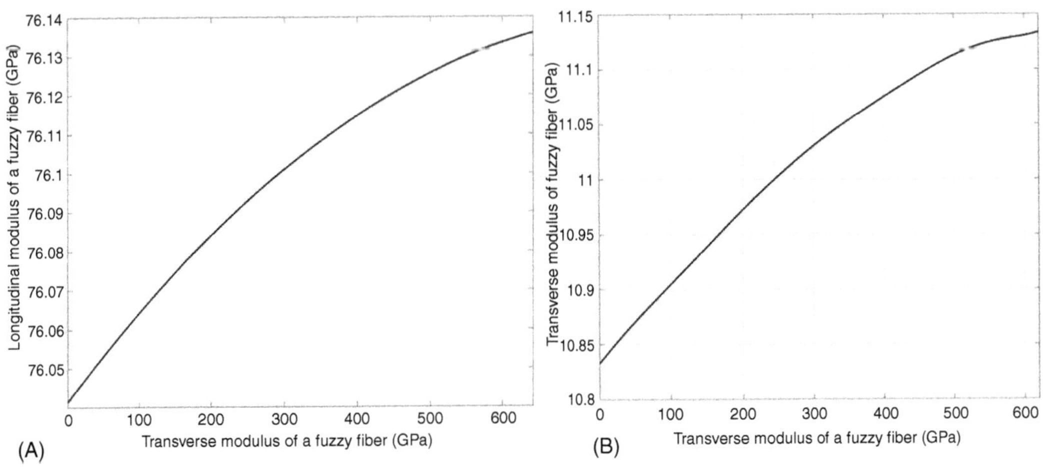

FIGURE 21.8 Effect of the transverse modulus of the equivalent fiber on the longitudinal (A) and transverse (B) moduli of a fuzzy fiber.

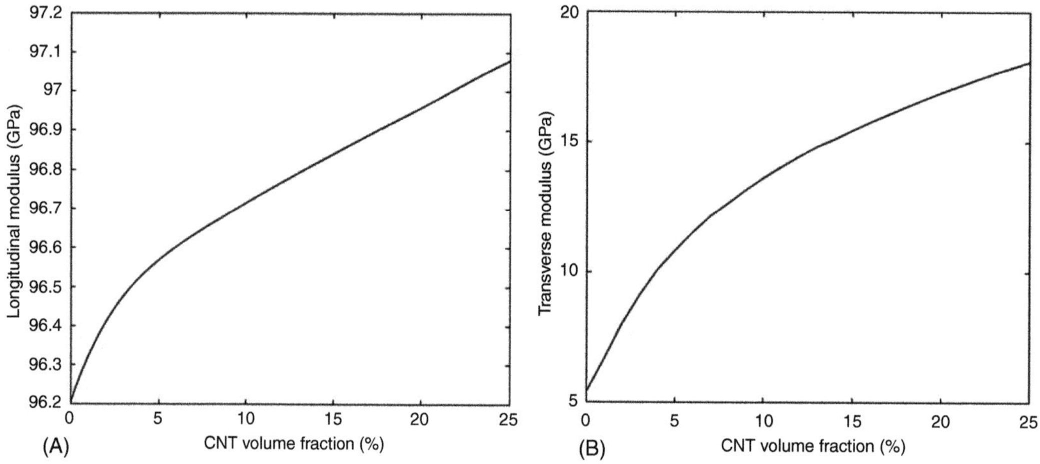

FIGURE 21.9 Influence of CNT contents on the longitudinal (A) and transverse (B) moduli of a fuzzy fiber–reinforced laminate.

Finally, the effect of CNT volume fraction on the longitudinal and transverse moduli of fuzzy fiber–reinforced laminate is investigated by increasing the CNT volume fraction up to 25%. A negligible improvement in longitudinal modulus is observed in Fig. 21.9, while a significant enhancement in the transverse direction is tangible.

21.8 Concluding Remarks

A stochastic multiscale modeling procedure is developed to predict the mechanical properties of both isolated fuzzy fibers and fuzzy fiber–reinforced polymers. The modeling procedure is able to consider both radially aligned and randomly oriented CNTs as two common configurations of fuzzy fibers. Processing-induced consistencies including CNT curvature and CNT volume fraction are treated as random parameters due to their inherent uncertain behavior. The modeling procedure is a full-range multiscale modeling covering all scales of nano, micro, meso, and macro. First, a top-down scanning is performed to identify the effective parameters of each scale defining appropriate RVEs. Then, a hierarchical modeling procedure is conducted. Stochastic implementation of the developed modeling is carried out taking into account random variables. The results are compared with published experimental observations and a very good agreement is observed for higher contents of CNTs. Finally, a parametric study is conducted to study the degree to which key parameters affect the estimated properties. The observations justifying the important role of CNT and CNT–polymer interactions can be summarized as follows:

- By increasing the CNT volume fraction, the transverse modulus of a fuzzy fiber is enhanced while the longitudinal modulus is even less than in the untreated carbon fibers.
- The effect of the longitudinal modulus of equivalent fibers (CNT and CNT–polymer interaction) on the transverse modulus of fuzzy fiber–reinforced laminate is

significantly higher than the influence of the transverse modulus of equivalent fiber on longitudinal modulus.
- The effect of CNT volume fraction on the transverse modulus of fuzzy fiber–reinforced composites is extremely higher compared with the longitudinal modulus of the fuzzy fiber.

References

[1] X. Reng, J. Burton, G.D. Seidel, K. Latifi, Computational multi-scale modeling and characterization of piezoresistivity in fuzzy fiber reinforced polymer composites, Int. J. Solid Struct. 54 (2015) 121–134.

[2] M.M. Shokrieh, R. Rafiee, A review of mechanical properties of isolated carbon nanotubes and carbon nanotube composites, Mech. Compos. Mater. 46 (2010) 155–172.

[3] R. Rafiee, R. Maleki Moghad, On the modeling of carbon nanotubes: a critical review, Compos Part B 56 (2014) 435–449.

[4] G.M. Odegard, T.S. Gates, K.E. Wise, C. Park, E.J. Siochi, Constitutive modeling of nanotube–reinforced polymer composites, Compos. Sci. Technol. 63 (2003) 1671–1687.

[5] K.I. Tserpes, P. Panikos, G. Labeas, S.G. Panterlakis, Multi-scale modeling of tensile behavior of carbon nanotube-reinforced composites, Theor. Appl. Fract. Mech. 49 (2008) 51–60.

[6] S.J.V. Frankland, V.M. Harik, G.M. Odegard, D.W. Brenner, T.S. Gates, The stress–strain behavior of polymer–nanotube composites from molecular dynamics simulation, Compos. Sci. Technol. 63 (2003) 1655–1661.

[7] R.G. Villoria, A. Miravete, Mechanical model to evaluate the effect of the dispersion in nanocomposites, Acta Mater. 55 (2007) 3025–3031.

[8] G.D. Seidel, D.C. Lagoudas, Micromechanical analysis of the effective elastic properties of carbon nanotube reinforced composites, Mech. Mater. 38 (2006) 884–907.

[9] J.L. Tsai, T.C. Lu, Investigating the load transfer efficiency in carbon nanotubes reinforced nanocomposites, Compos. Struct. 90 (2009) 172–179.

[10] C. Li, T.W. Chou, Multiscale modeling of carbon nanotube reinforced polymer composites, J. Nanosci. Nanotechnol. 3 (2003) 423–430.

[11] G.I. Giannopoulos, S.K. Georgantzinos, N.K. Anifantis, A semi-continuum finite element approach to evaluate the Young's modulus of single-walled carbon nanotube reinforced composites, Compos. Part B 41 (2010) 594–601.

[12] R. Rafiee, A. Fereidoon, M. Heidarhaei, Influence of non-bonded interphase on crack driving force in carbon nanotube reinforced polymer, Comput. Mater. Sci. 56 (2012) 25–28.

[13] R. Rafiee, R. Maleki Moghadam, Simulation of impact and post-impact behavior of carbon nanotube reinforced polymer using multi-scale finite element modeling, Comput. Mater. Sci. 63 (2012) 261–268.

[14] A. Freridoon, R. Rafiee, R. Maleki Moghadam, Modal analysis of carbon nanotube reinforced polymer using multi-scale finite element method, Mech. Compos. Mater. 14 (3) (2013) 1–8.

[15] J.M. Wernik, S.A. Meguid, Multiscale modeling of the nonlinear response of nano-reinforced polymers, Acta Mech. 217 (1–2) (2011) 1–16.

[16] C. Li, T.W. Chou, Multiscale modeling of compressive behavior of carbon nanotubes/polymer composites, Compos. Sci. Technol. 66 (2006) 2409–2414.

[17] R. Rafiee, R. Pourazizi, influence of CNT functionalization on the interphase region between CNT and polymer, Comp. Mater. Sci. 96 (2015) 573–578.

[18] M.M. Shokrieh, R. Rafiee, Prediction of mechanical properties of an embedded carbon nanotube in polymer matrix based on developing an equivalent long fiber, J. Mech. Res. Commun. 37 (2010) 235–240.

[19] H. Rafii-Tabar, Computational Physics of Carbon Nano Tube, Cambridge University Press, Cambridge, (2008).

[20] M.H. Sadd, Elasticity, Theory, Application and Numerics, Elsevier Butterworth-Heinemann, Oxford, (2005).

[21] S. Nemat-Nasser, M. Hori, Micromechanics: Overall Properties of Heterogenous Materials, North-Holland, Amsterdam, London, (1993).

[22] T. Mori, K. Tanaka, Average stress in matrix and average elastic energy of materials with misfitting inclusions, Acta Metall. Sin. 21 (1973) 71–575.

[23] J.D. Eshelby, The determination of the elastic field of an ellipsoidal inclusion, and related problems, Proc. R. Soc. Lond. Ser. A 241 (1226) (1957) 376–396.

[24] S. Kundalwal, M.C. Ray, Micromechanical analysis of fuzzy fiber reinforced composites, Int. J. Mech. Mater. 7 (2011) 149–166.

[25] M. Kulkarni, D. Carnahan, K. Kulkarni, D. Qian, J.L. Abot, Elastic response of a carbon nanotube fiber reinforced polymeric composite: a numerical and experimental study, Compos Part B 41 (2010) 414–421.

[26] J.M. Whitney, R.L. Mccullough, Micromechanical materials modeling Delaware Composite Design Encyclopedia, Technomics, (1990) p. 2.

[27] B.J. Sullivan, Z. Hashin. Determination of mechanical properties of interfacial region between fiber and matrix in organic matrix composites, 3rd International Conference on Composite Interfaces, 1990, Case Western Reserve, University of Cleveland, OH. 1990, pp. 520–537.

[28] J.C.H. Affdl, J.L. Kardos, The Halpin–Tsai equations: a review, Polym. Eng. Sci. 16 (5) (1976) 344–352.

[29] R. Rafiee, Influence of carbon nanotubes waviness on the stiffness reduction of CNT/polymer composites, Compos. Struct. 97 (2013) 304–309.

[30] R. Sager, P. Klein, D. Lagoudas, Q. Zhang, J. Liu, J. Baur, Effect of carbon nanotubes on the interfacial shear strength of T650 carbon fiber in an epoxy matrix, Compos. Sci. Technol. 69 (2009) 898–904.

[31] M.A. Shazed, A.R. Suraya, S. Rahmanian, M.A. Mohd Salleh, Effect of fiber coating and geometry on the tensile properties of hybrid carbon nanotubes coated carbon fiber reinforced composite, Mater. Des. 54 (2014) 660–669.

Index

Q

R

T

CPI Antony Rowe
Chippenham, UK
2017-10-03 18:43